THE HISTORY AND GEOGRAPHY OF HUMAN GENES

The History and Geography of Human Genes

ABRIDGED PAPERBACK EDITION

L. LUCA CAVALLI-SFORZA

PAOLO MENOZZI

ALBERTO PIAZZA

PRINCETON UNIVERSITY PRESS
PRINCETON, NEW JERSEY

Library of Congress Cataloging-in-Publication Data
Cavalli-Sforza, L. L. (Luigi Luca), 1922–
    The history and geography of human genes /
  Luigi Luca Cavalli-Sforza, Paolo Menozzi, Alberto Piazza.
        p. cm.
    Includes bibliographical references and index.
    ISBN 0-691-08750-4 (unabridged cloth ed.)
    ISBN 0-691-02905-9 (abridged pbk. ed.)
    1. Human population genetics—History. 2. Human evolution.
  3. Human geography. 4. Human population genetics—Research.
  I. Menozzi, Paolo, 1946– . II. Piazza, Alberto, 1941–
  III. Title.
  QH431.C395   1993
  573.21'5—dc20   93–19339

This book has been composed in Times Roman
Designed by Jan Lilly

The cover illustration is a map of the world showing four major ethnic
regions. Africans are yellow, Australians red, and Caucasians green.
Mongoloids show the greatest variation retaining some similarities with
Europeans on one side (a light brown greenish tinge in middle Siberia) and
with Australians on the other (a pinkish color in parts of America and on the
way to it). The extensive gradients due to admixtures between Africans and
Caucasoids in North Africa, and between Caucasoids and Mongoloids in
Middle Asia, are clearly visible. (See chapter two.)

Princeton University Press books are printed on acid-free paper and
meet the guidelines for permanence and durability of the Committee
on Production Guidelines for Book Longevity of the
Council on Library Resources

First printing of the abridged paperback edition, 1996

Printed in the United States of America

10 9 8 7 6 5 4 3

To our wives
Alba, Wallis, *and* Ada

CONTENTS

PREFACE TO THE PAPERBACK EDITION

THE FIRST edition of *The History and Geography of Human Genes* has been in print for less than two years and has received a favorable welcome by scientists and the public. In a friendly review that appeared in *Nature*, Jared Diamond compared it to a box of chocolates one keeps in the refrigerator as a source of precious morsels. He also noted that the volume is heavy to handle: the geographic maps of single genes occupy about half the volume and are, with their large format, responsible for its size. It seemed reasonable to print an abridged paperback edition, retaining only the unmodified text and its own bibliography and analytical index. The original edition, with its numerical tables and geographical maps, will still be available for those who prefer it, and for libraries where it can always be consulted.

Collecting data from the literature for the original book was begun in 1978 and ended in 1986, after which the genetic and statistical analysis was started, and the actual writing begun. During the long process of publication of a volume this size (the final proof correction took place in the summer of 1993, and publication in summer 1994), there were chances of updating the text. Inevitably, the need to avoid further delays restricted somewhat our taking advantage of these opportunities. We hope to put soon on the Internet the complete numerical data on which the book is based. They refer to almost 2,000 different populations and contain as many as 86,000 entries. The tables in the first edition of the book contain only a summary of this file.

In the last ten years new data have accumulated, enriching the information on "classical markers," and especially that on DNA markers. Quite a number of new experimental methods of molecular DNA analysis have been devised, generating new types of data, as well as new methods of evolutionary analysis. With the experience we accumulated, the existing organization and the available network of communications, updating the information and analysis should be much faster than the first collection and review of the existing data. Depending on the resources that will be available, old and new information could also be shared by interested readers much more promptly. The Human Genome Diversity Project aims at collecting a balanced and representative set of human samples across the world in order to have information on genetic differences of our species. In five to ten years it should accumulate an entirely new matrix of data, much more systematically and efficiently than has been possible so far. New evidence will probably resolve old problems and provide new clues and new problems worth considering.

For this paperback edition we still offer—to a wider audience, we hope—the same suggestions, proofs, and conclusions we have presented in the full volume. As we discuss more specifically in the Epilogue, which is reprinted from the first edition, many of our conclusions are tentative and are there to be tested, challenged, or confirmed. The reduced size of this volume does not reduce the importance we attribute to their message. What we have seen published since the writing of the manuscript has not suggested any dramatic change in our conclusions, but a new era is now beginning.

PREFACE

Thirty years ago the first effort was made to reconstruct the history of human differentiation by employing the genetic divergence observed among human groups. The data base comprised gene frequencies, that is, frequencies of alleles at polymorphic loci known to be clearly inherited. Observed frequencies are very stable and seem to be rather insensitive to short-term environmental change. There are, however, very few if any data from the past, and stability in time is inferred from the stability in space, essentially the regularity of gene-frequency distributions and the very small differences usually observed among populations that live in widely different environments. Fortunately, in very recent times, new developments in molecular technology have generated the hope of obtaining substantial information from individuals or populations that have been dead for a long time.

Data from physical anthropology (including skin color, body build, and facial traits) had previously been the only source of information. Some of these data, especially measurements on bones, have the great advantage of being readable in the fossil material. Unfortunately, data available for the past have shown conspicuous changes in the last 200 years, as, for instance, the trend to increase in stature and changes in other measurements observed in Europe. It is difficult to ascribe these observations to genetic causes, and it is more likely that they represent responses to recent environmental changes. They are therefore less suitable for the study of genetic history. Even so, major differences observed in the fossil material have been important for reconstructing the general lines of evolution of the genus *Homo*. More detailed conclusions are still controversial because of the rarity of informative specimens and of dating difficulties in the time range of greater interest. Some of these limitations are slowly being removed.

Genetic data about extant populations useful for our purposes are extremely numerous. Two of the first polymorphic loci discovered, the ABO and RH blood groups, had considerable clinical importance and were tested very widely. Many other markers with no clinical interest were nevertheless investigated in many populations because of the anthropological information they can provide. Unfortunately, the existing data vary widely in number and geographic distribution. If they had been collected more systematically according to a rational plan, as occurred for important markers like those of the HLA system, we would have a much more informative body of data. Today molecular genetics is providing us with enormously more powerful technology, but the data base thus generated is still minimal, and we should better organize our future efforts.

There is another important reason for starting a major program in analyzing human diversity now. While our potential skills for analyzing human evolution are increasing, social changes taking place in developing countries are rapidly destroying the identities—if not the very existence—of the most important aboriginal populations. Thus, organized research efforts to save this precious information about our past have acquired a new urgency. Fortunately, recent technical developments make the prospects very exciting, so that this is a good time for taking stock of available knowledge and using it as a guide for planning future research.

This book was started with the desire to analyze the geography of human genes, using new techniques we have developed for the purpose of studying ancient human migrations. While the very demanding work of computerizing the enormous data base in existence was proceeding, it became clear that there was a need to analyze the same information with other techniques, developed by us and by others, which can lead to conclusions of historical interest. But the challenging task of reconstructing the history of human evolution can hardly be entirely satisfactory using only evidence provided by the genetic data. Information from historical, linguistic, anthropological, and archaeological sources is also useful, and it should be compared with the genetic evidence if we wish to reach fully satisfactory conclusions.

Needless to say, all these sources have their own limitations. Relevant data from history are infrequent, far from quantitative, and do not usually probe deep enough in time. Archaeology says very little about the physical populations it studies, but it gives dates and some, however vague, information on demography, especially on numbers of people, that are important for predicting the rates of genetic evolution. But archaeologists often find it difficult to distinguish between the migration of people and the diffusion of artifacts or the techniques for making them. Linguistic change follows rules that are somewhat analogous to those of genetic evolution, except that it is much faster and the reconstruction of early stages is therefore especially difficult. Moreover, languages are sometimes replaced by others of totally different origin in a very short time, partially blurring the concordances. Physical anthropology can be misleading because certain physical traits observed in bones can sometimes change quickly with changing environmental conditions. Only genes almost always have the degree of permanence necessary for discussing fissions, fusions, and migrations of populations that took place during the history of our subspecies, which goes back for at least 100,000 years. A large fraction of the genetic variants we study appeared before that time. Their relative proportions have changed considerably since and can orient us in understanding population history.

Although population geneticists often summarize knowledge about the archaeology, history, and linguistics of the ethnic groups they have studied, there has been no comprehensive treatment or attempt at a global picture of our species from the points of view of general history that are relevant for genetics. We hope to fill this gap with the present volume. In the first chapter we give some general historical information on the subject, a discussion of the concept of race, its failure, and an elementary introduction to the major analytical techniques used for our purposes. We have tried to make the book readable to scientists of as many disciplines as possible, given that not only geneticists but also scholars from fields as diverse as archaeology, anthropology, history, geography, and linguistics have a potential interest in the subject. Most barriers to cross-disciplinary exchanges are the result of the specialized vocabularies of each field, and we have tried to counter this limitation as much as possible. This is tantamount to saying that lay readers could also understand this book, if they have the motivation necessary for going through a scientific analysis. Inevitably, the discussion is kept at an elementary level in each of these disciplines, and the language used is as simple as possible. Statistical methods and basic population genetics theory are explained in a qualitative way with economical use of scientific terms; all of which are defined at their first introduction.

The second chapter is dedicated to an analysis of the world data with the aim of understanding the general history of *Homo sapiens sapiens*. Trees of descent are reconstructed and compared with archaeological data and linguistic classifications. Other methods of analysis are applied to the global data for an evaluation of the genetic structure of the species as a whole.

The five chapters that follow are dedicated to the major geographic subdivisions of the inhabited Earth. We start with the continent where the genus and probably also the subspecies to which we belong have first developed, Africa, and then proceed with the other continents successively occupied, though not in the strict order of occupation: Asia, Europe, America, and Oceania. In each chapter we briefly discuss geography and ecology, and then history, starting with paleoanthropological and archaeological information. We pay special attention, when possible, to population numbers and densities, as well as to migrations that have special relevance for the evolutionary processes in which we are interested. Physical anthropology and linguistics follow. Then an analysis of the available genetic data is given for each continent in general and for its most important subsections. Geographic maps of genes for which there is enough information are given for the world and each continent at the end of the volume. "Synthetic" geographic maps derived from them are given in the text and show the major genetic patterns that can be abstracted from the total genetic "landscape" by suitable methods. There is not always enough genetic information to make full use of or to interpret all the historical and other nongenetic information given in the first sections of each chapter, nor is there enough of the latter to explain all details of the former, but we hope the unused information may be a stimulus for further research.

The last chapter is an epilogue that discusses generally our conclusions from a methodological point of view and the most urgent problems facing the continuation of research at this crucial time. We now have the tools for doing a much better job than has been done thus far at the level of both data collection and analysis. There is, of course, room for improvement in both, but the usefulness of living populations is being destroyed by a rapid increase in the rate at which human populations are vanishing. The mixing of formerly isolated groups is especially damaging for future research. This is a critical time for organizing our efforts before we lose a unique opportunity for understanding our genetic heritage.

As already mentioned, the second half of the book is dedicated to geographic maps for all genes for which the amount of data of aboriginal populations was deemed adequate. It is difficult to establish an objective criterion for deciding when data are sufficient for making a map, and the choice of alleles and continents represented in the maps was in part subjective. Gene frequencies from samples that were geographically too close had to be averaged before they were used in constructing maps. For different populations inhabiting the same region, we had to choose between discarding some of them or pooling them. In general, when there were both aboriginal populations and late settlers that could be easily distinguished, the former were chosen. The pooling of distinct populations living in the same narrow area generated local heterogeneities, which were systematically estimated and are shown on each map.

For satisfactory map construction, the regularity of the geographic distribution of the data is even more important than the total number of observations. Even for the most intensively studied genes, some areas are not well sampled. In order to give an idea of the strengths and weaknesses of each map from the point of view of the spatial distribution of data, we have indicated the locations at which data were available, as well as significant local heterogeneity, if any. No smoothing of the data could be perfect; we have therefore indicated where the calculated surface of gene frequencies departed significantly from the observations, and the direction of departure. A brief comment on the map of each gene is given in a special section of the appropriate chapter. These single-gene maps were used to generate the synthetic ones.

All gene frequencies obtained from the literature were used for building the geographic maps, but only a selection of populations tested for a greater number of genes was employed for tree analysis. The two methods, trees and geographic maps can be considered complementary descriptions of the same reality. The first stresses historical aspects, and the second the geographic ones. The historical interpretation of trees needs to be strengthened by tests of the validity of the hypotheses underlying them, which is sometimes possible. We usually find good agreement between genetic and nongenetic information, which encourages us further to believe in our conclusions. If nothing else, the presentation of clear hypotheses that can be tested is, we believe, a valuable contribution.

The gene frequencies of the population samples used for tree analysis in the various chapters and their sources are also given in the second part of the book (Appendixes 1 and 2). The bibliography of gene frequencies used for trees and maps is separate from that of works cited in the text (Appendix 3 and Appendix Bibliography). The largest part of gene-frequency data is also found in earlier tabulations that report the relevant sources, and we make specific reference to them population by population.

The task we set before ourselves was not an easy one, and we hope critical readers will recognize that the need to summarize a substantial amount of information of varied nature has inevitably generated the possibility of important omissions and errors. In particular, we apologize to authors who may feel their work has not been adequately considered. In many cases we have preferred to give our conclusions without comparing them with dissenting ones. Our excuse is that we wanted to present testable hypotheses and indicate the basis on which we have accepted or discarded them, without attempting to be fully comprehensive (a nearly impossible task). We are hopeful that our effort will help to spread knowledge and interest in human population genetics, and to recognize the usefulness of thinking in multidisciplinary terms. Much work is necessary for filling important gaps, for organizing future research more satisfactorily at an international level, and for making full use of the power of present techniques at this critical time, when crucial information is slipping out of our hands.

ACKNOWLEDGMENTS

OUR LIST of acknowledgments is long, and we hope we are not leaving out anyone. Many people helped us with computerizing our data and in many other ways. In particular, we want to thank Juliana Hwang for computer programming, and Mariangela Trentadue, Sharon Feingold, and the late Michelle Leo for data input. Most text figures were computer-generated by Megan Betz, and in part by Kim Ha. In addition to this effort, which was rather heroic considering the computer programs that were used at the time, they took part in the final preparation of the bibliography. Much of the statistical analysis was skillfully performed by Nazario Cappello, Eric Minch, Joanna Mountain, and Sabina Rendine. To all of them we express our heartfelt thanks.

The production of the geographic maps was an eight-year endeavor undertaken at the University of Parma under the auspices of its computer center, which made available unlimited amounts of computer time and untold miles of plotter paper and ink. We could not have carried out our project without the unlimited support and assistance of the computer center staff. Words cannot express our gratitude for Enzo Siri's wizard capacity for solving graphics programming problems and his biblical patience in producing the maps that appear in this book.

The consistently creative help of Eleonora Olivetti and Sabina Rendine, always over and above the demands of duty, in preparing the plots included in each map and the color synthetic maps played a critical role in completing the graphic part of our work.

Many people kindly read parts of the manuscript and suggested changes: Peter Bellwood, Jaume Bertranpetit, Anne Bowcock, Giacomo Giacobini, Barry Hewlett, Eric Minch, James V. Neel, Colin Renfrew, Philip Rightmire, S. M. Sirajuddin, Robert Sokal, Chris Stringer, Ken Weiss. Merritt Ruhlen read the entire manuscript. Mike Crawford helped greatly with data from Siberia.

Frank Livingstone, Arthur Mourant, and Don Tills provided us with proofs of their tabulations in advance of publication of their books. Franco Scudo made us aware of the Darwin citation used in chapter 8. Rosalba Guglielmino Matessi contributed in processing and interpreting the hemoglobin data included in this book. None of these helpers must be blamed for our mistakes, but we are grateful for their help and encouragement. Finally, Ed Tenner and Judy May of Princeton University Press provided the initial encouragement to extend our 1978 geographic analysis of Europe to the rest of the world, and the entire Press (in particular, Emily Wilkinson) have been remarkably helpful and patient as we missed deadline upon deadline for presentation of our final manuscript. Copy editing, an endless job, fell upon the shoulders of Teresa Carson, who contributed gracefully to improving the final product. Last but not least, Marilyn Anderson lightened our load by providing secretarial help with sensitivity and good spirit.

Grants-in-aid should be acknowledged. Without the National Institutes of Health (NIH) grants GM20467 and GM10452, this research would not have been possible. Financial help also came from the NIH National Library of Medicine (grant LM04106), the Lucille P. Markey Charitable Trust, the Wenner-Gren Foundation, and the National Science Foundation (Anthropology Division). On the other side of the Atlantic, the Italian CNR (Consiglio Nazionale delle Ricerche) Projects "Biotechnology and Bioinstrumentation," "Genetic Engineering," "Biological Archive," and MURST (Ministero Universita' Ricerca Scientifica Tecnologica) funds of 40% and 60% are gratefully acknowledged. Computer time for graphics (plots and color maps) was supported by grants and facilities of CSI-Piemonte (Consorzio per il Sistema Informativo, Torino, Italy). The help of their staff was especially appreciated.

THE HISTORY AND GEOGRAPHY OF HUMAN GENES

1 INTRODUCTION TO CONCEPTS, DATA, AND METHODS

1.1. INTRODUCTION

For some time, geneticists had been aware of a certain amount of genetic variation among the individuals forming a species, but the remarkable extent of this variation was not appreciated until about 25 years ago. Conspicuous human traits like hair and eye color clearly vary from one individual to the other in many populations; these differences are easily perceived by the layman, as are variation in height, weight, body build, and facial traits, which are also genetically determined to some extent. Their hereditary transmission, however, is complex, and these traits contribute little to our understanding of the extent of variation. The first example of clear-cut genetic variation, that of ABO blood groups, was described at the beginning of the century (Landsteiner 1901). Dissimilarities between individuals regarding ABO blood-group variation are due to small chemical differences between molecules found at the surface of red blood cells.

These studies were soon extended to other blood-group systems, and a body of data began to accumulate showing that different human populations have different proportions of blood groups. However, the first glimpse of the staggering magnitude of genetic variation came later—beginning in the 1950s and coming to full development in the 1960s—when individual differences for proteins could be systematically studied. A *protein* is a large molecule made of a linear sequence of components called *amino acids*; different proteins vary considerably in their amino-acid composition and serve very different functions. The relationship between structure

and function has been demonstrated for many proteins. The same protein may show small, strictly inherited differences between individuals. The first example was observed in the protein hemoglobin, in which the replacement of a specific amino acid by another was shown to determine a hereditary disease known as sickle-cell anemia. This first case of "molecular pathology" was detected by subjecting the protein to an electric field with a procedure called electrophoresis (Pauling et al. 1949; Ingram 1957). The amino-acid replacement involved in sickle-cell anemia causes a change in the electric charge of the hemoglobin molecule, which allows the separation of normal and sickle-cell hemoglobins. Electrophoretic analysis has since been further developed and has helped detect a great deal of variation in proteins. It is now known that the majority of the tens of thousands of different proteins found in an organism exist in more than one form, so that some individuals may have one form of the protein, whereas others may have another form.

Protein variation is still the tip of the iceberg. Only when the analysis could be carried out at the level of the hereditary material itself, deoxyribonucleic acid (DNA), could the full extent of individual genetic variation begin to emerge. This technique became widely available only in the 1980s, and although comparisons of segments of DNA in different individuals are still rare, they are becoming more common. They are, however, adequate to convince us that there is much more variation at the DNA level than was suspected when only proteins and blood groups could be analyzed.

Techniques of DNA analysis are still being developed rapidly, and the future will undoubtedly see more and more attention being paid to individual variation at the DNA level. Meanwhile, an enormous wealth of information has accumulated and keeps accumulating on individual variation studied with immunological techniques (as the blood groups are) or with electrophoresis of proteins.

If we know that there exist different genetic types of a specific protein or other strictly inherited character, we can count individuals carrying one type or the other and establish the proportions of that type in the population being examined. These proportions vary from one population to another because they change over time in each population in a relatively unpredictable manner. The change in proportions of these types over time is the *evolutionary process* itself. It proceeds slowly but incessantly over generations. The analysis of populations living today in different places gives us a cross section in time of this continuing process, which is inevitably diverse in the various parts of the inhabited Earth.

Our primary interest is in understanding this evolutionary process. The first task is to describe the existing variation, using a variety of techniques that lend themselves to this work and allow us to test the relevant evolutionary models. We restrict our interest to aboriginal populations, which we define as those already living in the area of study in A.D. 1492. After this time, geographic discoveries stimulated the expansion and migrations of the economically more advanced populations all over the planet. Some movement took place before A.D. 1492, but at a smaller scale. Ordinarily, populations that migrated after that date have mixed only partially with earlier residents and are easily recognizable on the basis of physical appearance and historical and social knowledge. They, and some populations that are highly isolated and/or have had a complex history—such as Samaritans, Jews, Gypsies, and several others—need special study and are not considered in this book. Samaritans, as well as many Jewish populations, have been the object of analysis by Batsheva Bonné-Tamir (1980; Bonné-Tamir et al. 1992). Several general articles and books have been dedicated to Jews (e.g., Mourant 1978; Carmelli and Cavalli-Sforza 1979; Karlin et al. 1979; Morton et al. 1982; Livshits et al. 1991).

One way of studying living populations is a geographic representation of the data. For this purpose we first consider each *gene* (a segment of DNA endowed with a specific function) by itself, and for each gene we separately analyze the different forms that we can recognize, the *alleles* of that gene. The proportion of a given allele in different populations is the raw material of this approach. It is well established that the proportion of an allele varies considerably from place to place, but usually there is little difference between neighboring populations so that the greatest variation is observed at large distances. It is thus possible to prepare geographic maps representing these proportions for a particular allele (also called *allele frequencies*, or simply, *gene frequencies*) when a sufficient number of populations have been tested. The standard procedure is to draw *isogenic curves* or lines connecting points of equal gene frequency.

Geographic maps of an allele are useful for understanding facts specific to that allele, including its evolutionary history and the effects of evolutionary factors like mutation and natural selection. The geographic distribution of a particular allele may give information on the place of origin of the genetic change (*mutation*) that generated it. Correlations of the distributions of gene frequencies with environmental parameters at the geographic level have been instrumental in the discovery of specific genetic adaptations. The sickle-cell anemia gene was the first example, because its geographic distribution showed a correlation with that of malaria (Haldane 1949). The hypothesis that this gene may confer resistance to malaria was later confirmed by more direct tests.

For a long time anthropologists tried to reconstruct evolutionary relationships and history on the basis of a single character or gene. A favorite for over 100 years was the cephalic index (the percentage of skull breadth to length) introduced shortly before the middle of the last century. However, with a single trait, two populations of different origin could well turn out to be more or less identical. Anthropometric traits of this kind also have another very serious drawback: there is no guarantee that the character is completely under the control of biological inheritance and the variations observed could be due to short-term response to environmental changes. This was shown by Boas (1940) at the beginning of the century, but this lesson was, and still is, usually forgotten. The main advantage offered by such traits, namely the availability of data from fossil bones, was therefore minimized because of the uncertain nature of the observed differences.

After the first blood-group system (ABO) was discovered, ABO gene frequencies soon became a favorite for classifying populations. The information thus obtained, however, is also inadequate, even if it escapes to a large extent the limitation of possible short-term changes under direct environmental effect. Every gene frequency varies over time in ways that can be considered, at least superficially, nearly random. Therefore, it is not surprising that populations having clearly different evolutionary histories may show similar gene frequencies. This drawback can be avoided if one cumulates the information from more than one gene. As one increases the number of genes considered simultaneously, the probability that a similar confusion takes place becomes more and more remote. In 1963 it was shown that even with as few as 20 alleles from five genes one could successfully attempt a reconstruction of human evolution (Cavalli-Sforza and Edwards 1964). Later experience proved that a larger number is desirable or even necessary.

Several methods allow us to combine the information from many genes into appropriate statistical indices. They are usually called *multivariate* to distinguish them from those using single traits or genes (*univariate*).

Multivariate analysis is especially useful for understanding evolutionary forces that tend to operate in a parallel fashion on all genes: migration and random genetic drift (the random fluctuation of gene frequencies in time, to be further explained later). These and other methods are applied to the existing data with the aim of extracting information of genetic and evolutionary interest.

The reconstruction of human evolution, including the fissions, the major migrations, and the understanding of the roles of mutation, drift, and natural selection is often difficult and challenging. There is clearly little hope of an experimental approach to our species, in which the evolutionary process could be repeated and interfered with in known ways. This, as well as the present almost total lack of fossil data on genetic variation (from populations living at earlier times), generates a strong desire for external evidence that can support the conclusions of genetic analysis. Fortunately, information from other sources can supply some clarification. The credibility of our conclusions can be greatly strengthened if these conclusions can be confirmed in the light of an interdisciplinary approach. Results from genetic data should be compared with relevant knowledge from other fields, in particular, paleoanthropology, prehistory, history, the geographic and ecological setting, and the cultural evidence that comes indirectly from linguistic studies. We have considered such feedback an essential part of our analysis, and we have designed our book in order to satisfy this requirement. The remainder of this chapter is dedicated to an introduction to specific concepts, data, models, and methods.

1.2. GENETIC DEFINITIONS

The purpose of this section is to provide some elementary definitions for readers who have no background in genetics. Genetic information is present in every cell of an organism in the form of chromosomes, of which there exist 23 pairs per cell in humans. Each pair is made up of one member of paternal and one of maternal origin that are morphologically indistinguishable but show subtle differences detectable at the chemical level. The main constituent of chromosomes, and the carrier of genetic information, is deoxyribonucleic acid (DNA), a long thread consisting of a linear sequence of relatively small molecules called *nucleotides*, or simply *bases*. Each nucleotide is chosen from four different ones, indicated by the symbols A,C,G,T. A short segment of DNA may look like a superficially random sequence of nucleotides, for example, TAACATGCCAT. . . .

The order of the nucleotides is actually responsible for the specific actions of DNA and is copied almost without error at reproduction of cells and individuals. Thus, the progeny contains DNA with a sequence essentially identical to that of the parent, and this is the mechanism that ensures the maintenance of the properties of living organisms.

The DNA thread is most probably continuous along a chromosome and is extremely long, since the average number of nucleotides per chromosome is over 100 million. In spite of its continuity, one can recognize shorter segments in the DNA that have specific functions and are called *genes*. The genes we recognize most easily are those that direct the structure and shape (and thus also the function) of proteins, complicated biological molecules that perform a great variety of specific activities in the cells. A chromosome may contain, on the average, many thousands of genes, each made of thousands of bases.

At cell division, DNA is replicated so that each of the two daughter cells generated by the division of one cell contains DNA that is practically identical to that of the parent cell, with very few errors in replication. Such errors are transmitted to the progeny because the new DNA is the master from which all future copies are made. Transmission error in the reproduction of DNA is called *mutation*. It may be the replacement of a nucleotide by another of the four, or the addition or deletion of nucleotides. Mutation may have trivial or serious consequences for the whole organism, depending on the alteration in the function of the specific gene to which the altered DNA belongs. Mutation in the dividing cell of an organism made up of many cells, like humans, may lead to alteration of part of the organism, but it is not transmitted to descendants unless it occurs in *germinal* cells, or *gametes*. Gametes are dedicated to the production of individuals of the next generation, and mutations occurring in them can be passed on to progeny and thus have evolutionary consequences. The male (sperm) and female (egg) gametes contain only 23 chromosomes. Reduction in number takes place by a very precise mechanism of random assortment of chromosomes so that each gamete receives only one member of each chromosome pair. The union of a sperm and an egg generates a new cell, a *zygote,* which again has 46 chromosomes, that is, 23 pairs in each of which one member is of paternal and the other of maternal origin (fig. 1.2.1).

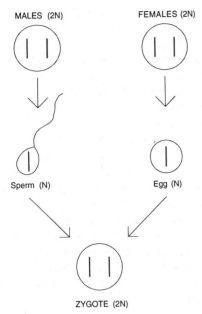

MALES (2N)   FEMALES (2N)

Sperm (N)   Egg (N)

ZYGOTE (2N)

Fig. 1.2.1 The mechanism of reduction of chromosomes by which gamete cells are formed. From the gamete cells, a fertilized egg cell (a zygote) develops. For simplicity, only one chromosome pair is represented. In humans, N, the number of chromosomes per cell, is 23.

Mutation results in a new gene that is slightly different from the old one; the two types are called *alleles* of that gene. In the first generation after mutation, there is only one individual in a given population carrying the new "mutant" allele. If the first individual carrying the mutation reaches the adult stage and has several offspring, there is a higher chance that the new allele originated by mutation will be found in later generations; however, many new mutations are lost in the first few generations. It is also possible that under the influence of evolutionary forces described later a new allele will become more and more frequent, on the average, in succeeding generations; and after many generations no copies of the old allele may be left in any of the individuals forming the population. This replacement of an old allele by a new one (the "fixation" of a new mutant) is the elementary process of evolution. It may take a great number of generations, on the average, tens or hundreds of thousands.

Even if mutation is rare, there will be at least some dozens of different mutations in any gamete, considering all the many different genes it contains. The total number of nucleotides in a gamete is very large (3 billion), and the mutation rate per generation may be of the order of 1 in 200 million nucleotides. Most mutations will happen in different genes, but over time the same gene may be hit again by another mutation. In this way, many alleles of the same gene can arise and coexist in a population.

When we detect the presence of two or more alleles of a gene in a population, we call the gene *polymorphic*. Polymorphisms produce the genetic markers used in all sorts of genetic studies, including evolution. The usual types of polymorphisms analyzed are summarized in the next section.

Because all the cells of every individual have one (and the same) gene of paternal and one of maternal origin for each type of gene, some individuals may have received different alleles of a polymorphic gene from their parents. They are called *heterozygotes*, whereas individuals that received the same alleles from both parents are called *homozygotes*. The percentage of individuals that are heterozygous at one gene is the *heterozygosity* of the gene, which is the simplest measure of the degree of polymorphism for that gene. It is equal to zero if the gene is not polymorphic.

Assuming that there are only two alleles, say M and m for a gene, an individual can be MM, mm (homozygous), or Mm (heterozygous). These are the three possible *genotypes,* and if the three can all be distinguished by direct observation or by laboratory tests, then it is easy to determine the gene frequency of M or m. In fact, it is enough to count alleles. If there are, for instance,

5 MM individuals,
6 Mm,
3 mm

on a total of 14 individuals then there are $2 \times 5 = 10$ M genes in MM homozygotes and 6 in heterozygotes for a total of $10 + 6 = 16$ M genes. There also are 6 m genes from heterozygotes and $2 \times 3 = 6$ m genes in mm homozygotes, for a total of $6 + 6 = 12$ m genes. Altogether there are $16 + 12 = 28$ genes, or twice the number of individuals counted. The *gene frequency* of M is $16/28 = 0.57$ (57%) and that of m is $12/28 = 0.43$ (43%). The sum of frequencies of all alleles of a gene is 1 (100%).

It may be impossible to count genes directly. If MM individuals are indistinguishable from Mm, but both are different from mm, one cannot separately count individuals that carry two M or only one M. This phenomenon is known as *dominance,* and the types that we can distinguish are called *phenotypes*, meaning distinguishable by appearance. Even if dominance makes it impossible to count MM and Mm individuals separately, under certain conditions one can still determine gene frequencies. The major assumption is that the choice of mates is random for that particular gene. Let us call p the gene frequency of M; that of m is $(1 - p)$ if there are only two alleles. Homozygotes for a given allele are expected to be present in a randomly mating population with frequency equal to the square of the gene frequency of that allele, that is, p^2 for MM, and $(1 - p)^2$ for mm. Heterozygotes are expected to be twice the product of the gene frequencies of the two alleles of which they are formed, in this case $2p(1-p)$. Thus, the three genotypes *MM, Mm, mm* should have frequencies p^2, $2p(1-p)$, $(1-p)^2$. This rule is named after its discoverers, Hardy-Weinberg. It is easily extended to more than two alleles. We often must have recourse to this rule, but its validity is restricted to populations and genes for which mating is random (as discussed in sec. 1.7). In this book we do not give frequencies of genotypes or phenotypes, but only gene frequencies calculated directly from them by gene counting or by application of the Hardy-Weinberg rule. For traits that are conspicuous—like hair, skin and eye color, or height—mating is not random, and hence this rule would not apply. In any case, their genetic determination is usually unclear or complex. For further reading on these topics see Cavalli-Sforza and Bodmer 1971a; Bodmer and Cavalli-Sforza 1976a; Christiansen and Feldman 1986.

Our current evolutionary thinking is mostly in terms of gene frequencies and their changes. Genes, unlike phenotypes, are essentially stable because mutation is rare; their frequencies are stable in time except for the evolutionary factors we consider later: mutation, migration, selection, and random drift. Thus, the genetic study of evolution is essentially an analysis of the role played by these factors in observed changes.

The study of gene frequencies restricts our analysis to the behavior of single points or very short segments of DNA. As we extend our knowledge, we will direct our attention more and more frequently to the structure of longer and longer segments. The longer a segment, the more polymorphisms for single points it can accommodate. The information will increase but so will its complexity. Long DNA sequences do not behave rigidly in evolution but can exchange segments. The ease with which the sequencing of DNA has recently become possible will make the direct study of long DNA segments more and more commonplace, but most of our present data allow us little more than a point-by-point approach.

1.3. Techniques for detection of polymorphic markers

Polymorphism refers to the presence of more than one allele of a gene in a population. Because we usually observe relatively small samples in terms of numbers of individuals, we tend to call a gene polymorphic if the rarer allele is not too rare—say, not less than 1%—so that there is a good chance of observing a polymorphism in a sample of 100 individuals or more because this is the order of magnitude of most sample sizes. Examined at the highest resolution (the DNA level), almost all genes are highly polymorphic, but this expectation is unfortunately based on few data. It is well known that some DNA regions are much more polymorphic than others. It is more difficult to estimate the average individual variation at the DNA level. Very roughly, 1 in every 500 nucleotides, on the average, differs in two chromosomes taken at random from a population, and therefore also in the two members of a chromosome pair of a random individual. Because genes are usually made up of many thousands of nucleotides, every gene is likely to be polymorphic if fully analyzed. We know, however, that some DNA segments—in particular, those coding for proteins—are much more highly conserved in evolution than others, and therefore we can expect them to show much less individual variation than others.

Polymorphic genes are caught in the middle of a transition from the first appearance of a mutation to its likely final event, fixation or extinction. This is a long process, and most of the time we cannot say which of two alleles is the older and which the newer (the "mutant," by definition). In the human species we can obtain some information on this point by looking at the presence of one or the other allele in the nearest species, chimpanzees or gorillas. Whether or not we know the remote history of polymorphisms, however, they provide us with pointers to the variation of specific chromosome segments. In this sense they are *genetic markers* and thus our door to the understanding and measurement of genetic variation.

The markers we analyze are conveniently classified by the technique used for detecting them and the tissues to which they apply. Polymorphisms can be found in almost any cell or biological fluid, but blood is by far the favorite because it is most easily obtained and gives the greatest opportunities for detecting them. The list of markers analyzed on a geographic basis sufficiently wide for our purpose is given in table 1.3.1. The most important categories of genetic markers are the following.

Blood groups. Blood groups are detected in red blood cells by immunological techniques. Substances at the surface of red cells act as *antigens*; that is, they determine the production of *antibodies* when injected in other individuals of the same or a different species. Antibodies are proteins (immunoglobulins), potentially produced by every individual, but only in large amounts when the organism is stimulated with the specific antigen. Antibodies also react specifically with antigens in test tubes in ways that can be easily made visible. The first system of "blood groups" discovered were called ABO, A and B being the antigens on the surface of red cells with which

Table 1.3.1. Genetic Markers Selected for Use in This Book because Adequate Data Are Available for Many Populations

Name of Locus	Symbol	Chromosome Location	Alleles Used
ABO blood group	ABO	9q34.1-34.2	A, B, O, A1, A2
Acid phosphatase 1	ACP1	2p25	A, B, C
Adenosine deaminase	ADA	20q13.11	1
Adenylate kinase 1	AK1	9q34.1-34.2	1, 2
Alkaline phosphatase placental	ALPP	2q37	S1, F1
α-1-antitrypsin	PI	14q32.1	M, F
β lipoprotein, Ag system	AG		X
β lipoprotein, Lp system	LPA	6q26-27	Lp(a+)
Ceruloplasmin	CP	3q23-25	A
Cholinesterase 1	CHE1	3q26-qter	U
Cholinesterase 2	CHE2	2q	+
Complement component 3	C3	19p13.3-13.2	S, F
Diego blood group	DI		A
Duffy blood group	FY	1q21-25	A, B, O
Esterase D	ESD	13q14.1-14.2	1
Glucose-6-phosphate dehydrogenase	G6PD	Xq28	A-, B-, def
Glutamic-pyruvate transaminase	GPT	8q24.2-qter	1
Glycine-rich β-glycoprotein; factor B	BF	6p21.3	S, F, F1, S0.7
Glyoxalase I	GLO1	6p21.3-21.1	1
Group-specific component	GC	4q12-13	1, 1F, 1S, 2
Haptoglobin	HP	16q22.1	1, 1F, 1S, 2
Hemoglobin, α	HBA	16p13.3	
Hemoglobin, β	HBB	11p15.5	
Hemoglobin, δ	HBD	11p15.5	
Hemoglobin, γ	HBG	11p15.5	
HLA-A histocompatibility type	HLAA	6p21.3	1, 2, 3, 9, 10, 11, 28, 29, 30, 31, 32, 33
HLA-B histocompatibility type	HLAB	6p21.3	5, 7, 8, 12, 13, 14, 15, 16, 17, 18, 21, 22, 27, 35, 37, 40, 41
Immunoglobulin* GM1; GM3	IGHG1G3	14q32.33	za;g zax;g f;b0b1b3b4b5 za;b0b1b3b4b5 za;b0b1c3c5 za;b0b1c3b4b5 za;b0stb3b5 fa;b0b1b3b4b5 zx;g za;b0sb3b5
Immunoglobulin KM (Inv)	IGKC, KM	2p12	1&1,2
Kell blood group	KEL		K, k, Kpa, Jsa
Kidd blood group	JK	18q11-12	A, B, O
Lactate dehydrogenase	LDH		A & B variants
Lewis blood group	LE	19	Le, le, Le(a+)
Lutheran blood group	LU	19q12-13	A
Malate dehydrogenase	MDH1	2p23	1
MNS blood group	MNS	4q28-31	M, N, S, s, S^u, MS, Ms, NS, Ns, He
P	P1	22q11.2-qter	1
Peptidase A	PEPA	18q23	1
Peptidase B	PEPB	12q21	1
Peptidase C	PEPC	1q42 or 1q25	1
Phenylthiocarbamide tasting	PTC		T
Phosphoglucomutase 1	PGM1	1p22.1	1
Phosphoglucomutase 2	PGM2	4p14-q12	1
Phosphogluconate dehydrogenase	PGD	1p36.2-36.13	A, C
Phosphoglycerate kinase 1	PGK1	Xq13	1, 2

(continued)

Table 1.3.1 *(continued)*

Rhesus blood group	*RH*	1p36.2-34	*D, C, E, Dᵘ, Cʷ, CDE, CDe,* *CdE, Cde, cDE, cDe,* *cdE, cde, V*
Secretor	*FUT2(SE)*	19q	*Se*
Superoxide dismutase 1	*SOD1*	21q 22.1	*1*
Transferrin	*TF*	3q21	*C, D*

* The GM markers also appear in the numeric notation. The correspondence between it and the alphanumeric notation we use will be found in Steinberg and Cook (1981).

anti-A and anti-B antibodies combine, respectively. An individual may carry the A antigen, the B antigen, both, or neither. Thus, four groups of individuals are defined and can be unequivocally tallied (fig. 1.3.1). Individuals of the same ABO blood group show the same reactions to the testing reagents employed and can also exchange red cells by transfusion without adverse consequences. The detection of the ABO system and its inheritance predates World War I. Many other blood-group systems were detected after ABO, of which only a few (especially RH; Landsteiner and Wiener 1940; Levine and Stetson 1939) are important in clinical practice. RH (formerly Rh) has a large number of alleles and is probably a family of adjacent genes. In addition to ABO and RH, MN blood groups and a few others are very widely studied.

Protein electrophoresis. Proteins, the main product of genes, move in an electric field with mobility that depends on their surface electric charge, which in turn depends on their chemical structure. The main polymorphisms studied are those of proteins present in the liquid part of blood (serum or plasma) or in red blood cells. The first detection of a blood protein polymorphism was that of hemoglobin (Pauling et al. 1949), showing the molecular nature of the mutation leading to sickle-cell anemia. Abundant serum proteins (e.g., haptoglobin) were found to be polymorphic soon thereafter. Proteins active as specific catalysts in biochemical reactions (enzymes) are usually present in the blood in small concentrations. When it was learned how to uncover enzymes through very sensitive and specific staining reactions, enzymes provided the first statistical evidence that polymorphisms are much more common than earlier believed. This was discovered simultaneously in humans and in *Drosophila* in 1966 (Harris 1966; Lewontin and Hubby 1966). An example of a two-allele variation detected by electrophoresis is given in figure 1.3.2.

Human lymphocyte antigens. A precious addition to the arsenal of genetic tools has been that of human lymphocyte antigens, HLA (in vertebrates known as MHC, major histocompatibility complex). These are proteins located on the surface of white blood cells, which participate in the formation of antibodies and have practical importance for organ transplants. Work started in the

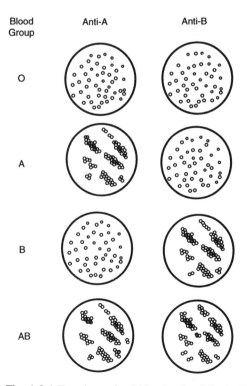

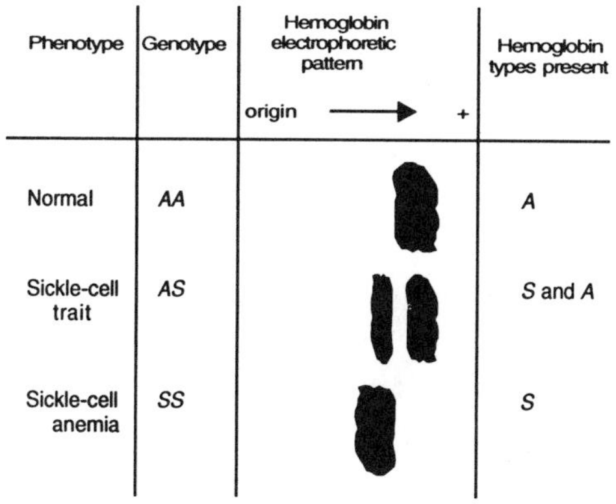

Fig. 1.3.1 Reactions of red blood cells of O, A, B, and AB individuals to anti-A and anti-B reagents.

Phenotype	Genotype	Hemoglobin electrophoretic pattern	Hemoglobin types present
		origin ⟶ +	
Normal	*AA*		*A*
Sickle-cell trait	*AS*		*S* and *A*
Sickle-cell anemia	*SS*		*S*

Fig. 1.3.2 Electrophoresis of a protein (hemoglobin) in which two alleles are detected as bands running at different speeds in the electric field. A homozygote (*AA* or *SS*) has only one band; the heterozygote (*AS*) has both.

early 1960s on this superfamily of genes has shown it to be almost as polymorphic as all the rest of non-DNA markers together and has generated the most informative single genetic system known today.

Immunoglobulins.   Variants of the immunoglobulins (*GM, KM, AM*, etc., formerly, *Gm, Km, Am*, etc.) found in plasma or serum are tested by special immunological techniques and provide a rich source of genetic variation. Immunoglobulins are the usual "antibodies." Other protein variants are also tested by similar immunological techniques.

Other polymorphisms.   There are a few other polymorphisms detected phenotypically by specific techniques (immunodiffusion, often done for lipoproteins, immunoelectrophoresis, autoradiography, etc.). A widely studied one, tested by tasting a substance called phenylthiocarbamide (PTC) has long been suspected to have irregular inheritance. A recent reanalysis (Reddy and Rao 1989) indicates penetrance of the heterozygote is incomplete and there might be a small contribution by polygenes. Gene frequencies of this widely studied marker may be slightly incorrect, therefore, but it would have been difficult to exclude it at this stage; we believe the approximation arising from its inclusion is likely to be negligible.

All the above markers reveal variation at the level of proteins or protein products. They were discovered some time ago and are the only ones for which information is abundant. In general, the longer the time since discovery, the more data at the geographic level are available (fig. 1.3.3), with exceptions tied to the practical difficulties of testing some of them.

DNA polymorphisms.   In the future one can expect a sharp rise of information on variation of polymorphisms detected by direct study of DNA. At the moment, data on DNA markers are minimal for many world populations. (They are reviewed briefly in sec. 2.4 in chap. 2.) A method used for DNA study at the population level is restriction analysis. Restriction enzymes cut DNA at specific sites, short segments defined by sequences of usually four, six, or (rarely) more nucleotides ("restriction sites"). A mutation in these sequences will prevent cutting; other mutations may generate new restriction sites. DNA fragments resulting from restriction of the DNA of an individual are electrophoresed and thus separated according to size. Those belonging to a region under study are revealed specifically by binding to DNA "probes," human DNA segments from the chromosome region of interest, cultivated in bacteria and especially prepared by attaching radioactive or other labels that allow their detection. A polymorphism appears as variation of labeled fragment sizes in different individuals (fig. 1.3.4) and is called RFLP, or restriction fragment length polymorphism. By this technique, hundreds of thousands, and perhaps millions, of polymorphisms can be detected; to

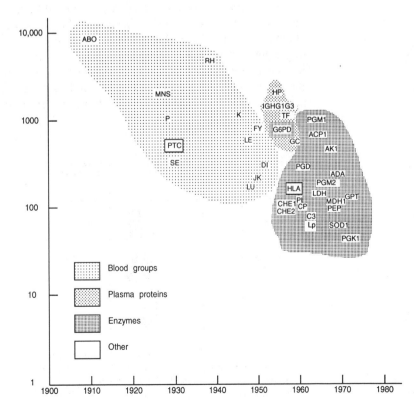

Fig. 1.3.3 Year of discovery of major genetic markers (abscissa) and number of observations in our data base.

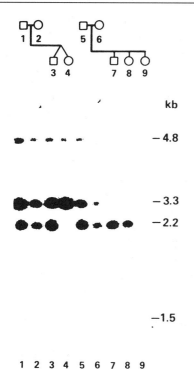

kb

— 4.8

— 3.3

— 2.2

—1.5

1 2 3 4 5 6 7 8 9

Fig. 1.3.4 DNA polymorphisms using restriction enzyme *DraI* of the *CD8* gene detected by restriction-fragment-length analysis. Alleles of 3.3 and 2.2 Kb of gene *CD8* behave as codominant Mendelian traits. DNA was digested with the restriction enzyme *DraI*, fractionated by agarose-gel electrophoresis, transferred to a nylon support by Southern's method (1975), and hybridized with a radiolabeled cDNA probe that detects the *CD8* gene.

date, more than 2,000 probes showing polymorphism are available.

DNA sequencing is the ultimate method of analysis, but existing sequences are almost always limited to one or few individuals. Powerful techniques (in particular, PCR, the polymerase chain reaction) (Erlich 1989) make the tests of known polymorphic nucleotide replacements and the sequencing of DNA segments much faster and more sensitive. PCR uses the capacity of DNA to be multiplied indefinitely by DNA polymerase, and a single DNA molecule can be amplified at will, usually with little error. One of the major novelties made possible by PCR is the analysis of very old samples in which small amounts of DNA are still left and are not too badly damaged. Some encouraging results have been described (Pääbo et al. 1989). Old samples have also been tested a number of times for non-DNA polymorphisms, in particular ABO; results are usually unsatisfactory, mostly because many individuals give uncertain reactions.

There are textbooks on blood-group polymorphisms (Race and Sanger 1975) and on protein and enzyme polymorphisms (Giblett 1969; Brock and Mayo 1978; Harris 1980). They were published a few years ago, but work on these lines has been slower in recent times. Summaries on HLA are found in a collection of symposia and workshops called *Histocompatibility Testing*. A list of all available DNA polymorphisms appears yearly in the publications of the Human Gene Mapping workshops.

1.4. THE EVOLUTION OF GENE FREQUENCIES

In this section we give a very elementary introduction to population genetics for readers who have no background in genetics. More complete introductions can be found elsewhere (Crow and Kimura 1970; Cavalli-Sforza and Bodmer 1971a; Bodmer and Cavalli-Sforza 1976a; Christiansen and Feldman 1986; Nei 1987; Hartl and Clark 1989; Weir 1989).

Gene frequencies change over time. Mutations supply the raw material by generating new alleles and even new genes, when whole regions are duplicated. Thus, *mutation* is a key ingredient of evolution. Without it, evolution would soon come to a standstill. But a specific mutation very rarely reoccurs in an individual other than the first, and thus the rate of recurrence of the same mutation has little effect on the overall rate of evolution of a particular mutation. Rather, its fate depends on the other three evolutionary forces, *migration, natural selection,* and *random genetic drift,* all of which can affect the gene frequency of an allele present in a popu-

lation. The first two drive gene frequencies in specific, and to some extent predictable, directions; of the two, natural selection has special importance in determining the future of a species. It is the only evolutionary factor that has direct *adaptive* consequences, because it is the automatic process sorting out and favoring useful mutations while eliminating deleterious ones. It thus makes the functional improvement of living organisms possible. Drift is nondirectional because it is the effect of random sampling of a gamete at each generation, and does not have any simple adaptive consequences. Like drift, mutation is random but may have different probabilities in different directions.

Natural selection is the automatic choice of "fitter" types, which can eventually make an initially very rare type, a single mutant, the most common in a population, provided it is advantageous to the individuals carrying it. The complex adaptations we observe in living organisms would have essentially zero probability of spreading to whole populations and species by mere chance. Natural selection is responsible for these extraordinary functional adaptations and the complex mechanisms responsible for them. Before Darwin, and after him for people who have not really grasped the power of natural selection, these adaptations have often appeared, understandably, as the product of design, and hence of intelligent creation. Under closer scrutiny, biological adaptations are wonderful but clumsy, like the result of "tinkering" (Jacob 1977), the accumulation of useful mechanisms not by design, but by trial and error, in a historical process, dictated by the chances of spontaneous mutations happening at particular times and places. When mutations offer acceptable solutions to the needs of organisms, they are adopted via natural selection. But they inevitably set later constraints on the further evolutionary process (see, e.g., Crick 1988).

Seen at the most elementary level, natural selection is simply the automatic enrichment of populations in genetic types that produce more descendants, and impoverishment in those that produce fewer. The rate of change under natural selection can be predicted on the basis of the numbers of descendants of each genetic type, strictly speaking, the number of children reaching sexual maturity. This number is called *Darwinian fitness* and is based on demographic parameters like survival and fertility. It is usually expressed in a relative scale, comparing two or more phenotypes or genotypes in the same population. On the basis of Darwinian fitness of two genetic types, one can predict which type, if any, will prevail in the end, and the rate of the process of change in gene frequencies, provided fitnesses do not change over time.

Natural selection acts directly only on phenotypes, but it acts on genotypes in an indirect fashion, depending on the extent to which the phenotype is determined by the genotype. Its genetic effect is thus dictated by the correspondence between the genotype and the phenotype. Phenotypes are chosen or discarded by natural selection, but the phenotypes on which selection acts are not necessarily the same that seem superficially "fitter" to us. Moreover, natural selection may differ in different environments. As an example, there are three phenotypes corresponding to the three genotypes for the sickle-cell gene: *AA*, *AS*, *SS*. The last of the three is severely sick and often dies from sickle-cell anemia. One would expect the gene *S* to be rapidly eliminated (in fact, it usually is, but not under all conditions). The situation is quite different in the presence of malaria, the great killer in tropical and subtropical environments.

By electrophoresis of hemoglobin or by DNA analysis, we can distinguish all three sickle-cell genotypes, whereas simpler laboratory tests separate only the first from the other two. In addition to ensuring the poor sur-

vival of the *SS* genotype, natural selection also acts at another level: in certain malarial environments, *AA* individuals survive less well than *AS*, whereas in nonmalarial environments no selective difference is noted between these two genotypes. This classic example shows that natural selection may well change in different environments. It also shows a peculiar evolutionary behavior in that, under environmental conditions favoring the spread of a certain type of malarial parasite (*Plasmodium falciparum*), the heterozygote is at a selective advantage over both homozygotes. The evolutionary outcome of natural selection in favor of the heterozygote is the stabilization of the frequencies of both alleles *A* and *S* at an equilibrium value usually near 90% *A* and 10% *S*; thus, neither allele becomes fixed, but the allele with higher fitness as a homozygote retains a correspondingly higher frequency. In the case of sickle cell, the anemia observed in the *SS* homozygote is devastating and is especially bad in the presence of malaria. Under such conditions, the *SS* genotype is about 10 times less viable than the *AA* genotype, and the allele frequencies at equilibrium in the presence of malaria tend to stay close to a similar ratio, 10 *A* versus 1 *S*. Thus, a heterozygous advantage determines a balanced or stable polymorphism. It is likely, however, that many other polymorphisms we observe are not stable, but their frequencies are slowly changing in time.

In the absence of heterozygous advantage, the allele favored by natural selection will eventually prevail and the other or others will be eliminated. The time taken by the selective process to reach this goal depends on the strength of selection, as measured by relative Darwinian fitnesses of the competing genotypes. These are often expressed as selective coefficients, s, which are percentage differences between the Darwinian fitness of a given genotype (or phenotype) and the Darwinian fitness of a genotype (or phenotype) taken as reference. If the heterozygote has a Darwinian fitness exactly between that of the two homozygotes, the formula for calculating the time t in generations necessary for an advantageous gene to increase from a gene frequency q_0 to $(1 - q_0)$ is especially simple:

$$t = \log[q_0/(1 - q_0)]/s,$$

where log is the natural logarithm. This formula assumes that selection remains at the same level all the time. Naturally, one can think of many other models in which selection varies in space or time, and in many other ways; the present model is the simplest.

Taking, for example, $q_0 = 0.01$ such that t is the time required to go from $q = 1\%$ to $q = 99\%$, and expressing the time in years (with 25 years per generation), the formula predicts that the spread to the population of the advantageous gene will take the following number of years, given the selection coefficient s: for $s = 0.1\%$, 115,000 years; 0.3%, 38,300 years; 1.0%, 11,500 years; 3.0%, 3800 years; 10.0%, 1150 years.

Unfortunately we know very little about the selection coefficients that prevail in human evolution, because it is difficult to measure them. It would be necessary to observe an impossibly large number of individuals, especially if s is small. A selection coefficient of 10% is very large, and values of this order of magnitude have been measured in the case of the advantage of the heterozygote for sickle-cell anemia versus the normal homozygote in the presence of malaria, as well as for other similar cases of genetic resistance to malaria, such as thalassemia. Another gene for which a selection coefficient of the advantageous type has been estimated is that for lactose tolerance, the capacity of an adult to digest the milk sugar, lactose (practically all young individuals digest lactose until 3–4 years of age). Its frequency varies considerably between populations, being very low where adults do not use fresh milk, and 50%–100% where they use large amounts. The evaluation of this selection coefficient is based on the length of time since bovines and ovines were domesticated and fresh milk thus became available for consumption.

In a previous estimation that did not take into account the interaction between cultural transmission of the custom of drinking milk as an adult and genetic transmission, the selection coefficient for lactose tolerance was estimated at minimally $1\frac{1}{2}\%$–3% (Bodmer and Cavalli-Sforza 1976a). Considering the cultural transmission of milk consumption, the selection coefficient must be at least 10% (Feldman and Cavalli-Sforza 1989).

Other genes also showing great differences between populations are those that control immunoglobulin types, perhaps because they involve differential resistance to infectious diseases, the incidence of which varies widely in different parts of the world. There is no direct estimate of selection coefficients for these genes. One can venture a guess since these genes are among those showing the greatest differences in frequencies between human populations, and these differences must have evolved over the last 50,000–100,000 years. From the data given above, it is likely that the selection intensities involved were expressed by s values very rarely greater than 1%. These are the highest selection coefficients likely to be affecting favorable genes in human populations. There is, however, very strong selection against deleterious genes determining serious diseases and early deaths.

Random genetic drift (also called *drift*) is the fluctuation of gene frequencies from one generation to another as a result of random sampling of gametes (sperm and egg cells). The transition from one generation to another is effected by gametes. The adults participating in the production of offspring, and the gametes made

by them and used for this purpose completely determine the gene frequency of the next generation (Cavalli-Sforza and Bodmer 1971a). When a population is small, the total number of gametes involved in forming the next generation is also small and subject to a large *sampling deviation* that depends on the total number of parents contributing to the next generation. Qualitatively, in a small population, the gene frequency may fluctuate wildly from one generation to the other; whereas in a large one, it will be more stable, and stability increases with population size. The effects of drift also accumulate in time, because the gene frequency of a generation is determined entirely by the gene frequency of the preceding generation, without memory of earlier gene frequencies. As a result, deviations caused by sampling increase over time, and the gene frequency of a population may eventually become 0% (extinction of the allele) or 100% (fixation). In fact, fixation or extinction are the inevitable fate of an allele if drift continues long enough, irrespective of population size; but the process takes much longer in large populations, on the average, in proportion to population size. Figure 1.4.1 shows computer simulations with three different numbers of individuals and three different initial gene frequencies. In the two cases with the smallest population sizes and in one with an intermediate one, the rarer allele was lost after a few generations.

The census size of a human population cannot be taken directly as an estimate of the N value in the simplified model used for figure 1.4.1, which, as in virtually all theoretical treatments, is that of a population reproducing synchronously. What matters in practice is the number of active parents, and because about only one-third of the individuals in a real population are of parenting age, the "effective population size" N_e, corresponding to that of a synchronously dividing population is approximately one-third of the census size. More accurate estimates may be made but are usually not necessary.

It is interesting to compare the consequences and to consider the interactions of drift with the other evolutionary factors. The effectiveness of drift depends on demographic factors: we have seen the importance of population size in figure 1.4.1. Migration also strongly affects and in general reduces the consequences of drift, and we consider it later. It is worth stressing that drift affects all genes in a quantitatively similar way, though it affects each randomly and independently of other genes, in the sense that a small population will have high drift fluctuations for *all* genes, whereas a large one will have very little drift, but again for *all* genes. By contrast, natural selection affects each gene in a unique way; we may anticipate, however, that many of the genes we study seem totally unaffected by selection, that is, they are *selectively neutral*. All genes are subject to the effect of drift, and even a mutation favored by

N=20 A

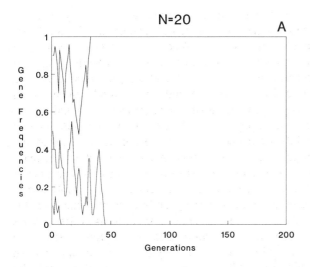

N=100 B

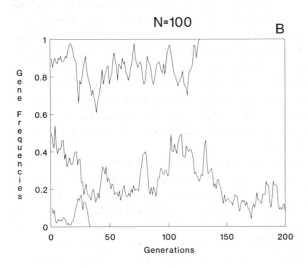

N=2500 C

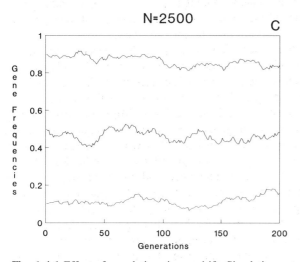

Fig. 1.4.1 Effect of population size on drift. Simulation experiments at three different initial gene frequencies (90%, 50%, 10%) with: A, $N = 20$ individuals; B, $N = 100$; C, $N = 2500$.

selection may be lost because of drift. This can happen with a fairly high probability in the first generations after its appearance by mutation while it is still rare; it is then more strongly exposed to chance effects, almost as strongly as a gene unaffected by selection. However, after an initial period, as soon as the favorable alleles have reached a higher frequency, drift will have little effect on the selective process of genes that are at an advantage.

A mutation that has just occurred is exposed to a fate that depends largely on its effect on the carrier. If this is unfavorable, it will be soon eliminated, together with the carrier or carriers; if favorable, it will increase and may ultimately become fixed, although at the beginning there may be some uncertainty on account of drift, as we have seen. If the mutation is selectively neutral, only drift will matter, and its final fate is either extinction or fixation. Extinction is much more likely, and there is a great probability that a new mutation will disappear in a few generations. But there is a small probability that it will survive and may eventually be fixed under drift alone, fixation taking, on the average, a long time that increases with increasing population size. One might superficially deduce that the pressure of new mutations will have very little effect, per se, on the process of evolutionary change. However, this conclusion, which was popular for some time, has been shown to be wrong. A population is made up of many individuals, all of which are exposed to mutation, and this counterbalances the rarity of mutations. In fact, it has been proved theoretically that the rate of neutral evolution—that is, under mutation and drift alone, without selection—equals the mutation rate (Kimura 1968, 1983).

The relative number of amino-acid differences of the same protein in two different species increases with the time of evolutionary separation between the two species (suggested by geological data). From such numbers one can calculate the rate of molecular evolution. For a given protein it is approximately constant, and there are reasonable explanations for the rate differences observed between proteins. Similar considerations can be extended to the evolutionary rates of DNA. Kimura noted that evolutionary rates thus calculated are comparable in order of magnitude to mutation rates, and this was one of the arguments for proposing that changes observed in molecular evolution are mostly or almost exclusively due to selectively neutral mutations. It is clear that disadvantageous mutations are not uncommon and are fairly rapidly eliminated, but they can be disregarded for this purpose at least at a first approximation; the real question is the relative importance of advantageous mutations versus neutral ones, because these are the only two types of mutations that are fixed in evolution. There must be favorable mutations, or adaptation would not occur, but the analysis suggests that they are indeed rare with respect to essentially neutral ones.

Kimura's theory that molecular evolution is mostly due to neutral mutations was initially met with strong skepticism by many, but evidence in its favor has accumulated (Kimura 1983). In DNA and protein regions of vital importance for function, one finds perfect—or almost perfect—conservation; that is, variation does not occur between individuals and occurs only rarely between species. This indicates strong selective control against changes that would be deleterious; it also shows that evolutionary improvement in this region is rare or absent. However, variation is quite common in chromosome regions that are not of vital importance. A clear example is the variation observed, for instance, in pseudogenes, which are derived from duplications of active genes but are completely inactive. The gene function is maintained by the active gene, but the inactive copy is not directly exposed to the action of natural selection. If we compare the variation of a pseudogene with that of the corresponding active gene, we find a great difference: unlike the active gene, which is under selective control, the pseudogene can freely accumulate all the variation that can be produced and fixed in evolution when there is no control by natural selection. Thus pseudogenes are under the influence of mutation and drift alone and can be observed to be about 5 times more variable between species than a functioning gene (Li and Graur 1991). Clearly, the active gene is evolutionarily more stable because natural selection weeds out all unfavorable mutations.

All considerations made thus far apply to *closed populations* that do not receive migrants. The species is, by definition, a population closed to migration from other species, but sections of a species can undergo cross-migration. In fact, practically all populations exchange individuals with other populations of the same species and such "migration" usually shows strong dependence on distance: the shorter the geographic distance, the higher the migration rate. Physical obstacles like mountains and rivers can further reduce migrational flow, and routes and means of transportation increase it. Because intercontinental travel began around A.D. 1500, we limited our study to *aboriginal* populations, those that were in place before that date; intercontinental transportation has subverted the earlier patterns of migrational flow.

It is useful to consider two different types of migration: (1) the relocation of individuals or small clusters, like families, leaving one group of individuals—a village, a town, a city—and moving to another, and (2) the relocation of groups, usually larger, moving to new, often uninhabited, territory (Cavalli-Sforza 1973). The first is very frequent and takes place mostly at short distances, generating a continuous internal mixing of populations. The second is rare, but can sometimes cover large distances. The second type was responsible for the occupation of large regions and whole continents. We call the first type *individual migration* and the second, *mass migration* or *colonization*, when the distinction is necessary.

The classical results of drift referred to above are given for closed *isolated* populations that receive no immigrants. The migratory exchange of individuals between these populations tends to buffer the effects of drift, the more strongly, the greater the fraction of immigrants received by a population per generation, m. Knowing effective population size N_e and m, assumed constant per generation, one can evaluate the reduction of drift caused by migration. The quantity $N_e m$ is an index of the degree of relative genetic isolation of a population made up of N_e individuals and receiving a proportion m of immigrants. It allows one to take into account the joint effects of drift and migration. The larger $N_e m$, the smaller the fluctuations of gene frequencies expected over time or over space, that is, between neighboring populations examined at the same time. Simple as it seems, this quantity is not always easy to estimate, especially for larger populations with a complex structure.

In the colonization of a new territory the "founders"—that is, the first colonizers—are sometimes few. They may therefore have an important effect on the subsequent history of the population, and the same will be true of every other demographic bottleneck that may occur at a later time. The phenomenon is especially prominent in the colonization of small islands, but is also observed in *isolates*, populations that do not mix with neighbors for social and/or geographic reasons. Extreme cases have been well documented in a few religious isolates like the Amish (Bachman 1961) and Hutterites (Hostetler and Huntington 1980) in the United States or the Samaritans in Israel (Bonné-Tamir, in prep.). Some investigators prefer to distinguish between drift and "founders' effect," as if they were two different phenomena (Bellwood 1979), but founders' effect is clearly only an episode of drift. Given the slow rate of human reproduction, an initially small population will remain relatively small for a certain number of generations thereafter, increasing the total drift effect over and above that resulting from the small number of founders. Many rare alleles are usually eliminated in these processes because they were either absent among the founders or were lost in later generations. If founders were few or there were strong bottlenecks, rare alleles that happen to be present at the beginning and survive until later will eventually be found at higher frequencies in comparison with other populations. It is especially easy under these conditions to find that one or more genetic diseases have become very frequent in the island or "isolate," whereas others have completely disappeared.

The colonization of large new territories, including those of uninhabited or scarcely inhabited continents or large regions, are of special importance for the history of human evolution. When conditions are favorable to the growth of the migrants and exchanges with the parental population are limited or absent, these colonizations are major examples of *fissions*. In this process a sample from the parental population generates, by a sort of budding, a new population that expands to occupy the space available to the degree of saturation compatible with the new environment and the available food technologies. When aborigines were present in sufficient numbers in the areas to which migrants were directed, there could be impingement, intermingling, or the formation of new population boundaries between earlier settlers and newcomers on account of linguistic, geographic, and ecological isolation. Often relics of more or less unmixed aborigines may remain for as long a time as isolates in refugia.

Clearly, mass migration offered many chances for the formation of new groups that became separated from the original population and had the opportunity to begin diverging from it. Often this has brought in contact groups showing substantial genetic differences, which may have maintained their individuality but may also have exchanged immigrants in one direction or both. When migration takes place prevalently in one direction (from one group to another) it is often referred to as *gene flow*.

It is characteristic of both individual and mass migration that all genes are equally interested in the exchange. It is also true of drift that all genes are affected with equal intensity, but each in an unpredictable direction, whereas in migratory exchanges all genes are affected in parallel, predictable ways. One can evaluate the extent of gene flow into a population when the gene frequencies of the parental populations are known or can be estimated with reasonable assurance.

Technological innovations have frequently determined local population growth. Sometimes the innovations (e.g., new food technologies or easily produced weapons) may have rapidly diffused to neighbors, affecting them in a similar way, and little if any genetic variation was determined. But if such innovations were complex and depended on specific social structures, their diffusion under exclusively cultural contacts may have been slow or impossible. Once the growth of the new population oversteps the limits of local population density compatible with the new conditions, it usually determines outside migration, often to the nearest possible place. Cycles of local demographic increase followed by migration can, in the long run, cause major geographic expansions of an initially small population. Such processes generate *population expansions* comparable to those that can take place in the occupation of uninhabited, or less densely inhabited, continents or large areas. All these processes have had an important role in the construction of the present genetic picture of populations and will be discussed in more detail below (sec. 2.7). Under these different mechanisms of evolution, gene frequencies have varied from population to population, and one of our tasks is to use modern geographic and ethnic variation for the purpose of reconstructing various aspects of this history. Using terms common in anthropology, the problem is that of inferring diachronic variation on the basis of the synchronic variation. Obviously, there are limits to the extent to which this is possible in the absence of fossil data. Continuous cross-checking with independent sources of evidence is the best insurance that conclusions are correct.

1.5. CLASSICAL ATTEMPTS TO DISTINGUISH HUMAN "RACES"

The study of human "races" dates to antiquity. The existence of conspicuous differences between humans of diverse geographic origins must have been a familiar sight to the first long-distance travelers. The father of Greek history, Herodotus (fifth century B.C.), gives the name, geographic location, customs, and physical appearance of a great number of people, mostly around the Mediterranean. He is the father not only of history, but also of anthropology (Myres 1953). Sometimes the information he gives us is clearly legendary or mixed with tales and superstitions, but at other times he has been vindicated by modern archaeology. When he lists the ethnic groups contacted by the Greek traveler Aristeas at the extremes of the Central Asian steppes, one is tempted to recognize proto-Mongolian nomads in some of them.

Egyptians and Phoenicians were certainly aware of the existence of sub-Saharan African populations. The Roman empire was in contact with Africans, Indians, and indirectly, by trade, with East Asians. The Roman naturalist Pliny the Elder (first century A.D.) had a naive explanation of the physical differences between Africans and Europeans, which he thought were a direct consequence of the climate. The Roman poet Lucretius (first century B.C.) had a more subtle approach to evolution that anticipated the idea of natural selection. But for Pliny, Africans are "burnt by the heat of the heavenly body near them, and are born with a scorched appearance, with curly beard and hair"; while in the north, being far from the sun, "the races have white frosty skins, with yellow hair that hangs straight" (Rackham 1979).

Although taxonomic ideas and examples, biological or not, go back mostly to Aristotle (fourth century B.C.), serious attempts at a classification of human races had to wait for substantial geographic knowledge. This became common only in the eighteenth century when interest in the classification of animals and plants was already flourishing. A definition of the *species* of living beings was given by John Ray (1627–1705) and is basically the same that we follow, namely, a group of individuals that can interbreed.

One of the first naturalists who discussed human variation, the Frenchman George Leclerc comte de Buffon (1707–1788), was a pioneer evolutionist whose work is said to have influenced Lamarck. Buffon used a definition of species very similar to Ray's, but he probably reached it independently. He clearly stated his conviction that humans are a single species and,

> after multiplying and spreading over the whole surface of the earth, they have undergone various changes by the influence of climate, food, mode of living, epidemic diseases, and the mixture of dissimilar individuals. At first these changes were not so conspicuous, and produced only individual varieties; these varieties became afterwards specific, because they were rendered more general, more strongly marked, and more permanent by the continual action of the same causes; they are transmitted from generation to generation, as deformities or diseases pass from parents to children.

Many citations like this one and the ones below are found in Count (1950).

Lists of races, or varieties as they were called by Linnaeus (1707–1778), appear in the eighteenth century with Linnaeus and with Kant (1724–1804), who also made various hypotheses on their mechanism of origin. Kant's hypotheses are unconvincing today, but the philosopher acknowledged their futility in the lack of adequate knowledge. J. F. Blumenbach (1752–1840), considered the father of physical anthropology, exercised great influence with his doctor of medicine thesis from the University of Göttingen (Blumenbach 1775). He stated that the human species is one, with five varieties: Caucasian (he might have been the first to use this term), Mongolian, Ethiopian (including all Africans), American, and Malay (including the islands of Southeast Asia and the part of Oceania then known). At that time skin color, the most conspicuous of all traits, had the dominant role it still has in the layman's mind. He defined Caucasians as we define Caucasoid today, including Europeans, North Africans, and people from the Near East and India. He did not, however, include Lapps and Finns, whom he assigned to Mongols. He stated that he chose the name of this variety from the Mount Caucasus, on the basis of what one might call a poetical motivation, because of the widespread belief that this region harbors the most beautiful people, like the Georgians who live in the southern part of the Caucasus. He also considered this area the likely origin of modern humans and followed Buffon in regarding white as the original color of the human species.

Trying to get away from the ever-present and obviously unsatisfactory criterion of skin color, the Swedish anatomist Anders Retzius (1796–1860) showed it was possible to generate a classification of races using craniometric criteria. Retzius invented the *cephalic index*; the ratio of the width to the length of the skull. This measurement enjoyed tremendous success in physical anthropology for a century, until the advent of multivariate analysis and genetic markers after World War II. Its popularity was tied to the simplicity of measurement both in living individuals and in skulls, including fossil ones, and to the superficial impression of precision it conveys. Ideas of biometry were introduced at that time by the Belgian scientist, Adolphe Quetelet (1796–1874). After World War II, interest in the cephalic index essentially disappeared because its heritability is probably low and because the index is sensitive to short-term environmental effects.

In the early nineteenth century, other means of distinguishing human races were suggested, and some contributors also argued against complete interfertility, challenging the idea that there is only one species. The summary of Charles Darwin (1809–1882) in *The Descent of Man, and Selection in Relation to Sex* (Darwin 1871) is especially illuminating. He noted arguments given by others for and against full interfertility in humans (we have no doubt today that there are no limitations in humans to interfertility). In the face of contrasting evidence existing at the time, Darwin concluded nevertheless that the species is likely to be one because all the races "graduate into each other"; moreover, "the races of man are not sufficiently distinct to inhabit the same country without fusion; and the absence of fusion affords the usual and best test of specific distinctness." He also specified that, however conspicuous the differences between races, they are mostly unimportant, because for most important traits, including mental ones, there is much similarity. In spite of the external difference between American aborigines, Negroes, and Europeans he was "incessantly struck . . . with the many little traits of character showing how similar their minds are to ours." Concerning classification problems, Darwin cited 12 authors who all disagreed on the numbers of races, giving numbers that vary from 2 to 63; he cited this disagreement as further evidence that "it is hardly possible to discover clear distinctive characters" between races, because they "graduate into each other."

As to the origin of variation, Darwin believed that "the external characteristic differences between the races of man cannot be accounted for in a satisfactory manner by the direct action of the conditions of life; the differences between the races of man, as in color, hairiness, form of features, etc., are of a kind which might have been expected to come under the influence of sexual selection." It is noteworthy and unfortunate that very little research has been done in humans on the evolutionary consequences of the choice of mates.

The American anthropologist Franz Boas (1858–1942) was among the first to throw considerable doubt on the evolutionary stability of quantitative phenotypic variation like stature, limb measurements, and in general most anthropometric traits. In a classic work (Boas 1940) that compared physical characteristics of children of immigrants to the United States with those of relatives who did not migrate, he showed the magnitude of short-term environmental effects. As was almost inevitable at that time, his work was statistically weak. In any case, confidence in anthropometry remained unshaken for a long time and may still be strong in the most conservative quarters. The magnitude of short-term environmental effects is well documented, and there also exist slow environmental changes, the physiological effects of which are difficult to test, but which throw considerable doubt on genetic interpretations of phenomena like the recent secular trend for an increase in stature in Europe and other parts of the world. Nonmetric variation of bones has recently become popular, but evidence that it is determined by genotype and is insensitive to short-term environmental change is still far from adequate.

We believe that the major breakthrough in the study of human variation has been the introduction of genetic markers, which are strictly inherited and basically immune to the problem of rapid changes induced by the environment. One should not expect, of course, that in the long term they show complete stability; otherwise, there would be no evolution. The nature and the dynamics of the major forces that mold the frequencies of genetic markers are well understood: natural selection (including also sexual selection), mutation, migration, and chance. Chance is effective in two ways: (1) because mutations are rare and random, the occurrence of a specific mutation at a particular point of time and space can be considered a chance event; (2) random genetic drift is another strictly indeterministic process.

The pioneers of this approach, Hirszfeld and Hirszfeld (1919), showed (fig. 1.5.1) the differences in frequencies of A and B blood antigens in different ethnic groups, sampled from the armies of World War I. They proposed a biochemical index to differentiate populations on the basis of the two antigens. Starting in the 1930s, American immunologist W. Boyd used information on gene frequencies of ABO and the other blood groups then known (MN and P; RH became known in 1940) for reconstructing the evolutionary history of humans and the differentiation of races (Boyd 1950). Boyd and others also started looking for ABO antigens in mummies, research that met with criticism because of the possible contamination with related bacterial antigens and destruction by specific bacterial enzymes.

The theoretical contributions of R. A. Fisher (1890–1962) to the understanding of the structure of RH (see Race and Sanger 1975) and his interest in evolutionary applications stirred considerable activity on blood-group studies in Great Britain. The person who emerged as the major student of human evolution through genetic markers was Arthur Mourant, who was instrumental in im-

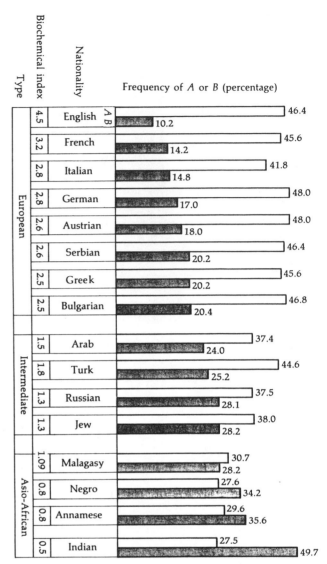

Fig. 1.5.1 This graph showing ethnic differences in *ABO* gene frequencies was the first use of genetic markers to study racial differences. The figures given are percentages of positive reactions with anti-A and anti-B reagents. The "biochemical index" is the ratio of *A* to *B*. (Taken from Hirszfeld and Hirszfeld [1919, pp. 505–537] by Bodmer and Cavalli-Sforza [1976, p. 576].)

proving the quality of research on populations, thanks to his expertise in genetic hematology, his typing of many interesting ethnic groups, and his publication of the first modern tabulation of gene-frequency data with an evolutionary interpretation (Mourant 1954). This book was followed by three more volumes, which have been a major source of information (see sec. 1.9). A tabulation of gene frequencies of immunoglobulin polymorphisms by Steinberg and Cook (1981) was a useful complement. In addition, a recent tabulation of genetic data from a selected number of human populations, with some geographic displays, has been published by Nei and Roychoudhury (1988). We closed our data collection in the summer of 1986 and were therefore unable to use new information from this book.

1.6. Scientific Failure of the Concept of Human Races

The classification into races has proved to be a futile exercise for reasons that were already clear to Darwin. Human races are still extremely unstable entities in the hands of modern taxonomists, who define from 3 to 60 or more races (Garn 1971). To some extent, this latitude depends on the personal preference of taxonomists, who may choose to be "lumpers" or "splitters." Although there is no doubt that there is only one human species, there are clearly no objective reasons for stopping at any particular level of taxonomic splitting. In fact, the analysis we carry out in chapter 2 for purposes of evolutionary study shows that the level at which we stop our classification is completely arbitrary. Explanations are statistical, geographic, and historical. Statistically, genetic variation within clusters is large compared with that between clusters (Lewontin 1972; Nei and Roychoudhury 1974). All populations or population clusters overlap when single genes are considered, and in almost all populations, all alleles are present but in different frequencies. No single gene is therefore sufficient for classifying human populations into systematic categories.

As one goes down the scale of the taxonomic hierarchy toward the lower and lower partitions, the boundaries between clusters become even less clear. The evolutionary explanation is simple. There is great genetic variation in all populations, even in small ones. This individual variation has accumulated over very long periods, because most polymorphisms observed in humans antedate the separation into continents, and perhaps even the origin of the species, less than half a million years ago. The same polymorphisms are found in most populations, but at different frequencies in each, because the geographic differentiation of humans is recent, having taken perhaps one-third or less of the time the species has been in existence. There has therefore been too little time for the accumulation of a substantial divergence. The difference between groups is therefore small when compared with that within the major groups, or even within a single population. In addition, our species and its immediate predecessor, *Homo erectus*, showed considerable migratory activity in all directions, some of which are likely to have resulted in admixtures between branches that had separated a long time before. Whatever genetic boundaries may have developed, given the strong mobility of human individuals and populations, there probably never were any sharp ones, or if there were, they were blurred by later movements. There may still exist weak genetic boundaries in some regions, but they only mean that there has been less local admixture across certain barriers. For instance, Barbujani and Sokal (1990; Sokal et al. 1988) have found a number of weak genetic boundaries in Europe linked with geographic, ecological, and linguistic differences (see chap. 5).

From a scientific point of view, the concept of race has failed to obtain any consensus; none is likely, given the gradual variation in existence. It may be objected that the racial stereotypes have a consistency that allows even the layman to classify individuals. However, the major stereotypes, all based on skin color, hair color and form, and facial traits, reflect superficial differences that are not confirmed by deeper analysis with more reliable genetic traits and whose origin dates from recent evolution mostly under the effect of climate and perhaps sexual selection. By means of painstaking multivariate analysis, we can identify "clusters" of populations and order them in a hierarchy that we believe represents the history of fissions in the expansion to the whole world of anatomically modern humans. At no level can clusters be identified with races, since every level of clustering would determine a different partition and there is no biological reason to prefer a particular one. The successive levels of clustering follow each other in a regular sequence, and there is no discontinuity that might tempt us to consider a certain level as a reasonable, though arbitrary, threshold for race distinction. Minor changes in the genes or methods used shift some populations from one cluster to the other. Only "core" populations, selected because they presumably underwent less admixture, confer greater compactness to the clusters and stability to the classification tree. Although the hope of producing a good taxonomy is a lost cause—a minor scientific loss—that of reconstructing evolutionary history retains full strength and has the advantage that hypotheses can be tested on the basis of other, independent sources of data. Greater confidence in the conclusions must come from agreement with external sources of relevant evidence rather than from internal analysis.

The word "race" is coupled in many parts of the world and strata of society with considerable prejudice, misunderstanding, and social problems. Xenophobia, political convenience, and a variety of motives totally unconnected with science are the basis of racism, the belief that some races are biologically superior to the others and that they have therefore an inherent right to dominate. Racism has existed from time immemorial but only in the nineteenth century were there attempts to justify it on the basis of scientific arguments. Among these, social Darwinism, mostly the brainchild of Herbert Spencer (1820–1903), was an unsuccessful attempt to justify unchecked social competition, class stratification, and even Anglo-Saxon imperialism. Not surprisingly, racism is often coupled with caste prejudice and has been invoked as motivation for condoning slavery, or even genocide. There is no scientific basis to the belief of genetically determined "superiority" of one population over another. None of the genes that we consider has any accepted connection with behavioral traits, the genetic determination of which is extremely difficult to study and presently based on soft evidence. The claims of a genetic basis for a general superiority of one population over another are not supported by any of our findings. Superiority is a political and socioeconomic concept, tied to events of recent political, military, and economic history and to cultural traditions of countries or groups. This superiority is rapidly transient, as history shows, whereas the average genotype does not change rapidly. But racial prejudice has an old tradition of its own and is not easy to eradicate.

1.7. IDENTIFYING POPULATION UNITS

Although, in principle, the individual can be the unit of evolutionary study, this requires testing every individual for a large number of genes, and the amount of information generated soon becomes prohibitive. This approach has been attempted thus far only in studies of mitochondrial DNA (Brown 1983; Johnson et al. 1983) (discussed in sec. 2.4). The development of simple sequencing techniques has permitted great expansion of this approach, but a considerable increase in the efficiency of present methods of statistical analysis is still required. Moreover, trees developed using this method give a different sort of information from the population trees that we examine in this book. They supply mutational histories and refer to events that are not the same as the populational events that we follow here.

A rigorous analysis should start with a definition of the "population" to be sampled; in practice, one deals with samples that have already been collected and tested so that one is limited to deciding whether a sample is acceptable. A "Mendelian population" is one formed by individuals who mate randomly (see, e.g., Cavalli-Sforza and Bodmer 1971a). The "Hardy-Weinberg" rule (HW), briefly outlined in a preceding section, predicts the distribution of observed phenotypes in a diploid organism like the human one, given random mating, gene frequencies, and dominance rules, if any. It is well known that HW is extremely robust and the great majority of observed samples satisfy it. Deviations usually arise from one of four sources:

1. The genetic model (postulated alleles and dominance relationships) is incorrect.
2. Laboratory procedures or testing reagents are not always adequate.
3. Natural selection eliminates some phenotypes preferentially (an example is the effect of malaria mortality when testing the distribution of phenotypes for sickle-cell anemia among adults).
4. The population sample is heterogeneous, being made of socioeconomic and geographic strata that do not mate randomly with each other and differ in their gene frequencies.

The test of deviation from HW is made by the statistical index of goodness of fit, χ^2. When this is greater than a predetermined amount, the deviation from HW is considered statistically significant, but the numerical value of χ^2 is proportional to the number of individuals forming the sample. If there is a true deviation, the larger the sample, the more likely the deviation is significant. Large samples are thus inevitably more likely to show deviations, and many smaller samples, which apparently satisfy HW, may be actually as unsatisfactory as the larger ones, but have a smaller chance that their deviation from HW is statistically significant. Very large samples seem less satisfactory from the point of view of fitting HW expectations because they are almost always made up of blood donors, large numbers of which have been tested for blood groups important for transfusion. These large samples are inevitably drawn from a large area and thus come from a population that is more likely to be heterogeneous. An evaluation of the heterogeneity between samples of the same geographic or ethnic origin can alert us to the existence of such problems. For this reason, we have in certain cases avoided the use of sample sizes as weights when calculating mean gene frequencies. Further information on the application of Hardy-Weinberg can be found elsewhere (Cavalli-Sforza and Bodmer 1971b; Bodmer and Cavalli-Sforza 1976b, c).

A deviation from HW can determine a systematic error in the estimation of gene frequencies in the case of dominant genes, for example, in the *ABO* and *RH* systems. This error may be relatively important when comparing populations that are genetically close but often becomes trivial when comparisons are between very different populations, as is practically always the case in our work. However, it is difficult to set a nonarbitrary, objective threshold at which to include or exclude a gene-frequency estimate, based on the relevant HW χ^2. We have therefore made only rare exclusions on this basis.

The definition of Mendelian populations based on *panmixia* (general random mating within a population)—and hence the test of HW equilibrium—are, in practice, of limited use for our purposes. If applied to whole populations rather than to population samples, the HW test would probably prove that many populations are heterogeneous when they come from towns of large size. It is difficult to guess at what level this would be perceptible, but one cannot, of course, set a size threshold for this phenomenon, since the chance of significant heterogeneity will increase continuously with town size. The redundancy internal to any population or sample that includes several members of the same family is one cause of increase of χ^2 above random expectation and is not easy to take rigorously into account. In addition, the tendency of marriages to occur at a short distance between places of residence or of birth tends to generate some geographic heterogeneity unless the area investigated is extremely small. Socioeconomic heterogeneity and other strong barriers to free interchange can also create stratifications almost within each population studied. Thus, validity of the HW test is not crucial to investigations making very broad comparisons. Most populations have some internal genetic heterogeneity, even if it is not detectable in every case, and it is of little importance for our purposes. In the geographic maps that we present, heterogeneity between populations occupying the same geographic locality is indicated and evaluated at an opportune significance level. (All the symbols used on the geographic maps are explained in table 1.14.1.) Local heterogeneity is tested under the hypothesis of binomial random sampling, and this test is very sensitive with large samples. It thus may respond to minor differences that are most frequently trivial compared with the genetic distances between populations that we are interested in studying. The explanation of these heterogeneities is usually sought in the coexistence, in the same small area, of different ethnic groups that do not undergo frequent genetic exchange.

These difficulties are sometimes not easily understood by people unfamiliar with human populations, their demography, and population genetics (Bateman et al. 1990a; Cavalli-Sforza et al. 1990). The term *deme* has been commonly employed to indicate a population unit that is panmictic and receives specified proportions of migrants from other specified populations. Except for a very few human populations, one cannot give an operationally useful definition of a deme, for reasons similar to those that make it difficult or impossible to define races. Demes are of course much smaller in size than races, but the continuity of the variation of gene frequencies and of mating distances at the geographic scale of demes is even more extreme than for races (Cavalli-Sforza 1958, 1963, 1986a, b).

It may be worth mentioning some of the reasons that make it difficult to define human demes. Candidates could be ethnographic units (e.g., tribes) or geographically defined clusters of people (villages, towns, cities). They are all usually endogamous to some degree and may come closer to the definition of a deme, but there are always many possible, embarrassing choices. Many tribes have undergone extraordinary demographic expansions (e.g., in Nigeria) and are subdivided in complex ways. In tribal as in modern society, the choice of mates is largely dictated by geographic, socioeconomic, religious, ethnic, and other constraints. The gene pool is therefore subdivided and stratified in very complex ways.

Moreover, especially in Africa, most villages are made of several ethnic groups, and even most smaller tribes are spread over many villages (for a rare demographic investigation of a farming population in Africa, the Ngbakas of the Central African Republic, see Thomas 1963). In modern society most of the farming population may be very sparse (e.g., in the Po Valley of northern Italy; also in agricultural regions of the United States), with many people living in isolated houses near the farms. The farming population, once 90%–95% of the population, has decreased enormously (to 10%–15% or less) with the modernization of the economy. At present, the population tends to conglomerate in big cities with extremely complex patterns of residential segregation. The effect of geographic distance on the probability of marriage is well known and applies, though in different degrees, to all populations but is usually an incomplete description of the patterns of population distribution. Changes in means of transportation and labor opportunities have of course altered profoundly marriage customs: Dahlberg used the term "breakdown of isolates" to indicate this phenomenon, which includes a fall in consanguineous marriages and an increase in geographic distance between birth places of mates (see also Cavalli-Sforza 1957, 1958, 1963, 1986b; Cavalli-Sforza and Bodmer 1971; Bodmer and Cavalli-Sforza 1976b; Cavalli-Sforza and Hewlett 1982; Wijsmann and Cavalli-Sforza 1984).

From a practical point of view, a definition that makes it possible, if necessary, to obtain another sample from the same population is the major requirement for statistical validity. The indication of the population sampled by an earlier research worker is usually sufficient for this purpose. Even in the case of the American Indian populations that were most thoroughly investigated, the Yanomama and the Makiritare, one cannot expect a new sample to satisfy conditions of a good statistical sample, that is, to be within binomial sampling error. The genetic heterogeneity between villages is very high (Ward and Neel 1970) and their genetic composition is unstable in time and space. The source of much of this extreme heterogeneity is that internal population rearrangements take place by partitions that tend to follow lines of kinship. In general, because of the internal genetic heterogeneity of every population, one might find some variation between different samples, but most of the time it will not be embarrassingly large as with the Makiritare or Yanomama.

The details of the geographic or ethnic origin of the published population data given in the original papers varies considerably and may pose some practical difficulties. Generally, a gene frequency refers to a population occupying a geographic area that is sometimes poorly defined. A number of published samples, fortunately small, is given with very little detail (for instance, the population of origin may be defined simply as Australian aborigines or Africans). When other, better-defined samples exist for the same genes, the poorly defined ones have been omitted. Even the indication of nationality—often the only information available—is not satisfactory, since many countries have substantial internal ethnic variation. When it was necessary to include any such poorly specified sample, it referred to the capital.

Populations defined as mixed, without giving details, were systematically excluded. When the original paper gave no information on this point, strong internal evidence of admixture with other widely different ethnic groups coming from the distribution of markers, was considered sufficient reason for discarding the population (for instance, the presence of nontrivial numbers of A blood-group individuals or of sickle-cell anemia in Central and South American Indian populations). We considered it important to be conservative in this respect, as one might easily bias the estimates if one were too extreme in applying this criterion. We retained populations that had up to 10% admixture, rarely more, but unfortunately there are few markers that allow a reliable diagnosis of admixtures from internal evidence.

Multivariate analysis posed more difficult problems. Here there is a need for finding the maximum possible number of genes for each population, and many authors tested only a small set of markers. It was therefore necessary to provide a reliable method of pooling populations. There was a smaller chance of being wrong when the same population name was used in different studies, even if some confounding is certainly introduced by pooling different samples of the same population collected by different authors. Especially with small tribes, however, there was frequently a necessity of using somewhat wider definitions of populations, allowing the pooling of tribes, in order to reach a satisfactory number of markers. Few populations have been tested systematically for a large number of markers.

Our database contains 76,676 gene frequencies. They correspond to 6633 samples with different geographic locations and to 1915 different population names. These numbers posed a gigantic problem of classification and there would be little to be gained from physical anthropology classifications if any existed.

Our main criterion in pooling populations for generating higher categories was geographic, but it was clear that, especially for populations from the developing world, the geographic criterion had to be supplemented with general anthropological information of some kind

because populations of widely different origins occasionally live only short distances from one another. We decided to resort to linguistics when other criteria failed since it is increasingly clear that there is a certain amount of parallelism between the linguistic and genetic evolution of populations. This parallelism is certainly incomplete, however, and there are many well known exceptions. On the one hand, the use of a linguistic code of classification of our populations offered the possibility of giving us a chance to test further the genetic-linguistic parallelism and the deviations from it, a problem of interest per se. On the other hand, the pooling of populations on the basis of linguistic association offered an additional criterion of grouping, which we used within groups defined by geographic and other classical ethnic criteria. It is important that linguistic classifications also usually follow geographic criteria, so that the two go hand-in-hand, but the linguistic criterion is usually finer and more often attuned to ethnic differences, especially in developing countries. In fact, tribal names are very often the same as those of languages. The linguistic code is discussed in the following section.

By this process of pooling, we reduced the initial 6633 samples, most of which had been tested for very few genes (alleles), to exactly 491 populations at the lowest level of clustering. This reduction involved both culling and pooling. These populations are listed in Appendixes 2 and 3, their gene frequencies and ethnic composition are given, and a bibliography is provided. Many of the 491 "populations" appear in the analyses of each region in chapters 3–7, which are dedicated to single continents. A further selection from the 491 populations was carried out for analyses at the level of whole continents, always excluding populations with the smaller number of genes. For the purpose of further increasing the average number of genes in the analysis of the whole world, a final reduction of the number of populations was made by culling populations that had too few genes and no affine groups with which they could be pooled, and pooling those that showed high affinity. With this second cycle of culling and pooling, the number of populations decreased to 42 (with 120 independent alleles), but still with a nonnegligible number of gaps. Although some sacrifices had to be made, the sample of 42 populations was still reasonably compatible with the desire to adequately represent the whole world. Their analysis is described in chapter 2. The gene-frequency data of the 42 populations appear in Appendix 1.

1.8. LINGUISTIC CLASSIFICATION

The most recent linguistic classification lists 4736 languages (Ruhlen 1987). This indicates a million speakers per language, on the average, but a handful of languages are spoken by hundreds of millions of people, the great majority by tens of thousands of individuals, and many by only a few hundred or less than one hundred. Languages in the last category are likely to become extinct in a few generations. The same fate has already befallen a number of others, some of which were studied before their disappearance. Except for the very few widely spoken languages, there tends to be a one-to-one correspondence of tribal names to language names. Thus, except in the case of large modern nations in which the identity of original tribes is usually—though not entirely—lost, languages offer a powerful ethnic guidebook, which is essentially complete, unlike strictly ethnographic information. Moreover, there exist phylogenetic classifications of languages, which in linguistics jargon are called "genetic," and they supply a partial taxonomic hierarchy from which we could build a linguistic numerical code. Naturally such a code should not be used automatically for a biological classification, but it was a convenient point of departure that was modified on the basis of other information.

An important consideration is that modern linguistic classifications recognize major groups or *phyla* (also called families). Leaving aside a few isolates, which, although reasonably well studied, cannot be classified in the existing taxonomy, the few unclassified languages, the recent hybrids (pidgins and creoles), and the invented languages like Esperanto, there are 17 major taxonomic groups listed by Ruhlen. A list of these 17 phyla is given in the next chapter. Thanks to Dr. Ruhlen, we had access to his classification while it was still unpublished. Books like *Classification and Index of the World's Languages*, and the *Ethnologue* (Voegelin and Voegelin 1977; Grimes 1984) were also very useful for synonymies and geographic information. Grimes was especially useful for demographic census data. Some information was also obtained from the *Encyclopaedia Britannica* (1974) as well as government publications for the USSR and China (*The USSR in Figures for 1986* and *China Handbook Editorial Committee* 1985).

The code we eventually adopted for classifying our populations is geographic–anthropological (physical)–linguistic–ethnographic, the order of the four words reflecting the average importance of each criterion in making decisions in uncertain cases. On the basis of this code, we classified all gene-frequency data into a four-tier hierarchy: the continent, major classes within the continent, and two lower hierarchical levels. The third tier included clusters of very unequal numerical importance and was designed to recognize major groups as well as special populations likely to deserve separate analysis. The fourth tier consisted of the 491 populations initially selected for multivariate analysis and consisted of tribes, countries, or regions of countries, depending on areas. Our code was therefore inspired by pragmatic considerations and does not necessarily correspond to subdivisions that eventually turned out to be entirely meaningful from a phylogenetic point of view,

but it made it possible to easily reconstruct linguistic phyla and their major subdivisions.

Tribes do not necessarily correspond to Mendelian populations as defined above; the discrepancy between the two is especially important when tribes are numerically large. In fact, in large tribes, social stratification and geographic differentiation may be very pronounced and even relatively small tribes, when carefully analyzed, have shown internal heterogeneity. This is especially marked, for instance, in the case of the Yanomama and the Makiritare (Ward and Neel 1970). Supratribal, national aggregates and their geographic subdivisions are even worse in this respect. Apart from the heterogeneities that may arise in various circumstances, however, tribes, when not too large numerically, are a reasonable approximation to a population unit for the purpose of genetic analysis; in any case, there is usually no better choice. Because the ideal of a Mendelian population is difficult to attain, we consider the smallest subdivision available as a genetic pool of individuals who mate randomly or nearly so for most practical purposes of genetic analysis, recognizing that this unit may be larger than a Mendelian population but still offers the best practical compromise.

It is reassuring to note that the patterns of linguistic variation in space parallel those of genetic and/or geographic variation, as shown in a number of detailed studies on small regions (Sardinia, Piazza et al., in press; Micronesia, Cavalli-Sforza and Wang 1986), as well as on a wider scale (North America, Spuhler 1979; Europe, Sokal et al. 1988; Barbujani and Sokal 1990; Central America, Barrantes et al. 1990). There are in fact good a priori reasons why cultural and genetic pools have close similarities: both genetic and cultural contacts take place by the same routes; they respond to the same geographic and ecological barriers; and they also can influence each other, in the sense of mutual reinforcement. For example, take a tribe as an imperfect example of a cultural pool; a tribe is frequently endogamous, a property that fits to some extent the idea of a genetic pool. At a more general level, the constitution of a genetic pool is determined by geographic factors, socioeconomic distance, and a variety of cultural factors (religious, linguistic, etc.), all of which also operate on cultural pools and affect them in a parallel way. Although investigations of the joint effects of all these variables would be very interesting, there do not seem to be any. It seems likely that two individuals have a higher probability of marrying if their distance in any of these scales is shorter.

Important correlations are thus created between genetic pools on one side and sociocultural pools on the other. There are limitations, however, to the parallelism of linguistic and genetic evolution. Languages evolve much faster than genes; two languages may become mutually unintelligible in a thousand years or less because of progressive differentiation. Formally, this is similar to the origin of two different species in biology. Speciation involves the loss of interfertility, in some measure the genetic equivalent of the loss of communication, but speciation takes on the order of a million years. Moreover, a language can be replaced by an entirely different one in as little as three generations as a result of political events leading to domination by a new people. By contrast, the genetic changes accompanying the replacement of a language, usually by invasion followed by imposition of the language of the new masters, may be difficult to detect genetically because the new masters are often numerically a small fraction of the whole population they dominate. It is also possible that extensive gene replacement has occurred through prolonged contact and gene flow from neigbors, without language change. One thus expects, and finds, minor and major inconsistencies in the comparison of genes and languages.

Linguistic analysis does not cease to be useful in the analysis of large national and supranational aggregates in which a common language is spoken. The distinction, however, may have to be drawn at the level of dialects and is inevitably more subtle. The simplest method of measuring the similarity of dialects, or of languages that are not too widely separated, consists of evaluating the proportion of words that clearly have a common origin, even though they may have undergone some phonological or semantic change. Such words are called *cognates*. The percentage of words that are cognate in two languages (or dialects) is a measure of the languages' similarity and also of the linguistic affinity of the corresponding populations. Attempts at correlating the similarity (or its converse, the distance) between two languages with their time separation (glottochronology) have been only partially successful.

Dialects of a language should show, on the average, smaller reciprocal distances than languages. The transition from dialect to language is, however, continuous and there is a gray area in which designating two forms of speech dialects or languages is arbitrary.

1.9. NATURE AND SOURCES OF THE DATA

We have confined our analysis to aboriginal populations that were in their present location at the end of the fifteenth century when the great European migrations began. We have thus excluded Black Americans and all the recent colonizations of Caucasoid, Chinese, and Indian origins. We have also excluded all manifestly mixed populations; those stated to have 25% or more external admixture; all populations from Israel, for reasons already stated in section 1.2; and various isolates, including migrant groups like Gypsies. We have also usually excluded populations classified as living "abroad," for which there is only a vague indication of origin.

All data from the major compilations cited in the Appendix that satisfy the above definitions have been included in the data bank, irrespective of sample size (excluding sizes below 50 individuals, with very few exceptions for geographic areas where data were extremely rare) and of Hardy-Weinberg χ^2. As already discussed, the rationale behind this last decision is that a usually small heterogeneity is expected in the majority of the samples, but is not usually found because the great majority of samples are of small size, and hence unlikely to show heterogeneity. The error in estimating the frequencies of dominant genes because of this heterogeneity is likely to be small and unbiased compared with the differences between populations that we study. It would be difficult to set an arbitrary general limit to χ^2 above which data are ignored.

Gene frequencies from populations with the same geographic coordinates were averaged (weighting by sample size), and heterogeneity χ^2's were calculated in each case. The indications of local heterogeneity that appear in geographic maps are based on these estimates.

Our joint work began as geographic analysis of genetic European data in 1977. After publication of our first paper on what we called synthetic geographic maps of Europe (Menozzi et al. 1978a), we decided to extend the work to the rest of the world. At that time, the extensive tabulation by Mourant et al. (1976a) had appeared, but it carried only data published up to 1972. An update was therefore necessary, and it was begun by a computer search using common key words and available retrieval software. In this way, we also identified journals most frequently employed for publishing articles of interest. These journals were systematically searched by hand.

When our work was fairly advanced, we became aware that Tills, Kopeć, and Tills were updating the tabulations by Mourant et al. (1976) from 1972 onward. These authors summarized the next 9–10 years of data, and eventually published it (Tills et al. 1983) in a format very similar to that used earlier by Mourant et al. (1976a), using the same numbers for tables of the same genes. These numbers are used in our own summaries for referring to data cited in these two books. Thanks to the courtesy of Drs. Mourant and Tills, we could obtain photocopies of Tills' updated tables several months in advance of their publication. We found that much of the material we had collected was duplicated in their work. Material for *GM* by Steinberg and Cook (1981), and for hemoglobins by Livingstone (1985) was also found to be partially duplicated.

Another important compilation of data by Roychoudhury and Nei (1988) lists 362 loci, including polymorphic and monomorphic ones, for a variable number of populations (a total of about 180 when data are available). They also include 50 world geographic distributions of gene frequencies and show their numerical frequency values. Unfortunately, this work was published too late for use in forming our data bank.

Various circumstances forced us to delay the beginning of our analytical work, and it therefore became necessary to update our own files. This was done by covering the period between the summers of 1982–1986 by systematic analysis of the journals that had proved richer in relevant articles in the previous search. The first search had used articles from a total of 136 journals, of which 12 provided 66% of all articles. In the second search, we used articles from the following 12 journals: *American Journal of Human Genetics*, *American Journal of Physical Anthropology*, *Annals of Human Biology*, *Annals of Human Genetics*, *Genetics* (Russian), *Human Biology*, *Human Genetics*, *Human Heredity*, *Japanese Journal of Human Genetics*, *Journal of Human Evolution*, *Tissue Antigens*, and *Vox Sanguinis*. This second period is therefore covered less thoroughly than the period before 1982. Altogether, more than 2900 articles were indexed by us, of which only 777 survived after eliminating duplications with other tabulations in books or reviews. These form the alphabetized list of references given in the Appendix, cited by numbers in the bibliography accompanying the list of populations.

An important question—but one that is difficult to answer precisely—is, how much information is missing from our data bank (or any other). It must vary considerably from country to country. The proportion missing may be high for several European countries, since many medical journals in languages other than English are not easily available in American or English libraries. For Russian—in particular, Siberian—data, Michael Crawford was especially helpful in personally requesting Russian colleagues, whom he visited before Perestroika, to send us reprints. Data from the Iberian peninsula were very thoroughly searched in Spanish and Portuguese journals by J. Bertranpetit; the number of these missing from our files was not small. However, data from Italy were easily available to us for obvious reasons, and the bias for this country is likely to be in the other direction.

Even if no files are complete, these are probably the most detailed available at the moment. They are limited to genes represented fairly widely. We hope to make the full files available to research workers in the future, in suitable computer form, if there are enough requests. It is also possible that the data and bibliography published in this volume may satisfy most needs. They contain the gene frequencies of the populations, sample sizes, names and locations of the populations as given by the author, and the latitude and longitude of the place of residence (or birth, when indicated) of the individuals forming the samples. The genotype and phenotype frequencies are not given in the files. When the place of origin indicated in the original paper is simply the country, the data were omitted if other data on the same gene and country were available with more precise places of origin. Otherwise, the data were included and assigned the geographic coordinates of the country's capital.

At the time of writing, we hope to update the data bank in the same format, although no specific plans have yet been made. In any case, we would appreciate it if errors and omissions were made known to any one of the authors.

1.10. METHODS OF ANALYSIS

The data were analyzed according to a variety of methods, which are briefly defined here, and explained in a little more detail in the following sections where key references will also be given. Table 1.10.1 collects a number of formulas employed in the analysis.

Genetic distances (sec. 1.11). Genetic distances are used to measure the global genetic difference between two populations. There are many genetic distances; they are all highly correlated. In this book we have used almost exclusively one measurement of distance, which is explained in detail in the next section.

Trees (sec. 1.12). Trees are the most common method of phylogenetic analysis. A dichotomous tree is constructed by using a matrix of distances between all the possible pairs of *n* populations. In the intention of the authors who first suggested this approach for evolutionary purposes (Cavalli-Sforza and Edwards 1964, 1967; Edwards and Cavalli-Sforza 1964), the tree should represent the history of fissions (splits, separations) having taken place in the human species or parts of it. Possible sources of error demand internal and external controls for the historical interpretation for a sequence of fissions to be acceptable. There are many methods for reconstructing trees; some give rooted trees (the root or origin is the earliest dichotomy), some give unrooted ones. A root can be added to an unrooted tree on the basis of evidence internal or external to it; the latter is likely to be more robust. Rooted-tree methods force all branches from the origin to the populations being analyzed to be equal in length. This is a necessary consequence of the hypothesis of constant evolutionary rates underlying such methods. Unrooted trees are not bound by these constraints, but

Table 1.10.1. Summary of Major Formulas Most Frequently Employed in the Numerical Analysis

A. Homozygosity, Heterozygosity, and F_{ST}

For a given locus with L alleles, where p_{ij} is the gene frequency of allele i in population j,

$$\sum_{i=1}^{L} p_{ij} = 1 .$$

The homozygosity of population j is

$$H_j = \sum_{i=1}^{L} p_{ij}^2 ;$$

and the heterozygosity of population j is

$$h_j = 1 - H_j .$$

The average gene frequency of allele i in a cluster of s populations is

$$\bar{p}_i = \sum_{j=1}^{s} p_{ij} / s .$$

Weighting by sample size,

$$\bar{p}_i = \sum_{j=1}^{s} n_j \, p_{ij} / \Sigma n_j ,$$

where n_j is the number of individuals in a sample of population j.

The heterozygosity of a population cluster is expressed as

$$h = 1 - \sum_{i=1}^{L} (\bar{p}_i)^2 , \tag{1}$$

where subdivisions in s populations are ignored.

The average heterozygosity of s populations is:

$$h_s = \sum_{j=1}^{s} h_j / s . \tag{2}$$

F_{ST}, a measure of variation of gene frequencies of populations, is calculated from

$$F_{ST} = (h - h_s) / h . \tag{3}$$

F_{ST} can be rewritten for a single allele i as

$$F_{STi} = \text{var} \, (p_i) / [\bar{p}_i \, (1 - \bar{p}_i)] , \tag{4}$$

where $\text{var} \, (p_i) = \sum_{j=1}^{s} (p_{ij} - \bar{p}_i)^2 / (s - 1) .$

Cumulatively, over all alleles at a locus,

$$F_{ST} = \frac{\sum_{i=1}^{L} \bar{p}_i \, (1 - \bar{p}_i) \cdot F_{STi}}{\sum_{i=1}^{L} \bar{p}_i \, (1 - \bar{p}_i)} .$$

Under drift alone, in a population of N individuals, F_{ST} increases with time as

$$F_{ST} = 1 - e^{-t/2N} . \tag{5}$$

Hence, $-\log (1 - F_{ST}) = t/2N .$ \tag{6}

If t is small with respect to $2N$,

$$F_{ST} \cong t/2N .$$

B. F_{ST} Unbiased Genetic Distance (Reynolds et al. 1983)

F_{ST} for s populations (for the distance between two populations, $s = 2$) is

$$b = 2 \sum_{j=1}^{s} n_j \, h_j / s \, (2\bar{n} - 1) , \tag{7}$$

(continued)

Table 1.10.1 *(continued)*

where n_j = sample size of population j, and

$$\bar{n} = \sum_{j=1}^{s} n_j / s .$$

$$a + b = \frac{\sum_{j=1}^{s} n_j \sum_{i=1}^{L} (p_{ij} - \bar{p}_i)^2}{n^* (s-1)} + \frac{(2n^* - 1)\, b}{2n^*} , \tag{8}$$

where

$$n^* = \frac{\bar{n}\, s}{s-1} - \frac{\sum_{j=1}^{s} n_j^2}{\bar{n}\, s\, (s-1)} .$$

$a + b$ is an unbiased estimate of total heterozygosity and b that of the within-population heterozygosity, in an analogue of a standard analysis of variance; the difference between the two gives a, the between-population heterozygosity, and the estimate of F_{ST} at the given locus is given by:

$$\hat{\Theta} = \frac{a}{a+b} . \tag{9}$$

Averaging over loci, it seems preferable to weight for the denominators by summing the numerators and the denominators separately.

The estimate of distance is given by

$$D = -\log_e (1 - \hat{\Theta}) . \tag{10}$$

Note some differences between formulas for calculating values of F_{ST} between two populations. If p in equation (4) is calculated from the gene frequencies of two populations, then

$$F_{ST1} = \frac{(p_1 - p_2)^2}{2\,\bar{p}\,(1 - \bar{p})} ,$$

which tends to be large and varies between 0 and 2;

whereas from equation (9), for very large samples,

$$F_{ST2} = \frac{(p_1 - p_2)^2}{2\,(\bar{p} - p_1 p_2)} ,$$

which always lies between 0 and 1.

C. Nei's Unbiased Genetic Distance (Nei 1978, 1987) Using the Ratio of "Kinships"

A "kinship" between two populations x and y is calculated by

$$J_{xy} = \sum_{i=1}^{L} \sum_{j \neq i}^{L} p_{xi}\, p_{yj} . \tag{11}$$

The standardized quantity,

$$I = \frac{J_{xy}}{(J_{xx}\, J_{yy})^{1/2}} , \tag{12}$$

is calculated from $J_{xx} = 1 - h_x$, $J_{yy} = 1 - h_y$,

where h_x and h_y are the heterozygosities of the two populations.

The distance can be calculated as

$$D_N = -\log_e I . \tag{13}$$

Unbiased estimates can be calculated from

$$I = \frac{1 - a - b}{1 - b} \tag{14}$$

for one locus (Reynolds et al. 1983). With many loci, numerator and denominator are independently summed over all loci before calculating I.

their meaning in evolutionary terms is less clear. Different tree methods may produce different results. Some variation in results is also generated by the use of different distances. Unless the number of characters (gene frequencies in our case) is high, the statistical error in tree building can be enormous. One way to evaluate its impact is to use the bootstrap resampling method (Efron 1982), and to compare trees obtained after resampling the set of genes employed in the analysis a sufficient number of times. We found the bootstrap to be a useful method of tree validation.

A major weakness of trees is that they do not easily take into account the existence of cross-connections between branches, because of admixture between populations. Methods to recognize cross-connections and to introduce them into evolutionary models are in their infancy.

Principal-component maps (sec. 1.13). Principal-component (PC) maps are almost as popular as trees for phylogenetic analysis. They permit us to collect in a few simple graphs a large fraction of the information contained in all the genes tested. Usually the data of over 100 gene frequencies can be summarized for a number of populations with an efficiency of 20%–40% by replacing the genetic data of one population with a single numerical value, the first PC. About 60%–80% of the information (i.e., variation) is thus lost, but a second PC can be calculated to recover as much of it as possible. This information is less than that summarized by the first PC but is still a valuable addition. One can continue with lower PCs. With only the first and second PC values, one can frequently obtain an excellent display of the relative genetic similarities of all the populations. This two-dimensional Cartesian diagram usually allows recovery of more than one third (sometimes more than a half) of the information contained in all the genes. The addition of a third PC is graphically possible but never as clear as the two-dimensional map.

A PC can be described as a weighted average of all the gene frequencies of that population, with weights calculated so as to maximize the amount of information about a population that can be condensed into a single metric value. Two- or three-dimensional PC maps thus obtained can usefully supplement trees, but they tend to generate conclusions similar to those obtained by trees and cannot be considered an entirely independent method of analysis. They are likely to be more faithful descriptors of the data than trees when there is considerable genetic exchange between close geographic neighbors. There are some variations in the procedures for obtaining PCs (sec. 1.13).

Geographic maps (sec. 1.14). Geographic maps of allele frequencies show the geographic distribution of an allele. The frequency of an allele is likely to be higher in the place of origin as well as in the regions where selective factors favor it. The distinction between these two possible explanations of a high frequency is not always straightforward. The geographic distribution of specific alleles has been historically very important in indicating which mutants confer resistance to malaria, a widespread and serious disease.

By constructing geographic maps of principal-component values, it has been possible to test hypotheses of early migrations and make similar hypotheses or hypotheses concerning selective factors. These maps have been called *synthetic maps* (see sec. 1.15) to distinguish them from the two-dimensional PC maps referred to above.

Isolation by distance (sec. 1.16). Practically all populations exchange migrants with geographic neighbors, and the accumulation of this phenomenon over generations provides a strong correlation between the genetic distance of two populations and their geographic distance. This seems to be a general rule in human populations. Of the several methods of testing it, we have used an especially simple one, extending it to much greater geographic distances than done in the past.

Admixtures (sec. 1.17). Fissions, followed by independent evolution in the two branches formed after each fission, must have had an important part in shaping the genetic history of humans. In many cases human populations that evolved, after fission, in different regions may also have exchanged migrants with new neighbors found in their new habitats. Genetic exchange probably became progressively more important with the increase of transportation, but it also must have had an impact in the early history of human evolution.

Predicting the consequences of genetic exchanges is especially easy for the simplest situation, in which a population is the result of admixture (recent or not) between two or more ancestral populations. Migratory exchange alters the distance matrix in ways that can be easily predicted in more complex situations if the migration patterns are well known, but this is rare. The fitting of usual dichotomous trees does not include admixture in the underlying evolutionary model, and there is still no completely satisfactory method of analysis for "networks" (trees complicated by interconnections between branches), except in simple cases. A complete analysis of *reticulate evolution* remains largely a task for the future, but a beginning has been made (for an example, see sec. 2.4).

1.11. Genetic distances

Genetic distances are a special example of "distances" between populations, an important chapter in statistical methodology. The many applications to anthropometric traits—beginning in the last century (Heincke 1898)—and the great number of existing formulas attest to the need for summarizing the information available from many characters in order to sharpen the analysis of population differentiation. The art of measuring distances for many quantitative traits culminated in the creation of Mahalanobis's (1936) "generalized distance," which is the natural measure for traits normally distributed and correlated when the matrix of intercorrelations between traits within a population is the same for all populations.

Gene frequencies, however, are not normally distributed. The matrix of correlations within populations is not easy to measure directly, but it can be inferred with little error using the expectations of binomial variances and covariances for alleles of the same gene, and setting equal to zero all correlations in a population between alleles of different, unlinked genes. Thus, the matrix within populations is made up almost entirely of zeros and differs from population to population, making the Mahalanobis approach less than optimal. A first attempt at generating a distance appropriate to the statistical properties of gene frequencies involved the use of the "angular transformation." This originated from a suggestion given personally to the senior author by R. A. Fisher and is equivalent, for a single locus, to a formula proposed earlier by Bhattacharrya (1946). Various slightly different formulas based on the angular transformation were proposed later by Cavalli-Sforza and Edwards (1964). The angular transformation is still used, but an alternative measurement proposed by Nei has become more popular. Nei reviews the whole field in a long chapter of his 1987 book, in which many relevant formulas and references can be found. For applications to human data, see also Jorde (1985). In table 1.10.1 we show the most frequently employed of Nei's formulas.

The simplest form of genetic distance between two populations for one biallelic gene would be the difference between their gene frequencies. If the gene frequencies are x and y, the distance would be simply $x - y$. Averaging over genes, one should of course neglect the sign of the difference, but, as in most statistical applications, the square of the difference, $(x - y)^2$, would be more appropriate. This is still unsatisfactory as a measure of distance, given that gene frequencies closer to 0% or 100% would then have a lesser weight than intermediate ones. Genetic theory shows that, especially if random genetic drift is the cause of population differentiation, the problem is largely eliminated by calculating

$$d = (x - y)^2 / [2P(1 - P)], \qquad (1.11.1)$$

where P is, strictly speaking, the unknown ancestral gene frequency common to all the populations and, in practice, the mean gene frequency calculated from all the populations being considered. A quantity F_{ST} first suggested by Wright (1951) (also called Wahlund variance; Cavalli-Sforza and Bodmer 1971a) is

$$F_{ST} = V_p / \bar{p}(1 - \bar{p}) \qquad (1.11.2)$$

where V_p is the variance between gene frequencies of a set of n populations, and $\bar{p}$ their average gene frequency (see also table 1.10.1). With two populations, equations (1.11.1) and (1.11.2) give the same result. Because of a property of the variance, the average of d values between all possible population pairs formed from n populations equals the F_{ST} between the n populations. Another useful property is that the distance between two clusters formed of n_1 and n_2 populations can be calculated by averaging the distances between the $n_1 \times n_2$ population pairs formed by all the two-by-two combinations of the members of the two clusters. A further advantage is that one obtains the same distance for a locus if one calculates it allele by allele and averages the results, or if one considers the whole locus at once with all its alleles. These convenient properties are only shared in part by Nei's distances.

When there are many genes, one averages the distances obtained with each gene, preferably weighting by the denominator of each distance. Special terms have to be added in order to correct for sampling error when sample sizes are small; the more elaborate formulas thus obtained by Weir and Cockerham (see Reynolds et al. 1983) are shown in table 1.10.1. The name given by these authors, *coancestry coefficient*, is unfortunately a misnomer, as it seems to indicate a measure of similarity, while it is really a measure of distance. We prefer to call this an F_{ST} distance, because it certainly belongs to the F_{ST} family (as the angular distance). Through extensive simulations with this and other distances, Reynolds et al. concluded that F_{ST} is more satisfactory than Nei's distance when only a few new mutations emerge in the evolutionary time period examined.

This is the situation for the period of evolution of modern humans, and we have therefore decided to use the F_{ST} distance consistently. In our 1988 paper (Cavalli-Sforza et al. 1988) we adopted a modified version of Nei's distances. Every allele was considered as an allele of a biallelic locus. Weighted averages over all alleles and loci were calculated, with weights given as in formula (14) of table 1.10.1. This was made necessary by the incompleteness of the data matrix as not all alleles and loci were represented in all populations. In every analysis of trees and PCs, we give the average number of independent alleles and its standard error. For complex loci, as, for example, RH, there was some duplication of haplotypes and allele data. In the 1988 paper we unfortunately did not mention the use of a modified Nei distance. Figure 2.3.3 will show that there is a very good correlation between F_{ST} and modified Nei distance.

Our major interest is to evaluate evolutionary time according to the basis of the genetic distance between populations. For this purpose, an F_{ST} value calculated according to equation (1.11.1) must be made proportional to evolutionary time. F_{ST} increases with time, under drift, as

$$d = F_{ST} = 1 - \exp(-t/2N) \qquad (1.11.3)$$

where exp is the exponential function, t is the time in generations, and N is the effective population size (see, e.g., Crow and Kimura 1970; Cavalli-Sforza and Bodmer 1971a). For simplicity, N is assumed to be the same for all populations.

The quantity,

$$D = -\log(1 - d) = t/2N, \qquad (1.11.4)$$

where log is the natural logarithm, is for $d << 1$ proportional to separation time t, and the proportionality

constant is $1/2N$. We refer to D as the *genetic distance*, and we use it in all phylogenetic analyses.

In general, distances calculated by different formulas are always highly correlated. A study of a very diverse group of distances (Karlin et al. 1979) showed that they tend to give similar results. Nevertheless, different distances used on the same set of data may give somewhat different results in tree analysis which may be very sensitive to small changes in the distance matrix. This is often perceived as embarrassing. But differences in results employing different methods of measuring distances tend to arise more often when the dissimilarities between populations are small (and probably trivial when compared with standard errors). Therefore, such discrepancies are often unimportant.

Although some formulas for calculating theoretical errors of genetic distances exist (Nei 1987), we prefer to estimate them by the bootstrap, a statistical method described in the next section (see also Pamilo 1990). Throughout this book we use F_{ST} distances (coancestry coefficients), and in a few cases we compare them with Nei's distances as calculated by formula (14) of table 1.10.1.

1.12. PHYLOGENETIC TREE ANALYSIS

1.12.a. DEFINITIONS

Here we examine in some depth the methods of reconstruction of trees and the problems that may arise in their interpretation. We do not specifically describe the details of the techniques, which can be found in the original publications or in available computer programs. The section may be of special interest to readers engaged in using trees for research problems. We hope other readers will get some idea of the major conclusions. To put them in a nutshell, trees are fallible friends. They are invaluable for summarizing extensive bodies of data, and there exist indirect ways of evaluating how complete and trustworthy is the summary they offer, or for identifying major loopholes. At the moment, trees form the only way of inferring evolutionary histories. They are no better than the data on which they are based, as is true of all statistical methods. The most common weakness of real data is the low number of genes available. We are aware that we have occasionally worked below the limits we consider acceptable, 50 or more independent alleles (which we also call, for simplicity, genes, i.e., the number of alleles at a locus minus one, summed over all loci); but in general our trees are based on more genes than those in most other publications. We have achieved this in part by allowing a somewhat larger fraction of missing items than is usually accepted, on the basis of considerations and safeguards explained below. The alternative is to trim down data matrices (populations × genes) by cutting populations and genes until gaps in

the matrix have disappeared or have been reduced below some very small fixed amount; missing items are then filled, for example, by averaging geographic neighbors. This procedure almost invariably generates data sets that have too few genes to be informative.

That all living beings have a common origin, and therefore the connections between the various forms in existence can be represented by trees of descent from a single ancestor, is a basic belief expressed in Darwin's *Origin of Species* (Darwin 1859). The revival of general interest in tree representations in recent years was stimulated by the development of numerical taxonomy (Sokal and Sneath 1963; Sneath and Sokal 1973), the analysis of relations between species by means of algorithms allowing the rapid and objective construction of taxonomies, which usually take the form of trees. The new discipline relied heavily on the use of computers, without which it would have been difficult to study adequate numbers of species (or more generally OTU, operational taxonomic units) and characters. Comparing the first and second editions of Sokal and Sneath's work, it is clear that they initially had no interest in evolutionary applications but aimed at searching objective classifications for "general purposes." Two major systematists of the eighteenth century, Carolus Linnaeus and Michael Adanson, created classifications of plants and animals before there was a belief in common origin and evolution of living beings. The second edition of *Numerical Taxonomy* however did contain a chapter dedicated to evolutionary analysis.

The definition of a taxonomy dedicated to general purposes does not necessarily cover evolutionary aspects, but in some cases may come close to it. In an attempt to specify more clearly purposes for developing a taxonomy, three types of definitions were considered by Edwards and Cavalli-Sforza (1964): (1) formal—the processing of information with purely descriptive aims; (2) applied or practical—based on special characters representing a specific interest; a classical example is the classification of dog breeds by their skills (see John Caius in Edwards and Cavalli-Sforza 1964); (3) evolutionary—reconstruction of phylogenetic history.

Different names have been suggested for trees reconstructed for the first and for the third aim: *dendrogram* and *cladogram*, respectively, but the use of these terms may be deceptive for it reflects more often the aims and hopes of the research worker than the result achieved. Another recently introduced term, *phenogram*, which is usually synonymous with dendrogram, is a misnomer when it refers to data on genotypes, such as those we employ. Should the trees we use be called *genograms?* *Tree* seems accurate and short, and if necessary it can be specified by the attribute "phylogenetic."

Recently zoologists and botanists (calling themselves "cladists") have been interested in developing robust methods for the reconstruction of evolution based on a careful choice of characters that are most informative for this goal (see Hennig 1966; Farris 1973; Sober 1988). These methods, however, are not easily extended to the study of human populations based on gene frequencies. Some of the relevant differences are the following: the processes studied by cladists reflect evolution taking place over time spans longer by two or more orders of magnitude; the degree of differentiation is correspondingly much greater, whereas in human populations it is often near the limit of resolution; unlike the often-distant species studied by systematists, human populations are likely to exchange individuals; perhaps most important of all, gene frequencies have statistical and evolutionary properties very different from those of the traits commonly used in cladistic analysis; moreover, cladists have decided to use exclusively "maximum parsimony" methods, to which we shall return. There seems to be less in common between the problems presented by the two types of traits used in ordinary cladistic analysis and in gene-frequency analysis than might be expected, considering that evolution (whether macro- or micro-) is always the topic being studied.

1.12.b. METHODS OF RECONSTRUCTING PHYLOGENETIC TREES FROM GENE-FREQUENCY DATA

Quantitative methods designed specifically for phylogenetic analysis were presented for the first time at the International Conference of Genetics at the Hague in 1963 (Cavalli-Sforza and Edwards 1964) for the purpose of analyzing data based on gene frequencies of human populations. Four methods were developed initially (in 1962–1963) and were described in more detail in 1967 (Cavalli-Sforza and Edwards 1967). They are summarized below as methods 1, 2, 3, and 5. Many other methods have since been added, some of which are described in Sneath and Sokal (1973). We will consider only two other methods (4 and 6), which are more widely employed for our specific purposes.

1. Method 1 is based on the analysis of variance, searching for the clusterings that maximize the ratio of variances between to the variances within the clusters resulting from each branching. Edwards and Cavalli-Sforza (1964) called it *cluster analysis*; a similar method was put forward independently (Ward 1963) and is available in the SAS statistical package (SAS Institute 1985 and later versions).

2. The second method estimates the amounts of evolution occurring in every segment of the tree, based on the hypothesis that the sum of the segments separating two populations in the tree must equal, within statistical error, the genetic distance observed between them. The best possible tree is chosen as that giving the least sum of squared differences between the estimated and observed distances of each pair of populations. This was called an *additive tree* or a *least-squares tree*, and the method was employed for the first tree published in the 1964 paper. The algorithm used was published by Cavalli-Sforza and Edwards (1967). A similar method was proposed by Fitch and Margoliash (1976), in which differences were weighted on the basis of their standard errors.

3. *Maximum likelihood* is a classical method of statistical inference developed for fitting theoretical models to observed data (the *least-squares method* is a special case of maximum likelihood, valid under certain restrictive assumptions). In the present case, the model was based on the assumption of random variation of gene frequencies at a constant rate in all segments formed after fissions. This classical method of statistical inference proved unusually difficult to apply for the tree problem (Cavalli-Sforza and Edwards 1967); it was only much later that it could be perfected (Felsenstein 1973; Thompson 1975).

4. *Average linkage* (UPGMA, Unweighted Pair-Group Method using arithmetic Averages), was suggested by Sokal and Michener (1958) for numerical taxonomy. It chooses the population pair with the smallest distance between the members and makes it the lowest split in the tree. The two populations are pooled, and the matrix of distances is thus decreased by one; the process is repeated, until only two populations are left. It was not immediately clear that this method had a number of interesting features, which were revealed essentially by simulation experiments. It is one of the most pop-

ular methods today and has been used consistently in this book; it will therefore be further discussed later in some more detail (sec. 1.12.g).

5. The *minimum-path* or *minimum-string model* is a method based on minimizing the sum of the distances between populations in the character space. This was developed and used to construct the unrooted tree projected on the world geographic map (Edwards and Cavalli-Sforza 1964). It was the first example of "minimum evolution," a principle similar but not identical to "maximum parsimony," to be discussed later. The Neel group used it extensively for studies of phylogenesis in South American Indians (see chap. 6).

6. Saitou and Nei (1987) suggested a method called *neighbor joining,* which could also be considered a method of minimum evolution. In spite of its newness, neighbor joining has become fairly popular. It gives results similar, but not identical, to those of the minimum-path method, primarily because it usually employs genetic distances other than Euclidean distance in the gene-frequency space, which is used in the minimum-path program. To our knowledge, a systematic comparison of the two methods has not been made.

All these methods give similar, but usually not identical, results when applied to the same data. However, criteria for choosing a theoretically optimal method are not easy to find. The majority of methods also present several practical limitations.

A major theoretical problem is the hypothesis or set of hypotheses underlying the method of reconstruction. This is entirely clear only for maximum likelihood, which assumes constant evolutionary rates and has been recently extended to include admixtures (see sec. 1.17). For all other methods, the underlying hypotheses are less clear, but average linkage and additive tree have been shown by simulation to give results close to those of maximum likelihood (sec. 1.12.g).

A major practical problem is the number of trees to be examined, which can be very large, as we see in the next subsection. With the exception of average linkage, all methods demand the analysis of all possible trees, or a large fraction of them, to determine which tree is best. Its simplicity and the similarity of its results to those obtained by maximum likelihood are the reasons we chose average linkage for our analysis.

1.12.c. The number of possible trees

The number of possible trees obtained by successive bifurcations grows extremely fast with the number of populations. To count all the possible trees, it is necessary to distinguish two types of trees, rooted and unrooted.

Unrooted trees give no indication of the origin, that is, the root, or initial split or branching. For instance, with three populations, there is only one unrooted tree;

with four populations (A,B,C,D), there are three possible trees:

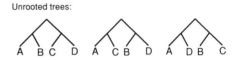

Table 1.12.1 gives the numbers of trees for higher population numbers.

Three populations result in three rooted trees:

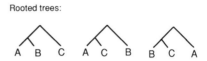

Adding a fourth population increases the number of trees to 15. For higher numbers, see table 1.12.1.

A *topology* is the form of a tree. Strictly speaking, a topology is independent of the populations attached to the branches (Cavalli-Sforza and Edwards 1967). For example, all rooted trees with three populations are of a single topology, and rooted trees of four exist in two topologies. In figure 1.12.1, the two topologies are the two trees at the extreme left and right. In practice, however, the number of topologies tends to be used synonymously with the number of trees.

The existence of a large number of possible trees is a serious obstacle. As mentioned above, almost all methods demand that all possible trees be tested in order to establish which is the best. Moreover, a few methods (in particular, maximum likelihood) demand nontrivial computer time, even for testing a single tree. With the most complicated methods, it may be very demanding or impossible to test exhaustively all the trees for as few as six or seven populations. In practice, one bypasses the problem of testing all possible trees by selecting with a simple method (e.g., the first or the fourth) an initial tree

Table 1.12.1. Enumeration of Possible Rooted and Unrooted Trees

No. of Populations	No. of Unrooted Trees	No. of Rooted Trees
n	$(2n-5)!!^{*}$	$(2n-3)!!^{*}$
3	1	3
4	3	15
5	15	105
6	105	945
8	10395	1.25×10^{5}
10	2.03×10^{6}	3.44×10^{7}
15	7.91×10^{12}	2.13×10^{14}
20	2.22×10^{20}	8.20×10^{21}

* For odd N: $N!! = 1 \times 3 \times 5 \times \ldots \times N$.

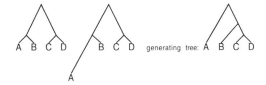

Fig. 1.12.1 *Left,* Four populations have been generated by three fissions and have undergone evolution at the same rate; hence, the segments representing the total amount of evolution from the common ancestor have the same total length. *Center,* After the second fission, more rapid evolution occurred in the branch leading to A. *Right,* Reconstruction of the tree of descent according to the hypothesis of constant evolutionary rates generates the tree at the right, in which the root is misplaced. (From Cavalli-Sforza and Piazza 1975.)

that gives a reasonable fit; then one tries to improve on the initial tree by modifying its structure and testing the goodness of fit of every new tree. The same can be done with an initial random tree or with several. In general, good trees near the best one show trivial differences in the goodness of fit. Especially when many populations are being examined, it is difficult, however, to avoid the nagging feeling that the best tree has not been identified and that there exists another, better tree that shows important differences from the one selected.

1.12.d. THE LOCATION OF THE ROOT

The location of the root is an important and difficult problem. When first confronted with it (in the paper published in 1964), Cavalli-Sforza and Edwards decided to place the root in the middle of the tree arc connecting the two most distant populations, Africans and Australians (see sec. 2.3). The root thus fell among Europeans and West Asians, a conclusion no longer considered correct by us and by many others. In principle, this method can generate errors; moreover, the small number of available genes may have contributed to the shift in root location in that first attempt (see sec. 2.3).

In general, the root can be located by using evidence internal to the tree or external to it. Internal evidence ordinarily uses the hypothesis of *constant evolutionary rates*. The rate of evolution is the amount of evolutionary change—usually measured as a genetic distance between an ancestor and a descendant—divided by the time in which it occurred.

Students of quantitative evolution have discussed the hypothesis of the constancy of evolutionary rates in comparisons between species. The accuracy of the molecular clock, the estimate of the time elapsed since the separation of two species based on genetic distance between them, depends on the validity of this hypothesis. A volume of the *Journal of Molecular Evolution* published in 1987 and edited by T. H. Jukes is dedicated to the molecular clock. One fairly widely accepted conclusion

(Goodman 1985) is that there has been a slowdown in the evolution of hominoids compared, for example, with that of rodents and equines. A reasonable explanation is that hominoids have unusually long generation times, causing lower mutation rates per unit of astronomical time. Comparisons between species, however, usually refer to long times, which differ by one or more orders of magnitude with those within the human species. They usually depend on numbers of mutational differences, which are independent of drift, whereas genetic distances based on gene-frequency differences, such as those we study, are largely determined by drift. Conclusions from species comparisons are therefore best kept separate from those based on intraspecific differences.

In chapter 2 we test the hypothesis of constant evolutionary rates in the case of human gene frequencies, with good results when comparing major populations occupying a whole continent or a large part of it. In other cases, the method may fail, essentially because of the vagaries of population sizes, as discussed later. In figure 1.12.1 we give a simple example with two trees of four populations, in which the length of each segment represents the actual amount of evolution since their origin. In the diagram at the left, the four populations have undergone the same amount of evolution. In the central tree, population A has undergone much more evolution than the others after it separated from B; hence its distance from its nearest ancestor is indicated as greater than that of any other tree segment. Whichever method of internal root reconstruction is used, the root in the second tree will be placed between A on one side, and B, C, and D on the other; this root placement is the same as that shown in the third tree in the figure, which has a different evolutionary history.

Three of the methods listed in section 1.12.b provide internal root evidence: cluster analysis, maximum likelihood, and average linkage. The last two are more clearly dependent on the validity of the hypothesis of constant evolutionary rates, and probably the first also depends on the populations available, a problem requiring further research.

The only really satisfactory methods for obtaining an unknown root or testing its validity use evidence external to the data set from which the tree has been reconstructed. Some of this evidence is genetic and some is not. Three examples are given below.

1. Fossil data for the traits used in the phylogenetic reconstruction, if they exist, can trace the position of "ancestors" in the tree; the similarities of these ancestors to living individuals or populations can help trace the root. Because such data are still absent or of unknown reliability for human gene frequencies, they are currently inadequate.
2. Knowledge from living "cousins"—for humans, the nearest anthropoid apes such as chimpanzees and gorillas—may help choose among various possible roots.

Table 1.12.2. Frequency of Four Alleles at the Locus *ERV3*, Enzyme *MspI*, in Primates and Several Human Populations, Indicating That the Origin of Allele *D* Is Late African, Whereas That of *C* Is Asian (Pacciarini et al. unpubl.)

Allele	Kb	C.A.R. Pygmies*	Zaire Pygmies	Mela- nesians	Chinese	Cauca- soids	Chimps
A	3.6	0.55	0.36	0.33	0.57	0.58	0.00
B	2.8	0.31	0.43	0.54	0.35	0.41	0.00
C	3.7	0.00	0.00	0.13	0.08	0.01	0.00
D	2.9	0.14	0.21	0.00	0.00	0.00	1.00
No. of chromosomes analyzed		42	28	24	88	176	16

* C.A.R., Central African Republic.

This method is also called the *outgroup* method. For example, for a particular gene (endogenous retrovirus ERV3) (Pacciarini et al., unpubl.), one can find external evidence for the root. Of the four alleles of the ERV3 polymorphism known in human populations, only one is found in nonhuman primates (gorilla, orangutan, chimpanzees), and it is the same as one of the three alleles found in Africa. This form is probably the ancestral form of the gene. Another allele is found only in human populations living outside Africa. This finding may support the idea that the most direct descendants of modern humans are Africans and that other human alleles found only outside Africa probably originated in Asia after the passage of modern humans from Africa to Asia (table 1.12.2). There are, however, other possible explanations of the phenomenon, and of course one gene is not enough to determine the position of the root for the whole species (see also Mountain et al. 1992).

3. The most important nongenetic source of information for locating the root comes from paleoanthropological and archaeological evidence on the origin of the species, when available and unambiguous. The relevant information for human evolution is discussed in chapter 2.

1.12.e. DNA SEQUENCES, MAXIMUM PARSIMONY AND MINIMUM EVOLUTION

The analysis of trees on the basis of DNA sequences could be carried out using as genetic distance the number of changes between two sequences and the methods developed for quantitative traits like gene frequencies. The current custom, however, is to employ methods of *maximum parsimony* suitable for qualitative traits; these methods choose the tree that minimizes the number of qualitative changes necessary for obtaining the observed tree (Farris 1972; Swofford 1989). These methods have been forcefully advocated by cladists for use in studies based on qualitative phenotypic traits and are believed to be especially useful for reconstructing phylogenies. (An excellent review and analysis of the field was published by Sober [1988]; see references therein.)

Counting the minimum number of mutations necessary for determining an observed difference between two

DNA sequences is the simplest way to determine the genetic distance between them. It does not necessarily give the number of evolutionary steps that have actually occurred, since single nucleotides may have undergone reversions or more complex cycles of mutation. One nucleotide difference does not necessarily mean one mutation, and the absence of nucleotide differences does not mean a lack of mutation. More complex mutational events like deletions, insertions, or transposition of several nucleotides are less likely to recur and hence have greater evolutionary weight. They are also rarer, and the majority of changes in sequences involve single nucleotides. The weight to give to different types of changes is a challenging problem.

Given the above, it does not necessarily follow that a method of tree reconstruction minimizing the number of mutations is the best or uses all the information contained in the sequences. The minimization of the number of mutations is intuitively attractive because we know that mutations are rare. There may be some confusion, however, between the advantage of minimizing the number of mutations and the sometimes invoked parallel of Ockham's razor (*entia non sunt praeter necessitatem multiplicanda*, meaning "entities should not be multiplied without necessity"), which was developed in the context of medieval theology. The extrapolation of Ockham's razor to the number of mutations in an evolutionary tree is hardly convincing.

The real weakness of the parsimony approach is that it does not contain a testable scientific hypothesis. Its use does not follow the standard scientific procedure of setting up a model that can be tested empirically, and rejected or modified if necessary. Specific evolutionary models like that of constant evolutionary rates, independence in evolution, etc. provide specific expectations that can be tested, although sometimes with difficulty. A more accurate analysis, therefore, shows that the intuitive attraction of maximum parsimony is not necessarily a guarantee that the approach is entirely rigorous. An attempt at presenting maximum parsimony as a model (Sober 1988) and reconciling it with maximum likelihood does not seem convincing.

Maximum parsimony methods do not provide a location for the root, which must be obtained by other criteria. Maximum likelihood provides a root and also has been developed for trees depending on discrete data like the number of mutational differences (Langley and Fitch 1973). It could be used for sequences but may be even more demanding of computer time than maximum parsimony. In principle, also, methods developed for gene frequencies could be used directly for matrices of genetic distances based on counts of differences for discrete traits like mutations. Simulations might be useful for testing their validity.

A method designed for gene frequencies, like the minimum-path method (see sec. 1.12.b), is parsimonious and is the first such method historically. But parsimony of the evolutionary path in the gene-frequency space is not the same as parsimony in the number of mutations. The number of mutations that actually occurred may be exactly equal to that counted by maximum parsimony, though the actual number of mutations generally tends to be higher, on the average, because of reversions. The probability that the evolutionary path in the gene-frequency space will be the same as the minimum path is practically zero. Evolutionary routes reconstructed by the minimum path are straight lines in the gene-frequency space, but under random drift, which has certainly played a role, the true evolutionary paths will be more similar to that of a particle under Brownian motion, and therefore extremely tortuous. Similar objections apply to the neighbor-joining method.

Some evidence shows that the minimum-path method is *not* the best for reconstructing a phylogeny on the basis of gene frequencies when the evolutionary rate is, on the average, constant. Simulations to test this point (Astolfi et al. 1981) compared the five methods listed in section 1.12.b (excluding neighbor joining, which did not exist at the time). Maximum likelihood was, not surprisingly, the best method, followed closely by least squares and average linkage, which were practically as good. Cluster analysis was definitely worse, but the worst of all was minimum path, which consistently reconstructed trees with more errors than all the other methods. This simulation does not tell us how the minimum-path method would perform if the hypotheses of constant evolutionary rate and of independence of evolution in the various branches were not satisfied. However, the deviations from such hypotheses could happen in many different ways, and this complex problem remains to be explored.

1.12.f. Statistical error in tree building

The test of goodness of fit of a tree is a delicate matter. Gene frequencies do not satisfy the requirements of normality with constant variance. The nature of a tree is so different from that of other statistical approaches that the usual concepts like standard errors or confidence intervals cannot be applied to its form. The most satisfactory method available today for testing the effects of sampling error on tree inference is the bootstrap analysis (Efron 1982). Its basis is the resampling with replacement of the original data matrix. By producing a random sample of the genes employed, one can obtain a new data matrix that can be considered an independent sampling realization of the same data. In practice, from the original data matrix (gene frequencies by populations), a random sample with replacement of genes is taken, generating a new matrix in which some of the genes appear only once, others two or more times, and quite a few (about one third) have completely disappeared. The total number of genes after each resampling of the matrix

is the same as the original, and the populations remain exactly the same in every bootstrap repetition. The number of repetitions of the bootstrapping procedure depends on available computer time; 50 or 100 are usually adequate. The tree analysis (or any other desired analysis) is repeated on each new matrix generated from the original one by a new resampling.

Felsenstein (1985) introduced the use of bootstrapping for testing the stability of tree nodes. The frequency with which a given node reappears in independent bootstrap samples is thus evaluated. The standard error of this frequency is easily obtained using the total number of bootstraps. With trees on many populations, however, one cannot expect this frequency to be very high, and it would be unrealistic to expect it to reach levels as high as those customarily employed in significance testing (e.g., 95% or 99%). Examples of applications to real trees are given in subsequent chapters.

Using the bootstrap analysis, one can calculate the standard errors of predetermined quantities like lengths of specific branches. One can then test whether these lengths differ significantly from zero, assuming a normal distribution with standard error calculated from the bootstrap. The nature of the resampling process tends to generate distributions for most bootstrapped values that are close to normal, even for quantities that are usually not normal. This makes the use of the standard probabilities from the table of the normal distribution adequate for testing significance in conjunction with the standard error derived from the bootstrap. All standard errors given in our tables, for segments or for genetic distances, were calculated by bootstrapping. Segments of small length compared with their standard error may be replaced by tri- or multifurcations. Several other significance tests that would be difficult to perform using normal procedures may be arranged in a simple way by bootstrap.

Bootstrapping also supplies useful information by noting whether a population tends to reassociate with a different cluster, when, on resampling, it leaves the cluster of populations with which it is normally associated. This reassociation indicates fairly close relationships with one cluster or another and may be indicative of earlier admixture, to be tested more directly with other procedures.

The introduction of the bootstrap has been a major addition in tree testing, since it has made a number of validity tests possible and easy by providing standard errors for the quantities usually calculated in tree reconstruction.

1.12.g. Treeness and the discovery of deviations from the simplest model of evolution

Constancy of evolutionary rates and independence of evolution of the segments of a tree, two important properties of the simplest evolutionary model, lead to a very characteristic property of the matrix of distances between all pairs of populations. In a population tree, such as in the example of seven populations shown in figure 1.12.2, one expects the matrix to have the block structure shown in the figure; that is, any two populations across the same node should have the same distance. For a simple visualization of this property, the populations in the distance matrix must be ordered so that populations belonging to the same node—that is, evolving from a common ancestral population—are adjacent to each other. In the distance matrix of figure 1.12.2, the populations are thus ordered; the blocks of distances expected to be identical are enclosed in rectangles.

The block structure of the matrix corresponds to the principle that a rooted tree is identified by $(n - 1)$ distance parameters, where n is the number of populations and $n - 1$ is the number of nodes. Empirical distances are affected by statistical error, but in a good empirical matrix the block structure is recognizable if rows and columns have been correctly ordered. Deviations from this expectation can, in principle, be tested by appropriate statistical methods, one of which was based on the maximum-likelihood method (Cavalli-Sforza and Piazza 1975), called the *treeness test*.

The treeness test we developed originally is difficult to apply, especially to large trees. In section 2.4, we give an analysis by bootstrap of a simple hypothesis in a five-population tree. This example can easily be extended to other situations. A simple, qualitative application of the treeness test is shown in various tables in chapters 3–7, where we show in rectangular boxes distances that are expected to be equal under a specific tree hypothesis. A real example is given in table 2.4.2.

There are two major sources of variation in evolutionary rates of gene frequencies that come to light by testing treeness: *true variation*, caused by different rates of random genetic drift and/or to the presence of natural selection in the various branches, and *false variation*, caused by population admixtures.

If only selectively neutral characters are used for the calculation of distances, natural selection does not distort the picture. It is likely that only some genes will be affected by strong natural selection and then only in some situations. The analysis of F_{ST} values, which we consider in chapter 2, provides some guidance. The presence of natural selection is not necessarily misleading for the purpose of reconstructing evolution. If it acts on all populations in a parallel way, it has no biasing effect (Cavalli-Sforza and Edwards 1967). If natural selection varies more or less randomly over time, space, or both (selective drift; Crow and Kimura 1970), its effects may be confounded with those of random genetic drift, described later. The forms of natural selection that can be especially misleading for our purposes are strong directional selection occurring only in some populations and environments, and selection for the heterozygote with strong differences in various environments and populations. Such situations do occur but may be relatively rare for the polymorphisms that we study most frequently. As long as they affect only a fraction of our genes, they may have little effect on distances.

Because drift is a function of population size, strong variations in population size—in particular, strong past *bottlenecks*—may cause a departure from a constant evolutionary rate. When population size is small (N_e, see sec. 1.4), one expects, on the average, greater changes in gene frequency under drift at every generation. Thus, when N_e is small, evolutionary rates caused by drift are, on the average, large. It is not easy to be aware of past bottlenecks since demographic history is usually unknown. It is easy to understand that isolates that are small today are most likely to have had strong drift effects. In principle, such populations tend to differ from other populations of the same geographic or ethnic origin, but they are not likely to show similarities with other geographically more distant and genetically unrelated populations if enough genetic markers are considered. They therefore tend to behave as outliers in a tree. This is not, however, the only mechanism by which outliers arise, and we discuss several situations of this kind.

Given that past population bottlenecks are often unknown, we usually avoid considering single small populations and often pool them with others known to be related for ethnic, anthropological (physical), or linguistic reasons that make a common origin likely. By forming population clusters and averaging their gene frequencies, one maximizes the chance that evolutionary rates are also averaged and thus differ less between clusters. However, we analyzed populations of special interest without pooling with others. The Yanomama are an example of a fairly small tribe that has drifted extensively from other South American aboriginal populations and has also been sampled completely, thus forming a nonnegligible proportion of all South Amerindian data. When they were averaged with neighboring South Amerindians they had a minor effect on the group means. When considered separately and analyzed with methods that allow us to observe individual branch lengths, like additive trees and minimum path, the effect of drift in producing long branches was clearly noticeable.

Another important cause of variation in branch length is an artifact resulting from the admixture of two populations. When two populations are geographically dis-

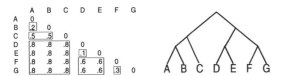

Fig. 1.12.2 A theoretical distance matrix for a tree of seven populations.

tant from one other, their admixture per generation from individual intermigration is usually small or negligible. There are, however, other important opportunities for admixture: major fusions between populations may take place in critical periods of the history of two tribes, in which the usual restrictions to exogamy between tribes are relaxed and mixed populations emerge. Admixture of two populations that were initially quite different genetically gives rise to a genetically intermediate population, and its properties stand out rather clearly in distance matrices and in trees reconstructed with methods not assuming constant evolutionary rates, because they have shorter branches. One can thus use genetic evidence internal to the tree to detect putative admixtures; sometimes historical or other types of information support the hypothesis. Admixtures and their consequences in tree reconstruction are discussed briefly in section 1.17.

1.12.h. THE RATIONALE FOR OUR CHOICES

The multiplicity of available methods of tree reconstruction was inevitably a source of uncertainty, but an important contribution to the choice came from simulation experiments (Astolfi et al. 1981), in which the efficiency of various methods for tree reconstruction was tested on a set of random trees generated on the assumptions of independent evolution, constant probability of branching, and constant evolutionary rates. In these simulations, the correct form of the tree was known and had to be inferred from analysis of the simulated data. We have already mentioned in sec. 1.12.b. that maximum likelihood gave the least error, as expected since the hypotheses on which it is based correspond exactly to those of the simulations. Fortunately, the simplest method, average linkage, was somewhat unexpectedly similar in behavior to maximum likelihood. Similar results were obtained in simulations by others (quoted by Nei 1987).

As mentioned above, average linkage also has the considerable advantage of leading directly to the best tree, and therefore—unlike nearly all other methods—does not require the testing of many other trees. Because, like maximum likelihood, it implicitly relies on the hypothesis of constant evolutionary rates, it directly produces a rooted tree. The root is valid if the hypothesis is verified.

It is somewhat ironic that trees obtained by average linkage—a method commonly used to produce dendrograms (subsection 1.12.a)—are good candidates for being considered cladograms. This, of course, is not tantamount to saying that a tree obtained by average linkage is necessarily correct or can be labeled a cladogram. One can rely on the fact that its results are close to those of the maximum-likelihood method, which is the best if evolutionary rate is constant. Average linkage is also the only usable method if the number of populations is large.

One weakness of average linkage is that ties arising when more than one population pair has the same

distance are broken by arbitrary choice. A change of distance method may cause changes in the pattern of distances between populations, which may alter the shape of the tree. When this happens, however, the discrepancy between the two trees obtained with the two distance methods is often trivial, because the differences between the two distance sets are probably small.

In any case, *one must realize that a tree with a large number of populations can hardly be without errors.* Even assuming that the evolutionary hypotheses are correct, it would take a large number of characters (genes) to make the chances of sampling error negligible (for estimates, see Astolfi et al. 1981). As an example, to maintain an average of one error per tree, 7 populations require 20 independent characters (equivalent to independent alleles), and 20 populations require 100 characters. Experience with the bootstrap is an excellent introduction to understanding the weaknesses of a particular tree and to acquiring the necessary degree of humility.

There would be no need to emphasize further the importance of using a large number of genes for reliable results, were it not for the serious problem of the incompleteness of most existing data. In order to have a complete populations-by-genes matrix, one should either collect the data, an ambitious endeavor, or sacrifice a great number of populations or genes. Filling empty cases with the general mean for the missing gene, or with data from neighboring populations, generates an arbitrary "smoothing" of the data, which has a systematic effect. It could be tolerated for only a few genes.

In principle, it is better to use exactly the same genes for every population; and in the past we have followed this rule very strictly. Given the need for many genes, it has been necessary to cut down considerably on the number of populations. The available data in the literature include a great number of genes and populations, but they have been collected by thousands of research workers with no preorganized plan, making it impossible to study all the most interesting populations with a sufficient number of genes. This is clearly a lesson for the future, now that new techniques are available for analyzing DNA and for generating cell lines from which any amount of DNA can be harvested. It will be possible to study any part of the genome on any population provided that the collection of samples from individuals, their storage, distribution, and analysis are adequately organized.

We have successfully tried another approach that permits us to use incomplete matrices of existing data to calculate distances between each pair of populations on the basis of the genes available in both members of the pair. This procedure is based on the principle of using a random sample of genes for every comparison between two populations, provided that the sample is large enough that no comparison is based on too few genes. By "genes" we mean independent alleles or total number of alleles at all loci minus the number of loci.

Matrices of populations by genes with gaps do in fact produce samples of genes that are somewhat different from population pair to population pair. This procedure allows us to increase considerably the number of genes used. We have been careful to eliminate all populations whose genes are poorly known, on the basis of the total number of genes available, because comparisons involving them would be especially weak. We have eliminated populations rather than genes, but in this way we could eliminate far fewer populations than if we had adhered to the rule of complete matrices. When possible or necessary, we have strengthened genetic knowledge of populations by increasing the number of geographic neighbors forming a "population." This is what we call the process of "culling and pooling" employed in forming the set of populations examined. Simulations have shown that, with 110–120 genes employed, a number of random gaps between 20% and 50% of the whole matrix has little effect on the conclusions.

Major difficulties with this approach might arise if the sample of genes were biased. In fact, the sample of genes available for study in these conditions is not exactly random, but the bias is in a favorable direction. A certain number of genes known for a longer time have been tested on more populations, as we have seen in section 1.3. Data on these genes tend to be always present, and comparisons for them are always available. This hard core of most frequently studied genes (mostly *ABO, MNS, RH, GM,* and others) helps to stabilize distances. We also see in chapter 2 that the variance of gene frequencies across populations is distributed as if drift were responsible for most differences, helping to avoid systematic biases. The greatest insurance against possible problems arising from the use of a random or semirandom sample of genes instead of a constant set of them is provided by the bootstrap, which calculates standard errors that take gaps into account, and generates a variable set of trees compatible with the data with given probability. As the bootstrap eliminates an important fraction of genes (more than one third) at every random resampling, and randomly increases the weight of those (a little less than one third) sampled two or more times in the resampling procedure, it also automatically tests for the effect of gaps. Allowing for gaps, one can thus include many more genes than if one eliminates all genes showing only one, or a few missing values.

The confidence that a tree will be invariant with the number of genes used obviously increases with their number. Even when the number is large enough to make sampling error negligible, there is no guarantee that the same tree would be obtained with a different method of reconstruction, or different formulas for distance, or other changes. The number of possible trees is so high that, with a fairly large number of populations, every change of data or methods, even if minimal, is likely to alter the tree somewhat. This is inevitable, and probably the best way to avoid it would be to limit dichotomies in the tree to those that are more strongly supported. We are not aware of a completely satisfactory routine that would generate such a result. Standard errors of segments generated by bootstrap helped in specific circumstances, but we found it unnecessary to constantly resort to polychotomies to which it is not easy to assign an accurate probability. It seems enough to be aware that all short segments are potentially weak and that one can evaluate their weakness by estimating their standard error.

1.12.i. Conclusions on the usefulness of trees

In essence, what can one expect from reconstructing a tree? A tree can be viewed minimally as a simplified description of a matrix of distances (or similarities). With n populations, the elements of a distance matrix are $n(n-1)/2$. The parameters that define a tree are the values of the nodes, which equal $n-1$ if the tree is rooted. Therefore, a tree permits substantial economy in describing a data set, cutting the number of parameters by the factor $n/2$. Inevitably, economy may mean loss of information; the loss may be more or less serious, depending on the nature of the data. If the distance matrix is compatible with a perfect tree (i.e., has perfect treeness), apart from sampling errors, then there is no loss of information when replacing a distance matrix with a tree.

The study of trees is the major technique for understanding the complex relationships between different populations. It offers a simple graphic aid for visualizing those relationships and a path to infer the possible evolutionary history behind them. It gives first help for interpreting a distance matrix, and when treeness is good, it replaces it completely. Otherwise, further analysis of the matrix will be important, but one should not think that the distance or the similarity matrix contain all the information. In calculating a distance or similarity matrix, the data of individual genes, and thus their variation and correlations, are lost. Evolutionary forces affecting different genes may be quite different, and the possibility of distinguishing them is thus also lost. However, the complexity of the data is usually so high that one must be prepared for some loss in order to be able to condense the information, at least at the beginning.

A tree specifies no order to the populations, other than that available in the nodes; but nodes can be rotated at will without changing the tree. In a perfect tree, rotation of the nodes should involve no loss of information. Gould (1989) used the metaphor that each tree node is like a point of suspension in a "mobile" to indicate this property. The following four trees are totally equivalent:

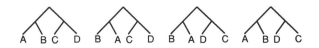

But if we know that B and D show greater similarity than the BC, AC, and AD pairs, then we may prefer the fourth tree. In so doing, we use information that the tree does not ordinarily express. We have not used this criterion except in very special circumstances because it has other disadvantages, but occasionally it might be useful.

Under the influence of the cladists, the trend has been toward the preferential use of maximum parsimony and, more recently, of minimum-evolution methods for evolutionary studies. As mentioned before, one difficulty with maximum parsimony is that it does not strictly correspond to the preferred scientific approach of testing specific evolutionary models and changing them if they do not fit the data. The maximum parsimony approach of choosing the tree that minimizes the number of changes is perhaps more appropriately described as a model-free procedure, which uses an acceptable postulate and may help to generate a reasonable tree from data. However, this tree probably does not correspond to that expected if evolutionary rates are constant. The parsimony tree is not ordinarily used for calculating expectations to be compared with observations or for testing alternative evolutionary hypotheses.

In our experience, trees are ordinarily useful summaries of potentially enormous data sets. They almost always contain inaccuracies, as a result of insufficient number of observations or faulty assumptions, but it may be difficult to locate errors, most probably because of insufficient information. It is essential to use as large a number of genes as possible. This was our reason for increasing the data base, even at the cost of using incomplete matrices. We might also be able to improve our trees considerably if we could give more attention to a second important source of error, admixtures, by extending trees into true "networks," that is, showing the most important interconnections between branches. Techniques for making this next step are in their infancy and demand considerably more work.

A tree is not the only method of analysis available and can be usefully complemented with others that are reviewed briefly in subsequent sections. In addition to many methodological papers of importance, Felsenstein (1982) has published a history of phylogenetic analysis and supplied a useful collection of programs called *Phylip*. The book on numerical taxonomy by Sneath and Sokal (1973) provides the rationale and technique for many classification methods, including some devoted more specifically to phylogenetic analysis.

1.13. ANALYSIS OF PRINCIPAL COMPONENTS (PCS) AND DERIVED METHODS

Principal components and related or derived methods—for example, principal-coordinate analysis, multidimensional analysis, factor analysis, biplot, and the analysis of correspondences—are procedures for simplifying multivariate data with minimum loss of information. Almost all sets of data formed by many populations and many gene frequencies contain some internal redundancy that is measured, for instance, by the correlations existing between genes in a pair of populations. Two identical populations would have a correlation of one; as they differentiate from one another in the course of evolution, the correlation decreases. Thus, correlation between populations can be said to measure the history of their common descent. A distance is the opposite of a correlation; it is zero for identical populations and increases with their differentiation. For certain formulas of distance, one can express the exact interdependence between distance and correlation and, in the simplest case, the relationship between them is $d = 1 - r$, where r is a correlation and d a distance measure (to be further transformed, however, if one wants to make it proportional to evolutionary time). Thus, the matrix of correlations and that of distances between population pairs are in a one-to-one relationship, but they have opposite meanings.

Principal components offer the simplest mode of analysis of a set of populations-by-gene frequency data. We

show a very simple example of application, using only five populations and five genes (table 1.13.1). In order to visualize the procedure more clearly, let us consider at the beginning only the first two of the five genes, and plot their frequencies as abscissa and ordinate of a Cartesian diagram. The five populations are represented by five points in the diagram (figure 1.13.1).

We draw a straight line as close as possible to the five points, using the criterion that the sum of the distances from the points to the line (dotted lines) must be a minimum. Incidentally, this criterion differs from the familiar one employed for calculating usual regression lines, where the segments (the sum of squares of which is minimized) are parallel to the abscissa or to the ordinate. Here the segments are orthogonal to the

Table 1.13.1. Gene-Frequency Data (%) on Five Populations and Five Genes

Gene	Population				
	Africa	Asia	Europe	America	Australia
*RH*D-*	20	15	36	2	0
*ABO*O*	69	60	65	90	76
*FY*A*	11	60	42	70	99
KM(1&1,2)*	34	19	8	35	30
*DI*A*	0	2	0	9	0

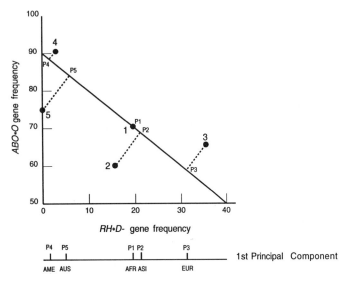

Fig. 1.13.1 A Cartesian diagram for five populations (1, Africa; 2, Asia; 3, Europe; 4, America; 5, Australia) and the first two gene frequencies in table 1.13.1, with the first principal component (PC) on the principal axis. Below the diagram, the first PC is shown as a horizontal line. The locations of the five populations are obtained by the projections of perpendiculars (normals) from each point to the principal axis.

line, which is called the (first) principal axis or component. The original values of the two gene frequencies can now be replaced with the five points P1, P2, P3, P4, and P5 on the principal axis, with an arbitrary scale (usually chosen so that the five points have zero mean and variance equal to 1). The value of each of the five points is formally a linear compound of the original gene frequencies, as below:

$$PC = a_1 \cdot x_1 + a_2 \cdot x_2 + \cdots$$

where $x_1, x_2, \ldots$ are the gene frequencies of a particular population and $a_1, a_2, \ldots$ are coefficients calculated in the process of minimization. The sum is extended to all the gene frequencies used; in figure 1.13.1 there are only two.

The resulting PC representation is shown at the bottom of the diagram. By replacing the original data with their first PC value, we have lost some information, but the process of minimization has reduced the loss and we have gained in simplicity because we have a straight line instead of a bidimensional display.

The real advantage occurs when there are many genes. A third gene added to figure 1.13.1 could be plotted on an axis orthogonal to the page, or, more clearly, in a three-dimensional model that could be photographed or drawn. Although it is mathematically easy to add more axes (as many as genes), our intuition and ability to make a spatial representation fail us after the third dimension.

One can add a second principal axis in figure 1.13.1, placed across the solid line in its center and orthogonal to it; but it would not help in this case because we have

considered only two genes and, with the second principal axis, we simply recreate the original diagram after relocating the origin and rotating the axes. If, however, we have more than two genes, as in table 1.13.1, then it is convenient to calculate a first principal axis in the five-dimensional space formed by all five gene frequencies, and then a second one orthogonal to the first and crossing it at its center, using the same criterion of minimizing distances of the points from the new line. Orthogonality guarantees that the second axis does not contain information already present in the first and is tantamount to saying that the two principal components have a correlation of zero. One can continue to add more axes with the same criteria. However, our major aim is to produce a visual aid to the overall variation, and we therefore usually stop at the second or at most at the third components, which produce an easily evaluated two- or three-dimensional picture. This procedure guarantees that the highest principal components contain as much information as possible and the fraction lost can be estimated as a proportion of the total variance. Further components will recover smaller and smaller amounts of the residual variation. Representing all the data from table 1.13.1 by the first two principal components, we obtain table 1.13.2 and figure 1.13.2. From table 1.13.2, giving the *eigenvalues* of the data matrix, one can calculate the percentages of information recovered by the various components. Here the first component expresses 64% of the total information (variance) and the second, 27%; the total for a two-dimensional representation is 91%, a substantial value with a relatively trivial loss (9%).

Displays of principal-component values like those in figure 1.13.2 will be called PC maps. The distances of the populations in the map mimic their genetic differences, but relationships can be more or less distorted, depending on the amount of information lost. Naturally, the first PC in figure 1.13.2 differs from that in figure 1.13.1. The first is based on five genes and the second on only two of them. In the PC maps the two axes are often scaled to express the relative importance of the principal components: the first is more important than the second, etc. The scale factor for each component is the square root of the corresponding eigenvalue.

Table 1.13.2. The Five Eigenvalues of the Matrix Shown in Table 1.13.1, to Which the Percentage of Variance Corresponding to the Five Principal Components Is Proportional

Principal Component	Eigenvalues	% of Variance	Cumulative % of Variance
1	3.195	63.9	63.9
2	1.365	27.3	91.2
3	0.276	5.5	96.7
4	0.164	3.3	100.0
5	0	0.0	100.0
Total	5		

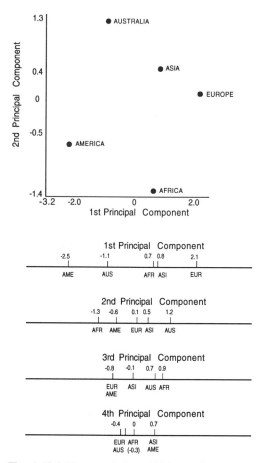

Fig. 1.13.2 Five populations (Africa, Asia, Europe, America, and Australia) with five genes (their frequencies are listed in table 1.13.1) represented by their first two principal components.

The real usefulness of principal components is manifest only for many more genes and populations than in these examples, which are given only to illustrate the method. It is often possible to retain a relatively high proportion of the total genetic information (40% or 50%) with the first two principal components, even when there are many dozens or even hundreds of genes. This allows an enormous simplification because the great number of variables is thus reduced to two or three arbitrary dimensions, with reasonable (and measurable) loss of information.

Although many consider the PC map a multivariate method of analysis unrelated to a tree, the connection between the two procedures is very close (Cavalli-Sforza and Piazza 1975). If the data behaved exactly according to a model of independent evolution, the first split in the tree would correspond to the separation of the populations operated by the first PC. If there is a substantial inequality in the number of populations separated by the first split of the tree, then the position of the zero on the PC axis may not exactly separate the same populations as the tree root, but the order of populations is such that some location of the first PC other than zero would tend

to reproduce the first tree split. The second split in the tree corresponds similarly to the second PC, and so on. The relations between close neighbors correspond to the lowest principal components, that is, to the lowest splits in the tree and can hardly be accurately represented in the usual PC maps that use only the first two or three PCs. Thus, information on closest neighbors is usually missing in PC maps, and the representation of similar populations may be misleading unless lower principal components are considered.

Genetic distances (see sec. 1.11) are sensitive to geographic distances between populations and tend to increase regularly with them. This is often visible in PC maps. If geographic distance were the only factor in genetic variation, one might expect the first two principal components to give a very close representation of the geographic map. The direction of principal axes and their origins are arbitrary and, in order to compare the two maps, it is usually necessary to rotate the axes and perhaps turn them around.

It is sometimes of interest to estimate the correlation of the PC map with the geographic map or with another PC map based on other variables (e.g., anthropometric or linguistic). This is done by a method that generates the best possible congruence of the two maps being compared. The procedure has sometimes been called Procrustean after Procrustes, a legendary bandit in classical Greek mythology, who employed brutal methods to adapt his prisoners to the length of his bed. The method of optimum congruence, developed by Schönemann and Carroll (1970) and programmed by Lalouel (1973) can be usefully employed to obtain various types of measures of agreement.

Other variations on the same theme are worth mention. Principal components can be calculated on genes instead of populations and populations and genes can be scored so as to represent both on the same map. This kind of "biplot" display (Gabriel 1971) shows for which gene or genes a population deviates most from the others.

These refinements were usually difficult to use in our PC maps because of the large number of populations appearing in them: we have usually concentrated on making it easy to recognize the position of populations in the PC maps.

Principal components are obtained in most applications by first calculating "Pearsonian" (i.e., standard) correlation coefficients between all pairs of characters. The original gene frequencies can be usefully transformed before calculating correlations. The subtraction of the mean gene frequencies on all populations is important because otherwise the first component mostly indicates the differences between mean gene frequencies, which are uninteresting from our point of view. A standardization of the gene frequencies by their average standard deviation, or better by dividing them by $\sqrt{[\bar{p}(1-\bar{p})]}$, where $\bar{p}$ is the mean gene frequency, can be expected to improve the recovery of information. The use of correlations of gene frequencies is, however, not entirely satisfactory for various reasons, one of which is that it is preferable to use measures of relationship proportional to times of evolution, like distances. The transformation of a distance matrix to a correlation matrix is necessary for calculating PCs. It is done, for instance, in multidimensional scaling (Torgenson 1958; Kruskal 1971) and in principal-coordinate analysis (Gower 1966). We have used the latter approach for generating all our bidimensional PC maps, starting from matrices of genetic distances, and all our PC maps are therefore, strictly speaking, principal-coordinates maps. This procedure generates a non-Euclidean space and, with it, a potential loss of information. It does, however, have advantages derived from the evolutionary meaning of genetic distances.

For synthetic maps, described in section 1.15, we have instead used ordinary principal components. There are therefore two sources of differences between results shown in PC maps as described in this section, and in synthetic maps, apart from the fundamental difference in the two modes of display: (1) In PC maps, the variation of gene frequencies is to some extent standardized because the divisor $\sqrt{[\bar{p}(1-\bar{p})]}$ used in the distance formulas tends to equalize them, without completely doing so. In synthetic maps, gene frequencies are taken at face value; (2) The other source of difference is the introduction of non-Euclidean variation in PC maps. Although the two approaches might have resulted in discrepancies

between conclusions, we have not observed any major ones.

In general, PC maps and synthetic maps are more useful the greater the fraction of total information supplied by the first two or three PCs. They can supplement trees by presenting data in a different form, which is especially convenient if clusters of populations are clearly visible. With perfect treeness, PC maps add little information to the tree, but observed matrices may deviate enough from this expectation that trees and PC maps often give different conclusions. PC maps are, in general, more flexible than trees, since they use a greater number of parameters; they are especially useful for comparing different sets of data because it is easier to compare two PC maps than two trees.

Principal-components analysis was first proposed by Hotelling (1933). Today, almost all statistical computer packages can calculate principal components, and all books of multivariate analysis give the basic theory. It was first applied to anthropometrics by Rao (1948) and to human gene frequencies by Cavalli-Sforza and Edwards (1964). The method gained popularity only after standard computer programs became widely available. PC maps are increasingly used in anthropometric and genetic analyses. It is hoped that the very elementary presentation given in this section may have clarified for novices some aspects of this widely used methodology.

We use PC for both principal components and principal coordinates, but we specify the full name when there is danger of confusion. Principal components are limited exclusively to synthetic maps and climate maps.

1.14. Geographic maps of gene frequencies

A geographic map of gene frequencies is comparable to a geographic map of altitudes—or any other quantity that varies from point to point on the earth—in which altitudes are represented by colors or shadings of varying intensity. Ordinarily, isopleths (in our case, lines of equal gene frequency) are first calculated for various levels of gene frequencies, for instance 10%, 20%, etc., or other suitable intervals, and the areas between two successive isopleths are filled with a color or shading of increasing intensity. In this way the map conveys immediate information on the location of relative maxima and minima (peaks and troughs) and gradients (clines) of the gene-frequency surface. Unless precise numbers are needed, one figure is worth a thousand numbers.

Mourant (1954; Mourant et al. 1976a) was the first to make extensive use of geographic maps of gene frequencies. The gene-frequency surface was constructed by sight (i.e., subjectively). Our specific interest was generated by the desire to build *synthetic maps* (described in sec. 1.15), and for this purpose it was considered important to use objective methods. The methodology for

automatic construction of geographic maps of gene frequencies was, at the time, an almost unexplored field, but now a variety of methods have been proposed for the interpolation or smoothing of geographic maps showing a number of variables (weather, mineral deposits, etc.). Each type of variable has special needs, but some aspects are common to all or most variables. In this section we summarize the general methodological problems (omitting some questions already discussed at length in Piazza et al. 1981a), the approaches already tried, and the final procedure adopted.

1.14.a. Choices to be made

The main decisions to be made are

1. *Selection of genes.* When are data sufficient for drawing a geographic gene-frequency map? There is no simple answer that allows an objective decision. Two variables are important: the total number of data points

available in the area to be mapped, and the regularity of their geographic distribution. Data points that have equal geographic coordinates were averaged after weighting for sample size. Their internal heterogeneity was calculated, and it was judged desirable to indicate graphically the presence of local heterogeneity on the final map (see map symbols, table 1.14.1). Because of the difficulties generated by the irregular geographic distribution of observed points, empirical decisions determined which maps would be included; in general, maps show at least 20 points (counted after averaging geographically close data).

We have considered it essential to indicate the location of the data points available for map building. Several areas are well studied; several others are exceedingly poorly represented. Some may have been examined in more detail in papers that appeared after the end of our survey, but they are probably relatively few. Leaving poorly represented areas blank could generate

Table 1.14.1. Symbols Used in Geographic Gene-Frequency Maps

 Presence of data points that are not outliers with respect to the map. If more than one data point with equal coordinates are available, they are averaged and indicated with the same symbol if not significantly heterogeneous.

 Data points are outliers, i.e., the deviation between the observed data point and its map value is large and reaches significance by the criteria indicated later in this section. Arrows show if the data point is greater or smaller than the map value.

 The data point is an average of more than one gene-frequency estimate with equal coordinates, and the estimates are heterogeneous as judged by χ^2 at a significance level of 5%.

 Combination of [symbol] and ↑ symbols

Combination of [symbol] and ↓ symbols

 When a single grid point belongs to a gene class different from all surrounding grid points, it is too small for graphical representation. When it does not include a data point, it is omitted. When it includes a data point, it is represented by one of the symbols given to the left, depending on whether the grid point is above or below the surrounding ones.

a false impression of a low gene frequency. Indicating the location of observations by the set of symbols shown in table 1.14.1 gives an immediate feeling for the validity of the areas interpolated between observations and for the need of further data collections.

2. *Choice between interpolation and smoothing.* Strictly speaking, *interpolation* is the fitting of a surface that goes exactly through the observed points; *smoothing* is the fitting of a surface that is close to the points but is not forced to go through them. Gene frequencies are affected by a fairly high sampling error, especially if they are based on a sample of a small number of individuals. Heterogeneous locations have an even greater standard error than expected on account of the total number of observations on which they are based. It is also clear that some points are true outliers; this is especially visible in areas of high concentration of data and is common in some particular regions of the world (e.g., India), for reasons discussed in chapter 4. Smoothing therefore seemed the right answer.

3. *Method of fitting the surface.* This is a major problem, to which the second part of this section is devoted.

4. *Choice of gene-frequency classes and patterns.* The classes of gene frequencies must be presented so that one receives a clear visual impression of the relative values of gene frequencies. Only in this way can the position of maxima, minima, and the regularity and steepness of gradients be appreciated. The maps in Mourant's books have excellent graphic quality. The increase in density of the shading is, however, not always regularly proportional to the gene frequency. In our first multivariate maps (Menozzi et al. 1978a) we used gray tones of varying intensity, an easy solution in terms of computer graphics. But the human eye has difficulty in clearly recognizing more than five tones, which are not sufficient in most cases. Another more subtle drawback is the tendency of the human eye to perceive tones in an altered way at the boundaries between two different tones. Finally, the print of gray tones extended to large surfaces may easily generate in print a moiré pattern which may be disturbing to the eye.

The use of patterns instead of gray tones also generates optical illusions that alter the perception of the average intensity of color or blackness. Even if patterns are ordered by the percentage of black per unit of area with a logarithmic coefficient of increase of density from interval to interval, the perceived order of intensity of the patterns is often incorrect or the comparison with the standard scale is insufficiently clear. We found that the use of parallel stripes at varying distance from one another offered the best solution (see Le Bras and Todd 1981), but with this approach appreciating the intensity differences of eight or more intervals was not easy. Alternating the direction of the shading (e.g., horizontal and vertical in successive intervals) obviated this difficulty, and made it easier to evaluate the gene frequency of an area of the map by comparison with the legend on the edge of each map. The device of alternating horizontal and vertical stripes should make it even easier to recognize the gene frequency of a particular area by comparison with the standard scale of gene frequencies given in each map. If not, copying the scale on transparent paper and placing it near the area of interest allows a completely unequivocal comparison.

Using discrete intervals causes a loss of information that we calculated as excess of variance produced (as well as loss of information measured in the manner of Shannon and Weaver 1972). We found the loss to be relatively small even with as few as four or five intervals. We also preferred using percentage intervals that were integer values from 1 to 10, thus avoiding fractional increments. An interesting alternative, the use of percentiles, has a disadvantage in that the evaluation of the actual gene frequencies requires a fastidious conversion. The shape of the distributions of gene frequencies (shown in each map) is also frequently complicated; most areas have a large number of outliers, usually making the distributions skewed and highly leptokurtic. We made sure the interval was, as often as possible, not smaller than the standard error resulting from sampling of the gene frequency. Smaller intervals would generate a false impression of precision.

5. *Goodness-of-fit tests of the surface.*   In previous work we used jackknifing (Piazza et al. 1981b) to test the goodness of fit of the surface. In the present applications we used a procedure described below (sec. 1.14.c).

6. *Relation of gene frequencies to areas.*   Because they are calculated from populations and not from single individuals, gene frequencies refer to areas, not to points on the surface. This problem might demand a special statistical treatment that would introduce several complications, but many populations cover a small area on the scale of our maps. At our geographic resolution, it is usually unnecessary to consider the actual area occupied by a population, which is, in any case, poorly known and not sampled homogeneously.

7. *Areas occupied by more than one population.* We selected, wherever possible, only aboriginal populations, for example, Lapps in northern Scandinavia, and Khoisanid populations in the deserts of Namibia and Botswana. In other cases, the situation is more complicated and we did not try to separate, for example, African Pygmies and African farmers in the areas in which they coexist. The degree of geographic resolution available on our maps would make such a distinc-

tion difficult. For genes that have different frequencies in Pygmies and African farmers, one will therefore find local genetic heterogeneity, as indicated by the appropriate symbols. These minority populations are treated with greater precision in the chapters dedicated to each continent.

8. *Areas in which aborigines have disappeared.*   Areas where there are no aborigines (e.g., the Carib islands and Tasmania) are not shown on the maps. Areas for which data on enough genes are not available (e.g., Madagascar) are also not shown.

9. *Choice of geographic projection.*   For maps of the whole world, we chose Mercator's projection, in which all meridians are straight lines parallel to each other. This causes a distortion that increases with increasing latitude but remains the same for a given latitude. For partial maps, we chose a method generating an equal-area projection (Odyssey World Atlas 1966), which does not distort the relative areas of the various parts of the map.

1.14.b. METHODS FOR FITTING SURFACES

In the remainder of this section, we discuss the major problem of fitting the gene-frequency surfaces. Some early attempts were made by fitting polynomial surfaces to the gene-frequency maps (Cavalli-Sforza and Bodmer 1971a; Cavalli-Sforza 1974), but the degree of the polynomials necessary for a good fit were prohibitively high, and where points do not exist, especially at the margins, the polynomial surface tends to curl and often generates absurd values. A second attempt led to a program by D. Schreiber for the maps published in the book *Genetics, Evolution and Man* (Bodmer and Cavalli-Sforza 1976a), using Shepard's formula (Shepard 1968) for calculating each point on the map as a weighted average of all data. The weight decreased with increasing distance between the point of the gene-frequency surface being calculated and the observed data. This method can be recommended for its extreme simplicity. Attempts made to choose a satisfactory function of distance for weights gave ambiguous results. The inverse of the square of the distance, plus a small constant, was finally adopted as weight (this agrees with the general scheme suggested in Shepard [1968]).

These methods can be called *global* in the sense that they simultaneously use all data points. *Local* methods use only local points chosen with some suitable criterion. They can also be made *adaptive* by a suitable change of parameters or functions when local conditions demand it.

Results shown in (Menozzi et al. 1978a) used a computer program (described in detail in Piazza et al. 1981a) that was local and partially adaptive. Each point was smoothed on the basis of data from the nearest populations selected in the four quadrants determined by latitude and longitude. The smoothing was carried out by Shepard's formula or by a quadratic polynomial, depending on the number of data points available in the neighborhood.

This computer program was tested by jackknifing and compared with a *kriging algorithm* (the name used for map smoothing in mineral engineering) implemented in France by Delfiner (1976) and based on Mathéron's theory (1971) of regionalized variables. This algorithm can be local or global, and also locally adaptive if necessary, in the sense that it can use a local or a general function connecting genetic distance and geographic distance. This function can be an empirical relation between the two types of distance, or any specified theoretical formula. In spite of its remarkable flexibility, the method did not show substantial advantages with respect to goodness of fit in comparison with our less sophisticated method. The surfaces it generated had a flaw similar to—though less extreme than—that already noted with polynomial fitting; namely, outside the field of observed points, even at a relatively short distance from the nearest point, the calculated surface tended to take extreme unjustified values (Piazza et al. 1981a). This is, of course, a general problem met in all extrapolation procedures.

Another elegant method we tried was developed by Wahba and co-workers (Wahba and Wendelberger 1980; Wahba 1982, 1984). It can be called global and adaptive. In this approach two functions are employed: one estimates a surface using polynomial splines or other more complicated expressions that minimize the difference between observed and expected values. A second function evaluates local smoothness (or its converse, roughness) by taking an average of the second derivatives over the surface. The requirements of minimizing the difference between observed and expected values for each point, and maximizing the smoothness of the surface, are reciprocally incompatible. Thus, a compromise must be reached by minimizing a weighted sum of the quantities evaluated by these two functions and using a cross-validation approach to calculate the relative weights to be given to the solutions provided by these two functions. One can call the method adaptive with respect to the average roughness of the region being mapped. By regulating the relative weight of the two functions, it usually reaches an optimum overall fit.

The main disadvantage of the method (at least of the early version we used) is that it is based on the *average* roughness of the map. We know by experience (and the examples shown for the region occupied by Caucasoids in Piazza et al. [1981a] show it clearly) that some areas are much rougher than others. Extending such examples to the whole world, one finds areas that are even rougher (e.g., the northern part of South America). Therefore, a region being smoothed might be divided arbitrarily into subareas believed to be of different roughness. Sewing together the results of fits obtained in different areas presented unusual mathematical difficulties. We abandoned this method with regret and are aware that it has been further improved upon since that time. We have, however, proceeded to develop other methods with greater local adaptiveness.

1.14.c. PROCEDURE ADOPTED FOR CALCULATING GENE-FREQUENCY SURFACES

The approach we finally adopted for constructing the geographic maps was developed by A. Piazza and Margareth Wright. "Expected" frequencies were calculated for each observed point, using only information from neighbors and not the observed point itself. To this aim, a plane was fitted to neighbor data, each of which was weighted with Shepard's formula (1968), including in each weight also the sample size. Neighbors were chosen sequentially, starting from the three nearest data points and adding more and more remote ones. At every addition, one calculates the difference between the observed point being interpolated and the expected value calculated on the basis of neighbors. The expected value changes with the addition of each nearest neighbor. The process of adding neighbors is stopped when the error of the interpolated point (the difference between the observed and the expected value) stops decreasing.

The number of neighbors thus considered depends on the local smoothness and could be used to express a measure of it. It depends also on the functions used for calculating the expected value. Two functions were tried and compared: a first-degree surface and an average of neighbors, weighted for distance (with Shepard's formula). There was little difference in the results of the two procedures, and the second was chosen. If the empirical point being tested is a real outlier, the goodness of fit may not improve at all after the first three nearest neighbors have been fitted.

The observed point was not used in the process of fitting an expected value to it, and the difference between the observed and expected gene frequency allowed calculation of a χ^2 value with one degree of freedom to indicate the local discrepancy of the map at each point. The χ^2 values calculated for neighboring points are not completely independent, but the probability of disagreement for a single point evaluated by the χ^2 distribution is valid. The probability of significance for this and every other test was chosen as one divided by the number of observed points on the map so that exactly one significant discrepancy per map is expected. More than one is found, on the average, but they are ordinarily distant from one another and therefore nearly independent. The indication of a discrepancy given by χ^2 is not, however, sufficient, as it depends on sample size, which varies enormously between samples. The distributions of sample sizes are given on maps. Large samples are more likely to show significant differences between observed and expected values because χ^2 is proportional to sample size. Thus, points for large sample sizes are likely to be significant, but the actual difference between the observed and the expected gene frequency may be trivial. The absolute deviation between the expected and observed gene frequency at each point was therefore also considered in the definition of outliers. The probability of each absolute deviation in a normal distribution was calculated and considered significant when less than one divided by the total number of data points in the map. *Outliers* were considered points significant according to both criteria simultaneously and indicated on the map by arrows pointing in the direction of the observed deviation.

In a second phase of the procedure, a lattice of regularly spaced points (lattice or grid nodes) was evaluated on the basis of the expected frequency values. For this purpose, the Voronoi neighbors (see fig. 1.14.1) were identified for each grid node and gene-frequency values for these nodes were calculated as weighted averages of the Voronoi neighbors. Weights were proportional to the

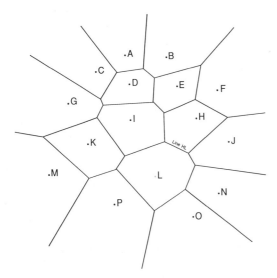

Fig. 1.14.1 Example of Voronoi polygons and Voronoi neighbors associated with 16 points A–P. Vertices of the Voronoi polygons are called *Voronoi points,* and the edges on the boundary are called *Voronoi edges.* The locus of a point closer to, say, point H than to point L is one of the two half-planes determined by the line (HL). The locus of a point closer to H than to any other point is then given by the intersection of all half-planes associated with H, that is, the H-associated *Voronoi polygon* (Lee 1982). In our procedure, the points A–P correspond to the "smoothed observed values" defined in the text. Every lattice point, which generally does not coincide with those values, is automatically associated with a set of smoothed observed values that form their Voronoi polygon.

Table 1.14.2. Number of Interpolated Grid Points per Map and the Average Distance in Miles between Contiguous Lattice Points in the Various Maps

	No. of Inter-polated Points	*Average Distance between Two Neighboring Points*
Africa	1604	92.0
Asia	2549	87.5
Europe	2379	47.5
Eurasia (prepared for world)	1601	125.5
Japan	79	56.3
America	1184	133.9
Australia	1221	54.1
New Guinea	477	29.8

sample size of each point and inversely proportional to the square of the geographic (great circle) distance between the node and the Voronoi neighbor. The resolution of the map depends on the fineness of the lattice, and table 1.14.2 gives the number of points evaluated per map. The final result of this phase of the procedure is a matrix of expected gene-frequency values calculated for each node of the lattice.

1.14.d. Isogenic lines (isopleths)

The last phase of the procedure is the calculation of the isogenic lines (isopleths, or lines of equal gene frequency for the chosen class intervals) from the matrix of expected gene-frequency values and the preparation by pen-plotter of the maps shown in the present book. (This set of procedures was prepared by P. Menozzi and E. Siri, also from University of Parma, Italy.)

The resolution level had to be increased to avoid discontinuities that, if not important in conveying the information, are certainly unattractive to the reader. Each of the original grid points was substituted by a 4×4 matrix, leaving the expected gene frequency unchanged in each of the 16 new points. Isopleths were then calculated on this blown-up grid. The step curve that would result was then further smoothed by bidimensional moving averages.

The problem of joining isopleths that are defined by the boundaries between two class intervals and the coast line was solved by dividing coast coordinates into groups that could be assigned to each node of the interpolated grid contiguous to the sea. The bookkeeping required during computer computations is quite complex and significantly increases the computer time required for map production, but to our knowledge, no alternative solution exists.

Further subroutines were required for shading the areas corresponding to various gene-frequency classes. The shading patterns we used were designed to guarantee a direct proportionality between the perceived density and the value of gene frequencies.

Each map was computed using the observed points represented with the symbols of table 1.14.1. For Pacific islands no interpolation was used and distributions (gene frequency, sample size, variogram) are not reported: they can be found in the maps of the continents themselves. World maps do not show the data points in Europe (too many data points crowd the area) except when the corresponding map of Europe has not been produced because of insufficient or irregularly distributed data points.

World and Pacific maps were prepared fitting each continent independently in order to avoid artificial continuity across continents. In order to avoid possible discontinuities in world maps, Eurasia was prepared anew rather than by joining maps of Europe and Asia. For some alleles, this may generate differences between the geographic gene-frequency maps of the world and those of the two continents, since a different interpolation grid and a joint data set were used for the world.

The isolated nature of islands had to be addressed, although there is no obvious way of measuring it. As a compromise, a group of islands—large, close, and genetically not too different from the continent—was treated differently from all the others. Greenland in America, Great Britain, Ireland, Sicily, and the islands between Denmark and Sweden in Europe, and the Japanese Archipelago in Asia were considered part of the continent. Japan, for which more information is available than for other Asian countries, was artificially enlarged. The grid of the interpolated continent points was extended to these islands when preparing the map of the continent. All other islands were considered isolated and their gene-frequency data were not used in interpolating the continental grid. Their geographic locations are shown in figures 1.14.2A–F. An average, weighted by sample size of all available data, was computed for each island and used to choose the shading for the whole island. In many cases, islands are too small for containing a perceivable shading, and a larger area surrounding them has been shaded. Even this larger area was often too small to contain at least two shading segments of the first two lightest shading classes: when only one horizontal (or vertical) line appears, it must be interpreted as the lowest class of shading represented by horizontal (or vertical) lines.

1.14.e. EVOLUTIONARY ASSUMPTIONS OF GENE-FREQUENCY SURFACES

What are the evolutionary assumptions or implications of gene-frequency maps? In general the calculation of a gene-frequency map assumes an underlying theoretical gene-frequency surface. It is interesting to consider whether this surface is in principle continuous or may have discontinuities. It is difficult to prove the existence of discontinuity in a rigorous way. Boundaries calculated by a recently introduced method (Barbujani et al. 1989), should not be taken to necessarily represent discontinuities. They may simply be zones, usually narrow, where gradients of gene frequencies are found to have higher slopes. Boundaries are calculated after a local interpolation of gene frequencies (by an algorithm different from the one we have used). It is not necessary to assume

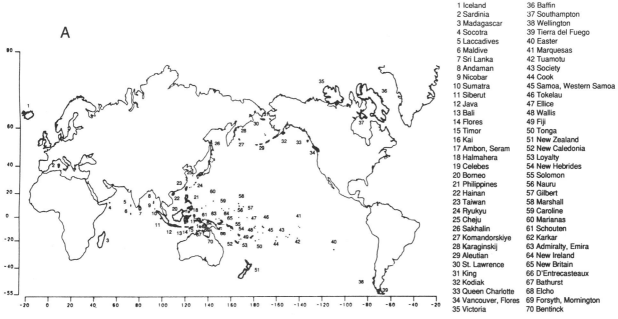

1 Iceland	36 Baffin
2 Sardinia	37 Southampton
3 Madagascar	38 Wellington
4 Socotra	39 Tierra del Fuego
5 Laccadives	40 Easter
6 Maldive	41 Marquesas
7 Sri Lanka	42 Tuamotu
8 Andaman	43 Society
9 Nicobar	44 Cook
10 Sumatra	45 Samoa, Western Samoa
11 Siberut	46 Tokelau
12 Java	47 Ellice
13 Bali	48 Wallis
14 Flores	49 Fiji
15 Timor	50 Tonga
16 Kai	51 New Zealand
17 Ambon, Seram	52 New Caledonia
18 Halmahera	53 Loyalty
19 Celebes	54 New Hebrides
20 Borneo	55 Solomon
21 Philippines	56 Nauru
22 Hainan	57 Gilbert
23 Taiwan	58 Marshall
24 Ryukyu	59 Caroline
25 Cheju	60 Marianas
26 Sakhalin	61 Schouten
27 Komandorskiye	62 Karkar
28 Karaginskij	63 Admiralty, Emira
29 Aleutian	64 New Ireland
30 St. Lawrence	65 New Britain
31 King	66 D'Entrecasteaux
32 Kodiak	67 Bathurst
33 Queen Charlotte	68 Elcho
34 Vancouver, Flores	69 Forsyth, Mornington
35 Victoria	70 Bentinck

Fig. 1.14.2.A Geographic locations and names of the islands represented in the maps of the world. Their mean gene frequencies are plotted but not used in the grid interpolation process.

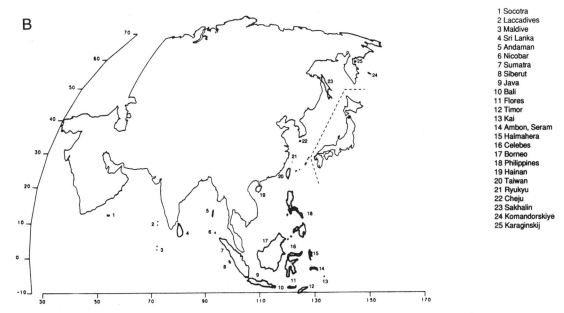

Fig. 1.14.2.B Geographic locations and names of the islands represented in the maps of Asia. Their mean gene frequencies are plotted but not used in the grid interpolation process.

discontinuities for using the method, nor was it claimed that boundaries represent true discontinuities of the gene-frequency surface. Barbujani et al. (1989) maintained that boundaries are distinct from gradients or clines, but they acknowledged that the distinction is difficult. Perhaps the main difference is that boundaries are likely to result from very little or zero local genetic exchange (because of physical, linguistic, social, or other barriers), whereas gradients can originate in many ways and local

exchange has a lesser role in their origin. Operationally, however, boundaries are defined as very steep clines and are usually observed for many genes simultaneously.

Some populations that are extreme outliers and are effectively isolated geographically, socially, or otherwise could, however, be conceived as local discontinuities, and some of our arrows in the maps may be examples of such situations. This is difficult to prove without further local analysis.

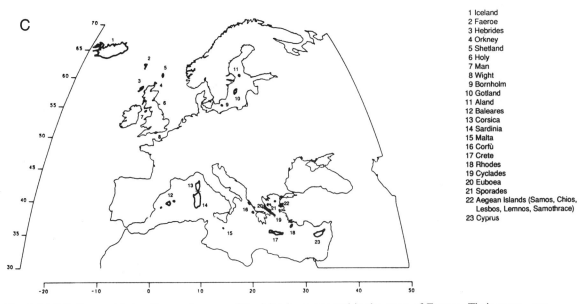

Fig. 1.14.2.C Geographic locations and names of the islands represented in the maps of Europe. Their mean gene frequencies are plotted but not used in the grid interpolation process.

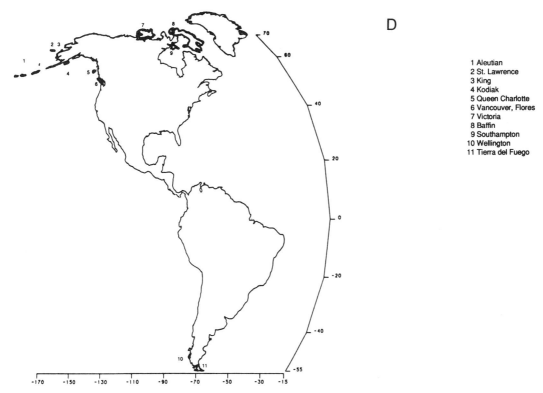

1 Aleutian
2 St. Lawrence
3 King
4 Kodiak
5 Queen Charlotte
6 Vancouver, Flores
7 Victoria
8 Baffin
9 Southampton
10 Wellington
11 Tierra del Fuego

Fig. 1.14.2.D Geographic locations and names of the islands represented in the maps of America. Their mean gene frequencies are plotted but not used in the grid interpolation process.

The gene-frequency gradients observed in maps could be due to selection or drift. The distinction should be based on external evidence. Local migration (genetic exchange between geographic neighbors) is almost always present and always tends to smooth down every gradient, but it is not very effective in some conditions (see secs. 2.7, 5.3, and Rendine et al. 1986). When migration tends toward one direction and is sustained for several generations, it can create gene-frequency gradients that may be relatively stable over time. Unlike gradients caused by selection or (recent) drift, such gradients have a historical origin (in the wide sense). This phe-

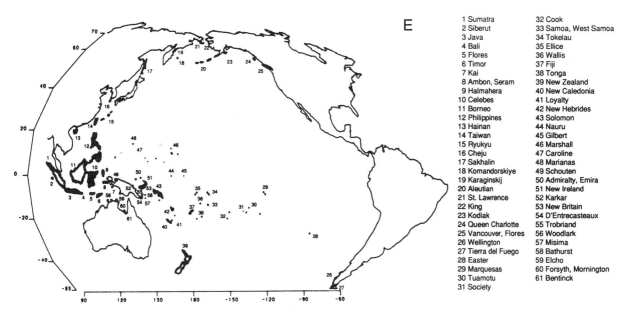

1 Sumatra	32 Cook
2 Siberut	33 Samoa, West Samoa
3 Java	34 Tokelau
4 Bali	35 Ellice
5 Flores	36 Wallis
6 Timor	37 Fiji
7 Kai	38 Tonga
8 Ambon, Seram	39 New Zealand
9 Halmahera	40 New Caledonia
10 Celebes	41 Loyalty
11 Borneo	42 New Hebrides
12 Philippines	43 Solomon
13 Hainan	44 Nauru
14 Taiwan	45 Gilbert
15 Ryukyu	46 Marshall
16 Cheju	47 Caroline
17 Sakhalin	48 Marianas
18 Komandorskiye	49 Schouten
19 Karaginskij	50 Admiralty, Emira
20 Aleutian	51 New Ireland
21 St. Lawrence	52 Karkar
22 King	53 New Britain
23 Kodiak	54 D'Entrecasteaux
24 Queen Charlotte	55 Trobriand
25 Vancouver, Flores	56 Woodlark
26 Wellington	57 Misima
27 Tierra del Fuego	58 Bathurst
28 Easter	59 Elcho
29 Marquesas	60 Forsyth, Mornington
30 Tuamotu	61 Bentinck
31 Society	

Fig. 1.14.2.E Geographic locations and the names of the islands represented in the maps of the Pacific. Their mean gene frequencies are plotted but not used in the grid interpolation process.

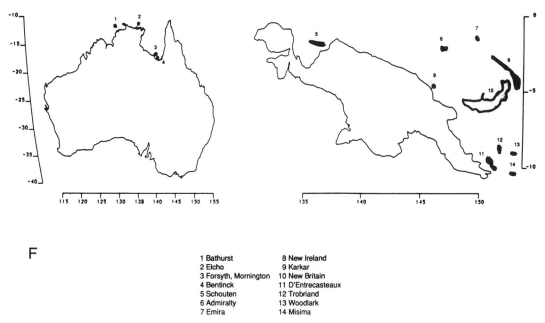

F

1 Bathurst	8 New Ireland
2 Elcho	9 Karkar
3 Forsyth, Mornington	10 New Britain
4 Bentinck	11 D'Entrecasteaux
5 Schouten	12 Trobriand
6 Admiralty	13 Woodlark
7 Emira	14 Misima

Fig. 1.14.2.F Geographic locations of the islands represented in the maps of Australia and New Guinea. Their mean gene frequencies are plotted but not used in the grid interpolation process.

nomenon is common in population expansions, and gradients caused by expansions may be distinguished from other types of gradients because they tend to affect many genes in a similar way (see secs. 2.7, 5.3). A relevant method is discussed in the next section; further discussion and examples are given in chapters 2, 3, and 5. All symbols used in maps are explained in table 1.14.1.

1.15. SYNTHETIC MAPS

The major stimulus for developing computer-designed geographic maps of gene frequencies came to us from the desire to calculate geographic maps of principal components (Menozzi et al. 1978a). The original purpose was to test the hypothesis that early farmers came to Europe from the Middle East. It was reasonable to assume that the migration of farmers into an area sparsely populated by foragers would have generated approximately circular gradients of gene frequencies around the region of the origin of the farmers, provided that later migrations did not destroy the pattern. A particular gene may or may not show the expected gradients, depending on whether there was an initial difference in gene frequencies between farmers and foragers, and depending on possible selective effects altering the pattern caused by the original migration of farmers.

Unlike selection, migration has the same effect on all genes. The gene frequency of a population that has received immigrants is changed in a very simple, easily predictable way, which depends on the difference between the gene frequencies of the immigrants and resi-

dents and on the number of migrants relative to that of the residents (sec. 1.17). The latter is clearly independent of the particular gene studied, and the effect of migration on different genes should therefore show a certain degree of congruence. An additional mathematical consideration is that both the effect of admixtures resulting from migration and the expression for calculating principal-component values are linear in gene frequencies. This will make principal components potentially very useful for the detection of migrations. In simple terms, a geographic map of a principal-component value may provide a synthetic view of all gene frequencies, abstracting from the data on all genes common patterns caused by single, earlier migrations. Building such *synthetic maps* of individual PCs may permit visualization of older migrations, as well as other phenomena affecting in some parallel way all, or many, genes. If some factor of strong natural selection (e.g., climate or an important disease) affected many genes in a systematic way, it might also be detected through the study of synthetic maps. In principle, one might thus hope that an individual PC could isolate the effect of individual factors of evolution like old mass migrations or strong selective factors, and perhaps others.

Our expectations were satisfied because we found genetic patterns in Europe that were in general agreement with those expected on the basis of the postulated early farmers' migrations and other archaeologically or historically known migrations. This method has now been extended to other continents and regions; results are discussed in the various chapters.

Once a method for calculating geographic maps of gene frequencies is available, such maps must be made for every allele for which there are sufficient data. As discussed in section 1.14, the adequacy of the coverage of the area is more important than the total number of data for determining whether an allele is usable. From such maps, values of gene frequencies are derived from an adequate number of lattice nodes (several hundred is the norm), principal components are calculated, and a geographic map is constructed for each principal component.

The geographic map of a PC has one or more peaks and troughs. Depending on the initial normalization of the calculation, positive and negative values of a PC are interchangeable and there is no difference in meaning between a peak and a trough: they both indicate the presence of a population whose gene frequencies, globally considered, take extreme values in the area being analyzed. They therefore correspond to populations that are unique in the sense of being most different from all the others that are present in the same area. A peak and a trough in the same map indicate opposite behavior. Two similar peaks (or two similar troughs) in different parts of the map may or may not indicate similar populations. Other PC values of the same populations should be considered simultaneously to judge whether the corresponding populations are genetically similar, because in this case they should be similar for all PCs, or at least for the highest ones.

Usually, but not necessarily, smooth gradients connect a peak and a trough and may indicate clines resulting from selection between two environments that have a different selective coefficient. Clines caused by migrational expansion from an area frequently, but not necessarily, generate approximately concentric gradients around the center of expansion, especially if migration can proceed unimpeded in all directions. When such a center of approximately circular gradients is clearly observable in a synthetic map, then the dominant direction of migration can be centrifugal, as is the case when a population grows and expands. These population expansions (also called radiations) are discussed in chapter 2 and later chapters. The reverse (a predominantly centripetal migration), although less likely, cannot be excluded immediately. For instance, the center of concentric gradients

may be an important capital, which has attracted immigrants for a long time. Some knowledge of history may be necessary for distinguishing such cases. Finally, physical or sociological barriers or favorite migration paths may be sources of complication, causing a distortion of an otherwise concentric pattern of gradients.

The meaning of peaks or troughs, in practice, is that a population differs from its neighbors for one of several reasons. For example, (1) at some previous time the population's members were settlers from a distant origin, or (2) initially the population did not differ genetically from its neighbors, but differences developed locally because of high isolation, small population size, and/or a long time spent under these conditions, which led, especially in combination, to substantial drift. Whichever of these factors or combination of factors is correct, one is more likely to find such extreme populations in mountain areas or otherwise isolated, impervious regions, areas jealously guarded by a highly endogamous population, or which for one reason or another have had no immigration for a long time.

Given the importance of recognizing differences between populations on the basis of simultaneous consideration of more than one PC, we have generated trichromic maps by superimposing three colors (usually green for the first, blue for the second, and red for the third PC). There are various limitations to the use of such maps: a major one is the cost of printing them and another is the difficulty of choosing the right combination of intensities of the primary colors. Alterations of intensity of one color may have important effects on the overall perception of patterns. We have tried to use a range of intensities proportional to the percentage of variation explained by each component.

The area chosen for PC analysis determines the patterns that will be observed. We have used entire continents, and repeated the analysis on the whole world. Our general experience suggests that the higher PCs are very unlikely to respond to extreme variation for one or very few genes. Selective effects are thus less likely to be observable by this technique, because a single selective factor rarely affects many genes at once. Even in cases like resistance to malaria, a major killer, it is rare for resistant mutants at more than one or a few loci to be found in the same population. The highest PCs ordinarily tend to show patterns that are found consistently, although in a less clear way, in 10%–40% of all the genes. These patterns are likely to occur for major, prolonged migratory phenomena involving populations that at the outset showed sufficient divergence from the other populations located in the area into which they expanded. In some cases, however, possible selective effects of climate have been detected.

1.16. Isolation by distance

The expression "isolation by distance," introduced by S. Wright (1943, 1946), indicates the tendency of populations to exchange genes with nearest neighbors, resulting in a greater genetic affinity between geographically closer groups and the likely occurrence of genetic differences between groups that are far apart because of genetic drift. Wright introduced the concept of distribution of individual migration distances, that is, the distances between the place of birth of one parent and that of his or her child. Collection of this kind of data is obviously much easier for humans than for other species, but unfortunately the actual distributions are usually highly asymmetric and often difficult to analyze. Mean and mean-square distances, however, provide measurements that, even if inaccurate, help in assessing orders of magnitudes. One finds that the mean migration distance per generation ranges from a few kilometers to some dozens, depending on populations. Estimates of standard deviations of migration distances in modern populations range from 1 to 300 km with a median value of 20 (Wijsman and Cavalli-Sforza 1984), but one has to remember that in more recent times there has been a substantial increase because of improved transportation.

The models developed by Wright have been conceptually influential but have had little practical application. More popular have been the models of isolation by distance developed by Malécot (1948, 1950, 1965, 1966, 1967) and by Kimura and Weiss (1964) in spaces of one or two dimensions, according to whether the migrating populations are supposed to live on a linear habitat (e.g., a group of islands placed along a line) or on a more common two-dimensional habitat. The approaches by Malécot and by Kimura and Weiss differ, but the qualitative predictions are alike. The unidimensional models are consistent in showing an exponential decrease of the genetic similarity between two *demes* (villages, or groups of people living in a defined geographical area) with their geographic distance. More precisely, such a relation can be written in the general form,

$$f(d) = a \exp(-bd)/d^c, \qquad (1.16.1)$$

where d is the geographic distance, and a, b, c are constants that do not depend on the distance. The parameters a, b, c, and the function $f(d)$ are given slightly different interpretations (and scales) according to the model of choice. The model by Kimura and Weiss assumes demes placed at discrete and equal distances scaled to unity ("stepping stone" model), with any distance being measured accordingly by integer numbers. In this model, $a = 1$, $f(d)$ is the correlation between gene frequencies of populations at distance d ($d = 1, 2, \ldots$), and c is 0 or 1/2, depending on whether the demes are placed on a one- or a two-dimensional habitat. The parameter b is expected to equal the square root of twice the ratio

M/m, where m is the migration per generation between two neighboring demes. M is a quantity that summarizes external sources of genetic input whose effect on gene frequencies is linear: it includes mutation—which is usually negligible—selection for heterozygotes near equilibrium, and gene flow from a large external gene pool, either located at a long distance (long-range migration) or unrelated to distance. A primary difficulty is that migration distances are distributed along a continuum and it is arbitrary to separate it into short (m) and long-range migration (which is probably the most important factor contributing to M).

After Wright's pioneering work, the mathematical theory of inbreeding under isolation by distance was greatly advanced by the probability treatment of Malécot, who introduced the concept of kinship and its measure, the coefficient of kinship. The *coefficient of kinship* is defined as the probability of identity by descent of two genes, one from each of the two populations under comparison, from a common "founder population." Malécot (1950 and later papers) showed that, in a continuous as well as in a discrete linear habitat, the variation of kinship with distance may be expressed by the general formula (1.16.1) with $c = 0$, $b = \sqrt{(2M/m)}$ (formally as in the Weiss-Kimura model)

$$a = 1/\left[1 + 4D\sqrt{(2mM)}\right], \qquad (1.16.2)$$

where in the continuous case, m is the mean-square migration distance per generation and D is the effective population (see sec. 1.4) density per generation; and in the discrete case, m is the migration between nearest neighbors per generation, and D is the average number of individuals per deme per generation; M has the same meaning as in the Kimura and Weiss model. Note that all these expectations are always valid at equilibrium.

In the case of a two-dimensional continuous habitat, formulas (1.16.1) and (1.16.2) are still valid but $c = 1/2$ according to Kimura and Weiss. The simpler approximation $c = 0$ holds in this case only for small d and M/m nearly equal to 1 (Imaizumi et al. 1970).

In conclusion, the model of isolation by distance predicts that kinship, as a measure of genetic similarity between populations, decreases exponentially with a population's distance from an initial value that measures the local kinship.

Morton and many other authors have applied the model of Malécot to a number of human populations; a summary of their results can be found in Morton (1982). From a qualitative point of view, the relationship predicted by equation (1.16.1) is almost always verified with $c = 0$, even for populations living in a supposedly two-dimensional habitat. The predictive power of this approach in obtaining demographic parameters like migration rates and population densities (eq. 1.16.2) is, however, limited. Two important reasons can be given. The first one is implicit in the simplifying assumptions of the model (equal population sizes for all demes, constant migration rates m and M, and, most important of all, lack of meaningful and empirically valid distinction between short- and long-range migration). Other problems arise from the exact meaning of the kinships used in Malécot's theory. Kinship is supposed to be a probability that depends on the gene frequencies of the *ancestral* founder populations. What we can actually measure is a covariance or correlation of the gene frequencies calculated on *contemporary* populations. Under these conditions, some of the calculated "kinships" are negative and the model (1.16.1) predicting kinship cannot be fitted without a correction. Morton et al. (1982), in their "theory of kinship bioassay," have proposed that an equation of the form

$$r(d) = (1 - L)f(d) + L \qquad (1.16.3)$$

best relates kinship estimated from contemporary populations ("conditional kinship," $r(d)$) to geographic distance.

The validity of this formulation and its capacity to cope with gene-frequency gradients unrelated to the equilibrium of migration and drift are still open problems.

The availability of gene-frequency data on a worldwide basis has allowed us to test the relation between gene-frequency variation and distance over ranges wider than the narrow ones tested by Morton and others. One of the main problems faced by the Malécot model in Morton's formulation (1.16.3) is that the simultaneous estimate of the parameters a, b, and L means that wide-range clines may affect gene-frequency variances measured at short distances. We introduced (Piazza et al. 1981a; Piazza and Menozzi 1983) a different parameterization, originally developed for geostatistic studies, that may avoid this problem by using the *variogram*, that is, a plot of gene-frequency variances at various distances. The isolation-by-distance model predicts that the variogram increases with the distance between pairs of populations and reaches a plateau at a distance at which the effects of gene flow compensate for those of drift.

THE VARIOGRAM

If the distribution x_i of a variable p (gene frequency in our case), specified at n geographic coordinates, is random and stationary (in genetic terms, equilibrium between mutation plus migration and drift is assumed), the expected value $E[p(x)]$ at any point x does not depend on x, and the variance between any two points $p(x_i)$ and $p(x_j = x_i + d)$ is independent of x and a function of the distance d only.

In the model by Kimura and Weiss and the one by Malécot, stationarity of the evolutionary process is al-

ways assumed: at least for short distances, the covariances of gene frequencies at any two points, x_i and x_j, depend only on the distance $d = x_i - x_j$. The *variogram* $v(d) = v(i, j)$ between gene frequencies $p(x_i)$ and $p(x_j)$ at locations x_i and x_j can be written as

$$v(i, j) = 1/2 \, \text{var} \, [p(x_j) - p(x_i)], \qquad (1.16.4)$$

and the isolation-by-distance model predicts (Piazza and Menozzi 1983)

$$v(i, j) = \frac{P(1 - P)}{1 + 4D \sqrt{(2mM)}}$$

$$\left\{ 1 - \exp\left[\sqrt{(2M/m)}(x_i - x_j) \right] \right\},$$

$$(1.16.5)$$

where $P = E[p(x)]$ is the expected value of the gene frequency assumed to be in equilibrium in the geographical region being considered (hypothesis of stationarity). The other variables have the same meaning as in equation (1.16.2). We use the standardization,

$$v(i, j)^* = v(i, j)/P(1 - P), \qquad (1.16.6)$$

(*standardized variogram*), in which the variogram is divided by the maximum theoretical value of the gene-frequency variance. This is analogous to transforming gene-frequency variances into F_{ST} values. This transformation is useful for comparing different alleles each with different equilibrium frequency $E[p(x)]$.

The variogram corresponding to the isolation-by-distance model is zero at the origin, is always increasing, and reaches the asymptote H,

$$H = v(\infty) = 1/[1 + 4D \sqrt{(2mM)}]. \qquad (1.16.7)$$

The behavior of the variogram at short distances is approximately linear with an initial slope, s, which can be calculated to be approximately

$$s = 1/[4mD + \sqrt{(m/2M)}]. \qquad (1.16.8)$$

The distance at which a variogram reaches an asymptote is referred to as the *range* of the variogram: it measures the size of the geographical area within which isolation by distance causes an exponential decrease of genetic similarity. The initial slope of the variogram calculated according to equation (1.16.8) depends only on the population structure (within the geographic area in which it is calculated) and should be the same for all alleles.

The variograms for each allele and for each continent analyzed in this book have been estimated by calculating equation (1.16.4) for distance classes of 200 miles (when available). These values, once transformed according to equation (1.16.6), have been smoothed by using the locally weighted regression and smoothing technique described in Cleveland (1979), which results in the almost continuous curves shown in the third of the three plots accompanying the gene-frequency maps.

The origin of the empirical variogram could differ from the expected zero: it may show a positive cutoff value determined by the average sampling variance of the populations. Moreover, the distributions of variances for given classes of geographic distance are highly skewed and robust techniques should therefore be used.

The initial slope of each variogram has also been calculated and is shown in each variogram plot. This estimate, however, is biased because the expected cutoff value of the variogram at distance zero is not zero as predicted by the model, but is inflated by the sampling variance.

An elementary analysis and interpretation of the variograms are given along with the description of single-gene maps. As far as we know, this is the first application to such an extensive data base. A preliminary and incomplete account of the approach has been given in Piazza and Menozzi (1983) and a more recent application (11 gene frequencies in European and Asian populations) can be found in Barbujani (1988).

In general, linear behavior has been confirmed for large intervals of geographic distance and for many genes. Some exceptions have usually simple and sometimes interesting explanations. The deviations from monotonic behavior (maxima and minima other than the final asymptote) are probably due to one of at least three possibilities: (1) data are numerically inadequate (the variogram is not given when this is unequivocally true); (2) equilibrium in the spread of a new mutation has not been reached; (3) there may be boundaries because of selection. We have not attempted any serious trial to test these possible explanations. The average behavior over all genes (sec. 2.9) is certainly clearer than the behavior of a single allele and agrees with the idea that geographic distance is a major factor of population differentiation.

Variogram versus correlogram

The study of correlation between gene-frequency differentiation and geographic distance can be treated as a problem of spatial autocorrelation. We can define spatial autocorrelation as the correlation of the values of one variable with the values of the same variable at all other points of a two-dimensional surface. The analysis of spatial autocorrelation is a complex subject covered in several books (for a review, see Cliff and Ord 1981). The application of this approach to gene frequencies has been specially advocated by Sokal and his associates in a series of papers (Sokal and Oden 1978; Sokal 1979; Sokal and Menozzi 1982; Sokal and Wartenburg 1983; Sokal et al. 1986; Barbujani 1987b; Sokal and Winkler 1987). A widely used coefficient of spatial autocorrelation is Moran's index, I (Moran 1950), which is a correlation coefficient restricted to population pairs whose geographic distance (estimated by any desirable definition or structure) is within a given interval, the central value of which is d. The plot of I values versus distance is referred to as the spatial *correlogram*.

It has been shown (Barbujani 1987a) that Moran's I, for any gene frequency and at any distance, d, is the ratio of the kinship, estimated for that allele and that distance, to F_{ST}:

$$I = f(d)/F_{ST}. \qquad (1.16.9)$$

Under isolation by distance, correlograms should decrease monotonically with distance as kinship does; at zero distance, $f(d = 0) = F_{ST}$, and then $I = 1$. An exponential decrease of the correlograms has actually been observed in the analysis of gene-frequency patterns when a model of isolation by distance has been simulated (Sokal and Wartenburg 1983). At large distances, I tends to L/F_{ST} (see eq. [1.16.3]), indicating that empirical correlograms—as in the Morton formulation of kinship—are more difficult to interpret than standardized variograms.

1.17. Admixtures, their estimation, and their effect on tree structure

The generation of diversity among modern humans is the result of a large number of separations of splinter groups that migrated to other territories. The separation need not have been abrupt every time: simple demographic expansion beyond the range usually covered in matrimonial migration, which is usually less than 100 km in a settled population, is sufficient to increase diversity. If the demographic expansion extends beyond physical barriers (like mountains, rivers, and seas that are difficult to pass), then some genetic discontinuity may eventually arise.

The movements of whole tribes, even those of farmers, were certainly not infrequent, even when demo-graphic expansion was limited or had completely ceased. People with more advanced agriculture are much more firmly settled, but even they may have been forced to move by long-term changes in climate, permanent loss of soil productivity, harassment by neighbors, or wars. Long-range movements cause the acquisition of new, sometimes very different, neighbors, and—in spite of the endogamous tendencies of most tribes—genetic exchange with them may occur and affect their genetic pool.

When two populations are geographically distant from one another, they tend to be rather different genetically, as discussed in the preceding section. Later movements may bring two such different populations geographically close to each other. Their mixture will generate a new population, intermediate between the two and probably unique. Ordinarily population mixtures do not occur in a "catastrophic" fashion, but are more likely to take place by the continuous slow infusion of individuals from one group into another neighboring group, usually in small amounts per generation. In certain circumstances the *gene flow*, as it is called, may occur only, or mostly, in one direction. This is the case for Black Americans, who, in the 300 or more years since they were forcibly taken from Africa to America have received genes from Caucasoid people in small proportions at every generation. In the 10 or more generations since African arrival, the accumulated gene flow is perceptible, as noted in the lighter skin colors of Black Americans compared with that of individuals from West, Central, or South Africa. The direction of gene flow can also be noted for all genes that have different gene frequencies in Caucasoids and Africans, for example, blood groups, RH, FY, or GM immunoglobulins. Glass and Li (1953) have used this information for estimating white ancestry in Black Americans. If the two parent populations, Caucasoid and African, have frequencies of a given gene q_A and q_B, and the proportion of the A type in the mixture is m of type A and $(1 - m)$ of type B, then the gene frequency of the mixed population, q_M, is given by

$$q_M = (1 - m)q_B + mq_A \qquad (1.17.1)$$

so that given q_A, q_B, and q_M one can estimate m. An important check of the hypothesis of admixture is that all genes should give the same estimate of m. Genes giving abnormal values of m may be affected by natural selection, an important test unfortunately not easy to carry out because gene frequencies of parent populations are known only approximately. Moreover, estimates of m are inaccurate for recessive genes and when the difference $q_A - q_B$ is small. New information on DNA polymorphisms, which are practically all codominant could change prospects of this type of research. In addition, research on mitochondrial DNA and Y-chromosome markers could distinguish the proportions of unions involving males and females of the two groups, which are likely to be very different.

In the case of Black Americans, the mixture was probably by continuous slow infusion. Assuming that the proportion of genes entering a group is constant at every generation and knowing the number of generations, one can estimate the admixture rate per generation. Today the Black American gene pool is 30% Caucasoid; this is an approximate average for the United States, with values ranging from 10% in the South to 50% in the North (T. Reed 1969). Probably some fraction of individuals with highly diluted African traits has passed the racial barrier, but the majority did not. After n generations, the proportion of admixture becomes $m(n)$

$$m(n) = 1 - (1 - m)^n. \qquad (1.17.2)$$

Assuming that 10 generations have elapsed since the peak rate deportation of slaves from Africa to the United States (1500–1880), the proportion m per generation can be estimated at 3.5%.

It is not often appreciated that, in the long run, gene flow may be very effective in almost completely replacing the gene pool. Although most tribes are substantially endogamous around the world, some degree of exogamy per generation is common. Assuming that a tribe receives a gene flow of 1%, 3%, or 10% from an external source, the genetic dilution thus resulting can be calculated as in table 1.17.1 (generation time of 25 years).

Table 1.17.1. Percentage of Residual Genotype After a Per-Generation Gene Flow m of 1%, 5%, 10%, and 20%, Using Formula (1.17.2)

Time		Gene Flow (%) per Generation			
Generations	Years	1	5	10	20
0	0	100	100	100	100
1	25	99	95	90	80
4	100	96	81	66	41
8	200	92	66	43	17
20	500	82	36	12	1.1
40	1000	67	13	1.5	–
80	2000	45	1.7	–	–
200	5000	13	–	–	–

The great majority of exogamous marriages are formed by one member from one tribe and one member from the other tribe, and in such cases the migration rate useful for calculating gene flow is one half the frequency of exogamous marriages.

When the residual genotype of the ethnic group that has received gene flow from an outside group is on the order of 10% or less, it is very difficult to estimate accurately, or even to detect. The table shows that even with a low rate of gene flow (1%–3%), it may be difficult to detect an admixture after 2,000–5000 years. Five hundred years are sufficient to cancel traces of the original population with a gene flow of 10%. These limitations must be kept in mind; gene replacement may in some circumstances be so nearly complete that it may be very difficult to reconstruct the genetic origin of a group. The important conclusion is that the genetic origin of groups that have been surrounded for a long time by populations of different genetic type can be recognized as different only if they have maintained a fairly rigid endogamy for most or all the period in which they have been in contact with other groups.

Many formulas and procedures have been suggested and used to estimate the cumulative degree of admixture. Much work has been dedicated to improve statistical methods (review in Chakraborty 1986), but they primarily take care of the sampling error of observed gene frequencies. These sources of error are usually trivial compared with those of obtaining satisfactory gene frequencies for the ancestral populations and making assumptions on the evolution of the mixed and the parental populations. This is especially true when we are interested, as we here are, in aboriginal populations whose admixtures often happened long ago and for whom the living descendants of the putative ancestral types are poorly known genetically. There is usually little if any knowledge of the demography of the parent and mixed populations in the intervening period after the admixture. The evolutionary changes that took place since admixture depend on the intensity of drift and other possible evolutionary factors that affected the populations whose data are used for the analyses. The best insurance against the various difficulties is to have data on more genes.

In some cases we have used more sophisticated methods (Wijsman 1984), but for rough estimates we have preferred to follow a simple approximate procedure using the F_{ST} distances we have employed throughout this

book. Figure 1.17.1 explains their meaning in terms of the gene-frequency space. One can easily realize that in the multidimensional space formed by all gene frequencies, a mixture between two populations is found on the line joining the two ancestral populations. This is the straight line shown in the first graph of figure 1.17.1. Since the admixture occurred, however, all gene frequencies of A, B (ancestral populations) and M (the admixture) have changed to new values, A', B', M', which are no longer on a straight line but form a triangle (see second graph of fig. 1.17.1). These conclusions do not change if we replace the original gene frequencies p with the transformation $p/\sqrt{(\bar{p}(1-\bar{p}))}$, where $\bar{p}$ is the general mean of frequencies of that gene. In such a space, the squares of the distances between points are the F_{ST} distances we have previously described. The actual genetic distances are a logarithmic transformation of F_{ST}, but for the purpose of calculating admixtures we return to the original F_{ST} values before the logarithmic transformation.

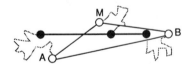

Once formed, a mixture M is on the straight line joining the two parental populations A and B, at a point determined by the proportions of A and B in the admixture. Here M is a mixture of 30% population A and 70% population B.

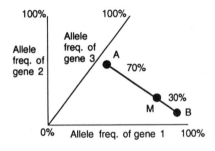

After some independent random evolution of A, B, and M, the straight line becomes a triangle.

Full circles: original values
Open circles: after some time

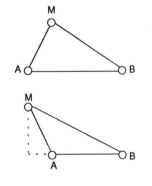

In older admixtures the triangle is less flat.

The admixture in this example is no longer recognizable as such.

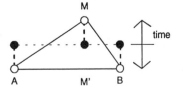

Admixture proportion and time of occurrence can be estimated assuming minimum evolution at a constant rate. The admixture proportions are estimated from AM', M'B relative to AB.

Fig. 1.17.1 Triangles expressing population mixtures.

If we could estimate the drift, or, more generally, the amount of evolution that has taken place since the admixture, we could obtain better estimates and tests of admixture. Usually, however, no information is available, and it is then simplest to assume that the amount of evolution accumulated since admixture in all three populations is equal (the hypothesis of constant evolutionary rates). In the last graph of figure 1.17.1, the amounts of evolution are the three vertical dotted segments of equal length. The proportion of genes contributed to M by A, m is then given by

$$m = 1/2 + (f_{BM} - f_{AM})/(2 * f_{AB}), \qquad (1.17.3)$$

where f_{AB} is the F_{ST} distance between ancestors and f_{AM}, f_{BM}, the distances between each of the two ancestors and the mixture.

The amount of evolution since admixture for A,B,M populations is then a function of half the height of the triangle. In Cavalli-Sforza and Piazza (1975), we gave formulas predicting genetic distances between the mixed population M and the ancestors. Here we use those formulas to calculate the time between the separation of the ancestors and the occurrence of the admixture, t, expressed as a fraction of the total since the ancestral separation. The time since admixture, always a fraction of the total time since the ancestors separated, is $1 - t$. From the formulas given in Cavalli-Sforza and Piazza (1975), one can derive the following formula for t, assuming constant evolutionary rates and using genetic distances, that is, in general, $d_{IJ} = -\log_e(1 - f_{IJ})$:

$$t = \frac{1 - (d_{AM} + d_{BM})/2d_{AB}}{1/2 + m(1 - m)}. \qquad (1.17.4)$$

The formula is not valid if the estimate of relative time obtained is greater than that of separation of the ancestors (i.e., if $d_{AM} + d_{BM} > 2d_{AB}$), in which case a postulated mixture is certainly suspect.

Acceptability of results depends on the validity of the hypothesis of constant evolutionary rates, but there is little to be gained by making more complicated hypotheses that would essentially rule out the possibility of making time predictions. Those made under the set of assumptions suggested here will be best tested against external evidence.

Graphic presentations of populations in the multidimensional space of gene frequencies, as seen in figure 1.17.1, show that recent admixtures lie on the line joining the two parental populations. This property is unchanged if we use principal components instead of gene frequencies as coordinates. Thus, a mixed population lies on the line joining the two parental populations even in maps of the first two or three PCs, but the reverse clearly is not true: populations that are on the line joining two others are not necessarily admixtures of those two. We have already seen that geographic maps of PCs are use-

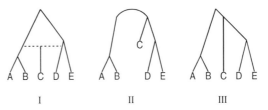

Fig. 1.17.2 Examples of distortion in tree reconstruction caused by admixture (horizontal dotted line in I). II, Unrooted tree reconstructed by various methods. III, Rooted tree reconstructed by average linkage.

ful for detecting migrations responsible for admixtures between immigrants and residents.

Further important consequences of admixture affect the interpretation of trees. The genetic exchange going on all the time between neighbors makes the use of trees less desirable for geographically close populations. It might seem that the tree would reflect the geographic relationships rather than the historical ones. This expectation is denied, however, by experience with the Makiritare (Wand and Neel 1970), in which the opposite seems to be true. The Makiritare villages, however, are often on the move and their geographic interrelationships may be less meaningful than those for truly settled farming communities.

When population samples come from widely scattered geographic locations, the effect of migration between neighbors is diluted enough that it may be negligible. Here ancient genetic exchanges between remote branches are especially important and interesting, and their discovery has special value.

One can easily predict the effect of hybridizations and fusions on the expectations of a tree. In figure 1.17.2 we have a theoretical example of a fusion between two populations by admixture in known proportions of two branches, followed by independent evolution. The relevant theory was given in Cavalli-Sforza and Piazza (1975) and later extended (unpublished), and it shows that admixture affects the distances of mixed populations in a number of ways that can often be detected by inspecting a distance matrix and the reconstructed trees.

1. A mixed population is expected to have a shorter branch than the parental populations from which it originates. The short branch is detectable, however, only with methods of tree reconstruction that do not postulate constant evolutionary rates. Shorter branches cannot be observed, therefore, with average linkage or maximum likelihood, but only with additive tree, minimum path or neighbor joining. Trees reconstructed with either of the first two methods are affected by the distortions summarized in the two points below.

2. If a population has a mixed origin, the characteristic block pattern of the distance matrix shown in figure 1.12.2 will be altered. A mixed population will show a characteristic distortion of all its distances, those with the parental populations being especially small, and the overall treeness will be affected (Astolfi et al. 1978).

Table 1.17.2. Genetic Distances (x 10,000) between Three East Asian and Four European Populations (the last row is an artificial admixture of 40% Ainu and 60% English)

	AIN	JPN	KOR	BAS	ENG	ITA	GRK
Ainu	0						
Japan	223	0					
Korea	358	138	0				
Basque	1153	1481	1063	0			
English	1220	1244	983	120	0		
Italian	1111	1145	937	141	51	0	
Greek	739	1175	904	232	204	77	0
Admixture	452	334	450	313	198	203	201

Note.– The upper triangle in the distance matrix includes all distances between East Asians. The lower triangle includes all distances between the four European populations. This is a reasonably good tree, as the rectangle formed by the East Asian x European distances are all high and only moderately variable.

3. In the ordinary type of bifurcating trees constructed on the assumption of constant evolutionary rate (in practice, average linkage), the mixed population will tend to be inserted in the tree in a position not far from the parental population that has given the greater genetic contribution. Its apparent branching time will tend to be earlier than that at which the mixture occurred.

In table 1.17.2 we represent the effect of admixture on a distance matrix and its corresponding tree. The table clearly shows two clusters: one of three East Asian populations, and the other of four European populations. The upper triangle in the distance matrix includes all distances between East Asians, varying from 138 to 358, while the lower triangle includes all distances between the four European populations, varying from 77 to 232. This is a reasonably good tree, as the distances between the East Asians and the Europeans (in the rectangle) are all high and only moderately variable (from 739 to 1481). A more exact analysis could be done, as in Cavalli-Sforza and Piazza (1975), but here we are interested only in a qualitative picture. The tree is shown in figure 1.17.3.

In order to introduce the effect of admixture on the distance matrix and tree, an artificially mixed population was created. The gene frequencies for this population were calculated arbitrarily, summing 40% of the gene frequency of the Ainu and 60% of that of the English.

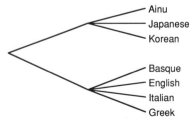

Fig. 1.17.3 The tree of seven populations shown in table 1.17.2.

The distances between the admixture and the real populations were then calculated (shown in the last row of table 1.17.2). Distances between the admixture and the three East Asian populations (first three values of the last row) are definitely smaller than those between Europeans and East Asians (forming a clearly distinguishable rectangle). The same is true for the distances between the admixture and the four European populations (last four values of the last row). Thus, the last row of the distance matrix shows that the admixture is a unique population that does not fit the pattern of a tree.

If a tree is nevertheless calculated from the distance matrix of table 1.17.2 by average linkage, the admixture attaches to the European cluster (fig. 1.17.4). This result is reasonable, as the admixture is predominantly European. In addition, the admixture's distinct identity is shown in that it is an outlier within the European cluster.

In theory it is possible to construct a tree with connections between the branches, and we show an example in section 2.4 that follows fairly closely that given in figure 1.17.2. As mentioned earlier, interconnected trees are called *networks*. In the language of graph theory, trees bifurcate or multifurcate, but their branches do not connect; in fact, in the simplest hypothesis, all branches should evolve independently. In a few papers, especially regarding the Yanomama and Makiritare, the word network has been used to define ordinary trees reconstructed from populations known to be historically interconnected. This usage seems confusing. It is preferable to avoid the word "network" when referring to ordinary trees. An attempt has been made at reconstructing true networks (Lathrop 1982). This method tries all possible mixtures between all possible pairs of ancestral populations presumed to have existed at a given time for all times and selects those that seem acceptable. Unfortunately, the method gave erroneous results in an application to American Indian data for which the history of major fusions is at least approximately known. The reason for the failure of this elegant method became clear by experimenting with simulations on the possi-

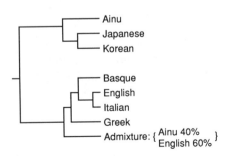

Fig. 1.17.4 Tree of the seven populations to which an artificial admixture made of 40% Ainu and 60% English was added, as in the last row of the distance matrix of table 1.17.2.

bility of recognizing only a few (one or two) specific admixtures introduced in a simulated tree. The investigation (P. Darlu, pers. comm.) showed that a tree with interconnections can be successfully reconstructed only with a large number of genes, much greater than those ordinarily available. For instance, with simulated trees of 7–8 populations, it was necessary to use 200–400 independent characters (or genes) for error-free detection of admixtures generating interconnections between branches. Even higher numbers of genes could be reached today by using the new batteries of polymorphisms known. Nevertheless, in searching mixed populations in the analysis of reticulate evolution, it is simpler to analyze potential admixtures directly by methods that test only the two parents and the putative mixture using simple methods like the one suggested above. Help in detecting potential admixtures may come from applying the bootstrap to the tree because mixed populations often tend to be attached to different clusters in different bootstrap trees. The instability of attachment of a population to its cluster under bootstrapping is especially noticeable for populations in which admixture is substantial. The tendency of a population to leave its cluster and join another in different bootstrap trees gives some cues on the clusters contributing to the admixture. The presence of mixed populations in standard tree reconstruction may sometimes alter the shape of the reconstructed tree. It is therefore good practice to try reconstruction with and without populations suspected of admixture, or to avoid including them. The full analysis of reticulate evolution remains an important task for the future.

2 GENETIC HISTORY OF WORLD POPULATIONS

2.1. PALEOANTHROPOLOGICAL BACKGROUND

2.1.a. THE GENUS *HOMO*

Africa has been the site of the most interesting recent discoveries of fossils in the human line and is therefore, unless new findings change the picture, also the cradle of the most ancient living beings that paleoanthropologists are willing to call *Homo*. According to the most commonly accepted recent views, the nearest ancestor from which the genus *Homo* separated is *Australopithecus afarensis*, specimens of which (including the famous "Lucy") were found in different locations in East Africa and dated between 3 and 4 million years ago. From *A. afarensis* descended four other species of australopithecines: *A. africanus* and *A. robustus* in southern Africa and *A. aethiopicus* and *A. boisei* in East Africa, as well as the genus *Homo*. The first species in our direct line of descent is *H. habilis*, followed by *H. erectus* and by *H. sapiens*, the latter being the only species living today. All australopithecines on the collateral lines are extinct.

The history of the genus *Homo* begins around 2.5 million years ago; and the divergence of modern humans among themselves may be only 100,000 years old, although several paleoanthropologists favor older dates. We choose to refer to these dates with the following symbols: my, million years; mya, million years ago; ky, thousand years; kya, thousand years ago. In general, for dates younger than 10,000 years ago (i.e., 10 kya), we prefer, for example, 6000 years, rather than 6 ky. "Years ago" corresponds approximately to years B.P. (years before the present). The approximation derives from the custom of referring to B.P. as the year A.D. 1950. Unless

otherwise specified, radiocarbon dates are not corrected by the tree-ring method. Historical dates are given as B.C. or A.D., as is usually done.

Until recently there was considerable uncertainty on the phylogeny of *australopithecines* and human ancestors (Johanson 1989). A consensus has now been reached in favor of the phylogenetic structure shown in figure 2.1.1 (Johanson, personal communication).

All australopithecines were small, with an average weight of about 50 kg but with extreme individual size variation within each species (probably due in part

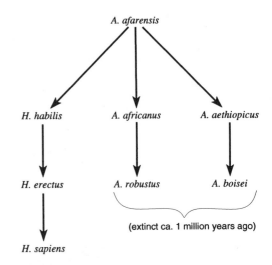

Fig. 2.1.1 Common part of four phylogenies proposed for the descent of humans from Australopithecines (after Johanson 1989).

to sexual dimorphism). From new reconstruction, the stature of early *afarensis* is about 1.1 m (3'6"). "Robust" refers to large tooth and masticatory apparatus, not to body size (Johanson 1989). They were all bipedal, but perhaps at the beginning not exactly like modern humans. Their brains were smaller than those of later humans, as calculated from endocranial capacity (range 380–520 cm^3; all braincase volumes from Klein 1989a). Although their brains were about as large in absolute values as those of our closest surviving relatives (chimpanzees and gorillas), they were definitely larger relative to general size.

In the human line we find three species, in order of time:

1. *Homo habilis.*   *Homo habilis* has been found with certainty only in Africa, where it must have originated from australopithecines; this species lived between ca. 2.5 and 1.5 mya (Klein 1989a). The brain size of these first humans is definitely larger than that of the australopithecines (range 510–750 cm^3), but their brains were still less than half our size, on the average. The skull shows a low cranial vault and heavy brow ridges. *Homo habilis* was capable of making tools; in fact, the oldest known stone tools are of African origin and are 2.4–2.7 my old, somewhat older than the oldest known specimens of *Homo habilis* (2.2–2.3 my). The first stone tools made by humans are of a type called the "Oldowan industrial complex" and are especially simple.

2. *Homo erectus.*   From 1.5 to 0.3 my, *H. erectus*, the first hominid clearly found outside of Africa, expanded from Africa to Europe, West and South Asia, and East Asia but not North Asia (which was perhaps too cold for the survival skills of this ancestor). It may have persisted in Java and China later than in Europe and Africa. The braincase is larger than that of *H. habilis*, with a range of 850–1100 cm^3. The name seems a misnomer since bipedalism is older. There has been considerable discussion over the phylogenetic position of *H. erectus*, but the sequence *habilis-erectus-sapiens* is now accepted by the majority of paleoanthropologists. A new style of stone artifacts, Acheulean, accompanied the development of *H. erectus* in Africa and is proof of its presence in Europe and West Asia where no human bones were found (Klein 1989a). These artifacts did not extend to East Asia, where tool kits always remained consistently poorer after this time, possibly because many tools were made from the widely available but highly perishable bamboo (Pope and Cronin 1984).

3. *Homo sapiens.*   The next human type to appear is that of our species; *Homo sapiens*, first found in Europe early at least 300 kya. By the time of its appearance the modern brain size, 1200–1500 cm^3, had been reached.

Table 2.1.1A. Geological Subdivisions of the Last 1.5 Million Years

Era	Time Period	Development of Humans
Pleistocene		
Early (or Lower)	1.7 m.y.a. – 700 k.y.a.	*Homo habilis – H. erectus*
Middle	700 k.y.a. – 130 k.y.a.	*H. erectus – H. sapiens*
Late (or Upper)	130 k.y.a. – 10 k.y.a.	anatomically modern humans (a. m. h. or a. m. m.) in Africa and Asia
Later	50 k.y.a. – 10 k.y.a.	a. m. h. in Australia, Europe (after Neanderthals), and America (radiating from Asia)
Holocene	10 k.y.a.	beginnings of agriculture in nuclear areas

At first *H. sapiens* showed no dramatic evidence of behavioral progress with respect to *H. erectus* (Klein 1989b).

Before continuing with the last part of human evolution, which is more accessible to genetic investigations, it may be useful to report schematically some standard names of the various periods defined for the last million years (table 2.1.1 A,B). Geological definitions include the Pleistocene (from 1.7 or 1.6 mya until 10 kya) and

Table 2.1.1B. Subdivision of Stone Industries over the Last 1.5 Million Years in Order of Appearance and Complexity, with Names of Some Major Cultures Used by Archaeologists

Stone Industry	Culture
I. Simple flakes and cores, choppers	Oldowan
II. Simple, formally shaped tools (e.g., hand axes); tools flaked by direct percussion	Acheulean
III. Flakes prepared by striking cores made in advance	Late Acheulean, Levalloisian, Mousterian, others
IV. Blades and burins	Many cultures of European and Near Eastern Upper Paleolithic
V. Microliths and composite tools	Many cultures of terminal Paleolithic, especially in Europe; also called Mesolithic
VI. Ground stone tools	Neolithic

Note.– Archaeologists use the terms "Paleolithic" and "Neolithic," which correspond only approximately to Pleistocene and Holocene. Paleolithic is subdivided into Lower (or Early), Middle, and Upper (or Late). These subdivisions are similar but not identical to those of the Pleistocene. In fact, the archaeological classification is based on a different criterion, the type and degree of development of stone industries. The sequence and periods of stone industries vary from place to place, as do the definitions of Paleolithic and its periods. To avoid this source of confusion, J. G. D. Clark (reported in Tattersall et al. 1988, p. 417) proposed the above subdivision of stone industries. The Paleolithic is followed by the Neolithic, which begins in some of the earliest areas with the beginning of the Holocene, but at other times in most areas. It has different local definitions: in Europe and West Asia it refers to the beginning of farming and lasts until the age of metals, whereas in other areas Neolithic is often marked by the presence of pottery and may be anterior to the development of agriculture.

Holocene Epochs (the last 10 ky). Archaeological definitions include the Paleolithic and Neolithic with similar, but not identical, durations and subdivisions.

2.1.b. THE SPECIES *HOMO SAPIENS*

The most ancient specimens of *H. sapiens*, referred to as *early* or *archaic*, are more robust than modern humans, showing thicker skull bones and bone crests usually not present in modern humans. The evolution from *erectus* to *sapiens* may have been somewhat different in different parts of the world, and later in the East. During the period before the origin of *anatomically modern humans* (a.m.h.) there appeared a subspecies of *H. sapiens*, first found 150–200 kya, *H. s. neanderthalensis*. More robust than modern humans, it developed in Europe from the *archaic* ancestor and was for a long time the only human type living in Europe and parts of the Near East until the appearance of modern humans. The expansion of Neanderthals to the Near East may have been secondary, having perhaps taken place between 100 and 50 kya. The nearly simultaneous disappearance of Neanderthals in various parts of Europe, coincided approximately with the first appearance of modern humans, between 40 and 30 kya. Neanderthal geographic distribution at its peak is shown in figure 2.1.2. The Neanderthal brain was a little larger than that of archaic *H. sapiens* and even of ours. There is no strong archaeological evidence of substantial behavioral progress, but the tools developed by Neanderthals, called Mousterian, were a little more advanced than Acheulean (*H. erectus* and archaic *H. sapiens*). Whether they had funeral rites is controversial.

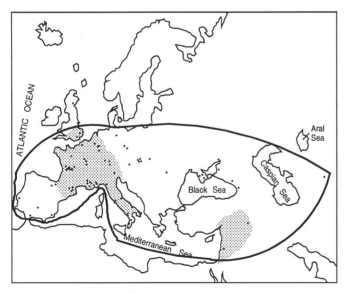

Fig. 2.1.2 Geographic distribution of *Homo sapiens neanderthalensis*. The shaded areas correspond to European and Near Eastern populations. (After Giacobini and Mallegni 1989; Vandermeersch 1989a.)

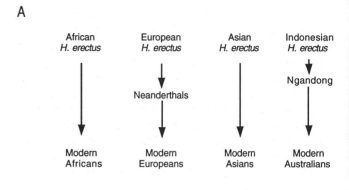

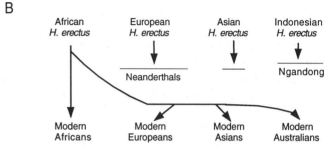

Fig. 2.1.3 The multiregional or polycentric (A) and rapid-replacement (B) models of the origin of modern humans (after Stringer 1989b).

The subspecies to which all a.m.h. belong is called *H. s. sapiens* and its most remote origins were, according to several paleoanthropoplogists, somewhat before 100 kya in East and South Africa. It shows rather clear-cut differences from all earlier humans, including the latest extinct representative, the Neanderthal. For instance, in modern humans, the angle formed by the axis of the skull and that of the face is close to 90°, whereas in Neanderthals, the face is projected forward and the vault of the skull is relatively lower with respect to the face; eyebrows are prominent in Neanderthals, the chin is rounded and not pointed, and there is an occipital bun. These and other differences usually allow one to classify skulls almost unambiguously into the two types. Although the earliest a.m.h. made Mousterian-like stone artifacts not dissimilar from those of Neanderthals, in the last 50 ky they developed a different, more advanced arsenal of tools, called Aurignacian in some important areas and periods, which showed considerable variation in time and space and is therefore given many different names depending on periods and areas. It was suggested (Isaac 1976) that this much increased variation of tool kits accompanies and parallels the use of modern language and its rapid differentiation in time and space.

How did modern humans originate? There are two different explanations (fig. 2.1.3). Weidenreich (1939) first suggested regional continuity, meaning that "racial" types observed today in four regions like the Near East, South and East Africa, North China, and Southeast Asia derived directly from locally found types of archaic *H. sapiens* and even *H. erectus*. Thus, he suggested not one but at least four centers of origin of modern humans. According to this polycentric hypothesis, Neanderthals were an intermediate stage in the transition from early *H. erectus* in Europe to modern Europeans, and modern Australians derived from Indonesian *H. erectus* also via intermediate stages, like the Ngandong skull from Java (a late *H. erectus* dated to perhaps 200 ky, Habgood 1989), and so on for the other regions. This polycentric or multiregional hypothesis was further elaborated upon and strongly defended by Coon (1963) and Wolpoff et al. (1984). The major difficulty of this hypothesis is that it requires parallel evolution over all of the Old World for a long period; the same changes from *H. erectus* to modern humans are postulated to have taken place to a large extent independently in regions far from one another. It is questionable from a genetic point of view whether parallel evolution can continue for a long period and eventually produce organisms very different from the original ones (e.g., in brain size and function, skull shape, etc.) but highly similar to each other (brain has increased, skull has changed shape, brain function has improved dramatically). There are local differences only in some details of skull shape, which seem to have been maintained across this long period of evolution, especially in East Asia. Other interpretations of these phenomena are possible, but we give a detailed discussion of the genetic aspects of the hypothesis later.

Other archaeological work has shown that modern humans appeared in East and South Africa (fig. 2.1.4) long before their appearance in the more remote parts of the world. The earliest African dates are around 100 kya, although with some degree of uncertainty (Border Cave, Klasies River Mouth; see Klein 1989a; Rightmire 1989). Finds at Laetoli, Omo, and others may be older and somewhat more primitive. In fact, in Africa the skull

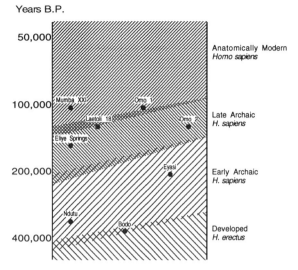

Fig. 2.1.5 Transition from *Homo erectus* to archaic and then to modern *H. sapiens* in East Africa, according to Brauer (1989a). Bodo in Ethiopia, Ndutu and Eyasi in Tanzania, Eliye Springs in Kenya, Laetoli 18 in Tanzania, Omo 2 and Omo 1 in Ethiopia, and Mumba XXI in Tanzania.

evolution from *H. erectus* to archaic *H. sapiens* to modern *H. sapiens* shows some continuity (fig. 2.1.5). The presence of these earlier African types further supports the evidence from initial dates that the first development of a.m.h. occurred in East and South Africa (Bräuer 1989a, b).

The hypothesis that the origin of modern humans took place in Africa is presented in figure 2.1.3b. It assumes that African a.m.h. expanded from Africa to Asia and the rest of the world, rapidly replacing the earlier human types living in these other regions. The replacement hypothesis has the advantage of not demanding a parallel evolution in many remote areas, but does not explain claims of regional continuity outside Africa.

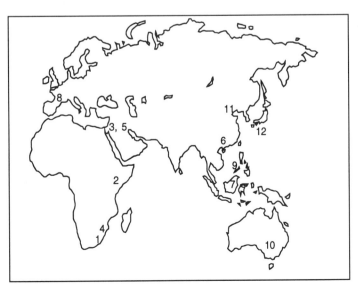

~ 100,000 years old
1. Klasies River Mouth cave
2. Omo
3. Qafzeh
4. Border cave
5. Skhul

probably >60,000 years old
6. Liujiang

probably >30,000 years old
7. Niah cave

about 30,000 years old
8. Cro-Magnon
9. Tabon
10. Mungo

probably >15,000 years old
11. Zhoukoudian upper cave
12. Minatagawa

Fig. 2.1.4 Sites where the earliest remains of *Homo sapiens sapiens* have been found (adapted from Stringer 1988a).

The sample of opinions of modern paleoanthropologists that appear in the most recent reviews (e.g., Giacobini 1989, and the more extensive Mellars and Stringer 1989) shows little consensus. Perhaps a small relative majority is in favor of the African origin and rapid replacement (Bräuer 1989a, b; Rightmire 1989; Stringer 1989a, b), but obviously the correctness of a view cannot be decided by ballot. Wu (1988), Wolpoff (1989) and to some extent Vandermeersch (1989b, c), maintained strong support of the multiregional hypothesis, while others preferred to discuss archaeological facts for and against both hypotheses (Groves 1989; Habgood 1989; Rouhani 1989).

The first appearance of a.m.h. outside Africa is in the Middle East. A number of finds of Neanderthals and of modern humans are in this region, sometimes at short distances from each other, as for example, at Mount Carmel caves. Until a short while ago, they were very poorly dated and believed to be much later (40–50 kya) than suggested by recent datings. Thermoluminescence dating of tools associated with some clearly modern human remains from a cave at Qafzeh, near Nazareth in Israel, provided an unexpectedly early date of 92 ± 5 kya (Valladas et al. 1988) followed by an estimate of 115 ± 15 kya obtained by electron spin resonance (ESR; Schwartz et al 1988). Another collection of modern humans from a nearby cave (Skhul in Mount Carmel) gives a similar date by ESR (Stringer et al. 1989). These dates do not differ greatly from the earliest African a.m.h. Taken in isolation, the Israel dates may raise questions concerning the validity of African origins for a.m.h. by providing a possible alternative origin in West Asia at a similar and perhaps earlier time. Support for an African origin, however, comes from the existence of earlier archaic specimens that seem to be in the direct line of descent to modern humans and that have not so far been found in the Middle East. It has been suggested that early a.m.h. occupation of the Middle East was only temporary. A Neanderthal date obtained by modern techniques from Kebara, 35 km east of Mount Carmel, is 60 kya (Valladas et al, 1987); this date is later by almost 40 ky than a.m.h. in the same area. This may indicate that Neanderthals replaced early a.m.h. settlers in the Middle East. However, the number of known samples of a.m.h. is so small and the areas investigated so few that new discoveries could dramatically alter the picture.

Outside Africa and the Middle East, dated remains of archaic *H. sapiens* and modern humans are definitely from a later period. The earliest evidence of modern humans in Europe is found between 35 and 40 kya (as we discuss in greater detail in chap. 5.); in China, perhaps as early as 67 kya at Liujiang (Brocks and Wood 1990); in Australia, 55 kya or earlier (see chap. 7); in America, at the earliest 35 kya (but according to many, only later, 15–20 kya. An important gap in the record, from 100 kya to 50 kya yields no information of events in most of Asia regarding a.m.h. (see fig. 2.1.4).

The gap might simply be due to the lack of archaeological or paleoanthropological findings, but it correlates with another phenomenon that demands an explanation: there is an important cultural difference between the earliest a.m.h. (ca. 100 kya) from Africa and from the Middle East and the a.m.h. of about 50 ky later. The arsenal of stone tools shows that a clear change occurred in these 50 ky: that of 100 kya is Mousterian, not clearly different from the tool kit of archaic *H. sapiens* or Neanderthals. That of 50 kya is Aurignacian or related, varies locally, and accompanies an intensely active population that is expanding to the whole world. The change of tool kit, its extensive spatial variation beginning at this time (Isaac 1976), and the migratory activities indicate that important behavioral changes took place in this black period between approximately 100 and 50 kya, and that "the appearance of the modern physical form [of *H. s. sapiens*, i.e., a.m.h.] preceded the appearance of fully modern behavior" (Klein 1989b).

2.1.c. TOTAL OR PARTIAL REPLACEMENT?

With all the caution that the current poverty of paleoanthropological information requires, it seems reasonable to assume that a.m.h. advanced geographically from Africa and West Asia toward East Asia, Europe, America, and Australia. Anatomically modern humans are first found in the west and seem to have gradually worked their way eastward to occupy all of the Old World, finally entering the New World from a western direction. Most of the eastward progress is poorly known, but at least the two continents entered last, America and Australia, were clearly occupied by expanding populations originally located in Northeast Asia and Southeast Asia, respectively. It seems reasonable to assume that such a geographic expansion of a.m.h. began in Africa and the western part of Asia and that its extension to America and Australia was simply the terminal phase of this radiation. Our understanding of the early phases of the expansion to Asia (especially its eastern regions) and to Europe is complicated by the fact that these regions were already populated by other human types. Therefore, important questions arise concerning possible replacements, admixtures, or more complex evolutionary phenomena involving the interaction of an expanding a.m.h. and preexisting types, in particular archaic *H. sapiens* and Neanderthals.

1. Archaic *H. sapiens* or earlier types in Asia seem to show superficially Mongoloid and, less clearly, Australoid features in some local fossils of East and Southeast Asia, respectively. This phenomenon suggests regional continuity and therefore a polycentric model; but, if one accepts the hypothesis of an eastward a.m.h. expansion, it could also be interpreted as the result of admixture between the immigrant and the resident populations.

2. Before their disappearance, Neanderthals may have established relations with modern Europeans and hence

may have contributed to the gene pool of modern Europeans. Three hypotheses have been made: (a) the Neanderthals transformed into modern Europeans; (b) a.m.h. from the Near East completely replaced Neanderthals; and (c) partial admixture took place in the expansion of modern a.m.h. from Asia (for fairly recent reviews, see F. H. Smith 1984; Stringer et al. 1984; Trinkaus 1984).

The admixture between expanding a.m.h. and earlier human types is a biological possibility. Barriers to fertility are usually slow to develop, and interspecific differences barring interfertility may take a long time, perhaps a million years or more in mammals, on the average. Barriers to interfertility of a cultural and social nature may be more important than the biological ones. One hypothesis made earlier (see, e.g., Cavalli-Sforza et al. 1988) is that a.m.h. may have owed its greater fitness, which conferred its undoubted capacity to expand demographically and geographically, to better communication—that is, to a higher level of language skills—which may have been the major process accompanying the formation of a.m.h. High-level language skills may have developed in the transition from archaic to modern humans, or in the period of maturation of a.m.h. between 100 and 50 kya mentioned above, or over both periods. This explanation is, of course, hypothetical and difficult to test, but it should not be easily dismissed. It is true that hypotheses of a lower level of linguistic ability in Neanderthals (Lieberman 1989; Marshall 1989) are difficult to reject or confirm, but it is also likely that communicative abilities must have evolved in the genus *Homo* starting at a very low level, comparable to that seen in nonhuman Primates. This long process must have gone through several steps. At the latest step, the most recent human types like a.m.h. are likely to have been provided with better means of communication than any earlier type, including most immediate predecessors like archaic *H. sapiens* or Neanderthals. If the speaking and linguistic skills of a.m.h. were truly different from those of other human types, the chance of genetic exchange between them may have been greatly reduced. Until the last century, humans with impaired possibilities of communication because of hearing or speech defects were essentially barred from reproduction, that is, they had a Darwinian fitness close to zero (Frazer 1976). If admixture with other human types surviving at the time of a.m.h. expansion was possible but limited, we can understand more easily why individuals with the characters we recognize as paleoanthropological markers of a.m.h. (skull-shape traits, the most important if not the only ones we have) spread so effectively across the world, and little, if anything, remained of previous types.

It is of interest to consider in more detail the evidence for admixture or other phenomena that may have occurred with respect to East Asians and/or to Neanderthals. For East Asia, Groves (1989) reanalyzed in some detail the traits that were proposed in favor of regional continuity in East Asia, like facial flatness, large bizygomatic breadth, small frontal sinuses, peculiar nose-root morphology, shovel-shaped upper incisors, smaller development or absence of third lower molars, high frequency of Inca bone (a supernumerary occipital bone), and others, as well as several traits considered characteristic of the Australoid region (discussed in chap. 7). Groves' concluded that the evidence is less strong than ordinarily believed, but did not dismiss it. In section 2.2 we see that multivariate craniometric analysis does not support regional continuity theories but does not falsify them. Other paleoanthropologists have strongly supported a degree of regional continuity in East Asia, and the problem should continue to be given serious consideration until a totally satisfactory answer can be obtained.

The case of the Neanderthals is more complex. Neanderthals were certainly the only occupants of a large area for a long time. They were not, however, homogeneous; European Neanderthals were somewhat more extreme than those in West Asia (Vandermeersch 1989a). In a relatively short time, however, their features disappear and are replaced in the fossil record by those of a.m.h.; in Europe, where data are most abundant, this happens between 40 and 30 kya. This period is far too short for assuming that this is the result of an evolutionary transformation. Moreover, at that time Europe was a mosaic, not only from the paleoanthropological point of view, but also from that of the archaeological record, which is more abundant. The Mousterian culture, typical of late European Neanderthals in this period alternates in different places with the Aurignacian, which is associated with modern humans. Even ignoring the genetic objections to a rapid direct transformation this picture is more in tune with a process of replacement, partial or total, than of transformation. In Howell's summary (1984), the situation with respect to replacement is unclear in eastern Europe, but at least in western Europe the hypothesis of a replacement of Neanderthals by modern humans cannot be ruled out. It would thus seem one should choose between the hypotheses of total or partial replacement, with the first more likely in western Europe and the second in eastern Europe. Partial replacement should be considered, of course, synonymous with admixture. In Klein's (1989a) view, total replacement seems more likely.

The first a.m.h. found in western Europe, the Cro-Magnon people, are typically modern. If they truly are the descendants of modern humans coming from West Asia, it would be surprising if they were unchanged after going through eastern Europe for a few millennia. In this area and period they would have been continuously exposed to potential admixture with Neanderthal. Cro-Magnon seems to emerge essentially unmixed, as far as we can say, and its morphology contrasts sharply with that of Neanderthals from neighboring French regions. In order to explain the lack of signs of admixture in the most western European a.m.h., and still accept the hypothesis that there was admixture in Eastern Europe, one would have to provide alternative explanations; for example, the Cro-Magnon people might have come via Gibraltar to France from North Africa where they might have originated from the Mechta-Afalou type, a strongly modern human (Clark 1972). This is entirely speculative; moreover, the coincidence of near-simultaneous arrivals of a.m.h. from different sources to the extreme west and the extreme east of Europe seems somewhat farfetched.

If one accepts the idea that Neanderthals were completely or almost completely replaced by a.m.h. coming from West Asia, the whole process was fairly rapid, lasting between 5 and 10,000 years, except for the possible Neanderthal survival in isolated pockets. It has been suggested that both Neanderthal and a.m.h. lived side by side, perhaps in somewhat different environmental niches, for a few thousand years, and that a.m.h. may have finally prevailed only because of faster population growth (Zubrow 1989). Evidence suggests that differences in the hunting and relocation customs of Neanderthals and a.m.h. may have given some advantage to a.m.h. Archaeological evidence of fighting has not been found, but it would be difficult to document.

Therefore, even if intermingling with Neanderthals cannot be entirely excluded in Europe, its extent, if any, remains unknown but is probably small. Intermingling might have been more likely in the Middle East, where biological differences with modern humans were perhaps less marked, where the cultural differences 100 kya may have been less important, as shown by the archaeological record, and where a.m.h. and Neanderthals lived in greater proximity. There is no evidence that Neanderthals and a.m.h. lived at the same time; in fact, in the Middle East, they are widely separated in time.

Following the hypothesis that a.m.h. replaced Neanderthals in Europe, the time necessary for a.m.h. expansion does not seem incompatible with other examples. Perhaps one of the most telling examples is the peopling of the Americas (see chap. 6) which, according to many, may have taken only 1000–3000 years from extreme north to extreme south. It is true that the expansion in the Americas took place much later than that of modern humans from Africa, perhaps 15–12 kya, and happened in a population vacuum, but it may have begun earlier. The expansion to Australia began at about the same time as that to Europe, or perhaps somewhat earlier, and may also have taken a similar length of time (see chap. 7) at least for occupation of the coastal regions.

If we look at the two hypotheses shown in figure 2.1.3, we conclude with a definite preference for rapid replacement, but we find it reasonable to leave open the possibility of complementing it by retrogressive hybridization (or gene flow) resulting from mixture with local types,

which may have differed at various places and times. Today modern humans are rather homogeneous in skull morphology (Howells 1989), and also (see sec. 2.3) in gene frequencies. These qualitative observations indicate that the contribution from more ancient, local human types was not a major one. It should, however, be sufficient to explain the examples of regional continuity that have been brought forward if they resist further critical quantitative analysis. Another limit to the acceptance of genetic contributions to modern humans by earlier human types through gene flow is set by observations on human mtDNA (discussed in sec. 2.4).

It might be useful to outline here why it is difficult to accept genetically the hypothesis that parallel evolution for a million years or more generated the present human aboriginal populations in four continents. If we first disregard genetic exchange (gene flow) and its possible influence on the process, we must justify the acquisition of a very similar external appearance (phenotype) by all modern humans, by independent evolutionary processes. There is no reason why independent processes should lead to essentially similar results. Artificial selection experiments for quantitative traits in animals show complex responses for a trait determined by many genes, even for different samples taken from one population (Mather 1949; Falconer 1960; Mather and Jinks 1977). There is evidence of the kind in humans for a common selective stimulus, malaria, which has led to different genetic responses in various parts of the world. If selective processes for the same or similar phenotypic response—for example, increase of brain size associated with improvement of certain brain functions—had occurred independently in many regions, genetic results would certainly have been quite different, even if perhaps similar at the phenotypic level. We do not know which genes would have been selected in these processes, but their linkage with nearby genes would have allowed them to "hitchhike" with these neighboring genes, generating considerable differences in the genetic backgrounds of various human groups. By contrast, data from a high number of genes prove that differences in genetic background of the most distant human populations are small compared with the variation among individuals of the same populations (Lewontin 1972; Nei and Roychoudhury 1972). Moreover, we shall see that the genetic variation in space observed today is comparable to what would be expected in the time available since the a.m.h. expansion. If the increase, say, in brain size and function were due to independent selective processes in the various continents over a million years or more, the results of these processes could also be expected to generate considerable genetic diversity in the genes that we observe.

These considerations seem fatal to a narrow polycentric theory à la Coon (1963, 1965), that is, without any exchange between continents. Wolpoff et al. (1984) extended the theory to include gene flow, referring specifically to a simulation by Weiss and Maruyama (1976), which showed that a regular exchange network (a stepping-stone model; sec. 1.16) covering the whole world could be very effective in altering estimates of time of fissions based on genetic distances. The applicability of such simulations to human expansions, which they were not designed to cover, is limited by the short time available to human evolution, in which it is unlikely that an equilibrium would be reached; by the complications of a complex system of physical barriers, and above all by the cascades of expansions determining new settlements, colonizations, invasions (see Weiss 1988, and sec. 2.7), and other phenomena. Simulations different from that of Weiss and Maruyama should be made to describe the problems we are considering here. Massive movements of populations have occurred in the last 50 ky, and a model that ignores these dynamic aspects is unlikely to be acceptable.

It might be useful to describe what is needed for a simple model to incorporate the hypothesis of polycentric origin and that of rapid replacement. This will also help to clarify the difference between them. It is essential, for this purpose, to supplement both of them with local gene exchange, so that all population units exchange genes with neighbors. In the current polycentric model, as defined by the Weiss and Maruyama simulation, the migratory exchange with neighbors is such that the same proportion of population is exchanged from population A to B as from B to A, and this is valid for all pairs of units. This is the standard assumption in stepping-stone models, and it also implies static demographic conditions, that is, stationary populations in all units. This is not true of the rapid-replacement model, supplemented by gene flow; here the western populations (the African ones) must be growing faster and perhaps also to a higher population density than the rest. This favors outmigration in the west-to-east direction. As a consequence, more migrants go in the west-to-east direction than vice versa, causing genetic replacement in the east. A similar model, specialized for describing another situation, was suggested for the agricultural expansion from the Middle East to Europe (see sec. 2.7 and chap. 5).

2.1.d. ONSET OF FOOD PRODUCTION

As we come closer to the present time, the passage from Pleistocene to Holocene is accompanied by one major change; the beginning of food production (by agriculture and animal breeding). Human population density must have slowly increased during the Pleistocene and may have been a potent factor favoring cultural innovation (Boserup 1965; Cohen 1977). Perhaps overkill of the fauna in the late Pleistocene, potentially a consequence of the same overpopulation factor, and climatic changes that may have required a difficult adaptation to the changing fauna and flora, were important additional stimuli to the development of the new technology that allowed the transition from foraging to food production. These climatic and demographic changes may well have occurred almost simultaneously worldwide, and therefore it is not too surprising that farming began independently and almost simultaneously in many parts of the world. Naturally, local crops and animals that were already used by the local foragers were employed for domestication and, with it, increased food production. Major areas in which farming originated were in the Middle East, Southeast Asia, Central America, and the northern portion of South America (see sec. 2.7 and later chapters). Not surprisingly, these were all temperate or subtropical zones that, for reasons of climatic eligibility, must have been, then as now, more heavily populated. The techniques of farming spread rather slowly around the areas of origin; usually the farmers themselves grew in numbers and started expanding geographically, as we discuss later. Agriculture certainly allowed a quantum jump in potential population density (the ecologist's "carrying capacity of the land") when compared with that possible by the previous economy of foraging (hunting-gathering), but it was inevitably limited to certain environments where food production was technically feasible. Arctic tundra, deserts, and tropical forests are inevitably closed to, or less suitable for, food production, except under special circumstances or for rare domesticates (e.g., reindeer for the tundra, camel for the desert). The increased carrying capacity of the land because of agriculture may have been 10- to a 100-fold, even for early forming. As the technology allowed increased population density, it must have stimulated the occupation of neighboring areas when the population grew above the new, higher limits of land saturation. In some particular areas, however, like the Pacific coast of North America (see Hassan 1975) and Japan (see Koyama 1978), foraging populations reached high population densities without agriculture. In the absence of a stimulus for technological change, farming did not develop spontaneously and diffused from the outside only much later.

Technologies of animal and plant breeding underwent rapid evolution and generated continuous increases in land carrying capacities. The rise in population density as a result of agriculture has been a major factor of change in human populations. It made possible the origin of cities and with them, civilization (from the Latin word *civitas* meaning city). Other technological changes have also been of great importance in determining growth and migration of people and, therefore, changes in the patterns of population structure.

2.1.e. POPULATION NUMBERS

The population of the Earth is estimated today from national censuses taken at regular intervals by every country. These censuses suffer from serious errors. Cen-

Table 2.1.2. World Population Estimates, in Millions, at Various Dates (Biraben 1980)

Population	B.C. 400	B.C. 1	A.D. 500	A.D. 1000	A.D. 1250	A.D. 1500	A.D. 1750	A.D. 1970
China	25	70	32	56	112	84	220	774
India, Pakistan, Bangladesh	30	46	33	40	83	95	165	667
Southwest Asia	43	49	43	36	25	27	29	118
Japan	1	2	5	4	9	10	26	104
Remainder of Asia (no USSR)	3	5	8	19	31	33	61	386
USSR	13	12	11	13	14	17	35	243
Europe (no USSR)	23	35	29	30	57	66	109	462
Northern Africa	10	14	11	9	9	9	10	87
Remainder of Africa	7	12	20	30	49	78	94	266
North America	1	1	2	2	3	3	3	228
Central and South America	5	8	11	14	23	34	15	283
Oceania	1	1	1	1	2	3	3	19
Total	162	255	206	254	417	459	770	3637

suses were rare in earlier times; perhaps the oldest censuses partially conserved today were taken in China at the time of the Han dynasty, almost 2000 years ago. Numbers for earlier periods are guesstimates that become inevitably more and more uncertain, the further back we go. There have been several attempts at calculating world population sizes for the late Paleolithic and they vary considerably. We use provisional estimates offered by Biraben (1980) who gives 400,000–800,000 for the period immediately before the expansion to Europe of a.m.h., increasing thereafter 3–5 million, until 8000 B.C., when the Neolithic revolution began to expand. In the next 8000 years the growth rate was high, although

certainly irregular; in A.D. 1, the population is estimated to have been around 250 million. For the last 2400 years, data begin to be more respectable so that a breakdown by major regions is meaningful. In table 2.1.2 we give an abstract of Biraben's estimates. A variance analysis of the full table showed that the local growth rate did not vary by regions, but there is very highly significant variation by periods. The numbers after A.D. 1500 also include immigrants from other continents, mostly Europe, but until recently they showed a decline mostly because of epidemic diseases imported by Europeans, slave raids, wars, and other major causes of population decline.

2.2. EARLIER QUANTITATIVE PHYLOGENETIC STUDIES

The first attempt at reconstructing human evolution on the basis of genetic data from living populations was undertaken in 1964 by Cavalli-Sforza and Edwards, who calculated genetic distances between pairs of selected populations for as many genes as knowledge at the time permitted and developed methods of reconstruction of trees of descent, based on simple genetic theories. The methods were briefly described in section 1.11. Fifteen populations were chosen, three from each continent; the genes were *ABO, MN, Rh, Diego,* and *Duffy,* with a total of 20 alleles. There is still something to be learned from this early attempt and therefore it will be briefly summarized. The results of two different methods are shown in figures 2.2.1 and 2.2.2. In the first figure the phylogenetic tree is given schematically, and in the second it is projected on the geographic map.

The evolutionary model behind the tree used in figure 2.2.1 assumes population fissions and then independent evolution of the branches. From a genetic point of view, *independence* is the expectation that there is

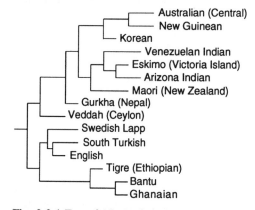

Fig. 2.2.1 Tree of 15 populations reconstructed on the basis of 20 alleles, using the method of additive evolution (after Edwards and Cavalli-Sforza 1964).

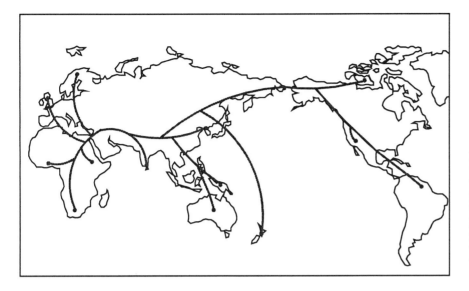

Fig. 2.2.2 The same data used in figure 2.2.1, with the unrooted tree reconstructed with minimum path and adjusted for display on the geographic world map (after Edwards and Cavalli-Sforza 1964, modified). The branch to the Maori was moved slightly with respect to the original.

no selective convergence or divergence of the genetic markers in some branches, and there are no important fusions or exchanges between the branches. These expectations are certainly not met completely, but minor deviations from these hypotheses would be practically inconsequential and hard to detect. Genes for which selective convergence in some populations could have been expected—for example, because they confer malarial resistance—were excluded. There are no other indications of selective discrepancies of a nature and intensity such that they could affect tree reconstruction. Some deviations resulting from major exchanges could be real, however (see sec. 1.17, 2.4, and various other places in later chapters). The method of analysis used in figure 2.2.1 (additive evolution) is faithful to the evolutionary model indicated above, because it assumes that evolutionary change in each branch is independent of the others so that the distance observed between any two populations is expected to equal the sum of the length of the tree branches between the two populations. Segment lengths were estimated by least squares for a specific tree topology. At first a reasonable topology was sought by the method of "cluster analysis" (Edwards and Cavalli-Sforza 1964) and improved upon by trying several similar topologies. Competing topologies were compared on the basis of their goodness of fit, estimated for each topology by the sum of squared deviations between expected and observed segment lengths. There is no assumption of a constant rate of evolution in the branches, and the method does not produce a root for the tree. In order to place a root in figure 2.2.1, it was assumed that the rate of evolution was constant in the path between the two most distant populations (Ghanaians in Africa and Australian aborigines), a method also used more recently by others (see, e.g., Cann et al. 1987). This method, however, does not consider population pairs less distant from the origin. In fact, in figure 2.2.2 there is a considerable heterogeneity of distances from the origin of the various populations, a fact that was perceived at

the time but was difficult to test statistically. Later experience has confirmed that several of these populations—for example, Europeans and Indians—always had shorter branches. The explanation for this was given only later in terms of admixtures (see secs. 1.17 and 2.4).

The method used in figure 2.2.2 offered a solution by a different approach: representing all populations in an *n*-dimensional space, where *n* is the number of gene frequencies, and calculating the length of a "string" connecting all populations and their ancestors in this space. The best tree is fitted by varying the topology and by calculating for each the ancestral positions that minimize the total string length; the topology showing minimum string (or minimum path) is chosen. Again, no root is produced by this method. To place the minimum-path tree on the world geographic map in figure 2.2.2, it was necessary to use an unusual world map having the Pacific Ocean in the middle, because the more common world map centered around the Atlantic Ocean does not allow one to represent the unbroken tree satisfactorily; we know, in fact, that the passage to America took place from Northeast Asia in the northern part of the Pacific. Can the minimum string path reproduce to some extent the actual paths of expansion of a.m.h.? The minimum-path tree corresponds to the shortest system of routes connecting the populations examined in the gene-frequency space, but not in the geographic space. Given the geographic positions of populations and connecting them with segments reproducing the tree, the lengths of the tree segments calculated in the gene-frequency space must inevitably be distorted in order to accommodate the tree on the geographic map, as was done in figure 2.2.2. Could the lengths of the geographic segments be used to calculate local rates of geographic expansion? Even if we knew all the necessary dates, the answer is negative, for various reasons. First of all, there is not enough information to place on the geographic map most points corresponding to ancestral tree nodes. For instance, the geographic position of the fission corresponding to Maoris was placed incorrectly in Alaska in the original paper. It could have been located in many other positions and still be formally compatible with the tree. We now know it should probably have been placed in some part of Southeast Asia and should have followed a devious route via Tonga and the Marquesas, as is true of most other Polynesian colonists (see chap. 7). Many populations may have moved considerably between the time of fission and the present, as Polynesians did, and the genetic tree usually does not provide enough information to reconstruct the entire path of the migrations.

The topologies obtained by the two methods are very similar, but not identical. As noted in chapter 1, different methods can, and often do, lead to different reconstructions. This should not be surprising, given the enormous number of possible tree topologies; there is, in fact, extremely little difference in the goodness of fit of the best topologies that can be calculated, and usually the best topology is found to give a better fit than the next-best ones by exceedingly trivial amounts. Very small changes in the original data, distances, or methods of tree reconstruction are bound to cause some variation of tree topology. Because of differences in results, however, one may find it difficult to choose the most satisfactory method for tree reconstruction. In our view, it is best to choose a method based on a specific evolutionary model that seems reasonable and can give rise to clear-cut expectations, so that the goodness of fit of the evolutionary model behind the tree can be tested and contrasted with other specific evolutionary models if necessary. This is the usual procedure in all scientific investigations. In the case of gene frequencies, the minimum-path method is not the most satisfactory because it does not correspond to a specific evolutionary model. It does, however, tend to give similar results to those obtained with other methods.

More genes, in particular, a few proteins that were studied in a sufficient number of populations (Kidd

1973) and HLA data (Piazza et al. 1975) confirmed the results obtained in the first investigation, including the position of the root we originally found. Extension of the work to a much larger number of genes, which were later discovered, has not confirmed the original location of the root. Nei (1978) extended considerably the number of markers employed in the analysis by studying the new enzyme polymorphisms that became available in large numbers in the late 1960s. These were originally investigated in only a few populations. Limiting the research to three more intensively studied ethnic groups—Europeans, Africans, and East Asians—he showed that Europeans were indeed closer to Africans than to East Asians for the blood groups then known (in agreement with our root), but were definitely closer to Japanese than to Africans for proteins and enzymes. The latter type of genes carried greater statistical weight so that in the overall analysis, the root was located between Africans on one side and Europeans and Orientals on the other. The location of the root between Africans and non-Africans was later confirmed on a larger sample of populations (Nei and Roychoudhury 1982). The most complete analysis on the three most highly investigated populations representative of the three major ethnic groups (mainly British for Europeans, mainly Japanese for East Asians, and mainly from Nigeria and Cameroon for Africans) was published recently (Nei and Livshits 1989) on a large number of polymorphic loci (see table 2.2.1). These authors concluded that they have for the first time shown that Europeans and Asians are *significantly* closer to each other than to Africans. There are some discrepancies between their results and ours, especially for HLA, but the overall conclusions are in agreement with our most recent ones, to be discussed in the next section. We must therefore consider the possible causes of the discrepancy between our early and our later results.

Table 2.2.1. Average Genetic Distances (Nei's) and Their Standard Errors among Three Major Ethnic Groups Based on 186 Polymorphic Loci (Nei and Livshits 1989)

Genetic Loci	No. of Loci	Europeans/Asians	Europeans/Africans	Asians/Africans
Standard genetic distance				
Proteins	84	0.028 ± 0.009	0.035 ± 0.009	0.048 ± 0.012
Blood groups	33	0.019 ± 0.010	0.059 ± 0.032	0.082 ± 0.041
HLA and immunoglobulins	8	0.329 ± 0.122	0.701 ± 0.341	0.386 ± 0.169
DNA markers	61	0.060 ± 0.012	0.081 ± 0.017	0.109 ± 0.025
Total	186	0.040 ± 0.007	0.063 ± 0.011	0.078 ± 0.013
Nucleotide (restriction-site) differences per site (enzyme)				
MtDNA[*]		0.02	0.05	0.04
MtDNA[†]		0.032 ± 0.048	0.045 ± 0.069	0.036 ± 0.069

[*] From Cann et al. 1987.
[†] From Brown 1980.

1. Do blood groups involve a bias? According to the first investigation by Nei (1978), it seemed as if the discrepancy between the positions of the root was a result of a disagreement between blood groups and proteins plus enzymes. Our early conclusions were based entirely on blood groups and have been confirmed recently by a careful and more complete study of Langaney's group (Excoffier et al. 1987; Sanchez-Mazas Cutanda 1990) with the most important blood groups we employed and with other immunological markers. But in very recent work, with a greater number of genes and populations, the difference between blood groups, proteins, and enzymes seems to have disappeared (Nei and Livshits 1989, and our own results, discussed in the next section). There is some indication that *HLA* may also favor the African-European connection (see sec. 2.10). The question of bias cannot be entirely excluded, at least for the blood groups we used at the beginning.

2. Was the initial sample of 20 (15 independent) alleles from 5 loci adequate? Most probably not. Had methods for estimating standard errors of segment lengths of trees been available at the beginning of this analysis, one would have most probably found no significant difference between the location of the root between Africans and non-Africans, and that between Africa + Europe and the rest. Our present analysis of all classical markers available on a larger number of populations is based on almost eight times more markers and confirms the position of the root between sub-Saharan Africa and the rest of the world. The same is true of our more recent work on 100 DNA polymorphisms (see sec. 2.4) based on a smaller number of populations.

3. Was there a special reason connected with the position of Europeans for missing the African/non-African split in 1963 and 1975? The answer is affirmative. We see in section 2.4 that Europeans (and other Caucasoids) have an intermediate position between Africans and the rest of the world: Asians + Americans and Oceanians. The shortness of the branches leading to Europeans and other Caucasoids in figure 2.2.1 is part of the same phenomenon, which was already discussed in section 1.17. Under these conditions, it is very likely that the root is more difficult to place.

4. How did this approach compare with knowledge from other sources, in particular, archaeology and physical anthropology? In spite of the small number of genes then available, the trees of figure 2.2.1 and figure 2.2.2 seemed quite satisfactory at the time they were obtained, at least by the criteria that they collected populations from the same continents together in clusters and grouped continents in agreement with their geographic proximity. There was no real external evidence—for example, from archaeology—with which to test this result. The only substantive difference between the two trees in figures 2.2.1 and 2.2.2 and the trees generated later is the placement of the root. It was possible, however, to exam-

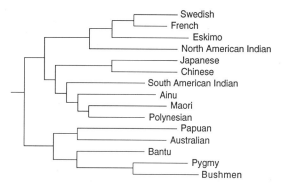

Fig. 2.2.3 Tree based on general anthropometrics (after Cavalli-Sforza and Edwards 1964).

ine independent evidence from anthropometric measurements. Data were available in the literature but had not been previously used to reconstruct a phylogenetic tree. Accordingly, at the time the first genetic trees were produced, we also constructed a tree from anthropometric characters, including measurements of the whole body and skin color (Cavalli-Sforza and Edwards 1964). This tree (fig. 2.2.3) showed marked differences from that obtained with genes; for instance, Australian aborigines and Africans were closely associated, whereas with genes these populations are the farthest apart. It seemed clear to us that the sensitivity of many anthropometric characters to climate was likely to bias the reconstruction of phylogenetic history. It has been well known since Darwin that adaptive traits are frequently not satisfactory for reconstructing phylogeny, because they express similarities of environments more easily than those of phylogenetic history. We concluded that the lack of agreement between the two types of trees was no cause of alarm, and that genes were more likely to reflect phylogenetic history. In fact, Africans, Australian aborigines, and New Guineans have been exposed to tropical climates for a very long time and are presumably highly adapted to them. The characters available for this first anthropometric investigation were essentially connected with body surface, in particular skin color and size measurements, which are known to be correlated with climate. Bergmann's rule (the correlation with external temperature of the ratio of body surface to weight) and Allen's rule (a similar correlation for the ratio of length of appendages to that of the trunk) are known to hold for animals and have been verified for humans (D.F. Roberts 1973). The correlation of skin color and climate is discussed further in section 2.13, where a world map of skin color is also given (fig. 2.13.4).

The anthropometric characters available for the 1963 attempt were, however, far from ideal; mean values had been taken from the data reported in Martin's classical handbook of anthropology (Martin 1957) and came from a number of investigations using insufficiently standardized techniques. Without collecting new data, there was

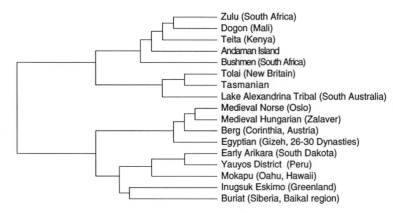

Zulu (South Africa)
Dogon (Mali)
Teita (Kenya)
Andaman Island
Bushmen (South Africa)
Tolai (New Britain)
Tasmanian
Lake Alexandrina Tribal (South Australia)
Medieval Norse (Oslo)
Medieval Hungarian (Zalaver)
Berg (Corinthia, Austria)
Egyptian (Gizeh, 26-30 Dynasties)
Early Arikara (South Dakota)
Yauyos District (Peru)
Mokapu (Oahu, Hawaii)
Inugsuk Eskimo (Greenland)
Buriat (Siberia, Baikal region)

Fig. 2.2.4 Tree based on skull metrics (after Howells 1973).

no hope of obtaining enough individual data to calculate the matrix of correlations between characters within populations, which is indispensable for using more refined statistical treatments like Mahalanobis's (1936) generalized distances. Theoretical considerations make it clear that correlations within populations need to be considered for anthropometric traits but not for genes unless they are very closely linked and could be neglected.

The careful craniometric analysis by Howells (1973) overcame the limitations of our 1963 study of anthropometrics. The author used a standard technique to measure a large number of traits on a large number of skulls from 17 world populations and carried out a discriminant analysis among them, thus eliminating the effect of intrapopulation correlations between the traits. Measurements in his study were made by a single observer, thus eliminating systematic errors that can be especially serious with craniometric traits. The tree that resulted (figure 2.2.4) was again in contrast with the genetic tree and rather similar to the anthropometric tree first described by us. But it was possible to prove (Guglielmino-Matessi et al. 1979) that part of the discrepancies between the tree based on the metric analysis of skulls and the genetic tree could be explained by a strong influence of climate. The first discriminant function of Howells, which was strongly associated with the root of the tree, was found to be very highly correlated (about 80%) with climatic variables, especially temperature, as well as with skin color, which is also highly correlated with climate. Howells' first discriminant was, in fact, determined to some extent by size measurements, which are inevitably correlated with general size, and these are known to be highly correlated with climate. By contrast, Howells' second discriminant was in reasonable agreement with the first dichotomy of the genetic tree. This is, we think, an important lesson on the problems encountered in the use of anthropometric measurements for evolutionary considerations. There may be ways of eliminating climate effects on anthropometric measurements and some preliminary work done by Guglielmino-Matessi et al. (1979) allowed a partial correction. Tests of the effects of climate on genes

(Piazza et al. 1981b) showed that although some genes were correlated with climate, on the average, the correlation with climate was much more pronounced for skull measurements and other anthropometrics (see sec. 2.13).

Using a different approach to further analyze the same material, Howells (1989) attempted to separate size from shape differences and construct trees based on shape differences, recognizing, however, that methods for separation of size and shape are not perfect. Trees for skull shape were very similar, but not identical to the earlier ones based on distances for the whole set of measurements. Africans and Oceanians (Australians and New Guineans) showed even greater similarity than in the former study and tended to form close clusters. It seems logical to conclude that shape, and not only size, are strongly influenced by natural selection by climate. Especially face shape is sensitive to climatic factors.

Using the same technique to compare all the human populations examined with some of the available fossil skulls yielded additional important results. Howells'(1989) findings on this point are summarized together with those of similar studies also using multiple craniometric measurements and multivariate analysis. They include especially contributions by Stringer (1978), Vark (1985), and several others also cited by Howells (1989), whose review we follow here and which should be consulted for further details and references.

Multivariate craniometric analysis showed two Neanderthal skulls to be completely distinct from any modern human skulls and to show no special resemblance to modern European skulls. Moreover, a.m.h. skulls—for example, Border Cave, Skhul, Kafzeh, and older African skulls—show no tendency to be associated with any particular modern group. A Zhoukoudian Upper Cave skull (near Beijing), dated from about 18 kya and decidedly modern (see chap. 4), was originally described by Weidenreich (1945) as Mongoloid, but shows no similarity to modern Mongoloids; by contrast, Chinese neolithics (7000 years old) can be recognized as similar to modern Chinese. A skull from the Cape, 12–13 ky old, shows a resemblance to modern Khoisan Bushmen. This seems to be the oldest South African human recognizably associated with Khoisan. Using present techniques, it seems difficult to ascertain reliable resemblances between skulls older than 10–12 ky and modern regional specimens from the same or a related area.

Multivariate craniometry has the great advantage of being more objective and comprehensive than inspection, qualitative description, or the small set of measurements which, until a short while ago, were the only procedures employed in physical anthropology. It is limited, however, by the need of specimens that are reasonably complete. Unfortunately fossil skulls that meet this requirement are rare. The method does not ordinarily include rare qualitative traits (e.g., the Inca bone) considered important in the problem of polycentric origins. The use of a great number of measurements may lose or dilute characters of shape that are more easily appreciated by direct inspection; in other words, there may also be a price to pay for objective measurement. Multivariate craniometry does not, therefore, completely replace the more conventional cranioscopic analysis on which older studies were based.

Another method of analysis that, like craniometry, has the advantage of being applicable to fossil material is the study of teeth, as practiced by Turner (1987, 1989). This interesting method has not yet been extended to the whole world, although it has already covered very wide areas. Results are described in chapters 4 and 6.

In the next section of this chapter, we discuss results obtained on the large collection of "classical markers" assembled for this book, that is, those that have become available from protein and immunological studies. We consider in section 2.4 the analysis of DNA markers, which adds independent information. Our results are also compared with archaeological data in order to study the constancy of evolutionary rates by calibrating genetic distances versus archaeological dates (sec. 2.5). Comparisons are made with the linguistic classification, which complements the information obtained with the genetic tree (sec. 2.6). One cannot emphasize enough the need for reaching agreement among independent approaches to human evolution. The problem is sufficiently complex that it is unlikely to be entirely solved with a single source of information.

2.3. ANALYSIS OF CLASSICAL MARKERS IN FORTY-TWO SELECTED POPULATIONS

2.3.a. TECHNICAL ASPECTS

Clearly the number of genes studied is of paramount importance in determining the confidence one can place in the results. At the same time, it is important to use a balanced sample of populations that are as close as possible to what may have been the (usually unknown) aboriginal set of populations.

The main practical problem is to sacrifice as little of the available information as possible in selecting populations and genes for our data base. Ideally, the genes-by-populations matrix should be without missing data. But the original matrix of data is very far from complete, and selections are therefore necessary. If one tries to keep the number of genes as high as possible, the number of populations is reduced excessively, and vice versa. Various strategies can be employed to obtain a representative set of populations without eliminating too much genetic information, but important help, in our experience, comes from some possibilities of coping with gaps.

Our populations were defined by the name given to them by the scientists who published the data. In order to take synonymies into account, we used two very useful sources, *Classification and Index of the World's Languages* (Voegelin and Voegelin 1977) and the *Ethnologue* (Grimes 1984). We also benefited from advice by M. Ruhlen. We pooled information from different authors referring to the same population. About 1950

populations differing by name were retrieved from the literature and used for geographic maps of single alleles. A first selection for multivariate analysis brought their number down to 491, by eliminating those too poorly known genetically and pooling populations that were geographically, ethnically, and linguistically reasonably homogeneous. The list of populations in Appendix 3 shows poolings, but not eliminations. The internal genetic variation was estimated at every step by calculating F_{ST} values. The different populations inevitably have variable genetic homogeneity, and the specifications available in the original papers do not always allow a precise ethnic and geographic characterization. Some of the populations that were less specifically characterized were labeled "unspecified," kept separate and employed only for higher-level pools. The 491 populations were the basis for the analysis of the detailed trees and PC maps discussed in chapters 3–7. Their gene frequencies are given in Appendix 2, and their ethnic composition and bibliographical references in Appendix 3, which is followed by bibliographic references. For the analysis at the world level in this chapter, it was necessary to reduce the number of populations by about an order of magnitude, and a second cycle of pooling and culling permitted a further increase in the total number of genes per population. The final sample used for multivariate analysis at the world level, to be discussed in this section, included 42 aboriginal populations and 120 alleles.

The list of populations and their gene frequencies are given in Appendix 1.

The process of pooling populations that are close to each other on the basis of geographic-ethnic-linguistic criteria has certainly generated some internal genetic heterogeneity, which is, however, minimal compared with that among the 42 populations. But, in addition to increasing the number of genes per population—an essential requirement for our purposes—it has also helped eliminate a certain amount of variation caused by drift, which is inevitable in small isolates that would otherwise be responsible for unwanted noise. Of the 42 populations, 7 were from Africa, 7 from Europe, 4 from Asian Caucasoids, 14 from other Asians, 5 from the Pacific Ocean, and 5 from America (see Appendix 1). In the section on linguistic comparisons 2.6, we have provided additional information on the 42 groups. The later chapters illustrate further the criteria of selection and grouping by discussing more specific analyses for each continent.

Among the groups selected for the purpose of reducing excessive drift and increasing the number of markers per group, a few relatively small isolated populations, which may have drifted more than average, were retained because of their intrinsic interest. Five such cases are discussed in more depth in the appropriate chapters: Mbuti Pygmies, !Kung San, Lapps, Samoyeds, and Sardinians. The first three today consist of 30,000–50,000 individuals each. The two important ethnic groups of African Pygmies and Khoisan were selected because they are least acculturated and have probably undergone the least admixture with neighbors. Mbuti Pygmies are from the Ituri forest in Zaire and the !Kung from Botswana-Namibia. They are most distinctive when compared with other Pygmies or Khoisan. Lapps are becoming largely acculturated but have not entirely lost their identity. Some Samoyed groups are much smaller than others but have been averaged with other more numerous ones. Sardinians number today about 1.5 million; they differ from neighbors probably because of drift to which they were exposed in the initial phases of settlement of this island.

Missing data ("gaps") account for 24% of the gene frequencies used for the 42 populations given in Appendix 1. The analysis was performed by first calculating a table of distances between pairs of populations, omitting genes not present in both populations, and averaging genetic distance (weighted by allele) for alleles present in both populations. F_{ST} genetic distances (see sec. 1.11) were used throughout unless otherwise stated.

The decision to use an incomplete data matrix needs detailed justification since it is unusual and may give rise to doubt. We believe the following reasons completely justify our decision.

1. The usual alternative of eliminating genes or populations so as to avoid *all* gaps would have caused an unacceptable reduction of the data base, making conclusions close to meaningless.

2. A method sometimes used, the filling of gaps by the interpolation from gene frequencies of geographic neighbors, may go too far in violating the rule of independent evolution. With populations as genetically remote from each other as our 42, geographic "neighbors," whose average would have to be used for replacing missing values, are fairly distant from one another.

3. A method sometimes employed, replacing missing data with mean values, would have introduced an unacceptable bias. All populations with large numbers of missing genes would have been shifted artificially toward the general mean.

4. A simulation with extensive resampling experiments by the bootstrap was made on a complete subset of 20 world populations (mostly representing Europe) for 89 genes. This complete data matrix was impoverished by the random deletion of various percentages of the data. The effect of gaps was studied on PC maps rather than on trees (J. Mountain et al., unpubl.). It is easier to compare PC maps to one another than to compare trees. Twenty-five percent gaps had practically no effect on the PC map; 50% gaps, however, altered somewhat the picture and introduced some discrepancies. The random elimination of genes for the simulation was made with uniform probability for each gene and with widely different probabilities for different genes. There were no important discrepancies between the two approaches.

In the case of the 42 populations, we are thus within the percentage of gaps considered safe. Probably the mean and variation of the number of genes represented in all population pairs are more important in determining the acceptability of a data matrix, and in critical cases we have made use of this further information. In general, some genes are tested much more frequently than others, and consequently the number of genes available for both populations when calculating a distance is definitely higher than if the gaps were completely random. In any case, whenever we excluded populations from calculations, we always proceeded by eliminating those tested with the lowest numbers of genes.

5. The analysis of F_{ST} for the various genes (alleles) shows that their general distribution is almost uniform, both at the world level and in the various continents (see sec. 2.10). The F_{ST} measuring the world genetic variation had a distribution among genes very similar for the sample of genes considered here and for another entirely independent sample of 100 DNA polymorphisms reported in section 2.4.

Table 2.3.1A. F_{ST} Genetic Distances between the 42 Populations, (×10,000) (lower triangle) and Their Standard Errors Calculated from the Bootstrap Analysis (upper triangle)

	BAN	EAF	NIL	WAF	SAN	BER	MBU	IND	IRA	NEA	URA	AIN	JPN	KOR	MNK	THA	DRA	MNG	TIB	INS	MAL	FIL	NTU	SCH	BAS	LAP	SAR	DAN	ENG	GRK	ITA	CAM	ESK	NAD	NAM	SAM	CHU	MEL	MIC	PLY	NGU	AUS
Bantu	0	149	36	43	290	227	450	487	394	747	706	643	469	626	739	450	993	484	758	562	633	619	720	358	402	677	410	533	449	474	536	474	589	565	586	530	508	775	581	557	721	622
E. African	658	0	202	196	776	534	1232	1078	1060	709	1690	1771	1345	1475	1538	1602	1066	1423	1126	1669	1216	1700	1600	1386	1664	1570	988	1600	909	1163	344	323	425	400	302	344	373	503	467	305	588	412
Nilo-Saharan	118	828	0	53	331	207	750	2073	2077	1823	2360	2161	2511	2368	2881	2978	2396	2519	2391	2703	1714	3033	2599	2302	2863	1570	1697	2599	1723	1767	626	690	540	637	560	613	497	871	909	623	899	822
W. African	188	697	53	0	284	186	801	1748	1796	1454	1921	2161	2252	1807	1951	2055	2480	2243	1717	1709	1365	2299	2062	2163	1958	1299	1328	2062	1459	1487	484	568	431	516	402	462	524	791	628	438	661	580
San	94	776	331	284	0	966	1642	2073	2077	1823	1921	2414	2511	2368	2881	2978	2396	2519	2391	2703	1714	3033	2599	2302	2863	1307	1697	2599	1723	1767	342	465	461	421	404	439	535	586	571	429	703	553
Berber	163	534	207	186	966	0	885	750	697	1002	1921	2161	2161	1475	1951	2161	2161	1761	1243	1709	1365	2062	2062	1958	1307	1299	1307	1459	892	892	51	341	347	314	275	280	376	520	528	192	650	570
Mbuti	714	1232	750	801	1642	885	0	1495	1748	1454	2360	2414	2252	2368	2766	2703	1851	1709	1365	2703	1216	3033	2599	2302	2863	1299	1697	2599	1723	1767	840	874	755	838	740	833	779	1040	1036	902	1032	1026
Indian	2202	1078	2073	1748	2073	750	1495	0	692	885	497	2663	2905	1411	1991	2064	1760	1434	1224	1548	849	1922	1442	1448	2231	1570	619	736	619	497	45	244	193	206	170	198	220	178	251	211	267	155
Iranian	2241	1060	2077	1796	2077	697	1748	2663	0	158	408	2588	2138	1950	1977	2480	1950	1717	2354	1625	1434	1268	1167	2302	1703	1299	408	619	273	273	33	334	282	278	293	315	367	271	361	308	348	233
Near Eastern	1779	709	1823	1454	1823	1002	1454	1246	158	0	263	2138	1977	1411	1991	1791	1760	1717	1224	1548	1434	1167	1448	1958	1703	1299	159	418	204	204	52	269	242	242	209	221	303	259	252	120	371	257
Uralic	2760	1690	2360	1921	2360	1921	2360	497	408	263	0	3283	3089	2996	2766	3872	2374	2568	2354	3125	1548	3776	2989	2373	3384	1309	1707	736	1707	418	258	199	233	264	150	191	324	472	646	522	392	256
Ainu	2769	1771	2161	2161	2414	2161	2414	2663	2588	2138	3283	0	763	681	866	852	279	509	337	418	459	872	638	847	2931	1309	1965	1940	293	272	175	106	131	180	264	296	374	276	384	355	255	259
Japanese	2361	1345	2511	2252	2511	2161	2252	2905	2138	1977	3089	763	0	79	718	763	911	681	549	459	261	449	293	272	1965	392	1723	1259	152	154	51	154	147	149	180	204	276	228	295	269	184	162
Korean	2668	1475	2368	1807	2368	1475	2368	1411	1950	1411	2996	681	79	0	154	3283	905	418	337	314	449	314	293	280	2547	1259	1723	1776	244	16	33	106	170	151	125	165	382	312	341	295	279	147
Mon Khmer	2446	1538	2881	1951	2881	1951	2766	1991	1977	1991	2766	866	718	154	0	158	279	509	549	261	681	638	293	700	3956	700	1776	3403	205	31	52	154	170	570	151	382	809	285	311	122	581	283
Thai	3364	1602	2978	2055	2978	2161	2703	2064	2480	1791	3872	852	763	3283	158	0	432	681	549	1282	866	821	293	918	261	918	368	3136	216	64	269	136	242	570	547	191	735	168	162	98	415	263
Dravidian	2112	1066	2396	2519	2396	1761	1851	1760	1950	1760	2374	279	911	905	279	432	0	107	549	432	549	549	246	272	2931	133	1965	1057	262	203	175	136	280	525	534	310	194	334	420	386	414	269
Mongol Tungus	2882	1423	2519	2243	2519	1709	1709	1434	1717	1717	2568	509	681	418	509	681	107	0	240	117	181	287	88	272	1703	261	391	1776	276	282	90	136	265	114	533	204	584	338	460	416	351	124
Tibetan	2320	1126	2391	1717	2391	1243	1365	1224	2354	1224	2354	337	549	337	549	549	549	240	0	126	240	250	368	186	2931	133	1776	1057	208	203	203	154	170	534	482	310	190	284	366	323	365	180
Indonesian	2908	1669	2703	1709	2703	1709	2703	1548	1625	1548	3125	418	459	314	261	1282	432	117	126	0	411	315	91	261	3125	392	1776	3403	238	133	238	461	539	534	523	540	811	115	87	121	394	273
Malaysian	1658	1216	1714	1365	1714	1365	1216	849	1434	1434	1548	459	261	449	681	866	549	181	240	411	0	1044	1024	315	1548	418	391	3956	255	510	510	611	470	469	482	523	674	186	175	156	385	233
Filipino	2913	1700	3033	2299	3033	2062	3033	1922	1268	1167	3776	872	449	314	638	821	681	287	250	315	1044	0	1109	315	3776	459	185	3275	272	553	537	523	537	449	379	335	536	106	163	211	272	363
N. Turkic	2760	1600	2599	2062	2599	2062	2599	1442	1167	1448	2989	638	293	293	293	293	549	88	368	91	1024	1109	0	271	2989	308	391	3275	224	291	290	292	418	537	449	335	536	163	384	294	429	339
S. Chinese	2486	1386	2302	2163	2302	1958	2302	1448	2302	1958	2373	847	983	306	337	918	246	272	186	261	315	315	271	0	3384	306	306	3136	346	321	321	156	166	368	339	350	612	119	164	164	293	234
Basque	1474	1664	2863	1958	2863	1307	2863	2231	1703	1703	3384	2931	1481	983	549	261	432	525	534	301	315	315	403	3384	0	106	55	34	321	50	25	393	321	344	322	348	400	417	503	386	509	406
Lapp	1904	1570	1299	1307	1570	1299	1299	1570	1299	1299	1309	392	481	392	392	459	357	194	142	246	418	423	306	306	106	0	113	59	71	49	53	279	210	219	255	279	344	344	434	394	274	225
Sardinian	2760	988	1697	1328	1697	1697	1697	619	408	159	1707	2547	1234	619	1411	1989	681	542	142	423	857	423	185	306	55	113	0	61	59	30	35	553	384	463	455	592	297	288	410	202	432	314
Danish	1708	1600	2599	2062	2599	2062	2599	736	619	418	736	1940	1965	3275	2989	638	549	246	250	314	449	329	238	828	34	59	61	0	7	30	14	290	270	260	264	268	359	313	410	358	335	236
English	2288	909	1723	1459	1723	892	1723	619	273	204	1707	293	152	244	205	216	179	197	208	236	314	238	236	306	321	71	59	7	0	67	10	311	241	265	253	242	342	405	252	187	357	234
Greek	1479	1163	1767	1487	1767	892	1767	497	273	204	418	272	154	16	31	64	175	203	203	133	510	553	291	321	50	49	30	7	67	0	34	302	159	397	402	396	490	380	452	366	317	215
Italian	1902	892	1534	1356	1534	51	1646	45	33	52	258	283	276	221	540	427	71	264	154	461	611	291	346	321	25	53	35	14	10	34	0	34	77	319	314	279	431	379	366	358	269	211
C. Amerind	1902	1374	1767	1767	1767	1374	1767	244	334	199	175	106	51	89	478	814	89	64	64	117	229	95	271	159	393	279	553	400	0	366	34	0	77	166	56	31	222	412	480	440	396	256
Eskimo	2292	1234	2138	1794	2138	1234	2138	1723	1776	1776	742	1118	802	936	931	1175	194	320	1024	411	341	341	403	366	25	210	384	344	255	0	166	0	140	143	197	279	344	434	459	394	385	256
Na-Dene	2237	1475	2701	2293	2701	1475	2293	1401	1470	1470	904	1173	1124	1446	1001	1175	455	251	320	1024	485	449	306	472	71	219	481	279	210	166	77	140	0	164	56	31	222	354	476	423	335	309
N. American	3251	2116	2598	2701	2511	2116	3099	1262	1364	1324	1063	888	1118	1017	264	789	455	551	411	908	485	481	403	619	240	140	197	260	131	164	319	166	164	0	738	180	197	313	410	476	335	434
S. American	2975	2083	3124	2754	2083	2598	2217	1234	1364	1198	933	1146	921	860	1463	682	303	531	455	828	268	463	368	304	688	143	180	265	0	131	314	56	140	738	0	197	268	405	252	366	317	210
Chukchi	2589	1358	2329	2011	1358	1358	2201	1199	1234	1204	869	1017	762	860	1100	589	331	535	301	316	316	264	316	159	472	56	143	131	54	211	400	31	56	180	197	0	555	396	443	386	609	247
Chukchi/Mel	2793	1654	2792	2502	2792	1654	2722	1234	1401	1262	1121	738	802	936	1482	1355	952	406	190	402	396	268	350	619	304	304	242	402	211	555	0	555	644	548	711	981	930
Melanesian	3375	1767	3109	2612	3109	1723	3109	1401	1429	1412	1055	751	1124	1074	1045	789	406	402	115	417	523	314	279	431	304	627	314	379	412	354	379	555	0	122	129	210	222
Micronesian	2518	1722	3120	2181	3120	1723	2722	1723	1364	1107	1412	807	936	1074	860	890	952	402	751	279	314	379	555	644	122	0	512	981	930	223
Polynesian	2649	1414	2491	1992	2491	1992	2181	1401	1751	1406	1063	1149	823	890	2240	1803	1992	814	1493	1974	1665	1300	1580	584	512	299	215
New Guinean	3372	2733	3324	2752	3269	2181	2612	1460	1751	1406	828	1539	1241	1420	2240	1803	1992	1493	1783	1974	1665	1300	1498	1413	711	981	1478	0	158
Australian	3272	2131	3557	2694	2705	1988	4287	1176	1546	1408	828	1246	821	850	1699	1314	1565	781	1055	1481	1665	1300	1580	1081	1949	1422	1868	1400	1534	1498	1413	1360	1230	1977	1264	1563	1736	1059	930	1145	974	0

Table 2.3.1B. Modified Nei's Genetic Distances between the 42 Populations (× 10,000) (lower triangle), and Their Standard Errors Calculated from the Bootstrap Analysis (upper triangle)

	BAN	EAF	NIL	WAF	SAN	BER	MBU	IND	IRA	NEA	URA	AIN	JPN	KOR	MNK	THA	DRA	MNG	TIB	INS	MAL	FIL	NTU	SCH	BAS	LAP	SAR	DAN	ENG	GRK	ITA	CAM	ESK	NAD	NAM	SAM	CHU	MEL	MIC	PLY	NGU	AUS
Bantu	0	25	4	8	38	70	32	142	133	106	130	153	134	157	128	193	102	125	106	160	86	164	116	172	62	83	155	95	140	72	159	106	164	157	139	156	66	212	144	144	190	172
E. African	87	0	31	31	29	20	50	55	51	34	68	83	58	76	65	79	45	66	46	78	54	81	57	85	35	40	72	44	61	41	65	56	95	89	65	66	49	98	78	61	114	86
Nilo-Saharan	12	87	0	7	34	72	23	109	102	93	83	103	123	102	127	149	97	83	94	115	81	128	84	125	56	65	124	83	84	70	121	111	100	116	88	105	66	153	144	116	144	131
W. African	29	90	19	0	38	73	32	117	106	96	69	93	135	93	97	137	91	68	97	86	69	125	89	101	59	71	130	81	82	72	128	113	110	122	87	123	68	164	130	108	133	128
San	121	99	91	108	0	37	58	62	60	47	75	96	81	86	88	87	74	56	70	84	72	87	53	97	61	59	65	58	60	52	60	113	82	81	72	75	53	107	91	71	119	90
Berber	192	58	175	180	83	0	79	19	14	10	42	72	68	57	67	62	35	50	37	57	52	49	40	65	10	59	18	58	8	12	9	78	64	72	81	50	35	65	71	55	84	67
Mbuti	88	165	61	84	179	193	0	148	137	114	145	167	153	162	127	181	117	127	116	160	97	184	153	176	103	101	169	92	136	97	160	145	160	174	145	167	91	201	159	160	202	181
Indian	471	177	304	351	180	67	429	0	9	8	27	54	53	38	55	83	34	21	11	61	63	37	29	44	16	18	21	10	12	14	13	47	48	48	37	45	31	50	54	39	68	41
Iranian	451	167	281	331	175	55	416	31	0	9	54	52	62	50	79	54	38	18	61	76	79	58	44	56	10	23	13	6	11	2	7	60	67	63	58	61	49	60	70	51	72	55
Near Eastern	344	116	241	270	121	37	354	47	31	0	40	59	59	53	55	54	18	22	36	50	56	45	36	50	7	20	13	10	9	6	10	52	62	62	45	51	40	63	53	32	81	61
Uralic	307	159	213	179	191	99	325	100	127	114	0	27	25	17	62	69	28	21	31	68	62	55	31	47	45	26	62	31	43	42	51	27	34	37	19	27	39	52	60	41	41	22
Ainu	439	256	271	289	304	191	421	171	174	185	80	0	9	14	61	66	53	32	41	69	75	55	22	41	42	28	63	57	61	28	64	33	44	41	53	53	40	52	61	59	59	39
Japanese	476	191	327	428	238	207	405	149	205	202	86	34	0	6	49	55	43	12	20	56	67	69	20	33	63	45	87	61	64	70	75	27	30	33	28	45	37	44	48	42	43	25
Korean	457	210	269	293	247	164	392	135	170	179	50	57	22	0	51	59	30	19	28	59	58	99	28	34	31	31	66	45	51	47	57	35	38	35	34	40	41	57	57	47	56	23
Mon Khmer	306	174	277	235	214	178	284	128	136	136	80	154	121	112	0	3	65	51	52	20	9	20	52	5	96	80	80	88	57	93	97	71	77	70	76	73	88	36	38	27	70	43
Thai	589	226	369	389	245	199	472	177	215	185	85	155	123	131	3	0	69	56	57	35	28	35	57	3	94	85	81	90	60	92	104	78	84	75	84	86	87	29	26	22	64	43
Dravidian	323	149	262	305	227	100	344	44	68	56	40	162	144	98	65	69	0	8	12	75	67	43	26	49	23	17	25	17	26	16	18	33	55	40	52	55	30	55	66	58	63	43
Mongol Tungus	259	143	156	145	116	108	221	57	74	93	36	32	12	16	51	56	8	0	7	12	54	34	29	14	38	29	47	37	37	2	33	23	14	17	15	22	25	32	40	39	32	13
Tibetan	324	148	248	309	200	131	302	52	83	82	40	69	20	45	52	57	12	7	0	60	52	17	27	30	16	16	26	31	31	22	29	23	30	33	16	43	25	45	50	43	51	24
Indonesian	428	215	263	218	211	156	372	169	211	163	69	83	56	38	20	35	75	60	60	0	11	18	56	14	73	80	49	51	51	69	89	82	81	84	84	82	90	24	17	24	59	49
Malaysian	228	160	188	181	157	159	190	193	238	181	79	163	142	145	9	28	67	52	113	11	0	54	156	140	79	80	80	64	64	87	93	79	75	78	78	77	71	32	30	30	63	49
Filipino	495	224	319	346	319	232	485	164	169	176	104	160	160	190	20	35	43	27	35	18	54	0	42	78	55	46	80	68	65	51	70	69	90	84	78	64	76	18	21	35	39	55
N. Turkic	329	195	235	293	156	174	439	99	129	112	95	197	176	190	26	13	103	95	113	40	79	42	0	40	30	14	53	41	44	49	53	26	29	31	21	24	19	51	54	43	64	45
S. Chinese	433	223	293	248	248	175	420	143	182	167	118	227	261	190	26	13	42	26	113	277	257	140	147	0	70	52	53	68	49	55	74	56	58	60	55	55	71	19	25	18	43	32
Basque	237	125	170	201	160	50	278	77	50	45	85	174	253	135	252	281	42	34	30	179	257	277	196	246	0	113	61	68	43	113	11	50	46	55	47	49	48	80	61	44	82	42
Lapp	330	141	199	294	177	91	298	92	79	79	43	238	176	197	34	41	199	113	33	140	216	184	84	162	113	0	14	14	13	5	11	82	86	94	83	80	71	66	53	46	88	72
Sardinian	492	240	329	358	183	76	486	56	60	60	61	238	292	237	158	236	77	128	112	153	221	175	171	201	43	61	0	14	0	6	2	58	60	55	51	54	58	66	54	44	76	54
Danish	319	145	223	260	237	43	216	33	11	44	31	186	217	164	221	246	75	70	118	123	267	123	219	221	113	58	61	0	1	6	58	60	55	58	51	51	56	81	58	54	72	54
English	462	188	236	274	170	38	376	40	47	47	43	227	261	188	159	227	92	99	142	208	213	208	196	213	130	109	114	6	0	11	59	58	63	66	56	56	47	82	52	44	76	59
Greek	214	122	183	201	129	48	237	11	20	20	51	110	197	135	207	225	57	52	90	110	227	102	135	207	67	52	67	34	34	0	5	71	72	68	67	68	62	71	67	50	62	50
Italian	524	202	311	359	175	43	477	58	27	42	150	212	256	190	218	293	66	101	121	227	275	149	227	225	200	163	197	200	200	176	0	197	264	221	218	264	197	240	192	171	226	208
C Amerind	324	183	300	323	237	175	489	159	186	229	85	118	94	102	175	185	128	83	77	56	217	197	26	38	50	82	38	58	59	176	197	0	129	59	11	4	27	69	46	57	57	35
Eskimo	651	318	475	429	277	244	485	172	245	254	108	186	217	164	210	248	183	77	84	60	245	231	29	56	55	60	60	55	63	72	264	129	0	115	30	31	24	75	81	78	88	59
Na-Dene	511	307	361	475	277	125	440	155	237	229	122	213	261	143	199	217	178	105	86	270	252	256	55	60	47	94	20	58	63	68	175	59	115	0	69	24	31	79	75	58	78	72
N. American	425	195	264	307	206	157	378	144	173	131	85	174	121	129	194	224	149	126	110	184	254	256	188	194	177	157	211	177	177	157	218	11	110	69	0	11	22	75	58	44	61	36
S. American	573	224	318	442	257	195	521	224	249	221	122	200	195	177	216	273	199	148	86	82	280	249	216	216	208	208	277	236	266	200	249	4	249	24	11	0	33	77	82	70	58	47
Chukchi	190	130	179	199	127	94	227	122	110	99	51	77	91	97	193	189	96	73	95	51	173	163	51	71	130	66	126	109	114	125	192	185	53	90	65	110	0	80	75	50	62	59
Melanesian	624	285	385	446	337	169	539	232	255	221	144	228	192	178	117	117	209	165	104	104	266	222	202	194	237	237	251	249	305	240	339	232	332	341	298	390	218	0	60	100	106	144
Micronesian	414	213	347	336	247	168	376	176	231	176	126	179	135	162	107	90	237	151	122	68	231	177	177	216	244	205	189	253	197	192	192	205	249	302	254	280	205	60	0	19	138	34
Polynesian	462	210	317	326	244	176	440	201	231	155	109	184	138	146	106	86	219	122	103	124	220	163	163	95	214	214	346	259	314	226	327	185	243	381	217	241	185	100	78	0	173	41
New Guinean	559	359	401	401	375	222	565	267	280	302	68	213	192	199	246	260	63	104	191	234	191	231	192	246	209	214	259	234	278	226	327	191	243	381	284	324	186	106	138	226	0	127
Australian	553	285	377	410	295	196	503	208	258	236	68	168	124	116	172	179	203	133	126	176	206	157	126	206	209	209	286	234	278	208	277	162	192	266	177	248	156	150	122	173	127	0

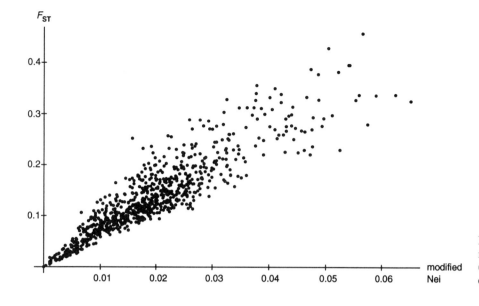

Fig. 2.3.1 Relationship between modified Nei's and F_{ST} distances (calculated according to Reynolds et al. 1983).

6. Extensive bootstrapping of our trees was carried out whenever it was especially important to test the validity of the clusters obtained or that of the statistical significance of the lengths of segments. Bootstrapping is a technique designed to evaluate statistical variation caused by random sampling when a whole experiment cannot be replicated. It randomly suppresses some genes and replicates others so as to keep the total number of genes constant in every bootstrap repeat. It thus tests the effect of missing data along with other effects of sampling.

One might summarize the rationale of our procedure for dealing with missing data by keeping in mind that when we measure the distance between two populations, we always use a sample of genes. In our approach we employ a somewhat different sample of genes for every population pair. This procedure is legitimate if the sample of genes is unbiased and numerically adequate. The difference in variation of individual genes is sufficient so that not using the same genes for all pairs of populations will inevitably result in loss of information, but the loss would be worse if one suppressed genes or populations in order to calculate distances on a complete matrix. Random resampling of genes by the bootstrap technique generates error estimates that also include errors resulting from incompleteness of the matrix. We believe this procedure should not be applied, however, unless the sample of genes is large.

Two types of genetic distances were calculated (table 2.3.1), Nei's distances modified as described in 1.11 to allow for missing allele frequencies and bootstrapping ("modified Nei") and F_{ST} or coefficient of coancestry (both unbiased; see sec. 1.11). The distances were highly correlated, but the relationship deviated from linearity, especially at the higher values, as shown in figures 2.3.1. From simulations (Reynolds et al. 1983), F_{ST} seems bet-

ter than Nei's distance for shorter evolutionary times, especially when the contribution from new mutations is not important (see sec. 1.11). The total evolutionary time in this case is short, and only a few new mutations have occurred in its course. Its duration is small compared with the more usual applications of Nei's distance (see Nei 1987).

Standard errors of the two distances are given in table 2.3.1. They were calculated by bootstrap. With the number of genes available here, the standard error of the F_{ST} distance is, on the average, 26.7% of the distance.

2.3.b. THE TREE FOR FORTY-TWO POPULATIONS

We present here (figures 2.3.2A, B) the world phylogenetic trees based on modified Nei and on F_{ST} distances; the tree using modified Nei's distances has already been published (Cavalli-Sforza et al. 1988). As happens whenever different distances or different methods of tree reconstruction are compared, there are some discrepancies in the results, which are discussed below. Both trees were calculated by average linkage; no other method corresponding to a specific evolutionary model for gene-frequency data would give the best tree for 42 populations within an acceptable computer time. In addition, this is the nearest substitute for maximum likelihood and therefore corresponds to an evaluation of the data by the model of constant evolutionary rates.

The first fission in figures 2.3.2A and B clearly separates Africans from non-Africans, with the exception of the Berbers in North Africa, who join the Caucasoid cluster. The separation of the Berbers from other Africans is certainly no cause for surprise, as North Africans have long been considered Caucasoid.

The two trees differ in only one important respect, the

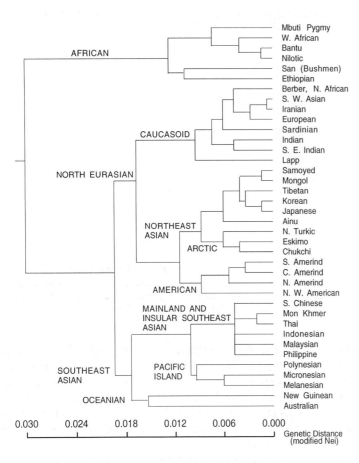

AFRICAN

NORTH EURASIAN

CAUCASOID

NORTHEAST ASIAN

ARCTIC

AMERICAN

MAINLAND AND INSULAR SOUTHEAST ASIAN

PACIFIC ISLAND

SOUTHEAST ASIAN

OCEANIAN

Mbuti Pygmy
W. African
Bantu
Nilotic
San (Bushmen)
Ethiopian
Berber, N. African
S. W. Asian
Iranian
European
Sardinian
Indian
S. E. Indian
Lapp
Samoyed
Mongol
Tibetan
Korean
Japanese
Ainu
N. Turkic
Eskimo
Chukchi
S. Amerind
C. Amerind
N. Amerind
N. W. American
S. Chinese
Mon Khmer
Thai
Indonesian
Malaysian
Philippine
Polynesian
Micronesian
Melanesian
New Guinean
Australian

0.030   0.024   0.018   0.012   0.006   0.000
Genetic Distance (modified Nei)

Fig. 2.3.2.A Average linkage tree for 42 populations, with 5 populations grouped as Europeans as in Cavalli-Sforza et al. (1988). The abscissa shows the genetic distances (modifed Nei) calculated on the basis of 120 allele frequencies from the following systems: *A1A2BO, MNS, RH, P1, LU, KEL, FY, JK, DI, HP, TF, GC, LE, LPA, PEPA, PEPB, PEPC, AG, HLAA* (12 alleles), *HLAB* (17 alleles), *PI, CP, ACP1, PGD, PGM1, MDH1, ADA, PTC, CHE1, SOD1, GPT, PGK1, C3, SE, ESD, GLO1, KM, BF, LDH, CHE2, IGHG1G3,* and *PGM2*.

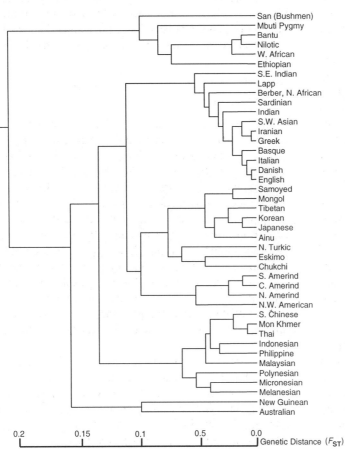

San (Bushmen)
Mbuti Pygmy
Bantu
Nilotic
W. African
Ethiopian
S.E. Indian
Lapp
Berber, N. African
Sardinian
Indian
S.W. Asian
Iranian
Greek
Basque
Italian
Danish
English
Samoyed
Mongol
Tibetan
Korean
Japanese
Ainu
N. Turkic
Eskimo
Chukchi
S. Amerind
C. Amerind
N. Amerind
N.W. American
S. Chinese
Mon Khmer
Thai
Indonesian
Philippine
Malaysian
Polynesian
Micronesian
Melanesian
New Guinean
Australian

0.2   0.15   0.1   0.5   0.0
Genetic Distance (F_{ST})

Fig. 2.3.2.B Average linkage tree for 42 populations. The abscissa shows the genetic distances (F_{ST}) calculated on the basis of 120 allele frequencies from the systems listed for Fig. 2.3.2.A. The five European populations form a single cluster and are pooled in Fig. 2.3.2.A. They are not clustered here and are kept separate.

order of the fissions of New Guineans and Australian and Pacific islanders as summarized in the following figure:

Tree obtained with Nei distances:

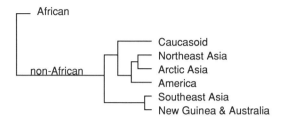

Tree obtained with F_{ST} distances:

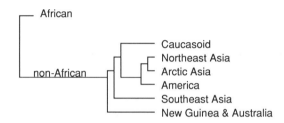

In other words, the cluster Southeast Asia + small Pacific islands has shifted in figure 2.3.2B to the immediately lower fission. This possibility was already noted when modified Nei's tree was tested by bootstrap (Cavalli-Sforza et al. 1988): "of special interest is the second bifurcation shown in figure 1, separating North Eurasians from Southeast Asians. This split occurs most often among bootstraps, but two alternative partitions are also fairly frequent: one separates Caucasoid from all Asian, Oceanian, and American populations, and the second separates New Guinea and Australia from all other non-African populations." This last partition is the one favored by the F_{ST} tree.

Summarizing bootstrap results concerning the major fissions in the F_{ST} tree, we find the first fission includes at least the four core African groups (Bantus, Nilo-Saharans, West Africans, and Mbuti Pygmy) in 83 of 100 bootstraps, very close to results with the modified Nei tree (84 of 100). Of these 83, 56 include all 6 African groups, 2 include all 4 core populations + San, 5 all 4 core + Ethiopians, and 15, only the core. There are also 5 trees in which one or two other populations are included with the Africans: Berbers twice, Malaysians once, Melanesians once, and Melanesians + Indonesians once. Thus, the African/non-African first fission is strongly confirmed.

In addition to the discrepancy between trees obtained with modified Nei and F_{ST} trees regarding the second fission (and the above-mentioned variation in bootstrap results of the modified Nei tree in the second fission), other evidence indicates that this second fission is weaker than the others. Bootstrapping the F_{ST} tree, one finds

the separation of the cluster New Guinea + Australia from all other non-Africans 35 times out of 100. In 6 of these, some populations associate with Australians and New Guineans: Melanesians alone twice, Micronesians + Melanesians three times, and Micronesians + Melanesians + Polynesians once. In 10 trees, New Guineans + Australians separate at the second fission from all other non-Africans, jointly with all of Southeast Asians including inhabitants of the small Pacific islands, as in the modified Nei tree. In 7 trees, New Guineans + Australians separate together from all other populations at the first fission. The residual 48 trees present a variety of different situations in which Australians and New Guineans often do not form a pair, or they associate with both Northeast Asians and Southeast Asians, or they form other combinations difficult to classify simply. The other clusters are strongly confirmed as being rather compact, as in the modified Nei tree, and there is no point in repeating details like those already mentioned for some of these clusters (Cavalli-Sforza et al. 1988).

2.3.c. ANALYSIS OF NINE CLUSTERS

In order to analyze further these results on a more compact set of data, we grouped the 42 populations in the following nine clusters:

Africans (sub-Saharan)
Caucasoids (European)
Caucasoids (extra-European)
Northern Mongoloids (excluding Arctic populations)
Northeast Asian Arctic populations
Southern Mongoloids (mainland and insular Southeast Asia)
New Guineans plus Australians
Inhabitants of minor Pacific islands
Americans

All these clusters are reasonably compact as shown by the bootstrap analysis briefly summarized before and commented on further in the original paper (Cavalli-Sforza et al. 1988).

The tree formed by these clusters by F_{ST} distance is shown in figure 2.3.3 and the distance matrix in table 2.3.2. The genetic difference between two clusters is aptly indicated by the average genetic distance between the two clusters. This can be calculated by reading the position of the node that separates two clusters on the abscissa given in figure 2.3.2B. It can also be calculated by averaging the distances of all possible population pairs formed by a population of the first cluster with a population of the second cluster. The nine clusters chosen differ in their genetic homogeneity, but we are interested in establishing history and not in generating a classification scheme. A criticism raised by Bateman et al. (1990a) on this point misses the difference between taxonomy and phylogenetic analysis. Even if we were interested in taxonomy, calibrating the homogeneity of

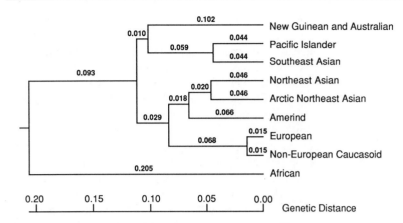

Fig. 2.3.3 Summary tree obtained after averaging the 42 populations in the nine clusters listed in the text. F_{ST} distances are used.

clusters on the basis of genetic distance in a tree would still generate an arbitrary classification that would inevitably depend on the sample of populations chosen. Lest there be no misunderstanding, we (Cavalli-Sforza et al. 1988; Cavalli-Sforza and Piazza 1990), unlike others (Bateman et al. 1990a, b) do not give to the clustering obtained in the tree of figures 2.3.2 or 2.3.3 any "racial" meaning, for reasons discussed in the first chapter. Clusters were formed for reducing the complexity of the data and were given specific names in order to simplify discussion.

The first fission is now found in 98% of 100 bootstraps, showing that even if the original African cluster has three outliers (San, Mbuti, Ethiopians) it is, on the average, quite distinct from non-Africans. The second fission is very weak and, somewhat surprisingly, corresponds to the modified Nei tree rather than to the F_{ST} tree. A more detailed analysis of the second fission helps to explain this reversal of the earlier conclusion. Among the 98 bootstrap trees in which the first fission separates Africans from the non-African clusters, one observes a variety of fissions, which can be classified as follows:

1. In 25 trees, New Guineans + Australians + inhabitants of small Pacific islands + Southeast Asians versus all other non-Africans;

2. In 24 trees, inhabitants of small Pacific islands + Southeast Asians versus all other non-Africans;
3. In 15 trees, New Guineans + Australians versus all other non-Africans;
4. In 8 trees, Europeans + extra-European Caucasoids versus all other non-Africans;
5. In 8 trees, Americans versus all other non-Africans.

The above 5 partitions account for a total of 80 trees; the residual 18 are difficult to describe synthetically, generally because subclusters disintegrate. For instance, in some of them, New Guineans and Australians part from each other, one of the two being separated from all others in the second fission.

A relative majority of trees is thus in agreement with that of figure 2.3.3, but Southeast Asians (with inhabitants of Pacific islands) show the greatest instability, clustering an almost equal number of times with New Guineans + Australians, or with other non-Africans, or staying separate from all other non-Africans. The first three partitions (represented in 25, 24 and 15 trees, respectively) do not differ significantly in frequency from each other ($\chi^2 = 2.8$, df $= 2$, $P > 0.20$); the first five are significant ($\chi^2 = 19.2$, df $= 4$, $P < 0.05$). A conservative conclusion would be to reduce the second fission to a trichotomy, separating the three subclusters New

Table 2.3.2. F_{ST} Distance Matrix of the Nine Clusters Shown in Figure 2.3.3 (x10,000 with standard errors obtained by bootstrap analysis)

	AFR	NEC	EUC	NEA	ANE	AME	SEA	PAI
Africans	0.0							
Non-European Caucasoids	1340.0 ± 301	0.0						
European Caucasoids	1655.6 ± 416	154.7 ± 29	0.0					
Northeast Asians	1979.1 ± 452	640.4 ± 134	938.2 ± 217	0.0				
Arctic Northeast Asians	2008.5 ± 387	708.2 ± 160	746.7 ± 210	459.7 ± 98	0.0			
Amerindians	2261.4 ± 434	955.5 ± 204	1038.2 ± 276	746.5 ± 183	577.4 ± 89	0.0		
Southeast Asians	2206.3 ± 529	939.6 ± 262	1240.4 ± 339	630.5 ± 299	1039.4 ± 326	1341.7 ± 418	0.0	
Pacific Islanders	2505.4 ± 648	953.7 ± 230	1344.7 ± 354	723.8 ± 262	1181.2 ± 331	1740.7 ± 544	436.7 ± 87	0.0
New Guineans and Australians	2472.0 ± 536	1179.1 ± 189	1345.7 ± 231	734.4 ± 118	1012.5 ± 257	1457.9 ± 283	1237.9 ± 277	808.7 ± 264

Guineans + Australians, Southeast Asians + small Pacific islands, and all other non-Africans, but this would not indicate the existence of a strong relationship between Southeast Asia and New Guineans + Australians. Altogether, the summary tree in figure 2.3.3 may remain the favored one, while we wait for more genetic data.

Among lower fissions, the triplet New Guineans + Australians, inhabitants of small Pacific islands, and Southeast Asians reappears in 29 trees. This may seem a low number, but there are at least 39 trees (plus a number of others more difficult to specify) in which this triplet could not appear because New Guineans + Australians or Southeast Asians have separated earlier at the second fission. Thus, the percentage of bootstrap trees at this node is greater than 29 in 61 (48%). The instability of the triplet New Guineans + Australians, inhabitants of small Pacific islands, Southeast Asians is in good part due to the intrusion of Northeast Asians. This quadruplet appears 17 times, mostly because of the tendency of Northeast Asians to pair with Southeast Asians (a total of 18 times). Northeast Asians pair much more frequently with the Northeast Asian Arctic populations (66 times), but it seems likely that there has been enough exchange between Northeast Asians and Southeast Asians, which are geographic neighbors, that the second fission is destabilized. The same reason seems responsible for the relatively low frequency of the triplet Northeast Asian Arctic populations + Northeast Asians + Americans (48 trees). The number of different triplets in a tree of 9 populations is 84 so that, in a binomial distribution with a probability of 1 in 84, the probability of observing by chance a given triplet 48 times in 98 is exceedingly low.

An elegant method by E. Minch (unpubl.) clearly confirms that there is an uncertainty if New Guineans and Australians truly join with Southeast Asians before they join with other non-Africans. The difference between the modified Nei and F_{ST} trees is due to an instability of present results, which will require further data and analysis. The source of the uncertainty is probably in the difficulty of distinguishing Southeast Asians from Northeast Asians. Their geographic proximity may be responsible for enough admixture that a dichotomous tree cannot give an entirely satisfactory representation.

In conclusion, there has probably been enough intermingling of the clusters that a network representation (i.e., a tree with interconnections between branches) would be highly desirable. But the tree in figure 2.3.3 is probably the best result that can be obtained using present methods, that is, a phylogenetic tree without interconnections. The way in which this tree was built, by first assembling all populations into clusters, is very different from the usual procedure in which one chooses a single, highly localized population from a vast region. This region might be equivalent to one of the nine clusters of figure 2.3.3, or to subclusters of them. This procedure was used, for instance, in generating all

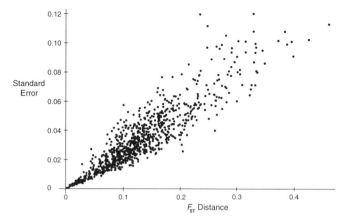

Fig. 2.3.4 Distances (F_{ST}) among 42 populations of table 2.3.1 (abcissa) versus standard error of distances (ordinate) as estimated by 100 bootstraps resamplings.

the trees represented in the figures of section 2.2, and in practically all papers published on human phylogenetic analysis, because this is the way in which data are most easily collected and found in the literature. This "atomistic" collection of population samples inevitably gives a clearer tree, since most geographically intermediate populations have been eliminated and links between branches are less likely to be observed. It is doubtful whether the sharpness thus acquired is real. Considering that the nine large clusters we have used represent large, geographically contiguous regions (but separated according to natural boundaries that tend to create some discontinuity), it may be almost surprising that the tree we obtained is reasonably reproducible in different bootstrap samples.

Bootstrap analysis has also been used to estimate the standard error of F_{ST} distances. Figure 2.3.4 shows the correlation between F_{ST} distances and their standard errors.

2.3.d. PRINCIPAL COORDINATES ANALYSIS

We have further examined the 42 populations by PC analysis. The first two PCs summarize 27% and 16% of the variation, respectively; the map is given in figure 2.3.5. Africans cluster clearly in the lower right quadrant of the map, with all Caucasoids in the upper right quadrant; Berbers are in between, but closer to Caucasoids. The lower left corner is occupied by Southeast Asians and the inhabitants of small Pacific islands; in the upper left corner are Northeast Asians, the Northeast Arctic populations, and the Americans. Between the Northeast and the Southeast Asian groups is the pair of New Guineans and Australians, somewhat closer to Southeast than to Northeast Asians. Thus, the PC map clearly epitomizes the conclusions reached by the tree analysis. The first principal component, however, does not correspond exactly to the first fission in the tree;

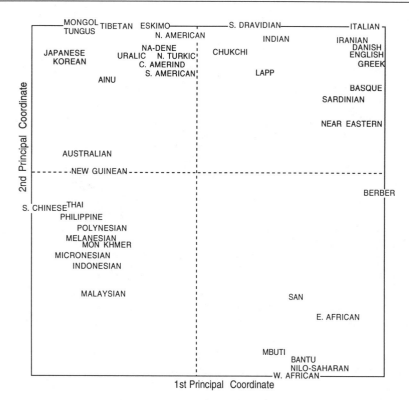

Fig. 2.3.5 Principal-component map of the 42 populations.

in fact, it groups Africans and Caucasoids versus the rest, as in the older trees discussed in section 2.2. In principle, the correspondence between the first PC and the first split in the tree may be blurred if the numbers of populations separated by the split are very unequal. This is true here, because there are 6 sub-Saharan Africans versus 36 non-Africans. In addition, there is a deviation from treeness (see sect. 1.12), that is, partial admixture between the products of the first fission, which also contributes to altering the correspondence between PC maps and trees. The separation of sub-Saharan Africans and Caucasoids is brought about clearly, however, by the second PC, which also separates northern Mongoloids from southern Mongoloids.

2.3.e. DIFFERENCES BETWEEN MARKERS

This brings us back to the problem of whether blood groups, which dominated the analysis in the earlier trees described in section 2.2 incorrectly favor the split of Africans + Caucasoids versus the remainder. In this study, there are more blood-group data than in our earlier investigations (in absolute numbers), but there is also a better balance between blood groups, HLA, proteins, and enzymes. The problem was attacked at the single allele-level by calculating for each allele two quantities, T_1 and T_2, for each of the two splits: T_1 compares Africans + Caucasoids versus the rest, and T_2 Africans versus the rest. The T_1 and T_2 values were differences between average gene frequencies of the clusters appearing in the two comparisons, divided by the standard de-

viation within populations represented in the two groups. One hundred and eight alleles were informative, that is, had a nontrivial number of populations in both comparisons. The difference between the absolute values of T_1 and T_2, T, was, if positive, in favor of the first split (Africans + Caucasoids vs. the rest), and if negative, in favor of the second split (Africans vs. the rest). The results obtained from classifying the genes into five types are shown in the table 2.3.3.

The T values were normally distributed so that the use of a Student's t-test for the difference of T from zero was appropriate. It gave Student's $t = -2.06$, with 107 degrees of freedom, significant at $P = 0.05$. The overall result favors the split of Africans versus non-

Table 2.3.3. Numbers of Systems of Polymorphic Markers (by gene type) Favoring the Split of Africans and Caucasoids Versus All Other Populations (negative T values) or the Split of Africans Versus All Other Populations (positive T values)

Gene Type	No. of T Values		
	+	−	Total
Blood groups	18	19	37
HLA	17	12	29
GM	2	5	7
Proteins	3	12	15
Enzymes	10	10	20
Total	50	58	108

Note.– χ^2 of this 2x5 table is 7.02 with 4 degrees of freedom and shows no significant heterogeneity.

Africans, but only protein genes taken in isolation show a significant deviation from equality. The early results (sec. 2.2) in favor of the split of African + European versus the others may have been caused by an insufficient sample of genes, rather than a tendency of different types of markers to give different answers.

2.3.f. BRIEF SUMMARY

The most important conclusion in this section is that the greatest difference within the human species is between Africans and non-Africans, but the inference that this was the earliest split rests on the assumption of constant evolutionary rates, which cannot be tested in full rigor from internal evidence alone. We consider this problem in the section on comparisons between genetic and archaeological data (sec. 2.5). The clusters we have formed are reasonably compact, but there are indications for some populations that they may have received genetic contributions from other clusters, as we see in more detail in chapters 3–6. The fissions after the first are less sharply defined. Correlations between Northeast

and Southeast Asians are probably responsible for the uncertainty about the second fission. The modified Nei tree and the F_{ST} tree on the nine clusters split Southeast Asians, New Guineans, and Australians from all other non-Africans. The F_{ST} tree on 42 populations separates New Guineans and Australians from all other non-Africans, and Southeast Asians split from the remainder at the next fission. Bootstraps show a slight advantage for the former partition. The cluster formed by Caucasoids, northern Mongoloids, and Amerinds is reasonably compact in all analyses. Uncertainty, if any, is rather connected with the extent of the similarities between Southeast Asians and Australians + New Guineans. Most probably, the early migration to Australia and New Guinea started from Southeast Asia, and enough ancestors of these migrants remained in Southeast Asia that a genetic similarity is still evident, but is complicated by later contacts between Northeast and Southeast Asia. Data on more populations and especially on more genes, in addition to other methods of analysis and independent sources of evidence, are needed for a sharper answer.

2.4. ANALYSIS OF DNA DATA

2.4.a. MITOCHONDRIAL DNA TYPES BY RESTRICTION ANALYSIS

The first DNA polymorphisms examined in humans for evolutionary purposes were from mitochondrial DNA (mtDNA). Mitochondria are self-reproducing units contained in all cells of higher organisms (eukaryotes, from fungi to Mammals), usually in many copies per cell (up to 10,000 or more). They are transmitted only by the mother but are present in both sexes. They are responsible for the type of cell respiration that produces most of the energy used by the cell and are determined genetically in part by DNA contained in the mitochondria themselves and independent of nuclear genes (*mitochondrial* DNA or *mtDNA*) and in part by nuclear genes. Human mtDNA was completely sequenced in 1981(Anderson et al. 1981); it is a closed circular molecule of 16,569 nucleotides and it is thus approximately 200,000 times shorter than the DNA contained in the nucleus.

The most general method of DNA analysis, the sequencing of whole segments was only very recently applied to the study of individual variation (see sec. 2.4.c). All the genetic work published before 1989 on mtDNA variation was done by electrophoretic analysis of restriction-fragment-length polymorphisms (RFLP, sec. 1.3) with one of two methods.

1. The usual way of detecting DNA fragments after restriction is known as "Southern blotting" (for a description of the technique in evolutionary applications to mtDNA, see Denaro et al. 1981; Johnson et al. 1983).
2. An alternative technique, called "end-labeling," permits a better resolution because it allows visualization of very short DNA fragments that usually escape detection in Southern blots. The latter method is easily applicable only if relatively large amounts of mtDNA are available; they are usually obtained from placentas. End-labeling (Brown 1980) was adopted in the early papers by Wilson's group (Brown 1983; Cann et al. 1987; Stoneking et al. 1990).

The first studies of five human populations analyzing RFLPs by Southern blots (Denaro et al. 1981; Johnson et al. 1983) showed considerable difference between two African groups, Bushmen (Khoisan) and Bantu, and the other populations: Caucasoids, Chinese, and Amerind. Bantus were in fact intermediate between Khoisan and the rest, and closer to the rest than to Khoisan. At face value, this result would indicate that the first fission was between Khoisans and the rest of the world (fig. 2.4.1A), but the number of polymorphisms studied was very limited, and this conclusion seemed at the time to disagree with other sources of information and was statistically insufficient to warrant a basic change of views.

By contrast, reconstruction of the history of mutations that led from one mtDNA type to the other gave a

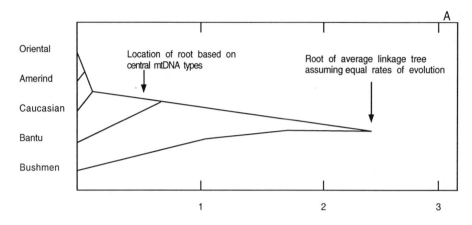

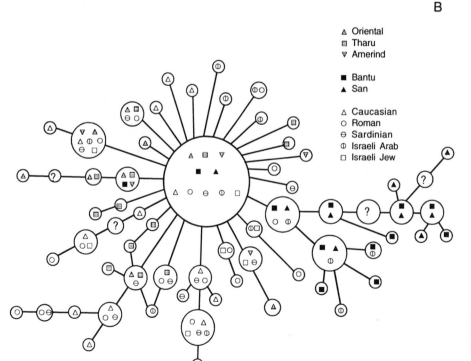

Fig. 2.4.1.A An early tree of five human populations (Oriental, Amerind, Caucasian, Bantu, and Bushmen) using mtDNA polymorphisms by Southern blots. The abscissa shows the number of mutational substitutions (after Johnson et al. 1983).

very different picture, closer to the earlier one obtained from classical polymorphisms. By far the most frequent mtDNA type was Asian, and from it seemed to derive all other types found elsewhere. Work with Southern blots has since been extended to a few more polymorphisms and to many other populations by many research workers. A recent summary of data in the literature (Excoffier and Langaney 1989; Excoffier 1990) has led to a picture (fig. 2.4.1B) which is still very similar to the original one (Johnson et al. 1983), and has been interpreted in a similar way, with an Asian origin. There are some indications that these data are not in agreement with neutral variation, but techniques used assume evolutionary equilibrium, which has probably not had the time to be established.

End-labeling gives a resolution about 10 times greater than that possible with Southern blots and thus gen-

erates many more mtDNA types, so that almost every individual—even from a sample of more than 100 individuals—was found to be unique. In the initial work on a number of human individuals from various parts of the world, the root of the human evolutionary tree was misplaced (Cann et al. 1982). Deletions later discovered were probably responsible for this early misunderstanding (Cann and Wilson 1983). Thus far, the most important set of data analyzed by this technique in Allan Wilson's laboratory (Cann et al. 1987; Stoneking and Cann 1989) includes 148 people from five geographic regions with 134 mtDNA types; a tree of all mtDNA types, reconstructed by a maximum parsimony approach, is shown in figure 2.4.1C and is rooted at the midpoint between the two most divergent types, a procedure assuming constant evolutionary rates. The root

Fig. 2.4.1.B A recent picture of mtDNA variation as tested by Southern blots (after Excoffier and Langaney 1989).

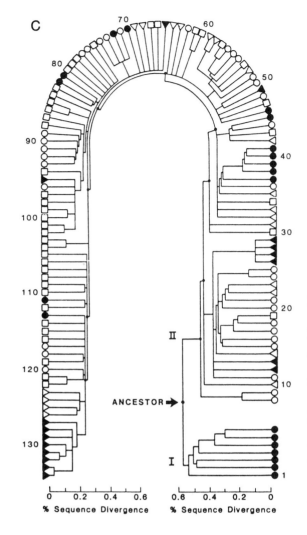

Fig. 2.4.1.C Genealogy of 134 mtDNA haplotypes found in 148 individuals from 5 different geographic regions (taken from Stoneking et al. 1989, p. 18, fig. 2.1). This tree consists of two primary branches (labeled I and II) and was constructed from restriction maps of about 370 cleavage sites per mtDNA, as detailed in Cann et al. (1987). The computer program PAUP (Swofford 1985) was used to relate mtDNA types in a branching network that minimized the total number of mutations; the resulting network was converted to a genealogy by placing the ancestor at the midpoint of the longest path connecting any two mtDNA types (i.e., midpoint rooting). Inferred ancestral nodes were then positioned approximately with respect to the scale of sequence divergence by averaging estimates of pairwise divergence (calculated from the restriction map by the method of Nei and Tajima 1983) for the appropriate descendant mtDNA types. *Closed circles,* Africa; *open circles,* Asia; *open triangles,* Australia; *closed triangles,* New Guinea; *open squares,* Europe.

separated about one third of all the Africans into one branch and the other Africans with all other populations into another. This was taken to be evidence in favor of an African/non-African first split. Africans are thus found in all major branches of the tree, whereas non-Africans are found in only one major branch, supporting the "out-of-Africa" hypothesis. Further analysis of the same data gave a time estimate of the first split of the tree in figure 2.4.1C. This result was made possible by the assumption of a constant evolutionary rate of mtDNA and an estimation of the rate based on a comparison of the divergence among humans with that between humans and chimpanzees. An estimated time for the human-chimpanzee split is based on other molecular information. Thus, the date of the first split in the mtDNA tree of figure 2.4.1C was set at about 200 kya, within an interval of 150–300. This is not based on standard error estimates, but on the possible fluctuation of the rates of mtDNA mutation estimated in primates.

Some weaknesses in these very important data need comment. "Africans" in the tree are almost all African Americans, who are known to have 10%–50% white ad-

mixture, on the average, depending on their origin (Reed 1969). This limitation is probably less serious than might otherwise appear, given that mtDNA is transmitted by the maternal line and that most white gene flow into African Americans has probably been from white males. More recent data obtained by using an independent technique have included better samples of Africans and are discussed in section 2.4.b. Repeating the tree analysis by alternative methods of tree reconstruction and testing the root location by bootstrap are desirable means of confirming the conclusions. Estimates of the number of mutations leading to each of the living individuals from the root, and tests to determine whether it is reasonably constant would be important. The analysis of 134 mtDNA types by maximum parsimony, a demanding task, does not guarantee finding the tree involving the minimum number of mutations, or a tree close to it. A further weakness is the estimate of the confidence interval of the date of 200 kya. The basis for estimating this interval is not explained in sufficient detail. This date depends on the validity of another date, that of the separation of chimpanzees and humans, which has been given as 5–7 mya on the basis of molecular data on protein immunology (Wilson and Sarich 1969; Sibley and Ahlquist 1984).

Clarification of these points would certainly strengthen mtDNA analysis, which is of considerable interest in that it provides a genetic date for human evolution independent of dates given earlier by other methods. A list of existing estimates, given by Weiss (1988) shows that they vary considerably but tend in general to be smaller. Conclusions drawn from mtDNA and from nuclear genes cannot be directly compared because they refer to different events. In principle, even the place of origin of nuclear genes and mtDNA genes might turn out to be different. Complementing the analysis with Y-chromosome markers (which, although now available, have not yet provided as good polymorphisms as desirable) would supply information on the paternal line of transmission, which is the missing half of the picture (see sec. 2.4.c).

There have been many misunderstandings among readers of these results on their exact meaning. We discuss three of them.

1. The reconstruction of a single woman, who may have lived 200 kya and carried an mtDNA type ancestral to all the types of mtDNA found in living human populations, has been misunderstood not only by laymen but also by a few distinguished colleagues, who have accepted it as evidence that 200 kya there lived a single woman from whom all living humans descend. The widespread use, especially in popular magazines, of the word "Eve" for naming the first mitochondrial ancestor of all mitochondria found in modern populations was probably responsible for generating this common misunderstanding. Earlier criticism of the use of Eve's name (Wainscoat 1987) is correct. There is absolutely no evidence from mtDNA work that the human population went through a bottleneck in which there was only one (or few) women.

This statement is apparently not intuitively grasped by all people who are exposed to trees of mtDNA mutations like that of figure 2.4.1C, and it may be useful to discuss it further. Unfortunately, the exact treatment of the problems of the genealogy of a given gene and its mutations is complicated, as is the new mathematical development to which they have given rise, called the theory of the *coalescent* (Kingman 1982a,b). MtDNA is transmitted by only one parent, the mother, in contrast with all nuclear genes (from chromosomes other than Y), which are transmitted by both parents. In this inheritance, called "unilinear," the mtDNA of all children is identical to that of their mother, barring mutation, and receives no contribution from the father. We can thus limit our consideration to women only. Moreover, since mitochondria originate from only one parent, recombination in the usual sense cannot occur as it does with nuclear genes, which can undergo exchange of segments between the paternal and the maternal chromosome at the time sperm and eggs are generated.

Unilinear inheritance has a property that is often misunderstood. At a given time, one can usually distinguish in a population many types that differ because of mutations that have arisen in their ancestry. If one traces their genealogy, one can always find a single common ancestor—a woman, in the case of mitochondria—that lived many generations before. This does not mean that the population at that time comprised of only one woman, but that the mitochondria of all the other women who lived at the same time and were different are extinct. To understand in an elementary way how this can happen, use

$$A, B, C, D, E, F, G, H, J, K$$

to label all the different mitochondrial types (DNA sequences) that exist in a particular generation. Some of them will be represented by one woman, others by two, three, etc. A fraction of women leaves no offspring. If a woman carries a type of mitochondrion, say A, not represented in other women, and has no children, then type A will be lost in the next generation. In addition, mutations may occur. For instance, a nucleotide at a given site of 16,569 forming the DNA sequence of E may have been replaced by another one because of mutation, but because all the rest of the sequence of the mutant (which we call E_1) is identical to E, the relationship of E and E_1 can be recognized. If there were several E women all of whom had children, and one of the children had the mutation E_1 in the next generation the mtDNA types might be:

$$B, C, D, E, E_1, F, G, H, J, K$$

Only 9 of the original 10 types survive, but a new type has formed by mutation. At every generation the process will be repeated; some of the original types are lost at every generation. Similarly, new types produced by mutation appear. A second mutation in E_1 causes a new type E_2, etc. Note that, if there were no mutations, drift would cause the whole population to be made eventually of only one type, irrespective of the number of individuals of which it is formed. Mutation continuously generates new types. Therefore, if the total number of individuals in a population remains constant, at some time, there will be a constant number of different mtDNA types in the population. This is called the *equilibrium* between the random extinction of old types by drift and the generation of new ones by mutation.

This process will continue and, as time goes by, the population will keep losing more and more of the old types and new mutants will arise. A possible result at some time is a population with

$$B, B_2, E_1, E_3, K_2, K_3, K_5$$

in which descendants from only 3 of the original 10 types survive (*B, E, K*). It is expected that, in every generation, a proportion of these types remains without descendants. If this process continues long enough, the probability that all individuals present are descendants from a single ancestral type increases and becomes closer and closer to unity, that is, certainty. Tracing the genealogy of mtDNA, we may thus find only one ancestor of a unilinearly transmitted set of markers, but of course the descendants of the original type will differ among themselves, because of the new mutations that will have accumulated. Thus, although all mtDNA present today can be traced to a single common ancestor, this is not evidence that the human population went through a period when only one woman was alive and reproducing. The same is true of the Y chromosome, the largest part of which is also unique (not represented in the X chromosome) and unilinearly transmitted, unless it undergoes some possible rare events like illegitimate recombination with other chromosomes.

2. The number of different mtDNA types found in a population has sometimes been taken to indicate a minimum number of founders of that population (e.g., for American Indians; Shurr et al. 1990). Especially in the Americas, but wherever drift was very strong, an analysis based on a few tribes provides only inadequate information on this point. The general statement that the minimum number of founders equals the minimum number of mtDNA types observed is correct, but this number is likely to be a very gross under-estimate of the number of founders, being an increasing function of the length of the segment of DNA considered and of the number of individuals. Presently, these data are based on a very small fraction of the total mtDNA. Moreover, the analysis is limited to one or a few populations of a continent or large region, which is also likely to cause serious underestimates. In the Johnson et al. (1983) paper, a Venezuelan tribe showed zero variation; should that have been taken to mean that it was founded by just one woman?

3. A frequent misunderstanding of the "birthdate of Eve," 200 kya, has been generated by the confusion between the date of an ancestral haplotype and the actual fission of populations (see fig. 2.4.2). The 200-kya date does not represent an event in the history of human *populations*, but only in the history of mutations leading to presently surviving haplotypes. It only sets an upper limit to the history of separations of human populations without giving much information on how much later the fission between Africans and non-Africans may have occurred. Archaeological information indicates that the population fission may have taken place perhaps as much as 100 ky later than the birth date of the oldest mtDNA ancestor whose descendants survive today. This may be still in reasonable agreement with the mtDNA data, but there is currently no theory for evaluating the agreement quantitatively. From a qualitative view, however, it is clear that the separation date of African and non-

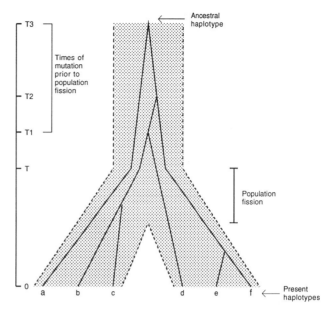

Fig. 2.4.2 Comparing the date of origin of the earliest mtDNA type and that of separation of populations; the second is later (after Nei 1987, p. 277, fig. 10.4).

African populations must be *later* than the date of origin of the mtDNA types common to all human groups. In section 2.5, we show that the date of fission of populations estimated from genetic distances obtained with nuclear markers agrees approximately with the 100 ky of separation of African and non-African populations. Therefore, the 200-ky mitochondrial date agrees qualitatively with the estimates of population separations around 100 ky derived from archaeology and confirmed by genetics.

The 200-ky date for the ancestral mtDNA of modern human populations is of considerable help in distinguishing between the rapid replacement and the multiregional hypotheses (figure 2.1.3). In fact (Stoneking and Cann 1989), it sets an upper limit to the common origin of the genome of a.m.h. that is much younger than that claimed by exponents of the multiregional hypothesis. Taken at face value, this is sufficient to discard the narrow multiregional hypothesis with no gene flow. The amount of gene flow acceptable with the results available and the numbers of individuals tested is to be determined, but is unlikely to be large. The main argument that is left to supporters of the polycentric origin is criticism of the validity of the 200-ky date, which depends on the calibration of the mitochondrial clock. These supporters therefore do not accept the hypothesis of constant evolutionary rates or the validity of the separation between chimps and humans around 5 mya. Further work on other regions of mtDNA or of the genome in general will be important for buttressing this result. But attempts at back dating the migration out of Africa to early *H.*

erectus, as supporters of the polycentric region desire, demand an error of calibration of mtDNA of almost a full order of magnitude, which seems unlikely.

2.4.b. SEQUENCING OF MTDNA

In the last few years, sequencing of DNA segments has become much simpler. Thanks to the method of DNA amplification by the polymerase chain reaction (PCR), one can use minute amounts of DNA, in principle, even a single molecule. This method was employed in very elegant research by Wilson's group (Vigilant et al. 1989) to study the sequence of 84 individuals for two segments of the control region of mtDNA, for a total of about 700 nucleotides. DNA samples from 14 !Kung (Khoisan) and 7 other individuals were obtained from hair roots. The tree obtained from the 84 sequences is shown in figure 2.4.3. The tree root was obtained using chimpanzees as an outgroup, and falls within the !Kung group, which is peripheral to the rest of the individuals. The general dates derived from this tree are in reasonable agreement with those from the tree of restriction analysis (fig. 2.4.2). This result would make the Khoisan the original human group from which the others have derived.

This very important conclusion agrees with figure 2.4.1A, and other observations of an entirely different nature discussed in chapter 3 regarding the origin of the Khoisan people. One would like to see it confirmed in many other ways, given its interest. Other lesser statements made in the Vigilant et al.(1989) paper on rates of movement of hunter-gatherers (African Pygmy populations) and the date of "fusion" of eastern and western Pygmies are difficult to accept, as may be gathered from the data in chapter 3. More importantly, the tree needs further quantitative analysis of the number of mutations among the various groups represented in the tree. This analysis is difficult with the data published up till now.

Further studies of the control-region sequence in other populations are briefly summarized elsewhere (for Sardinians and Middle Easterners, see chap. 5; for North American tribes, see chap. 6).

There has been considerable controversy on the validity of some of the statements made about mtDNA since the manuscript of this book went to press. For a review, see A. Gibbons, *Science*, 257 (1992): 873–75. Further work will be necessary for clarifying the root of the mtDNA tree and the estimate of its time of origin.

2.4.c. THE Y CHROMOSOME

Mitochondrial DNA is informative only with respect to the evolutionary history of the maternal line, but its informational power could be greatly increased if it were coupled with the analysis of genes in the portion of the Y chromosome that does not undergo recombination with the X chromosome, and is transmitted by the paternal

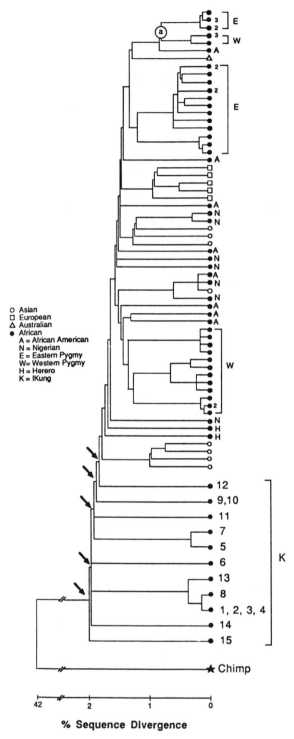

o Asian
□ European
△ Australian
● African
A = African American
N = Nigerian
E = Eastern Pygmy
W = Western Pygmy
H = Herero
K = !Kung

% Sequence Divergence

Fig. 2.4.3 Mutational tree of 84 individuals, mostly Africans, in which 700 bases of the control region of mtDNA were sequenced (taken from Vigilant et al. 1989, p. 9353, fig. 3).

line. The following notes on the Y chromosome are from an unpublished review by A. S. Santachiara-Benerecetti. Among the numerous sequences isolated from the Y chromosome, only a few have Y-specific polymorphisms (Casanova et al. 1985; Lucotte and Ngo 1985; Jakubiczka et al. 1989; Nakahari et al. 1989; Oakey and Tyler-Smith 1990). The most informative probe today, *p49f (Yq11.2;* Quack et al. 1988), reveals about 15 male-specific bands in *TaqI* digests, 6 of which can be present or absent (*A–D, E, F*) and are variable in size (*A* and *D*) (Ngo et al. 1986; Guerin et al. 1988). Other fragments (*G, H, N, O*) were later shown to be present or absent (Spurdle et al. 1989; Torroni et al. 1990). The molecular basis of the variation is not clearly understood, but the system is powerful in distinguishing ethnic origins. The A_1 band is virtually absent in 900 Caucasoids examined (Spurdle et al. 1989; Torroni et al. 1990; Lucotte et al. 1990; A. S. Santachiara-Benerecetti, unpubl.) and is present in 70%–80% of West and South Africans examined. A variant *D* band was found in all 22 populations examined. In studies on Mediterranean populations (Torroni et al. 1990; A. S. Santachiara-Benerecetti, unpubl.), several haplotypes showed unusually large variations in frequency among neighboring populations (Algerians, Tunisians, northern Italians, southern Italians and Sardinians). Clearly, the variation here is of considerable interest, and parallel studies with mtDNA and nuclear genes in the same populations might be especially informative.

2.4.d. DNA POLYMORPHISMS IN NUCLEAR (CHROMOSOMAL) GENES

In recent years, several population investigations of nuclear genes have used DNA, but they were usually limited to one or a few genes. The β chain of hemoglobin has been a favorite subject. In a region of approximately 35,000 nucleotides, about a dozen polymorphisms are known, and an analysis of a number of populations (Wainscoat et al. 1986) showed considerable distance between Africans and non-Africans (table 2.4.1). Analysis of a single gene—even a so-called supergene (at least five proteins are coded in this chromosome region)—is inevitably limited in the number of polymorphisms that can be studied, and conclusions must always be confirmed on many other chromosome regions to avoid the possibility that results may be idiosyncratic to the gene. This is especially true of the hemoglobin region, which is subject to intense natural selection in malarial areas where most samples investigated for hemoglobin originate. Taken at face value, however, the data show a considerable separation between Africans and all non-Africans, in good agreement with all other data.

Hemoglobin mutants specifically conferring resistance to malaria may well give biased results in terms of ge-

Table 2.4.1. Relative Frequencies (%) of Haplotypes of Five Restriction-Fragment-Length Polymorphisms in the β-Globin Region in Various Different Human Populations (modified from Wainscoat et al. 1986)

Haplotypes	Population (no. of chromosomes analyzed)					
	European (169)	Indian (111)	Thai (32)	Melanesian (173)	Polynesian (55)	African (61)
+ − − − −	63.3	52.2	90.6	67.6	78.2	1.8
− + − + +	26.0	25.2	0.0	16.7	10.9	9.8
− + + − +	8.9	13.5	6.2	1.7	0.0	1.6
− + + − −	1.8	2.7	0.0	0.0	0.0	0.0
+ − − + +	0.0	0.0	0.0	6.4	0.0	0.0
Several other haplotypes	0.0	6.3	3.1	4.0	3.6	0.0
− + − − −	0.0	0.0	0.0	0.0	0.0	3.3
− + − − +	0.0	0.0	0.0	2.3	7.3	19.7
− − − − +	0.0	0.0	0.0	1.2	0.0	60.7

netic geography; but they offer one advantage, namely, that natural selection can amplify the frequency of favorable mutants and thus reveal the presence of genetic markers otherwise undetectable because of their rarity. They make it possible to detect even a few migrants who carried some of these selected markers and, if a migrant population had a relatively high proportion of one or more of them because of previous exposure to malaria, it may leave a detectable trail. One can follow and to some extent confirm the migrations of Phoenicians and Greek colonizers in the Mediterranean, which are historically well known, by the geographic distribution of molecular types of thalassemia mutants (see sec. 2.14). In addition, the haplotypes for standard polymorphisms in the hemoglobin β-chain region, associated with these mutants, show a peculiar geographic distribution among patients and their families that is indicative of the geographic origin of the genes. Although probably biased for quantitative historical studies, markers favored by natural selection are useful for tracing qualitatively ancient migrations because their frequencies are amplified in the natural selection process.

A systematic analysis of DNA polymorphisms for nuclear genes (RFLPs) was begun in 1984 by one of us (L. Cavalli-Sforza) at Stanford in collaboration with Ken and Judy Kidd of Yale University with the aim of preparing a reference panel of populations of evolutionary interest, to be tested with a large number of DNA polymorphisms. From blood samples of a number of individuals (ideally, 50 unrelated) lymphoblastoid cell lines were grown in vitro in order to provide any desired amount of DNA from each individual, thus making it possible to test any number of RFLPs or other DNA markers. The number of available DNA polymorphisms is over 2000 and growing rapidly; every probe may give information about more than one polymorphism. The potential information is enormous, but a considerable amount of DNA is necessary for testing many probes. In vitro growth of human cells is therefore mandatory for the most important aboriginal populations. Although in vitro growth seems to cause no detectable variation in our experience—except for well-known rearrangements of chromosome regions involved in antibody production—it requires a process of "immortalization" by Epstein-Barr (EB) virus transformation. This procedure is not easy to carry out with remote populations, because blood lymphocytes begin dying quickly 1 or 2 days after blood collection. Our original plan (Cavalli-Sforza et al. 1987) was to collect 25 unrelated families (father, mother, two children) from each population, for the purpose of obtaining direct information on haplotypes. These can also be inferred from unrelated individuals, but with less accuracy. Actually, mortality of cells before they can be cultivated in the laboratory has substantially decreased the number of families and individuals available for populations located in remote regions (see also Bowcock and Cavalli-Sforza 1991).

In practice, the percentage of successful in vitro cultures has been low for the most remote populations, and therefore samples of individuals are often smaller than we intended. Families have frequently been broken up by lack of growth of random individuals, and haplotypes had to be calculated from the whole sample using information available from the few intact families. Population samples vary considerably in number. The Melanesian sample, the smallest, is made up of only 14 unrelated individuals (28 chromosomes). The largest samples have close to 100 individuals.

The following populations have been analyzed for each of 100 DNA polymorphisms (listed in Bowcock et al. 1987, 1991, 1992) of which 99 are RFLPs and 1 ($HLA\,Dq\alpha$) was tested by PCR.

1. Central African Republic (CAR) Pygmies are Western Pygmies, who, as tests with classical markers have shown (see chap. 3), have a 70%–75% admixture with other Black Africans, mostly of Nilo-Saharan or Bantu stock (collected by Cavalli-Sforza and by B. Hewlett).
2. Zaire (Ituri forest) Pygmies are Eastern Pygmies, who, as can be judged from their height (they are the shortest

population on earth), are the "purest" in the sense that they seem to have had the least admixture with neighbors. They also show the greatest distance from all other African populations (collected by Cavalli-Sforza and B. Hewlett).
3. Melanesians are from the island of Bougainville (collected by Jonathan Friedlander of Philadelphia).
4. Chinese are persons living in the San Francisco Bay Area, born in mainland China, both North and South, of Chinese parents (collected by L. Wang and L.Cavalli-Sforza at Stanford).
5. Caucasoids are local residents of the area of around Stanford and Yale Universities, or from data available in the literature. They are likely to be mostly northern Europeans.

See however Mountain et al. 1992.

The data currently available for 100 polymorphisms used a total of 74 probes (Bowcock et al. 1991, 1992). Most polymorphisms had two alleles. The F_{ST} distance matrix obtained is shown in table 2.4.2. All the following conclusions have been confirmed on Nei distances, which are closely proportional to F_{ST} distances. The standard errors calculated by the bootstrap method are appended to each distance; their average is 15.8% of the distance (median value).

The distance matrix shows that two pairs are made of rather similar populations: the two Pygmy populations have a very short distance from one another, and the Chinese and Caucasoid populations have a somewhat greater distance than that between the Pygmies. The distances of Melanesians from Chinese and Caucasoids are smaller than their distances from the two Pygmy populations. Therefore, Melanesians are closer to the Chinese and Caucasoid groups, and the root is located between the two African populations on one side, with the three non-African populations on the other. We have a miniature confirmation of the trees obtained with classical polymorphisms discussed in sections 2.3.b, 2.3.c.

The tree of figure 2.4.4 was obtained by average linkage and then by maximum likelihood. A second, somewhat different tree was obtained using the nearest-neighbor method (fig. 2.4.5). Minimum path and additive least squares also gave similar results. Unlike the methods used for figure 2.4.4, the last three do not re-

Table 2.4.2. Matrix of Genetic Distances of Five World Populations Tested for 100 DNA Polymorphisms (± standard errors)

	C.A.R. Pygmies	Zaire Pygmies	Melanesians	Chinese	Europeans
C.A.R. Pygmies [*]	–	0.023 ± 0.006	0.133 ± 0.020	0.144 ± 0.023	0.088 ± 0.014
Zaire Pygmies	0.043 ± 0.011	–	0.139 ± 0.021	0.139 ± 0.022	0.084 ± 0.011
Melanesians	0.242 ± 0.031	0.265 ± 0.034	–	0.094 ± 0.014	0.086 ± 0.013
Chinese	0.235 ± 0.032	0.235 ± 0.033	0.171 ± 0.019	–	0.058 ± 0.011
Europeans	0.141 ± 0.022	0.142 ± 0.017	0.148 ± 0.019	0.093 ± 0.016	–

Note.– The bottom left triangle corresponds to F_{ST} distances whereas the upper right triangle corresponds to Nei's distances. Boxes include distances that should be the same, within statistical error, if the tree of Figure 2.4.4 is correct. In the larger box, there is an unacceptable variation between the six distances, as explained in the text.
[*] C.A.R., Central African Republic.

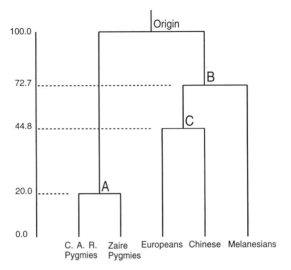

Fig. 2.4.4 A rooted tree obtained from the matrix of five populations (C. A. R. [Central African Republic] Pygmies, Zaire Pygmies, Europeans, Chinese, and Melanesians) and analyzed with 100 DNA polymorphisms using maximum likelihood reconstruction. Genetic distances among the populations are shown in table 2.4.2 (from Bowcock et al. 1991).

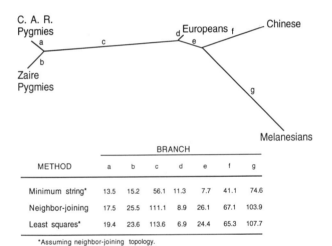

METHOD	BRANCH						
	a	b	c	d	e	f	g
Minimum string*	13.5	15.2	56.1	11.3	7.7	41.1	74.6
Neighbor-joining	17.5	25.5	111.1	8.9	26.1	67.1	103.9
Least squares*	19.4	23.6	113.6	6.9	24.4	65.3	107.7

*Assuming neighbor-joining topology.

Fig. 2.4.5 An unrooted tree obtained from the distances of table 2.4.2 by a variety of methods. The topology was obtained by the neighbor–joining method, and the lengths of segments a–g were obtained using the three methods shown in the table given below (from Bowcock et al. 1991).

quire equal total length for all segments. With all the methods, the European branch is clearly very short, for reasons discussed in this subsection.

In figure 2.4.6 we give the results of analyzing 100 bootstrap samples using average linkage. Clearly, the tree found by average linkage on the total data is also the most frequently occurring of all bootstrap trees: 84% of all bootstraps return the same answer. However, in 14 trees a different answer is obtained: the first split separates Europeans from all other populations, and the next split separates Africans from non-Africans. Two rare

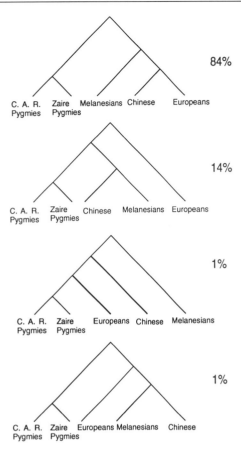

Fig. 2.4.6 One hundred bootstrap trees from the data set from which the F_{ST} distance matrix of table 2.4.1 was obtained. The first tree is the same as that obtained from a direct analysis of the distance matrix in table 2.4.2.

trees (1% of each) have somewhat different forms and can be considered random accidents.

One wonders if the tree has the internal consistency expected if the evolutionary rate is constant, a hypothesis under which the use of average linkage is satisfactory. If evolutionary rates are constant in every branch of the tree, the distances enclosed in a rectangular box of table 2.4.2 are expected to be equal (see sec. 1.11). The bootstrap estimate of the standard error can help test whether this expectation is satisfied. In the small rectangle the two distances 0.148 and 0.171 are very close within standard error. In the large rectangle, the six distances expected to be identical are rather heterogeneous. One notices two blocks, one made of the distances of the two African populations from Caucasoids: 0.141 and 0.142, which are homogeneous and smaller, whereas the other four distances (0.235, 0.235, 0.242, 0.265) are larger and also rather homogeneous. Because the distances of a matrix are intercorrelated, we tested the difference between the two blocks in the six-distances rectangle by bootstrapping. The quantity,

$$D = \frac{D_{ae} + D_{be}}{2} - \frac{D_{ac} + D_{ad} + D_{bc} + D_{bd}}{4},$$

(where a,b,c,d,e are the five populations of table 2.4.2, and D_{xy} is the distance between x and y, with x,y being any of the five populations a,b,c,d,e), was found to be much higher than its standard error, with a probability of 10^{-4}. Thus, there is significant heterogeneity among the six distances of the rectangle, and Europeans are closer to Africans than they should be; or, in other words, treeness is unsatisfactory. This is in agreement with the shortness of the European branch in figures 2.4.4 and 2.4.5.

There are three possible explanations for this phenomenon: a smaller evolutionary rate in the European branch, a bias resulting from the selection of all markers among European samples, or admixture between two populations of the tree leading to the formation of Europeans (see sec. 1.17). In the last 5000–6000 years, Europe was densely inhabited, and the resulting decreased drift may have lessened the evolutionary rate. Even if drift had been completely frozen in this period, it is unlikely that the shortening of the European branch by 80% (on the basis of the nearest-neighbor tree; see also Bowcock et al. 1991) was entirely caused by this phenomenon. There is little if any reason for supporting a slower European evolutionary rate of the magnitude observed. A selection bias cannot be entirely ruled out at present, but it is unlikely to be important, considering that heterozygosity in Europeans is only slightly higher than in Africans. However, it should be given serious attention when planning future studies, by using polymorphisms not selected in Europeans. There are good historical reasons for suspecting that admixture events may have happened repeatedly. In more recent times, modern Europeans also received an important genetic contribution from the Middle East at the time of farmers' expansions (9000–5000 years ago), as discussed in chapter 5. West and Central Asia, which are the probable origins of these and other migrations to Europe, are geographically in the middle between Africa and East Asia. An ancient population of West Asia, sandwiched between those of Africa and East Asia before the expansion of a.m.h. to Europe, may have become genetically intermediate between East Asia and Europe because of admixture between the groups. Similar, later events, including the above mentioned farmers' expansion, may have acted in the same direction.

We tested the hypothesis of the origin of Europeans by admixture by trying various admixture hypotheses (between African ancestors and ancestors of Melanesians, of Chinese, or of both). The model assumed that at a given time (t_m) there occurred an admixture of a proportion m of African ancestors, and a proportion $(1 - m)$ of ancestors from the non-African branches of the tree. The formulas used for the expected matrix under this admixture model are given by Cavalli-Sforza and Piazza (1975). The only hypothesis compatible with the data is that there was an admixture between African an-

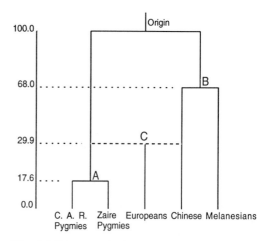

Fig. 2.4.7 A tree calculated by the maximum-likelihood method and showing that admixture between ancestral African and ancestral Chinese was responsible for the genesis of the European population (from Bowcock et al. 1991). C. A. R., Central African Republic.

cestors and Chinese ancestors (figure 2.4.7). The values $m = 0.35 \pm 0.09$ and $t_m = 29.9 \pm 6.0$ ky were estimated by maximum likelihood.

In a recent paper, Nei and Livshits (1989) examined the distances of Africans (mainly Bantus from Nigeria and Cameroon), Europeans (mainly from Great Britain), and Asians (mainly Japanese) for 186 loci (84 proteins, 33 blood groups, 8 HLA and Immunoglobulins, 61 DNA markers, including about one-half of those examined above). They showed that Europeans and Asians are significantly closer (Nei's distance with standard error, 0.040 ± 0.007) than Europeans and Africans (0.063 ± 0.011) or Asians and Africans (0.078 ± 0.013). A rooted tree cannot be built with as few as three populations; one can however, use a different model, that of admixture (sec. 1.17). The application is approximate because the method is entirely appropriate only for F_{ST} distances, but there is a fairly close proportionality in this range between the two types of distances. The estimate of admixture based on these data is 35% African and 65% Asian, in excellent agreement with that derived above using maximum likelihood on the full set of DNA markers.

The model of admixture we have employed is of necessity simple. It fits well, but it is likely that other models of admixture would also fit the data equally well. One possibility, already mentioned, is that there was more than one opportunity for admixture, in which case the times and proportions calculated are averages of various events (say, one before 35 kya and one after). The gene exchange between neighbors leading to admixture may have lasted for several thousands of years. In the model we used, admixture happens instantaneously, but it would be very difficult to distinguish the two situations from each other or from a clinal variation as in a stepping-stone-like model. There would be little point in trying to discriminate among many such similar models on the basis of so few data.

The earliest tree (fig. 2.2.1) obtained with classical markers, and analyzed by a method that allowed the detection of shorter branches, already showed this phenomenon. Figure 2.3.1 would not show short branches, because in a tree obtained by average linkage all segments from root to modern populations are forced by the hypothesis of constant evolutionary rates to take the same total length. The table of distances of the nine clusters (table 2.3.2), however, also shows that with that set of classical markers Caucasoids tend to be closer to Africans than expected. The phenomenon is therefore independent of the type of markers.

One would like to be able to extend this approach including fusions as well as fissions in evolutionary human history, but accurate analysis of a greater number of populations would most probably demand information on many more genes than are available.

Further discussion of the interpretation of the admixture was given in the original paper (Bowcock et al. 1991). It suffices here to conclude that DNA polymorphisms agree with data from classical markers; in addition, they allow one to hypothesize that the European population underwent substantial hybridization about 30 kya. The phenomenon may also have happened repeatedly at different times, before and after 30 kya. In general, it reflects the geographic intermediacy between East Asia and Africa of European ancestors, who probably originated in West Asia. The trees obtained by methods other than average linkage or maximum likelihood (e.g., fig. 2.4.5), show a slightly different structure; the shortness of the European branch is most probably the response of the method of tree reconstruction to admixture.

2.5. COMPARISON WITH ARCHAEOLOGICAL DATA

The most important difference in the human gene pool is clearly that between Africans and non-Africans, correcting earlier conclusions. This suggests that the split between Africans and non-Africans was the earliest in human evolutionary history, a suggestion subject to validation of the hypothesis that rates of evolution are constant. The genetic tree does not necessarily tell us whether the first humans were in Africa and expanded to Asia, or vice versa. We know human mobility can be high, thus reducing the strength of arguments based on the present residence of people, but we can assume that today they are not too far from where they once were, unless this conflicts with other evidence. We thus have two separate problems and turn now to archaeology in search of useful evidence.

Does this early African/non-African genetic split correspond to what is known of the history of human evolution? The earliest African a.m.h. with which we are familiar are dated about 100 kya in East Africa and South Africa. Today these areas are mostly dry and savannalike with a climate perhaps not very different from around 100 kya. Dryness is a prerequisite for the conservation of bones because they rapidly disappear in a humid forest. Therefore, we may never be able to determine whether these early a.m.h. also existed in the humid parts of Africa. In no other continent have comparable a.m.h. or their nearest predecessors been found in this period. The recent finds of a.m.h. in Israel have been very important in setting a date for the possible separation of Africans and non-Africans. The finds are so early that they have even generated some apprehension about the uniqueness of the African contribution to the genesis of a.m.h. We do not discuss this point further, given the limitations of the relevant evidence. We also do not include in our discussion the multiregional hypothesis, because we are not persuaded that there is yet enough evidence to make it the central event in the evolution of modern humans.

It is certainly encouraging that the greatest distance observed in the phylogenetic tree corresponds to the oldest separation accepted by many archaeologists. This leads us to research further the extent to which other archaeological dates provide support for the genetic tree. We have kept insisting on the importance of the constancy of evolutionary rates in the different branches of the tree, which is the condition for the usefulness of tree reconstruction. On what principle do we expect it to hold? The average genetic distance between two populations or population clusters will increase in proportion to the time of separation under random genetic drift, provided that (1) the population size remains approximately the same, on the average, in the two branches during the process of evolutionary differentiation, and (2) genetic exchange by migration between branches is not important, or at least it is proportional in the various branches. Neither expectation will ever be entirely satisfied, but it is enough if they hold approximately. Under certain conditions, error arising from these sources can be minimized. This brings us back to consider two possible approaches: the use of small populations chosen to represent a large area, and often located centrally in it (an *atomistic* approach); or the use of whole large regions as population units for tree analysis. Both methods have been employed, the first much more often than the second because it is more easily fitted to the usual structure of available data. The advantage of using small populations is that the various population units are all very distant one from the other, and cross-migration between them is extremely diluted. However, the populations being investigated may be very small and of different sizes and therefore very susceptible to drift of variable intensity. A population 10 times smaller than another one, if each mates randomly, will evolve 10 times faster than the larger one. The whole-large-region approach may seem more highly exposed to the problem of genetic exchange; if two large regions are neighbors, they inevitably have some exchange, unless they are separated by strong physical barriers. In fact, genetic exchange is not a difficulty for large areas, because it occurs primarily between immediate neighbors. Thus, exchange between two large regions is limited to their periphery. The effect on the whole group may be very small. If the large regions have similar population sizes, taking account of population density, they are likely to have similar amounts of drift.

It has seemed likely to us that the best conditions for comparing archaeological dates and genetic distances are found in the earliest splits, using large regions. There are three major early fissions or mass migrations for which we have approximate dates.

1. The movement from Africa to Asia (sec. 2.1) must have occurred after the origin of a.m.h. in Africa. Until a few years ago, it was believed that the first colonization outside Africa was in the Middle East around 40 kya. But these dates were very uncertain; unfortunately older sites are difficult to date with the conventional radiocarbon technique. New finds and new dating methods have provided evidence that a.m.h. existed in the Middle East in two different places around 100 kya. It is, of course, possible that this finding reflects a settlement later than the actual first arrival, or that this colonization was not successful. In any case, the appropriate node of the genetic tree with which to compare this time is the genetic distance between (sub-Saharan) Africans and non-Africans.

2. The arrival of a.m.h. in Australia is the next event for which we have dates supported by many later findings. A few years ago, the Australian date was believed to be 40 kya (or more) (Cavalli-Sforza et al. 1988). On the basis of new thermoluminescence data, recent work has suggested a date of 50 or 60 kya (Roberts et al. 1990). We use the date of 55 kya. The transit to Australia must have occurred from Southeast Asia, and possible routes are described in chapter 7. At least one part of the route required the use of boats or rafts for which we have no archaeological information. New Guinea and Australia were not separated by sea in this period, and New Guinea was probably occupied at the same time as Australia; the most likely route to Australia included a leg on New Guinea. We use as a genetic counterpart the average distance between New Guinea + Australia and Southeast Asia.

3. The entry into Europe occurred about 35 kya, corresponding with the disappearance of the Neanderthals (Cavalli-Sforza et al. 1988). Straus (1989) suggested an earlier date, around 43 kya. Using genetic data from Europeans is problematic because they have an abnormally short branch. It is probably more satisfactory to use all Caucasoids. Their nearest common ancestors are Northeast Asian + Americans, and their genetic distance with the whole cluster was taken for this separation.

If the evolutionary rate is constant, the genetic distance (G) divided by the time (T) should also be constant. These values are given in table 2.5.1 and are derived by averaging the relevant distances of table 2.3.2, after dividing by 10. For ease of comparison with the calculation (Cavalli-Sforza et al. 1988) previously performed with Nei's modified distances, which used a tree topology more similar to figure 2.3.3 than figure 2.3.2, we use the F_{ST} distances of table 2.3.2 that gave rise to figure 2.3.3.

Another advantage to using large regions (in practice, continents or large parts of them) is that they give some guarantee that one region will not diverge excessively from others in terms of the average population size since settlement. It is the average population size over long periods that counts, and the average census size over time to be used is the harmonic mean (see, e.g., Cavalli-Sforza and Bodmer 1971a), which is very sensitive to small values. All continents are likely to have been settled by a relatively small number of people at first, and thus no single continent will be favored by the procedure. Using large regions, the effects of fluctuations over time and over space are likely to be minimized. All continents seem to have been populated fairly rapidly, and, after an initial rapid increase, the total population may have increased only very slowly during the Paleolithic. Major differences in the density observed today usually developed much later. Dates of first settlement of continents, though approximate, are perhaps better known than most other dates. In the worst case, America, there is at most a factor of two between the extremes of the possible dates of occupation.

Table 2.5.1. Comparison of Genetic Distances (F_{ST} x 1000) from Figure 2.3.3 with Archaeological Separation Times

Clusters Defining Fission	G (genetic distance)	T (time)	G/T
African vs. non-African	205	100 k.y.a.	2.05
Southeast Asian vs. Australian and New Guinean	124	55 k.y.a.	2.25
Caucasoid vs. Northeast Asian and Amerind	84	43 k.y.a.	1.95

The three G/T values of table 2.5.1 have the average $G/T = 2.08 \pm 0.09$, and the low standard error of the joint estimate indicates good agreement.

As we see in chapter 6, there is considerable uncertainty about the first date of entry into America. In the 1988 paper, we used two different dates (15 and 35 kya) on which the preferences of different groups of archaeologists seem to concentrate. There was a slightly better agreement of genetic distance with the older date, but many considerations made a conclusion uncertain. Much movement of aboriginal Siberians has occurred in the last few centuries, and it is not clear which populations were in the extreme Northeast at the time or times of the passage to America, or how many founder populations or independent migrations there were. The earliest migrations, postulated to have given rise to Amerinds (chap. 6), may have comprised many different groups. Moreover, these populations may have diverged from other Asian populations for an unknown length of time before entering America. One can reverse the procedure and calculate the date of entry from the genetic distance. Using figure 2.3.3 and the data from table 2.5.1, the genetic distance between Americans and the cluster of Arctic Northeast + Northeast Asia is 66. With the G/T constant of 2.08 calculated above by averaging data from table 2.3.2, the date of entry to America is $66/2.08 = 32$ kya, closer to the older date. In chapter 6 we give another possible estimate, not very different from this one, based on the separation between the Asian ancestors of Arctic Americans and residual Americans.

We consider it safe to use the constant 2.08 (say 2.1) for large distances. Tests for shorter distances involve clusters that may give false results because of possible errors resulting from admixture or drift; applications to shorter distances should therefore be taken with caution. As mentioned before, admixture tends to artificially increase dates estimated in trees built with average linkage (sec. 1.15, 1.17), and the same happens with drift in small populations. Admixture may also artificially increase the time of separation by another mechanism. The sample of new settlers may come from a peripheral, splinter population that is geographically, and therefore also genetically, fairly distant from the average of the whole region that we accept as putative parent. In a low-density region such as Siberia is likely to have been at all times, this is certainly one of the possible scenarios. Taken at face value the colonization of America was early.

Three examples illustrate the use of the G/T constant for presumably shorter time separations and smaller populations.

1. *Lapps.* One calculates 26 ky of separation from other Caucasoids on the basis of their genetic distance. In fact, Lapps are a mixture of Northeast Asians and Caucasoids, with a slight majority of the latter. The date of their entry into Lappland is not known, but must be greater than 2000 years ago. They live in small groups and there is high variance between them, indicating high drift. Because they originated from a hybridization involving distant populations and because drift is high, the estimated date is most probably too early.

2. *Sardinians.* The date of 16 kya is in excess of the earliest radiocarbon date known (9120 ± 380 B.P.) (Spoor and Sondaar 1986), though not in serious disagreement given the rarity of archaeological data. However, there certainly were later arrivals in the Neolithic and even more clearly in the Bronze Age—in addition to the Phoenician colonization of southern Sardinia—which have added genes from many other sources, generating chances for admixtures. The bulk of the population was probably little influenced by these late migrations, and even less so by more recent ones (see chap. 5).

3. *Polynesians.* The genetic dating of the separation from Melanesians + Micronesians is around 26 kya, almost 10 times higher than the known dates of colonization of Polynesia. Here there was a genetic contribution from two distant populations, Southeast Asian and Melanesian. As in the case of the Lapps, the mixture is fairly distant from either ancestral population and artificially increases the date of separation. Polynesians have also had an enormous amount of drift as insular populations that underwent many extreme bottlenecks. Both admixture and drift must contribute to this high increase in the apparent date (see chap. 7).

Dating based on genetic distances has been criticized (Weiss 1988) because migration among differentiating populations may considerably reduce their genetic distance. This criticism would be correct if one used a priori formulas, for example, those given in section 1.11 for the dependence of genetic distance on time, knowing population size N. This approach would be useful only in exceptional circumstances. When genetic dating is made by archaeological calibration using genetic distances and archeological dates as above, the criticism is of lesser importance, as long as the amount of genetic exchange with neighbors remains within normal limits and in similar amounts in the various branches. We have seen in the case of Lapps and Polynesians, however, that genetic exchange may make these estimates unreliable, because the genetic distance becomes much greater than expected.

Several demographic factors other than migration and admixture affect genetic date estimates. When an area is colonized more than once, as usually happens, the genetic picture depends on the proportions of the ancestry coming from the different waves of colonization. These proportions are usually not known and may vary widely, depending on the numbers of descendants of the earlier colonizers present when the later colonizers arrive. The social relations established between earlier and later colonizers are of the utmost importance. These relations may vary from genocide (e.g., in the Antilles, in Australia and Tasmania; about the latter, see Diamond 1991), or slavery, to acceptance and admixture which may eventually cancel all traces of the hybridization process. Sometimes latecomers do not survive as a distinct population in the long run, but may well contribute to the gene pool of aborigines (as may have happened to Vikings in Greenland and Newfoundland).

The genetic distances employed here are based entirely on gene frequencies of classical polymorphisms. There is not yet enough information for a detailed comparison with DNA polymorphisms in equivalent populations, but whenever it was possible to set up comparisons for other problems, no major discrepancy between classical and DNA polymorphisms was found.

With the sequencing of highly mutable regions, as in the control regions of mtDNA, one has the opportunity to use an entirely different approach: the count of mutations differentiating individuals. If the mutation rate in a specific region is stable over time, the count of mutations is a better measure of genetic difference than those based on gene frequencies; the number of mutations is, in principle, independent of population size as long as there is no natural selection, and proportional only to evolutionary time (Kimura 1968, 1983). The standard error of these estimates may be high and affected by variation of the mutation rate at different nucleotide sites. Most of the experience in mtDNA dating seems to be restricted to the dating of the first ancestral haplotype (see sec. 2.4.a, b; see also chap. 5).

2.6. COMPARISON WITH LINGUISTIC CLASSIFICATIONS

2.6.a. PROBLEMS CONFRONTING LINGUISTIC CLASSIFICATIONS

Languages have very scarce "fossil" information, usually limited to situations in which writing was developed, taking us back at most about 5000 years. Toponymy may occasionally give older, but undated information. Whether human languages had a single or multiple origins is considered by most linguists as insoluble. The speculations on the origin of language were sufficiently wild in the last century that in 1866 the Linguistic Society of Paris forbade discussions of this topic at its meetings (Crystal 1987). The strength of this taboo was such that it still survives to a great extent. The real difficulty is that human language evolves so fast that the differentiation between presently extant languages is extreme, and it is difficult to establish similarities between them. Nevertheless, a small group of linguists, especially those interested in classification, have recently started a search for words possibly common to all language families. It is well known to linguists that some words are more highly conserved than others; they refer usually to body parts, kinship terms, personal pronouns, low numerals (one, two, three), and others. The first universal root (found in many families) was proposed by Greenberg (Lecture, Stanford University, 1976): "tik," meaning "one," "finger" (usually the index finger), or "hand." The semantic change is easily acceptable in this case, as is the phonological change from "tik" to "digit," for instance. Additional roots have now been proposed and the research is proceeding (Bengtson and Ruhlen 1993). There are two major traps in this research. One is the possibility of chance coincidence; this seems unlikely on the basis of qualitative considerations, but it can be tested quantitatively and should be examined more fully. A more subtle cause of error is the early borrowing of a word by another language. If, however, one has to systematically invoke borrowing for many roots of meanings known from other evidence to be highly conserved, the explanation of borrowing becomes suspect.

Fortunately, in recent years, the hypothesis of monogenesis of language is being considered seriously. Nevertheless, problems of classification remain and are made difficult by the absence of agreement on the criteria for considering two languages related. The lack of widespread understanding of statistical and probability concepts and methods in the linguistic community is a further barrier to consensus. Greenberg has recently published a new analysis of Amerindian languages (Greenberg 1987), considerably strengthening earlier proposals that most Amerindian languages belong to a single family, which he calls Amerind, and which consists of 11 families previously considered unrelated (Ruhlen 1975). Nevertheless, a symposium of Americanists meeting a few years ago (Campbell and Mithun 1979) declared that no less than 58 families of Amerind languages should be recognized for North America alone. One of Greenberg's responses is that if the criteria employed by his critics were extended to the well-accepted family of Indo-European languages, which has been recognized for 200 years, this family too would have to be rejected. More detail is given in chapter 6 in the discussion of the Americas, where the disagreement on the definition of families is particularly acute.

A linguistic family is a group of languages showing enough similarities for a common phylogenetic origin to seem likely (linguists use the word "genetic" for phylogenetic). This definition is mainly heuristic, and linguists find it difficult to agree on the exact composition of certain families. This is not too surprising, as even in zoology and botany there are still doubts about the acceptance of some phyla 200 years after Linnaeus gave the modern general classification of living beings. The original Linnean classification had well-known mistakes, like the inclusion of whales and dolphins among fishes, not to mention the acceptance of legendary humans in the genus *Homo*. But the linguistic classification has not reached the Linnean stage, in the sense that higher levels of classification (above the family) are few, incomplete, and usually rejected by many linguists. Thus, there is no tree of human languages. For such a tree to become possible, the monogenetic origin of languages would have to be accepted or, if refuted, specific independent origins would have to be spelled out.

The introduction of quantitative methods is still in its infancy, but could considerably improve the quality of linguistic classification if there were no objective difficulties in defining satisfactory criteria for measuring relationships. The evolutionary usefulness of characters varies considerably, with some being more highly conserved than others. A similar, though not identical, difficulty was met by the early taxonomists of plants and animals. It should be stressed that the problem would not be entirely resolved by simply applying existing methods of numerical taxonomy nor by the application of the prin-

ciples of cladistics (sec. 1.12). The extension of these concepts to intraspecific biological variation or to linguistics is not likely to be very constructive. Ascertaining a genetic relationship among languages may always require Pascal's "esprit de finesse" (the qualitative weighting of many different attributes and points of view by an expert) and is perhaps not ready for the use of the "esprit géométrique" (a fully deductive approach). The study of mechanisms of language evolution, which is underdeveloped, may be very helpful or, indeed, necessary before a rational "genetic" classification can be completed.

Even if a complete tree has not yet been constructed, a number of families has been recognized, and agreement is substantial for a subset of them. Others are in a state of flux. The history of linguistic classification is recorded by Ruhlen (1987), who is responsible for the most recent and complete guide to classification. With regard to the highest-level linguistic taxa both "phylum" and "family" are used, and we consider the two terms interchangeable. The 17 families specified by Ruhlen and their geographic distribution are shown in figure 2.6.1 and in

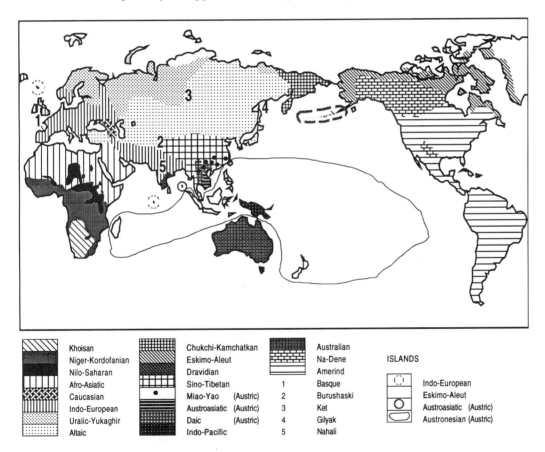

Fig. 2.6.1 Geographic distribution of the 17 linguistic families (from Ruhlen 1987, pp. 284–285, map 8.1): 1, Khoisan; 2, Niger-Kordofanian (made up of Kordofanian and Niger-Congo); 3, Nilo-Saharan; 4, Afro-Asiatic; 5, Caucasian (now considered two distinct families, North Caucasian and South Caucasian; Ruhlen, pers. comm.); 6, Indo-European (Indo-Hittite by Ruhlen); 7, Uralic-Yukaghir (Yukaghir is a single language long considered separate by some); 8, Altaic (some prefer to consider Korean-Japanese-Ainu separate from Altaic); 9, Chukchi-Kamchatkan; 10, Eskimo-Aleut; 11, Elamo-Dravidian (called Dravidian until now); 12, Sino-Tibetan; 13, Austric (this formerly includes Miao-Yao, Austroasiatic, and Austro-Tai, the latter consisting of Daic and Austronesian); 14, Indo-Pacific; 15, Australian; 16, Na-Dene (spoken in North America, mostly in the northwest); 17, Amerind (a new family proposed by Greenberg, which includes all Amerind languages of North and South America with the sole exception of those belonging to Na-Dene).

the color section. Ruhlen's first classification (1975) had 28 families, now reduced to 17, mostly because of the grouping of the Amerind and Austric families.

Some of these families were discovered long ago and are universally accepted today, like Indo-European (the first recognized) and others. For other families, differences of opinion exist as to whether some languages or subfamilies should be included in one or another family; a few of these differences were indicated in the legend for figure 2.6.1. The acceptance of the Amerind family has proved controversial, and even the 11 Amerind subfamilies that form it are not universally accepted by Americanists. A discussion of this controversy may be found in Ruhlen (1987). The methodological problems involved in language classification are of considerable scientific interest, and Greenberg's recent book on languages of America (Greenberg 1987) contains an excellent critical introduction to them (see also chap. 6). As of today, the Amerind family is the subject of a number of attacks (for an impartial commentary, see Diamond 1990), the heat of which surpasses the experience of the worst controversies in biology, unless one goes back to the diatribe against Darwinism in the last century. For people outside the field of linguistics, it may be useful to know that Greenberg is very highly respected by linguists for his work. His classification of African languages (Greenberg 1949, 1950, 1954, 1955, 1963) was attacked initially but is accepted today. It is difficult to predict when and how the new controversy will be resolved. The major need is for a discussion of methodology, although the opposing methodological points of view, summarized in Greenberg (1987), are so far from each other that it is difficult to see how an agreement can be reached (see chap. 6 for further discussion).

At the present stage of taxonomic efforts by linguists, only a few steps forward have been made beyond the description of the basic families and toward the possible final target of a complete hierarchic and truly phylogenetic classification of all linguistic families. As we stated previously, this is objectively a very difficult task, because the rate of linguistic evolution is so high that there is superficially almost no relationship between languages from very different families. The steps forward are the recognition of relationships between some of the families, which have led to the identification of superfamilies or superphyla. In this phase of the research, conclusions are still fluid and subject to criticism, but the following results are nevertheless worth mentioning. Two groups of researchers have independently recognized two superfamilies or superphyla that clearly overlap to an important extent.

1. *The Nostratic superfamily.* Following earlier suggestions by Pedersen (1931), the Russian linguists Illich-Svitych (1971) and Dolgopolsky (1970) independently recognized a superfamily made up of six families. Further contributions were made by Menges (1977).

Western: Indo-European
Afro-Asiatic
Caucasian (Kartvelian)
Eastern: Dravidian
Uralic
Altaic (including Korean)

2. *The Eurasiatic superfamily.* Greenberg has reconstructed this superfamily, which includes six families.
Indo-European
Uralic-Yukaghir
Altaic
Korean-Japanese-Ainu
Chukchi-Kamchatkan
Eskimo-Aleut

There is an obvious resemblance between these superfamilies: they share half of the linguistic families of Caucasoids and northern Mongoloids. In addition, making use of the recent work by Greenberg on Amerind, Shevoroshkin (1989) has recognized a similarity between Amerind and the Nostratic proto-language (ancestral reconstructed language). According to Ruhlen (1990), the discrepancies between the two superfamilies are due to differences in methodologies of the Russian and the American researchers. An important difference is that Russians rely on families for which a proto-language has been reconstructed; families without a proto-language were not considered. Greenberg does not believe in the need for using reconstructed proto-languages for interfamily comparisons. He left out Afro-Asiatic, Dravidian, and Caucasian because he believes they are less closely related than the language families he has included in Eurasiatic. These families would be included only at a higher level of classification. Ruhlen concluded therefore that Nostratic is not a valid taxon, because it omits families at a lower systematic level that have not been considered by the Russians.

A smaller superphylum, Austric, is already part of Ruhlen's classification (phylum no. 13) in figure 2.6.1. In more recent time, relationships already described long ago between Na-Dene and Sino-Tibetan languages, as well as Na-Dene and Caucasian languages, and between this superfamily and some language isolates (Basque, Etruscan, Sumerian, Hurrian, etc.) were given fresh attention (Starostin 1989; Gamkrelidze and Ivanov 1990).

2.6.b. COMPARISON OF GENETIC AND
LINGUISTIC TREES

Armed with this knowledge, we have proceeded to map these language families and superfamilies on the genetic tree of figure 2.3.2. Sixteen linguistic families could be mapped onto the genetic tree. We have no data on speakers of Caucasian languages among the 42 populations. Of the 42, Mbuti Pygmies are believed to have lost their original language, having borrowed those of neighbors. There was a clear correspondence between the 16 linguistic families and the 41 residual populations in that each linguistic family associated with either a single genetic population or with a few closely related ones, that is, to a genetic cluster that is low in the evolutionary hierarchy, and hence presumably appeared late. In figure 2.6.2, the genetic tree (see fig. 2.3.2A; Cavalli-Sforza et al. 1988) is arranged to show parallels and discrepancies more clearly.

The one-to-one correspondence between genetic clusters and linguistic families is remarkably high, but is not perfect. Two processes may cause exceptions: *language replacement,* and *gene replacement.* In some cases they can be readily identified on the basis of historical information. Genetic data on modern populations also help to distinguish between the two processes, or at least to assess the contribution of the second of the two. Gene replacement is more likely to be partial and tends to follow demographic history. Two neighboring people may mix by asymmetrical gene flow, with only one of them contributing a small number of individuals to the other in every generation. As we have seen in section 1.17, the continuation of this process over a sufficiently long time

may determine an almost complete gene substitution. In general, the gene pool tends to reflect rather faithfully the numerical contribution from the two parental groups. Thus, genetically intermediate populations can be generated, with all possible degrees of admixture. This process need not be accompanied by language change. Languages tend to behave more like a unit, and be replaced as a whole, if at all. One can, and usually does, notice contributions to the lexicon from neighbors, but the structure of language is more stable, and certain specific groups of words are more highly conserved. In certain cases, therefore, one can observe massive genetic contributions from an external source with little if any language change, and, in other cases, language substitutions with little genetic change.

Some cases of language replacement are historically documented; for instance, Latin spread to western Europe and other countries under Roman rule, and foreign invaders imposed their languages in Hungary and in Turkey during the Middle Ages. In certain situations language replacement was a massive phenomenon, as

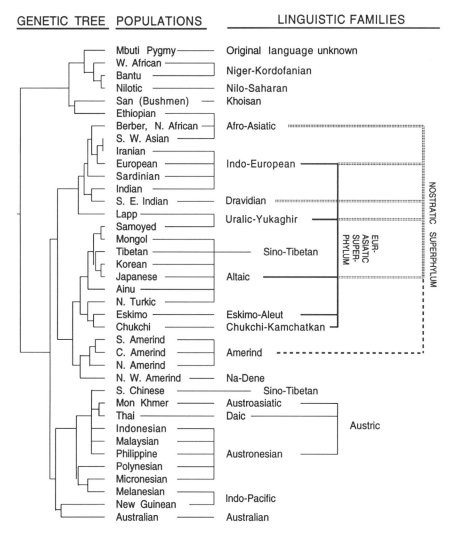

Fig. 2.6.2 The genetic tree comparing linguistic families and superfamilies published in Cavalli-Sforza et al. (1988). Populations pooled on the basis of linguistic classification belong to the following groups: Bantu, Niger-Kordofanian family; Nilotic, Nilo-Saharan family; Southeast Indian, Dravidian family; Samoyeds, Uralic family from Russia; North Turkic, branch of Altaic family; Northwest Amerind, Na-Dene family. The genetic tree was constructed by average linkage analysis of Nei's genetic distances and is the same as that of figure 2.3.2A.

shown by the spread of European languages to the Americas or Australia in modern times, or of Arabic to North Africa in the Middle Ages. The original languages may then appear only as *relics* in isolated places. This is believed to be the case for Basque, the only pre-Indo-European language surviving in Europe after the spread of Indo-European languages (see chap. 5). Similarly, the Dravidian family (see chap. 4) is mostly confined to southern India and a few more-northern places but was probably much more widespread in the past; in this case, the diffusion of Indo-European languages was also the cause of the partial disappearance of the languages of the earlier settlers. The processes leading to language replacement are described in some more detail later (sec. 2.6.d).

The exceptions to a full correlation between the genetic tree and the classification of languages by families in figure 2.6.2 are listed below.

1. Two members of the Afro-Asiatic language family correspond to different major branches of the genetic tree: East Africans (mostly Ethiopians), who are genetically African, and Berbers, North Africans who are genetically Caucasoid. Ethiopians are to various degrees mixtures of African and Caucasoid gene pools, with an African preponderance (on the average) so that they join Africans in the genetic tree. Most Ethiopians speak languages of Cushitic, a major branch of Afro-Asiatic, some speak Nilo-Saharan languages, but relatively recent and known historical events contributed to the spread to northern Ethiopia of south Arabian languages (the Semitic branch of the Afro-Asiatic family). Clearly there was in Ethiopia substantial gene replacement, which may be quite different in different places, and, at least in some cases, there was also language replacement (see chap. 3).

2. Two representatives of the Uralic family appear in the tree, Samoyeds (living east of the Urals) plus neighboring populations, and Saame (usually called Lapps) who live in northern Scandinavia. The first are genetically Mongoloids, and the second Caucasoids. Further genetic analysis, however (Guglielmino-Matessi et al. 1991; see also chap. 5) showed Lapps are a mixture of Caucasoids and people of Uralic origin who are Mongoloid. The contribution of the latter is less important and Lapps therefore associate in the genetic tree with Caucasoids. The simplest explanation is that gene flow going on for millennia from their neighbors of Caucasoid origin (see sec. 1.17 and chap. 5), has caused enough gene replacement in Lapps to produce their now predominantly European-looking gene pool without language replacement.

3. Tibetans and South Chinese speak languages of the same family, Sino-Tibetan, but belong genetically to northern and to southern Mongoloids respectively. Tibetans come from the north of China and their genetic association agrees with their geographic origin (chap. 4). Northern and southern Chinese are substantially different

genetically but Chinese languages were probably spread to southern China only in the first millennium B.C. when the rule of northern China was extended to the south (Wang 1991). Many ethnic minorities of southern China still speak native languages which do not belong to the Sino-Tibetan family.

4. Basque, as mentioned above, is considered an isolate language, very probably a relic of a pre-Indo-European one. All possible relatives of Basques are very distant from it linguistically and geographically. Genetically Basques are European but are sufficiently distinct that they can be recognized as an isolate, probably proto-European (Mourant 1954). They have remained endogamous and, in general, isolated enough from neighbors that they are recognized as linguistically and genetically distinct from them, but they certainly have received substantial gene flow over the millennia (see chap. 5).

5. Melanesians, and to some extent Micronesians and Polynesians, are mixed genetically. Some Melanesian tribes use Austronesian or other Indo-Pacific languages. A clear distinction of gene and language replacement is made difficult by the remoteness of contacts between earlier settlers (Melanesians, speaking Indo-Pacific languages) and later settlers (Indonesian and Southeast Asians, speaking Austronesian languages; see chap. 7).

The correlation between the genetic tree and the information from linguistic families is difficult to test statistically in a direct fashion, because the nature of a tree prevents the application to this problem of most if not all current statistical tests. An unusual measure of this correlation was, however, provided by an application of a coefficient (the index of consistency) developed for estimating the evolutionary congruence of traits observed in living organisms with the tree summarizing their phylogenetic history (Kluge and Farris 1969). The application of this measure has thus far been limited by the lack of a statistical test of significance and also by a strong dependence of the numerical value on the size of the sample tested (Archie 1989). The index was first calculated on our data by Bateman et al. (1990), but we disagreed with these authors on their method of calculation and their belief that it was impossible to calculate a statistical error (Cavalli-Sforza et al. 1990). If an adequate test of significance is constructed, the index can lend itself to test the congruence between the linguistic families and the genetic tree. A test of significance was applied to the problem by Cavalli-Sforza et al. 1992, by evaluating the index of consistency on the data from the Cavalli-Sforza et al. (1988) paper and also on 10,000 permutations of the linguistic families, which provide a distribution of the random values of the consistency index and permit a significance test. All values of consistency indices were calculated by the computer program MIX available in the Phylip package. The index for the observed data was 0.57 and in random permutations of linguistic families, such a high value would be found with a probability of 10^{-6}. The difference between the observed value and those obtained by chance is thus highly significant, and the statistical test confirms the congruence between genetic and linguistic differentiation.

There is also a remarkable correlation between the two linguistic superfamilies Nostratic/Eurasiatic and Austric and the genetic data. These superfamilies are the first important attempts at generating higher levels of hierarchy in the linguistic classification above the traditional families. Furthermore, Nostratic and Eurasiatic are similar enough that it is reasonable to consider them as classifications in progress that, when further work is done, will most probably converge and become one single superfamily. Taking the *union* (in the language of set theory, i.e., all elements present in one or the other superfamily) of Nostratic and Eurasiatic, plus Amerind, whose similarity to Nostratic/Eurasiatic has been suggested (Shevoroshkin 1989), we find an almost perfect correspondence with the whole north-Eurasian genetic cluster. There is only one exception to complete coincidence of the genetic and linguistic clusters, namely the presence in the genetic cluster of the Na-Dene, who are of uncertain origin but may belong to another, possibly older, Asian family called Dene-Caucasian (Starostin 1989). Na-Dene had considerable opportunities for admixture with Amerinds, making it more difficult to demonstrate their separate identity (see chap. 6). Moreover, the Austric superfamily shows a remarkable correspondence with a genetic cluster. These similarities at a higher hierarchic level are a dramatic confirmation of the existence of an important association between genetic and linguistic evolution. The superfamilies on which this last evidence is based are not recognized today by a number of linguists (Bateman et al. 1990a, b), and it is likely that reaching a consensus will take some time. Recent reviews in *Scientific American* and *Atlantic Monthly* tell the story of the fight between Nostraticists and their detractors, as well as the arguments concerning the "mother tongue" and the American languages (Ross 1991; Wright 1991). We hope that more satisfactory, quantitative methods of analysis of linguistic evolution will be developed and help to accelerate consensus (Diamond 1990). We suggest that such methods be first tested on material on which consensus has been reached (e.g., the Indo-European, and possibly other families).

Our findings demonstrate that the association between linguistic families and the genetic history of humans is far from random and the significance tests confirm our conclusion. Results of regional comparisons of languages and populations vary considerably. In a survey of 15 studies, not including ours (Barbujani 1991), 6 were found to be significantly in favor of association. This is a significant answer, since less than one study $(0.05 \times 16 = 0.80)$ would be expected to be significant by chance alone at a significance level of 5%. It is, however, unfortunate that most of the 15 studies reviewed by Barbujani have used linear correlation coefficients between genetic and linguistic distances. There is a serious drawback in testing the significance of such correlations because of the autocorrelation among the distances. Possible statistical solutions are given by the Mantel or the Clifford test (Mantel 1967; Clifford 1989) or, alternatively, by a direct comparison of trees as described above (see also Cavalli-Sforza et al. 1992).

2.6.c. WHY IS THERE A CLOSE SIMILARITY BETWEEN LINGUISTIC AND GENETIC TREES?

In subsequent chapters we examine many analyses at the regional level that reinforce the notion of a strong parallelism between linguistic and genetic evolution, as observed here at the world level. What explanations can one offer for this important correlation? The major explanation is the history of populations. The correlation is certainly not due to the effect of genes on languages; if anything, it is likely that there is a reverse influence in that linguistic barriers may strengthen the genetic isolation between groups speaking different languages. This effect of linguistic isolation on genetic isolation is observable at a level of linguistic difference much lower than that of language phyla, for example, between speakers of different branches of the Indo-European family (or even within languages of the Romance or Germanic branches; see chap. 5). It is crystal clear that all normal human beings have essentially the same skills in learning languages, and the native tongue of an individual is entirely determined by the social environment in which the cultural development of that individual has taken place. Because in most people these skills decrease considerably after puberty, it becomes difficult if not impossible after that age to attain native ability in a foreign language, especially in its phonetics. This phenomenon is probably true for all human groups, but it has not been properly investigated.

The explanation of the parallelism between genetic and linguistic trees is to be sought in the common effect of factors determining differentiation both at the genetic and at the linguistic level. The most important factors are events determining the separation of two groups. After fission and migration of one or both moieties to a different area, they are partially or completely isolated from each other. Reciprocal isolation causes both genetic and linguistic differentiation. It is reasonable to assume that both the genetic and the linguistic divergence thus determined will increase with the time since separation, although not necessarily as regularly as glottochronologists or some biological evolutionists would like them to. Thus, the history of major fissions in the process of occupation of continents or parts of them is likely to be mirrored in both genetic and linguistic evolution.

The historic process of change is also recognizable in the geographic patterns of variation, which reflect the pattern of past historical events, including genetic exchange. The model of isolation by distance (sec. 1.16) gives rise to the expectation that genetic distance will increase with geographic distance. We see in section 2.9 that genetic observations come rather close to expectation, at least within a range of 500–1000 miles. For wider ranges, genetic and geographic distances may also show a simple linear increase, but the relation may also be more complex, either because there has not been time to reach equilibrium, or because geographic barriers prevent its taking place. The situation for languages has not been examined in the same detail, but it may be similar to that of genes (Cavalli-Sforza and Wang 1986). At least for variation within a language, or a group of closely related languages, geographic distance and linguistic distance are reasonably correlated.

Factors other than separation may have similar effects on both types of evolution and contribute to the correlation. Two that come to mind are population size and migratory exchanges. Small population size favors faster genetic differentiation and may have the same effect on languages, but the mechanisms of linguistic evolution are too poorly known for this statement to be other than an interesting hypothesis. Migratory exchange between two populations favors both genetic and linguistic exchanges and therefore decreases or slows the divergence in both cases. For instance, in Ethiopians and Lapps, we have seen that admixture for languages and for genes are not parallel, because they are affected by different constraints.

2.6.d. Language replacements

Among the exceptions that may blur the picture of the correlation between languages and genes in figure 2.6.2, language replacement is likely to be most important and the easiest to study because there are many known historical examples. Therefore, we summarize here the mechanisms of language replacement. The archaeologist C. Renfrew (1987) has suggested three possible models. We have simplified and reduced them to two (pooling his second and third model, which he admits are similar). In both cases one language is replaced by another, but the genetic consequences are quite different. In the first case, a population replaces another and brings its own genes and its language. In the second, a small but powerful group imposes its rule (and its language) on a larger population. It is useful to consider both in more detail.

1. *Demic expansions* (see sec. 2.7). *Demic expansions* are the basis of Renfrew's *demographic-subsistence* models: expansions of people under demographic pressure to areas previously uninhabited, or inhabited by other ethnic groups living at a more primitive economic level and therefore at a definitely lower population density. Previous inhabitants are suppressed, enslaved, or absorbed. The expansion of farmers is one example that we explore in the next section and discuss in chapter

5 because the first such genetic model was advanced for Europe. Renfrew (1987a, 1989a) suggested that the spread of farmers from the Middle East, which he accepts as the explanation of the diffusion of agriculture from the area of origin, was responsible for the spread of Indo-European languages (see sec. 2.7 and chap. 5). Other competing theories about the spread of Indo-European languages are plentiful (Mallory 1989), as we show in subsequent chapters. One example of expansion of farmers accompanied by the spread of the farmers' languages is less controversial (see chap. 3). In the 3000 or more years it took for Bantu-language speakers to migrate from Cameroon/Nigeria to Central and South Africa, there was ample time for Bantu languages to diverge, but their original unity is easily appreciated. In general, two languages differentiate to the point that mutual intelligibility is lost in 1000 years (order of magnitude) and often less. Renfrew has more recently suggested (1992) that all agricultural developments causing demic diffusions have also spread the farmers' language.

2. *Conquest by a minority, including Renfrew's* elite-dominance *and* system-collapse *models.* Under conquest, a people, usually with strong social hierarchy and military organization, takes power in a country and imposes its language and usually much of its global cultural inheritance, retaining for itself positions of power and control of wealth. Conquerors, if well-organized, can be a small minority. Two such cases are the previously cited examples of Turkey and Hungary, which are well-known historically and are further studied in chapter 5. In both cases the genetic traces of the invaders are, at best, extremely modest since they were not sufficiently numerous to influence strongly the genetic pool of the previous inhabitants.

Conquest does not always involve language replacement. Several barbarian invasions after the fall of the Roman empire did not have a marked effect on local languages, although in some cases the original barbarians' dialect has been conserved to these days in certain small areas.

In Renfrew's terminology, *system collapse*, generating a power vacuum, may result in unusual circumstances, giving a chance to certain minorities to take control and impose their language. Two examples cited by Renfrew are the fall of the Roman Empire in Britain, after which Anglo-Saxon mercenaries, perhaps with the help of kin from abroad, acquired power; and the fall of the Mayan civilization around the tenth century A.D., about which much less is known. As acknowledged by Renfrew, this mechanism could be considered a special case of elite dominance.

Unlike Renfrew, who has chosen not to consider the genetic aspects of these phenomena, we are interested in the joint history. In the demographic-subsistence model, there is clearly replacement of both languages and, at least partially, also of genes. Most elite-dominance situations are likely to leave the genes largely or relatively intact, unless the size of the invading population is not negligible compared with that of the invaded country. In many cases the invaders generate the new aristocracy, and sometimes do not mix or mix only to a very limited extent. Even when admixture occurs, the aristocracy formed by the conquerors can remain relatively unmixed genetically for a long time. In the Americas one finds extremely varied situations from country to country, or city to city, with respect to the admixture between native Americans, Caucasoids, and Blacks.

In some situations, mechanisms other than those postulated by Renfrew seem to have operated, or the distinction between mechanisms is not easy. Especially in Africa, India, and New Guinea—but undoubtedly also in other parts of the world—there are tribes in which one finds an inconsistency between language and genes, but it is unclear whether it is due to language replacement or to gene replacement. One would expect language replacement to occur more easily, but in classical situations of asymmetric, modest gene flow, there may be extensive gene replacement of a population, provided contact and genetic exchange have lasted long enough (see sec. 1.17). African Pygmies have usually acquired languages from neighbors and retained them for some time; they may have gone through more than one such cycle, as they moved, or their neighbors changed. But Pygmies may also show evidence of partial, and sometimes advanced, gene replacement (Cavalli-Sforza 1986). These and many other cases show that language replacement may follow contact with other populations that are not at all at the level of social development required in the elite-dominance model. There is usually some economic dependence on the people whose language is acquired, but genetic exchange may be absent, moderate, or frequent, depending on circumstances. Although African Pygmies show examples of language replacement with or without gene replacement, in other groups extensive replacement of genes may have co-occurred with conservation of the language. This may be true for Lapps, who were in prolonged contact with neighbors of different ethnic and linguistic origin. We have seen in section 1.17 that gene replacement caused by prolonged gene flow can be fairly rapid. In certain cases, within 1000 years or less gene replacement can be complete enough to make the genetic origin of people very difficult to trace. In Africa the case of the Hadza and Sandawe is of special interest. These two Tanzanian populations, which speak Khoisan languages, show little if any evidence of Khoisan genes. This may be a case of gene replacement rather than language replacement; other similar situations are mentioned in chapter 3. We thus have a number of different situations, with or without genetic replacement, with or without language replacement, or without marked differences in social development. When history is well-known, the analysis of these phenomena is simple. A systematic study of language replacements,

of the historical conditions determining them, and its social and genetic correlates seems badly needed.

2.6.e. POSSIBLE TIMES OF ORIGIN OF LINGUISTIC FAMILIES

The genetic tree may help us set a time for the origin of linguistic families. There have been direct attempts (by Swadesh) at dating the origin of languages and their families by a method called *glottochronology* or *lexicostatistics*, which assumes (Swadesh 1971) a constant probability of replacement of words per unit of time with words of the same meaning but of recognizably different origin (root). This method, which is an exact parallel of the molecular clock in biology but had an independent origin, has received mixed acceptance among linguists. Swadesh recognized that different words have a different rate of replacement and tried to prepare a basic list of highly conserved words. Unfortunately, there are several difficulties: the replacement rates are not homogeneous even in the later revised word lists, and the curve of decrease of the percentage of cognates (words of same origin) is not a simple negative exponential (Kruskal et al. 1971; Sankoff 1970), as was expected under the original theory. As a consequence, glottochronological methods tend to underestimate older times of separation of languages. There are also difficulties at the semantic and phonological levels in establishing cognates. Further criticisms are illustrated by Embleton (1986), who proposed a mathematical method for taking account of the effects of words borrowed from neighbors. The expression "borrowing" is used by linguists to define an analogue of gene migration, but unlike the genetic phenomena, words are either borrowed or not, whereas the acquisition of genes from migrants is a completely gradual change of frequencies of genes, which proceeds in parallel for all genes.

The time of origin of the Indo-European family is usually given as 5000 years B.P. (Gimbutas 1966; Mallory 1989). Renfrew equates the spread of Indo-European with that of farming from the Middle East, which would put the origin, at the latest, at 10,000 years B.P. or 10 kya. On the basis of linguistic considerations, Dolgopolsky (1988) argues for a date between 10 and 20 kya, tentatively put at 15 kya. Genetic distances indicate a minimum date of 9 kya, the date corresponding to the average distance of Europeans, including Iranians and people in the Near East. An upper limit of 17 kya was obtained on the basis of the separation of all Indo-European speakers from Berbers, who speak a language from a different family. Indians were not included in this calculation of genetic distance for linguistic purposes, because they are highly hybridized with Southeast Indians who speak Dravidian and other languages. The existence of a disturbance in the apparent evolutionary rate of Europeans, discussed in section 2.4, makes these estimates less strong than one would like them to be.

The origin of Niger-Kordofanian is around 10 kya, but it is perplexing that Nilo-Saharan would seem to have a later origin according to the genetic tree (4 kya). The genetic data are probably giving an erroneous indication because Nilo-Saharans included in the tree are mostly Nilotics, coming from areas that must have had important contacts with Bantus at the time of their migrations (chap. 3). Information on other Nilo-Saharans is almost nil. The genetic divergence between Khoisan (!Kung) and other Africans on the basis of the tree is on the order of 50 kya, but a more complete discussion is given in chapter 3.

The divergence of Afro-Asiatic is based in this tree on the distance of Berbers from Indo-Europeans and is on the order of 15 kya. Data from Ethiopians are tainted by admixture problems, as repeatedly indicated above.

It is difficult to give a time depth for Dravidian using genetic data, as there is genetic admixture both in Indo-European and in Dravidian-speaking Indians. Taken at face value, the date would be 20 kya.

Excluding Lapps, who are mixed, Uralic appears to have 4000 years separation from Mongols (Altaic speakers). This seems short and may require reconsideration. For Altaic—excluding northern Turkic, which shows greater genetic affinity with Northeast Asians—the estimate is 19 ky. Eskimo and Chukchi have a separation of about 20 ky.

We have already discussed estimates of separation between native American and North Asians, which are on the order of 20–30 ky (see also chap. 6). This may also be the time for the separation of Amerind and Na-Dene, whereas that between Eskimo and Na-Dene may be only 10–15 ky.

Austric has a date around 21 kya. The data do not seem informative with regard to Sino-Tibetan.

The Polynesian fission from the tree seems unacceptably large and, as already mentioned, is likely to be greatly magnified by problems of initial admixture and of very high drift (see chap. 7).

Finally, the Indo-Pacific and Australian families are among the oldest, with a separation time around 45 kya.

The most interesting date is obtained from the correspondence between the Eurasian superfamilies (Nostratic + Eurasiatic) and the north Eurasian cluster of the genetic tree. The latter is dated at about 40 kya (see fig. 2.3.3, table 2.5.1). This is somewhat after the end of the "black period" before the archaeologically observed major expansions of a.m.h.

The dates indicated here for the origin of linguistic families are suggestions that should be taken with a ton of salt. A more specifically directed investigation is necessary. Rather than considering dates of origin for individual families, it is perhaps more meaningful to consider the global distribution of the dates of origin of linguistic families. The majority of families seems to have an age of 5 (more probably 7 or 8) to 20 ky, with some older ones in the range of 30–50 ky: Khoisan, Australian, Indo-Pacific. These values depend on the assumption that separation dates can be estimated accurately from genetic distances and that the date of separation for genes is also responsible for the fission of linguistic families.

Linguistic and genetic evolution can exchange useful information in another respect. Some research on the area of origin of linguistic families obviously can be of interest to genetics. Inevitably, the time and area of origin are closely connected (for a general analysis, see Sapir 1958). There are many possible sources of evidence, but they have to be used with caution. The question of the area of origin of Indo-European languages has been studied intensively, and there exist many different hypotheses (Mallory 1989): Northern Europe; Central Europe; the Balkans; a region north of the Black Sea, extending to the northern Caucasus and the western Asian steppes; Anatolia; and others (fig. 2.6.3). This list gives an idea of the difficulty of the problem.

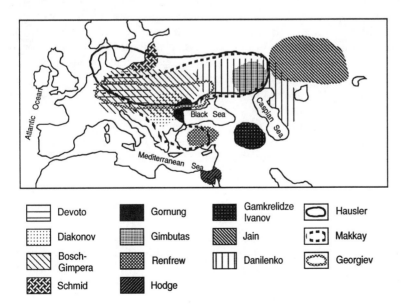

	Devoto		Gornung		Gamkrelidze Ivanov		Hausler
	Diakonov		Gimbutas		Jain		Makkay
	Bosch-Gimpera		Renfrew		Danilenko		Georgiev
	Schmid		Hodge				

Fig. 2.6.3 Postulated areas of origin for Indo-European languages (after Mallory 1989, p. 144).

Dolgopolsky (1988) favored an Anatolian origin for the Indo-European family, on the basis of ancient borrowings from proto-languages presumed to have been spoken near Asia Minor some 10–20 kya: proto-Semitic, in the Middle East, and proto-Kartvelian in the southern Caucasus. This hypothesis had also been suggested earlier and independently by many other authors on the basis of different considerations (Gamkrelidtze and Ivanov 1990); Renfrew (1987) has supported it with archaeological evidence (see chap. 5), disagreeing with earlier archaeological interpretations of the same problem.

Another contribution worth remembering, among the many in existence, is the area of origin of Bantu languages. These are spoken today over most of Central and South Africa; they are very similar to one another but are generally not mutually intelligible. They clearly must have evolved recently from a common origin. In contrast to earlier hypotheses, Greenberg (1955, 1963) showed convincingly that the non-Bantu languages most similar to the Bantu ones are spoken today in the area of confluence of the Benue and Niger rivers, around the border between Nigeria and Cameroon, and that the Bantu expansion must therefore have had its origin in this region. Greenberg's conclusion agrees with modern information from archaeology and history (chap. 3). Based on archaeological considerations, the beginning of the spread of Bantu-speaking farmers is located in Cameroon and dated at about 3000 years ago or earlier (see chap. 3).

2.7. IMPORTANCE OF EXPANSIONS IN HUMAN EVOLUTION

2.7.a. EXPANSIONS MAY HAVE PUNCTUATED HISTORY OF MODERN HUMANS

There have been, and still are, considerable individual and mass movements in human history. Individual (or local) movement is a major source of genetic and cultural exchange between groups, sedentary or not, and causes a slow but continuous genetic homogenization of populations countering differentiation caused by drift. When migration is not reciprocal in geographic direction, as is the case with urbanization, admixture is limited to the area of in-migration. Many towns and cities attract immigrants from a wide area and are thus complex gene pools. Until recently, cities, unlike rural regions, had negative net reproduction rates and were not reproducing themselves. Urban growth was thus maintained by immigration, and urban people typically represent a sample of the population of the whole area from which the immigrants have originated.

Migrations are determined by push and pull factors. Modern societies are highly sedentary, but migration is essentially a movement of individuals and of nuclear families. Pull factors are of an economic or social nature. Only in special circumstances such as war or famines do life and environment deteriorate to the point of acting as push factors. Traditional societies may have been forced to abandon their usual living place more frequently than modern societies. Many examples exist of long-term environmental changes that have caused whole populations to migrate: the desertification of the Sahara, a process continuing for the last 4000 years, caused most of its Neolithic inhabitants to leave for southern regions. In addition to natural phenomena, human deeds like invasions, wars, and plundering have helped to create hard times for local populations, not infrequently forcing them to leave.

These events have certainly caused mass migrations, but negative push factors are not their only cause: the history of the peopling of the earth is not simply a story of relocations caused by natural or man-made disasters. The growth of a population and its expansion to neighboring land must often have occurred under the stimulus of new, adaptive cultural developments permitting an increase in the carrying capacity of already inhabited lands, or the occupation of new regions and niches. The history of the peopling of the world must have been punctuated by a number of such *expansions* of which the occupations of continents like Europe, Australia, and America are outstanding, but not isolated, examples.

A reasonable scenario for the growth of human populations is that, for long periods, population numbers in a given region fluctuated around values not far from, but usually below, the limits imposed by the capacity to use the natural resources of the area. In other short periods there occurred relatively rapid growth and geographic expansion, determined by the chances of better use of existing resources offered by crucial innovations. It is therefore useful to consider the dynamics of such expansions, many of which brought people to new environmental niches and new regions and helped to generate the fissions that appear in the human evolutionary tree.

2.7.b. FROM FOOD COLLECTION TO FOOD PRODUCTION

The most important innovation allowing an increase in the carrying capacity of the land and the accompanying increase in population density was the transition from food collection (foraging) to food production through plant cultivation and animal breeding (Clark 1965; Harlan 1971; Flannery 1973; Cohen 1977; Reed 1977; Ammerman and Cavalli-Sforza 1984; Zohary and Hopf 1988).

Foraging (hunting, gathering, fishing) was replaced very gradually and never abandoned; even now, fishing remains an important source of food. Agricultural and breeding techniques underwent constant improvements over the millennia, and today the Earth can support densities several thousand times higher than it did at the end of the Paleolithic. Over time, technical improvements in plant cultivation and animal breeding have permitted a continuous increase in population numbers. We are presently in a critical period in which most populations living on the Earth have very recently decreased their mortality rates, but only those that did so earlier have learned to decrease their fertility rates. As a result, there is rapid exponential growth in many developing countries, leading the world toward demographic bankruptcy unless growth can be effectively curbed. Demography as a pure science of numbers of births, deaths, and populations can hardly explain why birth and death rates go up or down, or how they can be changed and directed. Historical records show that human populations respond rapidly with increased birth rates at the end of periods of high death rates resulting from epidemics, wars, or perhaps even famines; these disasters are relatively common occurrences, and the appropriate response is likely to be built-in, by genetic or cultural mechanisms or both. The response to overcrowding by a decrease in birth rates seems to be slower. Undoubtedly, some mechanism regulates population numbers downward when they are too high. Practically no human population—with the exception of a few religious isolates, which can accomplish the feat only for a short time—grows at the maximum rate allowed by human physiology. In most traditional societies, the growth rate seems to be adjusted to give a slight increase in numbers of people or none at all, when the population is near the limit of the carrying capacity of the land. When technical, economic, or social innovations permit rapid growth and local overcrowding is determined, two responses are possible: decrease of birth rates, and emigration. Emigration seems to have been more acceptable for modern societies (e.g., Europe after the Renaissance) and also for traditional societies. When reasonable opportunities for migration are available, it is obviously the immediately available solution. Reproductive customs are deeply rooted and, although their long-term regulation is poorly understood, historical experience in certain parts of Europe and the present experience in the Third World indicate that there is a considerable lag before birth rates decrease effectively. When overpopulation can be cured by emigration, there is little incentive to stopping local growth. Until regions of emigration are saturated, population growth will continue in the area of origin and in those of arrival. The combination of demographic growth with spread to other areas by migration is the essence of geographic expansions. A built-in mechanism favors the continuation of the expansion until total saturation is reached.

The availability of food has been the major brake on population growth, and its history is therefore essential for understanding expansions. It is convenient to consider four major economic systems, which tended to follow each other almost regularly, though with occasional exceptions in the sequence. The four systems therefore tend to correspond to four phases of food economy.

1. *Hunting-gathering*. Also called foraging or food collection, hunting-gathering was the only source of food until the Neolithic. In riverine, lacustrine, and marine environments it includes—or may even consist almost exclusively of—fishing. Foraging is historically the most primitive way of obtaining food; contrary to widespread beliefs, however, primitiveness of the economy does not involve high reproduction rates. Compared with populations that live in the same areas and have traditional economies but more advanced systems, hunter-gatherers have the lowest birth rates. Before the advent of agriculture, populations in most areas had been near saturation density. Because innovations determining increases in the carrying capacity of the land were few and relatively ineffective, most increases in total population were possible only by migration to new areas. Thus, in the Paleolithic, local growth must have been fairly slow, apart perhaps from periods of expansion. To obtain slow local growth, populations must have adjusted to a net population rate of increase near zero. A typical hunter-gatherer woman gives birth about every 4 years during her reproductive lifetime (Cavalli-Sforza 1986a), producing about five children. The resulting birth rate approximately matches their high death rate. Therefore, in terms of population growth, hunting-gathering populations are practically "stationary," that is, nearly constant in size. In exceptional conditions, as on the Pacific Northwest coast of North America, abundant fish harvests allowed the development of relatively high densities. Otherwise, population densities of foragers were, and are, mostly in the range of 0.01–0.1 inhabitants per square kilometer (Hassan 1975) and are higher in the tropics than in the Arctic.

2. *Primitive agriculture*. The development of agriculture is a recent phenomenon in human history, occurring entirely in the last 10 ky (the Neolithic, practically coinciding with Holocene; see sec. 2.1). It originated in a number of different areas, probably independently and under the action of a common stimulus, demographic pressure. In the beginning, agriculture was inevitably far from modern efficiency, but it did allow—and probably stimulated—population growth. Even though it never totally replaced the use of traditional sources of food obtained from nature, it increased food availability by introducing a new source, cultivated plants and animals, the abundance of which depended on the quality and

quantity of land, as well as on the working capacity and ingenuity of the individual. Food production has changed the world.

Even the earliest farming, at least in West and East Asia, included domesticated animals. This mixed economy was very successful in temperate areas, where it permitted increases in population density of 5 to 50 times that of hunter-gatherers within the first millennium of change. Populations that did not change to agriculture, but remained foragers, could continue their way of life to some extent, and some still do. The formation of a mosaic of food economies was also possible because macro- and micro-environments favored by early farmers were frequently different from those suitable for foraging. The fraction of foragers must have monotonically decreased, however, as more and more foragers were absorbed into the new society or their land was seized by farmers.

3. *Pastoral nomadism*. In other areas where ecological conditions did not permit people to obtain adequate support from plant breeding, some domesticated animals took on unexpected importance. In the grasslands of southeastern Europe and central Asia, pastoral nomadism developed into a complex culture. The domestication of the horse had a considerable impact on Eurasia, whereas reindeer herding spread widely in the Arctic. Other forms of pastoral nomadism (e.g., camel herding) also arose in arid regions where they are still of considerable importance.

4. *Complex agricultural systems*. Developments in areas especially favorable to agriculture led to considerable technical sophistication with the introduction of irrigation, fertilization (from natural sources), plowing, and a more and more elaborate and varied set of domesticates.

Enough food was produced that towns and cities could be built, with an early beginning in the Middle East. Given the primitive means of transportation, food could not come from great distances, limiting the size of communities; but cities of 50,000 inhabitants or more developed fairly quickly in the Middle East. Civilization (from *civitas*, Latin for city) is life in cities, etymologically, and the increasing needs of social organization made it necessary to create social hierarchies and stratifications. Technical developments in agriculture took millennia and continue today, allowing and causing increases in population densities up to 1000–10,000 times Paleolithic levels.

Agriculture originated independently in many places (see fig. 2.7.1), but probably first in the Middle East. The beginnings of the agricultural economy in the Middle East certainly took a long time and covered a fairly large region, the Fertile Crescent, extending from Israel to Syria, northern Iraq, and western Iran, with an extension to southern Anatolia. The beginnings were certainly early and preceded by local development of dependence on wild stalks of cereals (e.g., the Natufian culture in Israel). It seems a fairly natural development for a population foraging on wild cereals to stabilize fields of wild crops by cultivation. A long time was required before the agricultural economy was ready to move to other areas; it was only when several staple crops (wheat, barley) and animals (sheep, goats, cattle) were domesticated that the new economic complex could be successfully exported to nearby land. This expansion began around 9500 years ago and probably took place in all directions starting from the Middle East, but the archaeological information is adequate only in Europe.

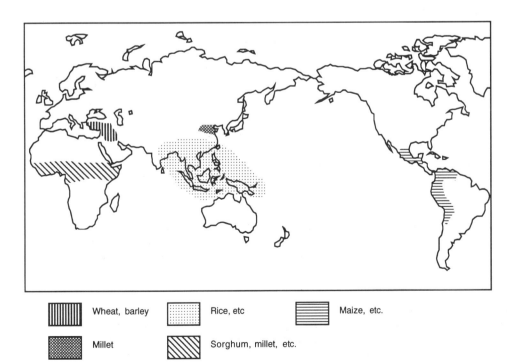

| | Wheat, barley | | Rice, etc | | Maize, etc. |

| | Millet | | Sorghum, millet, etc. |

Fig. 2.7.1 Areas of origin of agriculture (after Harlan 1971, p. 13, fig. 2.1).

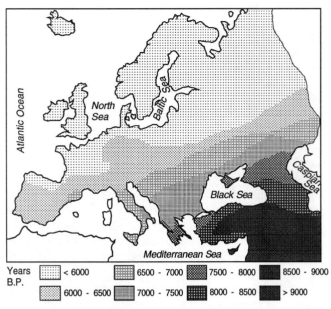

Years B.P.

< 6000	6500 - 7000	7500 - 8000	8500 - 9000
6000 - 6500	7000 - 7500	8000 - 8500	> 9000

Fig. 2.7.2 Timing of the expansion of early farming to Europe on the basis of radiocarbon dates (in years before the present [B.P.]) of archaeological sites (Ammerman and Cavalli-Sforza 1984, p. 59, fig. 4.5).

Figure 2.7.2 shows the times of expansion to Europe of agriculture, as measured by the initial appearance of wheat, which was first cultivated in the Fertile Crescent and was not available as a wild plant outside West Asia, with a few, probably insignificant exceptions. The rate of spread of wheat farming is very slow, on the average (about 1 km per year), and relatively uniform, with a slightly higher rate along the coasts, especially in the western Mediterranean, and in the eastern plains of Central Europe, where the diffusion took place along rivers.

2.7.c. DEMIC EXPANSION OR CULTURAL DIFFUSION?

It is important from a genetic point of view to address the question of whether the spread of agriculture was due to the farmers themselves or was simply the diffusion of a technology. In other words, was it a *demic expansion* of people, the farmers, or a *cultural diffusion*? The latter explanation would be tantamount to a diffusion of techniques; the expression "stimulus diffusion" is sometimes used by anthropologists to refer to an essentially similar phenomenon. There had been no serious attempt to ask or answer this question, because the trend of explanations in archaeological circles after World War II was biased in favor of cultural diffusion, and when the demic model was first put forward (Ammerman and Cavalli-Sforza 1973), it encountered very limited favor. The situation now seems to be changing, as we will discuss in chapters 4 and 5.

In the case of the expansion of farmers from the Middle East, four facts support the demic model.

1. The expansion of early farming was very slow, gradual and regular, and therefore more easily compatible with an expansion of people than of a technique.

2. Demographic knowledge of population growth and migration of early farmers (mostly from contemporary ethnographic observations) allows one to predict rates of expansion demonstrating that the diffusion of agriculture is compatible with a demic expansion.

3. From ethnographic observations, hunter-gatherers (e.g., African Pygmies) show little tendency to acculturate when in contact with farmers (Cavalli-Sforza 1986a). Some pygmies are now acquiring some simple cultivation techniques that are compatible with their nomadic way of life, but they rarely adopt a new and complex culture unless their habitat is totally destroyed or profoundly altered.

4. The study of the modern geographic distribution of genes in Europe strongly suggests a diffusion from a center of origin in the Middle East, as well as other less important migrations (Menozzi et al. 1978a, b). This pattern agrees with a model of diffusion of farmers and partial admixture with local hunter-gatherers, who were gradually absorbed by the much more numerous population of farmers, which could grow to higher population densities. In this way, a concentric cline of genes of the original farmers' population is created, which decreases with increasing distance from the center of origin. The possibility of separating such clines by principal-component analysis was proved by simulation (Rendine et al. 1986). These conclusions were confirmed by other independent approaches; further details are given in chapters 4 and 5.

2.7.d. THE WAVE-OF-ADVANCE MODEL

The kinetics of diffusion of a population from a center of origin, leading to the conclusion suggested in point 2 above is based on a mathematical theory by Fisher on the "wave of advance of an advantageous gene" (1937). Fisher's model can be applied to ecology, the diffusion of animals (Skellam 1951, 1973), and the spread of epidemics and rumors (Kendall 1965). A population that grows at a constant rate subject to local saturation (according to a logistic function as in fig. 2.7.3), and spreads at a constant rate of migration, randomly in all directions, tends to grow and move away from the center at a constant rate of advance or rate of radial diffusion. This rate can be calculated from Fisher's equation,

$$r = \sqrt{(2gm)}, \qquad (2.7.1)$$

where g is the growth rate (the initial growth rate in the logistic curve of population growth), and m is the migration rate per unit of time and space. More specific definitions are given in Ammermann and Cavalli-Sforza (1984).

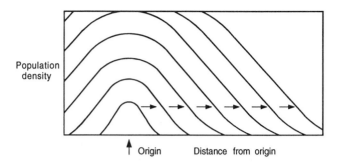

Fig. 2.7.3 The model of demic diffusion called the *wave of advance*. Abscissa, space (assumed unidimensional for simplicity); ordinate, local population density. The various curves indicate different times at regular intervals. Arrows indicate the passage of time. (After Ammerman and Cavalli-Sforza 1984, p. 69, fig. 5.2.)

As shown in figure 2.7.2, the spread of agriculture proceeded in approximately concentric equidistant circles from the area of origin. More accurately, the expansion is described by ellipses because of the greater rate of advance in the Mediterranean. The distance from the area of origin to southern England is about 3500 km, and it took approximately 3500 years to cover it, implying a rate of advance of 1 km per year. More accurate estimates, regressing the radiocarbon dates of the earliest sites for a total of about 100 sites (Ammerman and Cavalli-Sforza 1973, 1984) gave an average estimate of 1.1 km/year. Expression (2.7.1) allows one to test whether this estimate of the rate of spread of agriculture is compatible with the assumption that it is farmers who reproduce and migrate, in contrast with the hypothesis of cultural diffusion. The curves of figure 2.7.4 show the expected rates of advance, given the reproduction rate at the beginning of logistic growth (shown on the ordinate), and the migration rate shown on the abscissa.

It is very difficult, if not impossible, to estimate rates of population growth and migration from the archaeo-logical record. Ethnographic data, however, show that an agricultural population expanding in a practically uninhabited area grows at a rate of 3% per year. The demographic increases of French Canadians, or of Dutch colonists in South Africa, were both approximately a multiplication times a factor of 1000 in three centuries, practically without later migration. This is also a rate of population growth around 3% per year. Birdsell (1957) obtained a very similar estimate for two islands; Pitcairn and Tristan da Cunha. The curve in figure 2.7.4 shows that, for the rate of advance of 1 km per year and a growth rate of 3%, the migration rate must have been 200 km^2 per generation, which is less than the average observed in both modern and traditional agricultural populations. If the migration were 10 times as large, a value observed in a population of shifting cultivators in Ethiopia, the growth rate necessary for the observed rate of radial diffusion of 1 km per year would be 10 times smaller (0.3%). More details can be found in Ammerman and Cavalli-Sforza (1984).

2.7.e. Variety of expansions

The great power of expansion under agricultural developments is essentially due to the enormous increase in ecological carrying capacity; but the spread of agriculture was quite slow, as is expected for a demic diffusion. A true cultural diffusion could be much faster. For example, in the first 1000 years of agriculture, no pottery was made in the Middle East. Techniques for producing ceramics may have come from the Far East, where they were already present several thousand years before (see chap. 4) or may have been reinvented locally. Agriculture in Greece was initially aceramic. But when pottery was first introduced into the Middle East, it spread very rapidly to the European areas in which agriculture had previously expanded. From that time onward,

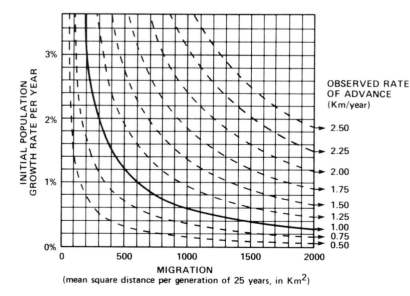

Fig. 2.7.4 Prediction of the rate of advance (radial diffusion) in Fisher's model of the wave of advance. Curves indicate rates of diffusion in kilometers per year, determined by the pair of values of migration and growth rates shown on abscissa and ordinate. The heavy curve represents the values that in combination would produce a rate of advance of 1 km per year. (Taken from Ammerman and Cavalli-Sforza 1984, p. 81, fig. 5.9.)

pottery diffused with agriculture (although in some cases it was acquired by Mesolithics in contact with farmers) and became a fairly reliable archaeological marker for the presence of farming at a site. Hunter-gatherers have limited use for ceramics. African Pygmies do not make pottery; they obtain some from local farmers.

Our understanding of expansions of farming is more advanced for Europe, where the archaeological and genetic records are more complete than for any other continent. Archaeological findings, however, show (fig. 2.7.1) several other areas of agricultural development, most of which were at least somewhat independent from the others, and one may ask about their genetic consequences.

From the Middle East, agriculture also expanded in other directions: toward Africa at about the same time or a little later as well as toward the East, (eastern Iran, Turkmenia and Central Asia, Afghanistan, Pakistan, India). The possible correlations between these expansions and the modern distribution of language families are discussed in chapter 4. The Sahara was not a desert at the time of the spread of agriculture, but farmers could not expand southward because cereals could not grow in the tropical belt. In chapter 3 we treat more thoroughly the local beginnings of agriculture just below the Sahara. Farmers speaking Bantu languages probably used cereals developed in the sub-Saharan area and were responsible for a major expansion starting from the Benue-Niger region, around the border between Cameroon and Nigeria (chap. 3). In perhaps 3000–3500 years Bantus occupied the largest part of central, eastern, and southern Africa, aided in a second phase by the use of iron tools. The maximum distance between the beginning and the end of the expansion is 4500–5000 km as the crow flies, giving an average rate 30%–70% higher than that of the European expansion.

Beginnings of agriculture in East Asia involve areas in North China, where millet was developed, and South China and Southeast Asia, where rice was developed (see also chap. 4). The northern and southern Chinese centers were probably independent developments. There was (and is) considerable genetic difference between the two populations, and although genetic information is still inadequate, what is available suggests little demographic exchange between the northern and southern centers of agricultural development in China. Climate and ecological differences may have been an important barrier to the exchange of crops cultivated in the north and south, and also to farmers.

In Mexico and in the northwestern part of South America, where maize and many other crops were domesticated, we observe a somewhat different behavior. Crops moved north of Mexico only late and then slowly, as discussed in chapter 6. There certainly were important cultural developments in the Andes, but the fragmentary nature of the genetic maps makes them of limited use in detecting expansions. Genetic drift has been very

powerful, especially in South America, and considerable genetic heterogeneity is observed in this continent.

Agricultural innovations can be expected to have had considerable genetic effects because of the large increases in population numbers they have caused. One may also suggest that they expanded more slowly in East Asia and the Americas because of the greater ecological heterogeneity of these regions. But the development of agriculture was not the only cause of expansions. In general, one is likely to find some technological innovations behind every expansion, but without important population growth, a detectable genetic correlate of the expansion is not likely. Populations that simply move from one place to another but do not undergo substantial demographic growth tend to change the genetic map in a less dramatic way. Events likely to have left more profound marks in the synthetic maps of section 2.11 and subsequent chapters are those in which a population genetically different from its neighbors undergoes important demographic growth. This may cause a demographic expansion and also eventually tends to freeze further drift effects in the group that has increased in size.

Innovations that involve increased food production are the most likely to entail demographic growth. One major innovation subsequent to agriculture that caused dramatic expansions was the domestication of the horse. This animal, which was probably domesticated in the western Eurasian steppe, was used not only for food but also, later, for transportation. This opened up the possibility of rapid movement of large population groups. As with all animal-breeding economies, a herd can be rapidly increased if better grazing grounds are available. The human population can then also grow, though not as rapidly, to a higher population density. Expansions of pastoral nomads may be stimulated in other ways, however. Pastoral life leaves much time free for other activities, like training for war and defense, and to some extent even requires skill in such activities. The long seasonal migrations of nomads, often carried out by thousands of people, pose logistic problems and defense needs that are very similar, or identical to military developments. In addition, the horse proved to be the greatest single military innovation before gunpowder.

Means of transportation and inventions of military or political importance may all contribute to expansions (fig. 2.7.5). If the population expands or moves to an area already occupied by earlier settlers, a genetic effect will be clearly noticeable, especially if the expanding or moving population undergoes substantial demographic growth. Sheer increase in mobility is not sufficient. In a Europe weakened by the crisis of the Roman Empire, the Visigoths rapidly moved from the Ukraine to Rumania and then to France, Spain, and Africa, but their kingdoms probably left few genetic or linguistic traces. The barbarian invasions that were responsible for the end of the Western Roman Empire probably had little genetic

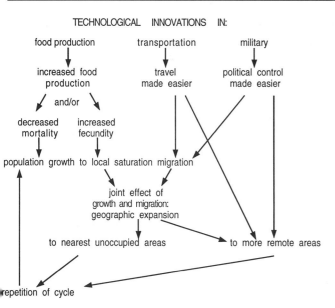

Fig. 2.7.5 A model for the genesis of demic expansions (Cavalli-Sforza 1986b, p. 24, fig. 1).

effect on the genetic structure of European populations. Their genetic impact is, however, difficult to distinguish from those of other earlier invasions of pastoral nomads from the steppes of Central Asia (chaps. 4, 5), since the barbarians are likely to have been mostly descendants of those same people.

Pastoral nomads were certainly responsible for a number of other expansions. From the western steppes, early invasions of Indo-European language speakers were successful in imposing new languages and, to some extent, genes, for instance in Iran and India. Later, nomads of the eastern Asian steppes expanded westward and southward in Asia, affecting the genetic picture and, even more, the linguistic one by spreading the languages of the Turkic subfamily.

Compared with these late expansions, the earliest one, that of *Homo sapiens sapiens*, is certainly poorly known but must have been the most important of all. It could have been a single episode or many. The occupation of continents from Asia took only a short time. We have already mentioned that a potential advantage of migrating a.m.h. might have been a language more sophisticated than that of earlier human types, making communication more efficient. Reconnaissance, scouting and search for camping sites must have been greatly aided. For the passage to Australia, innovations in sea transportation like rafts, logs, or boats were necessary (chap. 7) and were

probably also used for spreading along coasts, perhaps from Africa eastward along the southern coast of Asia.

The expansions to Australia and America were peculiar in that no previous inhabitants were found. In the archaeology and modern ethnography of hunter-gatherers, the minimum social group is a camp of 10–50 people, with an average of around 25–30 (for a summary and references, see Cavalli-Sforza 1986a) but these groups are usually exogamous and likely to have been combined into larger "tribes" of 300 or more (Wobst 1974). Models of budding and movement of the buds can explain the rapid occupation of large continents (for Australia, Birdsell 1977; for America, Cavalli-Sforza 1986b). Both major expansions to Europe, in Neanderthal times and at the time of early farmers, moved into areas of much lower density, at least in the case of the farming expansion. A model of movement of immigrants into an area more sparsely occupied by earlier residents has been developed and tested by a detailed simulation (Rendine et al. 1986; see also Ammerman and Cavalli-Sforza 1973, 1984). There were seven major variables in this simulation: the logistic growth rates and the saturation densities of early inhabitants (hunter-gatherers) and immigrants (farmers), the geographic migration rates of both groups, and the rate at which hunter-gatherers "acculturated," that is, joined the farming communities. One could envisage more complex acculturation processes, but that used in the model is not a true acculturation (adoption of agricultural customs by hunter-gatherers); it is rather the acceptance of individual hunter-gatherers into the farming society, which can generate gene flow. Actually, the transition might be more gradual if there were a first stage in which agricultural customs are acquired by hunter-gatherers, making it easier to establish a more lively cross-breeding later.

The expected increase in the population of farmers at various distances from the origin is shown in figure 2.7.3 and follows the predictions of equation (2.7.1). The formation of a gradient of gene frequencies similar to that observed (see chap. 5) depends especially on two critical factors, the ratio of the saturation densities of farmers to that of hunters, and the rate of acculturation (see also Sgaramella-Zonta and Cavalli-Sforza 1973).

A summary of some of the most important expansions punctuating human development is shown in table 2.7.1.

The importance of expansions in human evolution is further discussed in a paper by the same authors of this book in *Science* 259 (1993): 639–46.

2.8. EXTENT OF GENETIC VARIATION BY F_{ST} ANALYSIS

Here we discuss some simple properties of F_{ST}, the average F_{ST}s observed in our populations, the expected F_{ST} distribution in a simple tree and its comparison with the observed F_{ST} values of the DNA markers studied in section 2.4, the distribution of F_{ST}s in the 120 non-DNA polymorphisms, and its relevance for the study of natural selection.

Table 2.7.1. A Summary of Some Major Human Expansions (modified from Cavalli-Sforza 1986b) (y.a.=years ago; k.y.a.=kilo years ago)

| | | | | | Factors Favoring Expansion | |
| | | | | | Technical Innovations | Other Factors |
Period	People	From	To	Time		
Modern	Europeans	Europe	Americas	16th – 20th centuries	transatlantic travel	military and colonial conquest, religious persecutions, local poverty or famine, opportunistic factors
Modern	Europeans	Great Britain	Australia	19th – 20th centuries	discovery of Australia	
Historical	Greek	Greece	Central and West Mediterranean	10th – 6th centuries B.C.	nautical improvements	colonization and trade
Historical	Phoenicians	Lebanon	West Mediterranean	9th – 6th centuries B.C.	nautical improvements	colonization and trade
Prehistorical	Bantu-speakers	Nigeria – Cameroon	Central and South Africa	3 k.y.a. – 0 y.a.	agriculture, iron smelting	opportunities for farmers
Prehistorical	Nomadic	Eurasian steppes	Europe, South Asia	4 k.y.a. – 500 y.a.	horse breeding, military and social organization	exploitation of farming communities, plundering
Prehistorical	Polynesians	Samoan archipelago	distant Pacific islands	3 k.y.a. – 1 k.y.a.	Pacific navigation	
Prehistorical	Neolithic farmers	Middle East	Europe	10 – 6 k.y.a.	agriculture, boats	expanding and renewing fields, fast population growth
Prehistorical	*H. s. sapiens*	Northeast Asia	America	35 – 10 k.y.a.	superior	land bridge
Prehistorical	*H. s. sapiens*	West Asia	Europe	40 – 35 k.y.a.	language	ability to cross
Prehistorical	*H. s. sapiens*	Southeast Asia	Australia	60 – 40 k.y.a.	communication	sea tracts

2.8.a. SOME SIMPLE PROPERTIES OF F_{ST}

Measuring the variation of gene frequencies poses some special statistical problems that we summarize before proceeding to an analysis of the data. The standard measure of variation in a sample of n observations is the variance V,

$$V = S(x_i - M)^2/(n - 1), \qquad (2.8.1)$$

where S expresses the summation of all individual observations x_i from x_1 to x_n, and M is the mean, calculated as

$$M = S x_i/n. \qquad (2.8.2)$$

The variance calculated by equation (2.8.1) is an average of the squared deviations from the mean M. Elementary statistical knowledge shows that the sum of the squared deviations is not divided by n as for the usual arithmetic mean, but by $(n - 1)$ in order to remove a bias derived from the use in equation (2.8.1) of M, the mean of the sample.

When we study the genetic variation of the frequencies of a given gene in different populations, the natural measurement would therefore be the variance of the gene frequencies. If the mean gene frequency is very

small or very large (close to 0% or 100%), the variance of gene frequencies tends to be very small. One cannot therefore directly compare variances of gene frequencies of different genes without taking into account the mean gene frequency. This is one purpose of F_{ST}, which is the quantity,

$$F_{ST} = \frac{V_p}{\overline{p}(1 - \overline{p})} \qquad (2.8.3)$$

where V_p is the variance of gene frequencies, and $\overline{p}$ their mean (see also table 1.10.1).

A modified variance, F_{ST} has several special properties. Unlike the variance of usual metric characters, V_p has an upper limit; it cannot be greater than $\overline{p}(1 - \overline{p})$. Thus, F_{ST} can be considered the proportion of the observed variance of gene frequencies relative to the maximum value it can take. This also makes the observed variance to some extent (but not to all effects) independent of $\overline{p}$. F_{ST} varies between the limits of 0 and 1.

As explained in section 1.11, the genetic distance we employ throughout this book is identical to an F_{ST} between two populations, using $\overline{p}$ in the denominator as the mean gene frequency of all the populations being compared (not that of the two populations) and when the bias introduced by different sample sizes is not taken into account. Under these conditions, the F_{ST} of the group of populations being compared is equal to the average of all the F_{ST}'s between the pairs of populations forming the group.

Genetic theory (Wright 1943, 1946, 1951) has shown that F_{ST} can, under certain conditions, be equated to an inbreeding coefficient. It also tends to increase with the time of isolation between populations in direct proportion to evolutionary time, and in inverse proportion to effective population size. It is therefore of particular interest for the study of the evolutionary kinetics of diverging populations.

2.8.b. EFFECTS OF CLUSTERING ON F_{ST} VALUES, AND COMPARISON OF VARIOUS F_{ST} VALUES AT DIFFERENT LEVELS OF CLUSTERING

Values of F_{ST} are very sensitive to the mode of clustering of populations (Jorde 1980), and it has been shown theoretically that there is a simple relationship of F_{ST} with the degree of clustering, under conditions of equilibrium for a stepping-stone model of isolation by distance (Cavalli-Sforza and Feldman 1990). Figure 2.8.1 gives an example. Data on three blood groups (ABO, MN, Rh) from 37 villages in the Parma valley in Italy were grouped into 11 neighborhoods, which were grouped into 4 regions. The F_{ST} values obtained at the three clustering levels are shown in table 2.8.1. Theory has shown that F_{ST} can be expected to be approximately linearly related to the number of clusters, and table 2.8.1 confirms this expectation.

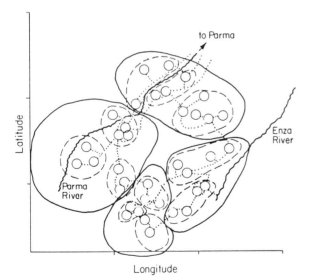

Fig. 2.8.1 Location of 37 villages from the upper Parma Valley and different clusterings adopted for computing the F_{ST} values given in table 2.8.1 (taken from Cavalli-Sforza and Feldman 1990, fig. 3, p. 10).

Table 2.8.1. F_{ST} Values (± standard errors) Obtained from the Three Levels of Clustering Shown in Figure 2.8.1 (from Cavalli-Sforza and Feldman 1990)

No. of Blocks (b)	No. of Villages per Block (n)	F_{ST} ± S.E.	Expected F_{ST} for Linearity
37	1	0.0261 ± 0.0059	0.02568
11	3.6	0.0080 ± 0.0019	0.00804
4	9.2	0.0032 ± 0.0013	0.00319

This possible source of error must be kept in mind when comparing F_{ST}s of different population groups that have been clustered in different ways. The difficulty is not tied to F_{ST} itself; any variance measurement is affected by clustering. In the present case, however, greater complexity is generated by the correlations existing between the gene frequencies of geographic neighbors.

The effect of clustering on F_{ST} for data in this book is far less dramatic than in the microgeographic example from Parma, but it is present and should be kept in mind. It is unfortunately less easy to study quantitatively because it is more difficult to form clusters of regularly increasing size.

2.8.c. THE THEORETICAL DISTRIBUTION OF INDIVIDUAL F_{ST}S, WITH AN APPLICATION TO DNA POLYMORPHISMS

When the variation of gene frequencies is determined by drift alone, as happens in neutral evolution, one can calculate the expected distribution of F_{ST}s. It is thus possible to test the hypothesis that only neutral variation is present, and natural selection is absent. In fact, when drift is the only evolutionary force causing population divergence, all genes of a given population are affected by drift of equal strength, because drift is determined entirely by the demographic properties of the population (which are the same for all genes) and not, as in natural selection, by the nature of the genes themselves.

When present, natural selection increases F_{ST} above that expected under drift alone if fitnesses of genotypes differ in different populations (disruptive selection), and decrease F_{ST} in the presence of heterozygous advantage (stabilizing selection), provided relative fitnesses of genotypes are constant in different populations. If we knew the demographic history of evolving *H. s. sapiens* populations, we could predict the F_{ST} expected under drift alone and recognize genes subject to disruptive or stabilizing selection (Cavalli-Sforza 1966). Unfortunately, we do not have this information, but we can still predict the overall F_{ST} distribution under neutrality on the basis of certain assumptions. This approach was tried on classical human polymorphisms (Lewontin and Krakauer 1974), but without taking into account the correlations between populations. These are generated by the common evolutionary history of populations as described by the tree, and Robertson (1975) showed that they must be considered. We now proceed to use tree information for evaluating the expected F_{ST} distribution.

The theoretical distribution of gene frequencies under drift alone is known (Kimura 1983). It depends on the initial gene frequency, the time since separation of two populations, and the effective population size N_e. No reciprocal gene flow is assumed after population fission, a reasonable assumption for populations separated by long geographic distances. The tree should provide information on the relative values of the times at which populations separated. Not knowing N_e, we assume that it is the same in the various branches. This assumption is not unreasonable because our gene-frequency samples represent mostly large populations, but it is truly difficult to replace it with realistic estimates, which do not exist. Actually, in a slightly more realistic and still extremely simple model, populations may slowly double in size between one fission and the next; but we are far from having reached, in our analysis, a level of precision at which this improvement of the model is really meaningful. Drift was certainly very important when population densities were much below those observed today in the same regions. Subsequent population growth has largely frozen drift effects. Therefore, when we consider large regions, the variance of gene frequencies observed today has probably been deeply affected by drift, and therefore by population sizes, existing until the late Paleolithic or early Neolithic. At our level of approximation population bottlenecks, even serious ones, do not have a strong effect on expectations, provided that they did not last very long. A further bias we are ignoring in this simulation is the nonrandomness of population splittings (Smouse et al. 1981). If it is important at the scale at which we are operating, this bias will effect a decrease in the estimate of population size obtained by this approach.

The problem is too complicated for a full analytical treatment and we must resort to simulation. Kimura's distribution of drift can be approximated by the β distribution (as done earlier by Lewontin and Krakauer [1974] and by Cavalli-Sforza and Piazza [1975]). The β distribution of gene frequencies after a given evolutionary time depends on the initial gene frequency p_0 and the F_{ST} expected at the end of that time. For example, consider the segment leading from the origin to the split of the proto-African branch in the tree in figure 2.4.4. We take a sample from an appropriate β distribution for calculating p_A, the gene frequency of the proto-African branch at the time it splits into the two African populations represented in the tree. This split takes place, as shown by the tree, after 80% of the total evolutionary time. The constants defining the β distribution are the initial gene frequency, which can be given any desired value, and the F_{ST} expected at this time, calculated as 80% of the total F_{ST}, 0.139. A slightly better approximation should take account of the

nonlinear relation between F_{ST} and time, but again this would be a trivial correction. The two values, p_0 and the partial F_{ST} determine the distribution of gene frequencies for the evolutionary segment beginning at p_0 and leading in the present case to the proto-African fission. Let us call p_A a random sample from the β distribution thus defined. One can similarly calculate the gene frequencies of the next segments, leading to the two living African populations resulting from the splitting of proto-Africans. These will be two random samples from a β distribution determined by the gene frequency p_A, and an F_{ST} equal to 20% of the total. All the gene frequencies of the other intermediate ancestors and living populations descending from them are calculated in a similar way. At the completion of one simulated evolutionary process, we have the gene frequencies of the five living populations, whose F_{ST} value can then be calculated. The simulated evolutionary process is repeated for an adequate number of times (100,000 in our case). In this way, the expected distribution of F_{ST} values can be obtained rather accurately for one particular p_0 value. Because of the dependence on p_0, many p_0 values must be tested: we examined the range from $p_0 = 1\%$ to 10% in increments of 1%, the range from 10% to 90% in increments of 5% and from 90% to 99% in increments of 1%; a total of 35 distributions, for a total of 35 initial gene-frequency values. The results show that the average F_{ST} reaches its maximum at 50% and then tends to decrease symmetrically as p_0 goes toward the extreme values, at first very slowly and then faster as p_0 approaches 0% and 100%. At the extremes of the range, $F_{ST} = 0$.

Figure 2.8.2 shows the F_{ST} distributions generated for two p_0 values. The 35 distributions obtained for all

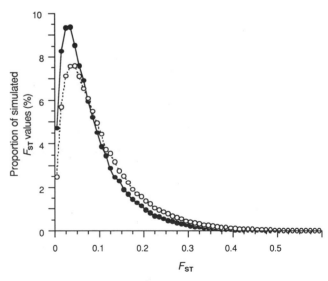

Fig. 2.8.2 Examples of F_{ST} distributions for two different initial values of gene frequency p_0, each obtained by 100,000 simulations of the evolutionary process by random genetic drift. *Open circles* $p_0 = 0.5$; *closed circles,* $p_0 = 0.1$ (from Bowcock et al. 1991).

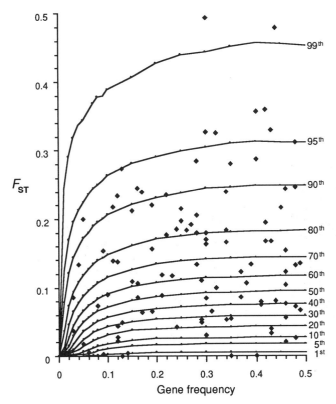

Fig. 2.8.3 Simulated distributions for F_{ST} expected in evolution under drift alone according to the tree in figure 2.4.4 are shown as percentiles from 1st to 99th. Distributions depend on their initial gene frequency; the average observed gene frequency is given on the abscissa as the estimate of the initial gene frequency. Percentiles from the distributions calculated for different initial gene frequencies were joined to form the percentile curves shown in the figures. Diamonds correspond to observed F_{ST} values for 100 DNA markers in five populations, as a function of the average gene frequency. The allele frequency nearest to 50% only is given for multiallelic markers. When the frequency of an allele, P, was greater than 50%, the value $100 - P$ is indicated in the graph (from Bowcock et al. 1991).

p_0 values tested were plotted in figure 2.8.3 by giving as ordinate values the F_{ST} cumulative percentile points of each distribution (1%, 5%, 10%, etc., to 99%) as a function of the p_0 value, which appears on the abscissa. Because of the symmetry of the distributions around $p_0 = 50\%$, only the p_0 range 0%–50% is represented. Percentile curves seen in figure 2.8.3 were built joining the percentile points of the 35 simulated distributions for the different initial gene frequencies tested.

Figure 2.8.3 also shows as diamonds the observed F_{ST}'s for the 100 DNA polymorphisms, using the average gene frequency of each gene as the abscissa. The initial gene frequency is unknown for the observed data, and we replaced it by the average gene frequency. In the evolutionary time considered, the initial gene frequency is not likely to have changed substantially, considering that the size of the world population probably increased

continuously. For multiallelic genes, only the frequency of the allele closest to 50% was shown. All observed values were indicated as having an average gene frequency $\overline{p}$ less than 50% by converting them to $100 - \overline{p}$ when greater than 50%.

This simulation was made for the hypothesis that there is no admixture between the branches. We have seen in section 2.4, however, that this is not likely, and an alternative hypothesis assuming admixture between proto-Africans and proto-Orientals gave an excellent fit. We repeated the simulation for the latter hypothesis, which gave a slightly lower variation of F_{ST} (see fig. 2.8.4).

Both figures 2.8.3 and 2.8.4 seem to indicate a slight excess of observed F_{ST} values at the high and at the low ends, as expected if some genes are exposed to disruptive selection and others to stabilizing selection. This indication is tested further in figure 2.8.5, where the number of genes in each percentile band (0%–1%, 1%–5% 5%–10%, 10%–20%, etc.) is shown against its expectation. If the observed distribution of F_{ST} values were in complete agreement with that expected for neutral genes, the bar diagram would be completely flat and follow the line drawn for an ordinate of one (equality of observed and expected numbers of genes). It shows instead a high, sharp peak for low F_{ST} values, and a broader one for high

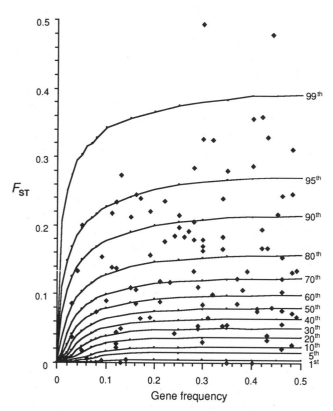

Fig. 2.8.4 Same as in figure 2.8.3, but the simulation follows the hypothesis underlying the tree in figure 2.4.7, that is, admixture between proto-Africans and proto-Orientals.

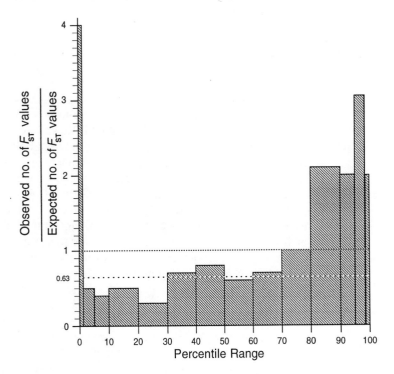

Fig. 2.8.5 Comparison of the observed and expected distributions of F_{ST} values shown in figure 2.8.4. If there were no deviation from the neutral hypothesis, the bar diagram would be flat along the upper horizontal line. The number of neutral genes that could be accepted on the basis of this analysis is that below the lower horizontal line, and the numbers of genes above it should be affected by stabilizing selection (at the extreme left of the figure) or by disruptive selection (at the right).

values. The deviation from the expected distribution is significant (all values given in Bowcock et al. 1991), indicating that at least some genes are not neutral. One can also estimate the proportion of neutral genes by drawing a lower horizontal line and counting how many genes must be discarded in each class to obtain a nonsignificant χ^2 between observed and expected. The lower, dotted horizontal line shown in the diagram corresponds to the hypothesis that about two-thirds of genes are neutral and about one-third are under selection. A few of them are under stabilizing selection; many are under disruptive selection.

The hypothesis under which this expected distribution is obtained assumes constant population size over time and in all branches, with no difference of evolutionary rates in different segments of the tree. If this simple hypothesis is incorrect, the distribution may be different, and the number of neutral genes is probably underestimated when a constant population size is assumed. In this case, the proportion of neutral genes is a minimum estimate.

Although we have no information about the possible mechanisms of selection in any of these genes, it is of interest that the four genes in the lowest percentile band of F_{ST} are closely linked with genes that can be suspected to be under stabilizing selection (two are near the gene for cystic fibrosis and two near that for another recessive genetic disease, Wilson's disease). They might have borrowed heterosis from these or other very close genes. It is more difficult to make suggestions about the genes

showing high F_{ST} values, because their selective behavior is not investigated. More information about their nature can be found in Bowcock et al. 1987, 1992.

We can evaluate the effective population size that could generate an average F_{ST} comparable with that observed. In independently evolving populations from a single origin, without intermigration, F_{ST} is expected to increase with time according to

$$F_{ST} = 1 - \exp(-t/2N_e), \qquad (2.8.4)$$

where N_e is the effective population size of each population, and t is the time in generations. Because 100 ky are equivalent to 4000 generations, for $F_{ST} = 0.139$, $N_e = 13,000$, corresponding to a census size of 39,000. This is not an estimate for the whole human species, but for the average branch of the evolutionary tree, and for the initial population of a.m.h., say 100 kya; this size is comparable to that of living Primates closest to humans (chimpanzees, gorillas). With the passage of time, the number of branches increases, and with a constant number of individuals per branch, the total population on Earth would also increase. Toward the end of growth and closer to the present, the number of branches and the total population would obviously be much higher than the five populations considered in our tree. At the end of the Paleolithic, when all the continents were occupied, the total population of the Earth was larger than the number calculated here for the beginnings of a.m.h. by a factor on the order of 100 (see sec. 2.1.e). These numbers confirm that, in spite of the many untestable hypotheses underlying these calculations, the overall picture cannot be too far from the truth.

2.8.d. VARIATION OF F_{ST} FOR THE NON-DNA POLYMORPHISMS

In this book, F_{ST} values for non-DNA polymorphisms were calculated in two different ways, which are not strictly comparable. (1) F_{ST} values given in the histograms of gene-frequency distributions in gene-frequency maps were calculated from the original values of individual populations after averaging over populations of the same location. The distributions include all the gene-frequency values available in the geographic region shown in the map, and used for map construction. (2) The second means of calculation used the gene frequencies for the 491 populations listed in Appendixes 2 and 3; these F_{ST}s are further illustrated in this paragraph and the following paragraphs of the present section. The individual values of these F_{ST}s are given in table 2.12.1, with other relevant information (average gene frequencies and numbers of populations for which the gene-frequency estimate was available). The average F_{ST} over all 120 genes obtained by this approach is 0.119 ± 0.010.

There are some discrepancies between the values found by the two approaches; usually values calculated by the first method tend to be larger than those by the second. In part, this is due to differences between the geographic regions used for calculations in the two methods. For instance, the F_{ST}s calculated for Africa in the gene-frequency maps also include North Africans, whereas values calculated for Africa from the 491 populations include only sub-Saharan Africans. The average over genes for Africa by the first approach is 0.068 ± 0.011, not surprisingly higher than that for the sub-Saharan Africans included in the 491 populations (0.035 ± 0.005). Even when the comparison is made between data from exactly the same region (e.g., for Europe), one does not find the same value, given the difference in the level of clustering in the two sets of calculations.

We note first that the F_{ST} (worldwide) averaged over all genes for DNA (0.139 ± 0.010) differs little from the F_{ST} averaged over F_{ST}s of non-DNA polymorphisms (0.119 ± 0.010). The latter value is calculated from the F_{ST} values of individual genes given in table 2.12.1. The difference is not significant when tested on the basis of observed standard errors.

The problem of the distribution of F_{ST}s for individual genes, and its comparison with that for selectively neutral genes, was attacked by the same method used for the 5-population tree and the DNA markers, but using the 42-population tree.

A simulation comparable to that for the DNA markers (fig. 2.8.3) has been constructed for the tree with 42 populations, and 120 non-DNA markers used in section 2.3 (fig. 2.8.6). Clearly, some polymorphisms

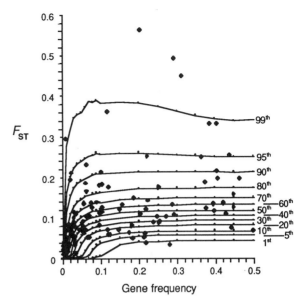

Fig. 2.8.6 Dots indicate F_{ST} values of 120 non-DNA polymorphisms among the 491 world populations given in Appendixes 2 and 3, versus the corresponding average gene frequencies shown on the abscissa. Curves correspond to the percentages of simulated distributions of F_{ST}s for a tree (as in Cavalli-Sforza et al. 1988), resulting in $F_{ST} = 0.119$ at the end of the evolutionary process, as observed for these genes.

have very high F_{ST} values, indicative of disruptive selection.

The comparison of expected and observed F_{ST} distributions for non-DNA polymorphisms (similar to the comparison in fig. 2.8.5) shows an excess at the upper and lower end of F_{ST} values. In addition, there is a hump to the right of center, which is more difficult to interpret. The sources of data are not entirely satisfactory, however, in that genes do not have a homogeneous distribution in the different areas, and many genes are represented in only a fraction of the 491 populations. The number of populations for which data for each gene are available are shown in table 2.12.1. Given these limitations of the data, we comment only on the genes showing extremely high or low F_{ST}s. Numerical values, including their percentile can be found in table 2.12.1.

Few, if any, markers show a low F_{ST}: $HP*1S$ (haptoglobin allele; but $HP*1$ is relatively high); $ALPP*S1$ (alkaline placental phosphatase); $GC*2$ (group specific component or vitamin-D-binding protein); $RH*cDE$, and $PGM1*1$ (phospho-gluco-mutase). These may show stabilizing selection. Genes for which there are potential explanations of their world variation are discussed in section 2.10.

The marker showing the highest variation at the world level is $FY*0$, which confers resistance to a malarial parasite (see sec. 2.10). Evidence of disruptive selection for Duffy (FY) seems indisputable. The two other alleles at this locus, $FY*A$ and $FY*B$, are also higher than average, but this results in part from their correlation with $FY*0$.

A system showing high values of F_{ST} is $GM(IGHG1G3)$, the immunoglobulin-G family of genes. The highest variation is shown by haplotype $fa;b0b1b3b4b5$, followed by $za;b0b1b3b4b5$, $f;b0b1b3b4b5$, $za;g$, $za;b0sb3b5$. Again because of the internal correlations between alleles or haplotypes it is difficult to see, on the basis of these data alone, whether some of these haplotypes showing lower F_{ST}s owe their F_{ST} to a direct effect of disruptive selection, or if they borrow it from those having higher F_{ST}s. It seems clear, however, that strong disruptive selection is present.

Two alleles of RH, cDe (very frequent in Africa, rare elsewhere) and CDe (omnipresent) show relatively high variation. Although it does not seem likely, the mechanism of selection of maternal-fetal incompatibility known for this system may be responsible for this finding.

$G6PD$ (glucose-6 phosphodehydrogenase) deficiency has a high F_{ST} with a low overall gene frequency. The evidence in favor of disruptive selection as a result of malaria is clear for this gene.

Other markers that stand out as having high F_{ST} values are $SOD1*1$ (superoxide dismutase 1), $PEPA*1$ (peptidase A1), S^U (allele of the MNS system), $JS*A$ (KELL blood group), $AG*X$ (lipoprotein), $ACP1*B$ (acid phosphatase), and $HLAB*22$.

This test would not detect a combination of disruptive and stabilizing selection—as might occur if a heterotic system showed a different ratio among the fitnesses of the homozygotes in different parts of the world—in which the two opposite selection patterns approximately balanced each other. We attempt to consider some related complications at the end of this section. The test could be strengthened if effective population sizes were known with some accuracy, because the average F_{ST} could then be predicted. We do not have reliable demographic information of this nature. We can only state that the order of magnitude of N_e calculated from observed F_{ST}s seems acceptable.

Geographic maps (sec. 2.10 and later chapters) confirm that markers with high F_{ST} show considerable local variation of gene frequencies in some areas, totally out of proportion to those seen for other genes in the same areas. This tends to exclude the idea that drift is the only cause of variation for these polymorphisms.

2.8.e. Averaging F_{ST} Values over all the Genes in the World and over Major Regions

The world genetic variation expressed by F_{ST}, averaged over the 122 (see table 2.12.1) classical genetic markers studied in this book, as we said, is such that the variance of these gene frequencies is about 12% of its potential maximum. Variation was calculated among the 491 populations defined in chapter 1 and Appendixes 2 and 3, and for some selected subregions. Gene frequencies, however, are not always available, or not in sufficient numbers for all subregions, so that results are uncertain in some cases.

All subregions were found, not surprisingly, to have F_{ST}s smaller than the world F_{ST}, an inevitable consequence of the existence of a strong positive correlation between the gene frequencies of geographic neighbors (as discussed in the next section).

1. America shows the greatest variation (0.070 ± 0.005), partly because it hosts several different ethnic groups (like Eskimo, the associated Na-Dene, and other Amerindians), but mostly because of the high variation found in South America (0.059 ± 0.006), where the greatest heterogeneity is observed.

2. The heterogeneity of Caucasoids, as a group embracing Europe, West Asia, India, and North Africa is next (0.043 ± 0.005), mostly because of the considerable variation among extra-European Caucasoids. Europe, by contrast, is the most homogeneous continent (0.016 ± 0.002).

3. Australia (0.019 ± 0.004), New Guinea (0.039 ± 0.007) and sub-Saharan Africa (0.035 ± 0.005) are intermediate, but Polynesia is more heterogeneous (0.049 ± 0.007), mostly because of extreme drift caused by the small size of the islands, as shown in chapter 7.

An approach that can further support indications of disruptive or stabilizing selection, or both, is the reproducibility of the behavior of F_{ST} in different regions. Data are insufficient to give firm conclusions, but the problem is sufficiently important that it is worth an approximate analysis. Table 2.8.2 shows high and low values of F_{ST} in seven regions, chosen for relative ethnic homogeneity and abundance of data. Only F_{ST} values definitely above the local mean (at least twice as high) or definitely below it (at most half as high) are shown in the table; the latter are marked by an asterisk. The corresponding mean gene frequency is also indicated, given that F_{ST} depends on it, and is progressively lower as gene frequencies approach 0% and 100%, especially in the 0%–10% and 90%–100% ranges.

We limit our discussion to genes for which there are abnormal results in more than one region. When data are available for several regions but values are abnormal in only one it is more likely that the result is due to chance. The majority of genes do not appear in this table, because their F_{ST} values are not sufficiently removed from the local averages. The mere fact that numerical values for a gene tend to recur in the table in different regions is an indication that the gene has an abnormal F_{ST} behavior. The indication that a given type of selection is operating is clearly stronger if a given gene has values all or most below normal (or above normal), but some genes may be under both disruptive and stabilizing selection, one or the other type being dominant in specific regions.

Table 2.8.2. F_{ST} Values (x 1000) and Average Gene Frequencies (x 100, in parentheses) in Seven World Regions. Only Values Definitely Above the Local Mean or Definitely Below It (the latter marked by asterisks) Are Shown. Empty Spaces Indicate That Alleles Were Not Present or That Data Were Not Available

Loci	Alleles or Haplotypes	Sub-Saharan Africa	East Asia	Southeast Asia	Asia (India)	Europe	South America	New Guinea
ABO	O	*10(69)	*6(55)		*10(58)			
	A				*7(19)			
	B	*8(14)	*9(20)	*12(17)	*9(23)			
ACP1	A		*10(22)					*16(22)
ALPP	F1				*2(72)			
	S1					*2(65)		
BF	S0.7					*3(1)		
C3	F					*1(20)		
ESD	1							
FUT2(SE)	+		69(56)				300(93)	190(70)
FY	O	322(93)			202(5)			
	A	102(61)	58(8)			21(41)		
	B	232(36)						
GC	1F		92(55)					
	2		*3(27)					
GLO1	1	*10(31)						
GPT	1	*10(84)	*8(57)					114(73)
HLAA	1		82(2)					
	2	*9(16)	*9(26)			*3(29)		51(2)
	9	*6(12)		100(33)		*4(11)	146(21)	210(66)
	11							275(15)
	28	*6(9)				*3(4)	124(14)	24(0.9)
	29			34(0.8)				
	31	71(49)						
	32		40(3)					
	33							112(3)
HLAB	5			*9(8)		26(7)	152(13)	
	13		41(2)					
	14	*6(3)						
	16						116(17)	
	18							66(1)
	22		6(6)					282(17)
	27							*12(6)
HP	1		*6(24)					*19(70)
IGHG1G3	fa;b0b1b3b4b5			78(79)	27(18)	36(0.9)		
	f;b0b1b3b4b5	111(0.9)			73(47)	23(71)	91(1)	
	za;b0b1b3b4b5	147(75)						
	za;b0sb3b5	187(10)		67(2)				
	za;g	65(3)						265(47)
	zax;g	30(0.9)	*9(16)			21(8)		
JK	O							
	A	111(67)	*7(64)				131(6)	
KEL	Jsa						190(5)	
	K		25(1)			*3(5)	56(0.7)	
KM	1&1,2		*7(30)				148(42)	285(32)
LU	A				54(1)			

(continued)

Table 2.8.2 (continued)

MNS	M		*10(46)		*3(57)			
	S		86(23)					
	S^u	132(12)			16(0.1)			
P1	1							
PEPA	1			309(90)	53(99)			
PGM1	1		*9(78)					
PGD	A			54(96)				
	C				185(1)			
PI	F			21(0.8)	58(1)			
	M				24(9)			
RH	CDE		54(6)					
	CDe		*7(61)					
	cdE							
	Cde				60(0.2)			
	cDe							
	cde							
TF	D			29(0.9)	148(2)	148(25)		
Mean F_{ST}		34.7	25.1	34.9	27.6	16.4	58.9	39.1

In most regions, ABO shows a tendency toward a relatively low F_{ST}. However, there are other indications, from correlations with infectious diseases, that it may be subject to disruptive selection. Both stabilizing and disruptive selection are probably present, but the stabilizing tendency dominates, except in America where this system has become largely monomorphic (see sec. 2.10).

Antitrypsin shows disruptive selection, especially for allele PI*F, in two of five regions for which data are available.

Duffy shows highly disruptive selection, not only because of the FY*0 allele in Africa (where it has a substantial advantage in certain malarial regions), but also because of alleles FY*A and FY*B. This is more clearly visible in areas where there is no FY*0 allele and therefore no indirect effect from it because of the negative correlation between alleles.

GPT*1 shows variable behavior.

In one region, GC*2 shows both stability and excess variation of GC*1F. It is possible that disruptive selection affects only the two suballeles of GC*1, but basic stabilizing selection affects GC*1 and GC*2.

The HLA supergene shows mostly stabilizing selection for major alleles like A*2 and perhaps A*9 and A*28, and mostly disruptive selection for some other alleles listed in the table. Again, the existence of stabilizing selection in some regions and disruptive selection in others is not surprising, given that the factors determining variable selection may not be present in all regions.

Immunoglobulins (IGHG1G3) show almost exclusively disruptive selection, of an intensity second only to Duffy.

Haptoglobin may be under stabilizing selection at least in some regions, but results from world data show that there may be also disruptive selection.

Kidd (JK) and Kell (K) show both stabilizing and disruptive selection. The same is true of MNS.

Among enzymes other than those already discussed PEPA shows disruptive selection, whereas others seem to be under stabilizing selection, observed, however, in only one region.

RH does not show clear selective effects, at least by this approach.

Secretor (SE+) seems to be under substantial disruptive selection in several regions, and the same is true of Transferrin D (TF*D).

Considering the total number of regions and alleles tested, the deviations from neutrality of classical markers detected by this test are not overwhelming. Results confirm the main observations already made at the world level, showing that Duffy and immunoglobulins are certainly subject to the strongest disruptive selection. Among blood groups, ABO seems mostly under stabilizing pressure, and to a lesser extent this may be true of the MN polymorphism. More observations are needed for enzymes that gave similar indications. This statistical analysis should be considered preliminary, but comparing F_{ST} behavior in different regions seems useful. This comparison apparently supports, at least in a qualitative fashion, the hypothesis that some genes are under stabilizing selection in some areas and disruptive selection in others. The simplest formal explanation for the coexistence of both types of selection is that the fitnesses of the two homozygotes, relative to that of the heterozygote, may be different in various environments.

2.8.f. POSSIBLE EFFECTS OF NATURAL SELECTION ON TREE RECONSTRUCTION

As was already clear to Darwin, neutral characters are best for reconstructing evolutionary history. From the above analysis, it would appear that most of the genes used by us for tree analysis are neutral, but there is some evidence of natural selection, especially of the disruptive type. For purposes of tree reconstruction, the nature of disruptive selection is important. If many genes showed intercorrelated responses to the various environments in which human evolution has occurred, then the tree might be distorted and would represent similarities in environments rather than evolutionary history. This phenomenon is likely to be especially important when analyzing polygenic traits; most anthropometric characters like stature and other body measurements are probably polygenic. Natural selection in favor of a specific phenotype in certain environments would bias the tree toward showing similarities of the environments affecting the phenotype. We have seen that adaptation of anthropometric traits to climatic conditions probably explains the discrepancy between trees obtained with anthropometrics and with gene frequencies (sec. 2.2).

If, however, natural selection varies more randomly with geographic environments, as in the evolutionary model described by Crow and Kimura (1970) as "selective drift," then the effect on tree distortion could be minimal or nonexistent. The same may be true of sets of genes whose selective coefficients are constant over time and space but are not correlated in their responses to different environments, so that each is selected differently from the others. If all genes respond to the same environmental stimulus, but the same stimulus has different intensity in different places, this would endanger the validity of tree reconstruction. Examples are genes conferring resistance to malaria and, in particular, to the most lethal malarial parasite, *Plasmodium falciparum*. For this reason, genes that cause sickle-cell anemia, thalassemia, and G6PD-deficiency have not been considered for tree reconstruction. It might be argued that, for reasons not completely understood, these genes almost never spread to the whole area where malaria is endemic and are therefore less misleading than might be feared. We have included *FY*O* in our analysis, however, which also confers resistance to malaria, but to a milder parasite, *Plasmodium vivax*, which for one reason or another is now rare where *FY*O* is found.

Some genes confer resistance or susceptibility to infections in which antigens change rapidly, for instance, in viral diseases and probably in many bacterial and some parasitic diseases as well. Immunoglobulins and *HLA* genes determine responses to pathogens that are likely to fall in this category and probably have irregular geographic distributions that may change over time. Their selective pattern may be closer to the model of selective drift (see Crow and Kimura 1970) and similar to drift resulting from finite population size. If this is true, then conclusions about evolutionary history are not severely affected. It is important, however, to keep these possible limitations in mind.

2.9. GENETIC VARIATION AND GEOGRAPHIC DISTANCE

Moving around is an important part of human activity. Whether in search of food, a mate, work, or entertainment, moving occupies a fair fraction of the day for most people. Until 150 years ago, most movements covered short distances, and only rarely did people venture out of the short range in which daily movement took place. Recent advances in means of travel have changed this custom to some extent. The distance between birthplaces of spouses has also increased considerably in the last 150 years, on the average, and Dahlberg (1943, 1947) has referred to this change as the breakdown of isolates, although it would be more accurate to refer to it as the widening of isolates.

Even though marriage within the social group (however defined) is the rule, in practically every part of the world, marriages with neighbors from other groups always occur, at a frequency dictated by customs and opportunities. This exogamy in a limited geographic range is sufficiently common that it has generated a very significant genetic similarity between geographic neighbors. Under this regime of ever-present short-range migration, new genes can spread over generations at a rate that depends on the intensity and range of this migration. The genetic expectations have been formalized in the theories of isolation by distance and the stepping-stone models (see sec. 1.16), showing that a regular decrease of genetic similarity with increasing distance is generated (Malécot 1950, Malécot 1969, Kimura and Weiss 1964). Application of these theoretical models by Morton and other geneticists to a number of real situations confirmed the general theory but was always limited to short ranges.

We prefer to use a special version of the original theory, which eliminates some uncertainties by using the *variogram*, which we define as the plot of genetic distance versus geographic distance. We have discussed the specific expectations in section 1.16, and we apply them here to the data of classical polymorphisms. In the gene maps found in the second part of the volume, we give the variogram of that gene in the region shown in the map, whenever data seem abundant enough. Individual variograms vary widely, for reasons discussed later, but in general there is widespread agreement showing that most genes exhibit an increase in genetic distance with geographic distance, even if in a capricious way, especially when data are sparse. The variograms are not expected to be monotonic when the gene-frequency

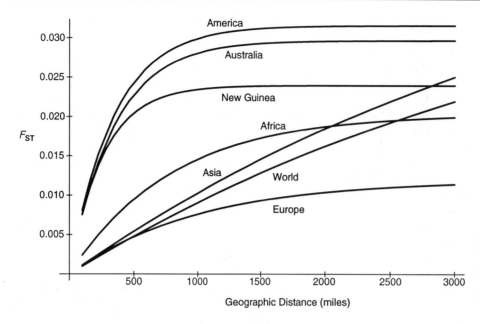

Fig. 2.9.1 The relationship between genetic distance (ordinate) and geographic distance (abscissa) expressed by the variogram for the world and continents or parts of them.

surface has marked peaks and troughs; in fact, maxima and minima of the variogram are determined by these peaks and troughs. When data from many genes are averaged, regularity increases and observed curves tend to a common shape that corresponds reasonably well with that expected at equilibrium for drift: an initial strong increase followed by a progressive flattening out of the curve. At long distances, the curve becomes—or tends to become—horizontal, forming an upper asymptote. Figure 2.9.1 shows the fitted theoretical curves of the world and of the continents, with the curves of Australia and New Guinea shown individually. The quantity used as the ordinate is F_{ST} for all pairs of populations, which introduces a standardization for different genes. Figure 2.9.2 also shows median values of real data for each curve. The fit is not very good at large distances, but the variation of the data is large, as the variograms of individual genes in the maps at the end of the volume show clearly. Medians were used because they are more robust than arithmetic means. The distributions of F_{ST}s are highly asymmetric (sec. 2.8) and therefore medians are systematically smaller than arithmetic means, a fact to be considered for comparison with other methods of estimation.

The theoretical curve is determined by two parameters, the initial slope S and the horizontal asymptote H. Rewriting equations (1.16.7) and (1.16.8) makes it easier to see their dependence on two constants A and B of genetic interest;

$$S = 1/(4A + B) \qquad (2.9.1)$$

and

$$H = B/(4A + B) \qquad (2.9.2)$$

with

$$A = mD \qquad (2.9.3)$$

$$B = \sqrt{(m/2M)}, \qquad (2.9.4)$$

where m is the short-range migration, expressed as variance of the distance between the birthplace of parent and progeny, and therefore in units of the squared distance (squared miles). M is a number incorporating mutation, heterosis, and long-range migration; it is expressed in the same time unit as m (year or generation). D is (effective) population density, or about one-third of the real density. Because we employed a continuous and unidimensional model of isolation by distance, D is expressed in miles$^{(-1)}$. There are difficulties in relating this measure to standard population density, which is expressed per mile or kilometer squared.

Estimates of A and B can be obtained from the parameters of the fitted curves, H and S, using

$$B = H/S, \quad A = (1/S - B)/4. \qquad (2.9.5)$$

Parameter B is the distance (in miles) at which the exponent of equation (1.16.5) is unity, and F_{ST} genetic distance therefore takes a constant value: $0.63/[1 + 4D\sqrt{(2mM)}]$. It might be useful for comparing geographic distances having a specific genetic effect.

The parameter A would be more useful if turned into the equivalent Nm of the discontinuous settlement model. Additional information would be necessary for this, and for estimating M from B and the variance, m, of the distribution of distances between parents and childrens' birthplaces, which is known in only a few instances and varies considerably from case to case. It was estimated to be between 100 and 1000 square miles per generation for early Neolithic (Ammerman et al. 1976), a range sufficiently wide that it may be valid also for the Paleolithic. The density D is even more difficult to evaluate since it is given for a unidimensional distribution. It may be of order one or lower for the late Paleolithic. It is clear that, in the absence or inadequacy of this auxiliary

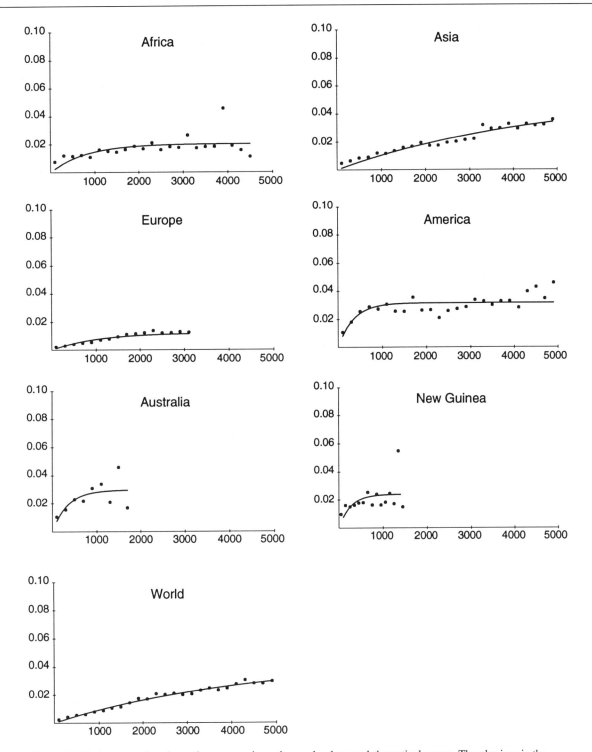

Fig. 2.9.2 Variograms of each continent or region: observed values and theoretical curve. The abscissa is the geographic distance (in miles), and the ordinate is the genetic distance.

information, only orders of magnitudes can be assessed and accepted with caution.

Table 2.9.1 shows the values of the initial slope and asymptote and the derived genetic quantities for the world, the continents, or parts of them. The B values show modest variation for Australia, New Guinea (the smallest), and the Americas; somewhat higher values for sub-Saharan Africa and Europe, and the highest for Asia. Not too surprisingly, the world value is dominated by the Asian component. Asia, the largest continent, must be farthest from equilibrium, which in any case could hardly have been reached even in the smallest region, New Guinea. The order of magnitude of B is especially influenced by the initial slope and not by the

Table 2.9.1. Parameters* of the Curves Fitted in Figure 2.9.1 and Calculations of Genetic Parameters

	Parameters of the Variogram		Genetic Parameters	
	Initial Slope S	Asymptote H	A=mD	B=(m/2M)$^{1/2}$
Africa	0.0000255	0.0204	9604	800
Asia	0.0000114	0.0511	20,809	4482
Europe	0.0000120	0.0120	20,583	1000
America	0.0000932	0.0315	2598	338
Australia	0.0000865	0.0297	2805	343
New Guinea	0.0000919	0.0240	2655	261
World	0.0000101	0.0440	23,663	4356

*Numerical values of constants calculated from the observed data with the theory indicated in the text.

asymptote, which is probably most poorly estimated for Asia and the world. These are most likely to be farthest away from equilibrium. Values for Australia, the Americas, and New Guinea are likely to be reasonably representative of the Paleolithic and early Neolithic. Taking $B = 300$ miles for such periods, and $m = 100 - 1000$ squared miles, M is on the order of 1/200 to 1/2000.

Values of A are very similar for Australia, the Americas, and New Guinea, highest for Europe and Asia, and intermediate for Africa. If division by m is adequate to transform A into D, this value would be between 3 and 30 for the Americas, Australia, and New Guinea, and higher for Africa, Europe, and Asia.

In conclusion, the variogram seems in general more attractive and shows a reasonably good fit to the data. We can conclude that the isolation-by-distance models hold for long distances as well as for short distances, and for large regions as well as for small and relatively isolated populations, such as those examined before. Our estimations of Nm are, however, tentative, like all the other ones found in the literature.

In the geographic gene maps that form the second half of this volume, variograms are given for each gene and region. The curve interpolated in each variogram is obtained by a procedure (see sec. 1.16) similar to the well-known moving-averages method using median values of the variograms. Medians were calculated after grouping genetic distances of all population pairs into classes of geographic distances at 200-mile intervals. When data points were scarce, the number of observations for many distance classes were inadequate, and no median was calculated for them. When points for calculating variogram curves were too few, the variogram was omitted from the gene maps.

Unlike the fitting procedure used in figure 2.9.1, the curves are not forced to pass through the origin, and the initial slopes are therefore much flatter than in figure 2.9.1. When there are too few points—and in some other circumstances—the initial slopes can be negative. This may happen especially if local genetic variation is high, that is, if many population pairs are very close geographically but genetically heterogeneous. When single genes are analyzed, erratic behavior of some variograms is unavoidable; and, in the process of averaging over many genes, the irregularities tend to disappear.

The slope of the initial linear segment was evaluated in order to compare different genes. The initial segment of the curve seems approximately linear in most cases for the first 500 or 1000 miles, or occasionally more (the length of the initial linear portion of the slope may, however, be in part an artifact of the moving averages method).

Table 2.9.2 shows the arithmetic mean (± standard error) and the median of the initial slopes for the various continents or regions, calculated from the individual variograms. The distributions have an excess of extreme values, and calculation of the medians offers a fairly robust procedure for their elimination. In a distribution

Table 2.9.2. Initial Slope of the Variograms for Six Regions and the World: Comparison of Two Methods of Calculation

	No. Polymorphic Alleles	Calculation of Initial Slopes			
		Averaging Values of Individual Variograms		Curves of Figure 2.9.1	
		Mean ± S.E.	Medians	Intercept not allowed	Intercept allowed
Africa	82	8.6 ± 2.6	4.7	25.5	9.1
Asia	97	13.6 ± 2.7	7.2	11.0	7.3
Europe	96	10.3 ± 3.7	4.3	12.0	12.0
America	65	11.7 ± 2.4	6.8	93	
Australia	17	19.9 ± 10.3	21.0	87	63
New Guinea	33	70.9 ± 32.6	24.1	92	
World	112	10.3 ± 1.0	6.6	10.0	9.4

Note.– The arithmetic means (± standard errors) and medians of the values are obtained from the variograms of individual genes, and the initial slopes of the curves shown in Figure 2.9.1 are calculated with and without intercept. All values x 10^6.

of this type, however, medians tend to be smaller than averages, as is seen here.

The table includes also the initial slopes of the curves calculated in figure 2.9.1 for whole continents, and those averaged from the individual variograms. The latter are regularly smaller for a good reason: in figure 2.9.1, as already mentioned, the curve is forced to pass through the origin. This is not the case in the individual variograms, in which, whenever acceptable, a straight line is forced to go through all points except the first, (i.e., for distance zero). This generates a systematic difference between the measurements, whereby the initial slopes calculated from the variograms are expected and found to be smaller.

One can relax the constraint that the curves in figure 2.9.1 must pass through the origin by interpolating a curve with three parameters: an intercept on the y-axis at the origin, and the other two constants already described. This procedure did not always converge to finite values, but it did provide statistically meaningful estimates in five of seven cases. Allowing for an intercept at the origin is reasonable, since all the differences between gene frequencies of population pairs include a sampling variance that inflates the variogram. In theory this could be estimated on the basis of the number of individuals in each population sample, but there are certainly other sources of error variance beyond sampling. The intercept thus calculated is not trivial; it is on the average, 15% of the asymptote. The initial slopes from the curves of figure 2.9.1 are given in table 2.9.2 before and after correction for the intercept. Correction somewhat improves the agreement with the data obtained by averaging the individual variogram slopes, but some important discrepancies remain.

The major discrepancy between the two data sets is that the initial slope is highest in the first set for New Guinea, and in the second for the Americas, Australia, and New Guinea. Even though we have no simple explanation for this discrepancy, and despite the rather erratic behavior of several individual variograms, interesting regularities are observed. The aim of the exercise was to test the variation in geographic behavior of different genes, as indicated by the initial slopes. Geographic mobility determining the slopes of genetic variation with distance should be a property of the individual, and hence should be the same for all genes. Other factors—for example, different selective effects—might, however, be idiosyncratic to each gene. The variation between genes has usually not been examined in published research on the correlation of genetic kinship and distance (Morton et al. 1982), but it is important to evaluate it. Given the differences between variograms of individual genes in different regions, we have averaged the initial slopes of all alleles at a locus or cluster of loci and limited the analysis to major multiallelic loci for which information was available in several regions. Medians were calculated for all alleles or haplotypes of each system. The median initial slopes were ($\times 10^6$):

ABO	4.15	*HLAA*	5.75
MNS	4.25	*HLAB*	6.78
RH	4.58	*GM*	11.43.

Despite not having the standard errors of these values, which would be of limited use with the skew distributions involved, it seems reasonable to conclude that *GM* has a higher variation with geographic distance than the other genetic systems. It is likely that the behavior of *GM* derives from variation at the geographic level of the selective factors affecting this system, as we have already hypothesized for explaining the interregional variation of F_{ST} for *GM* (sec. 2.8).

2.10. MAPS OF SINGLE GENES

World gene-frequency maps for 90 genes are in the second part of the book. For genes that have little geographic variation, or are polymorphic in only a small part of the world, no world maps were made. The construction of gene-frequency maps has been explained in section 1.14, where we have also explained the symbols used in the maps.

Gene-frequency maps are of interest for several reasons; we have already mentioned that they may permit visual detection or numerical calculation of correlations with possible factors of natural selection for which we know the geographic distribution. Naturally, the existence of a correlation at the geographic level is only a cue, which must be pursued with further research at the individual level; it is well known that correlations between two phenomena in space (or in time) could be generated trivially by other common causes. The classic, nontrivial example that we have already briefly mentioned is the correlation of sickle-cell anemia and malaria. There have been others, and the study of correlation between genotypes and diseases has attracted many investigators. It is of considerable interest not only from the point of view of pathogenesis, but also of natural selection and evolution. A summary of research on associations between polymorphisms and disease was published by Mourant et al. (1978), with a complete statistical analysis investigating the heterogeneity of results from different investigators and areas. All following references to statements on associations with disease are taken from this excellent summary.

Another valuable use of gene-frequency maps has not been explored in depth: the identification of the place

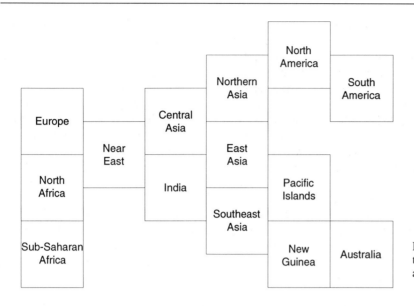

Fig. 2.10.1 A highly simplified scheme of the world showing the subdivision in regions and continents adopted for figure 2.10.2.

of origin of an allele, which may have been generated during the spread of modern humans, and is therefore found only in part of the world. In the absence of complications, the gene frequency of a neutral allele tends to be at a maximum at the place of origin, but it is admittedly not easy to distinguish this explanation of the maximum gene frequency from selective or migrational interpretations.

We thought it might be useful to add a simpler picture giving basic data of gene-frequency maps, like average frequencies by continent or by major region. To this aim, a graphic summary was prepared by computer from our data bank. The average gene frequency of each allele (including several for which no gene-frequency map was prepared) is shown by continent or major region, according to the general scheme specified in figure 2.10.1. The average gene frequencies of each region or continent are shown for 128 alleles in figure 2.10.2. An empty square indicates the absence of data.

Maps for *ABO* and *RH* do not display all the data symbols (geographic locations and statistical interpretation) as it was for all other maps. The location density of the collected samples is too high, at least in Europe, and it would have obscured the gene-frequency contour pattern. Therefore, in the world map, (1) the European data symbols are not represented (this information can be found in the corresponding map of Europe); (2) the statistical interpretation of the data points has been omitted for *ABO* (this information can be found in the maps of each continent).

The first polymorphism studied is *ABO*; its practical importance for blood transfusions made it the most widely investigated system. It is found all over the world, except for Central and South America where only *O* is present. A few exceptions occur, some of which are due to admixture with Africans or Caucasoids. In the northwestern part of North America, the frequency

of *A* is high, especially in Blackfoot and Blood Indians, who have the highest in the world. In East Asia, *B* is most prevalent. Except in northern Europe and Asia where a relatively small fraction of *A* is *A2, A* almost always includes only the *A1* subtype. One may add here a word of caution: when two subtypes of an allele are both shown, one cannot expect the sum of their frequencies to match exactly the frequency of the whole allele; in this case, the sum of the frequencies of *A1* and *A2* does not equal that of *A*. They come from independent samples and sets of data and cannot match precisely, here as in other similar situations for other genes.

The discovery of an association of *ABO* and stomach cancer (Aird et al. 1953) started an avalanche of investigations on *ABO* and disease. An introduction to the subject, statistical methods usually employed, and possible biases can be found in the work of Vogel and Motulsky (1986), who have dedicated special attention to the problem. Investigations were soon extended to other genes; an analysis and summary of published data was prepared by Mourant et al. (1978).

The possible associations of *ABO* and disease have been most widely studied. Of all other genes investigated, only *HLA* proved more interesting. The study of *ABO* associations was extended to many afflictions, ranging from a variety of infective and parasitic to neoplastic and degenerative diseases. High correlations have been found for syphilis, especially cerebrospinal (*A* more highly affected than *O*); better ascertained is the higher permanence of positivity of the Wasserman reaction (an immunological test for the disease) after chemotherapeutic treatment in *A* and *B* compared with *O* individuals. The possible higher resistance of *O* individuals to syphilis has considerable interest because it might explain the near absence of *A* and *B* individuals in the Americas, except for Eskimos and some northern Amerind groups. In fact, it has been believed for a long

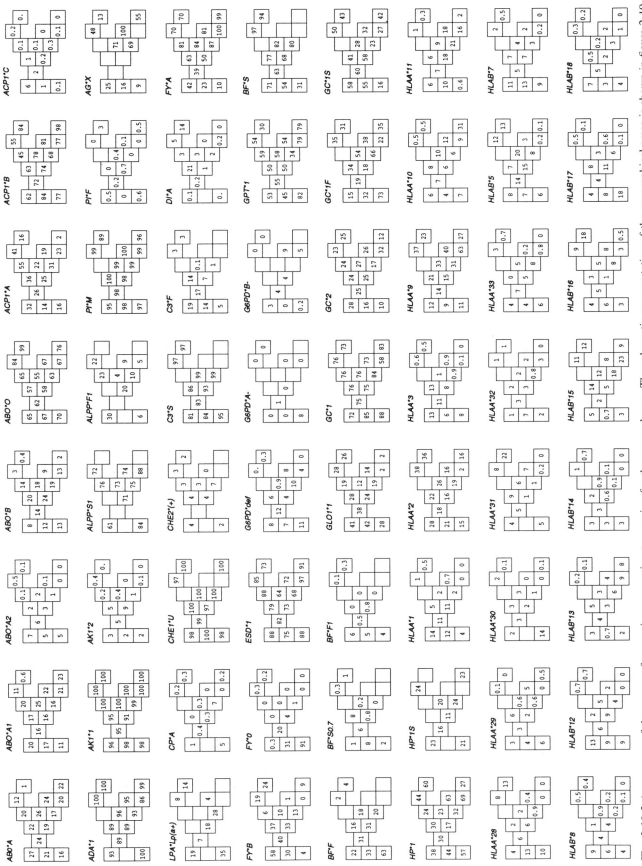

Fig. 2.10.2 A summary of the average frequencies per continent or region for the genes shown. The schematic representation of the geographical areas is shown in figure 2.10.1.

Fig. 2.10.2 (continued)

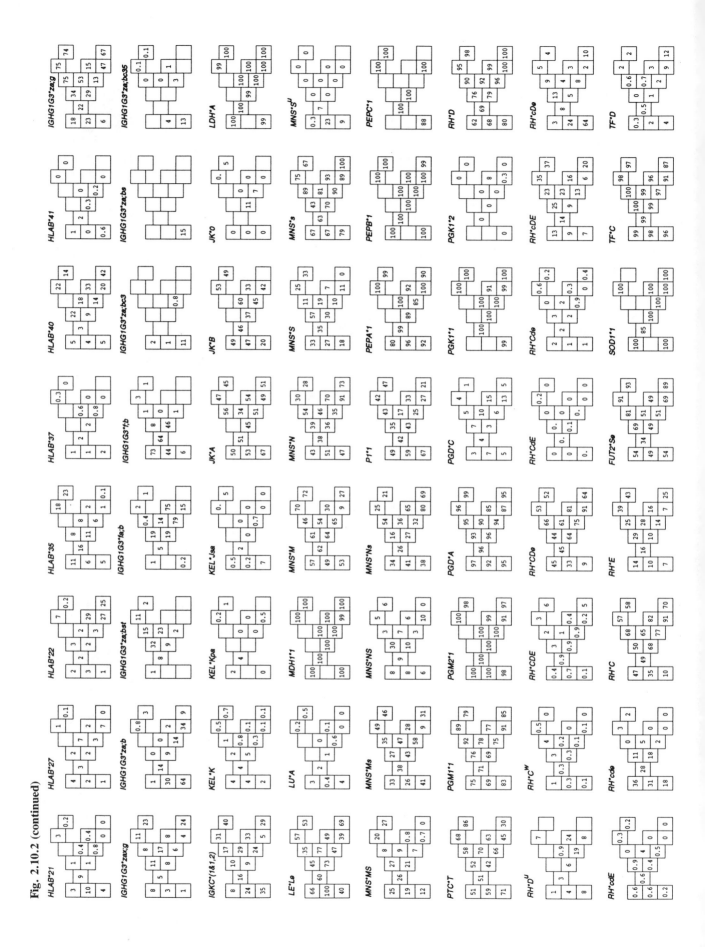

time that syphilis was imported to Europe and the rest of the world from America after Columbus's voyage in 1492. If it is true that *O* individuals are more resistant to syphilis and the disease was endemic in America for a long time, it may have eliminated most of the non-*O* individuals by natural selection. In more recent times, it was discovered that yaws, a disease widespread in sub-Saharan Africa, is due to a parasite that is almost undistinguishable serologically from syphilis; the theory was put forward that syphilis came from Europe to Africa (see discussion in McNeill 1976). Yaws is not a venereal disease and is transmitted by skin-to-skin contact, especially to and from babies. One cannot entirely exclude, however, that mutations in the African parasite caused it to spread by the venereal routes generating the syphilis syndrome.

Other important diseases like plague, cholera, and infant diarrhea caused by certain *E. coli* strains may also have favored group *O*. Smallpox was also said to have affected the geographic distribution of *ABO* alleles because *A* individuals were more susceptible. Enormous statistical heterogeneity between results of different investigators makes it difficult to reach a satisfactory conclusion; it is possible there are differences between local strains of pathogens. Tuberculosis (pulmonary) is also believed to be more virulent in *A* individuals than in *O* or *B*; heterogeneity between studies is not as high here. Influenza A2 virus shows a stronger preference for *O* than for *A* individuals (50% more), but there is no difference between *B* and *O*, and no heterogeneity. Malaria shows a preference for *A* individuals versus *O*, but with strong heterogeneity. The same is true for bilharzial hepatic fibrosis, with less heterogeneity.

Many neoplasms seem to have different attack rates for ABO phenotypes, and the numbers of individuals examined are sometimes so high that even small differences in incidences could appear statistically significant. There is perhaps less heterogeneity in neoplasms than in infectious diseases when more than one study is available. The effect seems to be fairly specific. For example, for malignant neoplasms of the salivary gland, esophagus, and pylorus, *O* individuals are more resistant than *A* or *B*, but the residual gastrointestinal tract is not affected differentially. Hypotheses have been put forward to explain the resistance of *O* individuals to adenocarcinomas (Yamamoto et al. 1990). Among other common diseases, diabetes has a slightly increased incidence (5% more) in *A* and *B* versus *O*, pernicious anemia a higher one (20%−25%); a rather high increase (2.5 times) in incidence of myopia is observed in *A* and *B* versus *O*. An average increase in incidence (20%−25% higher) is observed in *A*, and usually also in *B*, versus *O* individuals for rheumatic heart disease, thrombosis, and embolism, cirrhosis of liver, cholelithiasis; the reverse is true for rheumatoid arthritis. The most intensively investigated are gastric and duodenal ulcers, with a 15% higher incidence for gastric, and 30% for duodenal ulcer in *O* than in *A* or *B* individuals. All specific references to these data are in Mourant et al. (1978).

Except in the regions of the Americas, where there is an anomalous absence of *B* and, in part, of *A*, we have seen that the F_{ST} of *ABO* is often low. The differential sensitivity of *ABO* to a number of infectious diseases, which may have a variable geographic distribution, makes it likely that *ABO* alleles are subject to disruptive selection. However, the low F_{ST} seems to indicate a stabilizing selection. It is possible that both types of selection are present. The study of natural selection for *ABO* has been made more difficult by the inability of distinguishing genotypes with immunological techniques; phenotypically *A* individuals may show different reactions to disease depending on whether they are of genotype *AA* or *AO*, and the same can be said for *B* individuals, that is, that the phenotypic dominance we observe for the *A* and the *B* alleles does not extend to all possible manifestations. The molecular nature of the *ABO* gene has been discovered only very recently (Yamamoto et al. 1990) and may help to explain some of the mysteries of this classical blood group. All other genes are much less well known from the point of view of correlation with disease, with the exception of *HLA* for which associations much stronger than those observed with *ABO* have been found.

Several proteins or enzymes and a few blood groups are polymorphic as described in figure 2.10.2. In the following we comment on genes for which we can contribute some information or suggestion about their evolutionary history, especially in terms of their possible adaptive significance.

The blood group Diego, *DI*, allele *A* is found typically in the Americas but is present at low frequency in far East Asia where it probably arose. If this is correct, the geographic distribution of *DI*∗*A* is in disagreement with the hypothesis that an allele is at its maximum frequency at its place of origin. Figure 2.10.2 shows the highest frequency of *DI*∗*A* in Central Asia (21%)—it was found in only one sample of Turkomans from Iran. The authors of the original paper (Akbari et al. 1984) claim that this high frequency (unique in Eurasia) is due to the Mongol origin of the Turkomans.

Duffy, *FY*, is another blood group of special interest; in vitro experiments have shown that the *O* allele confers resistance to a particular malarial parasite, *Plasmodium vivax* (Miller et al. 1976). This parasite is rare in Africa, where *FY*∗*O* is very frequent, reaching 100% in some areas. We do not know which was the oldest *FY* allele in Africa. Two interpretations have been suggested. According to one, *FY*∗*O* appeared in Africa but was not the oldest allele, and because of the presence of the parasite, *FY*∗*O* underwent selection to reach the high frequencies it enjoys today. The near disappearance of susceptible individuals *FY*∗*A* and *FY*∗*B*, which were both originally present in Africa and are still found in areas where the endemicity of malaria is less strong or absent, has caused the near eradication of the parasite in this part of the world, but not outside Africa where $FY*O$ is rare or absent. According to the second interpretation, $FY*O$ was already present in Africa at very high frequencies before the appearance of the parasite, preventing its spread in Africa (Livingstone 1984). The $FY*A$ allele is present in all parts of the world, whereas the B allele is almost absent in Southeast Asia.

Glucose-6-phosphodehydrogenase, $G6PD$, a sex-linked gene, has produced a number of deficiency mutations (indicated by a minus sign) conferring resistance to malaria. Deficiency mutants are therefore quite common in many parts of the world, except the Americas and Australia. G6PD-deficiency was originally discovered because it determines sensitivity to a number of drugs, including some antimalarials, which may cause serious hemolytic attacks in deficient individuals. The ingestion of fava beans causes similar attacks (favism; see Vogel and Motulsky 1986). The normal gene exists in two electrophoretically different forms, A and B; deficiency mutants have arisen in both forms, but the $A-$ deficient mutants are found almost only in Africans, whereas the $B-$ forms are found on the other continents in which the deficiency is known. Most frequency determinations of deficiency are done by straight assay of the enzyme, which does not distinguish the two types (see also sec. 2.14).

Another gene with several alleles, connected with immune defense functions, BF (properdin), exists in two more common forms, and is polymorphic in most of the world but rare in the Americas. Two more rare alleles are found mostly among Caucasoids.

Glyoxylase-locus 1, $GLO1$ is regularly polymorphic all over the world except in Australia and New Guinea, where one of the two alleles is rare.

First called "Group-specific component," GC is a protein originally typed immunologically and then shown to be identical to vitamin-D–binding protein. The two major alleles show little variation in frequency; Mourant et al. (1976b) found a correlation with solar intensity, but we find (see later) only modest correlation with climate, which is highly correlated with solar intensity. There is today a great wealth of mutants described in addition to the original two (Constans et al. 1983), and it is possible that their geographic distribution, if it were adequately known, might be more informative. We show data only for two subtypes of $GC*1$, F and S, which do not seem to clarify the question of the correlation with solar intensity.

Haptoglobin, HP, is another plasma protein that binds hemoglobin and helps to dispose of it when it leaks from dead or damaged red cells. It might be of special importance in regions where diseases, such as malaria, which produce significant hemolysis are common. Two alleles are common (1 and 2); the first is subtyped into $1S$ and $1F$. Haptoglobin shows little variation, as shown by F_{ST}

analysis and might be under stabilizing selection such that it is stronger in some regions than in others; the same may be true of GC.

The most important system of markers in our collection, HLA, is represented by 12 A alleles and 17 B alleles. Their frequencies never sum to 100, because not all alleles are known, but in both A and B there are so-called blank alleles that are operationally defined as the sums of all the unknown ones. Blank frequencies are higher for populations other than Caucasoids, which have been less intensively investigated. Some of the conclusions that follow may be weakened by this fact.

Perusal of HLA maps, and of figure 2.10.2, shows that most HLA alleles can be classified according to one of four patterns.

1. A few alleles are essentially ubiquitous, in the sense that they are found at nontrivial frequencies all over the world. They are $A*2$, $A*9$, $B*15$, and $B*27$ (however, $B*27$ is nearly absent from South America and Australia).

2. Other alleles, the most numerous, are absent or, more usually, rare in East Asia and the Americas, Australia, New Guinea, and the Pacific islands and are therefore common only in Eurafrica (Africa, and Europe, and the part of Asia occupied by Caucasoids): $A*1$, $A*3$, $A*29$, $A*32$, $B*7$, $B*8$, $B*12$, $B*14$, $B*17$, $B*18$, $B*37$. One allele, $B*40$, has somewhat the opposite pattern, but it is not truly rare in Eurafrica, where it reaches frequencies of 3%–5% versus 14%–42% in the eastern part of the world.

3. Another relatively common pattern comprises alleles rare or absent in Australia and New Guinea: $A*28$, $A*30$, $A*31$, $A*33$ ($A*33$ is also absent from South America), $B*5$, $B*7$, $B*16$ (which is found, however, in New Guinea), $B*21$, and $B*35$. The opposite pattern is found in $B*22$, which is relatively rare in most of the world except in Australia, New Guinea, and the Pacific islands where it is very common.

4. Rare in the Americas are $A*10$, $A*11$, $B*13$, and $B*22$; they all reach, by contrast, rather high frequencies, sometimes the highest, in Australia. Allele $B*35$ exhibits a somewhat opposite behavior, reaching its highest frequencies in America.

Although there probably was a partial loss of alleles in the migrations to the two East Asian appendages, Australia and the Americas, the complete absence of an allele is rare. Later gene flow may have reintroduced genes that were initially lost but are now present at low frequencies. Drift, and possibly selection are considered the most likely causes for the decrease of certain alleles. One should resist the temptation to consider a founders' effect the exclusive factor. Drift is not expected to take place in only the first generation, that of the founders. Every generation determines the gene frequencies of the next ones, and drift effects are cumulative; moreover, population bottlenecks do not take place exclusively at the beginning. Another possibility to be considered is that some of the alleles of the second, third, and fourth patterns originated rather late and did not have time to diffuse to the most remote areas. It would seem, however, that the phenomenon is not observed with the same intensity for non-*HLA* markers. Moreover, other evidence, discussed in later chapters, shows extreme internal variation in America and Oceania, most likely caused by very strong drift.

The ubiquitous *HLA* alleles (the first pattern of geographic distribution) might be those for which there is more stabilizing selection. The second pattern is so common that it might suggest, taken at face value, that the first split in the tree took place between Caucasoids and Africans versus all the rest. This is the root that had been suggested in the earliest tree analyses (see sec. 2.2). It is difficult to choose between a historical and a selective explanation for this pattern.

Direct evidence for selective involvement of *HLA* is abundant and based mostly on associations of *HLA* and disease (Dausset and Svejgaard 1977; Tiwari and Terasaki 1985; Thompson 1988). *HLA* research is very active and many more genes other than *A* and *B* have been detected, including the *C* locus and a variety of *D* loci, in addition to many pseudogenes that are not affected by selection, at least today. Information on ethnic variation of selective patterns is scanty but sufficient to show that there are frequently differences of correlation between *HLA* and disease in different parts of the world. Most of the evidence for correlations between *HLA* and disease comes from association studies showing that patients with specific diseases are more frequently of a given *HLA* type than others. The gene actually responsible for the disease is extremely difficult to ascertain, because there are many different *HLA* genes in the complex region of the short arm of chromosome 6 where they are located and where some other genes are found. All *HLA* loci are closely linked, and association of a disease with a particular *HLA* gene may reflect "linkage disequilibrium": when two genes are very close on a chromosome, there is little genetic recombination (crossing-over) between them, and they tend to stay together for a very long evolutionary time. Therefore, diseases tend to show association with many neighboring genes. Usually the highest degree of association is with the genes closest to the gene determining the disease. The *D* loci are probably more often involved than *A* or *B*, and many of the associations found with *A* and *B* are now known to be a result of their close linkage with *D* loci, with which we do not concern ourselves here. *A* or *B* loci may, however, be directly responsible for diseases: *B*∗27 is perhaps the clearest example to date. There is some consensus that it is somehow directly responsible for ankylosing spondylitis, a chronic progressive form of arthritis affecting a number of joints, primarily of the spine. This disease is almost never found in non-*B*∗27 individuals, but *B*∗27 is certainly not the only factor since only a small fraction of *B*∗27 individuals develop ankylosing spondylitis. *B*∗27 is also associated with several other diseases.

The immunoglobulin genes, *GM* and *KM*, which produce antibodies and play an important role in defense against pathogens, have also shown some, though less clear-cut, associations with specific diseases.

GM is one of the major supergenes, the second in importance after *HLA*. It is located in chromosome 14 and has a very large number of alleles, but not all can be recognized in all populations, in part because of the scarcity of important reagents. Its four genes control the constant part of the heavy chain of four *G* immunoglobulins; variants are suspected of showing different resistance to some infectious diseases, although clear-cut examples are still rare. Only *G1* and *G3* genes are commonly typed and the locus has been called *IGHG1G3* for *IGHG1* and *IGHG3*. Their alleles are separated by a semi-colon.

The geographic distribution of the heavy-chain polymorphisms (*IGHG1G3*) shows that *za;g* and, to a lesser extent *zax;g* are almost homogeneously present in the world, but certainly are more frequent in East Asia and its appendages. *f;b0b1b3b4b5* is mostly Caucasoid; *fa;b0b1b3b4b5* is primarily East Asian, especially in the south and southeast, but almost absent in the Americas; *za;b0b1b3b4b5* and its derivatives are mostly African. The light-chain polymorphism, on the contrary, has a fairly constant frequency throughout the world. The real selective meaning of the immunoglobulin polymorphisms awaits analysis by molecular techniques simpler than those now available.

Another important and widely studied blood-group system, MNS, has been analyzed by allele and haplotype frequencies. The pair of alleles *M*, *N* is ubiquitous, with a reasonably regular geographic distribution, but *S* of the pair *S*, *s* becomes rare in Southeast Asia and disappears completely in Australia. The allele *S*ᵘ is almost exclusively African. Of the four haplotypes, two (*MS*, *NS*) are nonexistent in Australia, and *MS* is rare in New Guinea and the Pacific. Analysis of possible associations with a large number of diseases has reached statistical significance in very few cases; some such results might be expected to arise by chance when many significance tests are carried out. The few positive results have never received strong confirmation by studies in other areas (Mourant et al. 1978) so that one can consider the results of these investigations basically negative.

Phenylthiocarbamide tasting, *PTC* has been named an "honorary blood group" by Race and Sanger (1975), but is tested as the capacity to taste PTC, a substance extremely bitter to many and tasteless to a minority. As mentioned earlier, inheritance, or at least the testing procedure, is not perfect so that gene frequencies may be approximate. The world variation of the gene frequency is modest. The interest in this gene comes from the resemblance of PTC to antithyroid substances that are present in some widespread vegetables; *PTC* itself has antithyroid activity. Tasters might be prevented from ingesting too many antithyroid substances in their diet by their bitter taste. A correlation between taster's state and the occurrence of some thyroid diseases seems significant, tasters being less exposed than nontasters to nontoxic, nodular goiter, and to nodular goiter; but, somewhat surprisingly, patients with toxic, diffuse goiter are significantly more often tasters than normal controls. Again, somewhat surprisingly, hypothyroidism is more common in tasters than in nontasters. Other correlations with disease have been detected but are not easy to understand (Mourant et al. 1978).

Another gene that has been very widely studied is *RH*, a blood group of clinical importance because *RH*-positive children born to *RH*-negative mothers are exposed to the danger of antibodies produced by the mother, which may damage the fetus if they enter the fetal circulation. The first pregnancy is usually normal, but successive ones are progressively dangerous for the health and survival of the progeny. *RH*-negative children, as may be born to an *RH*-negative mother and a heterozygous father, are unharmed. This maternal-fetal incompatibility is also observed with other blood-group systems, but *RH* is responsible for 90% of the cases. *ABO* determines very early incompatibility of *O* mothers and *A* or *B* progeny, which often goes undetected. Incompatibility has an interesting evolutionary fate (Cavalli-Sforza and Bodmer 1971a): heterozygotes are at a disadvantage (only heterozygous children of *RH*-negative mothers die), and the polymorphism is unstable. According to the simplest hypothesis, there is a critical gene frequency at 50%; if a population has more than 50% *RH*-positive genes, the *RH*-negative type will be lost over the generations. If it has less than 50% *RH*-positive, the *RH*-positive type will be lost. Only one population on Earth, a group of Basques, has more than 50% RH-negative genes. We will see (sec. 5.6) that Basques were hypothesized to be an ancient European population that was predominantly or perhaps 100% *RH*-negative, and mixed with later settlers who were predominantly *RH*-positive. If there were no admixture between neighbors, Basques would evolve toward a full *RH* negativity, and the rest of the world toward full positivity. There is probably enough admixture of Basques with neighbors at low proportion of *RH*-negative genes, that *RH*-negative may be decreasing among Basques also. It is possible, however, that in addition to *RH*-incompatibility there is also some heterozygous advantage that helps maintain the RH gene polymorphism (Cavalli-Sforza and Bodmer 1971a).

RH was the first complex genetic system detected which may involve various loci. The *RH* theory proposed by Fisher (Feldman et al. 1969; Race and Sanger

1975) has not been falsified by later developments, but the number of alleles has grown considerably. Purely immunological methods of analysis have limitations that could be avoided only by molecular knowledge, when available. The *RH* polymorphism could be analyzed by immunological reagents at all three major loci postulated by Fisher, *C*, *D*, and *E*. Two alleles at each locus, *C*, *c*, *D*, *d* and *E*, *e*, were considered in the analysis by haplotype. Two more alleles, *C^w* and *D^u*, were also considered separately; *C^w* is highest in northern Asia, where it may have originated; *D^u* is high in Africa and in Southeast Asia. The *C*, *c* polymorphism is ubiquitous, with the highest values in New Guinea and Southeast Asia and lowest in Africa. The *D*, *d* polymorphism is most frequent in Europe; the distribution of the *d* allele (not given) is very close to that of *cde* haplotype, present in all Caucasoids and Africans, with a maximum in Europe, definitely more rare in East Asia and the Americas, and totally absent in Australia and New Guinea. The *E*, *e* polymorphism is ubiquitous with highest values in the Americas.

The haplotypes are a little more informative on aspects of the history of this polymorphism. According to Fisher's evolutionary model (see also Feldman et al. 1969), these are the main points of the process.

1. The oldest haplotype is *cDe*, which has its highest frequency in Africa but is also found in the rest of the world (when this model was developed in the 1940s, African origin had not been suggested).

2. From *cDe* arose, by *mutations* at the three loci, three haplotypes, all of which are common.

   2a. *CDe*, (from mutation *c* ⟶ *C*), may have arisen in the a.m.h. expansion to Asia and is frequent all over the world. It is rarest, but not too rare, in Africa (9%).

   2b. *cde* (the *RH*-negative "chromosome" or haplotype) must have arisen by mutation *D* ⟶ *d* during the expansion of a.m.h. toward the northwest from West Asia and the occupation of Europe (where the highest frequency is found, in the Basque region). It is not, however, rare in Africa, and it may have originated before the passage from Africa to West Asia or have flowed back.

   2c. *cDE*, which is especially frequent in Northeast Asia, where it may have originated by mutation *e* ⟶ *E*, and in the Americas, where it went from Northeast Asia; since it is more rare but not absent in Southeast Asia and Australia, some active exchange must have taken place between Northeast and Southeast Asia before migrations to the two appendages of East Asia.

3. In the next phase, apart from other mutations that we are not considering like *D^u*, *C^w* and rarer ones, crossing-over between the four types in existence that were listed above, could generate three more types.

   3a. A relatively rare type, *Cde*, has maximum frequency in North Asia and is common in Europe and West Asia, where it may have originated from crossing-over between two haplotypes like CDe and cde that are both common in that region.

3b. Rare in most places, with a maximum in East Asia, cdE could have originated from crossing-over between cDE and cde. This is most likely to have occurred in West Asia or North Asia on the basis of the frequencies of these types in these regions today. It could also have originated from crossing-over between other haplotypes.

3c. Also relatively rare, CDE, is most frequent in Northeast Asia and the Americas. It could have originated from crossing-over between CDe and cDE. On the basis of the frequencies of these two types, this most likely occurred in Northeast Asia, where this haplotype is frequent today.

From some indirect observations (for details and references, see Race and Sanger 1975) it has been postulated that the order of the three loci is DCE, and that the three crossover products above are all expected to be the result of single recombination events and therefore more frequent.

4. A crossing-over between two crossover products originated in the preceding phase could generate the eighth haplotype, CdE, which is, as might be expected, the rarest, a fact in agreement with the postulated order of the three loci.

In order to complete the picture and summarize it, one should add that three mutations are necessary for producing the first four types from which all the eight types can derive, and that once this has happened, it is more likely that recombination, rather than more mutations, produce the residual four haplotypes. A double recombination event is rare, but gene conversion may make it likely.

This hypothesis of the origin of the RH system is more than 40 years old but has not lost its capacity to explain at least some of the evolutionary history of a complex gene system. It also predicts that most of the events leading to the various haplotypes must have happened in Asia, though in different parts of it, in the 100–50 ky period that preceded the expansions to the other continents. A final assessment of this theory will become possible only when the molecular genetics of the RH region become sufficiently well known.

RH has been the blood-group system of greatest clinical importance after ABO, and it is not surprising that many data have also accumulated here on correlations between RH and diseases other than those resulting from materno-fetal incompatibility (Mourant et al. 1978). Conclusions are the same as for MNS; although there are a few significant results, a little above the expectations of a totally random phenomenon, differences in incidence with controls are always modest, and there is little hope that even one of the very few positive conclusions out of the several hundreds of published investigations might prove reproducible.

The secretor system, $FUT2(SE)$, is responsible for bringing into secretions substances responsible for A, B, and the related H substances that are normally found on the red cells of individuals and define their ABO status. Much research has been done on the association between gastric and duodenal ulcers and the secretor status. The susceptibility of secretors to gastric ulcers is only 59% that of nonsecretors, and that to duodenal ulcers 57%. This important difference is well ascertained. The possible mechanism of protection is that A and B substances are mucoid and may be protective of the linings of the gastric and duodenal mucosa if they are secreted in the stomach. Other findings have not had the support of as many independent investigations as these two diseases (Mourant et al. 1978).

2.11. SYNTHETIC MAPS OF THE WORLD

Observations made on single-gene maps can give only crude indications of evolutionary history, because gene frequencies are subject to random sources of variation. It is only by compounding the information from many genes that a clearer picture can emerge. We have described in chapter 1 how synthetic maps can be obtained by plotting principal components, on a geographic map, and drawing curves connecting equal values. We refer to principal components as PCs.

As mentioned in sections 1.13 and 1.15, PCs are linear combinations of all gene frequencies of a given population. They are therefore especially suitable for describing gradients of gene frequencies generated by migrations, because migration has a linear effect on gene frequencies (Menozzi et al. 1978a). They inevitably tend to smooth gradients, which therefore may appear more regularly linear than they actually are. Fortunately, we can give evidence from computer simulations that the distortions are not serious and that the technique usually can—as it is expected to—detect ancient migrations (Rendine et al. 1986). Moreover, a single PC map extracts only one source of information, and all PCs are independent from each other, so that each tells a unique story. PCs are ordered by their importance in extracting the information contained in the data, and thus the first PC uses the major discrepancies observed in the geographic area being studied; the second repeats this operation on the residual information, etc. In this section we show seven synthetic world maps corresponding to the first seven PCs. The relative amount of information expressed by each PC, or, in other words, the percentage of the total variance of gene frequencies explained by each PC, is given

Table 2.11.1. Percentage of Total Variance Explained by the First Seven Principal Components of World Data for Eighty-Two Genes

Principal Component	% of Total Variance	Principal Component	% of Total Variance
1	34.8	5	4.2
2	17.8	6	3.8
3	12.4	7	1.8
4	7.4		

in table 2.11.1. PCs lower than the first seven may still have useful information, but they tend to have very low correlations with individual genes, and were not mapped.

Each PC is the sum of the 82 gene frequencies, each multiplied by an appropriate coefficient. The 82 genes considered for this analysis are those for which world genetic maps (see Table of Genetic Maps) are shown in the second part of the book except for the following eight alleles: ABO*A2, G6PD*def, G6PD*A+, G6PD*B+, G6PD*B−, IGHG1G3*za;g, IGHG1G3*zag;b0stb3b5, IGHG1G3*f;b0b1b3b4b5. The correlations between a PC and the 82 genes express the relative importance of each gene in determining the PC. The genes having highest correlations with the first three PCs are shown in table 2.11.2. In general, one can expect that these genes tend to have a gene map most similar to the synthetic map, but a gene may contribute information to more than one PC, and therefore the pattern found in a single gene map may not always be similar to that of the corresponding synthetic map, unless the correlation between that PC and the particular gene is very high.

It is clear that correlations decrease, on the average because the PC is of lower and lower rank. This is inevitable, since the higher PCs (those calculated first) have extracted a greater amount of information. The lower the rank of a PC, the more important is "noise," or variation that cannot be attributed to any simple, regular genetic source.

For ease of visual recognition, we use eight classes of PC values in synthetic maps. The choice of an increasing or decreasing density of shading is totally arbitrary; it could be reversed if desired, without any loss of information. Intermediate classes are closest to the average PC, and the maximum and minimum values will sometimes be called "poles" of the PC. They indicate the populations that globally differ most from each other for the particular axis of reference formed by the PC being studied. Populations or regions that have very similar values of a particular PC need not, however, be similar between themselves, for they may be very different for another PC. The color maps are very useful in this respect for they cumulate the information from the first three PCs. They can also directly answer some puzzling questions that can arise when geographically and (supposedly) genetically remote populations turn out to be very similar for a particular PC. The information that

Table 2.11.2. Genes Showing the Highest Correlations with the Three Highest Principal Components of World Gene Frequencies

P.C.*	Range of Correlation Coefficient		Genes
1	1.00 − 0.90	(+)	—
		(−)	FY*A
	0.90 − 0.80	(+)	HLAB*18, HLAA*3, HLAB*14, RH*cde, HLAA*29, HLAB*12, HLAB*7, P1*1, GLO1*1, HLAA*19
		(−)	RH*D, HLAB*22, C3*S, RH*C, HLAB*40
	0.80 − 0.70	(+)	HLAA*30, HLAA*1, HLAB*17, HLAB*8, LU*A
		(−)	IGHG1G3*zax;g, IGHG1G3*za;g, FUT2(SE)*Se, HLAB*13, AG*X, RH*CDe
	0.70 − 0.60	(+)	HLAB*21, HLAA*33, MNS*S, RH*cDe, KEL*Jsa, PTC*T, HLAA*28, TF*C
		(−)	RH*Dᵘ, HLAA*9, RH*CDE, MNS*Ns, PGM1*1, HLAA*10, HLAB*15
	0.60 − 0.50	(+)	MNS*He, MNS*MS, KEL*K, MNS*M, RH*Cde
		(−)	ABO*A1
	0.50 − 0.40	(+)	MNS*Ms, IGHG1G3*za;b0b1b3b4b5, MNS*NS, ABO*B, HLAB*5
		(−)	RH*E, ACP1*B, AK1*1, BF*F, IGHG1G3*fa;b0sb3b5
	0.40 − 0.30	(+)	HP*1, HLAA*31, HLAB*35, ACP1*C, GC*1F
		(−)	RH*cDE
2	0.90 − 0.80	(+)	HLAA*2
		(−)	BF*F, GPT*1
	0.80 − 0.70	(+)	RH*E, RH*cDE, HLAB*16, DI*A, HLAB*35
		(−)	—
	0.70 − 0.60	(+)	HLAA*31, MNS*MS, MNS*M
		(−)	MNS*Ns, JK*A
	0.60 − 0.50	(+)	TF*C, PTC*T, IGHG1G3*za;g, MNS*S, HLAB*5, MNS*Ms
		(−)	MNS*He, HLAA*10, ABO*A, RH*cDe, IGHG1G3*za;b0b1b3b4b5, ESD*1
	0.50 − 0.40	(+)	PGD*A, ABO*O
		(−)	HLAA*30, HLAA*29, ABO*A1, HLAB*13, HLAB*22
	0.40 − 0.30	(+)	HLAB*27, FUT2(SE)*Se, GLO1*1, PEPA*1
		(−)	KEL*Jsa, LU*A, GC*1, ACP1*B, GC*1F, HLAB*17
3	0.90 − 0.80	(+)	ADA*1, KM*(1&1,2)
		(−)	—
	0.80 − 0.70	(+)	—
		(−)	—
	0.70 − 0.60	(+)	GC*1F, ABO*O
		(−)	ACP1*C, HLAA*11
	0.60 − 0.50	(+)	KEL*Jsa, MNS*He, RH*cDe, AK1*1
		(−)	HLAA*1, KEL*K
	0.50 − 0.40	(+)	HLAA*30, HLAA*28, RH*Dᵘ, JK*A, ACP1*B, GC*1

(continued)

Table 2.11.2 *(continued)*

	(−)	*RH∗CDe, HLAB∗27, ABO∗A, AG∗X, ABO∗B*
0.40 – 0.30	(+)	*PGD∗A, DI∗A, HP∗1, PGM1∗1, IGHG1G3∗za;b0b1b3b4b5, FUT2(SE)∗Se, RH∗D, P1∗1, C3∗S, PTC∗T, HLAA∗31, HLAA∗19*
	(−)	*ABO∗A1, RH∗cde, HLAB∗8, HLAA∗3, MNS∗NS, RH∗C*

Note.– Genes giving positive or negative correlation values are indicated by (+) or (−), respectively.
* P.C., Principal component.

can come from color maps is limited to the first three PCs and is therefore not complete, but in the case of the world, it summarizes 65% of the whole variation.

The first synthetic map (fig. 2.11.1) shows that the poles of the first PC are Africa and Australia, which are the maximum and the minimum respectively. The first PC tends to indicate the expansion from the south and east of Africa, toward the rest of the world, first north and east, then toward Australia and the Americas. But PCs can only indicate a static situation, and we add the movement to the picture from external evidence. Mongoloids are together with American natives at the center of the variation. Because of the west-east linear gradient imposed by the first PC, western Europe is similar to Africa, an anomaly corrected by lower PCs. Note that the least African of all inhabitants of Africa are North Africans, eastern Pygmies (Mbutis), and Bushmen. Again, the difference existing between Pygmies and Bushmen are stressed by later PCs. As is shown—for example, for Scotland, Ireland, and Spain whose values of the first component share the same class with sub-Saharan Africa—one component alone is not sufficient for drawing absolute inferences about genetic similarity. In general, with the inevitable approximation resulting from the linear representation by PCs, the first map corresponds to the first split of the tree, that between Africans and non-Africans, but the split described by the first PC considered in isolation is not as clear-cut as that of the tree. In summary, the first PC approximately describes the west-to-east movement of a.m.h.

The poles of the second synthetic map (fig. 2.11.2) are America on one side and Australia + New Guinea

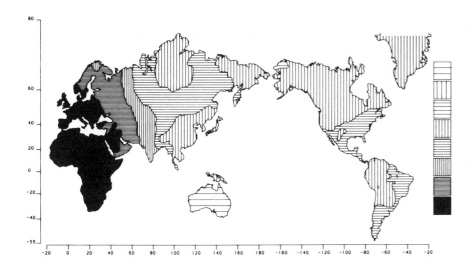

Fig. 2.11.1 A synthetic map of the world based on the first principal component (PC). Here, as in all subsequent maps (including the following chapters), the range between the maximum and minimum values of the PC has been divided into eight equal classes. The direction of increase of PC values is arbitrary.

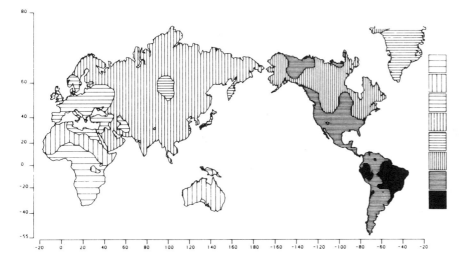

Fig. 2.11.2 A synthetic map of the world based on the second principal component.

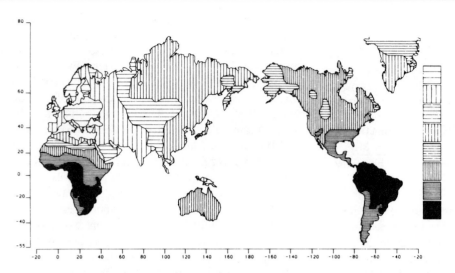

Fig. 2.11.3 A synthetic map of the world based on the third principal component.

on the other. Mongoloids are again in a central position. This is approximately equivalent to the second split of the tree, which separated Australia, New Guinea, and Southeast Asia from the rest of the non-Africans.

The maximum and minimum of the third map (fig. 2.11.3) are in sub-Saharan Africa and Europe, respectively. The intermediacy of North Africa and, to a lesser extent, East Africa between Africa and Europe is apparent.

The fourth map (fig. 2.11.4) stresses the separation of western Europe from Africa, South America, and East Asia, whereas the fifth map (fig. 2.11.5) contrasts western Europe with Southeast Asia alone.

The sixth map (fig. 2.11.6) emphasizes the difference between the American Arctic and Southeast Asia + South America, with North Asia and most of the rest of the world taking central values. If a climatic gradient is responsible for this, it shows independence from adaptations to the tropical forest in Africa.

The seventh map (fig. 2.11.7) shows the difference between Bushmen and other Africans, with the rest of the world intermediate.

This analysis of world synthetic maps corresponding to single PCs is only moderately rewarding. Each PC

repeats only approximately some of the conclusions from the tree. The combination of the first and most important PCs in a color map is more illuminating, as it is clearly easier to discern many of the clusters found by tree analysis at this higher synthetic level. Unfortunately, the human eye has only three pigments, and it is doubtful that the synthesis of more than three PCs by more than three elementary colors would yield a better picture. It has not been attempted. The color map of the world shows very distinctly the differences that we know exist among the continents: Africans (yellow), Caucasoids (green), Mongoloids, including American Indians (purple), and Australian Aborigenes (red). The map does not show well the strong Caucasoid component in northern Africa, but it does show the unity of the other Caucasoids from Europe, and in West, South, and much of Central Asia. The differences between North and South Africans are more visible in the Africa map.

There is one important limitation to the use of synthetic maps for describing the complex genetic patterns found when analyzing the whole world. It is well known that PCs can reveal a limited number of clusters: at a

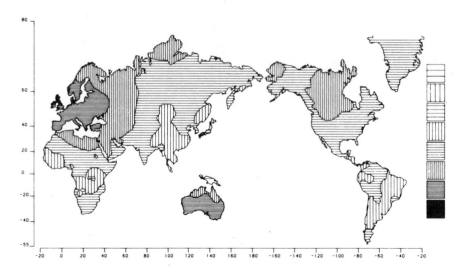

Fig. 2.11.4 A synthetic map of the world based on the fourth principal component.

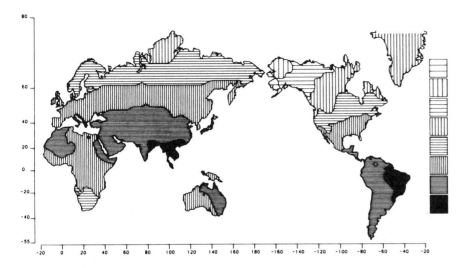

Fig. 2.11.5 A synthetic map of the world based on the fifth principal component.

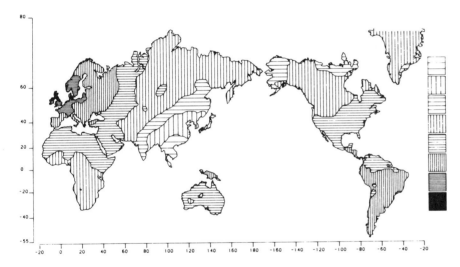

Fig. 2.11.6 A synthetic map of the world based on the sixth principal component.

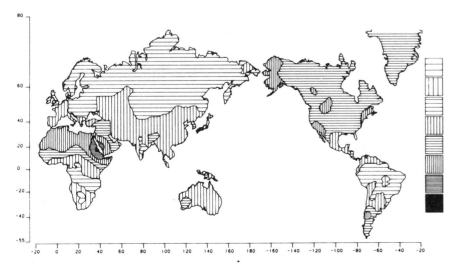

Fig. 2.11.7 A synthetic map of the world based on the seventh principal component.

minimum, the number of PCs used plus one. This principle is easily understood if one thinks of one PC, a straight line, which can only distinguish two extreme clusters. Adding a second PC allows us to distinguish at least one other cluster and so on. This limitation does not apply to clusters found in intermediate positions. Thus, a color presentation using three PCs, like ours, can in principle reveal only four clusters plus intermediate ones. We see, in fact, four major clusters and several intermediate ones. An important practical implication is that, in complex situations such as the world variation of gene frequencies, PCs may well change drastically when calculated on a fraction of the total area. The clusters appearing in the particular subregion examined will determine the weights of each gene frequency in each region. For this reason, the analysis was repeated for each continent and sometimes for each subcontinent in the subsequent chapters. Many details that cannot be perceived at the global level are visible in the partial maps in which PCs emphasize entirely different aspects. In general, one may add that PCs give an especially clear picture when there are not too many important factors or clusters to be considered; the world geographic distribution of gene frequencies is already a fairly complicated situation, and the analysis of partial areas is easier to understand.

The color picture of the world conveys 65% of the original variation and shows the areas occupied by four major ethnic groups: Africans are yellow, Australians red, Caucasoids green, and Mongoloids, showing the greatest variation, retain some of the similarities with Europeans on one side (a light-brown greenish tinge in central Siberia) and with Australians on the other (a pinkish color in parts of America and on the way to it). We thus have a good separation of Africans, Caucasoids, Australoids, and Mongoloids, but the first three components together do not seem to distinguish northern and southern Mongoloids. Three components are an incomplete representation of the global variation (in this case, they lose 35% of the variation). In addition to this disadvantage, the color picture also blurs to a large extent the gradients—that is, the major expansions—visible with single components. It provides, however, the best representation of global variation that we can give. The extensive gradients resulting from admixtures between Africans and Caucasoids in North Africa and between Caucasoids and Mongoloids in Central Asia are clearly visible.

2.12. HOMOZYGOSITY

The smaller a population and the fewer immigrants it receives, the more it tends to becoming homozygous for all of its genes. This is a standard result of drift, and also of inbreeding, but the two processes are not the same. In fact, most human populations avoid marriage between close relatives, and thus their degree of inbreeding tends to be smaller than if matings within a group were random. Expected homozygosity at a locus for random mating within a population can be easily calculated by summing the squares of the gene frequencies of all alleles:

$$H = S\, p_i^2, \qquad (2.12.1)$$

where the p_i are the relative frequencies of the m alleles at the locus, and the sum S is carried out over all m alleles. $S\, p_i$, the sum of the frequencies of all the m alleles is 1 (see also sec. 1.10).

Heterozygosity, the frequency of heterozygotes, is simply $h = 1 - H$. It is well known (Wahlund's principle; see Cavalli-Sforza and Bodmer 1971a) that if a population is made of many subpopulations that do not mate randomly with each other and have different gene frequencies, then heterozygosity is less than expected compared with that of a fully random-mating population.

If each subpopulation mates randomly, its heterozygosity can be computed from its gene frequencies and its homozygosity obtained using equation (2.12.1). The individual heterozygosities of s subpopulations (forming a larger one) can be averaged, giving average heterozygosity h_s, while h_t, the heterozygosity of the total population, can be calculated from the average gene frequency of the total population, assuming all its individuals mate randomly. Because of the incomplete interbreeding, h_s will be less than h_t, and the relation between the two provides another way of obtaining F_{ST} between the s subpopulations (see also sec. 1.10 and 1.11, and table 1.10.1):

$$F_{ST} = 1 - h_s/h_t \qquad (2.12.2)$$

Table 2.12.1 gives the h_s, h_t and F_{ST} values for all genes among the world populations selected in Appendixes 2 and 3.

We note that average F_{ST} values given in the single-gene maps are usually smaller than the corresponding ones given in table 2.12.1. Single-gene map F_{ST} values are calculated on all available samples, while F_{ST} values of table 2.12.1 are from the ethnic groups of Appendixes 2 and 3 and are averages of local values chosen to give a geographically and ethnically stratified sample of the world populations. The single-gene maps have a relatively larger representation of European data, which tend to take intermediate values in the world distribution of gene frequencies, and therefore their F_{ST} values are biased downward.

Table 2.12.1. Average Gene Frequencies (in increasing order), F_{ST} Values, Percentile Corresponding to Each F_{ST} in the Simulated F_{ST} Distributions, Heterozygosities h_s, h_t (from which F_{ST} was calculated), and Number of Populations Available among the 491 Populations (Appendix 2) Used for These Calculations

Gene	Frequency	F_{ST}	Percentile (%)	h_t	h_s	No. of Populations
RH*CdE	0.0004	0.0326	71	0.000753	0.000729	409
RH*cdE	0.0035	0.0411	73	0.007015	0.006727	409
HLAB*41	0.0051	0.0137	62	0.010104	0.009966	42
RH*C^w	0.0053	0.0233	67	0.010454	0.010210	158
PI*F	0.0067	0.0397	72	0.013199	0.012674	70
KEL*Kpa	0.0073	0.0331	71	0.014557	0.014075	65
HLAB*37	0.0087	0.0120	61	0.017287	0.017080	57
RH*Cde	0.0090	0.0311	70	0.017855	0.017300	409
ACP1*C	0.0094	0.0455	74	0.018566	0.017722	262
CP*A	0.0106	0.0531	75	0.020880	0.019771	78
G6PD*B-	0.0139	0.0714	76	0.027417	0.025459	66
HLAB*14	0.0149	0.0302	58	0.029301	0.028415	131
LU*A	0.0150	0.0314	58	0.029574	0.028645	134
PGK1*2	0.0153	0.0853	78	0.030128	0.027558	48
KEL*K	0.0170	0.0381	57	0.033327	0.032057	272
BF*S0.7	0.0178	0.0346	54	0.034940	0.033732	39
HLAA*29	0.0188	0.0382	54	0.036836	0.035429	111
RH*CDE	0.0192	0.0725	68	0.037574	0.034851	409
JK*0	0.0210	0.1946	94	0.040951	0.032981	52
MNS*S^u	0.0222	0.2122	95	0.043190	0.034025	174
HLAB*21	0.0223	0.0705	63	0.043551	0.040482	124
BF*F1	0.0224	0.0839	69	0.043250	0.039619	39
ABO*A2	0.0227	0.0614	58	0.044335	0.041614	362
HLAA*32	0.0238	0.0783	65	0.046427	0.042791	105
HLAB*18	0.0261	0.0558	52	0.050807	0.047972	125
KEL*Jsa	0.0287	0.1022	71	0.055574	0.049893	68
AK1*2	0.0291	0.0495	44	0.056544	0.053747	205
TF*D	0.0293	0.0833	61	0.056918	0.052178	331
HLAA*33	0.0296	0.0523	45	0.057459	0.054455	88
HLAA*30	0.0304	0.0894	63	0.058882	0.053618	90
HLAB*13	0.0306	0.0567	46	0.059332	0.055971	132
HLAB*27	0.0318	0.0427	38	0.061508	0.058881	130
HLAB*8	0.0321	0.0563	44	0.061987	0.058499	130
G6PD*A-	0.0354	0.1098	70	0.067951	0.060487	66
CHE2*(+)	0.0355	0.0149	21	0.068521	0.067497	59
HLAB*17	0.0432	0.0726	44	0.082622	0.076622	125
HLAA*28	0.0532	0.0928	52	0.100709	0.091360	125
HLAB*7	0.0534	0.0667	34	0.101009	0.094271	131
HLAA*3	0.0535	0.0804	43	0.101058	0.092933	132
HLAA*1	0.0543	0.0782	41	0.102642	0.094612	131
HLAB*12	0.0566	0.0638	30	0.106691	0.099884	131
PGD*C	0.0567	0.0703	35	0.106915	0.099401	241
DI*A	0.0567	0.1367	74	0.106917	0.092305	191
RH*D^u	0.0608	0.0715	33	0.114151	0.105988	91
HLAB*22	0.0608	0.1871	87	0.114229	0.092852	126
HLAB*16	0.0620	0.1040	56	0.116341	0.104247	114
HLAA*31	0.0638	0.1665	83	0.119330	0.099462	89
MNS*NS	0.0691	0.0527	18	0.128539	0.121762	345
HLAA*10	0.0695	0.1062	55	0.129279	0.115551	132
HLAA*11	0.0749	0.1380	72	0.138570	0.119451	132
IGHG1G3* za;b0b1c3c5	0.0763	0.1265	66	0.140032	0.122322	53

(continued)

Table 2.12.1 *(continued)*

IGHG1G3						
za;b0stb3b5	0.0880	0.1393	71	0.160144	0.137828	103
HLAB∗15	0.0955	0.1123	55	0.172627	0.153239	125
HLAB∗5	0.0956	0.0827	32	0.172811	0.158521	130
IGHG1G3						
za;b0sb3b5	0.0959	0.1809	84	0.171132	0.140175	17
IGHG1G3						
za;b0b1c3b4b5	0.1008	0.2138	91	0.179612	0.141216	24
IGHG1G3						
zax;g	0.1019	0.1162	57	0.182973	0.161718	241
RH∗cde	0.1026	0.1838	85	0.183858	0.150070	409
HLAB∗35	0.1028	0.1263	63	0.184285	0.161015	125
RH∗cDe	0.1173	0.3626	98	0.206253	0.131456	409
ABO∗B	0.1177	0.0722	20	0.207616	0.192616	477
C3∗F	0.1203	0.0611	13	0.211107	0.198212	56
ABO∗A1	0.1460	0.0724	18	0.249267	0.231221	362
MNS∗MS	0.1472	0.1220	57	0.250980	0.220372	345
HLAB∗40	0.1549	0.1328	63	0.261740	0.226977	132
ABO∗A	0.1731	0.0640	10	0.286166	0.267849	477
LPA∗Lp(a+)	0.1800	0.0822	23	0.294319	0.270130	21
PGM1∗1	0.1825	0.0510	4	0.297358	0.282184	29
RH∗cDE	0.1854	0.1495	71	0.301913	0.256771	409
FY∗0	0.2002	0.7831	99.9	0.313690	0.068033	97
IGHG1G3						
fa;b0b1b3b4b5	0.2031	0.5622	99.9	0.322648	0.141253	188
GC∗2	0.2121	0.0538	4	0.334026	0.316066	230
MNS∗S	0.2158	0.1226	55	0.338277	0.296821	351
HP∗1S	0.2167	0.0391	0	0.337930	0.324710	30
FY∗B	0.2201	0.2520	94	0.341163	0.255176	99
KM∗(1&1,2)	0.2218	0.1438	68	0.344899	0.295311	211
ACP1∗A	0.2362	0.0893	28	0.360587	0.328376	262
BF∗F	0.2444	0.1032	39	0.368053	0.330065	39
HLAA∗9	0.2486	0.1759	81	0.373347	0.307663	132
HLAA∗2	0.2502	0.0889	27	0.375001	0.341675	132
GLO1∗1	0.2687	0.1039	39	0.392672	0.351858	102
IGHG1G3						
za;b0b1b3b4b5	0.2910	0.4926	99.9	0.407946	0.206976	120
IGHG1G3						
f;b0b1b3b4b5	0.3118	0.4488	99.9	0.426996	0.235343	139
AG∗X	0.3639	0.2495	95	0.460605	0.345679	33
MNS∗Ms	0.3755	0.1246	54	0.468801	0.410409	345
IGHG1G3						
za;g	0.3854	0.3326	98	0.473171	0.315784	252
GC∗1S	0.3960	0.1182	50	0.476544	0.420208	87
MNS∗Ns	0.4032	0.2166	91	0.481048	0.376840	345
GC∗1F	0.4057	0.1983	86	0.478975	0.383980	87
P1∗1	0.4198	0.1370	62	0.486729	0.420050	337
JK∗B	0.4408	0.0760	16	0.491749	0.454395	63
HP∗1	0.4473	0.1579	73	0.494149	0.416144	373
JK∗A	0.4993	0.0688	10	0.499565	0.465205	184
MNS∗M	0.5379	0.1969	86	0.496992	0.399153	442
LE∗Le	0.5528	0.1266	56	0.493476	0.431004	148
RH∗CDe	0.5626	0.2539	95	0.491556	0.366730	409
PTC∗T	0.5726	0.1109	44	0.488759	0.434559	157
FY∗A	0.5970	0.3325	99	0.480143	0.320491	278
GPT∗1	0.6061	0.1348	61	0.476641	0.412374	105
SE∗Se	0.6278	0.1875	84	0.466832	0.379299	174
ABO∗O	0.7090	0.1088	43	0.412498	0.367637	477

(continued)

Table 2.12.1 *(continued)*

*BF*S*	0.7146	0.1334	61	0.405454	0.351356	39
*ALPP*S1*	0.7227	0.0357	0	0.399903	0.385609	29
*ACP1*B*	0.7488	0.0930	30	0.375902	0.340955	262
*PGM1*1*	0.7848	0.0624	8	0.337641	0.316559	278
*ESD*1*	0.7966	0.1103	46	0.323752	0.288048	172
*C3*S*	0.8767	0.0604	12	0.215678	0.202657	56
*RH*D*	0.8795	0.1776	83	0.211734	0.174128	443
*ADA*1*	0.9347	0.0540	20	0.122083	0.115487	162
*PEPA*1*	0.9363	0.2314	93	0.118232	0.090869	67
*PGD*A*	0.9416	0.0721	35	0.109931	0.102005	241
*TF*C*	0.9668	0.0792	56	0.064246	0.059156	338
*PI*M*	0.9680	0.0361	35	0.061888	0.059652	71
*PEPC*1*	0.9694	0.1149	74	0.058537	0.051814	29
*AK1*1*	0.9706	0.0488	43	0.056973	0.054191	205
*PGM2*1*	0.9802	0.0695	80	0.038722	0.036031	139
*PGK1*1*	0.9838	0.0804	82	0.031951	0.029383	48
*CHE1*U*	0.9876	0.0199	65	0.024468	0.023982	69
*SOD1*1*	0.9924	0.2968	99.9	0.015107	0.010623	47
*LDH*A*	0.9957	0.0197	65	0.008617	0.008447	125
*PEPB*1*	0.9976	0.0123	62	0.004732	0.004674	84
*MDH1*1*	0.9981	0.0242	67	0.003766	0.003675	116

Calculations of local homozygosity were also made for each locus and each map point from the gene-frequency maps, and averaged over all loci. A map of homozygosity was thus generated (fig. 2.12.1).

Average homozygosity is highest in the more isolated populations. A world map cannot show all the details. In some cases small or highly endogamous populations are located near others that are not small or highly endogamous, and the map cannot resolve the difference but merely shows the local heterogeneity. In general, as drift becomes more powerful, homozygosity increases. Hence, figure 2.12.1 is to a large extent, a map of drift, but other factors may also determine homozygosity.

Homozygosity varies in the world map from 0.63 to 0.79. The lowest values (and hence highest heterozygosities) are in the region that is geographically central to the history of population movements; the Middle East, extending to southern, central, and eastern Europe and western Asia. This is very reasonable, since the region of confluence of the diverse human populations must have the highest heterozygosity. There is also a possible source of bias that cannot be evaluated today but should not be forgotten; most human polymorphisms were found in Europeans, and results might be different if the initial search for polymorphisms had been carried out in another population. However, the majority of polymorphisms other than *HLA*—for example, most blood-group systems and electrophoretic polymorphisms of enzymes—were found by British scientists using samples primarily of the London population. The United Kingdom does not have the highest heterozygosity, and in fact is in the second homozygosity class with

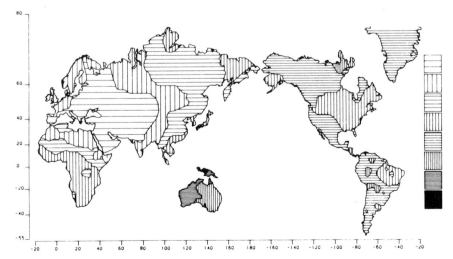

Fig. 2.12.1 A world map showing average genetic homozygosity.

the rest of northern Europe, most of India, and northern and eastern Africa. *HLA* has also been studied mostly in Europeans and people of European origin. The fractions of "blanks" (the sum of all unknown alleles at a locus, *A* or *B*), are expectedly lower for Europeans than for Orientals or Africans, but without a direct search of polymorphisms independent of ethnic origin, one cannot state which fraction of the differences is due to this bias and which is real. Since RFLPs are usually fully codominant, they may offer an answer to the question of blanks and of selection in a particular ethnic group. Unfortunately, all RFLPs examined until now have been selected almost entirely in Europeans. Heterozygosity for RFLPs is also higher in Europeans than in all other groups (Bowcock et al. 1987, 1991), but the difference is not too high and is quite comparable to that observed here. In order to avoid this bias in the future, new DNA polymorphisms must be selected from five or six different ethnic groups, ranging from Africans to Australians.

Northeast Asia has most of its homozygosity frequencies in the third class, increasing to the fourth in the extreme north and northeast (where Arctic living conditions generate considerable isolation) and in the more peripheral eastern and southeastern ends (Japan, South China, Southeast Asia). In New Guinea and Australia, there is a fairly abrupt transition to the highest homozygosities observed in the world, probably because of greater drift in the early settlements (fewer founders) and a smaller contribution from later migrations than in all other parts

of the world. America has higher homozygosity than East Asia, indicating that the migrations from Northeast Asia also involved a loss of heterozygosity because of the limited number of founders (or later effects of drifts), which was not, however, as marked as for Oceania. The slightly higher heterozygosity in the east-central region of North America is probably due to admixture with Caucasoids.

Homozygosity for all non-*HLA* loci is only slightly higher than the overall average given above; the map is not given, being almost superimposable on that of figure 2.12.1. The two *HLA* loci have much lower homozygosity, as expected, as they have many more alleles than the non-*HLA* loci but much greater variation is found in local *HLA* homozygosities: from 0.10 to 0.51. The general picture remains the same, but there is less difference between Africa and Europe + West Asia, or between the Americas and Australia.

The observation that heterozygosity is greatest among Caucasoids is certainly linked with the finding that Caucasoids are genetically intermediate between Africans and East Asians (see sec. 2.3 and 2.4) and are believed to derive from admixtures between ancestral populations giving rise to these ethnic groups. This hypothesis can explain both findings of a higher heterozygosity of Caucasoids and of their origin by admixture. The problem of bias resulting from selection of European polymorphisms, nevertheless, must be settled before one can put complete confidence in any conclusions.

2.13. CORRELATIONS WITH CLIMATE

In a previous study (Piazza et al. 1981b), we systematically analyzed the variation of gene frequencies and their principal components associated with latitude and longitude. Longitude showed a major effect, not surprisingly, given that most of human genetic variation is placed on an east-west axis that reflects the most important population movements. This contrasts with results from anthropometric measurements (see sec. 2.2). Some individually tested genes, however, were correlated with distance from the equator, indicating a climatic effect. Another study of correlation with climate was made for *HLA* (Piazza and Menozzi 1984) and for other markers (Piazza and Menozzi, unpubl.) separating temperature and humidity. This confirms an earlier finding by Piazza et al. (1980) who show that latitude plays a significant role, responsible for perhaps 10% of the *HLA* total genetic variation. The dissection of climate effects from history is not easy, but correlations with climate, if one can eliminate historical causes from them, are very likely to depend on selective effects.

We have repeated a similar analysis on the more abundant data collected in the present survey. Climatic data from 1473 localities (Great Britain Meteorological Office

1968) were subjected to a principal-components analysis. The 10 climatic variables employed are listed in table 2.13.1.

A principal-component analysis of the 10 climatic variables showed that 95% of the variation was explained by the first three PCs. Because January and July

Table 2.13.1. First Three Principal Components of Climate: the Percentage of Variance Explained and Correlations (x 100) with the T Climatic Variables

1. Maximum temperature in January	2. Minimum temperature in January
3. Maximum temperature in July	4. Minimum temperature in July
5. Maximum annual temperature	6. Minimum annual temperature
7. Maximum annual humidity	8. Minimum annual humidity
9. Average annual precipitation	10. Number of rainy days in a year

P.C.*	% of Variance Explained	Climatic Variables									
		1	2	3	4	5	6	7	8	9	10
1	51	91	86	92	91	98	92	-60	-66	14	-58
2	28	33	45	-20	13	18	37	70	67	93	70
3	16	-21	-16	23	36	-8	-2	16	20	4	-18
Total	95										

* P.C., Principal component.

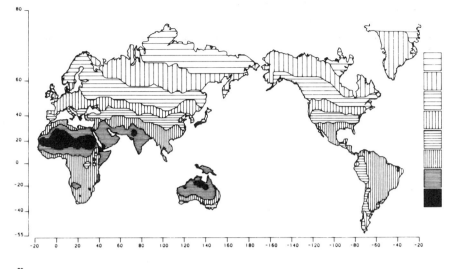

Fig. 2.13.1 Geographic distribution of the first principal component of climatic data. The range between the maximum and minimum values of the PC has been divided into eight equidistant classes. The direction of increase of PC values is arbitrary.

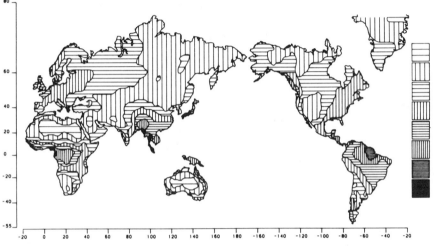

Fig. 2.13.2 Geographic distribution of the second principal component of climatic data. The range between the maximum and minimum values of the PC has been divided into eight equidistant classes. The direction of increase of PC values is arbitrary.

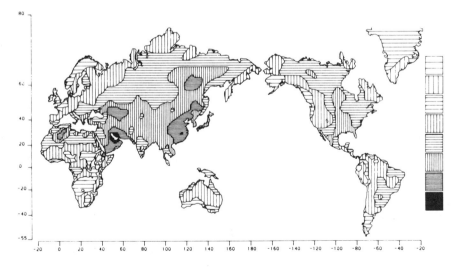

Fig. 2.13.3 Geographic distribution of the third principal component of climatic data. The range between the maximum and minimum values of the PC has been divided into eight equidistant classes. The direction of increase of PC values is arbitrary.

temperatures have opposite meanings north and south of the equator, they were exchanged when going from the northern to the southern hemisphere. The PC correlations with the 10 climatic variables are also given in table 2.13.1. The geographic distribution is shown in figures 2.13.1–2.13.3.

The first PC shows very strong correlations with ab-

solute temperatures, the second is most highly correlated with humidity, and the third is correlated with seasonal variation (the temperature difference between January and July).

The study of correlations between the gene frequencies of the 122 genetic markers and the climatic PCs has shown only 5 correlations above 0.50 for the first PC,

Table 2.13.2. Genes Showing the Highest Correlations with the First Three Principal Components of Climate

P.C.[*]	Range of Correlation Coefficient	Genes
1	0.70 – 0.60	(+) ACP1*B
		(–) RH*cDE
	0.60 – 0.50	(+) BF*F
		(–) HLAB*27, RH*E
	0.50 – 0.40	(+) IGHG1G3*za;b0b1b3b4b5, RH*Dᵘ, GPT*1
		(–) HLAB*5, MNS*M, HLAA*2
	0.40 – 0.30	(+) RH*cDe, HLAB*13, HLAA*10, MNS*Ns, HLAB*22
		(–) MNS*MS, MNS*Ms, MNS*S, TF*C
2	0.50 – 0.40	(+) HP*1
		(–) ABO*A, GC*1, LE*le
	0.40 – 0.30	(+) HLAB*15, DI*A
		(–) HLAA*10, ABO*A1
3	0.60 – 0.50	(+) HLAA*11
		(–) —
	0.50 – 0.40	(+) ABO*B
		(–) JK*A, ADA*1
	0.40 – 0.30	(+) AG*X, RH*CDe, HLAA*33, HLAB*5, MNS*NS, TF*C
		(–) AK1*1, GC*1, MNS*He, RH*cDe, KM*(1&1,2), PGD*A

Note.– Genes giving positive or negative correlation values are indicated by (+) or (–), respectively.
[*] P.C., Principal component.

Table 2.13.3. Correlations between Principal Components of Climate and of Gene Frequencies (x 1000)

Gene PC	Climate PC 1	2	3
1	-49	3	17
2	-483	126	166
3	313	48	-463
4	-110	-575	-249
5	137	125	-272
6	565	106	241
7	271	-224	145

none for the second, and 14 for the third. Altogether 19 correlations (of 366 studied) were above 0.30. More work is necessary to give a valid significance threshold, because the grid points (the interpolated gene-frequency values for each gene map) as well as the climatic PCs are all intercorrelated. We limit our consideration to the highest correlations, which are most likely to be highly significant by any criterion.

The genes giving correlation coefficients higher than 0.30 are listed in table 2.13.2. Several of them were found to be significant in the previous analyses (Piazza et al. 1981b; Piazza and Menozzi 1984 and unpubl.). One of the highest correlations with temperature observed is for Acid Phosphatase (ACP1), which has been reported previously (Ananthakrishnan and Walter 1972).

The correlation of the second climatic PC (tropical humid climate) with the Diego blood group (DI*A) confirms earlier findings by Wilson and Franklin (1968).

A study of the correlations between principal components of climate and those of gene frequencies shows (table 2.13.3) that the second PC of genes has an important correlation with the first PC of climate. The correlation is negative because the second PC of genes is lowest where the temperature is highest. In fact, the first PC of climate takes its highest values in northern Africa,

northern Australia, and northern India. The first climatic PC is also positively correlated with the sixth genetic PC, which has its lowest values in extreme northern regions. The second climatic PC, which is highest in regions with tropical forests and therefore corresponds to a hot, humid climate, correlates negatively with the fourth genetic PC, which is lowest in many hot and humid regions. Finally, the somewhat enigmatic third climatic PC, which peaks in some hot and humid regions, in particular the Persian Gulf, correlates negatively with the third genetic PC, which has some of its lowest values in humid areas. Other areas known to be hot and humid have high values for the third genetic PC, making the interpretation of this climatic pattern relatively obscure.

Some of the correlations between genes and climate are probably spurious, given that it is difficult to separate historical from ecological sources of association. Some evidence may come in the future from physiological information. Another method of analysis of potential effects of natural selection, F_{ST}, might be useful for confirming whether some of these genes have variation above that expected on a random basis. One cannot expect the agreement to be high, for a variety of reasons. Climate is not the only source of natural selection. Moreover, history may cause a discrepancy between the two approaches. For instance, $FY*O$ shows a very strong variation by F_{ST} and a dubious correlation with climate. The reason for the latter is that $FY*O$ is found at high frequencies in tropical Africa but not in other tropical areas of the world. The gene probably arose relatively recently in Africa after ethnic separation and did not spread elsewhere, or the resistance conferred against malaria does not extend to non-African areas. Other interpretations of the evolution of $FY*O$ have been offered (see sec. 2.10).

A systematic attempt was made at correlating the indication of disruptive selection available from F_{ST} values with that arising from correlation with climatic data. Because of the nonnormal nature of both distributions, the F_{ST} of a gene and its correlation with the first climatic PC (and, separately, its correlation with the second climatic PC) were replaced by their percentiles. A Pearsonian correlation was calculated between the per-

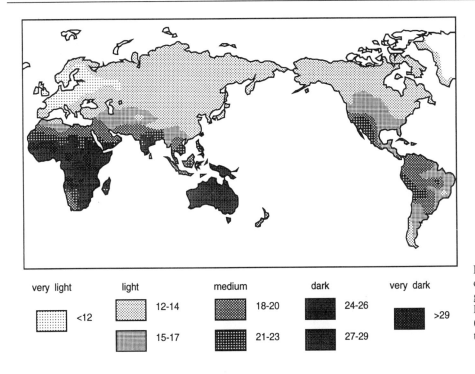

very light   light   medium   dark   very dark

<12   12-14   18-20   24-26   >29

15-17   21-23   27-29

Fig. 2.13.4 Geographic distribution of skin color, classified in eight grades according to the standard Luschan scale of skin darkness (after Biasutti 1959, p. 192, table 4).

centile corresponding to F_{ST} (given in table 2.12.1) and the percentile of the correlation for the same gene and a climatic PC. For both climatic PCs, the correlations did not differ significantly from zero. There is no indication, therefore, that the major factor determining high F_{ST}'s is climate.

Clearer correlations with climate originate for anthropometric traits. A classical trait is skin color, on which climate acts in many ways. By absorbing ultraviolet rays, a dark skin is advantageous in protecting against solar erythema and skin carcinomas, but it is a bar against production of vitamin D in the lower skin layers. Especially when the diet provides little or no vitamin D, a white skin is necessary in high latitudes. A diet especially poor in vitamin D may have been characteristic of the farmers who came to Europe from the Middle East and slowly made their way to Northern Europe, where they arrived about 5000 years ago. If this is true, then the white skin color of Northern Europeans evolved in

the last 5000 years from a light-brown color characteristic of Caucasoids from West Asia and North Africa. Skin color, however, is probably also under sexual selection which, according to Darwin, was of great importance in shaping the present looks of a.m.h. There is no possibility of investigating the tastes of people who have been dead for thousands of years, and there is little hope, if any, of being able to test satisfactorily Darwin's hypothesis, at least as far as the past is concerned. The present geographic distribution of skin color is shown in figure 2.13.4.

At least three or four different genes (Harrison and Owen 1964; Stern 1973) contribute to skin pigmentation. Such polygenic systems can be analyzed only superficially; there is also an interaction with customs, because skin color is affected by individual exposure to the sun. Molecular genetic approaches can, however, effect the resolution by linkage analysis or, better, by identifying the genes involved.

2.14. AREA AND TIME OF ORIGIN OF MAJOR MUTANTS, WITH SPECIAL ATTENTION TO HEMOGLOBINS

Our study has largely been restricted to major alleles, for which data were available in an adequate number of populations. The great majority of these alleles are present in all continents, with few exceptions. At least in some cases, the geographic distributions do suggest, however, possible places of origin of some alleles. In the absence of a study at the DNA level of the great majority of these alleles, there is potential confusion when phenotypic similarity hides genetic diversity. Even when DNA analysis shows identity, the possibility of repeated independent origin of the same mutation should be considered. With the number of individuals living today, and a mutation rate of 10^{-9} per nucleotide, the same mutation may occur in more than one individual every generation. But a few thousand years ago, the human population was much smaller than now. At the end of the Paleolithic, it is likely to have been one thousand times less. Recurrence of the same mutation can still be expected to have happened at these earlier times, but only if we consider the accumulation of the chance for mutations to occur over a period of a thousand generations.

A new mutation is not necessarily going to spread rapidly in a population, unless it has a strong selective advantage. Even so, it is subject to considerable drift in the first few generations, when there are only one or a few individuals carrying the new mutation. In fact, in this early period after its origin when the new mutation is exposed to strong drift, it may have approximately a one-third chance of being eliminated, even if it is very advantageous.

We dedicate special attention in this section to hemoglobin mutants and their origins. A full list is given by McKusick (1990). We did not include hemoglobin mutants among the markers used in our genetic analyses because, for most of them, their diffusion is clearly linked to the presence of malaria. They therefore mark the geographic distribution of this disease and only secondarily can be helpful in tracing historical origins. We report here maps of geographic distribution of the major alleles prepared by Guglielmino-Matessi et al. (unpubl.) on the basis of the data given in the tabulations by Livingstone (1985), to which we also refer.

We first consider the most important polymorphism, hemoglobin S, which was also the first example of heterozygous advantage of a single gene detected in humans or described in detail in any living organism. The original suggestion of the selective mechanism and its connection with malaria goes back to Haldane (1949) and is based on the parallel geographic diffusion of this parasite and the genetic disease. The first direct evidence was described by Allison (1954a, b). All hemoglobin S mutants found in the world are apparently identical not only at the protein level, but those investigated, at least, involve the same nucleotides. The protein change is due to substitution of the sixth amino acid of the β chain of hemoglobin, glutamic acid, by valine, which deter-

mines a change in the physicochemical properties of the hemoglobin molecules. Unlike normal hemoglobin (called A), hemoglobin S tends to crystallize at low oxygen tension, to which red cells are occasionally subjected, causing "sickling" of the red cell. This phenomenon is responsible in S homozygotes for the disease known as sickle-cell anemia. S homozygotes are not, however, protected against malaria. The AS heterozygote is normal clinically but, having about half hemoglobin S and half A, is partially protected against *Plasmodium falciparum*, the most dangerous malarial parasite. The precise mechanism by which the heterozygous state confers protection is not yet completely clear (Luzzatto et al. 1985), but the fact itself is well ascertained.

The hemoglobin S mutant can spread rapidly in the presence of malaria but tends to stabilize around the gene frequency of 10%. Individuals homozygous with the hemoglobin S mutation have a high probability of dying before maturity, and hence the S allele cannot replace the normal A allele. Although a balanced polymorphism will originate in the presence of malaria, in its absence, the mutant is eliminated or remains at a low frequency. Calculations have shown that the gene frequency observed—for example, in West Africa—is likely to be close to the equilibrium frequency, because it is precisely the stable value expected with the natural selection (the Darwinian fitness of the AA, AS, SS genotypes) observable today in malarial areas. The relative fitnesses of the three genotypes in the presence of malaria were calculated by comparing the frequencies of the genotypes in young children and in adults, that is, before and after selection. Starting from one mutation in a population of 50,000 individuals, the process for reaching the equilibrium value takes about 2000 years (Cavalli-Sforza and Bodmer 1971a). This is a minimum time, however, and genes for sickle-cell anemia may have been around much longer.

The map of the present gene-frequency distribution of the hemoglobin S gene is shown in figure 2.14.1. There

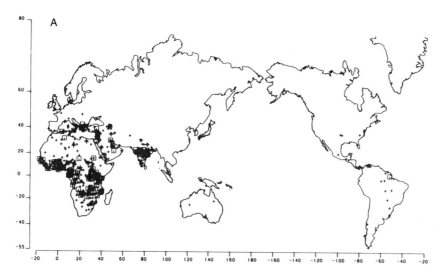

Fig. 2.14.1.A Geographic distribution of hemoglobin S available data points in the world. Symbols are described in table 1.14.1.

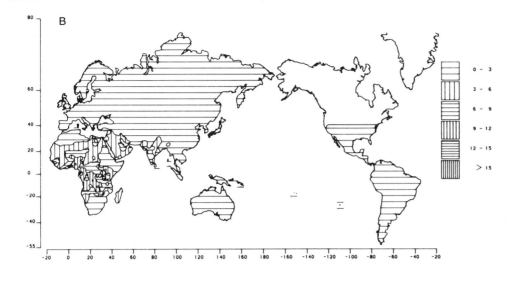

Fig. 2.14.1.B *Geographic distribution of hemoglobin S in the world. Numbers indicate the gene frequency (%).*

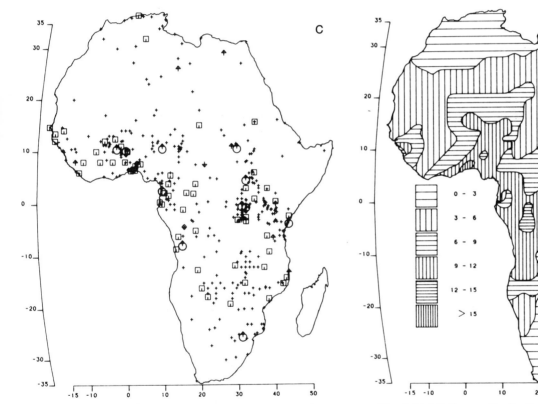

Fig. 2.14.1.C Geographic distribution of hemoglobin S data points in Africa. Symbols are described in table 1.14.1.

Fig. 2.14.1.D Geographic distribution of hemoglobin S in Africa. Numbers indicate the gene frequency (%).

are several peaks of gene frequencies on the geographic map. It is customary to distinguish at least four types called Bantu (East and Central Africa), Benin (Nigeria), Senegal (extreme West Africa), and Saudi (Saudi Arabia and India). The four types differ in the degree of severity, possibly because of other associated and closely linked mutations, because the mutations of the hemoglobin S gene are indistinguishable except for closely associated RFLP markers (Antonarakis et al. 1982, 1984a, b; Pagnier et al, 1984; Kuzolik et al. 1986; Chebloune et

al. 1988). The constellation of these markers in chromosomes carrying the S mutation—that is, the associated haplotype—is in general similar to that in chromosomes of the same geographic region not carrying the S gene. Local haplotypes may vary considerably from one population to another (see, e.g., table 2.4.1); S mutations are observed in more than one haplotype in the same region, indicating that the mutation arose some time ago. When S mutations from widely different locations are found in different haplotypes, usually those locally com-

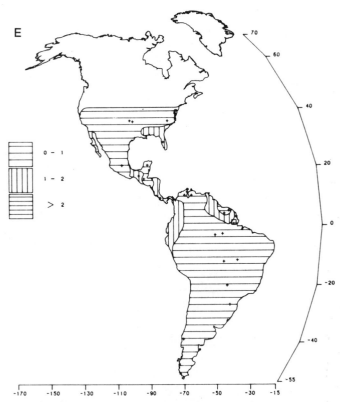

Fig. 2.14.1.E Geographic distribution of hemoglobin *S* in the Americas. Numbers indicate the gene frequency (%).

mon in non-*S* haplotypes, this is taken as evidence that they result from independent *S* mutations. This conclusion is reasonable, but weakened by the possibility that crossing-over, or more probably gene conversion, which may be more effective than ordinary crossing-over, has introduced an *S* mutation of foreign origin into a local haplotype.

Two thousand years ago, only a hundred million individuals or so might have lived in the malarial areas of the world (see table 2.1.2). Between, say, 3000 and 2000 years ago, there would have been a good chance that more than one mutation arose among these people even if the mutation rate were as low as 10^{-9}. The origin of some of these mutations may also be much older; 2000 years is a minimum age. There is no complete assurance from haplotype studies that separate *S* mutations have occurred, because crossing-over or gene conversion occurs more frequently than mutations and could bring the mutation introduced by an immigrant individual into the most common local haplotypes. A simulation (Livingstone 1989) indicates that enough time has lapsed for a single sickle-cell mutation that initially occurred in Africa to migrate to India. There is however, no adequate study of the time needed for intermingling at the haplotype level.

Another hemoglobin, C, occurs commonly in West Africa in part of the area of spread of one type of S hemoglobin (fig. 2.14.2). It involves replacement of the

same amino acid changed in hemoglobin S (the sixth of the β chain, glutamic acid), but to another amino acid (lysine). On the basis of RFLP haplotypes there was only one mutational origin of hemoglobin C. The question of its role in protection against malaria has been debated (Cavalli-Sforza and Bodmer 1971a; Livingstone 1976).

Hemoglobin D is polymorphic in Africa and in South Asia (fig. 2.14.3). The D label originates from the electrophoretic classification, and in this case as in others, it hides differences that were later detected at the protein level. At least two major hemoglobin D mutants have achieved polymorphic status in different parts of the world. Hemoglobin D Punjab is a replacement of glutamic acid with glutamine at position β 121, and is spread fairly widely in South Asia, with a probable center of origin in northwestern India. Hemoglobin D Ibadan involves replacement of threonine with lysine at β 87. The highest frequency is in West Africa and might mark the center of origin. A different hemoglobin D (Bushman) is centered in northern Namibia and involves the replacement of glycine with arginine at β 16. Other D hemoglobins have less well characterized origins. It is difficult to say if they were spread by malarial resistance. Current information on the prevalence of this disease may not correspond to the geographic distribution of earlier times. The coincidence of the area of spread of hemoglobin D in West Africa with a probable area

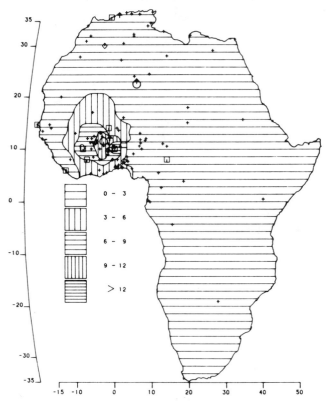

Fig. 2.14.2 Geographic distribution of hemoglobin *C* in Africa. Numbers indicate the gene frequency (%).

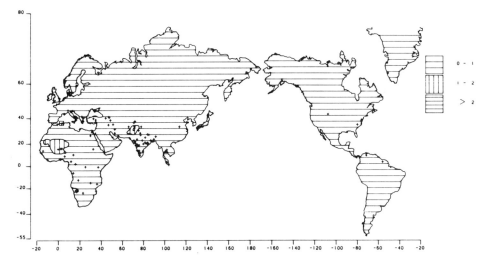

Fig. 2.14.3 Geographic distribution of hemoglobin *D* in the world. Numbers indicate the gene frequency (%).

of demic diffusion resulting from agricultural development in West Africa (discussed further in chap. 3), suggests another possible mechanism of diffusion. This polymorphism may have been common under drift (or selection) in the farmers' population that spread from this area in the early Neolithic. Some indication that may enable us to choose between the hypothesis of spread of a neutral or nearly neutral polymorphism and protection against a specific disease like malaria can come from study of the fitness of an adequate number of homozygotes for the mutant in the presence of malaria. But the numbers of individuals necessary for obtaining significant results are extremely large (Cavalli-Sforza and Bodmer 1971a).

Hemoglobin E is common in Southeast Asia (fig. 2.14.4). It is caused by replacement of glutamic acid by lysine at β 26. Two independent mutations are suggested by haplotype studies (Antonarakis 1982). Evidence in favor of malarial protection comes from the observation that there is a difference in frequency of hemoglobin *E*

in populations living short distances apart but in ecologically different areas: hilly and mountainous versus plains regions. In the Mediterranean, malaria is common in swampy coastal areas and rare in mountainous ones, but in Southeast Asia, *Anopheles minimus*, the vector of the malarial parasite, is adapted to forests in hilly and mountainous areas. Hemoglobin E is more frequent in these areas than in the plains, in parallel with the prevalence of malaria (Flatz 1967).

Another important group of hemoglobinopathies with a geographic distribution closely paralleling malaria is that of the serious hereditary anemias called *thalassemias*. The group name comes from the Greek word for sea (*thalassa*), because they were first found in people of Mediterranean origin. There is usually no structural change of the hemoglobin molecule. In the two most important classes of thalassemia, red blood cells have low or zero concentrations of one or the other of the two chains, α and β, forming the normal hemoglobin

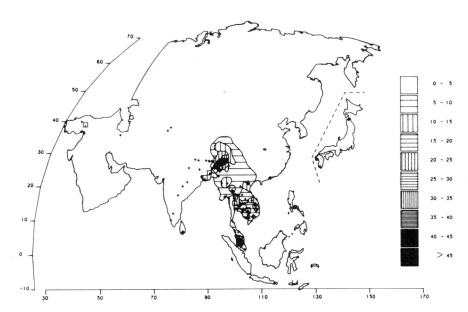

Fig. 2.14.4 Geographic distribution of hemoglobin *E* in Asia. Numbers indicate the gene frequency (%).

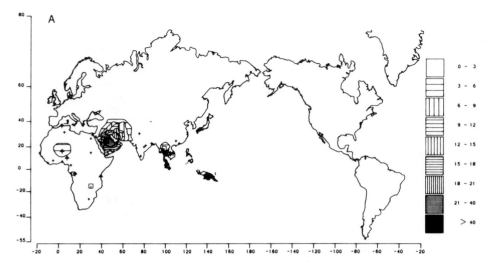

Fig. 2.14.5.A Geographic distribution of α-thalassemia in the world. Numbers indicate the gene frequency (%).

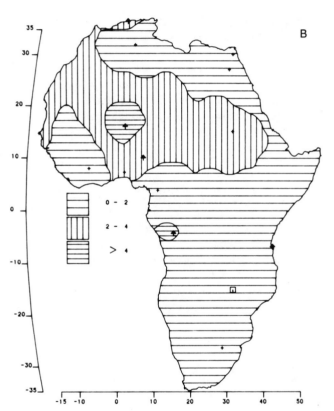

Fig. 2.14.5.B Geographic distribution of α-thalassemia in Africa. Numbers indicate the gene frequency (%).

molecule. The α-thalassemias (figure 2.14.5) are the result of a deletion of one or both of the α-genes, which are present in duplicate on normal chromosome 16. One of the two genes (α2) is more easily affected by polymorphic deletions. A normal individual has four α genes; a homozygote for the deletion of both α1 and α2 undergoes fetal death. The disease is usually observed in patients with a total of two α chains, for example, a homozygote for the deletion of one chain or a heterozygote for the double deletion. Ordinarily, an individual with three chains is clinically normal. Although the evidence is indirect, α thalassemia probably protects from malaria. In a south Nepalese population, the Tharus, an α-thalassemia mutation, has reached a frequency of 80%; it is estimated to confer a 10-fold resistance to malaria in the homozygote. This is by far the highest frequency ever observed in a population and became available too recently for inclusion in the map (G. Modiano et al., pers. comm.). α thalassemia is widespread in various parts of Africa, the Middle East, Arabia, India, and Southeast Asia.

Thalassemias of the β chain (chromosome 11) have been generated by a great variety of mutations (fig. 2.14.6 A–D). In some, called β^0, no β globin is made. One of these, $\beta^0 39$, is common in the Mediterranean and especially in Sardinia: the β-chain production is stopped by a terminator mutation. Other different mutations of β^0 type are found, for instance in India and China. Most mutations are of β^+ type and produce a certain amount of β globin. There is an extraordinary variety of β^+ mutants, all affecting in different ways the regulation of β globin production and leaving intact the globin structure. The clinical picture is less serious than that of the β^0 mutants. The overall geographic distribution resembles that of α, but β thalassemia is more common in Europe and Africa, and less common in East and Southeast Asia. There is usually a clear-cut parallelism between its geographic distribution and that of malaria, keeping in mind that this disease was once common but has been completely eradicated, for instance, in many parts of Europe. The geographic distribution in the Mediterranean, where the disease is frequent has been studied rather intensively. Figure 2.14.7 shows the main locations where mutants identified at the molecular level are found. There is a correlation between the early migrations of Phoenicians, Carthaginians (see sec. 5.2), and Greek colonists, which are known historically, and the distribution of some major molecular mutants. Haplotypes (not given) also help in the detection of major migrations. This correspondence sets the date of origin of most mutants, or at least of their arrival in

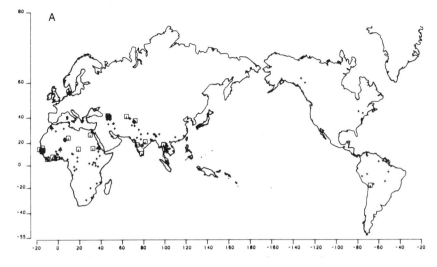

Fig. 2.14.6.A Geographic distribution of β-thalassemia data points in the world. Symbols are described in table 1.14.1.

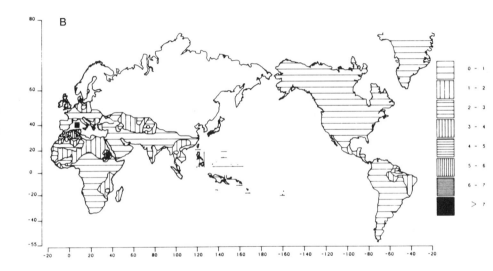

Fig. 2.14.6.B Geographic distribution of β-thalassemia in the world. Numbers indicate the gene frequency (%).

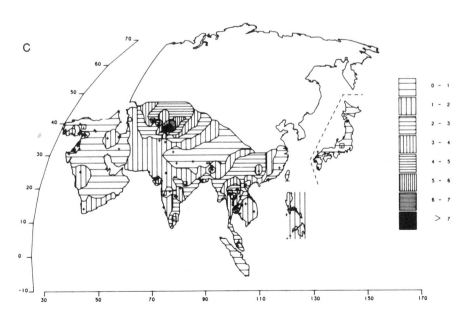

Fig. 2.14.6.C Geographic distribution of β-thalassemia in Asia. Numbers indicate the gene frequency (%).

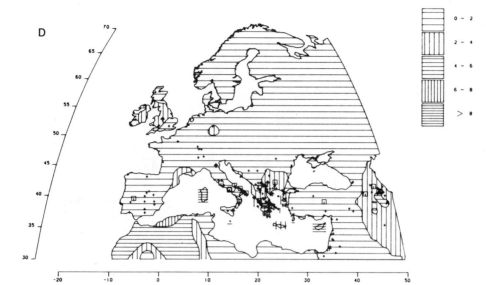

D

Fig. 2.14.6.D Geographic distribution of β-thalassemia in Europe. Numbers indicate the gene frequency (%).

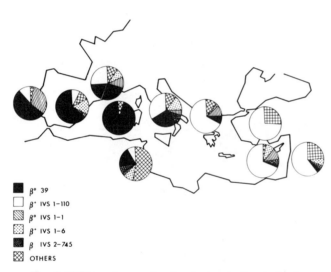

β° 39
β⁺ IVS 1-110
β° IVS 1-1
β⁺ IVS 1-6
β IVS 2-745
OTHERS

Fig. 2.14.7 Locations and molecular types of mutants of β-thalassemia in the Mediterranean (from Cao et al. 1989).

the colonies established by these people, at 2500–3000 years ago. Several mutations almost certainly originated earlier than 3000 years ago in the places of origin of the colonists.

The region between the β- and the δ-globin genes is responsible for the switch from the production of γ globin, contained in the fetal hemoglobin to that of β-globin, characteristic of the adult. The switch is gradual and occurs around the time of birth. Certain deletions of the β-δ region and other mutations in this region cause the persistence of fetal hemoglobin, which completely replaces adult hemoglobin in the homozygote. Heterozygotes have a proportion of fetal hemoglobin, which, in a

common β-δ deletion (HPFH or hereditary persistence of fetal hemoglobin) amounts to 17%–35% of the total in the adult. This condition, clinically benign, relieves some of the symptoms of sufferers of β-thalas-semia and other β-hemoglobin defects including sickle-cell anemia. The geographic distribution of elevated fetal hemoglobin is shown in figure 2.14.8, and it tends to mimic to some extent the distribution of β-thalas-semia.

Hemoglobin mutants all have restricted areas of distribution and usually confer a strong selective advantage on heterozygotes and a strong disadvantage on homozygotes. Their frequencies can therefore increase rapidly, but must stop as soon as a relatively low equilibrium frequency has been reached. The mutants are, of necessity, confined to areas of diffusion of the parasite against which the heterozygote is protected. This is not necessarily the case, however, for hemoglobin C; see fitness estimates in Cavalli-Sforza and Bodmer (1971a), and in vitro experiments cited by Luzzatto et al. (1985).

Another gene that was shown by in vitro experiments to give total protection against the malarial parasite *Plasmodium vivax* is Duffy, in particular, the *FY*0* allele, which we have already discussed in section 2.10. Here the resistant genotype is the homozygote, with a completely different dynamics.

Glucose-6-phosphate dehydrogenase (*G6PD*) has given rise to several partially deficient mutants (*G6PD*def*) which confer resistance to malaria (see also sec. 2.10). It is sex-linked, and therefore heterozygotes exist only in females. Males are either *G6PD* deficients or normal, but the *G6PD* deficiency state does not confer protection against malaria. Rather, *G6PD*

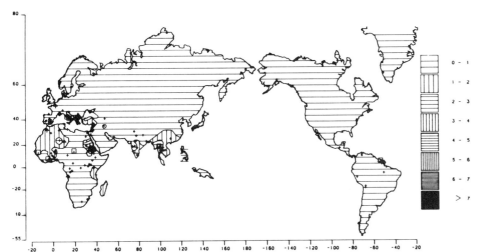

Fig. 2.14.8 Geographic distribution of elevated fetal hemoglobin in the world.

deficiency individuals may have lower fitness because they are prone to neonatal jaundice, and in regions where malaria is endemic have even higher parasite counts than do normal individuals; heterozygous women, however, have lower parasite counts and are protected against *P. falciparum* (Luzzatto et al. 1985).

Maps of *G6PD* deficiency are shown with those of other genes for the World, Africa, Asia, Europe. The deficiency is observed also in New Guinea and in insular Southeast Asia. On the northeastern coast of South America, *G6PD* was imported most probably by African slaves. The map of the geographic distribution of the *G6PD*∗*A−* is somewhat different from that of the *G6PD*∗*def* type, coming from independent data.

Many other genes may be involved in resistance to malaria, but those listed above are likely to be most important.

The malarial parasite must have been with humans for a long time, perhaps for the entire existence of the genus *Homo*, given that various species of malarial parasites are common in many other mammals including primates. The four species of malarial parasites in humans are not shared with other mammals. Various authors have suggested that the spread of agriculture has greatly favored the diffusion of malaria, because deforestation increases the chances of growth of malarial vectors. This can hardly be true, however, of Southeast Asia, where malaria persists in the forested areas. Another observation that shows that the diffusion of malaria in Central Africa must antedate the arrival of agriculture is that Mbuti Pygmies, who are hunter-gatherers and have only recently started practicing some agriculture, have one of the highest frequencies of hemoglobin S in the world (Cavalli-Sforza 1986). Malaria has been responsible for selecting a great number of mutants that confer resistance to it, not surprisingly because directly or indirectly it is probably the number one killer in tropical and subtropical areas. The present geographic distribution of malaria

is shown in figure 2.14.9. It should be noted that malaria was once more wide spread than it is today.

The situation is different for possible heterozygotus advantages observed in diseases unrelated to malaria. In the case of cystic fibrosis, heterozygous advantage has not been directly proved, but a possible mechanism for it has been suggested: the heterozygote may be more resistant to disturbances of saline metabolism caused by infantile diarrhea (Quinton 1982). Serious loss of sodium chloride is a major symptom of cystic fibrosis, and the cystic fibrosis gene is believed to be involved in chloride storage and elimination. The disease is common in all Europeans and perhaps other Caucasoids, and therefore may have originated at least 10 kya. One mutant allele (involving the deletion of three nucleotides

Fig. 2.14.9 Geographic distribution of malaria in the world before 1950 (after Allison 1961; data from Boyd 1949).

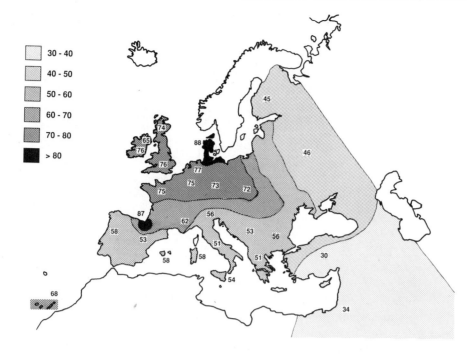

Fig. 2.14.10 Geographic distribution in Europe of the frequency (%) of Δ F-508 cystic fibrosis mutant, relative to other mutants determining the disease (data from European consortium on cystic fibrosis; modified from Devoto et al. 1990).

and therefore one amino acid) is very frequent, reaching approximately 70% of all known mutants determining the disease. Its distribution in Europe is shown in figure 2.14.10. The frequency at birth of the disease itself is less well known, but is found, on the average, in 1 of 2000 of all births in Europe. It varies somewhat from country to country, but the frequency of a genetic disease is not always easy to estimate accurately.

Very recent data indicate that the relative frequency of the common allele for cystic fibrosis (Δ F -508) is especially high in a relatively small sample of Basques (see fig 2.14.10). The geographic spread of the frequent allele is quite similar to the first synthetic map for Europe and therefore to that of the Mesolithic (pre-Neolithic) European gene pool. The conclusion would be that the most common cystic fibrosis allele in Europe is older than the spread of agriculture and must have been especially frequent among Mesolithics, while most other alleles come from the Middle East or originated in Europe after the Neolithic diffusion.

Many of the polymorphisms conferring resistance to malaria and other diseases may be relatively young; but the great majority of polymorphic alleles that we have studied in sections 2.10 and 2.13 are found in nearly all continents and therefore must have antedated the spread of a.m.h. Only a few are known at the DNA level, and it is impossible to say whether different mutations are hidden behind the same allele. The 100 RFLPs studied in section 2.4, however, are likely to have had a unique mutational origin, and the great majority of them are found in all continents (except the Americas which have not yet been studied for these polymorphisms). There are reasons to think that most human polymorphisms are much older than the date of spread from Africa. Theoretically, in the absence of selection, the average age of polymorphisms is comparable to that of the species. An estimate of the average time of appearance of the mutant allele of biallelic human RFLPs is 700,000 years (Mountain et al. unpubl.).

2.15. A BRIEF SUMMARY OF HUMAN EVOLUTION

The analysis of phylogenetic trees has used many different sets of data: nuclear polymorphisms tested by electrophoresis or immunological techniques, restriction fragment polymorphisms of nuclear genes, and mitochondrial DNA. All methods show a somewhat greater difference between Africans and non-Africans than between other human groups, and offer some information on dates supporting the interpretation that the origin of modern humans was in Africa, from which an expansion to the rest of the world started about 100 kya. The

interpretation rests on the assumption that evolutionary rates are reasonably constant. Some further comfort is derived from the consideration that the rates involved in the evolution of polymorphisms of nuclear genes and those of mtDNA are different; the first are determined mostly by differences in gene frequencies separating two populations, and the second by numbers of mutations separating two individuals. The dates to which they lead for the bifurcation of Africans and non-Africans are different, being of the order of 100 kya and 200 kya respectively. Given the ways in which these dates were obtained, this difference should be expected, with the second being greater than the first by an amount that has not been determined, but perhaps might be estimated on the basis of theoretical considerations. All in all, there is basic agreement between the trees obtained with mtDNA and with nuclear markers, the latter giving much more detail. It is worth remembering that naming the mitochondrial ancestor "Eve" has generated the false belief that there was a time when there was only one woman alive.

On the question of place of origin, the archaeological field is divided. A number of paleoanthropologists believe that modern humans originated in Africa, from which they spread to the rest of the world beginning about 100 kya. This is in agreement with the genetic data. A fairly large number of anthropologists reserve their opinion. Another group believes that the evolution of *Homo sapiens*, and perhaps even its predecessor *H. erectus* proceeded in parallel all over the Old World, and there was no expansion from Africa. The mitochondrial data are, at this point, the most useful in helping to reject this hypothesis, given that the origin of extant types of Asian mtDNA is more recent than this hypothesis would imply.

It is not yet possible, however, to exclude completely a partial participation of archaic *H. sapiens* from the Old World. New data and methods of analysis may help in this direction. What is very difficult to conceive is a parallel evolution over such a vast expanse of land, given the limited genetic exchange that could have occurred in earlier times. The capacity of the human genus to expand rapidly over a large fraction of the Earth's surface is more in tune with the idea of specific expansions from a nuclear area of origin. Such expansions must have been determined by some important advantage, biological or cultural. It is not difficult to accept the idea that the expansion of modern humans must have been strongly influenced by the possession of greater skills in communication by language. This increased ability to communicate is likely to have been extremely useful in favoring exploration and travel to unknown lands. Other technical improvements may have favored a trend to expansion. Although modern humans have now been found to have lived outside of Africa (in the Middle East) by about 100 kya, humans of this time in both Africa and the Middle East were biologically very similar to modern humans but culturally much less developed than at the time the real expansion began, perhaps 50 or 60 ky later. Many things may have happened in the meantime, in terms of cultural maturation and, perhaps, forward and backward movements between Africa and West Asia. Neanderthals are found in the Middle East after the earliest local appearance of modern humans in the same areas, and it has been suggested that they may have gained, or regained, lost ground in that period. The time between 100 and 50 kya (or, perhaps more exactly, between 90 and 60 kya) is currently a blank from an archaeological point of view. We hope that new discoveries will illuminate it. At the moment, the indications are that at the end of the blank period modern humans emerged with a new stone technology and started a radiation that took them to Europe, Australia and New Guinea, and America. Whether they partially mixed with or totally supplanted earlier inhabitants—for example, Neanderthals in Europe and archaic *H. sapiens* in East Asia—is difficult to state precisely on the basis of present knowledge.

Linguistic and cultural diversity increased conspicuously after that time, and the major linguistic families probably began less than 50 kya. Most of them are between 25 and 5 ky old. Genetic dating of linguistic families can only be approximate, but it agrees with ideas expressed by a few linguists. Moreover, the archaeological record shows increasing diversification, probably parallel with that of language.

An unsolved problem is determining the route by which the East was reached. Differences between East Asia and Southeast Asia make it reasonable to hypothesize that there might have been two routes, one through Central Asia and one through South Asia. Very little, if any, evidence of them exists today (fig. 2.15.1). The occupation of Australia and New Guinea was the major success story of the southern route, but it eventually led to an evolutionary cul-de-sac, as the separation between Oceania and Southeast Asia increased with the rising of the sea levels in the times after the last glaciation. It was only with the development of new nautical skills, 5000–6000 years ago, by South-East Asian populations who were also good farmers, that the Pacific routes were increasingly used. In the last 3000–3500 years, the expansion that generated the colonization of Polynesia began, most probably originating in a nuclear area in Southeast Asia.

There are two weaknesses in the present analysis, which will certainly require future work. One of them is the very short branch linking Caucasoids and, in particular, Europeans to the phylogenetic tree. One hypothesis is that they might have originated from an admixture between their southwestern and northeastern neighbors, Africans and Mongoloids, between which Europeans are sandwiched. One cannot completely exclude other hypotheses. Particularly serious is the possible bias resulting from the fact that almost all known genetic polymorphisms have been detected in Europeans. It will be important to remove this bias, especially in future data collections. Another area of doubt is the relationship between New Guineans + Australians, Southeast Asians and Northeast Asians. Our results have not settled this question unequivocally. It seems likely that the uncertainty arises because Southeast Asia is poorly known and may be heterogeneous, with some populations having an important genetic component in common with northern Mongoloids and others with people from Oceania. The

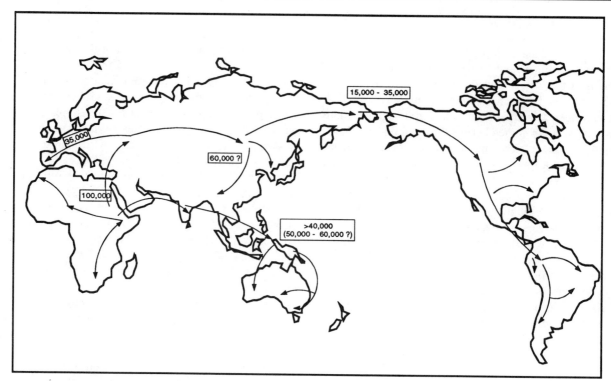

Fig. 2.15.1 Possible history and routes of expansion of modern humans in the last 100 ky.

heterogeneity may be in part due to ancient admixtures, and the arrow of northern Mongoloids pointing south in figure 2.15.1 express these considerations. There are also some undeniable physical similarities between northern and southern Mongoloids, leading one to wonder whether they have more in common than shown by the trees of sections 2.3 and 2.4. In other words, a fully dichotomous tree may be unsatisfactory in this part of the world, but more abundant and better evidence would be necessary for developing this explanation further.

The passage from Asia to America was later than that to Australia or Europe, perhaps because it first required a genetic and cultural adaptation to the more rigid climates of Northeast Asia. Genetic data, however, seem to agree with an early arrival, perhaps around 30 kya; possible uncertainties are discussed further in chapter 6.

Throughout the Paleolithic, population numbers remained small, leaving greater chance for random genetic drift to produce considerable diversification. Population size of a continental or subcontinental area at the beginning of expansion may have been on the order of 50,000–100,000 individuals. In the late Paleolithic, much of human action was in Asia, and the occupation of the rest of the world proceeded from this continent. Given the greater limitations on life in the north, Asia was like a relatively narrow, large landmass developed more in longitude than latitude. Because genetic divergence was subject more to random than selective forces, much of the gradient of the human gene pool goes from west to east. The first principal component therefore

extends in this direction and explains 35% of the total human variation, showing only moderate, if any, influence of climatic factors at the level of the nuclear genes investigated, but a greater influence on genetic factors involved in the adaptation of body build and bodily surface characteristics, which notoriously respond to climate. A dichotomy is thus observed between genetic data and observations based on the physical constitution, which is detectable also on modern and fossil bones. This explains the discrepancy between the evolutionary histories reconstructed from data on genes and on skulls (or, in general, anthropometric data).

Only in the last 10 ky, perhaps under increasing population pressure and climatic changes, did humans develop new food technologies, culminating in several different agricultural developments. These innovations caused the beginning of more rapid population growth, and in some cases of local expansions, which extended to ecologically similar areas, allowing the exploitation of domesticated plants and animals developed in the three major nuclear areas of agriculture. The consequent increases of population densities began a progressive freezing of drift effects. Farmers' expansions, followed by those of nomadic pastoralists, contributed in an important way to changing the patterns of gene geography. In spite of this, opportunities still remained for the survival of much local diversity, especially in refugia, few of which have been well examined.

A major conclusion is that linguistic and genetic evolution are closely related. In this chapter we have seen this relationship at the global level, but several investigations on specific regions or people that we examine in the following chapters have given similar results. The main reason for the relationship is that the evolution of both depend on the same historic and geographic factors. We have seen that discrepancies are not impossible, given that genes can be partially or even almost completely replaced under certain conditions, and languages can also be replaced. Language replacement is more likely to happen, perhaps, in recent history, and there are well-known examples of it. One can also express the necessity of a relationship between genetic and linguistic evolution (and, more generally, certain types of cultural evolution of which the evolution of language is a key example), considering the similarity of the relevant mechanisms of transmission. Genes are clearly transmitted from parents to children; in traditional societies, especially in the absence of schools, cultural transmission (unfortunately a poorly investigated subject; see, however, Cavalli-Sforza and Feldman 1981; Hewlett and Cavalli-Sforza 1986) also takes place mostly from parents to children, as does, presumably, the transmission of language from generation to generation. Two phenomena transmitted in basically the same way are bound to be strongly correlated.

In our original paper (Cavalli-Sforza et al. 1988), we expressed the strong conviction that language must have been a great asset that considerably helped modern humans in their expansion, and that it also may have limited or prevented admixture with other forms of humans that were less developed linguistically. The linguistic inferiority of the Neanderthals (Lieberman 1975, 1989) is controversial (Falk 1975). Nevertheless, the extreme complexity shared by all existing human languages seems likely to be a product of a final step in linguistic evolution, which peaked in a.m.h. and was spread by them to the whole world. An interesting relationship has been observed by Foley (1991). Using the genetic tree (Cavalli-Sforza et al. 1988) and information on the numbers of languages per family given by Ruhlen (1987), he has shown that there is a very strong linear relation ($r = 0.91$) between genetic distance between two groups separated by a node of the genetic tree and the number of languages spoken by the two groups together (see fig. 2.15.2). Although this evidence is indirect, and the correlation coefficient is biased upwards because the nodes of the tree are not independent, it adds to the persuasion that linguistic evolution goes hand-in-hand with the spread of modern humans.

The analysis of the genetics of human populations requires an enormous mass of information. Unfortunately,

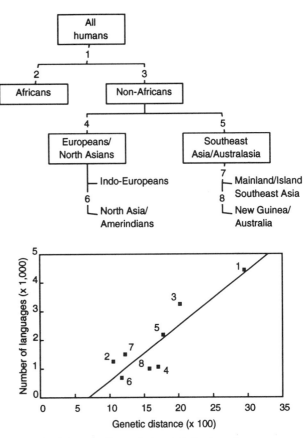

Fig. 2.15.2 According to Foley (1991), there is a strong relationship between human genetic variation and linguistic groups, showing that genes and languages may well diverge in similar ways. If genetic diversity found in each of the successive major groupings (top) of living human populations is plotted against the number of languages spoken by those groups, a strong linear relationship is obtained (bottom).

its retrieval has rarely been organized in an efficient way, and the data base available is the result of thousands of more or less haphazard collections and analyses of blood samples. An essential requirement of a sound analysis is that a large number of genes be thoroughly studied in parallel on all populations of interest. Today, there have been substantial advances in the techniques of analysis, unfortunately accompanied by nontrivial cost increases. The number of populations that can enlighten us on the past history of humanity is shrinking continuously. Only perhaps one or two decades remain in which we still have access to these populations. From the point of view of genetic history, we are an endangered species, and it is essential to avoid delay before taking the necessary steps to preserve this important knowledge about ourselves.

3 AFRICA

3.1. GEOGRAPHY AND ENVIRONMENT

Africa is the second largest continent, covering 20% of the total land surface of the world and hosting about 10% of the world population. Almost half of this is concentrated in the five most populous states, Nigeria, Egypt, Ethiopia, Zaire, and the Republic of South Africa. The climate varies from extremely arid to extremely humid. The humid area is concentrated in the lowlands, about 5° latitude north and south of the equator; rainfall is heavy and there is little variation in temperature (see table 3.1.1). This area is mostly covered by a tropical rain forest that extended more widely in the last 5000 years. Remnants are found on the northern coast of the Gulf of Guinea and in a few, mostly mountainous, areas. At increasing distance from the equator, humidity decreases fairly gradually, the vegetation changing accordingly. Both north and south of the forest belt are woodlands or wooded savannas with deciduous trees that lose their leaves during the dry seasons. At a lower humidity, the savanna or wooded grassland has fewer trees and a wet-dry cycle of seasons. The boundary of the forest is usually sharp, but there can be an area of savanna-forest mosaic between forest and savanna. With lower rainfall, thornbush vegetation prevails, with xeric trees like acacias and baobabs; with increasing dryness, the vegetation turns to subdesert scrubs and finally desert. Two major deserts are located almost symmetrically with respect to the equator. In the north the Sahara, the largest desert in the world, covers one quarter of Africa; inhabitants are few and limited to the major oases. In the south we find the Kalahari, located in Namibia and Botswana. The Somali desert in East Africa is smaller.

Figure 3.1.1 shows the vegetation zones of Africa, as determined by the climate, and table 3.1.1 summarizes data on the major climatic zones.

Very favorable climatic conditions are found in the valley of the Nile, which flows from its sources in Lake Victoria and Ethiopia to the Mediterranean, and also on a narrow coastal strip along the Mediterranean and near Capetown. The latter climates are called "subtropical dry" in figure 3.1.1, as contrasted with "subtropical moist" in the eastern portion of South Africa. In the highlands the climate is colder than in comparable latitudes at low altitude, and precipitation varies from area to area.

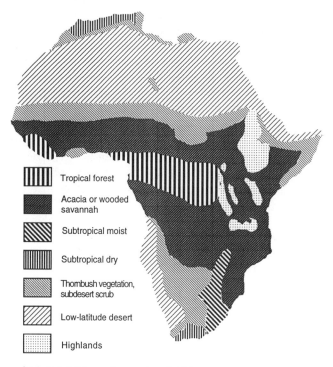

Tropical forest

Acacia or wooded savannah

Subtropical moist

Subtropical dry

Thornbush vegetation, subdesert scrub

Low-latitude desert

Highlands

Fig. 3.1.1 Vegetation zones of Africa.

Table 3.1.1. A Summary of Data on Major Ecological-Climatic Zones of Africa

Region	Location	Vegetation	Climate	Rainfall (inches/year)
Tropical forest	near equator: Zaire basin and mountains, north coast of Gulf of Guinea	canopy of high trees, plus dense lower layers	wet and warm	> 50
Woodland (or wooded savanna)	between 5° and 15° lat. north and south of tropical forest	trees deciduous in the dry season	hot, with variable seasons	35 – 55
Wooded grassland (acacia savanna)	belts between forest and deserts	occasional small trees, perennial tall grass	tropical wet and dry	20 – 30
Thornbrush vegetation	tracts in East Africa, from Sudan to Botswana	less grass, xeric and thorny trees	as above, but drier	12 – 20
Subdesert scrub	southern edge of the Sahara, much of Southwest Africa	sparse grasses, low shrubs	as above, but drier	5 – 12
Desert vegetation	Sahara in the north, Kalahari in the south	absence of, or rare, vegetation	very dry	< 5

The African deserts are homogeneously hot and arid except for a few oases and Lake Chad, which is very shallow and wide and, until a few thousand years ago, covered a much larger area. In Botswana the Okavango river, originating in Angola, forms a swampy delta in the Kalahari, generating a humid region in the middle of the desert.

The region immediately south of the Sahara is a dry strip of land that continues almost uninterrupted from the Atlantic Ocean to the Red Sea; it is called the Sudan, and the country by this name occupies only the extreme eastern part of it. In the west, a narrow strip of arid land south of the Sahara, forming the most northern part of the Sudan, is called the Sahel.

3.2. PREHISTORY AND HISTORY

3.2.a. PALEOLITHIC

There is reasonable consensus that the first human species, *Homo habilis*, originated in Africa (see chap. 2) from an earlier genus, *Australopithecus*, from which other branches also originated and died without leaving modern descendants, some less than 1 mya. It is believed that *H. habilis* originated about 2.5 mya, gave rise to *H. erectus* about 1.7 mya (Klein 1989a) and spread to Asia and Europe. Our species, *H. sapiens*, originated from *H. erectus* about 3–500 kya, perhaps in eastern Europe or West Asia, and spread to Asia, Europe, and Africa.

Homo sapiens sapiens (anatomically modern humans, a.m.h.) is a late-developing subspecies of *H. sapiens*. Complete consensus on its origins is lacking, but a certain amount of evidence indicates that modern humans appeared in East and South Africa around 100 kya (Klein 1989a; Rightmire 1989). Most important sites are shown in fig. 3.2.1. Unfortunately, all dates are weak and the material is scarce. Some of the presumably older finds (Omo, Broken Hill) may in fact be intermediate and thus establish a transition from the older, archaic *H. sapiens* type. Laetoli in Tanzania, dated at 120 ky, is also similar

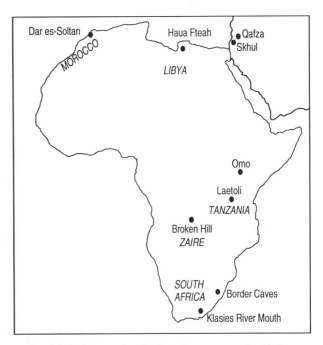

Fig. 3.2.1 Major archaeological sites connected with the origin of modern humans.

to modern humans but not quite in the modern range. The major modern finds are from the Border caves and the Klasies River Mouth (South Africa), dated at the earliest at 100 and 110 ky, respectively, but with wide and not well-known margins of error. The problem was discussed in section 2.1. The genetic distance between Africans and non-Africans indicates that the evolutionary split between these two groups is the most ancient in modern humans; it cannot argue per se for a geographic origin in Africa or in West Asia, because it is equally probable, other things being equal, that the genetic data indicate migration to Africa of a splinter group from a population originally located in West Asia, or vice versa. The question is more likely to be answered by archaeological information, such as, for example: "The African Middle Stone Age industries not only begin considerably earlier but also include more consistent evidence of technological sophistication (blades, geometric forms, fishing, ambush hunting)" (Tattersall et al. 1988). These more advanced industries are not associated exclusively with modern humans but also appear with more archaic types.

The question has been raised as to whether there is a similarity between these early examples of modern humans and the types presently living in Africa. Independent comparisons of craniometric data of the early South African modern humans with those of modern Khoisanids and Bantus living in South Africa gave contradictory results (Howells 1989; Rightmire 1989; see also sec. 2.1). It seems very unlikely that the difference observed today between modern Khoisanids and Bantus was recognizable more than 100 ky earlier, or that changes that occurred since that time were small enough to allow a meaningful comparison between old

and modern skulls. The uncertainty in the conclusions of different authors may simply mean that the distinction did not exist among the first modern humans and was generated later. It has also been noted that the collection of modern Khoisanid skulls used for the purpose of these comparisons is made up of many skulls that are not really representative or trustworthy (Morris 1986).

At the time of development of *H. s. sapiens* there may have been Neanderthals in the extreme north of Africa (Haua Fteah, see fig. 3.2.1). In the later Stone Age (35–8 kya), three major human groups are recognized in Africa on the basis of fossil remains (Phillipson 1982).

1. One group was ancestral to Khoisanids (Bushmen and Hottentots), who today live almost exclusively in southern Africa, but were once found in a wider area—probably including East Africa and perhaps also the Sudan and Egypt—and were therefore on the way of access to West Asia, as discussed below. According to some authors, fossil remains indicate these Khoisanids were there as early as 10 kya and perhaps earlier (fig. 3.2.2). It should be noted, however, that some authors (A. G. Morris, pers. comm.) have expressed disagreement on the identification of Khoisanid remains from Ethiopia and Egypt. The problem of Khoisanid origins is discussed in section 3.7.

2. A second group found in West Africa shows Negroid skeletal characteristics and is probably ancestral to all Negroid groups, who today live in the tropical forest and in much of eastern and southern Africa. Unfortunately, the soil of the tropical forest—in particular, the Zaire basin—is unfavorable for the preservation of human fossils, and no human remains have been found in this important area. The forest areas were not occupied in the early Paleolithic (Tattersall et al. 1988).

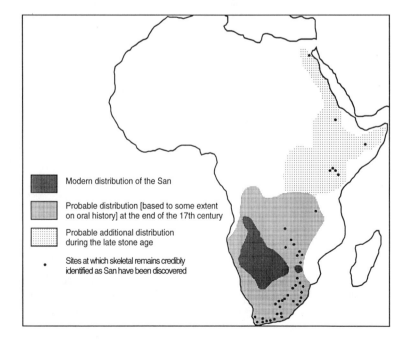

Modern distribution of the San

Probable distribution [based to some extent on oral history] at the end of the 17th century

Probable additional distribution during the late stone age

• Sites at which skeletal remains credibly identified as San have been discovered

Fig. 3.2.2 Area probably occupied by Khoisanids (Bushmen) in Africa around 10 kya at the end of the seventeenth century, and now. This illustration is a modification of figure 5.1 in Nurse et al. (1985), which is a modification of a figure in Tobias (1964).

3. Groups related to modern Caucasoid populations lived in North Africa. In the Maghreb, the western part of North Africa, there flourished from 20,000 to 7500 B.C. a late Paleolithic culture called Iberomarusian (Camps 1974) that extended from Spain to Morocco, Algeria, and Tunisia. Its people were hunters of the Barbary sheep and their sites were mostly within 100 km of the Mediterranean coast. Their type is sometimes referred to as "Mechta-Afalou." Skeletal evidence indicates that these people were of the Cro-Magnon type, that is, the same a.m.h. found in southwestern France and Spain.

3.2.b. THE NEOLITHIC IN THE NORTH

At the end of the late Stone Age, the Neolithic began in a few areas with the development of agriculture and spread slowly outward. The Middle East is the area of earliest agricultural origin nearest Africa. From there, the Neolithic spread to Egypt between 7500 B.C. and 5000 B.C. The newly domesticated Middle Eastern cereals, emmer wheat and barley, found a fertile soil in the Nile Valley, but the first farming culture known, Fayyum A, was not in the valley, though not far from it. As elsewhere, domesticated pigs and goats accompanied the first farming, and this new food-producing economy became the basis of the Egyptian civilization, which crossed the threshold of history with the introduction of writing about 3000 B.C.

After the disappearance of the Iberomarusian culture in the Maghreb, around 7500 B.C., a pre-Neolithic culture, the Capsian, possibly of local origin, took over and a related one appeared farther east in Libya. Capsians were hunters, gatherers, and fishermen and consumed large amounts of mollusks, characteristic heaps of which are found in the region (*escargotières*). Capsians later acquired pottery, sheep, and, to a lesser extent, other domesticated animals, but they always retained many of the earlier late-upper Paleolithic characteristics, including the local industries of stone tools and portable art objects like ostrich egg shells decorated with animal paintings.

3.2.c. NEOLITHIC DEVELOPMENT IN THE SAHARA

Farther toward the interior, in the mountainous parts of the Sahara in southern Algeria, Libya, and northern Chad, an incredibly rich collection of rock paintings and engravings offers a unique art gallery that allows us to reconstruct a series of local developments beginning with the late Paleolithic. Rock art has only rarely been dated, but to some extent it can be ordered in time sequence (Smith 1982).

1. The earliest phase, called *Large Wild Fauna* or *Bubalis* phase (from the name of the giant buffalo) is a period of hunters, and primarily rock engravings

are known. One such engraving has been dated to 9000–8000 B.C. (Mori 1974). Clearly, the climate was not arid at the time; in fact, the Sahara went through a phase of maximum wetness around 7000 B.C.

2. In parts of the Sahara this phase is followed by one characterized by paintings and engravings of large wild mammals and many human figures with round faceless heads, called the *Round Head* phase.

3. The next style is perhaps the best from an artistic point of view. It is called *Bovidian* because of the frequency with which cattle, clearly domesticated, appear in the paintings. Some of the painted cattle have coats of two or three colors, never observed in the wild. The earliest cattle paintings are from 3500 B.C., but domestication must have begun earlier. The paintings of the Bovidians sometimes show Caucasoid faces, sometimes Negroid. The latter are highly reminiscent both morphologically and culturally of a modern group of pastoral nomads called Fulani, now living in sub-Saharan West Africa. Cattle may have been domesticated locally, perhaps as early as 7000 years ago, or it may have come from the Middle East. Traces of cultivated plants are rare.

Progressive desiccation forced the Bovidians to leave by 1000 B.C., and they moved south to the savannas of the Sudan belt. They brought with them their cattle and pastoral customs, and most probably were ancestors of many pastoral nomadic tribes living today in West, north-central, and East Africa, and perhaps even farther south.

4. During the Bovidian phase, the Sahara started to undergo the desiccation that reduced it to a desert. Just before 1000 B.C., cattle is replaced in the rock paintings by an animal more resistant to dry heat, the horse, which was borrowed from the Middle East or Greece and used for driving chariots. These people, *Equidians*, were Caucasoid. In the sixth century B.C. Herodotus described, among other people living in Libya, the Garamantes, who drove horse-drawn chariots, probably employed for slave raids or war purposes. Even the horse was unable to endure the progressively drier climate of the Sahara, and in the first millennium B.C. was replaced by the camel, which had arrived in Egypt from Asia in 1600 B.C.

3.2.d. NEOLITHIC DEVELOPMENT SOUTH OF THE SAHARA

Middle Eastern cereals probably did not diffuse beyond Egypt, but in the meantime local developments of farming started in the Sudan and Ethiopia. Middle Eastern cereals did not grow in the sub-Saharan climate, and the presence of the tsetse fly made the survival of cattle impossible in humid areas. Several local grasses were domesticated, most probably in different parts of the Sahel, perhaps around 4000 B.C. (see fig. 3.2.3), the most important being sorghum or Guinea corn, and next in importance finger millet and pearl millet. In addition, cowpea (a pulse), the gourd, and the baobab were part of

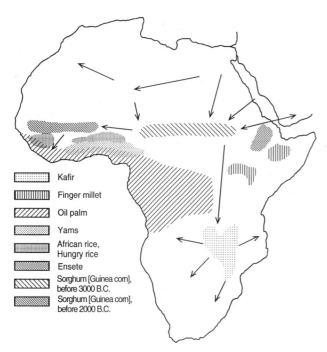

Fig. 3.2.3 Neolithic developments in Africa: cultivated local plants (a composite map using information primarily from Shaw 1980).

the "savanna complex." African rice was domesticated in West Africa somewhat to the south of the Sahel. In the more southern and humid part of West Africa, tuber yams, along with other tubers and oil palms, were major components of the "forest-margin" complex, an adaptation to an environment that was too humid for the savanna complex. Yams are still a major part of the diet in the tropical forest of the Guinea coast and were probably already known in 1700–1440 B.C., given that stone rasps useful for grating them were found in northern Ghanaian sites of that antiquity. The African banana (ensete) was domesticated in southwestern Ethiopia and is still the staple food of the Sidamo, whereas bananas, mango, taro, and plantains were imported later from Asia. The cereal teff was also domesticated in Ethiopia.

Sheep and goats may have spread to North Africa from the Middle East, together with cattle, which may have also been domesticated independently in the Sahara, as mentioned previously. However, when the Sahara became too dry, domestic animals had to withdraw south to the Sudan. There is evidence of cattle south of the Sahara in West Africa before 1500 B.C. and in Ethiopia before 1000 B.C. The only animals domesticated in Africa are the cat and the guineafowl (in Egypt) as well as the ass (Shaw 1980).

3.2.e. The Metal Age and the Bantu expansion

Copper was available only in parts of Africa that were settled relatively late and did not play a significant role in

the development of African civilizations (except Egypt). However, superficial iron ores are abundant in much of Africa. The iron-smelting technology was developed in West Asia around 1500 B.C. and came to Egypt and Nubia (northern Sudan) relatively late with the Assyrian invasion of the seventh century B.C. The first iron-smelting center is dated to the sixth century B.C. and located at Meroe in the Kush kingdom, which was in Nubia, the northern region of the present republic of Sudan. Kush was dominated by an Ethiopian dynasty that was initially under Egyptian control but in the eighth century B.C. took the lead and established its rule over all Egypt. It eventually withdrew to the south, with its capital at Meroe in Nubia (fig. 3.2.4). Iron rapidly supplanted stone, copper, and bronze in the making of tools. The new metal culture may have spread south from Meroe to the important iron center of Urewe, and perhaps even to Nigeria, where, however, it is more likely that development of iron smelting was indigenous.

The earliest large, well documented Nigerian iron-smelting center was at Nok, just north of the confluence of the Niger and Benue rivers. Here excavations have shown at least 13 furnaces for iron working dated between the fifth and the third centuries B.C. This culture is also known for attractive clay figurines, mostly human heads, found together with iron and stone tools.

Urewe is another large early center of iron smelting, just above Lake Victoria, with dates between 300 B.C. and A.D. The pottery is distinct from that found in the early Nok culture, indicating that the two developments

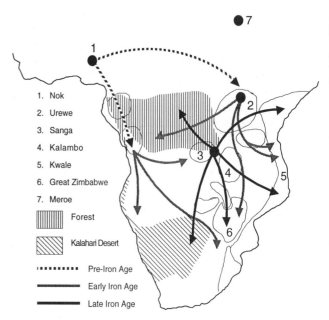

Fig. 3.2.4 Major iron-smelting centers in sub-Saharan Africa and Bantu expansions (information from Phillipson 1980; Denbow 1989). Some important archaeological areas are circled.

may have been independent, at least at the beginning. Tools used for iron smelting were also developed locally.

Both areas, Nok and Urewe, were most probably involved in the origin of the Bantu expansion, the linguistic evidence for which is discussed in the next section. Furthermore, both biological and archaeological evidence show that there were major expansions toward Central and South Africa that began *before* the Iron Age. Until recently, the archaeological information was almost entirely confined to the eastern and southern parts of the continent, and in Cameroon, Gabon, Angola, Zaire, and Namibia, data were limited or nonexistent. This situation is changing. We know today that there were two major streams directed toward the south, the western and the eastern one; the western one probably began earlier, before the Iron Age. Vansina (1984) has suggested that Bantu-speaking peoples were living in the western part of Central Africa by 1000 B.C. and that a western stream of expansion toward the south had more importance than previously believed. Recent archaeological evidence in Gabon (Digombe et al. 1988) shows that iron smelting in a heavily forested area of southeastern Gabon began around the second century B.C. Near the mouth of the Congo (Zaire) River, at Tchissanga, fourth century B.C. dates for iron smelting are the earliest in sub-Saharan Africa after those in Nigeria and Niger (Denbow 1989). At the same and other sites in the general area, earlier dates (between the tenth and fifth centuries B.C.) have been found for a pre-Iron Age farming technology. Thus, the first Bantu farmers of South Africa may have arrived with this western stream rather than with the eastern stream, in contradiction to earlier beliefs. In the Kalahari, fully developed pastoralism and metallurgy were already established by A.D. 500 and extensive grain agriculture by A.D. 800 (Denbow and Wilmsen 1986).

The eastern stream seems to have its starting point in the Urewe region. Connections between the origins of the western stream, probably in Nigeria and Cameroon, and the eastern stream in Urewe are unclear, mostly because of lack of adequate archaeological information from the intervening region. Linguistic evidence suggests that these connections were important. In Urewe, it is likely that Bantu immigrants also made contact with local and other cultures, possibly originating in the eastern Sahara. From Urewe, an expansion eastward reached the coast by the second century A.D. (Phillipson 1980) and continued rapidly toward the south in a narrow strip along the coast. These people grew millet and squash. A southwestern stream from Urewe pointed toward southern Zaire, where it may have connected with the western stream south of the forest, perhaps generating another, tertiary wave of expansion. From there, the expansion may have proceeded toward the central part of South Africa, reaching the northern borders of the present republic of South Africa in the fifth century A.D.

Developments after the fifth century A.D. showed the differentiation of local cultures that eventually gave rise to the late Iron Age expansion, perhaps from an area rich in metals—including copper—between Zaire and Zambia. The population density increased and centralized political systems emerged.

3.2.f. EARLY STATES IN SUB-SAHARAN AFRICA AND ETHIOPIA

Probably the first state to develop in sub-Saharan Africa was in Ethiopia. Here contact with Egypt and Sudan was early: Egyptians called Ethiopia the Land of Punt ("of God") because a source of the Nile was there. There is evidence from the sixth century B.C. onward of contacts with South Arabia and particularly with the two empires of the Sabeans and Minaeans. This is probably one source of the Ethiopian traditions of the visit of Queen Sheba with Solomon (who lived, however, several centuries earlier). Sabean influence on northern Ethiopia was profound until the first century A.D., and there were important migrations at the time and perhaps also earlier. It is likely (Encyclopaedia Britannica 1974) that the fusion of the immigrants from Arabia with Ethiopians of African origin, or of earlier immigrants from Arabia to Northern Ethiopia, created the people of the empire of Aksum, centered on the Tigre plateau. At that time, the political relations between Arabia and Ethiopia were reversed, with Ethiopia taking control of the Sabean and Minaean empires (Shinnie 1978). In the fourth century A.D., the Aksum empire extended to southern Arabia, the Beja territory in northeastern Sudan, and important parts of Ethiopia; it began declining in the seventh century A.D.

In the last two millennia, early states began to develop in many other parts of Africa. The earliest known is Ghana, in the western Sahel, the people being Mande farmers. In the extreme west of Africa, there were in the first millennium many tombs in the form of tumuli and megaliths, often with splendid ornaments. At the beginning of the present millennium, the Ghanaian state was destroyed by Arabs. In the Sahel there developed important trade centers like Timbuktu, Gao, and others, located along the caravan routes in Mauritania, Mali, Burkina-Faso (Upper Volta), and Niger, often in the form of twin cities, Arab and African, separate but close. There was trade in gold (mined in the Sudan), copper, and salt, found in the Sahara. South of the Sahel and closer to the coast, several populous states developed in the second millennium A.D.: the kingdoms of Ife (thirteenth to fifteenth century), Benin (fourteenth to eighteenth century), and Igbo Ukwu (with earlier beginnings, ninth century A.D.). The capitals of these states were all in the southern part of Nigeria (fig. 3.2.5).

The Arab expansion began soon after A.D. 632, the death of Muhammad, but was initially confined to Egypt

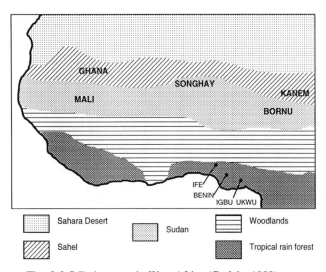

Sahara Desert

Sahel

Sudan

Woodlands

Tropical rain forest

Fig. 3.2.5 Early states in West Africa (Garlake 1980).

and the Mediterranean coast of Africa, extending with the Umayyads (Whitehouse 1980) to the Maghreb in the seventh and eighth centuries, and later farther south, both in the west and east. Moslem settlers are found in East Africa beginning in the ninth and tenth centuries. Kilwa in Tanzania and Manda in Kenya were East African cities, the first of which, erected in the fourteenth century, had the largest palace in sub-Saharan Africa. Trade flourished especially along the coast of the Indian Ocean; the language of traders, however, was not Arabic but Bantu (Swahili). Arabian Bedouins invaded North Africa repeatedly at the beginning of the present millennium.

There were also other Asian contacts, in particular with cultures speaking Austronesian languages from Indonesia, as shown by the presence of languages of this family in Madagascar. It is believed that the Austronesian arrival in Madagascar occurred in Iron Age times since no stone tools have been found on this island (Smith 1982). During this period, the banana was imported to East Africa from an unknown origin in Asia—probably not by Indonesians—and had a major impact on population growth.

Early states also developed in southern Africa during the second millennium. The major ruins are those of Zimbabwe, the capital of a Shona state (twelfth to fifteenth century). Once believed to be the result of the spread of Arabic influence, it is clear today that this was one of many local developments of organized states.

3.2.g. POPULATION NUMBERS

During the upper Paleolithic, Africa was a very populous continent, probably second only to Asia until 4000–5000 B.C., after which it was surpassed by Europe. It may have grown slowly to 50 million people by A.D. 1500, the most densely inhabited parts of the continent being then and now those in which agriculture and the use of iron developed first (especially Nigeria, Uganda, and Kenya). From A.D. 1500 on, growth slowed and came to a near standstill because of the slave trade, which lasted more or less until the end of the nineteenth century in the Americas and continued into this century in Arabia. Altogether more than 10 million people were shipped as slaves to the Americas, and 2.5 million to Arabia and Europe. The slave trade to the Americas was almost exclusively from sub-Saharan West Africa, Zaire, and Angola; that to Arabia, from the Central African Republic, Sudan, Kenya, Tanzania, and Mozambique (McEvedy and Jones 1978).

3.3. LINGUISTICS

Approximately 1400 languages are spoken today in Africa, about one-third of all the world's languages. According to the most widely accepted classifications, proposed by Greenberg (1949–1950, 1954, 1955, 1963), these languages belong to four major families (see Ruhlen 1987). Their geographic distribution is shown in figure 3.3.1.

3.3.a. GENERAL CLASSIFICATION

1. *Khoisan*. The Khoisan family contains about 30 languages and 120,000 speakers, all confined to the western part of South Africa (Bushmen and Hottentots), except for two languages spoken by the Hadza and Sandawe, two Tanzanian tribes. Khoisan languages are characterized by the use of "click" sounds as ordinary consonants, not generally found in any other family. Click sounds have been borrowed, however, by Bantu tribes of South Africa who were earlier in direct contact with Khoisan people. Of special interest is the presence of dental clicks in Kenyan populations north of Mombasa (Tucker 1969). Other indications of linguistic contacts between Ethiopians and Khoisan speakers are available (Ambrose 1982), and are especially interesting for reasons discussed in section 3.7. However, these interpretations of Ethiopian-Khoisan contacts have been criticized by others (Schepartz 1988).

Khoisanids were hunter-gatherers at the time of the arrival of the first farmers in southern Africa. Among the Khoisan living in South Africa today, the Khoikhoi or Khoi (also called Hottentots) are predominantly pastoral and are believed to have obtained cattle from the Bantu who had already arrived in the area 1500 years ago (Denbow 1984). The San are hunter-gatherers but are mostly acculturated today.

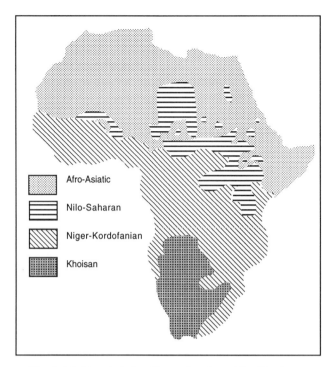

Fig. 3.3.1 Linguistic families and major subfamilies in Africa (after Ruhlen 1987).

2. *Niger-Kordofanian*. The Niger-Kordofanian phylum comprises a small branch spoken in the Sudan, Kordofanian, and the very large Niger-Congo branch, with more than 1000 languages and some 180 million speakers. To simplify the classification of the complex Niger-Congo subphylum, we list the branches, from west to east.

a) West Atlantic is spoken by the Fulani and related groups, originally nomadic pastors in the western Sahel, most probably from the Sahara; many of them are now settled farmers.

b) Mande is the language of a tribe of the same name; they are farmers.

c) North-Central Niger-Congo includes speakers of Kru, Gur, and Adamawa-Ubangian, who are also farmers; Adamawa-Ubangian is spoken in north-central Africa, just north of the Bantu languages.

d) South-Central Niger-Congo is by far the most important group in terms of the number of speakers: the western branch includes languages from Togo, and the eastern branch includes those of southern Nigeria and the Benue-Zambezi. This last cluster includes the Bantoid languages, from which Bantu languages developed, as well as the related groups non-Bantu, Bantoid, Bane, and Narrow Bantu. Bantu languages are spoken south of the third parallel north (approximately) in all of Central and South Africa, except for the area occupied by Khoisan-speaking tribes. A classification and tree of Bantu languages is given in section 3.3.b.

3. *Nilo-Saharan*. The Nilo-Saharan group is relatively small (140 languages, 11 million speakers), spreading from southern Libya to Kenya and Tanzania. It originated from a reelaboration of a group previously called East Sudanic and includes various people, pastoral nomads like the Nilotic (Dinka, Nuer, Shilluk in southern Sudan, and the Maasai in Kenya and northern Tanzania) and farmers like the Sara in southern Chad and the Mangbetu and Mangbutu in northeastern Zaire.

4. *Afro-Asiatic*. The Afro-Asiatic family consists of about 240 languages spoken in northern Africa from Morocco to Egypt (except for an area of southern Libya speaking Nilo-Saharan languages) and also in Eritrea, Ethiopia, and Somalia; it also includes all languages spoken in the Middle East and Arabia.

There are five major branches in addition to one extinct language (ancient Egyptian):

a) Berber, spoken in much of North Africa by pastoral nomads whose ancestors include the Caucasoid inhabitants of the Sahara;

b) Chadic, including Hausa in northern Nigeria, spoken by pastoral nomads and traders living in the Sahel;

c) Omotic, spoken in a small region of Ethiopia;

d) Cushitic, spoken in most of Ethiopia by settled farmers, and in the northeast of the republic of Sudan by the Beja, pastoralists;

e) Semitic, spoken in all of Arabia, all of the Middle East, northern Ethiopia (Eritrea, Tigray) and parts of North Africa.

3.3.b. BANTU LANGUAGES

The Bantu subgroup includes about 500 languages and 100 million speakers, close to a half of all the Niger-Kordofanian languages, and more than half of all the speakers (one-fourth of all Africans). Its unity was recognized in the eighteenth century, and it is immediately apparent that the family is young from an evolutionary point of view. In fact, the node leading to Narrow Bantu in Ruhlen's tree of Niger-Kordofanian is very low in the taxonomic hierarchy: the eleventh echelon, of 13. The success of Bantu is most probably due to the enormous expansion of its speakers, who advanced from their area of origin north of the equator to all of central and southern Africa, with the exception of the extreme southwest which is mostly desert and still occupied by Khoisanid speakers. The location of the Bantu homeland between Nigeria and Cameroon is based on linguistic considerations (Greenberg 1963), that is, the recognition of an area in which languages most similar to the Bantu ones are spoken today. It is in good geographic agreement with the archaeological discovery of an iron culture of about 500 B.C. at Nok (for a history of the recognition of the nature and origin of Bantu expansion see Flight 1988). As mentioned above, however, the Bantu spread from Nigeria

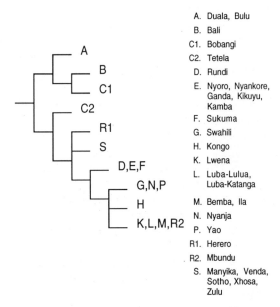

A. Duala, Bulu

B. Bali

C1. Bobangi

C2. Tetela

D. Rundi

E. Nyoro, Nyankore,
   Ganda, Kikuyu,
   Kamba

F. Sukuma

G. Swahili

H. Kongo

K. Lwena

L. Luba-Lulua,
   Luba-Katanga

M. Bemba, Ila

N. Nyanja

P. Yao

R1. Herero

R2. Mbundu

S. Manyika, Venda,
   Sotho, Xhosa,
   Zulu

Fig. 3.3.2 Henrici's (1973) linguistic tree of Bantu languages, simplified by condensing the 15 zones into 10 major clusters.

of spread of Bantu languages was located farther southeast, approximately between southern Zaire and northern Zambia (Guthrie 1967). Greenberg's center tends to coincide with the pre-Iron Age and early Iron Age Bantu expansion and is confirmed today by much archaeological research. Guthrie's area of origin corresponds perhaps with a tertiary center, that of the late Iron Age expansion.

Guthrie classified Bantu languages into 15 zones, named from A to S, and selected 28 better-known Bantu languages for the purpose of classification. The analysis of Guthrie's languages was repeated with taxometric methods by Henrici (1973), leading to a tree reported in a simplified form in figure 3.3.2.

Henrici's tree was plotted on a geographic map of the 15 zones (fig. 3.3.3) with the following externally im-

and Cameroon toward the east and south started even before the Iron Age, in Neolithic times (around 1000 B.C.). Using other criteria, Guthrie suggested that the center

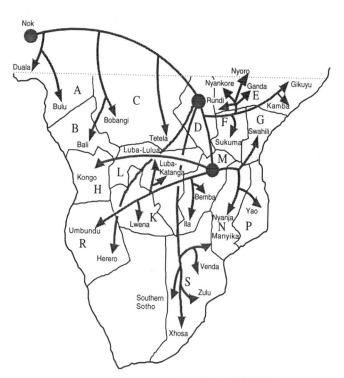

Fig. 3.3.3 A composite map of Guthrie's (1967) Bantu zones and Henrici's tree of linguistic divergence of Bantu languages. The three major nodes of the tree were plotted near the three major archaelogical areas that probably correspond to important centers of the Bantu expansion, chosen so as to maximize the correspondence with geography. (After Henrici 1973, figs. 1 and 5).

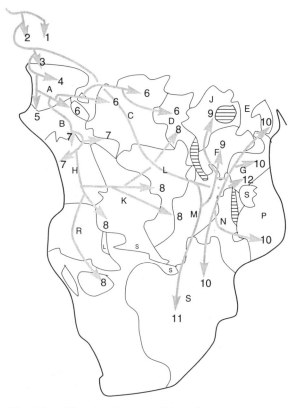

Fig. 3.3.4 The classification of Bantu languages by Bastin et al. (1983). Numbers refer to the language clusters the authors have established by numerical taxonomy. The original languages are too numerous to be plotted. Of 29 better-known languages referred to by Guthrie (1967) (fig. 3.3.3), 19 are considered by Bastin et al. (1983), in addition to many others. They belong to the following clusters (in parentheses Guthrie's zones): Bulu 3(A), Duala 3(A), Bobangi 6(C), Tetela 7(C), Kongo 7(H), Lewena 8(K), Luba-Lulua 8(L), Bemba 8(M), Herero 8(R), Nyoro 9(E), Nyankore 9(E), Ganda 9(E), Sukuma 9(F), Gikuyu 10(E), Kamba 10(E), Swahili 10(G), Nyanga 10(N), Yao 10(P), Zulu 11(S). Clusters 1 and 2 group Bantoid languages. The tree Bastin et al. (1983) have reconstructed is plotted on the geographic map as closely as possible to the language clusters.

posed constraints: (1) the first node or split was placed at the center of origin of Bantu languages according to Greenberg (in the area between the Benue and the Niger rivers); (2) the trichotomy leading to East and South African branches was placed in the important archaeological area centered at Urewe; (3) the next trichotomy was placed approximately in the region that Guthrie hypothesized was the primary area of origin. This picture derived from Henrici's tree minimizes the importance of expansion along the Atlantic coast, which we now know to be important. The problem could be corrected by modifying the second constraint cited above.

An independent analysis (Heine et al. 1977) generated a different picture, giving much greater emphasis to the western route of expansion to the south. This is in better agreement with another, more recent and complete analysis (Bastin et al. 1983) that used a much larger number of languages (over half of those in existence) and a standard taxometric technique (average linkage). It was the outcome of impressive field work lasting 30 years and followed by modern statistical analysis. The results favor

an area of origin of Bantu languages in a region between the Cross River, the Atlantic Ocean, and the Cameroon grasslands, very close to that suggested by Greenberg. As indicated earlier (Vansina et al. 1982; Vansina 1984), there are archaeological indications that a Neolithic culture was active in this area between 1600 and 700 B.C. so that most probably by 1000 B.C. the Bantu pre-Iron Age expansion was already under way.

The tree of linguistic evolution of Bantu languages by Bastin et al. is shown in figure 3.3.4, again plotted on the geographic map. The original languages are too numerous to be plotted individually; only their clusters, recognized by taxometric methods, are indicated (with numbers). Tree nodes are placed as close as possible to the clusters they subtend, but there is no attempt to constrain them near archaeological areas (unlike fig. 3.3.3). Both trees in figures 3.3.3 and 3.3.4 suffer from underrepresentation of some areas (the western one in figs. 3.3.2 and 3.3.3, and the eastern one in fig. 3.3.4), but the picture obtained in figure 3.3.4 is closer to the genetic data that we examine below.

3.4. PHYSICAL ANTHROPOLOGY OF MODERN AFRICANS

Today the continent is inhabited by two major aboriginal groups, Caucasoids in the north almost down to the southern border of the Sahara, and Negroids in subSaharan Africa. In the east, however, and especially in Ethiopia and Somalia, people have lighter skin and are considered Negroid by some, Caucasoid by others.

The Negroid type is not homogeneous: the classification by Hiernaux (1975) distinguished five groups (Bushmen, Pygmies and Pygmoids, elongated Africans, West African, and Bantu) and is summarized in table 3.4.1. As Hiernaux noted, it is not a biological, cultural, or linguistic classification, but in different cases, he used one or the other of these criteria for defining each group. Two groups stand out as peculiar, Pygmies and Bushmen (or Khoisan); references are found in the later sections dedicated to them. Until recently both of these groups were hunter-gatherers, but most of them have undergone heavy acculturation and also heavy admixture with Negroid neighbors. Nevertheless, there probably remain enough unmixed or less-mixed groups that one can form an opinion of their original type. Both Pygmies and Bushmen live in small social groups, believed to be similar to the archetypal societal structure of hunter-gatherers, and also have relations of servitude to local African farmers (usually Bantu). Both groups have been acculturating for some time, and only relatively small clusters have been less affected by this process.

Pygmies live in Central Africa, are mostly inhabitants of the tropical forest, and are adapted to a humid climate; they have a small stature that increases the ratio of surface to volume and allows them to take better ad-

vantage of the cooling effect of perspiration; in addition, their small mass means that their internal heat production is lower than that of larger people. The smallest Pygmy tribe has an average height of 145 cm in males and 137 cm in females (Mbuti, Ituri forest in northeast Zaire), but the head and, to a lesser extent, the chest have essentially the same size as those of people of average stature. Hunting-gathering is still an important activity, even if the great majority of Pygmies today live in part on food of agricultural origin, obtained either as compensation for work on Negroid farmers' plantations or from their own plantations in the most acculturated groups (Cavalli-Sforza 1986a).

Bushmen are somewhat taller than Pygmies but, after Pygmies, are the shortest African group. Those probably closest to the original type are the San, whereas the Khoi (Hottentots, Nama), who are herders, must have had greater contacts with Negroid pastoralists from whom they acquired the practice of herding. Among the several physical peculiarities that differentiate them from other Africans (Singer 1978, Nurse et al. 1985) are yellow skin, narrow and occasionally oblique eye slits reminiscent of Mongolian eyes (but they do not have typical Mongolian folds), female steatopygia or the tendency to accumulate fat high over the buttocks, and the tablier or Hottentot apron, as the elongated or hypertrophied labia minora are called. Like Pygmies, Bushmen have peppercorn hair, which is an extreme form of the typical African frizzy hair.

By far the greatest number of Africans are neither Pygmies nor Bushmen. The classification advanced by

Table 3.4.1. A Summary of Hiernaux's Classification (Hiernaux 1975) of Sub-Saharan African Peoples

Group	Country	Environment	Tribes	Economy	Physical Type	Languages
Khoisan	today mostly in Namibia & Botswana	usually dry	San (Bushmen): Kung, etc.	hunter-gatherers now mostly acculturated	females 150 cm males 160 cm brown-yellow skin color	Khoisanid (clicks)
			Khoi (Hottentot): Nama, etc.	pastoral nomads	eye folds, some steatopygia	
Pygmies & Pygmoids	N.E. Zaire (Ituri) N. Congo & S.W. Central African Republic S. Cameroon	tropical forest	Mbuti Biaka Baka	hunter-gatherers "symbiosis" (exchange food, etc. with local African farmers)	females 144 cm males 153 cm	many different languages from different Niger-Kordofanian stocks (Adamawan and Bantu)
	Uganda, Rwanda, S. Congo, etc.	drier environment	Pygmoid, Twa, etc.	various activities, cultivation, fishing, pottery, etc.	(155-160 cm)	
Elongated Africans and Nilotes	W. Africa	savanna	Moors Fulani	essentially pastoral nomads	tall (170-180 cm), thin, small head, small face, long limbs skin color from dark (Nilotes) to intermediate (E. Africa)	many different languages of Afro-Asiatic, Nilo-Saharan and Niger-Kordofanian stocks
	Sudan		Shilluk, Nuer, Beja, Dinka, etc.			
	Ethiopia Somalia Rwanda, Burundi		Afar, Galla Somali Tutsi			
	Kenya Namibia		Maasai Herero			
W. Sudan & Guinea Rain Forest	W. Africa	very dry savanna, just south of the Sahara, with transition to tropical forest on the Guinea Coast	many	sedentary farmers	intermediate stature, 165-175 cm dark skin color	Niger-Kordofanian other than Bantu
Bantu	C. & S. Africa	savanna & forest margins	many	sedentary farmers	intermediate stature dark skin color especially in the savanna, lighter near the forest	Bantu, a narrow section of Niger-Kordofanian

Hiernaux is in part based on a multivariate analysis using both anthropometric traits and a few genes. The most interesting novelty with respect to earlier classifications is the category of "elongated Africans," tall and thin people representing probably an adaptation to a hot and dry climate. They are practically all cattle herders, and they probably came from the Sahara, a hypothesis that explains both their dependence on cattle and their body build. They are located today south of the Sahara, some at a great distance from it.

The other two groups, West Africans and "Bantus,"

form the great majority of African people. "Bantu" is originally a linguistic definition and here is extended to the physical type that includes almost all people in central and southern Africa. The two groups are rather similar to each other.

The Malagasy, inhabiting Madagascar, are a group of mixed origin, as shown by languages, customs, and physical types. People of Austronesian origin came from south-central Borneo, where today Ma'anyan, the language most similar to Malagasy is spoken (Ruhlen 1987). The original Southeast Asian settlers seem to have mixed heavily with Africans; later Afro-Arab immigrants arrived from the east coast of Africa. They occupy the western coast, have clear-cut African features, and were responsible for introducing animal husbandry.

Unfortunately, the available genetic data do not support a reasonable analysis and therefore are not considered further, although the island appears in the maps of a few genes.

3.5. GENETIC ANALYSIS OF THE CONTINENT

Forty-nine populations were selected as having the most genetic information. Some of these are single tribes, whereas others are pools of populations. These latter populations are indicated by their countries or by linguistic groups (e.g., Nilotic, Cushitic), or in other ways (Pygmoid). When ethnic groups could be recognized, often across country boundaries (e.g., Tuareg, Berber, Bedouin), all data from these groups were pooled together and were not included in the country of origin. Pooling helped decrease sampling error or drift in very small samples or populations.

In spite of the exclusion of a number of populations that were tested for too few genes, the average number of genes in the 49 populations was only 47.6, with a standard deviation of 21.9. Sixteen populations (mostly from the north) had fewer than 35 genes, 21 between 35 and 59, and 12 above 60. The average number of genes in pairs of populations was 28.6. As far as can be judged from the similarity of the genetic and geographic or linguistic classifications, results were reasonably satisfactory, despite the limited genetic information.

The tree (fig. 3.5.1) shows a major cluster consisting exclusively of sub-Saharan populations, and a smaller one of Northern and Eastern populations, called Northeastern, where Northern includes both Mediterranean and Saharan. Nilotic-speaking populations from the south of the republic of Sudan do not belong to this cluster.

The Northeastern cluster splits almost exactly into two subclusters, North and East. The only North African population not found in the Northern cluster is the Algerians, who join the Eastern cluster; the only East African population not found in the Eastern cluster is the Somali, who join the sub-Saharan cluster.

The major cluster of sub-Saharan populations shows several outliers and minor clusters and two major subclusters, one of which consists of Bantu, Nilotics, and related populations, called the Central-Southern cluster, and the other, of almost all West Africans. The four outliers and minor clusters are listed in order of branching.

1. The Mbuti Pygmies are the most characteristic of all Pygmies (see sec. 3.7); two other Pygmy groups join other subclusters.

2. The second outlier is a cluster of Khoisanids including Somali. This may be an error, but there are other similarities between Khoisans and other West Asians.

3. The third outlier is a minor cluster of Wolof, Serer, and Peul (from Senegal); Peul is French for Fulani, but the Fulani from Nigeria join the West African cluster.

4. A minor cluster comprises the Sara, a Nilo-Saharan speaking population from Chad, and Biaka Pygmies from the Central African Republic.

5. Two more internally located outliers are the Funji and Bedik, speaking a Nilo-Saharan and a Niger-Congo language, respectively. While the first is poorly studied genetically (31 genes), the second is slightly better known (40 genes). The Senegal cluster is also poorly known (25–32 genes). Especially in the case of the Funji, the position as an outlier may be due to poverty of the genetic data. Mbuti Pygmies (82 genes), the Biaka Pyg-

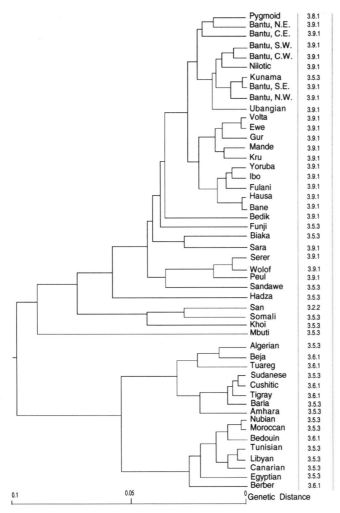

Fig. 3.5.1 Genetic tree of 49 African populations with the numbers of the figures where their geographical locations can be found.

Pygmoid	3.8.1
Bantu, N.E.	3.9.1
Bantu, C.E.	3.9.1
Bantu, S.W.	3.9.1
Bantu, C.W.	3.9.1
Nilotic	3.9.1
Kunama	3.5.3
Bantu, S.E.	3.9.1
Bantu, N.W.	3.9.1
Ubangian	3.9.1
Volta	3.9.1
Ewe	3.9.1
Gur	3.9.1
Mande	3.9.1
Kru	3.9.1
Yoruba	3.9.1
Ibo	3.9.1
Fulani	3.9.1
Hausa	3.9.1
Bane	3.9.1
Bedik	3.9.1
Funji	3.5.3
Biaka	3.5.3
Sara	3.9.1
Serer	3.9.1
Wolof	3.9.1
Peul	3.9.1
Sandawe	3.5.3
Hadza	3.5.3
San	3.2.2
Somali	3.5.3
Khoi	3.5.3
Mbuti	3.5.3
Algerian	3.5.3
Beja	3.6.1
Tuareg	3.6.1
Sudanese	3.5.3
Cushitic	3.6.1
Tigray	3.6.1
Baria	3.5.3
Amhara	3.5.3
Nubian	3.5.3
Moroccan	3.5.3
Bedouin	3.6.1
Tunisian	3.5.3
Libyan	3.5.3
Canarian	3.5.3
Egyptian	3.5.3
Berber	3.6.1

0.1   0.05   0 Genetic Distance

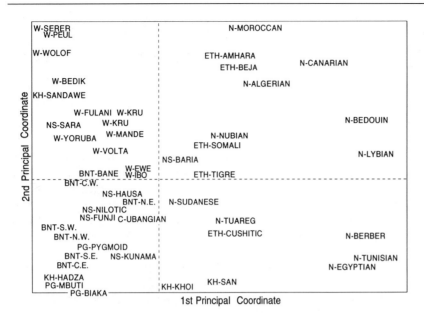

Fig. 3.5.2 Principal-component analysis of 49 African populations. BNT, Bantu and Bane; C, Central; ETH, Ethiopia; KH, Khoisan; N, North Africa; NS, Nilo-Saharan; PG, Pygmies; W, West Africa.

mies (97), and the Sara (50) are better known genetically (Cavalli-Sforza 1986a) and their outlying position is very likely to be real.

There are few deviations in this genetic clustering from a geographic, physical, or linguistic one. The North-eastern cluster includes only Afro-Asiatic languages, and nearly all of them, with the exception of Somali and Hausa. All Niger-Congo languages are found in the Central-Southern cluster, as are also Nilo-Saharan and Khoisan languages, which are, however, interspersed

Fig. 3.5.3 Geographic locations of some populations in the tree in figure 3.5.1.

among the various subclusters and minor clusters. Pygmies seem to have lost their original languages and speak those of present or past neighbors. Linguistic exceptions are considered further in the sections that follow.

Along with the genetic tree, we consider the results of a principal-coordinate (PC) analysis (fig. 3.5.2), which leads to more information and to a few methodological considerations. The geographic locations of the populations indicated in figures 3.5.1 and 3.5.2 are shown in figures 3.5.3, 3.6.1, and 3.9.1, and some of them in greater detail in figures 3.2.2 and 3.8.1 (see also the last column of figure 3.5.1). The first PC absorbs 30.4% of the information, and, as the picture clearly shows, separates Saharan and sub-Saharan Africans. All Ethiopian populations lie in the middle, and the San are at the very bottom of the diagram. These are the only Khoisan people who clearly share, with Ethiopians, this quality. The simplest interpretation for this behavior is that these two groups are genetically intermediate between Caucasoids, who all appear at the right side of the diagram, and the sub-Saharan Negroid Africans, who occupy the other side. This important cue is studied further in the next two sections using more direct techniques.

The second PC accounts for only 8.1% of the information but clearly separates West Africans (upper left quadrant) from Central and South Africans (lower left quadrant), with Nilo-Saharan speakers somewhat intermediate, but not clearly separated, from either group. Pygmies also are interspersed, but the most characteristic Pygmies, the Mbuti, are at the extreme bottom, a position they share with the San from whom they are separated by a substantial difference for the first PC.

The PC analysis thus confirms the observations made with the tree, without clearly separating Ethiopians into a compact cluster, but pointing out by their intermediate position that they may have arisen by admixture of Caucasoids and Negroids. It also separates West from Central-Southern better than the tree, which is unable to show the similarity between the minor and major clusters of West Africans. But it does not show as dramatically as the tree the outlying position of Mbuti Pygmies. This is inevitable because all these features are segregated in lower principal components and cannot appear in the representation of the first two. We have here a nice example of the rule that, even though PC maps and trees bring out the same major properties of the data, further probing of the individual distances may be useful. However, the distance matrix of 49 populations contains 1176 elements, a prohibitive number, especially since the genetic information is quantitatively modest. Therefore, in the following sections, we analyze appropriate submatrices in more detail by considering separately some major outliers and major clusters in the following order: Ethiopians and some North African isolates, Khoisans, Pygmies, and finally the largest cluster, formed by other sub-Saharan Africans.

The above conclusions are in reasonable agreement with results of other multivariate analyses (Hiernaux 1968; Nurse et al. 1985; Excoffier et al. 1987), in particular with that of Excoffier et al., who used single-gene systems (*RH, GM, HLA*) independently and restricted their study to the tribes in which data on each system were available and satisfied rigorous criteria of validity. Our clusters are inevitably sharper, because a greater number of genes and populations have been analyzed. The analysis, however, does not confirm the category of "elongated Africans," which seems to correspond to differences of a lower order compared with those considered here. Physical anthropology and culture set them aside from other sub-Saharan Africans, but, while radiating to many different parts of Africa, they may have had enough intermarriage with local groups that the gene picture is blurred. They may have maintained their peculiar elongated phenotype through sexual, and perhaps also natural, selection. One cannot exclude the possibility that more genetic information may bring out their ancient relationship.

3.6. ETHIOPIANS, SOME OF THEIR NEIGHBORS, AND NORTH AFRICANS

Genetic knowledge of Ethiopians and North African Caucasoids is limited. The best-investigated populations are used in figures 3.5.1 and 3.5.2. Figure 3.6.1 shows their modern geographic location, along with that of speakers of Berber languages and other populations of special genetic interest.

Table 3.6.1 shows the genetic distances used for building the tree in figure 3.5.1, limited to North and East Africans appearing in that tree, and a summary of the rest of the matrix. In table 3.6.1, populations are ordered so as to emphasize the split into the two major subclusters (1) North Africans, from which poorly known Morocco (15 genes, after exclusion of Berbers) and Algeria

(24 genes) have been omitted; and (2) East Africans. The three lines at the bottom collect three major sub-Saharan groups for comparison. The Tuareg, geographically in the North, associate genetically with the East subcluster. The two subclusters, North and East, are enclosed in triangles in table 3.6.1; the major rectangle collects cross relations between the two subclusters and is made up of uniformly high distances, as expected, considering the two clusters are also well separated geographically. The sub-Saharan populations shown in the last three lines have uniformly higher distances from all North and East Africans, though East–Sub-Saharan distances are smaller than the North–Sub-Saharan ones as expected if East Africans are a mixture of Caucasoids and Africans.

In the more developed countries of the Mediterranean, there is considerable urbanization and loss of ethnic identity. These populations are considered in the first paragraph that follows, separating them from populations of North Africa that have better kept their ethnic identity through geographic and social isolation (Berbers, the Tuareg, and the Beja).

1. *Countries of North Africa.* Populations appearing under the name of their country and not identified from an ethnic point of view include Algerians, Tunisians, Libyans, Egyptians, and Sudanese, and are usually from major cities; they must come from a fairly wide and unspecified area around these cities and are therefore least informative with respect to their more remote origins. The Berber contribution to these populations is

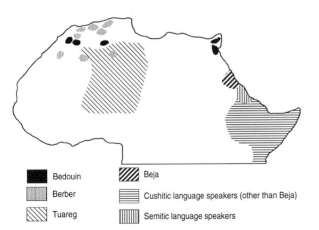

Bedouin

Berber

Tuareg

Beja

Cushitic language speakers (other than Beja)

Semitic language speakers

Fig. 3.6.1 Geographic locations of North and East African populations discussed in the text (after Murdock 1959, p. 112).

Table 3.6.1. Genetic Distances (x 10,000) Used for Building the Tree in Figure 3.5.1 Involving the Major Populations in North, East, and Sub-Saharan Africa

	EGY	LIB	TUN	BER	BED	TUA	BEJ	TIG	AMH	CUS	SUD	NIL	B&B	WAF
Egypt														
Libya	109													
Tunisia	162	37												
Berber	262	233	290											
Bedouin	204	170	141	212										
Tuareg	646	674	859	480	558									
Beja	539	605	538	320	477	135								
Tigri	442	548	455	561	516	320	220							
Amhara	811	507	899	725	517	278	258	46						
Cushitic	502	694	870	782	576	352	202	36	371					
Sudanese	359	625	621	689	488	240	290	47	184	20				
Nilotic	1393	1997	2664	1725	1840	766	536	394	449	478	381			
Bantu & Bantoid	1231	2146	1631	1576	1810	867	654	339	924	614	223	124		
West Africa	1408	1500	1736	1202	1476	472	500	397	511	675	340	236	139	

Note.– The three triangular boxes from upper left to bottom right include distances (1) between North African populations, excluding the Tuareg; (2) between East African populations, including the Tuareg; and (3) between Sub-Saharan African populations. The rectangular boxes include cross correlations among the three groups mentioned above. In the rectangular boxes, the variation between distances should be small if representation by a tree is satisfactory. Although this is usually true, there are some exceptions.

certainly important but is not the only one, especially where ancient colonies were known historically (Greeks in the eastern part, Phoenicians in the western part of the Mediterranean). We also know from historical sources (Murdock 1959) that Arab and Bedouin invasions of North Africa took place in the first and second millennium A.D., respectively, and are likely to have further contributed to the genetic background of these populations. The three reasonably well represented countries, Egypt, Libya, and Tunisia, show genetic distances in agreement with their geographic distances. The Berber and Bedouin backgrounds are obviously not negligible, but not overwhelming either. Genetic distances of the three countries from the Bedouin or Berbers are about equal, but definitely greater than those between the three countries, indicating that the urban population of these countries has undergone marked admixtures and perhaps received other contributions. Berbers and Bedouins are about as different from each other as they are from Egypt, Libya, and Tunisia. Bedouins are of recent Arabic origin and show relatively little difference from Arabs located today in the Arabian peninsula (see sec. 4.10).

2. *Berbers*. Berbers are located primarily in the northern regions of Algeria and Morocco, but somewhat to the interior, usually not far from the sea (see fig. 3.6.1). Bedouins in this analysis are found in these two countries as well as Egypt (especially the Sinai), but also in all parts of Arabia and Socotra. Berbers are believed to have their local ancestors among Capsian Mesolithics and their "Neolithic" descendants, possibly with genetic contributions from the important Neolithic migrations from the Near East. Similar radiations of Neolithic Middle Eastern farmers took place in other directions (toward Europe and Asia), where they contributed heavily to the local genetic background. It is reasonable to hypothesize

that the Berber (Afro-Asiatic) language was introduced by the Neolithic farmers.

Berbers were exposed to heavy pressure from Arabs and Bedouins after the seventh and eleventh centuries, respectively, and many Berbers have adopted the Arabic language. Berber languages and customs have survived only in smaller niches, mostly in the mountainous regions of northern Morocco and Algeria and in some northern oases of the Sahara. Berbers today are nomadic pastoralists or settled farmers. In Neolithic times, before the acquisition of the plow and advanced irrigation techniques by Berbers, the Canary Islands were occupied by a Neolithic Berber population, the Guanche, discovered by the Spaniards when they occupied the Canaries late in the fifteenth century. The Guanches and the related Canarios had lost contact with the continent after they stopped building adequate boats and were living an early Neolithic type of life at the time of Spanish discovery, with barley as a staple, and goats, sheep, pigs, and dogs. They made butter and cheese, pottery, did little hunting or fishing, and built rectangular stone houses. Physically, they had blue or gray eyes, brown skin, and blondish hair (Murdock 1959). There is only modest genetic information about them, and their original language is extinct. Their genetic relationship is highest with northern Africa, from where they probably traveled to the Canary Islands.

3. *The Tuareg*. Among speakers of Berber languages are the Tuareg, who inhabit mainly southern Algeria and northern Niger, with fewer in neighboring countries (Libya); there are also some representatives from Mali in the data. In the tree in figure 3.5.1, the Tuareg associate with the East African cluster, Algerians, and the Beja; Algerians are poorly known (24 genes); Berbers, the Tuareg, and the Beja are better known (72, 62, and

48 genes, respectively), and therefore the relations of the last three populations deserve greater attention than those of the Algerian data. We are confronted with the new finding that, although the Tuareg speak a Berber language, they show a closer genetic relationship with the Beja. The origin of the Tuareg is not fully understood. There are indications that they may have moved to the center of the Sahara in order to escape Arab attacks in the seventh and eleventh centuries A.D., but their earlier whereabouts are not known. They have always been a very mobile people, being pastoralists who may have been among the first in Africa to make extensive use of horses and, later, camels. Up to modern times, the Tuareg have been the major camel traders on the traditional caravan routes across the Sahara. They have a developed literature and use an ancient Libyan alphabet called *tif-finagh*, which suggest a more eastern and northern origin at some earlier time. The Tuareg have dark-skinned slaves, and there is considerable variation in skin color from the nobility to the servants. It is possible that their genetic similarity to the Beja and Ethiopians is a consequence of their having some black admixture; but Tuareg slaves have usually not been sampled, or were sampled separately. Although the Tuareg are classified with Berbers on linguistic grounds, it is clear that they show important ethnic distinctions from Berbers and have a developed culture of their own.

4. *The Beja.*   The Beja (or Bega) are a group of nomadic pastoralists speaking a Cushitic language and may have long inhabited the area that they now occupy in northern Sudan, between the Nile and the Red Sea. Egyptian records indicate they have been in this region at least since 2700 B.C. as independent pastoralists (Murdock 1959). They may have actually been in place since 4000 B.C. There are records of fights between the Beja and Meroe in the first millennium B.C., and later with Arabs. They were part of the Aksum empire. Even today their economy is almost exclusively pastoral, and they mostly subsist on milk, butter, and meat provided by sheep and goats, cattle, and camels.

The relation between the Beja and the Tuareg is quite interesting. Both are pastoralists and both speak Afro-Asiatic languages, though from two different branches. Because nothing is known about the more remote history of either group, the genetic similarity shown in table 3.6.1—one of the highest in the whole table—is unexpected and demands explanation. The matrix of table 3.6.1 clearly shows that the Tuareg do not belong genetically to the Berber-North African group, but rather to that of Ethiopians, to which Beja also belongs, although there is no marked similarity of either the Tuareg or the Beja with Ethiopian groups. Both the Tuareg and the Beja are numerically conspicuous (360,000 and 900,000, respectively; Grimes 1984) The Tuareg are widely dispersed over a large area and are very mobile;

the gene frequencies we used are averages of many groups. Hence, the data we use for both the Tuareg and the Beja may have been little affected by drift. At face value, their genetic difference may correspond to a genetic separation a little less than half as long as that from Ethiopians. Their time separation could be over 5000 years if the Beja were already in the area in which they are now some 5000 years ago and the Tuareg had already separated by that time. One can explain the similarities of the Beja and the Tuareg in other ways, but at this stage the hypothesis of their common origin should be considered a serious possibility. If the Beja and the Tuareg were once truly one population, their present linguistic divergence may be due to either the influence of the Berber milieu that the Tuareg may have joined after their separation from the Beja, or by the Beja language being a consequence of the political domination of Ethiopians. Today the Tuareg are by far the more mobile of the two; if this has been true for a long time, it is more likely that the Tuareg separated from the Beja and migrated westward 5000 years ago.

5. *Ethiopia and Somalia (East Africa in the narrow sense).*   The Tigre region in Eritrea and neighboring regions of northern Ethiopia were the center of the Aksum empire. Tigre and Tigrinya are two Semitic languages spoken in the north and south of the region, respectively. They are related to Geez, which, though now extinct, is still used for literary and liturgical purposes. These languages, together with Amharic and Gurage, all belong to the southern branch of West Semitic, originally spoken in southern Arabia. Amharic is today the official language of Ethiopia and has the most speakers, but it is confined to North-Central Ethiopia. The presence of Semitic languages must reflect the historical Arabian contacts.

Other populations that are less well known genetically include the speakers of the Cushitic branch of Afro-Asiatic, who are spread over most of Ethiopia and Somalia. The Beja form a branch of Cushitic by themselves. The major branch, Cushitic proper, includes Somali (2–3 million people), Billen (32,000, Eritrea), Falasha (20,000 near Lake Tana, of Jewish faith but of Ethiopian origin, now in Israel), Sidamo (farmers, about 900,000, in the south), Galla (pastoralists, 5 million), Afar (pastoralists, 200,000, in the east), and many others. The first three are the best known genetically. The speakers of two branches of Afro-Asiatic languages, one spoken in western Ethiopia (Omotic), the other in the area of Lake Chad (Chadic), are very poorly known genetically.

The Barya (more correctly known as Nara; Barya means slaves) number 25,000, and their neighbors, Kunama (also from Eritrea), 45,000–70,000. These relatively small groups speak Nilo-Saharan languages as do the Funji or Gule, who live in Sudan, on the Blue Nile near the Ethiopian border.

In the matrix summarized in table 3.6.1, we show among sub-Saharans, the averages of Nilotic speakers (mostly from southern Sudan, but also from Uganda and Kenya), of the major Bantu-speaking groups, and of West Africans, in order to illustrate the considerable difference of both North Africans and East Africans, but particularly North Africans, from all the rest.

In summary, the information available on individual groups in Ethiopia and North Africa is fairly limited but sufficient to show that they are all separate from sub-Saharan Africans and that North Africans and East Africans (Ethiopian and neighbors) are also clearly separate. Estimation of admixture by standard methods (Guglielmino-Matessi et al., in prep.) has given values of about 60% African and 40% Caucasoid, using sub-Saharan Africans as African "parents" and Southwest Asians as Caucasoid parents. Because very similar results are obtained using North Africans as Caucasoid parents, it is difficult to tell whether Southwest Asians or North Africans contributed the Caucasoid genes. Perhaps both did. Using the simple F_{ST} approach discussed in chapter 1 for calculating admixtures, average gene frequencies from Nilotic speakers as prototypes of African ancestors, as well as gene frequencies of North Africans averaged for the five groups of table 3.6.1 as Caucasoid ancestors, one obtains 53% of African (and 47% Caucasoid) contribution for Tigre, 57% for Amhara, 56% for Cushitic.

It is of interest, however, that there are relic populations in Ethiopia that may show greater similarity to the San or to Southwest Asian populations. One group from the mountain forest in Ethiopia near Bonga (in the Kaffa region) has been examined for markers on a small sample of individuals. Especially interesting were the GM markers that indicated a possible Asian origin (Nijenhuis and Hendrikse 1986). Little is known anthropologically about these populations.

The simplest conclusion is that most Ethiopians come from an admixture in which a slightly smaller fraction, of Caucasoid origin, may have come in part from northeast Africa and in part from Arabia, but ultimately mostly from the Middle East, considering that Neolithic Middle Eastern migrants must have contributed in an important way to North African genes. The slightly larger fraction of Negroid genes may have been in situ in early Neolithic times, perhaps having come to Ethiopia from the west, southwest or south, before the beginning of Egyptian civilization. A contribution from the San people is considered possible or even likely, if we accept their presence in Ethiopia 10 kya or earlier. On the basis of present data, it is difficult to distinguish from the later Caucasoid contribution coming from Arabia at the times of the Sabean and Axum empires and other likely similar contacts, earlier and later, with Caucasoids from West Asia. Originally, languages may have been Cushitic and have been replaced by Semitic languages in the north of Ethiopia under the influence of South Arabia. Knowledge of Omotic-speaking populations, at the moment extremely limited, would also be of interest, especially since Omotic is generally considered the most divergent branch of Afro-Asiatic. In the next section, we consider Khoisanids (for whom a problem of admixture similar to that posed by Ethiopians arises), compare the two admixtures, and examine possible times of origin.

3.7. KHOISANIDS

When white settlers in South Africa first made contact with the local aborigines in the seventeenth century they found two types of people: the Khoi (Hottentots and others), who were cattle-herders, and the San, who were hunter-gatherers. The two people are closely related biologically, as shown by their having in common several physical characteristics discussed above (e.g., steatopygia and the tablier; see Nurse et al. 1985, Tobias 1978). They also spoke closely related languages, as shown by, among other things, the presence of unique sounds, clicks. Clicks are also used by a few Bantu tribes that had extensive contact with Khoisanids (e.g., Zulu, Xhosa). We have already mentioned the Khoisan languages spoken by two groups in Tanzania, the Hadza and Sandawe, further considered below. Furthermore, there are also click sounds in groups living in Kenya. The latter, however, are controversial. We will consider the Khoi and San separately, since they show many genetic differences.

There is substantial genetic information on the Khoisan (for summaries, see Tobias 1978; Nurse et al. 1985) and in particular on the San. Total numbers for the latter may be of the order of 55,000, of which perhaps only 10% are relatively unacculturated. The San are divided geographically and linguistically into northern, central, and southern; they have all been pooled to reduce the effects of drift in the comparisons with other populations.

In the tree analysis of the 42 world populations discussed in chapter 2, the San associated with other Africans and showed somewhat greater similarity with Ethiopians, with which they form a loose pair. On bootstrapping, however, the association between the San (or Ethiopians) and the other four sub-Saharan African populations is loose and only sufficient to guarantee that all six sub-Saharan African populations form a single cluster in barely more than 50% of all bootstraps. In the other bootstraps, the San associate with Caucasoids, or, more rarely, with Orientals, and the same is true of Ethiopians.

Table 3.7.1. Genetic Distances (x 10,000) between San, Ethiopians, and Representative Groups of Sub-Saharan Africans and Asians

	BAN	NIL	WAF	MBP	SAN	ETH	NEE	IRAN	IND	SEA
Bantu	0									
Nilo-Saharan	119	0								
West African	188	196	0							
Mbuti Pygmy	715	750	801	0						
San	944	1002	885	1496	0					
Ethiopian	659	829	697	1232	777	0				
Near East	1779	1824	1455	2139	880	709	0			
Iran	2242	2077	1796	2589	1267	1060	158	0		
India	2202	2074	1748	2664	1246	1079	229	155	0	
Southeast Asia	2829	2590	2114	3499	1819	1534	1050	1119	844	0

Note.– The rectangular boxes show the cross distances between three major groups of populations. Bantu, Nilo-Saharan, West African, and Mbuti Pygmies are from Sub-Saharan Africa. The last four populations are all Asian. San and Ethiopians seem intermediate between Sub-Saharan Africans and Asians and are close or closer to West Asian than to other Africans. Homogeneity of distances in the rectangular boxes is not entirely satisfactory, and a modification of the observed tree might be advisable but is not easily carried out.

In the tree in figure 3.5.1, the San are the second outliers of the sub-Saharan group. In table 3.7.1 we give an excerpt of the F_{ST} distance matrix of the 42 world populations, showing the relationships between all African and a selection of Asian populations. The Near East includes all populations from Southwest Asia, from Turkey to Arabia.

From table 3.7.1, a clear pattern emerges. The San differ from other sub-Saharan Africans 0.1082 ± 0.0140, that is, more than any sub-Saharan group differs from any other (the mean distance is 0.0461, the maximum 0.0801). While the other sub-Saharan groups differ from the three geographically closer Asian groups 0.1966 ± 0.0146, the San differ from them only 0.1131 ± 0.0126, a highly significant difference. These three Asian groups genetically closer to the San are all Caucasoid, but Ethiopian groups show no indication of close relatedness to the San. An approximate test of the hypothesis that the San are an admixture between African and Caucasoid can be made by comparing the three distances (retransformed back to F_{ST}s, see formulas 1.11.3, 1.11.4):

	distance	F_{ST}
sub-Saharan–San	0.1082	0.1026
San–Caucasoid	0.1131	0.1069
sub-Saharan–Caucasoid	0.1966	0.1785

Using formula (1.17.3) these values give $m = 0.49$ as the contribution of Caucasoid, and $1 - m = 0.51$ as that of Africans.

Figure 3.7.1 shows a graphical analysis of admixtures, in which Near Eastern data are used in lieu of Caucasoid because they are closer geographically. The numbers are therefore slightly different. Note that the edges of the triangles are the square roots of distances. The contribution of Near Eastern is somewhat greater than that of Africans, the estimate of admixture being just above half ($m = 0.56$). A more thorough test of admixture gave similar results. Ethiopians showed a

slight preponderance of the African component, using Caucasoids as parents, and $m = 0.54$ using Near Eastern data, but they also have considerable genetic distance from the San (0.0777); thus, the San and the Ethiopians are likely to have been independent admixtures at rather different times, though with similar initial components. Even if the initial components were geographically and ethnically the same, they would have evolved between

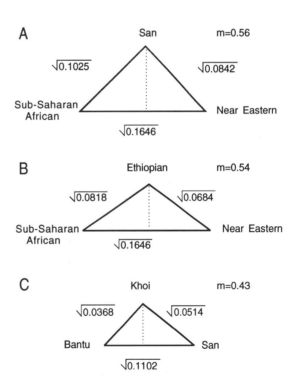

Fig. 3.7.1 Graphical analysis of admixtures involving San, Ethiopian, and Khoi. Triangles are built by using the square roots of the distances of population pairs as lengths of the sides. The presumed parental populations are the two corners of the base of the triangle. The presumed mixture is the top corner. For the calculation of percentage of admixture m, see formula (1.17.3).

the first and second admixture. On the basis of historical considerations, the mixture resulting in Ethiopians is likely to be later.

The map of the early distribution of Khoisanids (fig. 3.2.2) shows the sites of presumed skeletal remains of the San. The proximity of East and North Africa to Southwest Asia makes it extremely likely that there was admixture between Africa and West Asia.

An alternative hypothesis should be considered. Some peculiar external characteristics of Khoisans, and the uniqueness of clicks, have struck the imagination of many anthropologists to the point that some scholars have considered the Khoisan a separate race of very remote origin (Coon 1963). In line with this, some linguists have seen the clicks as primordial sounds of human language, preserved only in Khoisan. Data from the analysis of an approximately 700-nucleotide region of mtDNA, already discussed in section 2.4 (Vigilant et al. 1989), seem to give some weight to the idea that Khoisan are direct descendants of primitive human ancestors. Our analysis of gene frequencies on the basis of the admixture hypothesis leads to the opposite conclusion, namely, that Khoisans are the result of a relatively early admixture between Africans and Asians. It is not easy, however, to distinguish between the two hypotheses that Khoisans are the root of all humans or the result of an admixture, for in many respects these two hypotheses give the same expectations, especially looking at gene frequencies. At this time and until further data accumulate and other analyses are made, it may be useful to consider both hypotheses possible and wait for further elements that may help in distinguishing them.

The Khoi behave somewhat differently from the San; in the tree in figure 3.5.1, they are outliers of a minor cluster formed by the San and Somali. This minor cluster is now easier to explain on the basis of the above considerations. Distances between Khoi, Bantus, and San are given in table 3.7.2. Clearly, the Khoi are closer to all Bantu groups than to the San. They are also darker in skin color, and they have certainly had extensive contacts with Bantus since they acquired cattle-breeding practices from them. The hypothesis of admixture therefore seems quite reasonable (see fig. 3.7.1). It is not clear (Nurse et al. 1985) whether the contact happened north of the Zambesi (in what is now the country of Zimbabwe) or in

the Kalahari Desert. One cannot exclude the possibility that the contact may have occurred even farther north. There are, however, some linguistic considerations supporting the hypothesis that the contact happened in the Kalahari between an advance party of Bantus with cattle who may have fused with central Bushmen, whose dialect is most similar to that of the Khoi. We have already cited work by Denbow and Wilmsen (1986) and others suggesting that an important contact between Bantus and Khoisan occurred in the Kalahari 1500 years ago.

Today the Khoi (more than 100,000) comprise three major groups: the Nama in Namibia, the !Ora (Griqua) in Orange, and a number of individuals of varying degrees of admixture who have mostly lost identity. In cities they are classified with the "Coloreds," an officially designated group including individuals of very diverse ethnic origins. Because of pastoral customs, the Khoi have acquired the gene for lactose absorption, which is not found in populations outside Europe other than African pastoralists who use milk as adults. Gene frequencies (Nurse et al. 1985) that are close to zero in San and most Bantu groups reach as high as 29% in Namas, corresponding to about 50% absorbers (because the gene is dominant). This is a fairly high figure and implies a reasonably long period of pastoralism unless the proportion of non-Khoi in the initial admixture was high. Cattle did, however, arrive in South Africa earlier than was previously thought, and we have seen that first contacts between the Khoi and Bantus also occurred early.

Gene frequencies show the similarity of Khoi is highest with Southeast Bantus, their present neighbors, but next highest with Central-Eastern Bantus (table 3.7.2). Assuming that a mixture between Southeastern Bantus and San generated the Khoi, the average mixture proportions are 57% Bantu and 43% San. Figures would be slightly different using Central-Eastern Bantus as parents.

The contact of the Khoi, and perhaps also the San, with Bantu immigrants led to gene flow not only into the Khoi, but also into the Bantus, as suggested by the high figures of GM markers typical of Khoisan among Bantu tribes. The figures are especially high in South Africa, Zambia, and Mozambique. The admixture with San is estimated (Jenkins et al. 1970; Nurse et al. 1985) to be as high as 60% in the Xhosa, the most southern Bantu population in South Africa. This result needs con-

Table 3.7.2. Genetic Distances (x 10,000) Comparing Bantus and Khoisanids

| Khoisanids | Bantu* | | | | | | | San |
	NE	CE	SE	NW	CW	SW	Mean ± S.E.[†]	
San	996	1029	943	1339	1362	1338	1168 ± 81	
Khoi	304	309	256	441	420	413	375 ± 31	528
Sandawe	245	396	398	452	182	336	335 ± 42	1794
Hadza	418	440	670	381	461	406	463 ± 43	1444

* Bantus are grouped in six geographic clusters described in a later section (see Table 3.9.3).
[†] S.E., Standard error.

firmation from other genetic markers before it can be accepted. Results of tree analysis (fig. 3.5.1) suggest that the estimate should be lower, but for accuracy tree analysis should be repeated with the single tribes. A marker like GM, belonging to the immunoglobulin system, may respond to natural selection resulting from infectious diseases, substantially altering its gene frequencies. The excessively high degree of admixture suggested by GM may indicate that the Khoisan GM markers confer (or conferred) higher resistance to local diseases. Accurate analysis of the heterogeneity of admixtures calculated for single genes might help to assess the effects of natural selection for these genes. The new DNA polymorphisms may be very useful for resolving this discrepancy.

Although it may not have been as high as the GM estimates would indicate, there clearly was gene flow from the Khoisan into the southern Bantu populations, which was paralleled by the linguistic flow ("borrowing" in linguistic terminology) of clicks into a few Bantu languages. Some Negroid populations in South Africa do not have specific Khoisan genes but speak a Khoisan language (Dama; see Nurse et al. 1985).

Two populations speak Khoisan languages but are located far north of the Khoi and San, in Tanzania, the Hadza and the Sandawe. The Hadza are a relatively small group of nomadic hunter-gatherers (about 2000), living southeast of Lake Victoria, at a short distance from the Sandawe (about 40,000), who were hunter-gatherers but have recently acculturated. Surprisingly, the two languages have almost no reciprocal similarity, even though they both show similarity to the Khoisan languages (Greenberg 1963). These two groups appear in the genetic tree as outliers of the sub-Saharan African cluster. The Sandawe are at an average distance of 335 ± 42 (table 3.7.2) from all Bantus and have an especially small distance from Central-Western Bantu. If this greater similarity can be taken at face value, Sandawe may have hybridized while these Bantus were on their way toward the southeast; but it is possible that the Sandawe also moved considerably before reaching their present location. The genetic distance between Hadza and Sandawe is 426, of the same order of magnitude as those of the Hadza and the Sandawe with a group including all other sub-Saharan Africans. The Hadza, however, are a little more distant from other Bantus than Sandawe, perhaps because they are a smaller population and hence more subject to drift. They show no particular similarity with any Bantu subgroup.

Both the Hadza and the Sandawe have retained very little, if any, genetic similarity with the San; they show the same similarity with the Khoi as they do with Bantu subgroups or with each other. The lack of special genetic similarity between these two displaced Khoisan-speaking people is in agreement with the low level of linguistic similarity. The case of the Hadza and Sandawe may be similar to that of the Dama, a Negroid tribe speaking a Khoisan language, who are usually considered with the Khoi. Obviously, in several instances, an almost complete replacement of genes has occurred while the language was maintained, or vice versa. The second alternative may seem less likely, if it is true that for imposing one's language, a group must have an adequate social, political, and military organization (Renfrew 1989a). It has been suggested, however, that the Dama were originally brought to Namibia as slaves of the Khoi and then freed themselves (Nurse et al. 1985). This would indicate a higher social and political organization of Bushmen groups than present today. The case of the Hadza and Sandawe is different, however, from that of the Dama, because today they live very far from any Khoisan-speaking groups. Moreover, unlike the Dama, their languages have changed enough that their acquisition of the Khoisan language, or, more probably their loss of contact with other Khoisanids and the beginning of their genetic assimilation by neighboring groups, must be relatively ancient. The possibility may also be entertained that their genetic similarity with Khoisans never existed, and their language was acquired because of political imposition or social influence in Khoisan milieu in a fairly distant past, as hypothesized for the Dama.

3.8. PYGMIES

A recent publication on Pygmies (Cavalli-Sforza 1986) is briefly summarized below. References given here are limited to later publications.

Many groups of Pygmies live in Central Africa, all of which have several characteristics in common. Small stature is the most important one, since they—or at least one particular group, the Mbuti—are the smallest people on earth. In general, populations living in tropical forests are small the world round (see sec. 2.13), in agreement with the idea that small stature represents an adaptation to warm and humid environments. Although it is not completely clear how the height of African Pygmies is determined genetically, some information indicates that it is determined by a codominant single locus, perhaps, but not necessarily, the same in different Pygmy populations. Pygmies have normal levels of growth hormone, but lowered levels of growth-hormone receptor (GHR) and insulinlike growth factor I (IGFI); the latter remains low during adolescence, while in non-Pygmy individuals, IGFI increases about threefold. The major difference in the growth curves of Pygmies and non-Pygmies is that the former do not show the adolescent growth spurt, indicating that GHR and IGFI increase at puberty is responsible for this spurt and its lack in Pygmies determines their small stature (Merimee et al. 1987).

Facts and causes of the small stature of other non-African "Pygmies" around the world are less clear. They could well represent independent adaptations, given that a decrease in stature is very likely to be a major factor of adaptation to a hot and humid climate. There is in general a reasonable correlation between the stature of African Pygmies and the degree of admixture with other local populations. It should be noted, however, that non-Pygmy populations living in the forest are also smaller, on the average, than populations from the savanna (Hiernaux 1968), in a few cases because of Pygmy admixture. Pygmies are definitely smaller, however, than other populations from a similar environment.

The Pygmies' life in the forest is made possible by an economy of hunting and gathering, which many still practice. They are, however, progressively shifting to a mixed economy with some agriculture of their own, but more frequently to partial dependence on settled farmers, with whom they exchange game and labor for food like manioc and bananas, alcoholic drinks, and items that they do not produce like iron tools, tobacco, and pottery. Where the forest has been destroyed, Pygmies have been forced to change their economy; for instance, pottery making (in Rwanda and Burundi) and fishing (in Zambia) have become major occupations. Forest destruction has usually resulted in a loss of identity and faster admixture. Another major feature common to almost all Pygmy groups is the association with a local group of settled farmers, whose language they have acquired. In this association, Pygmies are always the lower caste, being the farmers' hereditary servants (the word slaves would not be accurate in this case). Pygmies have repeatedly decided to move to new forest regions, in which case an association with new farming masters is generated. Pygmies may continue to use the language of the old masters, which then becomes their "own," while they also acquire that of the new masters. An important fraction of their vocabulary, relating to forest flora and fauna, is not found in the farmers' language and may be more ancient (S. Bahuchet, unpubl.).

Genetic exchange between farmers and Pygmies is limited, but is more frequent in some villages and groups than in others. There are some rules that are almost always observed.

1. Farmers may acquire Pygmy wives, but not vice versa; the bride price to be paid is lower for a Pygmy wife, and Pygmy women are credited with higher fertility. In the case of the Rwanda Tutsi, there have been a few Pygmy women who married Tutsi kings and became queens (Desmarais 1977). This custom varies greatly between farmers' tribes and also between villages of the same tribe.

2. The children of the mixed couple are raised in the farming culture, although they often have lower status.

For this reason, gene flow is from Pygmy to Bantu or other farming group. We therefore expect to find Pygmy mtDNA among farmers who accept Pygmy wives.

3. Reverse gene flow (farmer-to-Pygmy) can also take place. This may happen in individual instances, probably infrequently, because of illegitimate unions, or in cases in which divorced Pygmy wives who married a Bantu farmer take their children back to the Pygmy group of origin.

These conclusions come from modern, limited field observations. It is necessary to conclude, however, that there are, or were in the past, occasions of more substantial gene flow from farmer to Pygmies. Assuming the Mbuti are the ancestral pygmy type, and Bantu the other ancestors (Wijsman 1984), it was estimated that Biaka Pygmies of the Central African Republic are only 18.5%–31% Pygmy (significantly different from zero), the different values depending on various assumptions made. The Pygmy component would probably be higher if non-Pygmy ancestors used in the estimation were tribes of origin closer to that believed to be correct, for which, however, there are not enough good data. No significant evidence could be found for a Pygmy component in Cameroon Pygmies; estimates for Rwanda Pygmies varied from 19% to 25% according to hypotheses (significantly different from zero). Pygmoid populations can therefore be indistinguishable, genetically, or hardly distinguishable, from some groups of farmers. Their stature remains on the low side, however, perhaps because of natural selection of the hot and humid climate of the forest in which they usually live, or lived until a short while ago, or because of sexual selection. Pygmoids have often maintained strong Pygmy traditions, although one cannot entirely exclude the possibility that in some cases they were acquired. Physical characteristics (e.g., stature) and Pygmy customs are usually the major criteria employed for the Pygmy classification.

We have already discussed (sec. 1.17) the possibility that the gene pool can be almost entirely replaced by continuous gene flow over hundreds or thousands of years. Although Pygmies are relatively protected against gene flow from other populations, many Pygmy tribes show extensive gene replacement and few have remained relatively unaltered. One example of prolonged contact between Pygmies and farmers, which has been investigated at the historical level by ethnographic and linguistic information, is that of the Biaka Pygmies in the Central African Republic (Thomas and Bouquiaux 1976). They clearly show extensive hybridization with non-Pygmies when compared with the Mbuti Pygmies, as will be explained later, but today there are very few examples that interbreeding is going on at present between the groups of Biaka studied and local farmers. Examples are limited to a few villages and take place in the forms described above under points 1–3, which can generate only minor farmer-to-Pygmy gene flow.

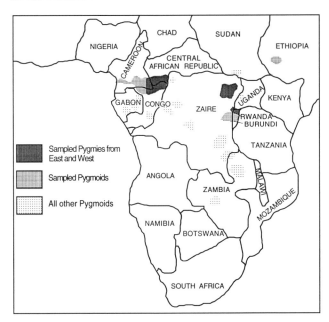

Fig. 3.8.1 Geographic location of major Pygmy groups and of those studied in this survey (Cavalli-Sforza 1986a).

Like all hunter-gatherers, Pygmies live in small groups (camps), typically averaging 30 individuals, which are highly mobile but move only within a clearly defined geographic area. A number of camps may be associated in mostly patrilinear bands. Pygmies choose mates at some distance from the group (on the average, 30–50 km), so their endogamy is not high. As they become sedentary, the distance between birth places of mates decreases. Populations intended as social (mating) groups are not small and hence drift is not too important, but there may be important genetic differences between populations located at some distance from one another and reciprocally isolated.

We have formed three geographic groups for the purpose of this analysis (fig. 3.8.1), Eastern Pygmies, Western Pygmies, and Pygmoids.

1. Mbuti or Eastern Pygmies, live in the Ituri forest, northeastern Zaire, almost directly on the equator. They speak languages of the Nilo-Saharan family. Cross-exogamy between groups speaking different languages is not known quantitatively. Mbutis are the smallest Pygmies (sec. 3.4). The total number may be around 30,000 or more.

2. Aka (also Biaka and Babinga) or Western Pygmies live in the forest between the Central African Republic, the eastern part of Cameroon, and the northern part of the Popular Republic of the Congo. They number about 30,000 and speak languages belonging to two different branches of the Niger-Kordofanian family, the Adamawa-Ubangian, and the Bantu group C. They are somewhat taller (8–9 cm more) than Mbuti Pygmies, but are otherwise fairly similar sociologically. There is no genetic exchange with Mbutis, who live at a considerable geographic distance, and in addition to a sub-

stantial average genetic difference from them, there are also different "private" genes in both groups; hence the reciprocal isolation may be ancient. It is also difficult to tell how private these alleles are because the immediate neighbors with whom some exchange takes place are usually not known genetically.

3. We have pooled all other Pygmy groups, from southwestern Cameroon, Rwanda, and northwestern Zaire, into "Pygmoids." They are usually somewhat taller and more widely acculturated than the other two. They are from very different regions and form an heterogeneous group.

The Pygmies that are most different genetically from the remainder of the Pygmies and other Africans are also the smallest in stature, the Mbuti. They are unquestionably the most representative Pygmies in most respects and were therefore kept separate from the other Pygmies. They are reasonably well known genetically (82 genes), though less well known than the Biaka (97 genes) (Cavalli-Sforza 1986a), who are more easily accessible. Pygmoids are less well known (55 genes, mostly due to lack of data on HLA). Mbuti Pygmies are clear outliers of the sub-Saharan cluster in figure 3.5.1, their average distance from the rest of the sub-Saharan cluster being almost as large (874) as that between Northeast Africans and Caucasoids (964). With the constant (genetic distance/time ratio) calculated in chapter 2, their separation time would be about 18 kya. This estimate of separation time is to be taken with great caution because it is certainly deeply affected by admixture. Table 3.8.1 shows the F_{ST} distances of Pygmies among themselves and with the most representative groups of other sub-Saharan Africans.

Mbutis are clearly the most distant from other Africans, the Biaka are intermediate but closer to the average sub-Saharan African than to Mbutis, and Pygmoids are closest to the average African. At least for Biaka Pygmies, linguistic and ethnographic analysis indicate that they were probably in the southwest of the republic of Sudan before the Arab invasion of Sudan (about 1000 years ago) and traveled from there in association with, and probably scouting for, the Ngbaka, an Ubangian-language-speaking tribe from that region. They are believed to have first traveled south toward northeast Zaire, where an Ngbaka tribe still exists and is perhaps related to the Zairian one. There is some genetic information on the Zairian Ngbaka and, in terms of genetic distance, they are closer to the Biakas than any of the tribes studied so far (distance data in Cavalli-Sforza 1986a). Biaka Pygmies, and probably a fraction of the Ngbaka, continued westward and then to the north, reaching their present location in the extreme southwest of the Central African Republic. There they associate today with various local tribes, one of which is also called Ngbaka (presumably a splinter of the Zairian tribe by the same name) and lives close to the Biaka, but unfortunately has not been typed genetically.

Table 3.8.1. Genetic Distances (x 10,000) of Pygmies among Themselves and between the Most Representative Groups of Other Sub-Saharan Africans

Pygmies	Bantu*						Nilotic	Sara	Ubangian	San	Pygmies		
	NE	CE	SE	NW	CW	SW					Mbuti	Biaka	Pygmoid
Mbuti	714	765	850	698	933	849	1064	800	849	851	0		
Biaka	363	295	310	256	389	339	456	248	339	311	465	0	
Pygmoid	131	132	223	216	262	266	407	311	226	224	385	179	0

* Bantus are grouped in six geographic clusters described in a later section (see Table 3.9.3).

Pygmies were probably once more widespread than they are today. Egyptian documents indicate that in 2300 B.C. the Egyptians were in contact with Pygmies called Aka, which is still the name of one major Mbuti group and is also the name of the western Pygmy group (Biaka is simply the plural form of Aka). The discovery of the first Egyptian document about the Pygmies goes back to the end of the nineteenth century. Although it was asserted that the document referred to chondrodystrophic dwarfs, Egyptologist Dawson (1936) showed that the Egyptian words for chondrodystrophic dwarf and for Pygmy are different. Pygmies were valued by the pharaohs because of their dancing skills, which are famous to this day in much of Central Africa, and probably also for their knowledge of herbal medicine. For this reason, a general was sent to capture them "in the Land of Punt" (probably, the sources of the Nile; it must have been the White Nile). The contact with Egyptians may have taken place in the southern part of the republic of Sudan, perhaps not far from the upper Nile Valley and most probably in a more northern location than that in which Pygmies are found today. They may have later left those areas, perhaps under Arab pressure, as probably happened with the Biaka.

The San show no special association with Pygmies greater than that with other sub-Saharan Africans. The Sara are somewhat closer than Nilotics to Pygmies and the fact, probably accidental, that their genetic distance to the Biaka is slightly smaller than that of the nearest Bantu subgroup is responsible for the association between Sara and Biaka in a minor cluster of figure 3.5.1. The distance between Biaka and Northwest Bantus (256) is slightly greater than that between Biaka and Sara (248), but the difference is negligible when compared with standard error. Under these conditions, average linkage will, however, associate Biaka and Sara in the tree. Tree building may be very sensitive to small random oscillations of distances; PC maps are less sensitive, but they are even less efficient at faithfully portraying the similarity between close neighbors.

Because of the lack of fossils in the acid African soils—which dissolve the main bone component, calcium phosphate—it is impossible to specify exactly the geographic distribution of Pygmies at earlier times. There is practically no Pygmy fossil record from dry areas where bones would have been preserved, confirming the belief that Pygmies represent a long-term adaptation to the tropical forest.

3.9. BLACK SUB-SAHARAN AFRICANS

Of the 49 populations studied in figure 3.5.1, those that are located in sub-Saharan Africa (excluding Khoisanid and Pygmies, already discussed) include West Africans, Bantus living in southern Africa (south of the third parallel north), and populations living in Central Sudan (more specifically, in a rectangle delimited by the Sahel at the north, the third parallel north at the south, Nigeria to the west, and Ethiopia to the east). In the northern section of this region, languages belong in part to the Niger-Kordofanian phylum (to which all West African and Bantu languages belong) and in part to the Nilo-Saharan phylum. We have seen that Bantu people tend to cluster in the PC map and do so more loosely in the tree; the same is true of West Africans.

Figure 3.9.1 shows the location of the tribes and groups of tribes compared in the tables of genetic distances; figure 3.9.2 gives a PC analysis limited to Niger-Congo and Nilo-Saharan speakers, after leaving out Pyg-

mies and Khoisan. This map is based on an average of 45.5 genes (they vary from 23 to 105) and accounts for 36.4 of the original genetic variation. Here Bantus are definitely more clustered than West Africans, with Nilo-Saharan still dispersed, but closer to Bantus. We see later that this greater Bantu homogeneity is confirmed by an analysis of distance matrices.

We now consider in more detail each of these two groups, as well as their relations with other people from Central Africa, making direct recourse to the distance matrix, which is not affected by the possible distortions of trees and PC maps. Table 3.9.1 includes F_{ST} distances between West African tribes and tribal groups, and table 3.9.2 F_{ST} distances between Bantu groups and their neighbors.

Considering West Africans first, we have ordered populations approximately from west to east: the Bedik were placed last in spite of their western position

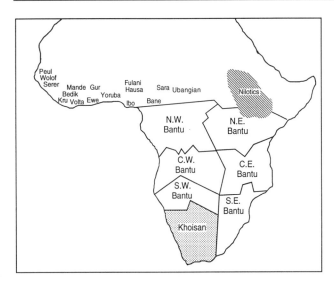

Fig. 3.9.1 Geographic location of African tribes and groups of tribes discussed in the text.

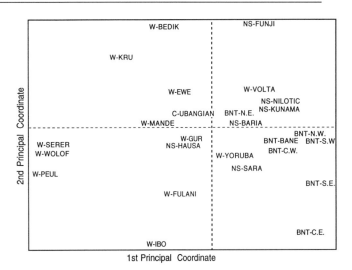

Fig. 3.9.2 Principal-component map of 26 Niger-Congo and Nilo-Saharan–speaking populations. BNT, Bantu and Bane; C, Central; NS, Nilo-Saharan; W, West.

Table 3.9.1. Matrix of Genetic Distances (x 10,000) for West Africans

	WOL	PEU	SER	KRU	MAN	GUR	VOL	EWE	YOR	HAU	IBO	FUL	BED
Wolof	0												
Peul	79	0											
Serer	42	155	0										
Kru	124	209	151	0									
Mande	208	185	232	79	0								
Gur	394	491	364	98	165	0							
Volta	494	573	381	108	148	83	0						
Ewe	175	173	115	119	75	103	13	0					
Yoruba	398	335	151	81	171	186	30	175	0				
Hausa	147	292	217	119	265	127	117	237	211	0			
Ibo	64	65	62	420	266	165	221	164	50	164	0		
Fulani	371	265	443	151	232	364	156	367	83	135	66	0	
Bedik	360	294	337	197	320	290	117	358	211	283	164	367	0

Table 3.9.2. Matrix of Genetic Distances (x 10,000) for Central and South Africa, Excluding Khoisanids and Pygmies

	Bantu									
	NE	CE	SE	NW	CW	SW	NIL	UBA	SAR	BAN
Bantu										
Northeast	0									
Central East	228	0								
Southeast	237	143	0							
Northwest	138	184	88	0						
Central West	167	189	107	173	0					
Southwest	193	263	167	138	52	0				
Nilotic	128	237	193	144	94	103	0			
Ubangian	152	326	130	126	132	115	127	0		
Sara	387	413	331	269	279	291	318	276	0	
Bane	153	301	118	53	115	91	222	123	269	0

because of their nature of outliers in the tree. All West Africans speak Niger-Congo languages, the major branch of the Niger-Kordofanian family, with the exception of one group, the Hausa, who speak an Afro-Asiatic language belonging to the western subgroup of the Chadic branch (the only representative of this branch). When data were insufficient, we grouped tribes by linguistic criteria, and we specify the tribes that have contributed the most genetic information to these groups.

The first three populations (Wolof, Peul, and Serer) are located in Senegal. Wolof (2 million) and Serer (650,000) are farmers and speak related languages. The Peul (900,000 in Senegal) speak a language of another Niger-Kordofanian branch, Fulani. Peul is the French word for Fulani, a large and heterogeneous group found in a narrow band located approximately in the Sahel, extending from the Atlantic coast to northern Nigeria and Chad. They are, according to Murdock (1959), either pastoralists, and Caucasoid-like (of Berber origin), or settled farmers of Negroid type. One group that was sampled, but is not considered here, the Tukolor, has elements of both Negroid Fulani and Berbers. Nomadic and settled groups have important social and ethnic differences, but the Fulani sampled here seem to belong to the second category, at least as far as we can tell from their known genes. The group later indicated by the name "Fulani" comes almost exclusively from northern Nigeria and is genetically different from the Senegal Fulani, who are more similar to the other Senegalese populations, the Wolof and the Serer.

The next five groups are fairly similar to each other and rather different from the Senegal populations. They appear to form a single cluster in the tree of figure 3.5.1; their most informative components are listed below.

The Kru, linguistically a branch of North-Central Niger-Congo, are represented primarily by eight groups: Bakwe (7000), Bassa (200,000), Bete (400,000), Gagu (40,000), Guere (200,000), Krahn (70,000), Kru proper (120,000), and Wobe (subgroup of Bete).

The Mande are, linguistically, a divergent branch of Niger-Congo, with 29 languages (the other branch, called Niger-Congo proper includes all the other 1003 languages). The Mande shown here include the Bambara (2 million), Bandi (40,000), Bozo (less than 90,000), Diula (250,000), Kpelle (500,000), Mandinka (1.4 million), and others. The Gur or Voltaic (linguistically a group of North-Central Niger-Congo) include Frafra (300), Kurumba (90), Lizzara, Mossi (2 million), and Tiefo (6–10,000).

The Volta-Comoe or Akan (a branch of South-Central Niger-Congo) include the Agni (400,000), Akposso (50,000), Ashanti (1.4 million), Baule (1.2 million), and Nzema (300,000).

The Ewe (another branch of South-Central Niger-Congo) include the Ewe (1.8 million) and the Fon (1 million).

The last four groups are all from Nigeria: Yoruba (15 million), Hausa (6 million in Nigeria; 25 million total, including second-language speakers), Ibo (12 million), and Fulani (6 million in Nigeria alone) and form the most important ethnic divisions of this very populous nation. The Yoruba and the Ibo belong linguistically to the eastern group of South-Central Niger-Congo, whereas the Fulani and the Hausa have different linguistic origins, as mentioned above. Nigeria has a long history, with early agricultural developments that may have attracted immigrants and perhaps favored admixtures, as seems likely on inspection of the matrix. It is worth noting that the Ibo have a strong similarity with both the Yoruba and the Nigerian Fulani. But the Ibo and the Yoruba are similar to the three populations from Senegal that are somewhat similar to the Kru and the Volta. Such genetic relationships indicate that major migrations and admixtures occurred within West Africa in earlier times. In addition, the Bane (a group shown in table 3.9.2) are highly similar to the Hausa, in part because of geographic propinquity. We discuss the Bane with the Bantu, to whom they are linguistically related.

The final population in table 3.9.1, the Bedik, is a very small group located in Senegal. They are outliers in the tree with respect to all West African and Bantu populations. This may be due to drift, given their small population size (5000). They do show some similarity to populations of the Kru-Volta group, and a less marked similarity to Nigeria, indicating that they may have originally come from a more eastern location than the one they now occupy. Linguistically, however, they belong to the West Atlantic branch of Niger-Congo proper.

In summary, West African populations can be described as showing three major clusters: Senegalese peoples; a central group formed by Mande, Kru, Gur, Volta, Ewe; and a group of four Nigerian populations. The last group shows great similarity to the Bane (see fig. 3.9.1), located at the boundary between Nigeria and Cameroon, and linguistically closely related to the Bantu.

The remainder of sub-Saharan Africans (excluding the Khoisan discussed in earlier sections) speak languages of two different families, Niger-Kordofanian (mostly the Bantu branch) and Nilo-Saharan. We have no genetic data for Kordofanian speakers (the branch of the Niger-Kordofanian family spoken in Sudan). Only a few groups of Nilo-Saharan speakers have been investigated genetically in a satisfactory way. The best known are the Sara (Majinga), numbering about 47,000 people in southern Chad (55 genes); the Nilotics, a linguistic group formed mostly by Shilluk (less than 200,000), Mabaan (25,000), Luo (2 million), Nuer (700,000), Dinka (250,000), Karamojo (250,000), Maasai (400,000), and Turkana (250,000). They are "elongated" pastoralists and live mostly in the southern Sudan, Uganda and Kenya (42 genes). Two tribes from very different locations are less well studied (26–31 genes): the Funji or Gule in east-central Sudan and the Nara or Barya (spelled also Baryia or Baria) in northern Ethiopia (western Eritrea), south of the Kunama. The last two tribes are of medium size (25,000 and 40,000, respectively).

Genetically, Nilo-Saharan speakers are very close to Bantu when considered at the world level (sec. 2.3). In the tree in figure 3.5.1 they show no internal similarity. Nilotics are, somewhat perplexingly, associated with Central-Western and Southwestern Bantu; a possible explanation is considered in the discussion about Bantu. The Kunama associate with Southeastern Bantu, again a perplexing result. The Barya are found in the northern Ethiopian group, presumably because of Caucasoid admixture. The Sara associate with the Biaka, an even more perplexing result. As we can see from table 3.9.2, they show a low, but constant similarity with all populations of Central and South Africa; the same is true with respect to West Africans (not shown in the table), but it is worth underlining that the population to which they show greatest resemblance are the Hausa (distance, 184). They do not associate with the Hausa in the tree because they had already formed a cluster with other Nigerian populations, but they do associate with the Biaka in the tree because, like the Biaka, they do not pair with any other population or cluster until they pair together, the distance between them (249) being relatively low. We cannot form a final opinion on the Nilo-Saharans because the most characteristic populations from the western and central Saharan groups are not represented at all. The similarity between Nilotics and Bantus, incidentally, disposes of the hope that the "elongated Africans" might correspond to a clearly separate genetic group. Nilotics contain most of the examples of this group. Much better data would be needed to test the hypothesis more satisfactorily.

Table 3.9.2 contains distances between the Bantu and their neighbors. The tree in figure 3.5.1 already showed that the Bantu are a relatively homogeneous group, about as homogeneous as West Africans, giving more weight to the hypothesis that the Bantu expanded relatively rapidly to central and southern Africa after the development of agriculture and iron tools. Four major populations infiltrated the Bantu cluster.

1. *Pygmoids.* The Pygmoids must have very heavily hybridized with the Bantu and other neighbors. They associate specifically with Northeast and Central-Eastern Bantus in the tree. This area is one in which the concentration of Pygmies is high and may well have been their area of origin. Pygmoids include Twa from Rwanda (20,000), Baka from south-central Cameroon (perhaps 20,000) and Badjelli in southwestern Cameroon (2000), as well as Pygmies from north-central Zaire who may be very numerous.

2. *Nilotics.* Most Nilotics are located a short distance north and east of one of the putative secondary centers of Bantu diffusion, Urewe. If the Bantu migrated from their area of origin between Benue and Niger toward the east, in the direction of Urewe, they may have crossed regions inhabited mainly by Nilo- Saharan speakers, as was most probably the Urewe region itself. Hence, there must have been considerable opportunity for intercrossing between speakers of these two groups. The more northern Nilo-Saharan speakers, still inhabiting the south-central part of the Sahara, are genetically unknown.

3. *Ubangian.* The Ubangian are part of the Adamawa-Ubangian branch of North-Central Niger-Congo. The group includes tribes like the Bangandu (2000; variable spelling), Ngbaka (750,000), Ngbundu (Mbugu), and Nzakara (3000).

4. *Bane.* The Bane include the Bamileke (a large group of tribes living in Cameroon), Bamun (200,000), and Banjun (a subgroup of Bamileke). Linguistically they belong to the "broad Bantu" group, nearest to the Bantu in the Bantoid branch of the South-Central linguistic branch of Niger-Congo according to Ruhlen. We have also added to this group the Tiv, (about 1 million) living mostly in Nigeria; linguistically they belong to the non-Bantu subgroup of the Bantoid branch.

The Bantu have been grouped according to the six classical geographic-linguistic areas (N.W., N.E., C.W., C.E., S.W., and S.E.). The tribes forming the six Bantu areas are listed in table 3.9.3.

Only a few Bantu tribes are well-known genetically, and the pooling into the six zones has considerably increased the number of genes available (62–105 for the six Bantu groups; 43 for the Bane, 38 for the Ubangian, and 55 for the Sara who are also given in table 3.9.2). The tree in figure 3.5.1 remains valid even after eliminating the Nilo-Saharan intruders. It is not as clear as the linguistic tree, which is based on more detailed information, but it adds some interesting novelty to it. The six Bantu zones, the Nilotics, Ubangian, and Bane are all genetically related. There is greater similarity between some neighbors (e.g., between CW and SW, distance 52, the smallest in the table; between Bane and NW, distance 53, the next smallest; between Nilotics and NE, distance 128), but perhaps not as much as one might expect. The pattern of expansion reconstructed by archaeologists and linguists is largely confirmed. The extraordinary genetic closeness of NW and SE Bantu might seem perplexing, but the most recent archaeological (see Denbow 1984, 1989; Denbow and Wilmsen 1986) and linguistic research (Bastin et al. 1983) confirm that the western stream of Bantu migration gave an early and important contribution to the formation of southern Bantu people, via the Central Western area. Another confirmation of the importance of the western route of the Bantu expansion is provided by the low genetic distances between Bane and all the western Bantu groups (compared

Table 3.9.3. Bantu Tribes Included in the Six Geographic-Linguistic Zones

Geographic-Linguistic Zone	Corresponding Guthrie's Zones cf. Figure 3.3.3	Bantu Tribes	From Africa	Bantu
Northwest	A, B, C	Bagandu, Balundu, Basakata, Batswa, Beti, Bushong, Busu, Combe, Duala, Eton, Ewondo, Lissongo, Maka, Mongo, Nkundo, Nkundo Maka, Ntomba, Pahouin, Pamue, So, Teke, Yengono, Wambissa	17	
Central West	H, L	Ambundo, Ankonde, Bangala, Kabende, Kaonde, Njinga, Quimbundo, Sambo, Uba, Umbundo	4 17	Angola
Southwest	K, R	Ambo, Bailundo, Cuangare, Diriku, Ganguela, Gciriku, Herero, Huambo, Kavango, Kwangali, Lunda, Lunda Tchokwe, Mbkushu, Mbunda, Mbunza Sambyu, Ovahimba, Ovambo, Ovimbundo, Quioco, Quipungo, Sambio	7 10 14	Lunda Ovambo Southwest
Northeast	D, E, F, J	Akamba, Babira, Bahuta, Baeundian, Embu, Fulero, Ganda, Gishu, Ha, Havu, Hima Ugandan, Humu, Hunde, Hutu, Ikoma, Kamba, Kiga, Kikuyu, Mabudu, Meru Gusii, Nkole, Nyamwezi, Nyanga, Nyaturu, Nyoro, Rega, Rwandan, Shi, Soga, Swaga, Tembo, Toro, Tutsi, Wabira, Wanande, Wanyali	5 6 9 15 16 18 17	Hutu Kenya Nyaturu Tutsi Uganda Zaire
Central East	G, M, N	Ankonde, Bemba, Bisa, Bondei, Chikunda, Gogo, Ila, Kisii, Kissi, Lamba, Luano, Nhugue, Sambaa, Sena, Tonga, Tumbuka, Nyanja, Wuzikili, Zambian, Zaramo, Zigua	19 17	Zambia
Southeast	P, S	Aiaua, Alolu, Bacca, Baka Hlubi, Barotse, Basuto, Bitonga, Bujga, Bunji, Changana, Chawa, Chopi, Fingo Pondo Xosa, Karanga, Kgaladadi, Koni, Korekore, Lozi, Maconde, Macua, Makocca, Mangwe, Manyika, Ndau, Ndebele, Ngoni, Nguru, Pedi, Pondo, Ronga, Rotse, Shangaan, Shangana, Shona, Sotho, Swazi, Tswana, Venda, Xhosa, Zezuru, Zulu	8 11 13 14 21 17	Maconde Shangana South Zimbabwe Zulu

Note.– The last two columns correspond to the population groups listed in Appendix 3.

with the respective eastern groups). Figure 3.9.3 is a proposed pattern of expansion based on the genetic data alone, in which the direction of arrows, which cannot be deduced from the genetic data, is suggested by the archaeological information. A more satisfactory picture may have to await more information, but the agreement between archaeology, linguistics, and genetics, though not perfect, seems nonetheless remarkable.

We can confirm with a direct analysis of distances that the Bantu (including Bane) are somewhat more closely clustered than West Africans: the F_{ST} between Bantu subgroups, including Bane, calculated from the average of F_{ST}s between pairs of Bantu subgroups in table 3.9.2 is 157 ± 14, whereas between the West African groups of table 3.9.1 one finds 211 ± 17. The difference is significant, showing that Bantus are more tightly clustered than West Africans, as might be expected, considering that the Bantu expansion took place in the last 2500

or at most 3000 years, while the differentiation among West Africans must have started earlier. The meaning of the comparison between the two F_{ST}s is complicated, however, by the fact that hybridizations with other external populations (e.g., Pygmies, Nilo-Saharans, and Khoisanids) contributed to the heterogeneity between Bantu subgroups. As for the heterogeneity among West Africans, it must be much older, as the linguistic evidence clearly indicates. Many West African people have been in place for a long time. They have received immigrants from the Sahara and had their own expansions under the impulse of the first agricultural development, but under the constraints of a fairly heterogeneous environment. Nevertheless, the major cluster of West Africans seen in the tree may be the outcome of a single, major agricultural expansion, probably earlier than the Bantu radiation. The Bantu expansion may to some extent have been a late continuation of it, directed toward the east

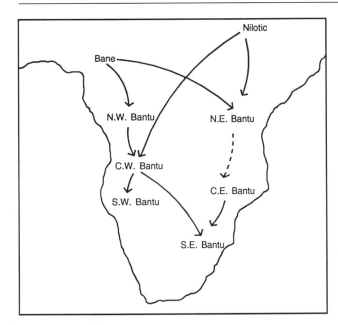

Fig. 3.9.3 Pattern of Bantu expansion, as suggested by the genetic data, with direction of arrows based on archaelogical information.

and the south because these areas were not yet occupied by farmers at the time. The Bantu expansion took place almost in a vacuum, because only hunter-gatherers occupied Central and South Africa at the time, and is thus comparable to the Neolithic expansion in Europe. At the

time of European colonization, the Bantu expansion was not yet complete; while West Africa was already heavily settled by farmers, whose further movements were frozen by the colonial powers, South Africa was occupied mostly by Khoi and San.

The hypothesis of a single expansion of West Africans mentioned above might be contrasted with the alternative one that three major clusters, one in Senegal, one in Nigeria, and one including the Mande, Gur, Ewe, Volta, and Kru (fig. 3.5.1) correspond to three independent agricultural expansions that started at similar times and proceeded with little reciprocal interaction. A finer analysis of genetic-linguistic-archaeological data might allow one to choose between these hypotheses, or make others. Only the most eastern group, in Nigeria, had the chance to develop beyond the boundaries of West Africa, settling in Central and South Africa. In the course of this final expansion, there was also some blending with populations and cultures of the eastern Sudan.

A final consideration concerns the use of the word *Bantu*. Clearly, Bantu is originally a linguistic term, and frequently concern has been expressed when it was used to define populations studied from a biological point of view. But, as already emphasized by Hiernaux (1975), the geographic expansion of people speaking Bantu languages has generated a group of populations that are genetically similar enough that the word "Bantu" also covers the biological reality, a conclusion we are able to confirm.

3.10. STUDIES OF SINGLE GENES

An apparent anomaly of the African maps is that Madagascar is shown only in some maps, that is, for the few genes for which data were available. The location of data collections was not indicated since the map reports the average gene frequency for the whole island.

ABO, the most thoroughly studied gene, shows a relative wealth of *B* (Africa being second only to Asia), mostly at the expense of *A*. *O* is about the same in Africa and Europe, on the average. *A* is higher in North Africa than in the center because of ancient Caucasoid flow from the Near East and perhaps Europe, caused by *A1* rather than *A2*. There is a slightly greater concentration of *A* in the south than at the center, but the data for the *A* antigen and for *A1 + A2* are inconsistent; one does not find, in the latter, an explanation for the relative maximum of *A* on the southwestern coast, between Angola and Namibia, but we did not find *A1 + A2* data in that region. *A2* has its relative maxima in East Africa and a low frequency in the west. *B* has its maxima at the extreme northeast and the extreme south, areas in which *A* also tends to be high and

O has its minima. We cannot offer any explanation for this behavior. The variograms show a fairly regular linear increase for 1500 miles or more. *A* shows conspicuous irregularities at more than 3000 miles, because of the low frequencies at the extreme northeast and variation in the south, and *B* shows a relative minimum at more than 2000 miles. The variograms of *A1* and *A2* are flatter.

Acid phosphatase (*ACP1*) has an interesting maximum of allele *B* in a region approximately between Mali, Niger, and the Upper Volta. This peak, which is confirmed by other markers, could have been an area of early agricultural development. The West African populations, with the exception of the Senegalese and others, have a central position on the PC map of African populations (fig. 3.9.2: Mande, Gur, Ewe) and may have been near the origin of a first agricultural expansion. That origin probably antedated the Bantu expansion, given that it is today in fairly arid land, and was perhaps an important center among those that made all of West Africa very populous. Another gene peaking in this region is Hemoglobin C (sec. 2.14), which may have or have had a function of resistance to a malarial parasite. We see that several other alleles peak in this area, however, and have never been associated with malarial resistance. *ACP∗B* is known to vary with climate in a way probably unrelated to malaria, and the hypothesis explaining this peak on the basis of an agricultural expansion of a local isolate may be more likely.

Adenylate kinase (*AK1∗1*) appears in two anomalous regions, probably not related historically to one another: the western Sahara (possibly because of the West African expansion hypothesized above) and the Khoisan area. Wealth of minority alleles in the Khoisan may be ancestral, whereas that in the western Sahara may be due to more recent local accidents.

The *M* allele of α-1 antitrypsin (*PI*), has a modest north-south gradient and a rather flat variogram.

Cholinesterase 1 (*CHE1∗U*) shows a very simple gradient northwest to southeast, but the poverty of data and the high gene frequency make the map less likely to reveal any complicated behavior of the gene-frequency surface. The variogram is approximately linear with a reasonably high slope.

Esterase D (*ESD∗1*) shows a gradient at almost 90° to that of *CHE1∗U*. The minimum is in the Suez region and the maximum in southern Africa. Only historical explanations seem possible for this cline, which has a fairly high slope as shown by the variogram and an even higher slope at long distances.

The Duffy blood-group system (*FY*) has a peculiar geographic distribution and an interesting selective explanation. It has been shown by in vitro experiments (Miller et al. 1976) that *FY∗O* red cells cannot be invaded by a particular malarial parasite, *Plasmodium vivax*. This may explain why *Plasmodium vivax* is absent over an area across West and tropical Africa. It is well known that malarial vectors grow especially well in an agricultural environment. They are not necessarily confined to them; for instance, Mbuti Pygmies, who are hunter-gatherers in a tropical forest and have been least exposed to agricultural development, have one of the highest frequencies of hemoglobin S and also have a 100% frequency of allele *FY∗O*. Outside Africa, *FY∗O* is rare or absent. Proof of its complete absence is difficult, because the gene is recessive. Therefore, even if no Duffy-negative individuals are observed, one cannot exclude a low frequency of a *FY∗O* allele, or of any other recessive *FY* allele different from those that can be demonstrated antigenically, *A* and *B*. However, the *FY∗O* allele as defined now is rarely found outside Africa.

As far as we can tell at present, the dark band of high *FY∗O* gene frequency may represent the region where the *P. vivax* parasite is absent. In fact, if susceptible human individuals become rare under natural selection for resistance to it, and the parasite has no other intermediate host in which to undergo its sexual cycle, the parasite will disappear. It was therefore not surprising that among Pygmies, who are 100% *FY*-negative, *P. vivax* infections have never been found (Pampiglione and Ricciardi 1986). Probably as a consequence of the high concentration of *FY∗O* in central Africa, variograms are rather flat with irregularities. The other two alleles, *FY∗A* and *FY∗B*, show a somewhat parallel behavior, probably because they have both been progressively replaced by *FY∗O* under selective pressure. It is likely that *FY∗O* originated in Africa and perhaps spread rapidly in areas rich in malarial parasites, to which the new mutants showed a degree of resistance. An opposite hypothesis has, however, been suggested (see sec. 2.14).

Glucose-6-phosphate dehydrogenase (*G6PD*), a sex-linked gene, is, like hemoglobins, under natural selection since *G6PD* deficiency involves resistance to malaria. Hundreds of different deficient mutants have been identified and the overall pattern of geographic distribution is probably unaffected by mutational history, but rather represents the partial geographic distribution of the malarial parasite(s) to which it confers resistance. *A* and *B* are electrophoretic variants polymorphic in Africa. *A+* has practically normal enzyme activity; *A−* has activity reduced to about 15% in males. The geographic distribution of *A+* shows a major peak in the northeastern part of the Gulf of Guinea, with a variogram that increases regularly up to 2000 miles. *A−* shows two peaks, the major one in West Africa where the first agricultural development arose and where *Plasmodium falciparum* malaria transmission is very high, constant throughout the year and not seasonal. The variogram is linear, with a high slope, for more than 2000 miles.

G6PD∗def is the sum of existing mutants; it seems to show various peaks. The selection coefficients involved (Cavalli-Sforza and Bodmer 1971a) may determine whether it increases to high levels of gene frequency in a few thousand years. The gene is recessive and, because it is sex-linked, is exposed to more powerful adverse selection in males. Heterozygous females are believed to be at an advantage over both homozygotes, normal and deficient. Heterozygous females have about 50% of red cells with, and 50% without the enzyme (or lower levels of it). The parasite can also prosper in red cells that are devoid of the *G6PD* enzyme, but it must adapt to this host cell environment before it develops at a normal rate (Luzzatto et al. 1985). In section 2.14, we have discussed present knowledge.

Glutamate-pyruvate transaminase (*GPT*) allele 1 has limited genetic information, generating a simple map with a south-center to north-east slow cline. The variogram is linear for almost 2000 miles.

Glyoxalase 1 (*GLO1*) is poorly studied. The north-south gradient of allele 1 could be of historical origin, with a pattern determined by Caucasoid genes from northwestern Africa. The variogram is linear for almost 3000 miles.

Group-specific component (GC) is a protein that binds vitamin D. As already explained (sec. 2.14), it has been suggested that its frequency varies with latitude as a function of the intensity of solar radiation and the need for producing endogenous vitamin D. The African pattern does not show a clear agreement with this hypothesis, but perhaps the variation of solar intensity in Africa is insufficient to cause a clear-cut selective difference. The lowest frequencies of GC*1 are observed in the savannas of East Africa where solar radiation is not low. This gene has many alleles, but only a few have adequate geographic information. A variant of GC*1, GC*1F, has a modest amount of data, but shows increased incidences in African Pygmies; another allele (GC*1S) has a higher frequency among Bushmen (Wang and Cavalli-Sforza 1986). Its map is not shown because of the poverty of the data.

Haptoglobin (HP) is a protein that binds hemoglobin and may be under selective control, especially where diseases causing profound hemolysis, like malaria, are present. It is difficult, however, to understand the presence of both high and low peaks of HP*1 in highly malarial regions, for example, West and Central Africa. A technical problem is worth remembering. In areas and periods of the year in which malarial infection is especially heavy, there is no or very little haptoglobin in the blood in a large fraction of individuals, because HP protein bound to hemoglobin is eliminated from circulation. It cannot be excluded that the phenomenon is more frequent in some genotypes, altering the evaluation of gene frequencies (see a discussion in Bernini et al. 1966; Cavalli-Sforza 1986) and simulating the genetic absence of HP caused by a rare HP*0 allele. The variogram of HP*1 is almost linear for about 3000 miles.

The HLA data come from relatively few areas, but their geographic distribution is more satisfactory than for any other gene, since their collection was organized by HLA workshops and maps are worth considering even if the data are few.

HLAA*1 is rare in Africa, except in the Caucasoid regions; it is especially frequent in Europe. A*2 follows a similar pattern. A*3 has a distribution indicating a greater frequency in West Asian Caucasoids; A*9 has a distribution somewhat opposite to that of A*3. A*11 shows again a northern Caucasoid type of distribution. A*19 shows higher values in the south than in the Caucasoid areas. A*28 has a peak in the area of Upper Volta, and a presumably unrelated one near the Red Sea. A*29 has a peak in the Northwest Bantu area.

HLAB*5 is infrequent in the south; B*7 has a peak in the Upper Volta region. B*8, a typical Caucasoid marker, has a fairly uninformative distribution. B*12 shows a different behavior in northern and eastern Caucasoids and a minor peak in the eastern tropical forest region. B*13 is highest in the eastern Caucasoid region and in the south; B*14 has a weak east-west gradient, and B*15 a weak north-south gradient, plus a certain amount of variation especially in the south. B*17 is very variable. Sub-Saharan Africa shows higher frequencies. Interestingly, the San are highest, whereas the Khoi form a local trough. B*18 has a minor peak in the eastern Caucasoid area, where B*21 has a more pronounced peak. B*22 and B*27 have a modest north-south difference. B*35 is highest in West Africa, and B*40 is higher (> 0.06) along the western stream of the Bantu expansion.

Most of the variation of HLA in Africa seems to mark the difference between Caucasoid and Black African areas, but many alleles show conspicuous differences between northwestern and eastern Caucasoid areas, indicating at least two doors of entry of Caucasoids into Africa; in the latter area, however, only one population was tested. Peaks or troughs occur near the Mediterranean coast opposite Gibraltar and in Libya. Most variograms have regular initial increases and remain linear for 1500–3000 miles, but at least six have a very flat or negative initial slope. It would be desirable to redo these maps with more data.

The Caucasoid alleles za;g and zax;g of IGHG1G3 of course show higher frequencies in the Caucasoid areas of Africa. An even more typical Caucasoid allele is f;b0b1b3b4b5. The first peaks on the northwestern Atlantic coast and the other two on the central and eastern coast of the Mediterranean. In all sub-Saharan populations, which carry instead the Z and the A antigens, f;b0b1b3b4b5 is rare or absent.

The za;b0b1b3b4b5 allele shows two peaks, most probably unrelated historically and perhaps also immunologically, on Madagascar and in the Ituri forest where Mbuti Pygmies are located.

Just as za;b0b1c3b4b5 shows a peculiar peak in northern Ethiopia, za;b0sb3b5 shows another such peak in the Khoisan area; za;b0b1c3c5 is more evenly distributed.

All variograms of GM alleles have initial positive slopes and are linear between 500 and 1500 miles.

Another immunoglobulin gene located on chromosome 2, KM, controls the constant portion of the κ light chain. The allele (1 & 1,2) is rare, as expected, in the northern Caucasoid region, particularly in the western zone. The variogram is linear for 1500 miles.

The Kell blood-group system (KEL) includes the K allele, which has most of its higher values in the Caucasoid areas, and Jsa, which has a peak in central Africa with a regular centripetal gradient. This may indicate the origin of the allele. Kidd (JK) is another blood-group system. Allele JK*A has a behavior difficult to characterize. Variograms are regular.

The Lewis blood group (*LE*) has the highest values of the allele *Le* on the Mediterranean shores and shows a regular north-south gradient. The allele *A* of Lutheran, another blood group, shows a difference between the northern and eastern Caucasoid zones. Variograms show regular increase, especially for Lewis.

MNS is a blood-group system with several alleles, of unknown function. The *M* allele has a peak in the Khoisan region and, subordinately, near the Red Sea; the Niger-Mali region shows a minimum. Another allele, *S*, at a close but different locus within the *MNS* system, is lowest in West Africa and highest near the Mediterranean coast. The two loci form four alleles in strong linkage disequilibrium: *MS, Ms, NS, Ns*. The first is highest in the eastern Caucasoid region and lowest in West Africa; *Ms* has an almost regular north-south gradient with a peak in the Khoisan region. *NS* has a strong peak opposite Gibraltar, but its gene frequency does not continue as high on the European side. *Ns* has a high peak on the Nigerian coast and a secondary one near Gibraltar. Because *NS* and *Ns* should drift independently, this is a weak indication of selection favoring the *N* gene. Another antigen that is especially common in Africa and rare or absent elsewhere, S^u is highest in Mbuti Pygmies, but also frequent among non-Pygmies extending toward the northwest and elsewhere. The S^u mutation probably originated in Africa. Variograms show regular increases except for S^u and *Ns*.

Another blood group, *P1*, shows the highest frequencies of allele *1* in sub-Saharan Africa and in the Khoisan region, with a peak on its border. The variogram is fairly regular.

Peptidase A (*PEPA*) allele *1* has an east-to-west gradient, with a low value approximately centered in Nigeria. Phenylthiocarbamide (*PTC*) tasting shows a north-south gradient of allele *T* (taster) with an east-west difference. One of the phosphoglucomutase genes, *PGM1*1*, has a north-south difference with various peaks and troughs. The most pronounced trough corresponds to the Mali-Niger region. *PGM2*1* has a distribution suggestive of spread with the western stream of the Bantu expansion. *PGD*, or 6-phosphogluconate dehydrogenase shows a high frequency of allele *A* except in the north and east, with four pockets of frequency close to 100%. In the Niger-Mali region, however, the frequency is especially low. All these enzymes show regular variograms, with initial portions that increase linearly for 800–2000 miles; *PEPA*1*, however, is irregular.

Rhesus *RH* is the most polymorphic blood-group system. It can be examined in terms of each of three genes, *C*, *D*, and *E* and also in terms of some variants, as well as of major haplotypes. The *C* gene shows a clear north-south gradient, being frequent among Caucasoids and almost absent in sub-Saharan Africa; there are peaks in Egypt and the northwestern Sahara. *D* has a more irregular pattern, as shown by several peaks and troughs, but also has a north-west gradient and is very frequent

in sub-Saharan Africa. The north-south gradient is most probably determined by the influx of *d* (the *RH*-negative allele) from Europe. The third gene, *E*, shows a peak in Libya and various troughs. Another marker of the *RH* system, *V*, shows a surface with a minimum in the Khoisan region and a maximum in Ethiopia. The D^u allele of *D* has an irregular surface with a considerable number of peaks and troughs. It is not an easy marker to test, and some of the irregularities may conceal variation in the testing techniques. It is most probably of African origin. Although the variograms of the other genes are regular, D^u has a strong negative gradient for 3000 miles, probably because of conspicuous irregularities of its surface.

Among the common *RH* haplotypes, *CDe* (also called *R1*) has a clear north-south gradient with peaks and troughs, somewhat opposite to that of *D*. Another major haplotype *cDE* (also called *R2*) has a maximum in the northeast and several minima. The standard *RH*-negative haplotype, which is of European origin, *cde* (also called *r*), has a maximum in the north; the peak in Libya may be due to drift or to contact via Sicily. The surface is especially complicated in the south. The most important haplotype in Africa, *cDe* (also called *Ro*), most probably originated in the south and decreases regularly toward the north. The high frequency of the *cDe* haplotype in Africa is in good agreement with the hypothesis of African origins of modern humans: *cDe* is also the original haplotype in the evolutionary genesis of *RH* haplotypes according to Fisher's hypothesis (see sec. 2.14) (Feldman et al. 1969). The variograms of all the *RH* haplotypes show regular initial increases for the first 1000–2000 miles.

Secretor (*SE*), a gene controlling the secretion of *ABO* substances in various fluids, has a largely flat surface except in the south. The initial slope of the variogram is negative for the first 1000 miles.

Transferrin (*TF*) has high frequencies in the Caucasoid areas, which have almost exclusively the common allele *C*; West Africa and the western-stream Bantu have the highest frequencies of the (almost) complementary *D* allele, which probably has a local origin. The variogram has a regular positive initial increase for almost 2000 miles.

A summary of these descriptions gene by gene is, of course, available in the principal-component maps, which are given in the next section. At the single-gene level, one finds a variety of patterns, of which six are the most important.

1. The maxima or minima are in the general Caucasoid area, including the north and the east (above the equator).

2. Variants of this pattern show maxima or minima (a) in the western Caucasoid area, perhaps referring to an ancient entry of populations from western Europe via Gibraltar (after 30 kya); (b) in the northeastern area, where exchange probably occurred at many different times (including very early ones), an important late one occurring with the agricultural Neolithic expansion from

the Near East, and the latest one with the Arab invasion; (c) in the eastern area, from Arabia across the Red Sea or via the Indian Ocean. This third contact may have happened repeatedly at many different times. A further possible area of contact is via Sicily.

3. The south is mainly occupied by aborigines of Khoisan origin who usually show some genetic difference from neighbors.

4. Agricultural expansions from West Africa are perhaps reflected in various pockets of high or low gene frequencies in various parts of West Africa, with Senegal and Niger-Mali being the most important ones. Expansion from Niger-Mali is often visibly connected with the western Bantu expansion more than with the eastern one. The western expansion (from Senegal) was probably earlier, and the more eastern one (Niger-Mali) did not go as far south but included rather strong participation by people from central and western Sudan.

5. Pygmies are responsible for some peaks and troughs of gene frequencies in the eastern or western tropical forest.

6. It is difficult to separate climatic effects from historical effects since the areas that have more characteristic climates have also been populated by people of non-African origin like Caucasoids, or people of uncertain origin like the Khoisanids, who had a mostly unknown but probably complex history before settling in the areas that they now occupy exclusively. In general, however, a large fraction of the genetic differences observed in Africa probably are historical in origin. Special climatic zones like the Mediterranean are limited to a narrow strip along the coast, and Caucasoid genes seem to occupy a much wider territory. Khoisanids are likely to have lived in less arid areas than those they now occupy, since they were most probably pushed back into marginal environments by invaders.

3.11. SYNTHETIC MAPS OF AFRICA

The synthetic maps have been obtained by calculating PCs from 79 genes for which data seemed sufficient to allow the construction of single-gene maps. The 79 genes considered for this analysis are the 83 for which genetic maps of Africa (see Table of Genetic Maps) are shown in the second part of the book except for the 4 alleles $G6PD$ $*def$, $G6PD*A^-$, $G6PD*A^+$, $G6PD*B^+$. The fractions of variance explained by the first seven components are shown in table 3.11.1, which altogether explain 82% of the total variation. The genes that account for most of the variation are listed in table 3.11.2, indicating the highest correlations between the genes and the individual components.

The first synthetic map, accounting for more than one-third of the total variation, strongly confirms the importance of the north-south difference, which shows clearly in the first principal component. The gradient is clearly rooted in the relatively ancient presence of Caucasoids in a northern strip along the Mediterranean and in additions from West Asia, which are visible in the second and third components. The present population of the Sahara is Caucasoid in the extreme north, with a fairly gradual

increase of Negroid component as one goes south. Traveling through the Sahara on the Algiers-Lagos axis (in southern Nigeria) by the classical route that goes through Tamanrasset in southern Algeria, one can observe the genetic gradient with the naked eye as a skin-color gradient, obviously with an intensity that is the inverse of that shown in the map (fig. 3.11.1). In the Sahara, however, the population is very sparse and concentrated in a few oases that are hundreds of miles from each other, making the gradient seem more abrupt. Moreover, the population is segmented in different tribes and ethnic groups that show skin-color differences and, presumably, parallel variations in genetic Caucasoid-Black admixture. The situation may be more complex on the central route across the Sahara, going from Libya through Tibesti and Chad to the Central African Republic, which has been closed to traffic for political and military reasons for more than 30 years. Here, in the south-central Sahara are many interesting populations, but this area has been closed to foreigners for a long time, and the genetic information is close to zero.

It is of interest to compare the next two PCs, the second (fig. 3.11.2) and the third (fig. 3.11.3). Both show a maximum in the northern part of Ethiopia, but the relations between the center and the rest of Africa are very different. Although the third PC shows no other similarities between the Ethiopian pole and other African areas, the second PC shows that the northern Ethiopian peak has a strong similarity with the Khoisan area and has the most marked contrast with West Africa. Thus, the second synthetic map reinforces the hypothesis that there was originally a population ancestral to the Khoisans, located in an area not far from Ethiopia (see sec. 3.7).

Table 3.11.1. Percentage of Total Variance Explained by the First Seven Principal Components of African Gene Frequencies

Principal Component	% of Total Variance	Principal Component	% of Total Variance
1	34.6	5	4.4
2	18.6	6	3.6
3	10.3	7	3.2
4	7.0		

Table 3.11.2. Genes Showing the Highest Correlations with the
First Six Principal Components of African Gene Frequencies

P.C.*	Range of Correlation Coefficient	Genes
1	> 0.90	(+) HLAA*1, IGHG1G3*f;b0b1b3b4b5, RH*CDe, RH*C, LE*le
		(−) RH*cDe
	0.90 − 0.80	(+) FY*B, HLAB*27, IGHG1G3*za;g, PGM2*1, GLO1*1, HLAA*2
		(−) KM*(1&1,2), MNS*Ms, GC*1F, FY*O
	0.80 − 0.70	(+) IGHG1G3*zax;g, RH*cde, PI*M
		(−) GPT*1, HLAA*10, IGHG1G3*za;b0b1b3b4b5, JK*A, RH*D, PTC*T
2	> 0.90	(+) G6PD*B+
		(−) —
	0.90 − 0.80	(+) HLAB*13
		(−) G6PD*A-, G6PD*A+, IGHG1G3*za;b0b1b3b4b5
	0.80 − 0.70	(+) PEPA*1, MNS*MS, MNS*M, IGHG1G3*za;b0b1c3b4b5, HLAB*18
		(−) HP*1
	0.70 − 0.60	(+) HLAB*14, HLAB*15
		(−) G6PD*def, HLAB*35
3	1.00 − 0.80	(+) RH*V
		(−) HLAA*9
	0.80 − 0.70	(+) HLAA*3
		(−) —
	0.70 − 0.60	(+) HLAA*28, HLAB*21
		(−) —
	0.60 − 0.50	(+) ACP1*B, AK1*1, LU*A
		(−) ESD*1, IGHG1G3*za;b0stb3b5
4	> 0.60	(+) HLAB*12
		(−) —
	0.60 − 0.50	(+) HLAA*29
		(−) HLAA*28
	0.050 − 0.40	(+) ABO*A1, AK1*1, HLAB*40, KEL*Jsa
		(−) GC*1, HLAB*18, HLAB*7
5	0.70 − 0.60	(+) HLAB*40, HLAB*7
		(−) —
	0.60 − 0.50	(+) FUT2(SE)*Se
		(−) AK1*1, HLAB*14
	0.50 − 0.40	(+) JK*A
		(−) —
6	0.60 − 0.50	(+) —
		(−) AK1*1
	0.50 − 0.40	(+) RH*cDE, MNS*MS
		(−) HLAB*35, G6PD*A-
	0.40 − 0.30	(+) HLAB*17, HLAB*7
		(−) HLAB*5

Note.− Genes giving positive or negative correlation values are indicated by
(+) or (−), respectively.
* P.C., Principal component.

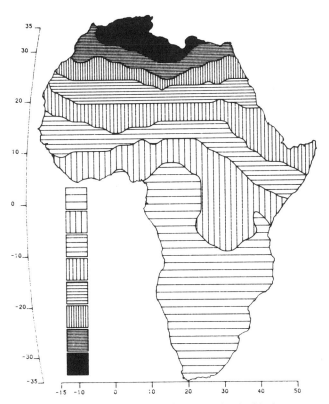

Fig. 3.11.1 Synthetic map of Africa obtained with the first PC.

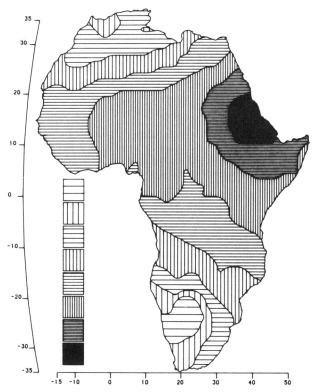

Fig. 3.11.2 Synthetic map of Africa obtained with the third PC.

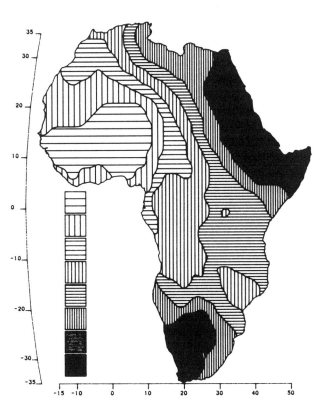

Fig. 3.11.3 Synthetic map of Africa obtained with the second PC.

Because the region between Ethiopia and South Africa was heavily settled in the last two millennia by the Bantu coming from the west, there is no visible genetic connection between the two peaks.

In the second synthetic map (fig. 3.11.2), the pole opposite East Africa lies in West Africa, the region probably ancestral to all the agricultural expansions to Central and South Africa that took place since the beginnings of the development of agriculture 4000 or 5000 years ago, including the last and best-known one, that of the Bantus, which started around 3000 years ago or somewhat earlier and seems, in this map, a continuation of the population expansion begun earlier, west of Nigeria. The Bantu expansion itself is more visible in the fourth synthetic map given later.

The third synthetic map (fig. 3.11.3) shows maximum divergence between the Ethiopian peak and the Khoisanid area, in which one observes the lowest values of the third PC. This PC emphasizes the differences between Ethiopians and Khoisans, whereas the second PC shows their similarity. The two PCs elaborate further the observation made in the tree (sec. 2.3), that Ethiopians and Khoisans differ from the other "core" Africans in a similar direction: they are both similar to Caucasoids but are otherwise very different.

In the fourth map (fig. 3.11.4), we note two opposite poles, one approximately in southern Cameroon and one in the Mali–Niger (and Burkina-Faso) region. The

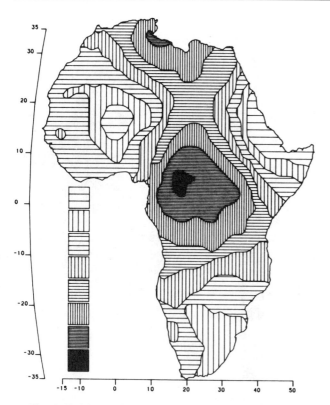

Fig. 3.11.4 Synthetic map of Africa obtained with the fourth PC.

first may be considered a major expression of the Bantu expansion, and the second of an earlier agricultural expansion in West Africa. Recent archaeological discoveries have given considerable importance to Cameroon and Gabon as nuclear areas of the western Bantu stream (Vansina 1984). Although the archaeological area that has been investigated is somewhat to the west of the darkest area in the synthetic map, the density of genetic data and that of archeological data may be insufficient to pinpoint exactly the areas of origin. As for the area of expansion at the boundary of Niger and Mali, it tends to coincide with maxima or minima of several genes (sec. 3.10) and of hemoglobins *C* and *D* (sec. 2.14). This area is repeated in more or less greater detail in other components (fifth and sixth, not given here), and a comparison with archeological data would be worthwhile.

The color picture conveys 63.5% of the original variation and shows clearly that there are three major ethnic zones in Africa. From north to south one notices the green band that represents the Caucasoid northern African; red in the west, center, and part of the south

corresponds to the characteristic Black African type; and blue in the south, the Khoisan ethnic type (Bushmen and Hottentots), whose skin is not so dark and who has some features in common with Asians, as already discussed, and may have been present in East Africa at an earlier time. The relative importance of the divergence of the Khoisanids from other Africans may be exaggerated by the intense blue color. It is also interesting that there is evidence of important Khoisan gene flow into the Bantu tribes in the area.

There also appear four other minor centers of ethnic differentiation. An approximately circular area in the western Sahel includes the eastern half of Burkina Faso (formerly Upper Volta), much of the southeast of Mali, the western part of Niger, the northwestern corner of Nigeria, and a northern slice of Ghana, Togo, and Benin. We have mentioned that this might be the result of an early agricultural expansion following the retreat from the Sahara. In addition to the markers used for building these synthetic maps, there is another interesting anomaly in this region: a high frequency of hemoglobins C and D. They may or may not have played a role in resistance to malaria, perhaps at an earlier time. This ethnic zone seems to continue to the southeast into a somewhat differently colored area centered in the Cameroons, which may represent the nuclear area of the beginning of the Bantu expansion to the south and the east. It is not a coincidence that this area is in part connected with the Burkina Faso-Mali area described above.

An altogether different brown color is observed in the area immediately to the west of the great lakes, Victoria and Tanganyika, in particular. This region—politically corresponding to Rwanda, Burundi, Uganda, and western Tanzania—has been and is still to a large extent occupied by Pygmies, and the difference in color from other neighboring zones is probably due to their presence.

A fourth ethnic area that differs from neighboring ones is near the southern coast of the Red Sea. The nuclear area includes the north of Ethiopia, Eritrea, and extends to most of Ethiopia and Sudan. It probably expresses the genetic peculiarities of Nubians, Beja, and northern Ethiopians. The possible importance of both the agricultural expansion that took place in Ethiopia and the genetic contributions from southern Arabia are worth keeping in mind.

This information is really not different from that already discussed on the basis of the monochromatic, single-component maps, but is presented in a more compact way.

3.12. SUMMARY OF THE GENETIC HISTORY OF AFRICA

If modern humans originated in Africa, Africans must have migrated to Asia at an early time, through Suez, but possibly also further south, from Ethiopia. There clearly were important human developments in West Asia, after which the expansion eastward and northward may have occurred between 60 and 40 kya; at that early time there may have already been retrograde flows toward Africa. Perhaps the most exciting finding is the intermediacy of the Khoisan, which could be described as part African and part West Asian. A possible and simple explanation is that earlier Khoisans inhabited another region, closer to Asia, perhaps East Africa or Arabia or the Middle East, where hybridization occurred. This hypothesis is very speculative at this stage and, as stated earlier, we give little credence to the hybridization date of about 20 kya derived from the data, except that it shows that whatever relationship exists must be rather old. It is nevertheless very interesting to compare this hypothesis with the superficially opposite explanation derived from mtDNA data (Vigilant et al. 1989; see sec. 2.4 and 3.7), that the Khoisans are the closest descendants of the original human types. One realizes that gene-frequency information is unlikely to distinguish between the two hypotheses, whereas mtDNA can; but present mtDNA information is too limited, and analysis of the existing data is not sufficiently complete for full acceptance. Confirmation of these observations with data from another region of mtDNA would also be very important. When they can be carried out, archaeological studies of Arabia and further archaeological and genetic studies of Ethiopia may add considerable information.

Signs of a possible early Khoisan presence in Ethiopia and East Africa come from archaeological finds (disputed by some), linguistic cues (also hotly disputed by others), and a genetic indication, the presence of Asiatic genes among Ethiopian Pygmoids, a clear-cut testimony (see van Loghem et al., as well as Nijenhuis and Hendriksen in Cavalli-Sforza 1986a). Despite the smallness of the sample and the population, it is remarkable that more than one gene shows Asian alleles not found elsewhere in Africa. Other populations in the same area of western Ethiopia, in particular the patch of tropical forest near Bonga, should be investigated further, because this area has the potential for harboring relic populations. In the rest of Ethiopia, major historical developments are likely to have annulled or absorbed most prehistoric populations; thus, one can hope to find some "living fossils" only in the least accessible areas.

If East Africa (or earlier, Arabia) was the place of origin of the ancestors of the Khoisan, they must have later moved south. The southward expansion of the Khoisan may have occurred early and was probably reinforced by later migrations under the push of Bantu expansions. According to A. Morris (pers. comm.), however, there are no recognized Bushmen remains in South Africa before 12 kya, but this may also be due to the difficulties of recognition. The presence of Caucasoids in northern Africa is attested to by the archaeological information. Caucasoids arrived in the western part of North Africa from the Iberian peninsula at an early time, perhaps 20 kya or more. Migrations of Neolithic farmers from the Middle East are likely to have been numerically important

in the last 10 ky, but there must have also been earlier migrations. Suez, like Panama, is a narrow funnel, but it must have been crossed many times in prehistory.

In the Neolithic period, the northern and central part of the Sahara was probably populated by Caucasoids and the central and southern part by Negroid peoples. The few Berbers who are left in the western Sahara may have originated in part from Capsian Mesolithic and in part from early Middle Eastern Neolithic ancestors, both Caucasoids. Only much later did Arabs and Bedouins enter the picture; most of them mingled with the local Berber population, although a few (e.g., the Mzab, near the oasis of Ghardaia in Algeria) may have remained largely unmixed. There may have been some Negroid admixture of early Saharan populations, perhaps from slaves raided farther south or from Black Africans who were living in the Sahara in the early Neolithic. Some, perhaps many, Negroids from the Sahara were pastoral nomads when the Sahara started drying up and were forced to migrate to the south, settling first in the Sahel. A fraction of these people have continued to practice cattle herding to the present day and have developed high frequencies of lactose tolerance as a consequence of fresh milk consumption (Flatz 1987). Their cattle is perhaps of Middle Eastern origin but may also have been domesticated independently in the Sahara at about the same time or possibly even earlier. Other Saharan Negroids who migrated to the Sahel from the Sahara developed new domesticates from local wild plants, especially cereals, which were already adapted to the new climatic conditions. New crop development took place in a narrow strip below the Sahara from the Atlantic almost to the Red Sea, and several new domesticates appeared. West Africa may have been the first part of the continent to experience an important population increase from farming. The genetic data support at least two independent expansions: one in western Senegal and one starting in the Niger–Mali–Burkina-Faso region, the latter probably giving rise to the modern Gur, Mande, and Ewe, who are today located farther south. For settling in the more humid regions closer to the Atlantic coast of Guinea, the southern part of West Africa, further developments of crops that could better withstand the humid climate were necessary. These new domesticates may have also helped in the later spread to Central and South Africa. Nigeria, on the eastern border of West Africa, has reasonable genetic similarity with populations to its west and east. From Nigeria, the expansion could only be eastward and southward since agriculture developed in the west at the same time or perhaps even earlier. The development of iron smelting in the sixth century B.C. or before certainly favored expansion from Nigeria, but when this early Iron Age started, there had already been important pre-Iron Age expansions toward Cameroon and Gabon. Genetic data strengthen the recent linguistic and archaeological findings that there was an important western stream of Bantu speakers from Cameroon down the coast, and probably also more toward the interior.

The early connections between the origins of the western and eastern Bantu streams are more obscure. It seems likely that Bantu and Nilotic speakers intermingled, probably in the region around Lake Victoria where Urewe was located. This area may have been a melting pot, sometimes genetic and sometimes cultural, with Bantu languages being more often successful, but still allowing the diffusion and survival of Nilo-Saharan languages in East Africa.

Ethiopians, the Beja, and perhaps the Tuareg—who might be, contrary to the expectation generated by their Berber language, an offshoot of the Beja (or vice versa)—have a mixture of African and Caucasoid genes quantitatively similar to that of the Khoisan. The two mixtures are likely to have had a different historical origin but probably the same geographic cause, propinquity to Southwest Asia. If one tries to attribute the origin of the mixture of the Beja and Ethiopians to the same event, the hybridization date calculated from the genetic data for Ethiopians (3000 years ago) is too late. There may have been repeated genetic contacts between Arabia and East Africa during the last 5000 or 6000 years. One cannot exclude, however, the possibility that the Beja were of Middle Eastern origin, entered Africa from Suez, and acquired Negroid genes on their way south to their present location. If the common origin of Beja and Tuareg is confirmed, the westward migration and the acquisition of a Berber language by the Tuareg are likely to have been secondary events. Alternatively, it may be the Beja whose language was replaced.

The Bantu expansion is recognizable, at the genetic level, by the slightly greater internal homogeneity of Bantus compared with West Africans; Bantus are not, however, totally homogeneous. Genetics tends to separate the western and eastern streams and emphasizes the importance of the first. It also suggests considerable admixture of northeastern Bantu with Nilo-Saharans (at least with Nilotics, the principal representative here), which may have occurred in the region around Urewe. The nuclear area of the western stream is not located, genetically, in the Benue-Niger area, as was suggested by linguistic studies, but somewhat farther southeast. This should not be seen as a real disagreement with the hypothesis of Bantu origins in the Benue-Niger area. Genetic differences relevant to the problem are small and hardly significant. Minor population movements are common, and the power of geographic resolution of genetics and of linguistics is limited. The genetic map shows where populations differ enough from the background that they tend to take values of PCs sharply deviating from the mean. It is perfectly possible that the late Neolithic and early Iron Age developments in Cameroon-

Gabon were demographically important enough that a distinctive population arose locally and spread further south, although the actual beginnings were a little to the northwest.

It remains difficult to pinpoint an ancient place of origin for the Negroid type, which includes all West, Central, and South Africans. Contrary to many earlier opinions, modern Pygmies and Khoisans are not good candidates for a proto-African population. Khoisanids cannot be considered Negroid, if we accept that San are less mixed with Black Africans than Hottentots, in agreement with genetic evidence. But there is no such bar for Pygmies. Their skin is only somewhat lighter than that of local farmers (at least when they have not been exposed to the sun but live in the forest). Pygmies have extremely high frequencies for many typically African genes such as $RH*cDe$, $FY*O$ and others, and on this basis it is tempting to consider them as Proto-African, or at least less mixed with other latecomers than are other Africans. Unfortunately, the language of the Pygmies cannot help us, the original one having been almost certainly lost. However, the extreme frequencies of some alleles among Pygmies may also be interpreted on a strictly selective basis. It is also true that there are many indications of a high genetic isolation of Pygmies, especially but not only, the smallest ones (the Mbutis). This may explain their position as extreme outliers in the tree and the presence in both Biaka and Mbutis of many "private" or, more often, "semiprivate" polymorphisms: alleles reaching relatively high frequencies, and present in one, the other, or both Pygmy groups, but rare or absent in other populations (Cavalli-Sforza 1986a).

Altogether, Pygmies, Bantu, West Africans, and Nilo-Saharan peoples are all potential representatives of the modern African type, apart from possible minor Caucasoid admixture (most probably rare or absent in Pygmies) acquired through contact with Caucasoids in the Sahara or from Khoisanid admixture. For different reasons, Khoisanids and Pygmies are, however, outliers with respect to the main African type.

One group to which we have dedicated less attention because of unsatisfactory genetic knowledge is the pastoral nomads, most of whom may have originated in the Sahara, but probably in different parts of it: the west and center for Peuls, the center for Hausa, and the east for Nilotics. Their typical "elongated" phenotype makes it tempting to search for a common genetic origin. When the genes responsible for their remarkable body build are known, it will be possible to test whether they are shared by all or most elongated Africans, supplying some indication about a potential common origin. It is unlikely that the other genes will give a clue unless substantial population numbers and many genes are tested, because differences between most sub-Saharan Africans other than Khoisan and Pygmies seem rather small.

4 ASIA

4.1. GENERAL INTRODUCTION, GEOGRAPHY, AND ENVIRONMENT

Asia is the largest of the continents (nearly 17 million square miles), and its size and complexity make it the most difficult to study. Prehistory and history are very different in the various geographic regions, which will therefore be considered separately. In this introduction, we first give some general considerations, followed by a summary of the conventional geographic subdivisions.

Asia is really at the center of the world both geographically and historically. The geographically central position of Asia is shown by its direct connection with Africa at Suez and by the proximity of East Africa; by the direct connection, and indeed continuity, of Asia with Europe that would make it more reasonable to speak of Eurasia rather than of two distinct continents; by the direct connection with North America, through a strip of land called Beringia that existed at the time of the major flux from Asia to America; and by its proximity to both Australia and New Guinea. Thus, Asia is, or was in recent times, directly connected or extremely close to all the other continents.

The origin of modern humans is now believed to have taken place outside Asia, in Africa. However, one important area in which some of the earliest modern humans were found, East Africa, was very close to Asia but probably not physically connected with it during part of hominid evolution. Early contacts between Africa and Asia at Suez are also very likely. Once Asia was colonized, the path to expansion to the whole world was clearly open, but unfortunately we know very little of the actual routes by which modern humans spread for

the first time from the western part of Asia to its northern, northwestern, northeastern, southern, and southeastern parts. We have discussed in chapter 2 the possibility that there may have been two major routes of expansion from Africa to Asia, one through North Africa and one through Ethiopia; but clear genetic traces of the hypothetical southern route are detectable today only in southeast Asia, and perhaps, as we shall see, in South India. Also important are suggestions that an East Asian population hybridized with modern humans coming from the west (see sec. 2.1). In the western part of Eurasia, there was another human predecessor of *Homo sapiens sapiens* whose possible contribution to our subspecies is the subject of much discussion, *H. sapiens neanderthalensis*.

In any case, modern humans in Asia clearly underwent considerable demographic development that favored their extension to the rest of the world, even if the routes and timing of the expansion inside Asia are nebulous. We have approximate times only for the entry to the major continents. Later expansions are better understood, being tied to major technological development in food production that left clear archaeological marks. There were at least two major, and presumably independent, agricultural expansions, one in the Middle East (West Asia) and one in East Asia. They both started around 10 kya and the former expanded more widely. It originated in a zone with temperate climate in the middle of a large temperate area: southern and central Europe, North Africa, and Central Asia. Far Eastern agriculture originated in China, and here one can clearly speak of several independent developments that were, however, constrained by deserts and steppes in the west and north, by the highest mountains of the world in the southwestern direction, and by tropical conditions toward the south. When these two agricultural expansions had practically ended, the development of pastoral nomadism, a process initiated as a secondary and later adaptation of food production in the Central Asian steppes, was extremely successful in originating new expansions that continued for more than three thousand years, almost until the present. All these expansions have deeply affected the history and the linguistic and genetic geography of Eurasia.

Asia is not only the largest but also the highest continent. Mountains and plateaus occupy three quarters of the area. In the west, there is essentially no discontinuity with Europe. The geographic boundary is set somewhat arbitrarily at the Ural Mountains, but there is no effective barrier to communication, especially in the southern part. Longitude varies from 60° E at the Urals (25° E in Turkey) to 170° W at the Bering Sea (145° E in Japan). Latitude varies from 9° S of the equator in Indonesia (1° N in Singapore) to more than 75° N. Accordingly, temperature varies from tropical to arctic conditions.

It is common to distinguish seven large geographic regions that are ecologically very different (fig. 4.1.1).

1. *North Asia.* North Asia is covered along the Arctic coast by tundra (permanently frozen soil, no trees) that extends somewhat farther south in the mountain-ous parts of North Asia; south of the tundra belt is the taiga zone, with coniferous trees mixed with deciduous trees like aspen and birch; grass and shrub are found in drier parts. The North Asian environment extends from the Urals to the Pacific Ocean and continues west of the Urals into Europe to include major parts of northern Russia and Scandinavia. The northeastern part was connected with North America until the end of the last glaciation, when the sea level rose and submerged the flat land connecting Asia and America. The separation of the two continents was complete around 10 kya.

2. *Central Asia.* Central Asia, below North Asia, is a strip of almost homogeneous environment from the Urals (and west of them) to Korea (not included). This is the steppe, mostly unwooded grassland interrupted by deserts.

3. *Middle Asia.* Directly below Central Asia, Middle Asia extends from the Caspian Sea (including the Iranian plateau) to the northwestern part of China. From north to south, this region makes a more or less gradual transition from the steppe to a semidesert (brush, saltbrush, etc.) or desert with oases.

Middle Asia is separated from South Asia by mountain ranges. The Pamirs are in the west; and three major chains are in the east: the Tien Shan (north), the Kun Lun (central), and the Himalayas (south). Between the latter two lie the Tibetan Highlands and the climate is strongly affected in all these regions by altitude.

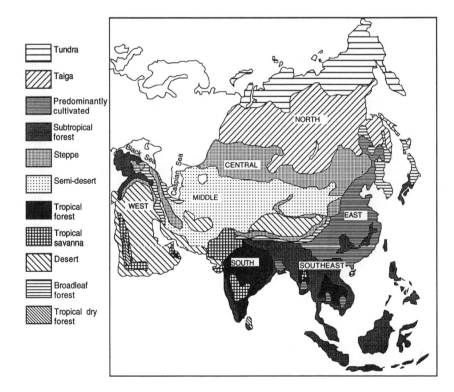

Fig. 4.1.1 Map of ecological and geographic regions of Asia.

4. *South Asia.* In the narrow sense, South Asia is the Indo-Gangetic plain corresponding politically to Pakistan and the north of India, plus the Deccan plateau of the Indian peninsula (the south of India). The western part is mostly tropical savanna. The climate in the north-central and peninsular area is mostly humid tropical, and the eastern area is mainly a humid, evergreen forest.

5. *East Asia.* East Asia is represented by the eastern part of China, Korea, and Japan. Japan was connected to the continent until the end of the last glaciation. It is primarily a cultivated land; the climate is that of a temperate zone, colder and drier in the north, under the influence of cold air masses from Mongolia, and warmer and humid in the south, under the influence of hot Pacific air masses.

6. *Southeast Asia.* Southeast Asia includes the peninsula of Indochina, the Malay peninsula, and the islands of Indonesia, Borneo, and the Philippines. This was a single land mass until the end of the last glaciation. It is covered by rain forest, evergreen, or deciduous humid tropical forest.

7. *West Asia.* West Asia is sometimes distinguished as West Asia proper, including the Anatolian plateau (Asia Minor) and the Armenian and Iranian Highlands, and Southwest Asia, made up of Arabia and Mesopotamia. It is a predominantly arid region, becoming a true desert in the Arabian peninsula and in much of the Middle East. The Mediterranean and Black sea coasts have a mediterranean climate.

Because of the great extent of Asia, and its consequent heterogeneity, we have chosen to treat each main region by itself from the point of view of prehistory and of history. Although geographically and ecologically distinct, Central and Middle Asia have enough common developments that they were pooled for our purposes. The next six sections deal with each of these geographic regions in sequence. A general summary of the linguistic picture of Asia is given in section 4.8, and of physical anthropology in section 4.9. The general genetic picture is discussed in section 4.10, with the next seven sections dedicated to Asian regions according to a breakdown suggested by the genetic analysis, which is similar, but not identical, to the standard classification used for the geographical and historical part given in the first seven sections.

4.2. PREHISTORY AND HISTORY IN NORTH ASIA

In the middle Paleolithic, North Asia was as cold as it is now. There is no evidence of human occupation of Siberia until 35,000 B.P. (Klein 1980), when signs of human presence begin to appear (at least as far as known today), showing people and cultures similar to those seen in Europe at the beginning of this period. Both Europeans and North Asians probably came from the Middle East. An important unanswered question is, when did a.m.h. first enter the Americas? The problem is more fully discussed in the chapter on the Americas, but one may anticipate different opinions, ranging between 35 kya and 15 kya (Fagan 1987).

Two primary upper Paleolithic cultures existed in Siberia: one in the Far East, where the climate was milder and there was a mixed forest (Amur maritime; Chard 1974); and one in the interior, which includes the Mal'ta culture, near Irkutsk, the Yenisei Paleolithic, near the Yenisei River, and the culture of the Diuktai cave (see fig. 4.2.1). There is undisputed evidence of Alaskan occupation, beginning 11,000 B.P. Beringia was then still above sea level, and Northeast Asia and Alaska were joined by land, making the passage to America easier. At the time, Northeast Asia offered two types of lithic technologies, both evolved from prototypes in Mongolia: one characterized by "pebble-type" tools, and the other by use of cores for making stone tools. There were also

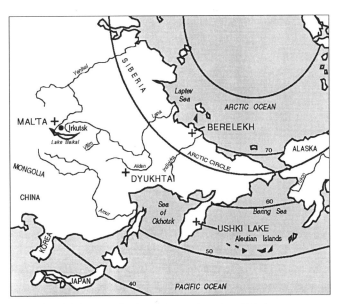

Fig. 4.2.1 Map of cultures in North Asia (Whitehouse and Whitehouse 1975).

two major funnels, through which people could have crossed to Alaska. One of them, the northern funnel, was entirely tundra, and would have been open to people from North Asia adapted to this harsh environment. The other funnel was along the Pacific coast, and people

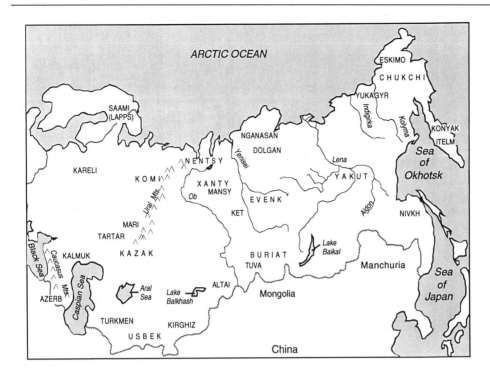

Fig. 4.2.2 Present location of ethnic groups in Siberia.

from the Amur basin, Japan, Manchuria, Korea, and northern China would have used it, but probably at times of low sea level; many lands that are now submerged may harbor the most interesting archaeological remains.

In the period between 9000 and 3000 B.C., the Siberian climate was warmer and may have stimulated immigration from the south; it then became cooler. From 7000 B.C., pottery is found, hence the name Neolithic, although there is no trace of agriculture. Again, the cultures of the Pacific coast and those of the interior are clearly distinct. The Bronze Age began late in the first millennium B.C., and the Iron Age in the next millennium.

In recent centuries, Siberian people have undergone substantial migrations and acculturations. Those located in extreme northern regions are probably the least acculturated; thus, only the southern Nganasan have become reindeer breeders, while in the extreme north they are still, or were until a short time ago, hunters of reindeer (fig. 4.2.2).

For a more extensive general introduction to the main topics treated here, see chapter 49 in the Cambridge Encyclopedia of Archaeology (Sherratt 1980). It has been a partial source of information for us and contains further references.

4.3. PREHISTORY AND HISTORY IN MIDDLE AND CENTRAL ASIA

In Middle and Central Asia, Neanderthal people left their mark, and the upper Paleolithic is represented in the region. So too is the Mesolithic, which shows strong similarities to that found in northern Iran. Major, better-known prehistorical and historical developments of possible genetic significance, occurred in the Neolithic.

Because extensive contacts between Central and Middle Asia have almost always existed, we discuss them together. Central Asia is a strip of land made of steppes and deserts that extends west into Europe, in the region between the Volga and the Urals, north of the Caucasus and the Black Sea. Middle Asia is a dry region east of the Caspian sea, which has had important exchanges

northward with the people of the steppes of Central Asia and southward with the Neolithic cultures of Iran and the Indus Valley and the later urban civilizations that developed in those areas. Nevertheless, Middle Asia, and in particular its western part, which is often referred to as Turkmenia, also had developments of its own.

4.3.a. NEOLITHIC IN CENTRAL ASIA AND NORTHERN IRAN

In Middle Asia, the Neolithic appeared in the sixth and fifth millennia B.C.. The Djeitun culture (fig. 4.3.1) was clearly imported from the Iraqi-Iranian region. It was a culture of settled farmers that, in the fifth and fourth

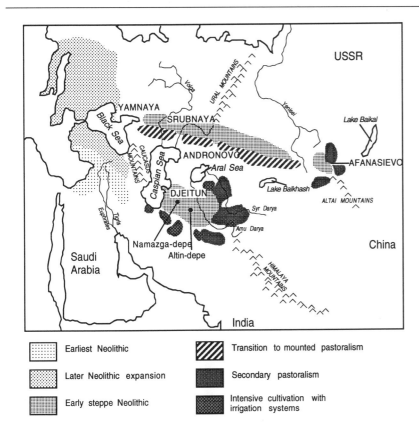

Earliest Neolithic

Later Neolithic expansion

Early steppe Neolithic

Transition to mounted pastoralism

Secondary pastoralism

Intensive cultivation with irrigation systems

Fig. 4.3.1 Map of middle and central Asia in the Neolithic and Bronze Age (Mallory 1989; Sherratt 1980; Zvelebil 1980).

millennia, reached a certain level of prosperity in spite of the aridity of the soil that confined most of the development to oases. In the fourth millennium, early Chalcolithic cultures are observed. In the first half of the third millennium, settlements increase in size and have incipient specialized structures, fields show beginnings of irrigation, and one observes signs of important contacts with Iran. During the period 2500–2000 B.C. there is an early Bronze Age and proto-urban development. Altin-depe (fig. 4.3.1.; Zvelebil 1980) is more than 60 acres in size, with monumental architecture, many multiroomed houses, and use of seals. Throughout southern Turkmenia, culture seems to be largely unified, and in the next 400 years (2000–1600 B.C.), a truly urban development takes place (e.g., Namazga V). Writing is borrowed from neighbors. But between 1600 and 1500 B.C. a mysterious decadence sets in. Altin-depe and Namazga-depe are abandoned, and agricultural oases also decay. Abandonment is gradual; there are no signs of catastrophic events, but the possibility exists that decadence is due to incursions of stock-breeding tribes. There are in fact indications of contact with northern tribes from the steppes of Central Asia: sherds of Andronovo, a steppe culture, as well as symbols popular in the steppes, like swastikas, and burials typical of some steppe tribes (Masson and Sarianidi 1972). This period is approximately contemporary with that of the decadence of the more southern Harappan civilization (see South Asia), which may have been caused by related events.

4.3.b. THE DEVELOPMENT OF PASTORAL NOMADISM IN THE CENTRAL ASIAN STEPPES

The steppes of Central Asia, just north of Turkmenia, were a difficult environment for agriculture, but the grasslands lent themselves to husbandry. Goats and sheep and cattle bones are found from 4000 B.C. (Zvelebil 1980). In the third millennium B.C., the camel and horse came into use. The dominance of pastoralism over agriculture probably developed slowly but consistently. Animals, including their milk, soon became the most important source of food, but everything was used, from skin to excrement (as a source of energy). The next important developments are wheeled transportation, with two- and four-wheeled carts (middle of the third millennium B.C. in the western steppes, above the Black Sea; somewhat later in Central Asia), and horse riding, with horse-handling gear. Burial mounds called Kurgans, believed to be typical of this culture, spread to the whole steppe between 2500 and 1500 B.C. The Central Asian steppe is an excellent environment for pastoral nomadism, which probably originated between the Dnieper and Ural rivers (e.g., the Yamnaya culture) about 2500 B.C. and spread east to Asia, beyond the Urals, about 2000 B.C. Around 2000 B.C., the steppe Bronze Age began to flourish thanks to the rich mineral ore deposits in the Urals, and, later, in the Altai Mountains. Farming and irrigation techniques borrowed from Turkmenia helped complete the economic and technological picture. Two cultures descended from earlier Yamnaya: the Srubnaya and Andronovo, which were responsible for some later developments. All these cultures spoke Indo-European languages, and their members are believed to have been Caucasoid in physical appearance. Only later did pastoral nomadism spread eastward, to Mongolia and Mongoloid people.

Pastoral nomadism was the most important factor in the development of Central Asia, with the horse most probably playing a major role in subsistence, transportation, and eventually military operations. From there, it probably spread farther south, to the Middle East and Arabia, and later to Tibet.

Pure pastoral nomadism (i.e., totally without agriculture) is rare. More common is seminomadic pastoralism, in which there is some agriculture in a supplementary capacity, practiced, for instance, by women, who remain sedentary, or by some groups within the society (Khazanov 1984). There are all possible degrees of relative importance of agriculture and pastoralism. Pastoral nomadism, however, is characterized by mobility, essentially migrations accompanying the herds. The migrations, often seasonal, can range from 20 to 1000 miles or more and may last many months. They often follow traditional routes that have not changed over millennia. Very frequently, pastoralists engage in important economic exchanges with sedentary farmers, with relations varying from symbiosis to predation. Nomads occupy enclaves that usually intercalate with those occupied by farmers and are not useful for agriculture. When they occupied a politically dominant position, nomads did sometimes take farmlands and turn them into pastures.

In general nomads used more than one animal species, sheep, goats, horse or cattle but, especially in marginal environments, there is only one domesticated animal: for example, the reindeer in the Arctic, and the camel in the southern deserts. Like most economies based on a few—or sometimes only one—sources of food, pastoralism is unstable. Under favorable meteorological conditions, herds may be expanded, which takes a relatively small number of years; this may cause cyclical variation when changes in conditions occur. Several anthropologists (Khazanov 1984) believe that there is a trend to increase the size of herds, which, if successfully carried through, may allow an increase in the human population size. When herd size must be reduced, human birth control is sometimes practiced. It has been stated that "in the long term the pastoral economy is doomed to stagnation" (Khazanov 1984). When nomads could effectively exploit and control farmers, they had, however, additional chances of increasing their power and their numbers.

Other characteristics of nomadic life are also important for understanding their history. One man, even one child, can control large herds. Nomads work little to provide food and have time for other activities. On the other hand, nomadic life is challenging in terms of dangers and need for defense; it requires difficult decisions, excellent knowledge of the environment and animal breeding, and skills in the use of weapons (Turri 1983). It leaves time for exercising military skills, which are sometimes needed, and also for engaging in leisure and artistic activities such as weaving, which is well suited for the nomadic type of life, given the availability of raw materials like wool and vegetable pigments and the transportability of small, lightweight looms. Nomads can form relatively large social groups, which may be advantageous for attack and defense; there has thus been provided an excellent opportunity for generating structured societies with rigid classes and chieftains, hereditary or elected. The innovations in horse riding and the use of chariots enormously increased mobility and opportunities for developing military skills (Khazanov 1984). Even moving from summer to winter pastures and vice versa (transhumance), a group of many thousand people may travel 1000 km a year or even more and therefore need logistical skills similar to those of an army. Chariots and horses developed for use in war gave the potential for conquest of remote lands and the subjection of large, peaceful farming peoples.

Nomads despise sedentaries, thinking their own way of life is superior. The imprinting of a mobile and free life in attractive environments by exposure to it from the youngest ages may make it very difficult for the nomad to become sedentary; they rarely do so. Europe provides a classic example of nomads who refuse to settle and have maintained their identity for almost a millennium: the Gypsies.

The potential for rapid population growth following the increase of herds, the advantage of life in the open with frequent change of environment (which may decrease the chances of epidemics and parasites), the ability to move rapidly and form large bands of well-trained warriors, who can be collected from neighboring tribes if necessary, the development of very effective weapons and tactics, all of these factors have given pastoral nomads excellent chances of expanding, plundering and subduing sedentary populations, and occupying desirable areas. All these preadaptations have made it likely that strong groups of pastoral nomads could move to invade and control rich agricultural regions, even though they were, in terms of relative numbers, small minorities with respect to earlier settlers of agricultural lands. Especially in earliest times, farming societies had probably not developed strong military organizations. Their conquest was thus relatively simple and left few archaeological traces.

In the second millennium B.C., there may have started some true population explosions of the people of the steppes, which continued to generate repeated population waves until the Middle Ages, first of European-looking and Indo-European speaking individuals, then, begin-

ning in the third century B.C., of oriental-looking nomads, speaking languages of the Altaic family. It is likely
that, in the majority of the cases, the relative numbers of
conquering nomads to earlier settlers was rather small,
having therefore little genetic impact. But, by forming
local aristocracies that kept their grip over the earlier
settlers for centuries or millennia, the nomads may have
determined a gene flow disproportionate to their original
numbers, as well as a considerable class stratification
with differential gene flow according to class. A similar situation applies, for instance, to Mexico, where the
Spanish contribution was initially very modest but there
has been some hispanization at the genetic as well as at
the cultural and linguistic levels, with strong geographic
and social differentials (Lisker 1981). A similar situation
applies to India, with a further complication introduced
by the caste structure (see South Asia).

The archaeological record is uncertain, and therefore the "Indo-European" migrations to Europe from the
Asian steppes are not as clear as they were in the minds
of earlier prehistorians (Renfrew 1973, 1987). Those to
Iran, Pakistan, and India are perhaps more clear. The
penetration of Turkmenia, the Indus Valley, and India by
Aryans, probably from the Andronovo and Srubnaya cultures in the first half of the second millennium B.C., may
have been among the first expansions of pastoral nomads
in Asia (Zvelebil 1980). The culture described in the oldest Aryan texts is very similar to that of the steppe nomads. The Hurri and Mitanni kingdoms in Mesopotamia
(1500–1300 B.C.) probably had a similar origin (Moorey
1980). The early migrations of Indo-European speakers from the Volga steppes, which—according to older
prehistorical teaching—have led to the occupation of
Greece, Italy, and central Europe (the Celts) were dated
to the last part of the second millennium B.C. There is
little, if any, unambiguous archaeological trace of such
migrations, and many archaeologists today are somewhat
skeptical that such migrations of pastoral nomads ever
took place (see, e.g., Renfrew 1987). The steppes of
Central Asia also generated many later waves of nomads
described in the following paragraphs. At first these came
from the western moiety of the steppes and then from
the eastern part of the steppes toward the west and south.
There is historical confirmation for these later migrations
(Zvelebil 1980).

It is believed that the cultures of Srubnaya and Andronovo were also responsible for the origin of a strong
group of pastoral nomads called Scythians, a Caucasoid
group whose settlements, starting perhaps with the tenth
century B.C., extended from the northern coast of the
Black Sea to the Altai region. Their movements in war are
told in considerable detail by Herodotus, who described
the war waged, and lost, by the Persian king Darius
to the Scythians that inhabited at the time (sixth century B.C.) the northern coasts of the Black Sea. The in-
credible customs and ornaments of the Scythians related

by Herodotus (1964, see also Rice 1957) were not believed for many centuries until the discovery of Scythian-
like tombs in the Russian steppes vindicated his writings.
Scythians left a highly developed art that included gold
sculptures and carpets, of which a beautiful sixth century B.C. sample was conserved in good condition in the
frozen Pazyrik tombs of eastern Siberia. Among military innovations, they developed special skills in shoot-
ing with bows from horseback, perhaps because of the
introduction of stirrups, which gave them an advantage
over the military use of chariots.

Scythians generated several secondary nuclei in the
east, like Sacians (or Sacae) and Massagetes, which
may be different names given to Scythians by different people. Under pressure from other steppe nomads, another cluster of descendants from Andronovo
deprived the royal Scythians of power: the Sarmatians
or Alans, speaking an Indo-European language of which
there remain corrupted relics in Ossetic (southern Caucasus). Sarmatians also mixed in part with Scythians and
reached Europe. Their advantage over the Scythians was
due to the use of heavy armor and long lances on horseback.

Sarmatians (Zvelebil 1980; also Sulimirski 1970) were
later defeated by the Huns (perhaps the same people
as the Hsiung-nu referred to in Chinese annals), whose
bows could penetrate Sarmatian armor. The Huns were
probably the first Mongolian nomads, who dominated
the Asian steppes from 300 B.C. to A.D. 500 and formed
a confederation of nomadic tribes from Manchuria to
the Pamir. The Huns were even more mobile than earlier nomads, having given up cattle breeding. They were
very skillful mounted archers, and their excellent strategy gave them a series of victories. Their attacks on
northern China were the main stimulus for the construction of the Great Wall of China by the Qin (pronounced
Ch'in, which is also the earlier spelling) dynasty, at the
end of the third century B.C. The succeeding Chinese
dynasty (the Han) fought the Hsiung-nu vigorously and
eventually absorbed part of them, while a western horde,
the Huns, were driven into Central Asia. They invaded
Europe in A.D. 370, overcame the Alans (living between
the Volga and the Don), defeated the Ostrogoths and
Visigoths, and extended their power to Germany and the
Balkans. Finally, under Attila, they invaded Italy. After
the death of Attila and a major defeat, they disappeared
from history.

The Hsiung-nu had customs in common with the
Turks. Shortly after the disappearance of the Hsiung-nu
from the chronicles, the T'u-chueh, who claimed descen-
dance from them, became the most powerful nomads of
the Mongolian steppe. They are usually identified with
the Turks, being sometimes called Orkhon Turks, and
rose to power in Mongolia in the seventh century A.D.
They expanded westward and established several kingdoms, the most important one being that of Timur Lenk

(Tamerlane), who created in the last part of the four-teenth century a vast empire, stretching from China to the Mediterranean. He died while he was preparing for the final conquest of China. At different times, other Turkic people formed a number of empires. The Khazar in the Ukraine (seventh to twelfth century) originated from a confederation of Turkic people and Iranians. The Turkmen Oguz nomads occupied eastern and central Anatolia in the twelfth century, fighting Byzantines and Mongols. They founded the Ottoman Empire, which reached its peak in the sixteenth century, at which time it occupied most of the southern and eastern Mediterranean, and lasted until the twentieth century.

The last wave of nomads from Central Asia were Mongols. A tribe known as "All The Mongols" entered the chronicles in the eleventh century; other powerful tribes in the region at that time included the Tungus and Tatars. Between the twelfth and the thirteenth centuries, the Mongol Genghis Khan conquered a vast empire that was further enlarged by his successors and at the time of Kublai Khan included Mongolia, China, Tibet, Siberia, Iran, Iraq, the Caucasus, and southern Russia (the Golden Horde). The Mongols lost power to the Ming dynasty of China in A.D. 1367.

It is easy to believe that all these conquests by nomads had effects on the geographic distribution of genes. It is possible, however, that in earlier times the genetic contribution of nomad invasions may have been more important numerically, especially when large groups of nomads settled in new lands and prospered. As one comes closer to more recent times, the population density of farmers was higher, and hence the relative genetic contribution of invading tribes was likely to have been less important. At the beginning, when invading nomads were followed by their families and farmers had little military power (e.g., in the Indus Valley), the genetic contribution of invaders might have had a greater impact. In more recent times, the invaders had to face strong and organized resistance, which might be incompatible with being followed at close quarters by women and children. At the end of the Western Roman Empire, however, Goths were still traveling with families and obtaining from Roman emperors the right to settle in Roman lands. Although modern Turkey was repeatedly invaded by Turks from the late eleventh century A.D. onward, what little genetic evidence is available does not seem to support a major genetic contribution by the invaders.

For a more extensive general introduction to the main topics treated in this section, see chapter 38 in the Cambridge Encyclopedia of Archaeology (1980). It has been a partial source of information for us and contains further references.

4.4. PREHISTORY AND HISTORY IN EAST ASIA

4.4.a. JAPAN AND KOREA: ARCHAEOLOGY AND THE BEGINNING OF HISTORY

Japan was not an island in the period in which the first human occupation was detected; it was connected entirely or almost entirely with the mainland both to the south (through southern Korea) and to the north (through Hokkaido and Sakhalin). There existed an internal sea between Japan and the internal coast (Sea of Japan; see fig. 4.4.1). Connections with the mainland disappeared with the increasing sea level between 10,000 and 7000 years ago. The various islands forming Japan were uninterrupted land before that time (Glover 1980).

Extensive archaeological investigations of Japan point to signs of human occupation dating back to 30 kya near Tokyo, where one edge-ground axe was found at the Sanrizuka site (Glover 1980). The population in southern Japan increased by the year 20,000. Koyama (unpubl.) estimated the maximum number of Paleolithic inhabitants at 3000 individuals, on the basis of comparisons with the population density of Australian aborigines. This estimate may be low, considering that the climate was probably more favorable. The first pottery (the oldest in the world) is dated from 12.7 kya. The style of pottery changes somewhat around 10 kya, marking the beginning of the period called Jomon (or Japanese Neolithic) that lasted until the third century B.C.. One should note here the frequent confusion generated by the fact that the name Neolithic is given to the presence of pottery, whereas in other parts of the world, for example, Europe, it indicates the presence of agriculture.

During the Japanese Neolithic, there is no clear plant cultivation until perhaps the middle of the first millennium B.C. (millet, buckwheat, hemp). Domesticated dogs appeared in 7000 B.C.. Fishing was common, especially on the east coast; boats and paddles are found from 5000 to 4000 B.C.. There was hunting of pigs, deer, birds, and small mammals. Pottery may have been used for the collection and storage of acorns, and for boiling food.

Throughout the Jomon period, there was a marked population growth lasting until about 4000 B.P. According to Koyama (unpubl.), a survey of almost 28,000 archaeological sites in Japan belonging to the Jomon period (Koyama 1978), dated by pottery style but in part by radiocarbon, shows that from 10,000 to about 4500 B.P. population growth was regular and stable, at the rate of around 1% per year. As the maximum population density, reached around 4000 B.P., there were about 260,000

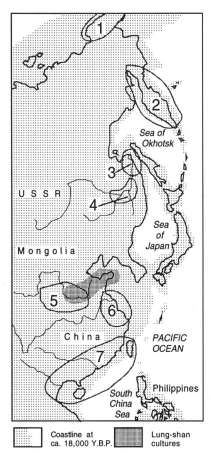

Fig. 4.4.1 Traditional Neolithic areas in China (Glover 1980). See the text for details. 1, Chukchi Peninsula; 2, Kamchatka Peninsula; 3, Lower Amur Valley; 4, Middle Amur Valley; 5, Yang-Shao; 6, Qing-Lein-Kang; 7, Ta-Pen-Keng.

tween 100 B.C. and A.D. 100 (Middle Yayoi). The number of Yayoi sites increased, and river valleys were occupied. Population growth has been estimated at 1.7% per year (Koyama 1978). One encounters the first mention of Japan in Chinese chronicles from A.D. 57, when an envoy from the state of Nu in northern Kyushu visited the Han court.

The beginning of history in Japan coincides with the Kofun period, fourth to the seventh century A.D.. The development of stone tombs of dolmen type here as well as in Korea indicates that a foreign influence is likely; warriors were buried in their armor, and saddlery is found in tombs. Social stratification is observed, and evidence of Korean and Chinese contacts increases. Some archaeologists suggest an invasion of a militant pastoral tribe from northeastern Asia in late Kofun times (Chard 1974).

Among anthropological problems in Japan are the relations between the Ainu, living in Hokkaido and farther north, and modern Japanese, and those between Japanese of the Jomon and Yayoi period. A discriminant analysis of a limited number of skulls (Howells 1986) found differences between the Ainu and the Japanese. The latter clustered in two major groups, one of which consisted mostly of southern and the other of northern Japanese, whereas most Jomon skulls clustered in a third group with Ainus. Modern Japanese skulls are markedly similar to Chinese ones. Similar results are obtained by Hanihara (1985). The analysis of a series of East Asian skulls with nine measurements suggested that in Jomon times Japan was inhabited by Ainus, and that a heavy infiltration of migrants from the Asian continent via the Korean peninsula took place in Yayoi times.

4.4.b. CHINA: FROM PALEOLITHIC TO HISTORY

China is one of the few regions of the earth that shows signs of human habitation in the lower Paleolithic. *Homo erectus* fossils and artifacts have been found in abundance in the lower caves of Zhoukoudian (Choukoutien in earlier writings) near Beijing. (Recent information can be found in Brooks and Wood [1990].) The date assigned to this, once called "Peking Man," is between 230 and 500 kya, possibly similar to the date for other famous human remains from Java. Finds made recently at Lantian (100 km southwest of Zhoukoudian) are earlier, perhaps 600–1000 kya. No Neanderthals have been found in China, the nearest one having been found in Uzbekistan.

The earliest specimens of *H. sapiens* are from Dali (near Xian, 180–230 kya), from Jinniushan (northeast of Zhoukoudian, 210–300 kya) and from Maba (in the southern province of Guangdong, 130 kya). The oldest modern human is from Liujiang (Guangxi) and has an approximate (uranium) date of 67 kya.

Modern human fossils have also been found in the upper Zhoukoudian cave. This upper Paleolithic material is dated to 11–18 kya. Morphometric analysis (How-

people in Japan (0.89 inhabitants per km^2) with considerable variation from region to region. Areas most densely populated (up to 3 inhabitants per km^2) were in Kanto and Chubu. After 4000 B.P., there was a cooling of the climate and concomitant reduction in population growth.

Contacts with Hokkaido were few. On the northeast end of this island, there is today an ethnic group that differs from the modern Japanese, the Ainu. Pottery developed in Hokkaido only 5000 years after its beginnings in Japan (Chard 1974). The archaeological situation in southern Korea is less well known, but pottery is found from around 4000 B.C., and there is evidence of contact with Japan. The Jomon population may have been of the Ainu type south of the island of Hokkaido (see below).

A dramatic change began in Japan in the period 400–300 B.C., initiating the Yayoi period. Farming, including cultivated rice with irrigation, iron tools, and weaving is first found in the southwest of Japan, presumably arriving there from or via Korea. There are no signs of massive immigration at this time; however, about 200 B.C. a rapid spread of the Yayoi cultural pattern began along the main axis of Japan, reaching its maximum be-

In the figure legend area:
Coastline at ca. 18,000 Y.B.P.   Lung-shan cultures

ells 1981) found the skulls to be more similar to those of American Indians than to those of modern Chinese. There is no great similarity between the upper cave Zhoukoudian and Liujiang; the latter is more similar to three skulls found at Minatakawa (Okinawa, about midway between Taiwan and Japan, dated to 28 kya) and to remains from the Niah Cave in Southeast Asia. Analyses of fossil material are usually based on very few skulls and may therefore be affected by gross sampling errors. Prima facie evidence, however, suggests a difference between northern and southern early modern Chinese humans, which continues later. A longer list of dates is found in Brooks and Wood (1990).

We should mention a strong belief expressed in some Chinese publications (e.g., Wu 1988) that there is a continuity in the human lineage in China from *H. erectus* to early *H. sapiens* to modern specimens, which suggests a parallel but independent evolution in West and East Asia. This idea, based entirely on paleoanthropological observations, is similar to that expressed by western supporters of polycentric human evolution. It is difficult to reconcile this hypothesis with the genetic data, but it is possible to hypothesize that there was admixture of expanding modern humans and local archaic *H. sapiens* groups, as previously discussed in chapter 2.

A different approach has been used by Turner (1989), who has analyzed extensive collections of ancient and modern teeth from large regions of the world. The most interesting results from dental studies to date have made it possible to distinguish different migrations from northeastern Asia, but some of the results are also relevant to the distinction between northern and southern East Asia. This approach is discussed in some detail in chapter 6 on the Americas, to which these studies have made the most outstanding contribution.

Chinese archaeology has recently blossomed, and one may expect a number of current ideas to be upset by discoveries that are unpublished or still to come. The new discoveries have made a major impact on the Chinese Neolithic: the beginning of farming has been moved back by about 2000 years. Until recently, the record of early farming started around 5000 B.C. in three areas, and the cultures recognized at that time showed well-developed villages, making it hard to believe they were truly the initial ones. Earlier sites with less-developed cultures have now been found, contributing to our understanding of the transition from the upper Paleolithic to the early Neolithic (Chang 1977). Figure 4.4.1 shows the location of traditional Neolithic areas. Information on the Chinese Neolithic is practically unavailable in non-Chinese publications, and a summary of the information obtained from Professor Gong of Xian during a recent journey to China is given below in more detail.

1. *Middle cultural area.* The Middle cultural area, the earliest, is found in the provinces of Shaanxi, Shanxi,

Hunan, and Hebei, and corresponds approximately to the area indicated as Yang-Shao in the map. The early stage (also called pre-Yang-Shao) starts around 8500 B.P.; millet was the major crop (up to 100 tons were found in one village). The pig and dog were domesticated. More than 60 villages were found, dated between 6500 and 8500 B.P., and occupying an area of 1000–20,000 square meters. The middle stage, corresponding to the classical Yang-Shao period, lasted from 6500 to 5000 B.P. in Shaanxi and Shanxi and from 7800 to 5000 B.P. in Hebei and Hunan, depending on areas. The celebrated village of Banpo, in Shaanxi, and the even better known nearby village of Zhangzhai belong to this period. At Zhangzhai, the whole village was excavated, revealing about 100 houses made of clay and a canal around the village. The houses were clearly grouped into five areas, each with a bigger house at the center, and presumed to correspond to different clans. Agriculture was definitely more important than foraging, with millet still the major crop; sheep, cows, and pigs were bred and kept in pens. Cemeteries were outside the village, one per clan, and tombs with female skeletons were much richer than those of males, suggesting a matriarchal culture that continued in China, at least in part, until the first millennium B.C.. Yang-Shao villages used written signs, of which more than 200 specimens have been found (Glover 1980).

After 4800 B.P. began the late Neolithic, or Lungshan stage, which is fairly different in different regions. Class distinctions appeared, and the end of the Lungshan coincided with the first dynasty, Xia (Hshia in earlier spelling). This time is still before the historical period, because writing began with the Shang dynasty, a few hundred years later. During the Xia dynasty, there were writings on scapular bones for divination purposes.

The Yang-Shao culture was centered on the middle portion of the Huang Ho (the Yellow River). Agriculture started later in the upper (5500 B.P.; Gansu and Qinghai provinces), and lower (7000 B.P.; Shandong province) areas of the river. Both areas also used millet. In the northeast (Liaoning, extending to northeastern Beijing and northeastern Mongolia), the Neolithic began in the western part (7000 B.P.), always with millet and pigs.

2. *Yangtze area.* By contrast with the Yellow River, all the cultures that developed in the Yangtze area (Qing-Lein-Kang) cultivated rice rather than millet. The earliest is in the lower Yangtze River valley (Homutu culture), beginning in 7000 B.P.; rice was present in large amounts. In addition, pigs were domesticated. The middle Yangtze was a little later (beginning 6000 B.P.); it was also based on rice and pigs, as was the upper Yangtze (Sichuan, Yunnan, eastern part of Tibet) beginning around 5000 B.P.

3. *South coastal cultures.* The south coastal cultures (Ta-Pen-Keng), which began in caves, were also based on rice and pigs. The earliest is in Guangxi (6000 B.P.),

later extending to Guangdong (5500 B.P.), Fujian, and Taiwan.

The distinction into northern and southern Neolithic reflects modern genetic differences, as we shall see. Around 1800 B.C. we find the emergence of *civilization*, as marked by the beginning of city life, metallurgy (bronze is rarely found in late Lungshan), writing, and excellent art. This corresponds approximately to the Shang dynasty (1750–1150 B.C.), which overthrew the dynasty traditionally considered the first, Xia (Hsia). The Shang dynasty was centered in Honan, but their power extended to much of northern China. The center of a Shang City had a wall of pounded earth and was inhabited by the aristocracy. Outside of it were industries and farmers' houses.

The end of the Shang dynasty came around 1120 B.C. at the hands of the Chou, the next dynasty, which reigned for about nine centuries. The Chou established a long-lasting feudal system. They successfully fought the nomads of the north and, at least for some centuries, guaranteed stability. Cast iron appeared for the first time in the capital, Shensi.

During this period, wheat started replacing millet in importance, and rice was spread to the north by the development of techniques of dry cultivation. Soybeans acquired great importance, crop rotation and organic fertilizers became common, and irrigation became necessary.

Population increases were presumably the major stimulus to technical improvements in agriculture. With the end of the feudal system, farmers started becoming owners of the land, paying tax in proportion to the land owned. Furthermore, merchants and artisans became independent of the lords. A new elite class emerged from the old aristocracy, providing administrators, teachers, and philosophers: among them was Confucius (551–479 B.C.), whose teaching was very influential in reshaping Chinese society and civilization.

In the fifth to third centuries B.C. (the period of the warring states), northern China was divided into a number of states in constant battle. The most western of these was controlled by the Qin dynasty. Extensive reforms generated power and wealth; the nobility was deprived of its privileges, and a new taxation system that favored the nuclear family, and thereby caused the breakdown of the traditional extended family, encouraged farmers to settle in Qin from neighboring states, or to migrate to new places. Eventually, Qin was successful in unifying almost all of China (221 B.C.). In order to strengthen the north against invasions of Mongolian nomads, the first Qin emperor built the Great Wall, but the dynasty did not survive the impact of the Mongols. Revolt soon restored power (202 B.C.) to the hands of the Han Chinese dynasty, which lasted until A.D. 220. Many technological advances were already well developed by this time. Cast iron, the ox-drawn plow, irrigation systems like pedal-operated chains of buckets or water-driven wheels, the opening of canals and a road system, and coinage (disks with holes appeared in 300 B.C.) were among the many innovations that enriched Chinese life before the Han dynasty.

After the end of the Han dynasty, three regional kingdoms arose. The northern one was exposed to warfare and aggression from the Turkic Hsiung-nu, Mongol tribes, and Tibetans. The north, which until then had been the most prosperous part of China, underwent progressive depopulation: from having three quarters of the people of China it decreased to less than half in the tenth century. Often northern tribal chiefs took power, and important northern Chinese families became related to Turkic chieftains from the steppes.

In the sixth century, the royal house of Sui reunited China and reconstructed the Great Wall; a related dynasty, the T'ang, had to fight Tibetans and the Turkic Uighurs. These were nomads who succeeded Turks in the steppes but were driven out by the Kirghiz, finally settling in the northwest of the modern Chinese territory (Xinjiang). The capital at that time, Qang-an, approached a million people, but was destroyed in the rebellions of the ninth century. Power was largely decentralized.

At the end of the tenth century, the Sung dynasty took over and brought prosperity, except to the north, which eventually fell to the Mongols. Genghis Khan conquered the north at the beginning of the thirteenth century, and his grandson, Kublai Khan, conquered all of China, establishing a Mongol dynasty. This was the time of Marco Polo's travel to China and of the first semiofficial contacts between West and East.

In Mongol China, four classes were established (in order of importance and privileges): (1) the restricted Mongol elite; (2) Turks, Near Eastern Moslems, and other confederates of the Mongols; (3) Han, the northern Chinese; and (4) the southern Chinese, who were considered the "southern barbarians."

In the middle fourteenth century, the Ming rebelled against the Mongols and were successful in driving them out. They established a new dynasty that favored the repopulation of the north by colonists from the Yangtze and by soldiers with their families. American crops were introduced in the sixteenth century (maize, sweet potatoes, peanuts).

Beijing had become the capital under Mongol rule; the Ming dynasty first moved the capital to Nanking, then returned it to Beijing. They forbade private slavery and restored independent ownership of the land by the farmer, breaking up the great land estates. In the seventeenth century, the Ming dynasty lost to the Ching, a dynasty originating from nomadic tribes of Manchuria that had already founded a Chinese dynasty several centuries before. The Ching dynasty lasted into the present century. It still had to fight nomads from the north and was successful in destroying the Dzungars (a tribe in western

Mongolia). It established good relations with Tibetans and Mongols.

4.4.c. POPULATION NUMBERS IN CHINA

In the middle of the Han period, around A.D. 1, the Chinese population, in an area similar to the present one, was close to 60 million. This number, the result of an official census, was probably an underestimate because there were advantages in not being counted. Eight hundred years before, there were about 14 million. In the first millennium A.D., the north was largely depopulated as a consequence of invasions of northern nomads, and part of the population migrated to southern China. Population numbers remained around the 60 million figure for a long time (with ups and downs), but started growing with regularity in the sixteenth century, reaching 200 million in 1760, 300 million shortly before 1800, 500 million at the beginning of 1900, and one billion in 1982 (Murphy 1969).

4.4.d. TIBET

Tibetans occupy a high plateau surrounded by very high mountains. A homogeneous group from the point of view of linguistics and religion, they have received contributions to their ethnic background from various neighbors to the southwest, southeast, and north. Northern contributions may have been the most important, given their present genetic similarity to northern Chinese (see sec. 4.12). Tibetans were, and still are to a large extent, pastoral nomads. We owe to professor Du Ruofu

of Academia Sinica, Beijing the following summary of prehistorical and historical knowledge on the origin of Tibetans. There is evidence of human habitation as old as 4000–5000 years ago. "Aboriginal" groups called "Bo" raised yaks and lived in Shannan (South of the Mountain) when the Xi-Qiang migrated to Tibet from the northeast. The earliest record about the Xi-Qiang is in Hou Han Shu (Later Han annals, early fifth century A.D.) and tells of the migration of one tribe of this group toward the south and then to the west for several thousand li (1 li = 0.5 km), after crossing the Si Zhi River (the segment of today's Yellow River in Qiang-hai province). The Xin Tang Shu (New Tang Annals, eleventh century A.D.) report on the Xi Qiang, who lived in the area bounded by the Yellow River, the Huang Shui, the Yangtze River, and the Min Shan mountains. They did not communicate with China, living west of the Si Zhi river. The ancestor of the Xi Chiang had the name Gu Ti Bo Xi Ye, and the name Bo reappears repeatedly in his descendants.

A Tibetan kingdom entered the historical record in the sixth century A.D. and dissolved in the ninth century. Apart from a period in which it was under Mongol rule, the country was often under the local control of feudal lords, with the Buddhist religion assuming more and more importance in politics until the recent annexation of Tibet by China.

For a more extensive general introduction to the main topics treated in this section, see chapters 21, 23, 37, and 41 in the Cambridge Encyclopedia of Archaeology (Sherratt 1980). They have been a partial source of information for us and contain further references.

4.5. PREHISTORY AND HISTORY IN SOUTHEAST ASIA

4.5.a. PALEOANTHROPOLOGY

East Asia and Southeast Asia were the seats of development of some of the most primitive hominids, belonging to *Homo erectus* according to one of the current paleoanthropological classifications. Several specimens in China, including Peking man (from the lower cave at Zhoukoudian near Beijing) and the Jetis specimens from Java, are rather archaic; other samples of *H. erectus* in Java are younger. In the line leading to modern humans, there are only the remains of Solo River at Ngandong, Java, which, according to some paleoanthropologists, dated at 60–100 kya. Others have expressed very different opinions, considering Solo much older and perhaps belonging to a different species or subspecies or *H. sapiens soloensis* (Bellwood 1979).

What is the connection between modern humans in Southeast Asia and *H. erectus*? This unsolved question has had very different answers. Weidenreich (1945), fol-

lowed by Coon (1963), suggested a direct line of inheritance, but others view replacement by modern humans coming from the Middle East as more satisfactory. Still others suggest partial continuity rather than replacement. The main support is not so much from human fossils but from the persistence of stone technology. There is a claim of a Mongoloid/Australoid north-south cline at a later time, during the Neolithic, (Bellwood 1979), with the Mongoloid component resulting from expansion from northern China, and the Australoid one representing remnants of the population that had earlier expanded to Oceania, with darker skin and frizzy hair. But the evidence is little more than anecdotal.

For an understanding of the early developments of modern humans in Southeast Asia, it is essential to remember that the islands now forming the insular part of Southeast Asia, belonging to Indonesia, Borneo, and the Philippines, were all connected with each other and with the Asian Mainland around 100 kya. In the middle

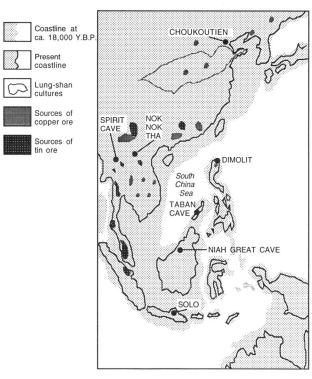

Fig. 4.5.1 The coastline in Southeast Asia at about 18,000 years B.P. (Glover 1980; Freeman 1980; Thorne 1980).

of the last glacial period (18 kya) Borneo, Indonesia, and the mainland were all connected, with the Philippines forming one big island (fig. 4.5.1). From 18,000 to 5000 B.P. the land was reduced considerably, to about 60% of the initial area. The present coast lines may have been formed between 7000 and 5000 B.P.

4.5.b. TERMINAL PALEOLITHIC AND EARLY NEOLITHIC

In spite of a land connection at different times, mainland and insular Southeast Asia show two different groups of cultures.

1. On the mainland and northern Sumatra (the largest Indonesian island) is a technological complex known as the Hoabinhian, from the first description of finds made in the province of Hoa-Binh, southwest of Hanoi (Vietnam). The Hoabinhian extends over a very large area, from southern China (as far north as the middle Yangtze) to all of Southeast Asia, and northeastern Sumatra.

The Hoabinhian is characterized by a pebble-and-flake tool industry that typically dates from the period 11,000–2000 B.C., but may perhaps represent the evolution of earlier local forms. Most sites are close to streams. One of the most interesting sites, Spirit Cave, is in northwestern Thailand. Until about 6800 B.C., there were people living on a foraging economy with a great variety of animals and plants; after this date, one finds ceramics (cord-marked and incised pottery), knives, and a differ-

ent type of adze, which may indicate the emergence or arrival of a farming culture but for which strong evidence is still lacking. Other sites have shown perhaps stronger evidence of hill rice agriculture, domestic cattle, and a sedentary life-style around 4500 B.C.

2. Island Southeast Asia began separating geographically from the mainland between 14 and 10 kya. At that time, the Hoabinhian cultures were flourishing on mainland Southeast Asia, and the islands had flake industries with some variation between Java, Borneo, and the Philippines. Several archaeological sequences lasting for long periods are found at Niah Cave in Borneo and Taban Cave in Palawan, a long, narrow island between Borneo and the Philippines. In general, the appearance of ceramics and of food production is late, especially in these two caves (Bayard 1980; Glover 1980).

4.5.c. LATE NEOLITHIC AND BRONZE AGE

The Neolithic period has not been well studied, but one can cite several unexpected findings, such as the early appearance of metallurgy. Two major sites in northeast Thailand (Ban Chiang and Non Nok Tha) have shown a somewhat similar development, both beginning around 4500 B.C. and including in their earliest layers early rice cultivation (from indirect cues) and cast bronze ornaments. In Ban Chiang, water buffalo and wet rice cultivation seem to appear around 2000 B.C., as does iron, which is found here earlier than in China. A little before 1000 B.C. the introduction of iron tools and of water buffaloes for plowing allowed a population to settle in the alluvial plains of the Red and Mekong rivers. While this was happening in some areas, life on most of the hills seemed to continue with the same late foraging economies; in fact, some of them continue today, like that of the Negritos (see sec. 4.9) and many other foraging cultures on the mainland and in the islands (Bayard 1980).

The spread of bronze and iron to the islands of Southeast Asia took place in the period between 1000 and 1 B.C.; rice cultivation spread late to the islands, and its diffusion is not complete even today.

4.5.d. FOREST EMPIRES

Paradoxes are numerous in the late development of Southeast Asia, if one compares it with the familiar succession of events in other civilizations. In the bronze age there was already a remarkable contrast between two cultures. On one side, sophisticated farming communities of the alluvial plains, which developed into urban cultures, were responsible for the early development of metallurgy and showed considerable interest in bronze art objects. Other areas were inhabited by isolated tribal communities that took a very long time to abandon primitive tools, techniques, and economies; today several groups are still foragers or slash-and-burn farmers. The tropical forest environment must somehow have contributed to this dualism in development. There was no lack of contact between the different cultures; on the contrary, there were and continued to be exchanges with probably mutual advantage (Presland 1980). A similar separation between tribal and urban cultures, with side-by-side survival, is also observed in India, with limited commercial but practically no technological exchange.

Another surprise is the lateness of development of urban civilizations in Southeast Asia, in spite of early local developments of metallurgy and signs of exchange with external civilizations by sea routes as early as 2000 years ago. Rome was a good market for many Southeast Asian forest products (spices, etc.), and archaeological evidence points to Roman trade centers in India, Malay, Sumatra, and Vietnam, and to exchange with India and China.

The earliest urban developments of the area are in northern Vietnam, where they may date from the third century B.C.; they were certainly influenced by the Chinese conquest under the Han dynasty. By the sixth to eighth centuries, there is evidence of towns and cities appearing in good numbers in Thailand, Burma, Cambodia, and Java. These population centers were not under Chinese, but Indian influence, beginning around 400–500 A.D., which explains why the monumental architecture that started flourishing after the seventh century in many different parts of Southeast Asia is clearly reminiscent of Indian art. The major center was in Angkor, Cambodia, which was almost entirely covered by the forest when it was discovered after French colonization in the last century. Angkor became the capital of the Khmer empire in the ninth century A.D. and was destroyed by Thai armies in the fifteenth century. Other major architectural developments patterned on the Indian model occurred in Burma and Java.

4.5.e. TRIBAL GROUPS

Hunter-gatherers still persist in the forests of the mountains and hills, for example, the Senoi of the Malay peninsula. There are also slash-and-burn agriculturists like the Semai. Sumatran groups like the Kabu (Mamak and others) alternate between seasons of hunting-gathering and of harvesting crops planted before leaving for the hunt. In the Philippines, the Tasaday were said to be still using stone tools, but doubts on the authenticity of these findings have been voiced. Other highly isolated groups also exist (Presland 1980).

Negritos are small, dark-skinned, frizzy-haired forest dwellers, usually living in isolates scattered throughout Southeast Asia (see sec. 4.9). The Semang of Malay are considered Negritos and, together with the aborigines of the Andaman Islands, they have extremely frizzy hair. They also have a relatively high frequency of Mongoloid inner epicanthic eyefold (Bowels 1977). Agta are scattered Philippine Negritos; the word Agta means "black" (Presland 1980). They are hunter-gatherers and part-time slash-and-burn farmers; their trade relations with neighbors are similar to those of African Pygmies and other forest dwellers. Some groups are highly acculturated.

For a more extensive general introduction to the main topics treated in this section, see chapters 40 and 42 in the Cambridge Encyclopedia of Archaeology (Sherratt 1980). They have been a partial source of information for us and contain further references.

4.6. PREHISTORY AND HISTORY IN SOUTH ASIA

There is a dearth of archaeological information about South India; a summary of archaeological dates is given by Dutta (1984). Mesolithic cultures are found from about 10 kya. The origin of agriculture in India is obscure (Vishnu-Mittre 1977). There are claims of animal domestication in central India in 5500 B.C., without concomitant agriculture. As explained in more detail in the next subsection, wheat, barley, and millet entered the area from the west, and an important agricultural civilization (Harappan) flourished in the Indus Valley. Rice may have been an original crop in the Gangetic plains, where it rapidly became successful; it later spread to the south, arriving in Maharashtra around 3500 years ago and in Tamil Nadu and Karnataka 500 years later. The geography of Indian provinces and the distribution of modern tribal Indians are shown in figure 4.6.1.

A multivariate analysis of prehistoric South Asian crania (Kennedy et al. 1984) showed three clusters. A major cluster includes upper Paleolithic skulls from Europe and all hunting-gathering populations of South Asia. A sharply delimited cluster includes all Harappan skulls of the mature period. A smaller cluster, somewhat intermediate between the two just mentioned, includes skulls from late foraging or from agricultural communities from central and southern India.

4.6.a. THE NEOLITHIC AND THE INDUS CIVILIZATION

The Indus Valley was the scene of urban developments that were not as early as those of Mesopotamia, but still retain an indigenous flavor despite some testimony to

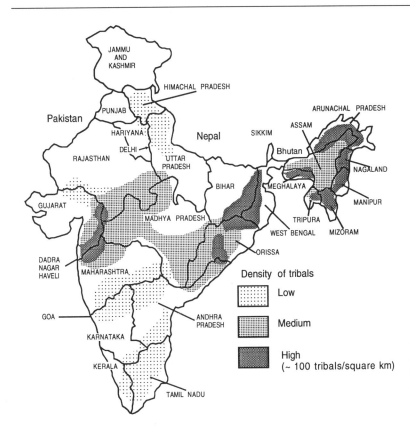

Fig. 4.6.1 The provinces of India with approximate local densities (km²) of tribals (people who are outside the caste system). Highest densities are around 100 tribals per km² (Vidyarthi and Rai 1985).

trade between the two cultures. The exploitation of the floods of the Indus River, which deposit alluvial silt, made it easy to cultivate wheat and barley without need of manuring and irrigating or even plowing (although a plow was probably known to the Harappans). Other domesticated plants found include cotton, leguminous plants, rice (not necessarily cultivated), and many others. Cattle, goats, sheep, buffalo, the Indian pig, camel, donkey, dog, and cat were domesticated. Terra cotta models show the existence of carts for transport (probably in addition to pack-transport); carts with wooden wheels practically identical to those seen in Harappan models are still in use in the Indus Valley today. Copper and bronze were extensively used for tools and ornaments, and most of the pots were wheel-turned. Many seals were found, which carry short inscriptions, probably in a syllabic alphabet.

The most important city was Moenjodaro, which, at its peak, may have had 40,000 inhabitants or more (cited in Chakraborti 1980). The name of the civilization comes from the second largest city, Harappa. There are over 200 known sites between 2300 and 1750 B.C. (see fig. 4.6.2).

This complex civilization did not mushroom unexpectedly. There was a long period of in situ development that seems to have led directly to the Indus culture. Both at the lowest levels of Harappan cities and at other sites, a pre-Harappan culture was found that seems to antedate in many respects the mature Harappan culture. Among the pre-Harappan sites, Amri is dated to 3320–3600 B.C. (two corrected radiocarbon dates from the middle level; Du et al. 1991); Kot-Diji has upper and lower levels of 3155 and 2599 B.C. and has pottery that is identical to that of the early levels of Moenjodaro.

Fig. 4.6.2 Map of India and Pakistan with primary early Neolithic sites (Chakraborti 1980).

The anthropology of Harappan skeletons has shown a substantial similarity with modern populations from a nearby area, as well as with an ancient series from Iran and one Egyptian population (less similarity, however, with other Egyptian populations; Dutta 1984).

The Harappan civilization disappeared between 1750 and 1500 B.C. Disappearance was ascribed some time ago to the Aryan conquest. The word "Aryan" indicates pastoral nomads from Central Asia but is probably a misnomer; it does not seem to refer to an original population, but it means noble, from the upper castes, or even foreigner. It probably reflects the social condition of pastoral nomads after they arrived in the Indus Valley from the north and established their authority over the Harappan farming community. There is no clear evidence of violence, however, and some suggest that the decay of the Harappan culture was due to a disaster—like a sudden major flood, or a change in the course of the Indus river (a not infrequent event)—responsible for diverting the necessary flood waters from the agricultural lands. A new explanation of the disappearance of the Harappan urban centers has been recently suggested by evidence derived from satellite photography, showing that the Saraswati River dried up when the continuing uplift of the Himalayas caused a shift of rivers, one of which— the Yamuna, originally a tributary of the Indus—shifted eastward to join the Ganges. The ensuing aridity is also documented by paleobotanical information (see Gadgil and Thapar 1990 and references therein).

In agreement with the indigenist trend in Anglo-American archaeology, it has been suggested that the Aryan migration is a total invention (see Shaffer 1984). However, as briefly discussed in section 4.3, events in the Neolithic cultures of Turkmenia, northwest of the Indus Valley, are well explained by assuming a migration of pastoral nomads from the north at about the same time; the end here was also not abrupt or violent. Linking the two series of events in Turkmenia and the Indus Valley, it seems very likely that both were due to the takeover of power by Aryan pastoral nomads who came from the steppes of Central Asia, spoke an Indo-European language, and used iron and horses (Masson and Sarianidi 1972). More about their origin was given in section 4.3. A possible archaeological marker of the first arrival of Aryan people is Painted Grey Ware (Thapar 1980). The events accompanying the arrival of the Aryan pastoral nomads from the oases of Central Asia are probably better described as a migration rather than a conquest or invasion, but pastoral nomads clearly introduced a social stratification and segmentation of all India based on religion and politics, in which they took the upper classes.

The Harappan civilization probably did not disappear completely but is likely to have contributed to the development of modern Indian culture. Shorty after the end of the mature Harappan phase, several regional cultures arose, sometimes called post-Harappan. In any case, the Harappans were socially, culturally, and probably linguistically, different from the Aryans, who made a major contribution to the new culture. It is difficult to attribute an exact date to the Aryan takeover. Most information comes from the Vedic literature, the oldest texts written in Sanskrit, which are still the heart of modern Indian civilization. The social restructuring that took place with the Aryan invasion involved the creation of a strictly stratified society that still exists (castes) and contrasts sharply with the apparent absence of class stratification or structure in Harappan society. The Aryans of the early Vedic texts (the Rig Veda) were largely pastoral, and the horse and iron were important components of their military strength.

The earliest Vedic texts refer to an area that is geographically included in the Indus system. The post-Harappan culture found here, characterized by a new type of pottery (Painted Grey Ware), is associated with the first appearance of iron, the domestic horse (Lal 1954; Thapar 1980), and a mixed pastoral and agrarian economy. The Painted Grey Ware culture later spread to the Gangetic region and with it appears also social stratification, in agreement with the idea of an Aryan eastward expansion.

4.6.b. INDIA

India has a population almost as large as that of China and is currently multiplying faster than China. A map of Indian states is shown in figure 4.6.1.

A northern Neolithic culture is found in the Vale of Kashmir (Burzahom, 3000–1600 B.C., fig. 4.6.2), with stone tools showing affinities with China (Chakraborti 1980). In the third and second millennium B.C. south of Maharashtra, there are sites like Brahmagiri and several others that indicate a prevalence of pastoralism over farming. Ahar in southeast Rajasthan is the oldest and largest of a number of nearby cultures of settled farmers who adopted copper at an early time. Iron is first found in the Indus Valley, then in the first millennium in the upper Gangetic valley (associated with Painted Grey Ware). In the southern part of India, iron is often associated with megalithic monuments (usually stone tombs like dolmens etc., that are also found in northern Europe, the Mediterranean area, and Japan). No simple connection has been established between very remote reappearances of megalithic monuments, and coincidence is not excluded.

In the middle of the first millennium B.C., a number of kingdoms and republics (i.e., confederations of tribes) existed in India. Fifteen states were listed in Buddhist scripts of the times. The successful campaign of Alexander the Great (325 B.C.) had a profound influence on Iran, Afghanistan, and northwestern Pakistan, with the creation of Greek-style cities and the diffusion of Greek art. The establishment of the Mauryan empire in India followed shortly thereafter and was responsible for the first unification of almost all India (except the extreme south); this empire lasted a hundred years and included most of South Asia. It marked the entrance of India into history (Thapar 1980).

At the end of the Mauryan rule, several small kingdoms arose and various external invasions are on record, by Indo-Greek rulers (from Bactria, Iran) and by Scythians (the Sacas, who settled in western India). The caste system (see below) was sufficiently rigid that there were problems in accepting foreign individuals because they could not be fitted into any endogamous group or into the hierarchy dictated by the Hindu religion. Entire foreign groups were somewhat easier to fit into the social structure as new castes. In the fourth to sixth century A.D., northern India was under the rule of the Guptas, a dynasty that was able to check an invasion of a Turkic group, the White Huns, from the northwest. The White Huns gave rise to migrations from Middle Asia west to the Volga and Europe, and southward, to Iran and northern India. After a rapid but largely unsuccessful campaign as far as Madhya Pradesh, the Huns established a short-lived reign in Kashmir. Other invaders included the Gurjaras, supposedly from the Kazars of Central Asia. Some Rajput dynasties may have had Central Asian origins. An Arab invasion took place between the seventh and the eighth centuries, via Kabul, but was only temporarily successful. Turks were active in northern India in the eleventh century and later. In the late twelfth century, a Turkish dynasty from Ghazni in Afghanistan occupied the Punjab. Rival Moslem Turks from Ghur were able to take power both in Afghanistan and northern India and conquered the Gangetic plain, establishing a sultanate that was for a while the most powerful state in northern India. Timur's invasion in 1398 left northern India without a strong ruler for some time.

When the sultanate of Delhi was in power, Moslem rule was not imposed in any rigid way. A descendant of Timur, Babur, who also originated from Genghis Khan through his mother, was able to conquer first the Punjab, then Delhi and the whole of northern India. He founded the Mogul empire, which lasted for 250 years, until the second part of the eighteenth century. The Mogul rulers were not rigid in applying Moslem rules, and the third emperor, Akbar, established religious freedom.

Europeans settled in India, beginning with the Portuguese, in the early sixteenth century. With a very small metropolitan population, the Portuguese had to man fortresses and ports in many areas of the globe. Perhaps for this reason they traditionally encouraged, in India as elsewhere, marriage with local women, a practice that has continued to this day, but was usually discouraged by other European colonizers. Goa was their base in India, and a mixed Portuguese-Indian population (of Roman Catholic religion) was thus established.

The Dutch, British, and French also participated in the colonial effort. The competition between the French and English was resolved in favor of the latter, and in the course of the late eighteenth and nineteenth centuries the British acquired political control of India, which lasted until independence (1947).

4.6.c. CASTES

Most of India has a unique social system, castes, most probably introduced by the Aryan conquerors. Since it is of considerable importance from the genetic point of view, we give a short summary of the history and development of the caste system, leaving genetic details for section 4.14.

According to the Greek historian Megasthenes, in what is perhaps the oldest historical mention of castes, seven endogamous castes defined hereditary professions in the third century B.C.: priests ("philosophers"), farmers, herdsmen, soldiers, magistrates, councilors, and artisans. The latter three lived mostly in towns.

Castes are endogamous groups conveying a series of other obligations. Belonging to a caste implies strict transmission not only of caste, but also of profession; purity of caste is essential, and violation of the endogamy rule involves pollution and punishment, usually by degradation of the culprits to a lower caste. The hierarchical order of castes and occupations is basic to Indian society and is fixed by religion; it is not the same but is similar in all states. All over India the highest caste is that of Brahmins. Two concepts indicating caste are *jati* (endogamous social strata as defined below) and the four *varnas* (castes in a more general, wider meaning), the origin of which is described in a late hymn of the Rig Veda. The hymn says that the four varnas arose from parts of the god Prajapati: from the mouth, *Brahmins* (who possess Brahma, magic or divine knowledge; these were the priests), from the arms, *Ksathriyas* (who possess Ksatra, power or sovereignty; these were the aristocracy, owning the land and playing the role of military leaders), from the thighs, *Vaysyas* (from vis, the settlement; these were businessmen, subservient to the higher castes), and from the feet, *Sudras* (a word of unclear origin; they are probably non-Aryan cultivators).

Until recently little was known about the numbers of the caste system, for lack of a proper survey. A governmental project called "People of India" has started such a survey that is now near completion; it will eventually be published in many volumes over a period of 20 years. The following information originates from a paper given by Sri K. S. Singh, director general of the Anthropological Survey of India, at a February 1991 meeting at the Bangalore Indian Institute of Science. The survey is based on a sample of 4400 communities (endogamous segments in which Indian society is stratified, which are defined geographically, professionally, and occupy a specific echelon in the social hierarchy fixed by religion). They all have a specified territory, which is usually fairly small; 96% of these communities live within the boundaries of one state, giving an indication of their geographic limitation. Eighty percent take their name from a specific occupation, which is passed from father to son, stressing the importance of professional stratification. Occupations are closely related to resources available. Of the communities, 87% are strictly endogamous; others have different rules, and 5% in particular follow hypergamy, whereby a woman is allowed to marry in a caste higher than that to which she belongs by birth.

Today an increasing proportion of marriages take place without regard to the caste system, especially in large cities; but in rural India the system is still largely unchallenged. The total number of endogamous communities today is around 43,000, and was on the order of 75,000 at its peak. The average number of members in a community is several thousand individuals. A caste can be made of exogamous lineages (*gotra*), within which marriage is forbidden, to avoid excessive inbreeding. Caste is, however, a slightly ambiguous term, meaning both *varna* and *jati*: the endogamous community defined in hierarchical, professional and geographic terms corresponds to the *jati*.

4.6.d. TRIBALS

Not all the Indian population is organized in castes. Moslems are outside the caste system, which is defined by the Hindu religion. Foreign minorities, even if they have lived in India for a long time, are not tied to the system, but their acceptance as a group has been made easier by their belonging to a group large enough that it can be endogamous and thus does not disturb or threaten the endogamy of traditional Hindu castes.

Tribes living in isolated areas have not been given caste status, and the people are referred to as "tribals." They ordinarily live outside towns, though not necessarily far from them; their economies range from foraging to slash-and-burn cultivation and more sophisticated farming. Tribals are generally aboriginal groups who have not been absorbed in the caste system imposed by Aryans (Vidyarthi 1983).

While some tribal groups like Bhils and Gonds in the central province of India are farmers who use plowing, many small groups are still hunters-gatherers, like the Malapandaram in the extreme south of India, the Chenchu from Andhra Pradesh, and the Veddas of Ceylon. Other forest-dwelling tribals who practice simple farming techniques, essentially slash-and-burn agriculture, live in Andhra Pradesh and supplement their diet with wild plants. The geographic distribution of tribals is shown in figure 4.6.1 (Vidyarthi and Rai 1985).

The official designation of tribals is "scheduled tribes." In the 1971 census (Vidyarthi 1983), the total number of individuals thus designated was 38 million (or 7% of the population). In that census, 427 tribes were listed, 6 of which (the Gonds of Madhya Pradesh, Maharashtra, and Andhra Pradesh; the Bhils of Rajasthan, Gujarat, Maharash-tra, and Madhyia Pradesh; the Santhals of Bihar, Orissa, and West Bengal; the Minas of Rajasthan; and the Mundaris and the Oraons of Bihar) had populations of more than one million individuals. At the other extreme, there are many very small tribes: the most interesting Andamanese tribe was made of 96 individuals at the latest count (1990). Recent information on aboriginal inhabitants of the Andaman islands is given below.

A subdivision of tribal groups taking into account a number of factors (geographic, social, etc.) finds four groupings (Vidyarthi 1983).

1. The Himalayan region (4.7 million or 12.5% of tribal India and 45% of the people in the region) includes the northeastern (Assam, Meghalaya, the mountainous region of West Bengal), central (Uttar Pradesh), and northwestern (Himachal Pradesh) Himalayan regions.
2. Middle India (21 million or 55% of the tribal population, 14% of the people in the region) includes Bihar, West Bengal, Orissa, and Madhya Pradesh.
3. Western India (10 million or 26% of the tribal population and 27% of the people in the region), includes Rahjasthan, Maharashtra, and Gujarat.
4. South India (2.5 million or 6% of the tribal population and 1.7% of people in the region) includes Karnataka, Andhra Pradesh, Tamil Nadu, and Kerala.

In the most eastern islands under Indian jurisdiction, the Andaman and Nicobar islands, most inhabitants are acculturated and of recent foreign origin, but there still survive—especially in the Andamans—a few very small tribes of hunter-gatherers, classified as Negritos. The Andaman Islands remained free of external influence until very recently, in part, because of the fierce and usually uncompromising customs of some of their aborigines toward all foreigners. Last-century contacts of the British with the Andamanese gave rise to extremely bloody confrontations. Of the four Andamanese groups the only one that accepted a peaceful contact at the beginning, the Ariotos of the Great Andaman, was rapidly destroyed by disease and alcoholism. Today, only 29 survive of perhaps 3500 that lived at the establishment of the Port Blair Penal settlement in 1858. The other three groups are also much reduced in numbers, but they were in contact with outsiders for a shorter time, and the Indian government is trying to be helpful toward them. They are the Jarawa on South and Middle Andaman, the Onge on Little Andaman, and a third group on Sentinel Island. In 1952 the Onge were the first of the three to accept an anthropologist for a long term, and there is an excellent report of that investigation (Cipriani 1966). The Jarawa were first visited in 1974, and the Sentinel islanders accepted contact for the first time in January

1991. There are about 200 Jarawa alive, 96 Onge, and perhaps 80 Sentinellese. They speak mutually unintelligible but related languages of the Indo-Pacific family. At least the Jarawa and Onge have very low fertility, possibly because of extreme inbreeding. Physically they have small stature, very dark skin, and peppercorn hair; the women have fairly high steatopygia. The most interesting aspect of the Andamanese is that they probably had the least admixture compared with other Negritoes, and perhaps represent relics of the human bridge that may have existed, perhaps 60 or 70 kya, between Africa and Australia. The few genetic data available show remarkable genetic homogeneity for 11 red-cell and enzyme proteins, according to a report by Pandit and Chattopadhayay (1991). A complete genetic investigation of these groups with modern techniques is very important, since they are probably the most interesting Negrito population, have undergone the least admixture, and are rapidly disappearing. The tendency to homogeneity is obviously a consequence of strong drift, but if very many genes are tested, especially from highly polymorphic systems, the information collected may still be useful and may determine whether these populations represent a missing link between Africa and Australia. Geographically the Andaman Islands belong to Southeast Asia, but politically to India. Other tribals in the Indian subcontinent have characters (mostly hair and facial traits) vaguely resembling either Africans or Australians. The information is extremely vague, and most probably these populations are highly mixed with other Indian ethnic groups. In some cases, there may have been recent admixture with Africans (e.g., among the Kadar of Kerala; Saha et al. 1974). Careful collaborations between Bombay and Canberra (e.g., Saha et al. 1974, 1976; Ghosh et al. 1977), in which R. L. Kirk tested a high number of protein markers, failed to find any close, uncontaminated resemblance of these tribals with Africans or Australians. However, a number of other tribes in many parts of India are potentially interesting for identifying pre-Dravidian features. *Dravidian* refers to languages spoken especially in southern India, but also farther north, that are believed to have once been spoken over a much larger area. They probably originated farther west and were suppressed in Iran and most of the northern and central Indian continent by Indo-European languages, whose speakers arrived later. Dravidian languages probably arrived first in Iran before moving to India with early farmers; it is unknown which languages were spoken before, but it seems reasonable to assume that the people speaking pre-Dravidian languages in India shared some of the physical traits found in some Indian tribals, which show the least similarity with traits of the present inhabitants of the Indian continent. Among the most interesting pre-Dravidian people are the Vedda of Sri Lanka, a Negrito group now reduced to very small numbers and largely acculturated (they speak an Indo-European language).

It is clear that Indian tribals form a large and heterogeneous population, made up mostly of relic populations, but also of later arrivals that were never totally absorbed in the Indian culture. Among tribals, one sees all stages of acculturation, from groups accepting and in part practicing Hindu customs,˙to groups who are positive toward them but do not practice them, and finally to others who are negative or indifferent. Linguistics is a partial help. Indo-European languages, which were the latest to arrive in India, are also perhaps the most common among tribals and are spreading rapidly. Dravidian languages, which were perhaps more common earlier, are limited to the southern and central-eastern subcontinent. The same linguistic picture is valid for the nontribals, with few exceptions. Two other linguistic families are spoken by tribals. In East Bengal and in the eastern Himalaya region, Tibeto-Burman languages (of the Sino-Tibetan family) are the most common. Two groups of languages from the Austroasiatic family are spoken in different pockets of India: the Munda (both Santhals and Mundaris), in different regions of the strip separating the North from the peninsula, and the Mon-Khmer (the Khasi), in the northeast. Although they speak a language of eastern origin, the Munda's physical appearance is not Oriental.

There clearly were many lesser migrations, in addition to the major ones believed to have brought Dravidian and Indo-European languages to South Asia. The endogamous caste system spread by Hinduism does not usually extend to tribals, who have had more free intermarriage outside the caste system, but the great variety of the origins of tribals, of the environments they occupy, and of their customs and local adaptations have helped to keep South Asia very heterogeneous genetically.

For a more extensive general introduction to the main topics treated in this section, see chapters 22 and 39 in the Cambridge Encyclopedia of Archaeology (Sherratt 1980). They have been a partial source of information for us and contain further references. Gadgil and Thapar's (1990) article is an excellent summary of the history of India seen from an ecological point of view.

4.7. Prehistory and history in West Asia

There are differences in the definitions of the expressions Levant, Middle East, Near East, and West Asia, which tend to indicate a progressively wider region, but there is little agreement in their usage. We use them somewhat interchangeably, reserving the name *Levant* for the Mediterranean region between Turkey and Egypt.

West Asia shows remains of both *Homo sapiens nean-derthalensis* and *H. s. sapiens* in many areas, and even side by side in the same caves—for example, in Mount Carmel, northern Israel—but the remains are probably from very different times. In section 2.1, we have summarized the present situation and, in particular, the discoveries in the Middle East and Africa of modern human remains dated from approximately the same period, about 100 kya.

For a long time the Middle East was tacitly considered the place where modern humans first appeared and from which they spread to other parts of the earth, probably as an extrapolation from biblical lore. In any case, since the Middle East is a region through which migrations between Africa and Asia probably took place, it would not be surprising if it played a central role in human evolution.

In more recent times, at the end of the Paleolithic and the beginning of the Neolithic, the Middle East, along with East Asia and Central America, was a major area in which agriculture originated. Present knowledge indicates that it was also the first of these major nuclear areas, but the difference in the dates of inception for these three areas of origin keeps shrinking as information increases on the latter two, which are still somewhat less fully investigated than the Middle East.

4.7.a. THE RISE OF AGRICULTURE IN THE NEAR EAST

The development of plant and animal breeding had the most dramatic effects on population growth and expansion, and it is remarkable that it took place almost simultaneously in different parts of the world, at the end of the Pleistocene and the beginning of the Holocene (see chap. 2). The distances between these areas of origin were far too great for there to be a direct cultural influence among these centers. Moreover, plants and animals that were domesticated were available locally and were therefore quite different in different places. Long-distance transfer of domesticated animals and plants took place only much later. The three centers of the origin of agriculture each gave rise to one of the three crops that are the major staples today: wheat in the Middle East, rice in East Asia (perhaps also in Southeast and South Asia), and maize in Central America. Almost all important animals like cattle, sheep, goats, and pigs seem to have been domesticated first in the Middle East, but other areas of origin are not excluded. The transition from the hunting-gathering economy to farming was never abrupt, and partial dependence on hunting-gathering continued for millennia, shrinking only very gradually.

Adaptation to the use of wild cereals for food seems to have developed in a culture, the Natufian (14,000–11,000 B.C. in Israel; Sherratt 1980), that predates the later local farming economies. As the warmer climate

at the end of the last glacial age favored the growth of grasses, the regions of the Middle East, Turkey, and Iran developed rich stands of wild cereals like wheat and barley, progenitors of the modern cultivated varieties. Since these are rarely, if ever, found outside these regions, it is necessary to accept that cultivated wheats and barleys spread from this area of origin. Moreover, direct archaeological evidence shows that the cultivated varieties, which are different from the wild ones, originated in the Near East.

The Natufians probably did not cultivate wild cereals but exploited wild stands near their camps and villages. They used flints to make sickle blades for harvesting and mortars and pestles for grinding seeds. Grains were in sufficient abundance to offer adequate food, and the proteins they produce (like those of rice but unlike those of maize) supply the dietary requirements without being complemented by other products derived from hunting and gathering. The latter activities were, however, continued. Moreover, harvested cereal grains can be stored and thus provide food for the whole year. The necessity of storing the harvest probably favored their becoming sedentary, as suggested by the fact that Natufians, unlike other contemporary populations, built stone houses and villages. The need to assure a more permanent, guaranteed supply of grains, reasonably close to the permanent homes, may have been among the stimuli to domesticate cereals.

Study of natural and cultivated varieties of cereals in the Middle East showed that cultivation caused substantial genetic changes in wild cereals. For example, wild varieties have shattering seeds; domesticated cereals do not. This may have been a direct result of selection from the use of sickles, which would favor the retention of nonshattering variants as soon as deliberate sowing was started. Other properties of cultivated varieties are more likely to have been the result of intentional selection. Depending on locations, wild cereals were diploid or tetraploid, the tetraploids themselves being the result of spontaneous crosses between different diploid species leading to the formation of allopolyploids. Cultivated cereals are diploid (Einkorn wheat), tetraploid (durum, Emmer wheats), hexaploid (e.g., spelt, bread wheats, from crosses of wild diploid and tetraploid).

The appearance of the new wheats—for example, at Jericho by 10 kya—provides further proof of the development of agriculture. Concomitantly, the settlements became larger and more clearly permanent. Mud-brick houses became common. Pottery did not exist here at this time; it would be found later; therefore, the Neolithic (the period in which agriculture first appeared in West Asia and Europe) in this part of the world and in its earliest diffusion to Europe is called "pre-ceramic." In East Asia, pottery appeared before agriculture.

The first appearance of agriculture is not localized in Israel alone but is observed over a vast area, the "Fer-

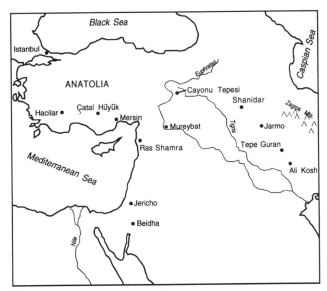

Fig. 4.7.1 Major archaeological sites of the Fertile Crescent and the Anatolian extension (Whitehouse and Whitehouse 1975).

tile Crescent" (fig. 4.7.1). The western horn of the crescent starts in Israel (Jericho and other sites), goes north through Syria (sites of Ras Shamra on the coast, Mureybat on the upper Euphrates), up to southeastern Turkey (Cayonu, where sheep, most probably domesticated, appear together with domesticated cereals). The eastern horn of the fertile crescent descends toward the Persian Gulf east of the Tigris. In the northern part of Iraq, just south of the Zagros Mountains, is Shanidar, with early domestication of sheep and goats; farther south Jarmo, probably with domesticated swine; farthest south in Iran is Ali Kosh, with goats and cultivated cereals. An extension from the central part of the crescent runs toward southwestern Turkey, where wild cereals also became common very early. Here we find important sites like Çatal Hüyük, the first agricultural village that reached the size of a "town," and others.

Obsidian from the Zagros Mountains, used mostly for blades (Renfrew 1969), was probably traded from village to village as far south as Ali Kosh. In later stages of the agricultural expansion, obsidian was obtained from islands in the Aegean and the Tyrrhenian seas. The first maritime expansions may have taken place from the coastal sites of northern Syria, like Ras Shamra. Cyprus, Crete, then Greece and Macedonia have the first farming sites known outside Asia.

In the ninth millennium B.P., pottery appeared and rapidly spread throughout the farming area then existing in Asia and Europe. From that time on, pottery and agriculture tended to expand together in Europe. While the focus of development seems to have spread from the southern part of the Levant toward the coastal areas of Syria in the north, the south tended to become pastoral. In southwestern Turkey, the development of Çatal Hüyük

continued and abundant use of cattle is clear from wall paintings. Cattle is also seen in the ninth millennium B.P. in lowland areas of eastern Syria (at Bouqras, on the middle Euphrates) and northern Iraq (Oates 1980). In highland areas, sheep and goats were prevalent.

In the eighth millennium B.P., we witness continuing development and extension of agricultural areas. New tools appeared (stone hoes); pieces of hammered copper, and possibly smelted ones, were used (hence the period is called Chalcolithic); trade began to prosper; and, most important of all, irrigation extended the areas brought under agriculture. Farming expanded in both eastern (Iran, Turkmenia) and western (Europe, Egypt) directions. The pattern of farming in Mesopotamia and even that of building simple houses with clay bricks remained largely unchanged for many millennia. Houses were built on top of old collapsed ones and many "hills" seen in the Near East (called "tells") and Anatolia are actually the result of the accumulation of layers of clay from earlier houses. Some of these tells are still inhabited today after 9000–10,000 years of accumulation of clay houses (e.g., Arbil in northern Iraq).

4.7.b. IRRIGATION AND CITY GROWTH

About 8000 years ago (the Samarra phase; Oates 1980) irrigation began and soon became common in southern Mesopotamia. The increase in farming potential under irrigation generated or emphasized disparity in the value of land, depending on the availability of water. The need for irrigation was a stimulus both for cooperation as well as for the development of social classes and hierarchical structures. Not long after this, the first "temples" appeared. These might have been centers for social and organizational meetings as well as religious practices, which might have been invoked as insurance against natural calamities (floods, droughts, etc.). The greater farming potential made possible the formation of towns and their continuous growth in size and also favored the development of trade, which extended as far as southern Arabia and the Indus Valley.

Mesopotamia in the eighth and seventh millennia B.P. thus became ready for the first development of a strong urban civilization associated with the Sumerians (fig. 4.7.2). One of the first Sumerian cities, Uruk, reached a considerable extent by 6000 B.P., with mass-produced pottery. It may have attained its maximum size in little more than 1000 years. From late Sumerian times, around 5000 B.P., comes the world's first written material. Only a few words of this script are understood, and the Sumerian language itself is of unknown origin. According to some linguists (Gamkrelidze and Ivanov 1990), it shows similarities with Caucasian languages. A sexagesimal system of numerical notation was used.

Uruk was not the only Sumerian city, and Sumerians established trade and economic and intellectual ties with

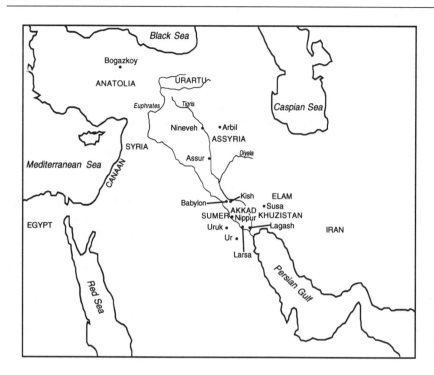

Fig. 4.7.2 Major sites and people in Mesopotamia and neighboring regions (modified from Oates 1980).

Syria and southwestern Iran. In southwestern Iran, in the area known as Khuzistan, lies Susa, the oldest site and later the capital of the proto-Elamite culture. These developments outside Sumer were certainly under strong Sumerian influence, which extended to most of the Middle East. Elamites were probably politically dependent on Sumer but had a certain degree of cultural independence, including a different language belonging to the Dravidian family.

4.7.c. BRONZE AGE IN THE NEAR EAST

Sumerians.   Early cast copper is present but rare in Mesopotamian towns, but a number of copper axes came from early Susa (Oates 1980). As the Bronze Age developed, city-states emerged, with kings who may have been elective and often competed with the authority of temples. Ur, Nippur, and Lagash were Sumerian city-states that showed a tendency for political expansion (see fig. 4.7.2.). This period, called Early Dynastic, is believed to have begun about 5000 years ago, roughly at the time of the first Egyptian Dynasty (3100 B.C.). Social stratification increased and aristocratic ruling families had their land property managed by clients, who may have been in a state not far from slavery. The older system of kin control over shared land was present, but probably on its way to disappearance. Kings tried to wrestle land property from temples, often successfully.

1. *Akkadians* (Moorey 1980).   Similar urban developments just north of Sumer fell slowly into the hands of Akkadians, people speaking a Semitic language; the city of Akkad was not far from Babylon, but its exact site is not known. Sargon, an Akkadian from the city of Kish, conquered Uruk around 2350 B.C. and expanded his domination to Syria, Anatolia, Iran, and the Indus Valley. The Akkadian empire did not last long and was overthrown by tribes from the north, west, and east; kings of the Sumerian city of Ur took power and extended it far and wide, in the footprints of the Akkadian emperors. This third dynasty of Ur was to fall around 2000 B.C. to the Elamites (themselves under Akkadian rule after 2300 B.C.). The Sumerian cities of Larsa and Isin contended for power, and Sumerian rule was again active in lower Mesopotamia until Hammurabi established a new dynasty with its capital in Babylon, shortly after 1800 B.C.. Hammurabi was a member of the Amorites, a Semitic-speaking group of nomads probably from the Syrian desert. Babylon later fell to the Hittites, but the Kassites, already in the area, took power around 1600 B.C. and remained in control of Babylonia until the twelfth century B.C., when Elamites and Assyrians (the latter from northern Mesopotamia) ended Kassite power.

2. *Hittites.*   Records from the twentieth century B.C. and somewhat later indicate that Hittites, an Indo-European-speaking people, were settled in central Anatolia. They took power and established a capital, Hattusha, at modern Boghazkoy. The Hittites were responsible for ending the dynasty of Hammurabi in Babylonia, and they fought with Egyptians for control of trade in the Middle East, and also with Hurrians, a people from Armenia that may have come from farther North. The Hurri spoke a non-Semitic non-Indo-European language (of the Caucasian family; Gamkrelidze and Ivanov 1990) from which the Urartian language may have originated. They

fell under the control of an Indo-European-speaking tribe coming from the outside and between 1500 and 1350 B.C. founded a short-lived kingdom called Mitanni (Moorey 1980).

3. *Sea People*. The fall of the Hittite empire in 1200 B.C. is credited to the Sea People, (Parr 1980), a diverse collection of ethnic groups from the Mediterranean islands, but the destruction that followed the conquests by these pirates in the Near East caused a break in ancient records, making historical reconstruction difficult. Egyptians resisted the Sea People fairly successfully, and the names of many ethnic groups are reported in Egyptian inscriptions as belonging to them: Acheans (Greeks), Tyrrhenians from western Anatolia, possibly ancestors of Etruscans, and also Lycians from the same geographic region; Sicilians, Sardinians, and Peleset (Philistines in the biblical tradition). The last are perhaps the only large tribe of Sea People who settled permanently on the southern coast of the Middle East, giving their name to the geographical region of Palestine.

4.7.d. THE IRON AGE IN THE LEVANT

In the turmoil generated in the Middle East by the attacks of the Sea People and the fall of the Hittite Empire in 1200 B.C., several new populations (in addition to the Philistines) established themselves on the Mediterranean coast or nearby (fig. 4.7.3).

1. *Arameans*. Pastoral tribes from inland Syria, the Arameans, speaking a Semitic language, were early users of the camel, which allowed them to strengthen trade across desert routes. They were also early adopters of an alphabetic script used by Phoenicians.

2. *Canaanites*. Canaanites, traders already located on the coast near Mount Carmel, were able to keep their culture and were the ancestors of Phoenicians, who started an expansion by sea in the tenth century (Parr 1980). Canaanites, and later Phoenicians, developed the alphabetic script that gave rise to all European alphabets. Very able traders, they founded a number of trading posts on the coast of North Africa, western Sicily, Sardinia, Corsica, Spain, and beyond Gibraltar. They are believed to have circumnavigated Africa. Many of their trading posts developed into successful colonies, like Carthage, not far from modern Tunis, and founded daughter settlements.

3. *Israelites*. Israelites had earlier settled in the Levant; they were originally one of many Semitic-speaking pastoral tribes, probably of Aramean or Canaanite stock. Among their earliest whereabouts that can be reconstructed from the Bible is their travel from Ur, which is associated with Abraham's name and began around 2000

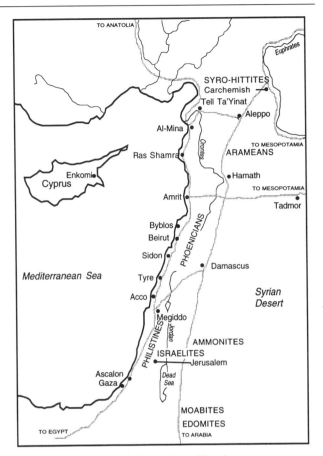

Fig. 4.7.3 The Levant in the first millennium B.C. (modified from Parr 1980).

B.C. in a northern direction up the Euphrates, toward the land of Canaan. Hebrews probably settled in Israel by the fourteenth century B.C., and tribes were encouraged to unify under the monarch Saul (1020 B.C.) by pressure from hostile neighbors like the Philistines, whom they fought successfully. Later the kingdom split into northern and southern realms, Israel and Judah, both of which eventually fell under foreign domination: Israel under the Assyrians (722 B.C., with deportation of the inhabitants), and Judah under the Babylonians (capture of Jerusalem, 597 B.C., with exile of the Jews to Babylonia). The major and last diaspora in many directions occurred in A.D. 70 after Titus's conquest of Jerusalem and destruction of the temple.

4. *Urartians*. The Urartians inhabited northwestern Turkey, and their name may have come from Mount Ararat. They had a common ethnic background with Hurrians, who were from the same region, and had a similar language, also of poorly understood origin (perhaps Caucasian; Gamkrelidze and Ivanov 1990). In Assyrian times (see later), Urartians had a kingdom north of Assyria that was under cultural and, to some extent, political Assyrian influence. The kingdom was invaded by the Cimmerians, a nomadic people from the Caucasus, at

the end of the eighth century. The Urartians lost their independence when power in the area went to the *Armenians*, an Indo-European speaking population. The name is probably a Greek misnomer caused by confusion with Arameans; they call themselves Hayk. The Armenians survived occupations and destructions by many armies, of which the Medes, Persians, Greeks, Romans, Byzantines, Arabs, crusaders, Turkmen, and Turks are a sample.

5. *Assyrians.*   Assyrians owe their name to the city of Assur, in northern Iraq, known since 2400 B.C. (Postgate 1980). They spoke a Semitic language similar to, but distinct from, Akkadian (in use farther south) and were traders with central Anatolia.

The Assyrians generated an empire that extended more or less to the whole area occupied by the fertile crescent and, with ups and downs, lasted until 605 B.C.. The Assyrians used cuneiform, script, and a large number of their documents have been preserved, allowing their history to be reconstructed with a fair degree of accuracy. Cuneiform first appeared a little before 3000 B.C. in Sumerian documents, replacing the first and older Sumerian hieroglyphic script. It was used by Akkadians, Elamites, Kassites, Persians, Mitanni, Hurrians, Hittites and people under their control, Babylonians, and Assyrians. Aramaic, which was soon written in the Phoenician alphabet, replaced Assyrian and became a lingua franca. It is still spoken today, for instance, by the people of northern Iraq, who call themselves Assyrian and live between Mossul and Arbil in northern Iraq, not far from the ruins of the Assyrian capital, Nineveh. They are Christian and are possibly bona fide descendants of their namesakes.

4.7.e. IRAN

As part of the Fertile Crescent, northwestern Iran participated in the agricultural Neolithic revolution. The first historical developments in southwestern Iran included Elam, which had close contact with Mesopotamia. Iran became exposed to incursions of pastoral nomads from the Central Asian steppes before 2000 B.C.. Among these were the *Medes*, first named in historical documents in the eighth century B.C., who settled near other Iranian tribes believed to be closely related in central-eastern Zagros. They built a strongly fortified capital, Agbatana (modern Hamadan, in west-central Iran), which was inhabited between the eighth and sixth centuries B.C.. The Medes, in alliance with the Babylonians, were responsible for the fall of the Assyrian empire. Shortly after that time, the Achaemenian dynasty from Parsa, south of Media and east of Elam, started gaining strength. The Persian king, Cyrus the Great, overthrew the last Median king and conquered an empire larger than the Assyrian one, extending from Anatolia to the Indus River. Among later Persian kings, Darius conquered Egypt, but was

unable to take Greece; Cyrus created a network of good roads with regular stations, whose nodes were administrative centers. Agriculture was favored, and Mesopotamian irrigation techniques spread to Central Asia. In addition, the 1000-year-old "qanat" system of irrigation (using underground channels where water was moved by gravity) was borrowed from Elam and Mesopotamia and widely adopted. Rice and sugar cane were probably imported from India; the peach and the apricot, originally from China, were also introduced (Stronach 1980).

The Achaemenian dynasty lasted two centuries and was suppressed by Alexander the Great, who conquered the Persian empire during the 13 years of his rule toward the end of the fourth century B.C.. He thus brought Greek civilization to Western Asia, but was careful to save most of the existing administration. His successors, the Seleucids, continued the Hellenization of the area.

The Seleucid period is followed by the Parthian (240 B.C.–A.D. 224) and the Sasanian (A.D. 225–641). The first Parthian rulers may have been nomads in central Asia; the Parthian empire was, however, heavily Hellenized and continued Alexander's policy of extending urbanization and agriculture by founding new cities, opening new canals, and reorganizing water supplies. The Sasanians took power but continued the policy of expanding state-controlled irrigation and massive urban growth. Populations of conquered cities and vanquished armies were forced to resettle so that local population growth was not all natural but reached density levels that were never equaled later. Organization was regional, with feudal lords taking care of administration. Taxation was, however, so high that it exhausted and angered the people. After Arab takeover, adequate maintenance of the irrigation system became impossible and led to the collapse of the system.

4.7.f. THE ARABIAN PENINSULA

The Arabian peninsula (fig. 4.7.4) is arid, and most of the population is concentrated on the coasts in the east and southwest.

In northern Arabia, south of the Dead Sea, is an area formerly occupied by Edomites, who fought Israel and later moved to Judea. From the fourth century B.C. comes the first historical reference to the new occupants of this area, the Nabataeans, who developed new techniques of water conservation in the desert that allowed them to build cities and temples in the southern desert. They owed their wealth to the control of the trade of incense from southern Arabia to the Middle East, since the caravan routes from Arabia passed through Nabataean land.

The Eastern coast of Arabia showed two major occupation zones (see fig. 4.7.4).

1. The federation of Dilmun occupied the northwestern coast on the Persian Gulf between Sumer and Bahrain

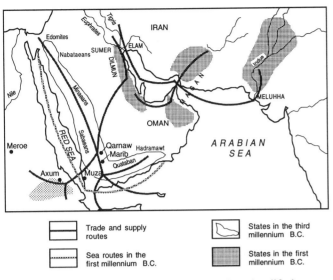

Fig. 4.7.4 Arabia, Ethiopia, and southern Iran (modified from Doe 1980).

Island. This was a trade center forwarding material to Sumer, and later to the Akkadians, from Magan (described below) and from the Harappan civilization in the Indus Valley. Copper, stone for carving, and wood were the major goods traded by seafaring vessels.

2. Magan is an area located partly on the Arabian peninsula, on the coast of the United Arab Emirates (Abu Dhabi) and on the eastern coast of Oman, and partly on the opposite Persian coast across the strait of Hormuz. On both coasts occupation extended inland. The architecture of tombs in the Persian and Arabic areas that go under the name of Magan indicate contact between Persia and Arabia.

The southwestern coast of Arabia developed a trade in myrrh and incense, which were grown locally. There was trade with Israel, during Solomon's times, via the Red Sea and also by land via a caravan route parallel to the west coast of Arabia that reached Petra in Edomite and, later, Nabataean territory. Another caravan route crossed the desert toward Dilmun. Several Arab peoples are historically known, among whom were the Sabaeans, a federation of tribes in modern Yemen, and the Minaeans, farther north along the caravan route. Both these peoples had important exchanges with Ethiopia, as discussed in chapter 3.

Arab trade flourished in Roman times and continued later under Arab rule. When the Portuguese sailed around Africa to India at the end of the fifteenth century A.D., they found prosperous Arab trade from the Persian Gulf to China.

4.7.g. ARAB EXPANSIONS

The city of Mecca (fig. 4.7.5), which became the center of Islamic faith, was an important node in the camel

trade between southern Arabia and the Middle East. It was the city of birth of Muhammad, who created the religion of Islam and unified much of Arabia before he died in 632. Under his successors, expansion continued at a rapid rate: Syria, Iraq, and Egypt were conquered next and in less than 80 years all of North Africa and part of Spain were occupied. The capital was moved first to Baghdad, with an independent western capital established at Cordoba in Spain (see fig. 4.7.5).

The conquest of northern Africa by Arabs in the seventh century A.D. was followed by a second invasion, the "Hilalian," of Bedouin nomads from central Arabia beginning around A.D. 1045 (Murdock 1959). Bedouins form about one-tenth of the population of the Arab Middle East and Arabia, being largely pastoral nomads even today. The Hilalian invasion forced many Berbers into servitude or migration to the interior of the Sahara and further degraded environmental conditions in North Africa through indiscriminate grazing.

Arab rule considerably changed the distribution of populations in western Asia; many earlier prosperous cities were abandoned, while new ones were founded. Arab rule was interrupted by the invasion of Turkic people, also Moslem. Among these, the Seljuqs established an empire in Iran in the eleventh century, occupying Baghdad, and then most of Anatolia. Baghdad was sacked again in the thirteenth century by another horde of pastoral nomads, the Mongols. Arab rule was replaced over an important part of Arab lands by Ottoman rule, which began with Osman, a Turkic chief in Anatolia (died A.D. 1320). After taking all of Anatolia and ending the Roman Empire of Byzantium in A.D. 1453, in succes-

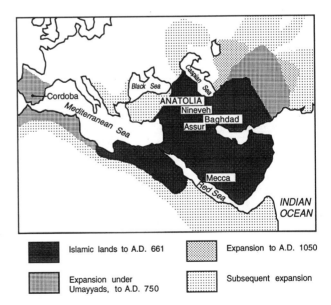

Fig. 4.7.5 Arab conquests (modified from Whitehouse 1980).

sive centuries the Ottoman Turks extended their rule to the Balkans, then to Syria, Egypt, Mesopotamia, Tripolitania, Tunisia, Algeria, Hungary, and much of Arabia. The dissolution of the Ottoman empire began in the last century and was completed with World War I.

4.8. LINGUISTICS

According to Ruhlen's classification, 10 language phyla are spoken in Asia (fig. 4.8.1):

1. Uralic-Yukaghir (northern Asia and northern Europe);
2. Chukchi-Kamchatkan in the extreme northeast;
3. Altaic, the family spread over the widest area: Northeast, East, Middle, and Central Asia and Turkey;
4. Sino-Tibetan languages: spoken in China, Tibet, and part of Southeast Asia;
5. Austric, including Austronesian, Daic, Austroasiatic, and Miao-Yao, spoken in mainland and insular Southeast Asia and the Pacific islands;
6. Dravidian, almost exclusively in southeastern peninsular India;
7. Indo-European, shared with Europe;
8. Afro-Asiatic, shared with Africa;
9,10. Caucasian, two families spoken in the northern and southern Caucasus. They were considered one family in Ruhlen (1987) and two families in Ruhlen (1991).

There are also four language isolates, currently not classified in any phylum:

1. Burushaski, spoken by the Hunza in the northern Pakistan mountains;

For a more extensive general introduction, see chapters 15–19, 26, 28, 30, 31, 43 and 45 in the Cambridge Encyclopedia of Archaeology (Sherratt 1980). They have been a partial source of information for us and contain further references.

2. Ket, spoken in central Asia at the border of Uralic and Altaic speakers;
3. Gilyak (or Nivkh), spoken on the lower course of the Amur river in East Siberia and on northern Sakhalin Island;
4. Nahali, in central India, which, according to some, belongs to the Munda (Austroasiatic) languages.

1. *Uralic-Yukaghir.* The Uralic-Yukaghir family comprises two subfamilies, one made up of Yukaghir (and two extinct languages from northeastern Siberia) and another that includes an eastern group called Samoyed—about 2 million speakers—divided into northern (Nenets, Enets, Nganasan) and southern (Selkup, Kamas) parts, and a western, or Finno-Ugric, group spoken in different parts of Europe (discussed in sec. 5.6).

2. *Altaic.* The Altaic family includes three subfamilies, Turkic, Mongolian and Tungus, and, according to some linguists, three isolated languages: Korean, Ainu, and Japanese (with Ryukyuan).

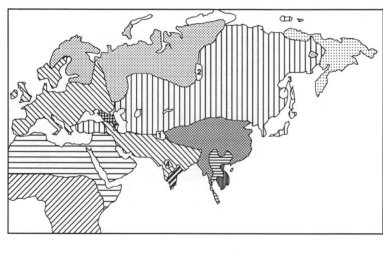

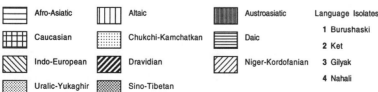

Afro-Asiatic	Altaic	Austroasiatic	Language Isolates
Caucasian	Chukchi-Kamchatkan	Daic	1 Burushaski
Indo-European	Dravidian	Niger-Kordofanian	2 Ket
Uralic-Yukaghir	Sino-Tibetan		3 Gilyak
			4 Nahali

Fig. 4.8.1 Geograpic map of the distribution of linguistic families spoken in Eurasia and North Africa (Ruhlen 1987).

Some 30 Turkic languages were spread widely by the pastoral nomads of the eastern Asian steppes. They are subdivided geographically into five groups:

1. eastern (Uighur, Uzbek, etc.);
2. western (Bashkir, Tatar, Kumyk, etc.);
3. central (Kazakh, Kirghiz, etc.);
4. northern (Yakut, Tuva, Altai, Dolgan, etc.);
5. southern (Turkish, Turkmen, Azerbaijani, etc.).

There are about a dozen Mongolian languages, including Kalmyk and Buryat, and 16 Tungus languages, including Even, Evenki, and Manchu. Altogether the Altaic family comprises 60 languages and 250 million speakers.

3. *Chuckchi-Kamchatkan*.   The Chuckchi-Kamchatkan family occupies the extreme northeastern corner of Asia and is subdivided into two subfamilies: the northern one includes Koryak, Chuckchi, and two other languages, whereas Kamchadal, the language spoken on the Kamchatka peninsula, is the sole member of the southern subfamily. Altogether there are 23,000 speakers. This family shows some similarities to the Eskimo-Aleut languages, spoken on the other side of the Bering Strait, which are discussed in the chapter on North America. There are also a few Eskimos west of the Bering Strait.

4. *Sino-Tibetan*.   The Sino-Tibetan family includes the Sinitic subfamily (12 languages, one billion speakers) and Tibeto-Karen (246 languages, but a smaller number of speakers). Tibeto-Karen branches into Karen, a single language, and Tibeto-Burman. According to some linguists, Tibeto-Karen is related to the Na-Dene family of northwestern America, to the Yeniseian family of Central Siberia, and to the North Caucasian family in a Dene-Caucasian phylum.

5. *Austric*.   Austric is a phylum that includes several other previously classified families.

1. *Miao-Yao* includes four languages and many dialects, spoken by a large number of fairly isolated populations in Thailand, Laos, Vietnam, and South China (7 million people).
2. *Austroasiatic* includes around 150 languages. Seventeen Munda languages are spoken in at least four different areas in northeastern India (6 million speakers). The others, called Mon-Khmer, are spoken by the majority of people in Vietnam, Cambodia (Khmer), Laos, one group in northeast India, and the inhabitants of the Nicobar Islands (total 50 million speakers).
3. *Daic* embraces 57 languages (50 million people) spoken in southern China and Southeast Asia, including Thai and Lao.
4. *Austronesian* includes almost 1000 languages (180 million speakers). There are three major branches:
   a. Formosan: the aboriginal languages of Taiwan, today numbering about a dozen, with 350,000 speakers;
   b. Western Malayo-Polynesian: about half of the languages, spoken in Indonesia, the Philippines, southern Vietnam, and the African island of Madagascar;
   c. Central-Eastern Malayo-Polynesian: includes the remaining languages, which are spoken in Melanesia, Micronesia, Polynesia, and also on the northern coast of New Guinea. The eastern half constitutes less than 1% of all the speakers of Austronesian languages.

6. *Dravidian*.   The Dravidian languages (28 in number) are, for the most part, found in southern India and Sri Lanka and number 145 million speakers. An isolated Dravidian language, Brahui, is spoken in southwestern Pakistan. Harappan inscriptions have been interpreted by some as belonging to a Dravidian language. A more thorough analysis is at the base of the relationship postulated by McAlpin (1981) between Elamite and Dravidian, thanks to the existence of documents in Elamite written in cuneiform characters from about 2500 B.C. to the fourth century B.C.. Elam is the biblical name for a region of southwestern Iran that had extensive contacts with Mesopotamia (see sec. 4.7). McAlpin has reconstructed proto-Elamo-Dravidian, and the Dravidian family has been renamed Elamo-Dravidian (Ruhlen 1987). Speakers of this family probably once extended continuously from Mesopotamia to India and thus may well have included Harappans. Two authors (Cavalli-Sforza 1988; Renfrew 1989a,b,c) have independently hypothesized that it was spread eastward by farmers originating in the Iranian horn of the Fertile Crescent. Despite its name, the Fertile Crescent may be approximately described as having three horns, one pointing toward Suez, one toward western Iran, and one extending into southwestern Anatolia. It is possible that a different language was spoken in each of the three horns some 10 kya, at the time of the spread of agriculture, and each of the three was carried by the Near Eastern farmers in the three directions in which they spread: North Africa/Arabia, Europe, and Iran/India, respectively (fig. 4.8.2). If this is correct, then the three languages may have been ancestral to the Afro-Asiatic (at least its Semitic branch), Indo-European, and Dravidian families, respectively. Today in Iran and most of India, Indo-European languages are spoken, but the most widely accepted interpretation is that they were brought there at the beginning of the second millennium B.C. by pastoral nomads originating in the western steppes of Asia or of southeastern Europe (sec. 4.6).

Initially, Renfrew (1987) assumed that farmers from the Fertile Crescent spread Indo-European languages and that the arrival of Afro-Asiatic in the Middle East was a later phenomenon. He left open, however, the more traditional option that Indo-European languages arrived in Iran, Pakistan, and India from the Asian steppes. In his later papers (Renfrew 1989a,b,c, 1992), Renfrew abandoned the hypothesis that Indo-European languages accompanied the diffusion of Neolithic farmers to Iran and India and, as mentioned above, favored the idea that the

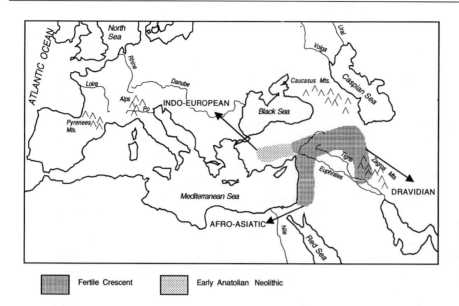

Fertile Crescent   Early Anatolian Neolithic

Fig. 4.8.2 The spread of Indo-European, Dravidian, and Semitic languages by Middle Eastern farmers (a hypothesis; modified from Cavalli-Sforza 1988, Renfrew 1989b).

language spread eastward by Middle Eastern Neolithic farmers was proto-Dravidian, ancestral to the modern Dravidian family. It is worth noting that the union of Indic and Iranian or Indo-Iranian is one of the few subgroupings within Indo-European that is universally accepted, indicating that it probably represents a very late split. Indo-Iranian languages are thus far too similar to have been spread from the Middle East 10 kya. In fact, the migration from the Central Asian steppes that led to the settlement in Iran and India may be only about 4000 years old or less.

7. *Indo-European.*   The Indo-Iranian branch of Indo-European (Indo-Hittite; Ruhlen 1987) is formed by 93 languages, of which 48 are Indic and 40 Iranian, with a total of 700 million speakers.

8. *Afro-Asiatic.*   The Afro-Asiatic language family has six branches, (discussed in chap. 3). The branch with the most people (121 million of 175), but not the most languages (only 19 of 241) is Semitic, spoken in North Africa, northern Ethiopia, Arabia, and the Middle East (Arabic, Hebrew, and Aramaic).

9,10. *Caucasian.*   Caucasian languages are very diverse and form two distinct families, North Caucasian and South Caucasian (or Kartvelian). It is of considerable interest that some authors (Gamkrelidze and Ivanov 1990) find similarities pointing to a common origin of Basque, Sumerian, Urartian, Hurrian, Etruscan, and North Caucasian languages. This very exciting hypothesis may be difficult to support in ways that will obtain wide recognition: the majority of these languages are extinct, some incompletely known, and the relationship must be very ancient, greatly reducing the similarities. The time of an expansion to such a large area of people speaking these languages must antedate the Indo-European, Afro-Asiatic, and Dravidian expansions, namely the development of agriculture in the Near East beginning 10 kya. The only expansion for which we have some evidence is dated to 40–30 kya and corresponds to the replacement of Neanderthals in Europe. The common characteristics of languages that diverged such a long time ago must obviously be very thin. Alternatively, of course, there might have been an expansion at an intermediate date between 35 kya and 10 kya. Progress is slowly being made, however, towards recognizing etymologies common to all or almost all language families (see chap. 2), whose common origin must be even more ancient. This brings hope that hypotheses of this kind can be tested more rigorously than has been so far possible.

4.9. PHYSICAL ANTHROPOLOGY

The layman's perception of characteristics of Asian people derives especially from facial traits, which notoriously vary from Caucasoid to Mongoloid as one goes from west to east. In extreme cases, it is easy to distinguish individuals of these two groups by simple inspection, and also to note that there are various degrees of admixture, especially in Central Asia. The morphological traits vary more in Caucasoids than in Mongoloids: for instance, there is a wider range of skin color from extreme white in northern Europe to black in southern India than there is in Mongoloids, who vary less from north to south. Hair in Caucasoids varies from blond to black and can be straight or wavy, whereas it is almost invariably black, straight, and coarse in Mongoloids. Mongoloids also have scarce facial and body hair, and graying is delayed, whereas hair is abundant in Caucasoids and graying is almost the rule. Eyes are brown in mongoloids with epicanthic eyefolds and fat-padded lids; they usually appear "slanted," a characteristic also found to some degree in the African Khoisan. The Mongoloid face is flat, with a flat nose and high cheekbones. Another well-known Mongoloid feature, the Mongolian spot, is not regularly present. Mongolian characteristics are most marked in southern Siberia, as we discuss more fully in the chapter on America, given the importance of these populations in the peopling of Americas. In Southeast Asia, though present, these characteristics are less marked; other physical characters derived from body build distinguish northern and southern Mongoloids. We show below that genetic data confirm this distinction and set a boundary corresponding to that of the basins of two major rivers, the Huang-Ho (Yellow River) and the Yangtze. The physical separation goes back to the Paleolithic and is maintained through the Neolithic, as already mentioned in earlier sections. Chang (1977) summarized the ancient picture as follows: "North China was populated throughout by people manufacturing stone implements of great complexity, displaying different features at different sites, from the Mongolian steppes to the Manchurian and northern Chinese forests. The North China stone implements were microflakes and microblades, quite different from those manufactured in South

China, where there were no blades or microlithic industry, but the 'Hoabinhian complex' of Southeast Asia, characterized by chipped pebble tools including hand axes and scrapers." Moreover, skeletons differ in North and South China (1977), the northern ones being "Mongoloid" and the southern ones more similar to contemporary Southeast Asian types.

A collection of Asian anthropometric data by Bowles (1977) allows an elementary multivariate analysis of nine standard measurements in 33 Asian populations. The variables are: stature, head length and breadth, forehead breadth, face breadth and height, lower jaw breadth, nose height and breadth. The geographic distribution of populations analyzed is shown in figure 4.9.1.

The analysis of Bowles' data collection by first and second principal components (PCs) is shown in figure 4.9.2 and that by second and third component in figure 4.9.3.

Even with the limited choice of anthropometrics available in this data set, one can recognize that the first component is especially sensitive to general size and approximately separates south (at the right of the diagram in fig. 4.9.2) from north (at left). South Asia (India) occupies the extreme right of the diagram. The first and second PCs segregate Southeast Asia in an upper cluster, and Northeast Asia almost entirely in the lower left quadrant. Two PCs alone cannot do justice to the great number of clusters (we have seen that in principle, n PCs separate a minimum of $n + 1$ clusters). Thus, it is not too surprising that the first two PCs do not succeed in

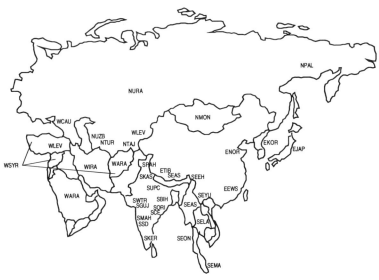

WARA	Arabia & South Afghan	
WSYR	Syria & Central Afghan	
WIRA	Iraq, Persian Gulf, Iran	
WCAU	Caucasus, Ossetes	
WLEV	Levant, Turkey, North Afghan	
NMON	Mongol, Kazakh, Kirghiz	
NUZB	Uzbek & Karakalpak	
NTAJ	Tajik	
NTUR	Turkmen	
NURA	Uralic & Tungus	
NPAL	Paleoasiatic & Eskimo	

ETIB	Tibetan & Ch'iang
EEWS	East, West & South Chinese
ENOR	North Chinese
EKOR	Korean
EJAP	Japanese
SEEH	East Himalayan & North Burma
SEAS	Assam, Burma, S.W. Chinese
SEYU	Yunnan, Kweichow, Vietnamese
SELA	Laotian, Thai, Cambodian
SEMA	Malayan
SEON	Onge Andamanese

SKAS	Kashmir, Punjabi, Pahari
SPAH	Pahari Botia
SUPC	Uttar Pradesh Caste & Tribal
SBIH	Bihar & Bengal
SORI	Orissa Caste
SCE	C. & E. Tribal Gond, Oraon, Munda
SSD	South Deccan Tribal & Vedda
SGUJ	Gujarat & Konkan Coast
SMAH	Maharashtra Caste
SWTR	West Tribal & Lower Caste
SKER	Kerala, Madras, Sri Lankan

Fig. 4.9.1 Geographic distribution of populations listed by Bowles (1977).

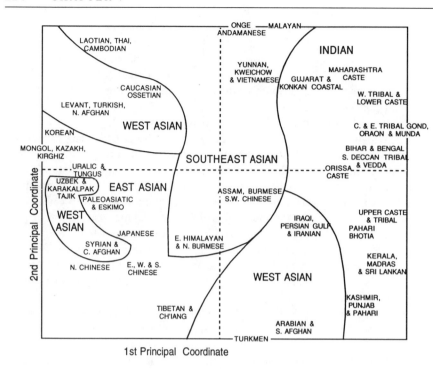

Fig. 4.9.2 Physical anthropology of Asia, based on Bowles' (1977) summary. First (horizontal) and second (vertical) principal coordinates. N., North; S., South; S. W., South West; C., Central; E., East; W., West.

clustering the populations from West Asia, which seem to form three clusters; but the third PC clusters them successfully (fig. 4.9.3).

Looking at individual measurements, stature is lowest in South and Southeast Asia, as might be expected because of the well-known negative correlation of stature and temperature (Roberts 1973). In agreement with this, the first principal component is especially sensitive to overall size. By contrast, face and nose breadth, indicators of Caucasoid/Mongoloid admixture, show an east-west gradient.

The east-west gradient is especially high in the speakers of Turkic (Altaic) languages. This is partly explained by the history of migrations in and from the central steppes. The first migrations moved from the western steppes of the southern Volga toward East Asia and South Asia (as well as Europe) in the third to first millennia B.C. until Altaic-speaking peoples (Hsiung-nu, Turks, and Mongols) reversed the dominant direction of gene flow in the steppes, making east-to-west flow dominant. These people also moved toward the south at about the same time. The very high geographic mobility of the

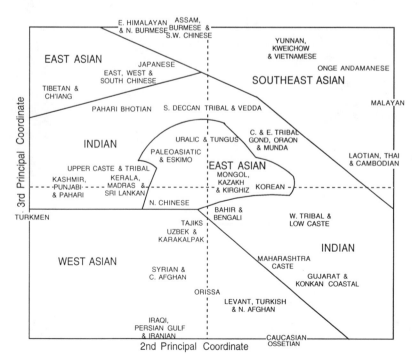

Fig. 4.9.3 Physical anthropology of Asia, based on Bowles' (1977) summary. Second (horizontal) and third (vertical) principal coordinates. N., North; S., South; S. W., South West; C., Central; E., East; W., West.

steppe people, both western and eastern, is explained by their economy of pastoral nomadism. The technologies of animal breeding and exploitation that they developed and perfected over millennia made them more and more mobile. As briefly indicated in section 4.3, there were not only relocations, but also chances for demographic expansion extending the range of their gene pool and, even more, the range of their languages. When mobility became very high, however, the chance of influencing the local gene pools of the invaded countries decreased considerably. Small and efficient armies could rapidly conquer large countries, and there was no time for invaders to multiply fast enough for their contribution to the local gene pool to be easily discernible, especially if the invaded countries were highly developed agriculturally and had a high population density. The chance of influencing culture and language, however, is much greater than that of influencing genes. A powerful elite

of conquerors can—even if an absolute minority—impose its rule and, with it, its language and customs, but is much more limited in extending its genes widely and rapidly.

In addition to Caucasoids and northern and southern Mongoloids, one finds in Southeast Asia a few scattered groups of Negritos (Omoto 1985). This population is characterized by small stature, brown skin, curly to frizzy hair, large teeth, a projecting jaw, and broad nose. Various Negrito groups are found in the Philippines (Luzon, Mindanao, Palawan), in the Andaman Islands, and in Malaysia (Semang). They show some physical similarity to Australian aborigines and Melanesians. Some Indian aboriginal groups (Vedda of Sri Lanka, Kadar of Kerala, and other tribals) bear a certain physical resemblance to Negritos (see also sec. 4.6). The relationships of Ainus with modern Japanese on the basis of skull data have been discussed in section 4.4.a.

4.10. General genetic picture of Asia

The tree constructed from 39 Asian populations or population pools is given in fig. 4.10.1. The genetic documentation includes 68.6 genes, on average. There are 2 major clusters: a smaller one, consisting of 7 populations from Southeast Asia, and a larger one including all the rest. North China and South China belong to

different major clusters, thus confirming the suspicion that, despite millennia of common history and many migrations, a profound initial genetic difference between these two regions has been in part maintained.

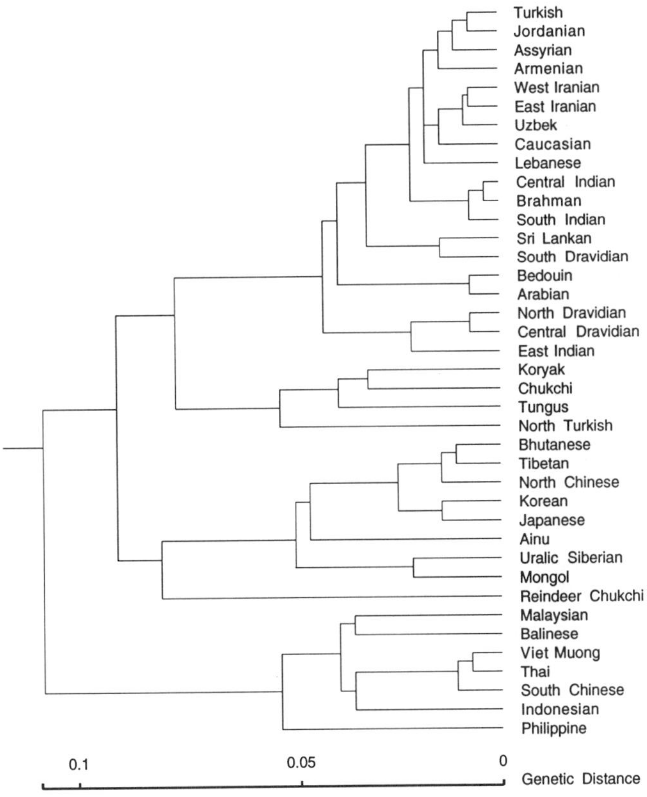

The larger cluster contains three subclusters.

1. A small subcluster comprises four populations from the extreme northeast: Koryak, Chukchi, Tungus (Even), and North Turkic. North Turkic speakers include the Yakut, Tuva, Altai, Dolgan, and others, as explained in more detail in section 4.11. Other northern groups, like Uralic-speaking Siberians, associate with the next subcluster. Reindeer Chukchi, instead of associating with other Chukchis, are an outlier of the next cluster.

2. The northern Mongoloid subcluster includes six populations from East Asia and three from the northeast (Uralic-language speaking Siberian, Mongol, and Reindeer Chukchi) that are missing in the previous subcluster. The limit between North and South China was drawn in the middle of China and, as we have seen, South Chinese join Southeast Asians, while the North Chinese associate with Koreans, Japanese, Ainu, Bhutanese, and Tibetans. We may surmise that the last two originally came from the north, thus explaining the association (see sec. 4.4).

3. The largest subcluster, consisting of 19 populations, is sharply separated from the others and includes all West Asians and South Asians in the sample. It is clearly Caucasoid, though one population from Central Asia, the Uzbek, may be more mixed with Mongoloids than the others. The two Arabian groups pair together and form a small outlying cluster by themselves. Indians tend to divide into two groups and are examined in greater detail below.

The tree thus confirms the major features of the world tree, while it shows some new features with the introduction of more populations into the analysis. The division into two main clusters of Caucasoid and Mongoloid

Fig. 4.10.1 Genetic tree of 39 Asian populations.

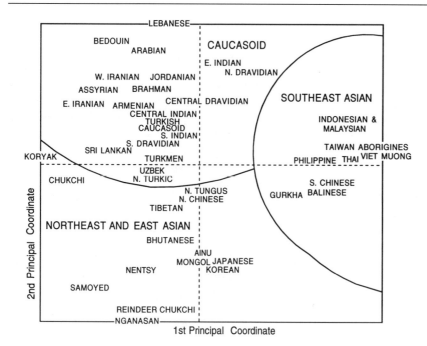

Fig. 4.10.2 Principal-component map of 42 Asian populations.

may exaggerate some differences, forcing Turkic speakers like the Uzbek to fall into one of the two clusters, although the variation is probably more gradual.

Unlike the data from physical anthropology, but not in real disagreement with them, the tree shows a major distinction between Southeast Asia and the rest. The association between East Asians and Northeast Asians and the clustering of all Asian Caucasoids are clear.

The choice of populations was based, as usual, on the number of genes. From a longer original list, three populations that had fewer markers than the 39 eventually chosen (and behaved as outliers in the tree) were retained for PC analysis: Taiwan aborigines, Turkoman, and Gurkhas. Siberian Uralics, used as a block in the tree, were split into their components: Nentsy, Samoyed, and Nganasan. The PC map is therefore based on 44 populations (fig. 4.10.2) with an average of 65.1 ± 6.2 genes and accounts for 39% of the original genetic variation.

The PC map gives conclusions very similar to those of the tree; the first PC separates Southeast Asia from the rest, like the first fission in the tree of Asia. The second principal component helps separate the rest of the populations into two major clusters clearly visible in the left part of the diagram: in the upper part all the Caucasoids (i.e., West and South Asia; tree subcluster 3 indicated above), and in the lower part all of Northeast Asia and East Asia (tree subclusters 1 and 2, defined above).

The PC representation gives only an approximate idea of the subclusters expected in this vast continent and shown by the tree, but it cannot be expected to resolve more than three clusters since it is based on only two components. It is therefore not too surprising that the separation of the clusters in the upper left and lower left quadrants is not as sharp as that of Southeast Asia from the rest. The two groups intermediate between the Caucasoid and Northeast Asia–East Asia clusters, Turkmen and Uzbek (who are also close to the center of the figure), may be mixtures between Caucasoid and northern Mongoloid. The Gurkhas, who are mostly located in northern India and Nepal, show some genetic association with Southeast Asia, but are probably not a clear ethnic entity, the name being that of a royal Nepalese dynasty that was given loosely to British soldiers of a Nepalese corps. They seem to differ, however, from other Nepalese, who associate more closely with Tibetans and Bhutanese. Taiwan aborigines associate closely with Southeast Asia, in agreement with earlier observations (e.g., Chen et al. 1985).

On the basis of these results, we focus our attention on the following groups:

The Arctic (including other non-Asian Arctic populations);
Northeast and Central Asia (including some neighbors);
Southeast Asia, mainland and insular;
South Asia (the Indian subcontinent);
West Asia, including Iran and the Caucasus.

4.11. Genetics of the Arctic

Arctic populations show considerable variation, as might be expected because of their low population densities and the small overall sizes of their groups. They are highly mobile, especially in the winter. For the purpose of a direct comparison with potentially ancestral Asian groups, we have included Eskimos even though Asian

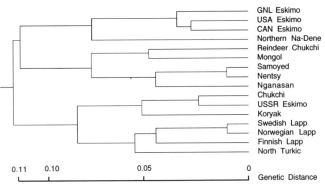

Fig. 4.11.1 Genetic tree of 15 Arctic populations and Northern Na-Dene. GNL Eskimo, Greenland Eskimo; USA Eskimo, Alaskan Eskimo; CAN Eskimo, Canadian Eskimo; USSR Eskimo, Soviet Union Eskimo.

Eskimos are very few and the vast majority are found in North America. We have also included another American group, the Na-Dene, which are somewhat similar to Eskimos. For the former USSR we use population size estimates from the USSR Central Statistical Board (1987) and we retain the old political designation in the following.

Unfortunately, few markers are known for most of the individual ethnic groups. Because tree analysis of a matrix of 26 single, unpooled populations with 64% gaps gave erratic results, we left out or pooled some populations as specified below. From genetic distances based on 60 genes, on the average (and 41.3 per population pair), the tree of 16 populations in figure 4.11.1 was obtained. In the PC map (fig. 4.11.2), which is perhaps less sensitive than trees to sampling error, the larger

matrix of 26 populations was used (average number of genes, 43.6). This map accounts for 35% of the original genetic variation. The four populations with fewer genes (not included in the tree) are: USSR Lapps, a very small group (1900 in 1979), Yupik Eskimos from western Alaska (population 17,000; too few genes), and two large groups, the Komi and Mari (population 47,800 and 622,000, respectively), who speak languages of the Finnic subfamily of the Uralic family, which also includes Lappic languages. The Komi and Mari are located around the north-central part of the Ural mountains.

The speakers of some Altaic languages, who were also tested for rather few genes, were pooled on a linguistic basis. One of the pooled groups is the North Turkic and its best-known members, genetically, are the Tuva (166,000), located in southern Siberia just northwest of the People's Republic of Mongolia; the Yakut (328,000), occupying the central part of Siberia up to the northern coast; the Dolgan, a small group (5100) now located near the Nganasan (see below) but living earlier in a more southern region; and the Altai (60,000) near the western border of Mongolia with the USSR and China.

Another pooled group is Mongol-Tungus, which includes the Even (12,500), located between the Yakut and the Chukchi on the Kamchatka peninsula; the Evenki (27,000), located mostly at the eastern border of Mongolia with China and the USSR; and the Buriat (353,000), located east of Lake Baikal and also in Mongolia and China.

Asian Uralics include the Nganasan, a very small population (900) that lives in or below the Taymir peninsula (they still hunt wild reindeer, but are increasingly becoming reindeer breeders); the Nentsy (30,000), also

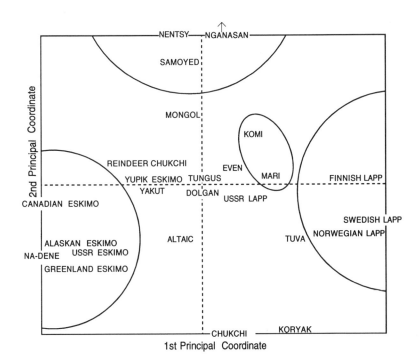

Fig. 4.11.2 Principal-component map of 26 populations in the Arctic region, and some neighbors.

Table 4.11.1. Genetic Distances (in the lower left triangle of the matrix) and Their Standard Errors (in the upper right triangle of the matrix) between Arctic Populations (all values x 10,000)

	FLP	NLP	SLP	EDR	GES	IES	USA	USSR	NN-D	MGT	NTU	CHK	KCHK	RCHK	NEUR	NGUR	SUR
Finnish Lapps	0	103	99	271	263	384	356	261	291	122	225	288	344	352	235	475	321
Norwegian Lapps	418	0	25	208	255	237	162	186	299	209	85	197	265	342	169	266	178
Swedish Lapps	465	96	0	273	375	405	194	196	337	205	119	170	234	354	344	488	271
Elamo-Dravidian	1158	1014	1343	0	75	50	76	66	99	92	225	240	275	145	98	289	241
Greenland Eskimos	920	1090	1534	302	0	66	79	173	205	162	223	185	248	259	192	488	370
Inupik Eskimos	1670	1217	1878	326	278	0	91	167	296	174	339	360	620	303	244	599	465
USA Eskimos	1275	734	905	259	330	417	0	125	120	190	177	219	495	102	273	325	156
USSR Eskimos	1336	803	1073	407	677	799	351	0	182	166	224	72	169	212	257	322	121
North Na-Dene	1748	1420	1748	645	690	872	595	746	0	107	258	390	457	350	322	521	429
Mongol Tungus	723	818	852	478	903	853	728	566	743	0	271	389	449	109	221	290	148
North Turkic	739	402	533	895	966	1261	686	837	1157	689	0	161	187	377	277	477	349
Chukchi	1061	502	559	658	682	1370	562	246	1136	807	505	0	157	511	303	454	238
Koryak Chukchi	1199	858	700	1070	1232	2015	1227	680	1516	962	658	335	0	666	399	574	304
Reindeer Chukchi	1659	1087	1013	621	1067	1494	496	583	1236	466	1398	980	1414	0	251	311	150
Nentsy Uralic	854	894	1219	619	776	946	844	780	1165	512	787	1028	1440	864	0	198	30
Nganasan Uralic	1927	1673	1861	1309	2125	2157	1451	1361	2232	904	1457	1874	2136	1321	631	0	65
Samoyen Uralic	1269	907	1022	773	1345	1508	501	380	1217	370	852	840	1036	498	114	234	0

called Yurak, Nenets, or Yenisei; and others. Other people were grouped as (other) Samoyen or Samoyeds, a generic name for the whole group.

The Chukchi (14,000) live on the Chukchi peninsula and are divided into two groups: maritime Chukchi on the Arctic and Bering coasts of the peninsula, and Reindeer Chukchi, reindeer herders who inhabit the interior of the peninsula. The Koryak (8000) live in the Kamchatka peninsula and also speak languages of the Chukchi-Kamchatkan family.

The Lapps (their correct name is Same or Saame; Lapp is derogatory in some countries, but we retain it because it is much more widely known) are divided here according to political geography. There are 20,000 Lapps in northern Norway, 8000 in northern Sweden, 2000 in Finland, and almost 2000 in the USSR. They speak at least three different Lappic languages. Coast, forest, and mountain Lapps have different economies, substantially transformed in recent times. Coastal Lapps were hunters of wild reindeer and sea mammals, with seasonal migration (i.e., seminomadic); mountain Lapps were fully nomadic reindeer herders; forest Lapps were seminomadic wild reindeer hunters and freshwater fishermen.

Eskimos are described more fully in the chapter on America. (Their correct name is Inuit. Eskimo is sometimes perceived as derogatory, but it is more widely known.) There are only about 1500 USSR Eskimos, near the Bering Strait, and they speak the Yupik language. Genetic information on the Asian Eskimos is limited. They seem to cluster with the neighboring Chukchi, and the linguistic families of the two groups show some similarities. Another group of Chukchi, the Reindeer Chukchi, are located farther south. Northern Na-Dene are from the northwestern part of North America (see chap. 6) and are included with Eskimos to examine their possible affinities with people from Arctic Asia.

The tree shows three major clusters, Eskimos (joined by their American neighbors, the Na-Dene), Lapps (with some Chukchi), and a group of Siberian populations. The PC map shows some more detail. Lapps and Eski-

mos are the clusters farthest apart, forming the poles of the first PC. They are followed by northern (Siberian) Uralics and Chukchi at the poles of the second PC. The two western Uralic groups, Komi and Mari, are somewhat intermediate between Lapps and northern Uralic groups, approximately corresponding to their relative geographic positions.

Nganasan are outliers of the Samoyed group in the tree and also in the PC map, where, however, this detail is not clearly visible. For the purpose of graphic display, they were shifted in the PC graph so as to take the same value of the second PC as the Nentsy; their real value would place them at almost twice the distance from the center as the Nentsy. Their position as an outlier must be a consequence of their very small population size (900), which must have been responsible for extreme drift. In spite of this, however, they still cluster with the other two Uralic-speaking groups who are their geographic neighbors and who belong to the same linguistic branch of the Uralic family (North Samoyed).

In conclusion, there has been considerable genetic differentiation in the Arctic region, especially among populations living in the extreme north, and here drift must have been especially powerful given the low population density and small tribal numbers. Geography retains some importance, as shown by PCs, but with exceptions: Lapps and Eskimos are farthest apart in the PC map and in geography, but Lapps show similarity, though not marked, to some Chukchi. Populations that are central in the PC map are mostly located in southern Siberia and do not diverge as much as the more northern populations. One group of Chukchi (Reindeer Chukchi) is widely separated from the other Chukchi and closer to the Tungus. It could be hypothesized that they originally spoke another language and acquired a new one, along with reindeer-breeding skills, from the Chukchi. Russian Eskimos are probably more mixed with other Asian populations and diverge from North American Eskimos. The F_{ST} genetic distances used in these analyses are given with their standard errors in table 4.11.1.

4.12. GENETICS OF EAST AND CENTRAL ASIA

The analysis included 21 populations from East Asia and some neighbors (F_{ST} distances and standard errors in table 4.12.1). The tree (fig. 4.12.1) shows a distant outlier, the South Chinese, who are more closely related to Southeast Asia than to Northeast Asia (see chap. 2). As already mentioned with regard to physical anthropology in section 4.9, there is a strong difference between North China and South China. The tree shows one large cluster that includes populations living in Japan, Korea, Bhutan, and Tibet and another large cluster made up of three smaller ones. Of the three small clusters,

the first includes Turkic-speaking populations of Central and West Asia (Turkomen, Uzbek, and Turk) who have Caucasoid components of various importance; the second includes Altai, Northern Chinese, and Nepalese (excluding Sherpas, who are kept separate); the third comprises two Altaic-speaking populations, already described in section 4.11, and also—perhaps surprisingly—Sherpas (85,000, living mostly in Nepal).

The PC map agrees largely with the tree (fig. 4.12.2). It is based on an average of 55.7 ± 5.4 genes and accounts for 39% of the original genetic variation. South

Table 4.12.1. Genetic Distances (in the lower left triangle of the matrix) and Their Standard Errors (in the upper right triangle of the matrix) between Populations in Eastern and Central Asia (all values x 10,000)

	AIN	HOK	HOC	HOKA	HOKI	SWHO	KYU	RYU	KOR	NCH	SCH	NNO	SHE	TIB	BHU	EUZ	ALT	TUV	YAK	TURK	TURO
Ainu	0	46	46	93	50	138	125	33	104	104	247	309	95	182	103	133	289	122	195	250	612
Hokkaido	127	0	54	41	84	61	43	25	11	45	72	92	87	32	109	121	140	112	157	261	108
Honshu Chubu	250	129	0	66	100	19	33	28	51	110	296	39	94	77	90	135	116	118	152	275	83
Honshu Kanto	300	84	120	0	94	12	245	39	63	90	271	85	63	63	79	158	156	233	169	214	168
Honshu Kinki	164	146	160	183	0	170	17	59	43	157	319	511	134	42	118	127	170	73	159	295	809
S.W. Honshu	304	128	59	44	237	0	141	72	58	97	128	118	90	50	90	104	121	92	120	87	213
Kyushu	319	94	60	322	46	149	0	179	241	135	400	310	106	165	165	184	136	163	107	249	132
Ryukyu	152	66	121	116	119	119	277	0	41	80	233	213	99	52	66	78	137	195	120	195	397
Korean	358	36	170	153	134	190	381	123	0	50	210	191	51	53	60	92	234	205	126	224	382
North China	462	143	326	290	447	242	413	262	208	0	152	53	48	67	50	86	77	120	180	146	63
South China	828	208	760	601	1004	377	1304	663	498	328	0	116	58	301	261	216	296	249	364	297	185
Nepal Nosherpa	819	256	126	338	899	252	602	477	475	90	507	0	121	142	63	121	65	238	236	189	102
Sherpa	513	414	562	376	699	528	527	420	282	149	225	386	0	72	111	103	51	98	77	179	192
Tibetan	390	116	206	200	85	112	322	147	174	234	830	367	190	0	38	119	206	123	140	168	156
Bhutanese	480	284	279	340	321	239	444	287	269	108	848	161	227	78	0	138	134	235	99	195	135
East Uzbek	512	412	492	523	587	301	726	374	370	347	721	442	447	447	547	0	150	96	111	67	25
Altai	863	349	314	415	490	278	349	365	576	73	727	272	200	446	420	353	0	172	123	125	53
Tuva	625	417	425	942	422	307	467	719	824	377	1148	1025	315	610	867	311	401	0	113	132	298
Yakut	555	592	516	731	515	448	453	429	519	399	1263	759	133	364	353	226	204	228	0	193	93
Turk	796	563	743	778	674	282	675	747	794	421	1084	752	586	620	811	167	453	636	735	0	55
Turkoman	1032	320	268	578	1519	554	317	693	709	105	622	456	418	406	543	51	146	618	217	104	0

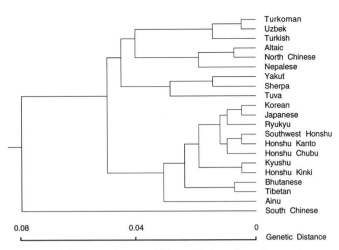

Fig. 4.12.1 Genetic tree of 21 populations from East and Central Asia.

Chinese are again a strong outlier. The first PC separates Japan and Korea from all others. Three Tibetan groups—from Tibet and from Bhutan, and the Sherpas from Nepal—are reasonably close, whereas Nepal is fairly distant (but close to Bhutan in the third PC, not shown). Tibetan groups take a central position but have apparently some affinities with Japan and Korea.

Not shown here are the Balti (population about 40,000) who are pastoral nomads, living in North Pakistan, who speak a Tibetan language; on the basis of the few genetic data available, they show some affinity with the Tibetans.

As already mentioned, Tibetan populations are historically of northern Chinese origin, which explains their clustering with Altaic and northern Chinese. The two Nepalese populations included in the analysis diverge between themselves, and the genetic evidence taken at

face value would indicate they have different origins in Northeast Asia. Tibetans were originally nomadic pastors who came from the North, as already mentioned (sec. 4.4); a large fraction of Tibetans are still nomadic pastors. Data from table 4.12.1 show that Tibetans have the shortest distance from Butanese and from Japanese. Tibetans are the most loosely connected population of the Northeast Asian cluster. In bootstraps (chap. 2) they separate from this cluster 24 times of 100, twice as often as the next loose member, the Ainu, and then they usually join South Dravidian members of the Caucasoid group, from whom they have a fairly small but poorly investigated genetic difference.

The Ainu have always attracted great anthropological interest and were considered Caucasoid by early (European) anthropologists. Those least acculturated or mixed are very few (Omoto 1972, 1973) and live on Hokkaido, the most northern Japanese island, with a few on Southern Sakhalin (USSR, 1500). They were in Japan before the arrival of modern Japanese, and there may have been reciprocal gene flow. The major physical characteristic that differentiates them from other Northeast Asian people, with whom they are tied by genetic and linguistic similarities, is their hairiness as well as the hair form. This was probably the major reason for thinking of them as having a Caucasoid origin, but there are also some other isolated Mongoloid groups other than the Ainu who show hairiness (Alexseev 1979). The Caucasoid origin is still a popular suggestion in classical anthropology, but other hypotheses have been advanced, for example, that they are related to Australian aborigines, or even that they are an independent "race" whose genetic similarity to Japanese is due to extensive admixture. Direct estimates of "purity" by analysis of pedigrees of the last three to four generations indicated an overall non-Ainu

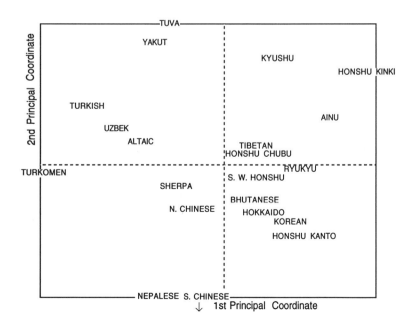

Fig. 4.12.2 Principal-component map of 21 populations from East and Central Asia.

component of 40% in a sample studied by Omoto (1972, 1973). It is not clear whether it is necessary to invoke sexual selection to explain the survival of a few, most probably genetic, traits characteristic of the Ainu like hairiness or hair shape. For the rest, the Ainu show no clear trace of Caucasoid ancestry. Omoto has also shown that the Ainu are Mongoloid, and not Caucasoid, on the basis of fingerprints and dental morphology. It is also possible that the Eta, the outcaste untouchables of Japan, are related to the Ainu (at least as judged from hairiness). Etas were strictly endogamous by law and are, or were, by profession butchers, tanners, executioners, and sweepers; the caste has not disappeared to this day and deserves study. Ryukyuans and Atayal aborigines from Taiwan were also thought to be related to the Ainu. The neighbors from whom the Ainu show the smallest genetic distance are the Hokkaido Japanese. This is not surprising because they live on the same island; the next closest neighbors are the Ryukyuans (152 ± 33). In the tree, however, the Ainu are outliers in the East Asian cluster.

In the world tree (chap. 2), the Ainu show shortest distances from Tungus, Japanese, and Koreans; their distance from Australian aborigines is greater (though not significantly) than that of Japanese or Koreans from Australians; their distance from Caucasoids is perfectly comparable to that of Japanese or Koreans. On bootstrap trees, the Ainu leave the Northeast Asian cluster 11 times of 100, second to Tibetans (who leave it 22 times). When the Ainu are not with the cluster of Northeast Asian populations, they are only slightly external to it, as outliers of a group including other Eastern populations, but, unlike Tibetans, they never join a Caucasoid group. It seems reasonable to discard the myth of a Caucasoid origin of the Ainu.

Most probably, the Ainu lived all over the present Japanese archipelago, perhaps as early as Jomon times, and were largely replaced by invaders of Korean or related origin in the first millennium B.C. and the following millennium. In most respects, they are northern Mongoloids and fairly closely related to all populations from Northeast Asia. They probably owe their outlying position in the northern Mongoloid cluster (chap. 2) to their being of fairly ancient insular origin, but the genetic effects of their ancient isolation may have been reduced by recent admixture with the Japanese.

Koreans show a difference of some magnitude from the Ainu and the Ryukyuans but are quite similar to the Japanese, in agreement with the idea that Korea was the origin of several waves of invaders of Japan.

Japan has been reasonably well studied and the data have been subdivided geographically. They all cluster fairly closely together and with Korea. The area of Kinki, the region around Kyoto, the old capital, is an outlier together with Kyushu, the southernmost of the four big islands forming the Japanese archipelago. Examination of the distance matrix (table 4.12.1) shows a

complex situation. As mentioned previously, the Ainu are least distant from the Hokkaido Japanese (127 ± 46); it is unclear whether the gene flow has been in one or both directions, from Hokkaido Japanese to the Ainu or also vice versa. The next closest to the Ainu are the Ryukyuans, (152 ± 33), who are in a geographic location diametrically opposed to Hokkaido with respect to Japan. The Ryukyu archipelago (55 islands south of Japan in the direction of Taiwan, including Okinawa) shows the smallest distance with the Hokkaido Japanese (66 ± 25). Hokkaido is the most northern island and, somewhat surprisingly from the geographic point of view, is most similar genetically to Korea. There were few Japanese in Hokkaido until 120 years ago and this northern island still has the lowest population density. A possible interpretation is that, before the major invasions of Japan from 2300–1600 years ago, the population was mostly similar to the Ainu throughout the Japanese and Ryukyu archipelagoes.

The tree clearly shows that northern and southern Chinese have different genetic backgrounds. Here, as in other investigations discussed later, the northern group always associates with Mongols or in general with speakers of Altaic languages, and the southern group with Southeast Asia. Modern China is a country of more than one billion people, and, as we saw before, has been densely populated for millennia. Clearly, internal migration has not been sufficient to create homogeneity; thus, the initial peopling must have been from two different sources, north and south. The northern and the southern parts of East Asia have different histories, as indicated by the world tree. China was most probably continuously populated from the upper Paleolithic, and some archaeologists believe that continuous occupation extends back even to *Homo erectus* (see, e.g., Wolpoff 1989). In spite of considerable time for migration, the difference has not been canceled and the genetic gradient is especially high in central China. The pressure by the pastoral nomads from the north has been strong throughout the last 2300 years and has certainly contributed to maintaining the gradient of gene frequencies, but the difference between north and south most probably antedates the nomads' expansion. We mentioned in section 4.9 that there are both archaeological and physical anthropological differences between the north and south that go back to the Paleolithic. The northern and southern Neolithic of China maintained the differences, and the substantial population growth from agricultural development must have strengthened them. Changes in the composition of the population because of the invasion of the northern nomads probably helped maintain the differences. The building of the Great Wall (third century B.C.), as mentioned earlier, was a response to incursions by nomads but was not sufficient to repel them. The north lost an estimated 35 million people, a third of the whole population of China and three quarters of that

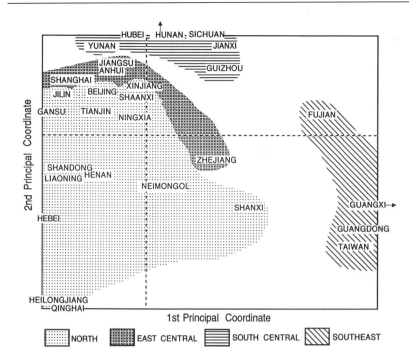

Fig. 4.12.3 Geographic display of a genetic tree of 28 provinces of China plus Taiwan (Han people only), based on *ABO, MN, RH* (Du et al. 1991).

Legend: NORTH — EAST CENTRAL — SOUTH CENTRAL — SOUTHEAST

of the northern provinces, because of massacres and systematic genocide by the Mongols of Genghis Khan and his successors in the thirteenth century (McEvedy 1978; but different sources on Chinese historical demography are somewhat contradictory). At the time, some northern Chinese migrated to the south, but these migrations may have been partially reversed later.

Today at least 52 "minorities" or isolated ethnic groups in China are in the process of being studied genetically. Some of these are known to be late arrivals (e.g., the Uighurs of the most western province of China), as the wealth of historical documentation existing in China can prove. Almost half of these ethnic groups live in the Yunnan, a southern province, and may be enclaves of original inhabitants. The study of these minorities can add greatly to our understanding of the history of China and of all of East Asia. By far the majority of modern Chinese, however, call themselves Han, the dominant group. A tree of the Han from the Chinese provinces, based on *ABO, RH,* and *MN* (Du et al. 1991) is shown in figure 4.12.3 in the form of a map joining provinces that cluster together at the first fissions. The separation is clearly between north and south, with the east (the lower Yangtze) being somewhat intermediate. Data on *GM* (Schanfield et al. 1979) were the first clear indication of a strong difference between north and south. Recent data on *HLA* (too late for inclusion in our analysis) also showed the existence of a strong genetic difference between north and south (Saitou, pers. comm.). The genetic tree shown in figure 4.12.3 is inevitably affected, especially in the lower fissions, by a strong statistical error, given the small number of markers.

Unambiguous results have been obtained by an analysis of a stratified random sample of about 540,000 Han surnames from the 1982 census (Du et al. 1991). Surnames are transmitted only by the male line, are inevitably younger than genetic polymorphisms, and are subject to historically known mutations. Nevertheless, their frequency distribution originates a tree (fig. 4.12.4) that is in substantial agreement with the genetic one and is much more detailed, thanks to the large sample size and the large number of alleles (i.e., surnames, which are over 1000). Chinese surnames are very ancient (some go back to late Neolithic), their historical origins are frequently known, and the great majority antedates the major northern invasions. In more ancient times, surnames were transmitted matrilinearly, a custom changed by emperor's rulings in the middle of the first millennium B.C.. The surname and the genetic maps are reminiscent of the locations of the three classical Neolithic areas. This is not surprising, since the major demographic increases that took place in the Neolithic must have blocked to a large extent subsequent genetic differentiation by drift and buffered the effects of later migrations. All the Han speak Sino-Tibetan languages, but genetically the northern Han are closely related to Mongolian and Japanese people (i.e., northern Mongoloids), and the southern Han to the Vietnamese and Mon-Khmer (see sec. 4.13), who belong to the Southeast Asian or southern Mongoloid cluster. The ancient origin of the Sino-Tibetan languages is obscure, but more recent historical evidence indicates that the diffusion of Sino-Tibetan languages to all of mainland China (with the exception of ethnic minorities) is a consequence of the early political union of the whole country under northern rule at the time of the Qin and Han dynasties over 2000 years ago (Wang 1991).

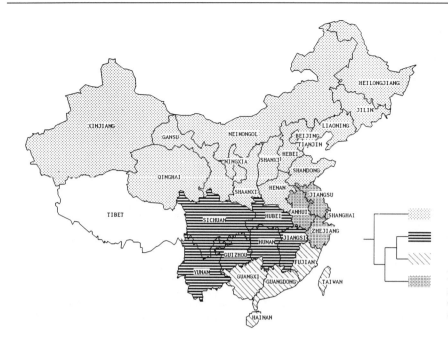

Fig. 4.12.4 Geographic display of a genetic tree based on Han surname distribution in 28 provinces of China (Du et al. 1991).

4.13. GENETICS OF SOUTHEAST ASIA

The 25 populations to which this section refers all come from the Southeast Asian cluster, insular and mainland, with the addition of some neighbors, of which the South Chinese are numerically the most important. Taiwan aborigines and Indian and Micronesian populations that are linguistically or otherwise related have also been included. The tree is shown in figure 4.13.1 and the PC map in figure 4.13.2. The PC map is based on an average of 50 ± 4.7 genes and accounts for 41% of the original genetic variation.

In the tree in figure 4.13.1, based only on 31 genes, the first two fissions distinguish three clusters that we call

A, B, and C. Cluster A is large and composed mostly of mainland Southeast Asian populations, whereas B includes insular Southeast Asia and one tribal Indian group. A and B are more closely related to each other than either is to the third, which consists mostly of Pacific island populations.

1. Cluster A. Cluster A is formed by two subclusters, A1 and A2, with the Filipinos being outliers. The first consists of island people from Sumatra and Bali and three of the four Taiwan aborigines studied, to be discussed below. It also includes the Sea Dayak or Iban (population 300,000) in northwestern Borneo who are, despite their name, riverine populations. Dayaks are believed to have come to Borneo before the Malays, as discussed later. Sumatra is a big island, genetically fairly heterogeneous.

The A2 cluster includes as an outlier the fourth Taiwan aboriginal group studied (Ami). The major group in A2 is very large in terms of population numbers, comprising of most mainland Southeast Asian populations: the South Chinese, who number in the hundreds of millions; the Thai, also a large group (30 million plus the Tai, 10 million); the Laotians (15 million); and the Khmer (7 million in Kampuchea and many in Vietnam). The Khasi, outliers in the cluster, are mostly in eastern India, in Assam and the Jaintia Hills district in the state of Meghalaya (450,000) north of Bangladesh; some of them are slash-and-burn farmers. They speak a Mon-Khmer (Austroasiatic) language.

The major component of this cluster is the South Chinese. The boundary between northern and southern China has been fixed approximately for the purpose of

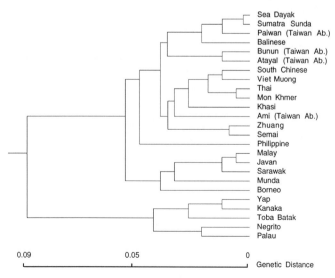

Fig. 4.13.1 A genetic tree of 25 populations in Southeast Asia. Taiwan Ab., Taiwan Aboriginal.

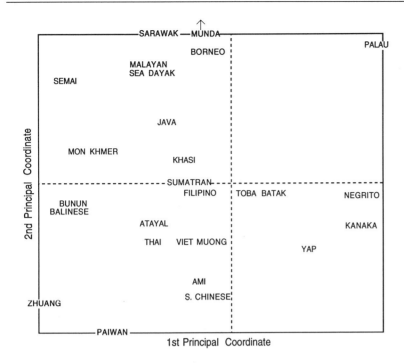

Fig. 4.13.2 Principal-component map of 25 populations from Southeast Asia.

this distinction, as a line intermediate between the two major rivers, the Yangtze and the Yellow River (Huang Ho): see figs. 4.12.3, 4.12.4. Clearly, despite possible gradients established by migration in this highly heterogeneous country, southern China clusters with Southeast Asia, and northern China with a more northern group, discussed above under East Asia. It is of some interest that the Viet Muong (370,000), a group of Vietnamese living mostly in the mountains of central Vietnam and speaking a Mon-Khmer (Austroasiatic) language, also cluster with the South Chinese. A reasonable genetic taxonomy of the people of southern China and mainland Southeast Asia awaits a better set of data from a more representative group of all the minorities present in this part of the world. Our sample is far too small for going beyond the statement that South Chinese classifying themselves as Han, and the few populations tested in the eastern part of mainland Southeast Asia, form a fairly compact genetic cluster. We can include in it the Zhuang (Chang), a tribal group or an ethnic minority in the official Chinese designation (13 million), inhabiting southeastern China (the lowlands of the province of Guangxi and neighboring provinces) and speaking a Daic language (northern Tai). The Zhuang belong to subcluster A2 and associate genetically with the Semai.

The Semai (Senoi) are a small group (18,000) of farmers in Malaysia, somewhat remote from the geographic center of gravity of this cluster. They speak a Mon-Khmer language, which may help in understanding their genetic association with this group.

Taiwan had an early Neolithic development (Bayard 1980). One of the most important Neolithic centers of the Yueh' coastal culture in southeast China was found at Tapenkeng, not far from the Taiwan capital, Taipei.

Much later, Taiwan was settled by three waves of mainland Chinese. The most important one was from Fujian (southeastern China), a migration starting in the eighteenth century, followed by the Hakka (from central and southern China, in the nineteenth century), and a smaller group, originating from many parts of mainland China, which followed Chiang Kai-shek around 1950. The Taiwan Chinese of these recent migrations are very similar to the mainland Chinese from whom they separated, as clearly shown by surname studies (Du et al. 1991). The analysis of surnames of Chinese settlers in Taiwan follows very closely the history of settlement of the island, which is well known (Chen and Cavalli-Sforza 1983).

The Taiwan aborigines, who inhabited the whole island before these recent migrations, are now mostly confined to the chain of mountains rising near the eastern coast. They speak at least 10 different languages. One group, the Ami (100,000 people) is the largest and lives on the southeastern coast. Nine other groups total approximately 100,000 individuals (Bunun, Paiwan, and Atayal number 18,000, 20,000 and 20,000, respectively). All languages are Austronesian, belonging to two (or more) independent branches; of those represented in our sample, the Ami, Bunun, and Paiwan are more closely related, whereas the Atayal are separate. In the genetic tree, the four groups are all in the same subcluster, but the Ami are somewhat removed from the others, as they are also in the PC map. Genetic differences between the four linguistic groups are not striking, however, as can be seen in the F_{ST} distance matrix (table 4.13.1). They are higher between the Ami and the others, ranging from 301 to 375 with high standard errors, whereas among the other three groups distances range from 104 to 227. Given the very few genes examined

Table 4.13.1. Genetic Distances (in the lower left triangle of the matrix) and Their Standard Errors (in the upper right triangle of the matrix) between Populations in Southeast Asia (all values x 10,000)

	KHA	MKH	SEM	VMU	MUN	CHG	THA	SCH	BAL	BOR	JAV	SSU	TBA	KAN	MAL	SRW	SDA	PAL	NEG	PHI	YAP	AMI	ATA	BUN	PAI
Khasi	0	90	144	51	174	161	87	147	226	143	214	27	156	318	93	118	79	289	250	108	257	168	100	53	195
Mon Khmer	182	0	89	142	114	161	17	87	47	172	46	35	243	351	33	64	111	418	360	80	362	165	80	152	98
Semai	494	162	0	124	98	54	53	157	130	86	266	86	267	282	94	86	86	445	295	266	336	218	202	192	292
Viet Muong	180	227	370	0	684	80	31	31	222	168	92	131	79	117	171	195	32	346	275	145	135	85	137	111	132
Munda	341	243	336	968	0	560	381	498	198	140	28	152	211	313	168	166	165	419	271	230	414	885	393	90	1374
Chuang	355	282	87	111	914	0	64	83	102	342	55	175	331	392	451	314	1180	650	371	169	373	101	272	59	56
Thai	240	59	183	81	755	209	0	21	92	211	226	45	119	169	132	146	345	313	291	74	190	60	226	68	87
South China	372	288	551	114	1073	252	105	0	111	216	165	68	126	182	178	228	408	497	135	82	80	153	170	130	136
Bali	567	143	161	466	550	268	247	380	0	216	147	28	160	267	171	210	58	415	462	75	270	165	186	76	62
Borneo	457	531	342	472	529	738	670	697	839	0	179	157	84	126	65	55	123	107	186	217	78	219	262	160	288
Java	405	146	434	142	116	65	494	378	408	377	0	121	130	209	27	32	114	267	419	37	215	208	37	99	83
Sumatra Sunda	144	111	250	260	264	269	143	241	127	589	234	0	69	143	111	102	23	297	114	47	201	139	164	82	53
Toba Batak	368	612	759	171	742	1093	364	320	431	404	304	107	0	107	146	119	581	110	132	134	54	136	162	343	217
Kanaka	624	981	1370	364	1222	1767	774	607	1252	438	633	383	436	0	240	210	133	270	129	80	41	143	372	822	450
Malay	279	106	169	326	396	512	348	510	387	266	50	233	436	1043	0	89	70	354	153	137	251	118	299	179	346
Sarawak	433	212	219	416	316	473	423	522	592	205	122	291	449	940	89	0	27	355	228	150	246	129	303	136	375
Sea Dayak	375	311	205	101	439	1735	547	614	178	290	286	18	966	881	155	50	0	215	320	263	169	176	202	238	32
Palau	961	1469	1353	880	1471	2085	1270	1026	2216	318	647	797	329	447	941	1003	1200	0	160	355	295	330	990	909	1179
Negrito	533	953	1369	699	1068	1620	918	584	1236	868	736	586	479	102	672	1079	1246	178	0	79	204	296	245	406	293
Philippine	462	304	613	342	602	542	334	241	341	636	151	147	395	360	364	362	323	937	345	0	98	267	157	294	328
Yap	594	1130	1439	278	1499	1431	578	365	1031	573	642	451	197	95	950	1098	725	565	264	350	0	232	305	599	320
Ami	436	303	562	207	1424	159	193	283	480	358	463	255	384	302	373	305	298	855	872	386	503	0	209	129	151
Atayal	288	198	480	232	765	267	506	511	499	468	130	327	505	804	380	637	423	1406	730	378	711	375	0	38	86
Bunun	340	370	543	323	350	116	306	440	332	535	324	188	707	1111	402	577	391	1821	1236	641	1095	301	104	0	52
Paiwan	607	354	562	390	1817	118	315	326	218	671	162	118	614	888	655	735	77	1641	924	598	899	375	227	165	0

(among the lowest in the Southeast Asian sample) and the chances of drift, there is little point in discussing the imperfect agreement with the linguistic tree.

In an earlier study (Chen et al. 1985) comparing one small group of Taiwan aborigines with Southeast Asian populations, it was shown that there is a regular gradient of gene frequencies across the Southeast Asian islands and that the Filipinos were genetically the nearest population (they are also geographically close). However, some Taiwan aborigines may have settled in the island in late Paleolithic or early Neolithic times before the island became separated from mainland China. A genetic contribution from other immigrants by sea, and by gene flow from Taiwan Chinese, are also possible and may blur the picture.

The Filipino appear as an outlier in subcluster A. They inhabit the archipelago of the Philippines, over 7000 islands northeast of Borneo with a population of about 38 million. Unlike the rest of Southeast Asia, these islands were colonized by Spain, which led to developments not found in most other colonies in the area. The first inhabitants of the Philippines were most probably Negritos, of which there remain scattered enclaves in these islands, in other parts of Southeast Asia, and beyond (Omoto 1985). Later occupants were probably Malays, who perhaps came to the Philippines from Indonesia; much more recently, starting in the tenth century, Chinese immigrants formed a fairly large group of Filipino-Chinese.

2. *Cluster B.*   Cluster B includes Malays (5 million people in Malaysia, over 3 million in Indonesia), Javanese (48 million), people from Sarawak, a region of northwestern Borneo, which is part of East Malaysia, and from Kalimantan (the greatest part of the island of Borneo, under Indonesian rule). The population of Borneo is about 4 million in the Indonesian part, and that of Sarawak about 1 million. As in all of Indonesia, there are many ethnic groups: in Borneo, Malays are found mostly on the coast, while Dayaks are mostly in the interior.

In cluster B one also finds the Munda, who have a very different geographic location, India. They speak Austroasiatic languages and form 10 distinct tribal groups (5 million total). The Munda inhabit hilly and forested areas, fairly poor for agriculture, of the three most eastern states of India: West Bengal, Orissa, and Madhya Pradesh. Another group speaking a Munda language is located farther to the west, in central India.

3. *Cluster C.*   Cluster C, which is fairly distant genetically from A and B, contains only insular populations, including the Yapese (5000) from the western Caroline Islands; the Kanaka (somewhat over 1000) from Saipan in the Mariana Islands (where they moved from the Caroline Islands a few centuries ago after the island was left by the Chamorros); the Toba Batak, like other Batak a large group from the northern part of Sumatra

(2 million). Other Sumatrans are in cluster A1. We have no specific information that might help to explain their presence in this cluster.

Palau islanders and Negritos (which in our sample come mostly from the Philippines) form a fairly loose pair in this cluster. Palau and Yap are small islands in the western Carolines, in the Pacific Ocean. One reason for studying them here is that Yapese, Palauan, and Chamorro languages are three isolated branches of Western Malayo-Polynesian, like other populations studied here, whereas Pacific island populations speak languages of another branch of Austronesian (central-eastern Malayo-Polynesian). Chamorros were not considered because they are known to have hybridized fairly heavily with Spanish colonizers and with Chinese.

Negritos' physical anthropology shows short stature, a broad nose, and yellowish-brown skin, all characteristic traits of populations from the tropical forest, as well as Australoid features like curly to frizzy hair, large teeth, and a projecting jaw. Women are frequently steatopygous. They are found mostly in the Philippines (where our samples come from), but also in Malaysia (Semang) and the Andaman Islands. We have already discussed the Andamanese, which are unfortunately too poorly known genetically. Negritos show similarities with Micronesians and Melanesians and also with inhabitants of several islands south of the Philippines, which are, however, poorly known genetically. They may be a relic of some of the earliest immigrants to Southeast Asia, who were once much more widespread. Their association in the tree with Palauans may be due to poor genetic knowledge of this last population; but it is interesting to note that Palau islanders have dark skin, which has been attributed to contacts with New Guinea (Bellwood 1979). Clearly, there is, in Southeast Asia and Oceania, a dark-skinned genetic substrate, probably resulting from earlier immigrations. Some of these immigrants, living in marginal environments, remained relatively unmixed, especially where later migrations (e.g., of Malayo-Polynesians) have had a lesser genetic impact. In highly populated areas, contact was unavoidable and often sufficient to cause language replacements. Moreover, some Negritos, especially in the Philippines, are relatively highly acculturated even though they remain reasonably endogamous.

The PC map (fig. 4.13.2) shows similarities with the tree in the sense that the first principal component separates cluster C from A and B, that is, the small Pacific islands plus Negritos from the rest; the second component less clearly separates A and B, with some discrepancies. Most probably this is another indication of the weakness of genetic data from Southeast Asia, which have, on the average, only 31 genes. The Munda are extreme outliers according to the second component, indicating that they should probably not be analyzed with these populations. The Toba Batak seem to associate more easily with

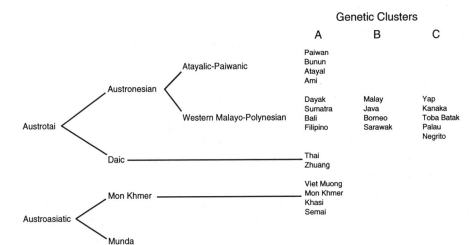

Genetic Clusters

A   B   C

Atayalic-Paiwanic — Paiwan, Bunun, Atayal, Ami

Western Malayo-Polynesian — Dayak, Sumatra, Bali, Filipino | Malay, Java, Borneo, Sarawak | Yap, Kanaka, Toba Batak, Palau, Negrito

Daic — Thai, Zhuang

Mon Khmer — Viet Muong, Mon Khmer, Khasi, Semai

Munda

Fig. 4.13.3 Genetic and linguistic classification of Southeast Asian populations.

cluster A than with C. The Sea Dayak and Semai show greater similarity to the B cluster than to A1. Semai and Zhuang separate widely according to the second PC.

The Southeast Asian populations form a fairly compact cluster (see chap. 2) when compared with other major human genetic clusters, but we feel less comfortable about their internal subdivisions. Our attempt to conduct a more detailed analysis within the group has certainly suffered from the considerable poverty of genetic markers, in spite of our limiting the analysis to the better-known populations. In addition, the degree of admixture of the majority of local people is often such that many of them can hardly be considered "aboriginal" in a narrow geographic sense, even though recent admixtures with external groups like Chinese or Caucasoids are probably not the source of the problem. The migrations of Malays, and perhaps other groups, may have generated considerable genetic homogenization in the area. The homogenizing effect, however, may have been greater for clusters A and B, which are therefore more similar to each other, than for cluster C, which may have preserved a little more closely the genetic characteristics of earlier occupants.

The clusters indicated by the genetic tree retain some correlation with the linguistic classification, despite numerous language replacements. A possible interpretation of the association in figure 4.13.3 is that Western

Malayo-Polynesian languages have been adopted by several populations of all genetic clusters, again as a consequence of the extensive migrations of these people. The speakers of these languages are part of a culture that was, as is well known, made up of exceedingly enterprising colonizers and navigators. In other words, the hypothesis suggested by the table is that Western Malayo-Polynesian languages were adopted by several populations in the area together with the whole culture, but with different degrees of genetic admixture, and that this admixture was least important for cluster C. The Munda constitute a special case. It seems likely that they retained their original language after migration to India but did mix to some extent with the local populations.

Not included in the count in figure 4.13.3 are southern Chinese, who speak Sino-Tibetan languages most probably because of the influence of northern Chinese for the last two millennia. Also excluded are Negritos, whose likely greater antiquity in the area and lower level of independent development make it probable that their original language was lost. Whatever the interpretation, the genetic data for Southeast Asia are weak enough that the detailed analysis done above should be considered of uncertain validity. It has at least the merit of showing that the number of genes that we have been able to use here is probably inadequate for a rigorous analysis of the internal differentiation within Southeast Asia.

4.14. GENETICS OF SOUTH ASIA (THE INDIAN SUBCONTINENT)

South Asia, more specifically the Indian subcontinent, includes the countries of India, Pakistan, Sri Lanka, and Bangladesh. It is a highly heterogeneous region from a historical point of view. The population is large and growing rapidly (over 800 million). The major division is linguistic: the most numerous are the Indo-European speakers, followed by Dravidian (150 million) in the south, and a smaller number of speakers of Austroasiatic and Sino-Tibetan languages. Most Austroasiatic

speakers are Mundas, living in the forests of central and northern India, whom we already discussed in Southeast Asia. They are agriculturists, some of them slash-and-burn farmers, except for a few (e.g., the Birhor, population 590) who are hunter-gatherers. Sino-Tibetan languages are spoken only in the northeastern part.

An important group of populations in India are the tribals, officially called Scheduled Tribes in the 1951 census (Vidyarthi 1983), who are spread over many re-

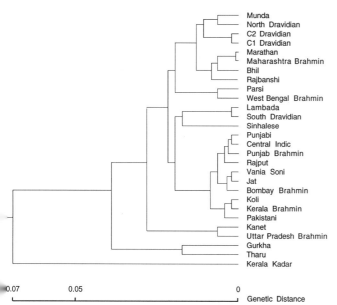

	Munda
	North Dravidian
	C2 Dravidian
	C1 Dravidian
	Marathan
	Maharashtra Brahmin
	Bhil
	Rajbanshi
	Parsi
	West Bengal Brahmin
	Lambada
	South Dravidian
	Sinhalese
	Punjabi
	Central Indic
	Punjab Brahmin
	Rajput
	Vania Soni
	Jat
	Bombay Brahmin
	Koli
	Kerala Brahmin
	Pakistani
	Kanet
	Uttar Pradesh Brahmin
	Gurkha
	Tharu
	Kerala Kadar

0.07   0.05   0

Genetic Distance

Fig. 4.14.1 Genetic tree of 28 South Asian populations. *C1 Dravidian* is the group comprising Brahmin Telugu, Chenchu, Gondi, Konda, and Javara Telugu. *C2 Dravidian* comprises Kolami Naiki and Parji Gadaba.

gions in the north and the south but are also found in a wide central band of territory. Their geographic distribution is shown in figure 4.6.1. Only a few have been studied with an adequate number of markers. Tribes are often fragmented in different small groups occupying different locations, spread over many states. Different groups of the same tribe may have different gene frequencies, because of various degrees of admixture with neighboring populations and drift in the smallest groups. Therefore, even when more extensive data become available, they may not be representative of the whole tribe. Even if they are, when tribes are very small, the data may have been altered by extreme drift. Tribals may represent relic populations of unknown origin but potentially great genetic interest, or intrusive populations whose origin is often known to some extent. Language may help in distinguishing the two cases, as for instance with the Munda, who are probably intrusive, given that the family to which their language belongs has a relatively large area of distribution, but they occupy a peripheral position in it.

There are in the subcontinent other populations of even greater interest, like the Hunzas, whose language, Burushaski, is an isolate regarded by most linguists as unrelated to any other language or family (but perhaps related to Caucasian; Gamkrelidze and Ivanov 1990). Unfortunately, the Hunza (40,000) are located in a mountainous region of northern Pakistan and have not been given adequate genetic attention. Another linguistic isolate in India, the Nahali, is a small group located in Madhya Pradesh (1200). Extremely limited genetic knowledge, totally inadequate for our aims, exists for the only Dravidian-speaking enclave in Pakistan (south-central re-

gion), the Brahui, who number about 225,000 people. They may not differ much genetically from other lowland Pakistanis; but their language suggests they had fewer contacts, and perhaps less admixture with incoming Indo-European speakers.

The genetic tree (fig. 4.14.1) collects 28 large, better-studied groups of Dravidian and non-Dravidian-speaking people and a few reasonably well analyzed smaller groups. In spite of choosing the best-known populations, genetic knowledge remains rather limited (see also Roychoudhury 1974, 1977, 1983), and in spite of considerable elimination of some peoples and pooling of others, in order to increase the number of genes per population, we have only 47 genes, on the average.

The Kadar, a small tribal group (about 1000 individuals) in Kerala, is a major outlier. This may be due to drift but is interesting because, morphologically, the Kadar are considered an Australoid group in India (Vidyarthi 1983). Extreme types have some Negrito characters—that is, frizzy hair instead of straight or wavy hair and especially dark skin—but it has been suggested that some observed examples of frizzy hair are due to rare admixture with Africans (Saha et al. 1974). The Kadar are nomadic; they do not farm but do not like to live as foragers. They work as laborers or as specialized collectors of commercial tropical plants.

Less extreme outliers are the Gurkhas (6 million in Uttar Pradesh–speaking Indo-European languages), from Nepal, who have been discussed previously, and in this tree show affinity with the Tharu (600,000 in Nepal). The next outliers are a pair, the Kanet (about 26,000) in Himachal Pradesh and Gujarat, probably with Tibetan admixture, and the Uttar Pradesh Brahmin, an unexplained association.

There are two major clusters in the tree, A and B. Cluster A is made of 10 populations and is linguistically heterogeneous. A subcluster is formed by three Dravidian-speaking groups (one northern and two central Dravidian groups, C1 and C2) and the Austroasiatic speakers, the Munda. We have already discussed the Munda in the section on Southeast Asia; they are spread over various central and northeastern states. The C1 Dravidian group includes the Chenchu-Reddi (25,000), the Konda (16,000), the Koya (210,000), the Gondi (1.5 million), and others, all found in many central and central-eastern states, though most data come from only one or a few locations. The C2 Dravidian group includes the Kolami-Naiki (67,000), the Parji (44,000), and others; they are located centrally, a little more to the west. North Dravidian speakers are the Oraon (23 million), who overlap geographically with some of the above groups but are located in a more easterly and northerly direction.

The rest of the first cluster includes only Indo-European speakers: one subcluster is made up of the Maratha (50 million, in Maharashtra and neighboring states), known champions of Hinduism. They have the distinction of being the most similar to the Brahmins from the same region. The Bhil (1.2 million) are found mostly in mountainous areas, in the central zone of India. The Rajbanshi live in West Bengal, Bangladesh, and Nepal. A small subcluster that is an outlier of the first cluster is formed by Brahmins of West Bengal and Parsi. The latter are Zoroastrian followers from Persia who migrated to India in the seventh century A.D. and now speak a Gujarati dialect. The association found in this small subcluster is also difficult to understand.

The second major cluster, B, contains a minor subcluster B1 formed by Sinhalese, Lambada, and South Dravidian speakers. Sinhalese is the language most widely spoken in Sri Lanka, where Tamil speakers number only 3 million, and the most likely aborigines, the Veddahs (numbering 411 in 1963) speak another Indo-European language that is being rapidly replaced by Sinhalese. The South Dravidian group includes a number of small tribes like the Irula (5300) in several southern states but especially Madras, the Izhava in Kerala, the Kurumba (8000) in Madras, the Nayar in Kerala, the Toda (765), and the Kota (860 in 1971) in the Nilgiri Hills in Madras (Saha et al. 1976). Large groups include the Malayaraya (22 million), various other populations from Kerala, and the Tamil (45 million) in Madras and in Tamil Nadu.

The second subcluster, B2, comprises 10 Indo-European speaking groups, 3 of which are Brahmins. Kerala Brahmin associate with the tribal Koli (750,000; Papiha et al. 1980) spread over many states and with the sample of Pakistani, forming a small subcluster. A second small subcluster includes the Vania Soni from Gujarat (Undevia et al. 1978); the Jat (linguistically a group of Western Hindi, from north-central India; Das et al. 1978); and people from Bombay of unspecified origin. A third subcluster includes unspecified people from Punjab, Brahmins from Punjab, a number of Indo-European speakers of Central Indic languages not included in the groups mentioned before, and the Rajputs. The Rajputs (11 million) live in northern and central India; they are landowners and their name comes from the royal dynasty that occupied Rajasthana in the eighth century A.D..

Data pertaining to Brahmins of various geographic origins were separated from the others in order to test whether this, the highest caste, shows similarities among subcastes of different geographic locations. At least in this tree analysis, the result has been negative: the six Brahmin groups do not tend to associate among themselves. A similar conclusion was obtained by Karve and Malhotra (1968). Only in two cases do they show affinity with local people of other castes (Punjab, Maharashtra). If there is any genetic kinship among Brahmins of different geographic origins, it must have been blurred, probably by drift following geographic isolation. Because marriage outside the caste is not accepted, endogamy is

likely to be especially strict among Brahmins, the highest caste. It is clear, however, that Brahmin marriages are subject to geographic constraints, and their caste is far from being one in all of India; there are probably a great many different Brahmin *jatis* that are totally segregated genetically. If they had a single origin, they are now highly differentiated. On an entirely different basis, other scientists (Chakraborty and Roychoudhury 1978) came to the same conclusion, that drift is the major factor responsible for genetic differentiation in India. Although tribals do not usually accept the caste structure, social customs, geographic isolation, the small size of some tribes, and the extensive geographic fragmentation of practically all tribes are likely to have contributed to genetic diversity. Tribals are more open to external marriage, however, and similarities between them and caste Indians could thus arise.

The linguistic differences account for much of the genetic diversity, considering that the three major Dravidian groups from central and southern India are reasonably similar, but the pattern of genetic and linguistic differentiation is complex; there must also be a geographic component, which is difficult to separate in this case.

The Kerala Kadar are remote genetically from the other Dravidians, but they are also different anthroposcopically (Saha et al. 1974). Moreover, if this group has been as small as it is now for a long time, its gene frequencies would be expected to have scattered in all directions. We later compare the Kadar in more detail with other groups. Dravidians, although clustering together, do not separate from Indo-European speakers at the first or second fission. This indicates considerable admixture. Although Indo-European speakers are found in the whole subcontinent, their genes must have been considerably diluted as they went through the center and south.

The PC map (fig. 4.14.2) of the first two components, based on 46.6 ± 4.6 genes on average, accounts for 33% of the variation. The first PC tends to separate, though imperfectly, north and south. Dravidians are mostly located at the center of the first PC and are all of the same sign with the third PC (not shown), but are not the only ones with this property. Brahmins tend to cluster together toward the center but do not form a close cluster.

Geographic maps of single genes, as exemplified in Piazza et al. (1981a), have shown a peculiarity of India: great microgeographic variation making the fit of gene-frequency surfaces especially difficult. The simplest explanation is the accumulation of the effects of a widespread net of tribals and those of the caste structure, both of which create a multitude of endogamous pockets. Considerable random genetic variation at a relatively short distance is thus generated. Theoretically there are only four Indian castes (and two among Dravidians, who do not have warriors' and traders' castes)

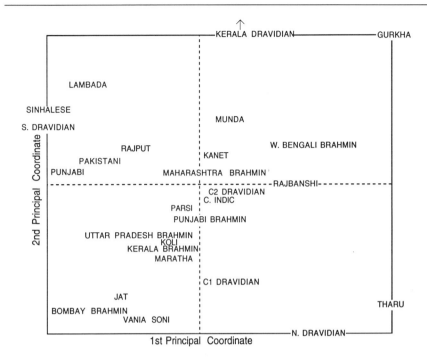

Fig. 4.14.2 Principal-component map of 28 Indian populations. *C1 Dravidian* is the group comprising Brahmin Telugu, Chenchu, Gondi, Konda, Koya, and Savara Telugu. *C2 Dravidian* comprises Kolami Naiki and Parji Gadaba.

In practice, there are many more than four because of territorial isolation, which is always present and often especially strong (see sec. 4.6). Thus, the gene-frequency surface is rich in peaks and troughs, impossible to fit except with very complicated surfaces that are well beyond the sophistication of ordinary fitting methods. When the goodness-of-fit of the surfaces interpolated for the most extensively investigated genes (Piazza et al. 1981a) was tested, considerable statistical heterogeneity (i.e., error beyond random sampling variation) was detected. The analysis of geographic variation is therefore more difficult in India than in most other parts of the world.

It is reasonable to agree with other authors (e.g., Vidyarthi 1983) that there are at least four major components of the genetic structure of India, which we interpret in the following way:

1. The first component (Australoid or Veddoid) is an older substrate of Paleolithic occupants, perhaps represented today by a few tribals, but probably almost extinct or largely covered by successive waves and presumably leaving no linguistic relics, except perhaps for the Hunza and Nahali. There seems to be no linguistic trace of the Australoid-Negrito language but Andamanese speak languages of the Indo-Pacific family. This may or may not be their original language.

2. The second is a major migration from western Iran that began in early Neolithic times and consisted of the spread of early farmers of the eastern horn of the Fertile Crescent. These people were responsible for most of the genetic background of India; they were Caucasoid and most probably spoke proto-Dravidian languages. These languages are now confined mostly, but not exclusively, to the south because of the later arrivals of speakers of Indo European languages who imposed their domination on most of the subcontinent, especially the northern and central-western part. But the persistence of a very large number of speakers of Dravidian languages in the center and south is an indirect indication that their genetic identity has not been profoundly altered by later events.

3. The most important later arrival was that of Indo-European speakers, the Aryans, who, about 3500 years ago entered the Indian subcontinent from their original location north of the Caspian Sea, via Turkmenia and northern Iran, Afghanistan, and Pakistan (see sec. 4.6).

   Even if the Aryans spread to the south of India in a fragmentary and incomplete way, they did arrive as far south as the island of Sri Lanka, where only 20% of the population speaks Dravidian languages today and where even the few surviving Australoid-looking aborigines, the Veddahs, now speak the dominant Indo-European language. Indo-European speakers were also successful in spreading north as far as Nepal, where the official language, Nepalese, is spoken by about half of the population and is Indo-European.

4. In the northeast and in the center, the many populations speaking Austroasiatic and Sino-Tibetan languages are a witness to other major migrations and infiltrations, mostly from the east and northeast. These are even less well known than the other three components and probably more diverse. In the case of the Munda, their genetic similarity with Dravidians indicates that their migration may have taken place before the Aryan expansion to the eastern part of India.

The presence (in very small numbers) of anthroposcopically Australoid types in southern India has always attracted much attention. Groups other than those already mentioned in India have some of the classic Australoid characteristics such as darker skin, sometimes frizzy hair, and small stature. The Negritos of the lesser Andaman Islands (politically Indian), the Semang of Malaya, and the Negritos of the Philippines have had less admixture. In the Indian mainland, less adulterated survivors may not exist but several minor tribes are worth examining. The disappearance of the original languages is not surprising or unprecedented; the languages of African Pygmies have also become extinct, but the African tropical forest may have permitted greater genetic conservation of its inhabitants. Almost all reputed survivors of this hypothetical Australoid population are on islands that were mostly connected with the mainland in earlier times. It is tempting to speculate that the migration from Africa

to Southeast Asia and Australia happened mostly by the way of the coast. Boats or rafts had to be used by the earliest Australian aborigines in order to enter that continent, as has already been mentioned, and it would not be too surprising if all or most of the migration of a.m.h. to Southeast Asia and Australia from Africa had taken place along the coast. Genetic traces of this migration may no longer exist, or may be very difficult to find. However, as has been repeatedly emphasized, Dravidian populations are not genetically similar to Australian populations (Kirk and Thorne 1976), as an analysis of the world data (sec. 2.3) has confirmed.

4.15. Genetics of West Asia

The region of West Asia as we define it here includes Iran, Afghanistan, Arabia, the Middle East, Turkey, and the Caucasus. It thus includes essentially all extra-European Caucasoids living in Asia, other than South Asia (see sec. 4.7).

The genetic tree (fig. 4.15.1) of 18 populations (average number of genes 53.7 ± 5.1) distinguishes a minor cluster made up of Arabs from the Arab peninsula and a major one with all the other populations. The small cluster includes all South Arabs except the Kuwaitis. It was necessary to define Saudis and Yemenites according to countries. Bedouins, however, could be kept separate. They are nomadic herders (their name means nomads) and are found in all Arabic countries, being about a tenth of the total population. Bedouins spend the winter in the desert and the summer in less dry, cultivated parts of the country, and maintain a tribal organization. As mentioned in chapter 3, they participated in at least one and probably two major invasions of northern Africa in historical time. Arabs separate in the tree from other populations of the Near East.

The major cluster shows two subclusters: the smaller one consists of inhabitants of the northern region of the Caucasus, Armenians, Pathans, and the Hazara Tajiki. The Tajiki speak an Indo-European language of the Iranian subgroup; they are believed to be descendants of sedentary farmers of central Asia who spread north and east from Iran in Neolithic times. They live mostly in the USSR (3 million) and were even more numerous in Afghanistan. The Hazara Tajiki are from northern Pakistan. Armenians (about 5 million) live in the Armenian Republic of the USSR, in Northern Iran, and in many other countries. Large numbers once resided in eastern Turkey, but the recent genocide of Armenians there (denied, however, by the Turkish government) is said to have eliminated or dispersed perhaps 95% of them. Armenians have an ancient civilization and a highly developed culture. Probably, their ancient name was Urartians as was that of the non-Indo-European language they spoke earlier (see fig. 4.7.2). The Pathan, whose Indo-European language is called Pushtu or Pashtun, are believed to be originally from Afghanistan, where 8 million of them live today, with an additional 7 million in northern Pakistan. The Pathan migration to Pakistan is believed to have occurred between the thirteenth and sixteenth centuries A.D.. Over 3 million recent Afghan refugees to Pakistan are Pathan herdsmen, farmers, or warriors. The language is classified as East Scythian and is likely to have come from the north with the invasion of pastoral nomads from the western steppes.

The larger subcluster has the Kuwaitis as an outlier. Kuwait is a small Arab country, extremely close geographically to a region traditionally occupied by Sumerians, the oldest civilization in Mesopotamia. Until a few years ago, a distinctive group—who, until the recent Iran-Iraq war, built huts in the shape characteristic of those of the Sumerians—in the marshes of southern Iraq. They were probably dispersed or destroyed by the fighting. The Kuwaitis were geographically very close and might show similarity to them. In any case, they are

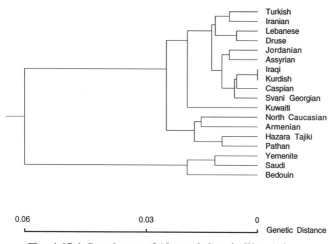

Turkish
Iranian
Lebanese
Druse
Jordanian
Assyrian
Iraqi
Kurdish
Caspian
Svani Georgian
Kuwaiti
North Caucasian
Armenian
Hazara Tajiki
Pathan
Yemenite
Saudi
Bedouin

0.06   0.03   0
Genetic Distance

Fig. 4.15.1 Genetic tree of 18 populations in West Asia.

somewhat distinct from other people of the Arab peninsula. The rest of the subcluster contains nine populations that form two groups, not clearly distinguishable from each other geographically or linguistically.

1. The Svani (35,000) speak a South Caucasian language and are from Georgia in the south Caucasus. They associate with the Caspians.
2. The Caspians are from northern Iran near the Caspian Sea (more than 3 million).
3. The Kurds, from western Iran, total about 6.5 million. They live in a region at the border between Turkey, Iran, and Iraq, but also in neighboring countries, and speak a distinctive Iranian (Indo-European) language. Some groups are still nomadic or seminomadic and often retain a tribal structure. They pair very closely with the Iraqis.
4. The Iraqis are defined by the country of origin, after exclusion of Iraqi Kurds.
5. The Assyrians are a fairly homogeneous group of people, believed to originate from the land of old Assyria in northern Iraq. A capital of Assyria, Nineveh, was not far from the modern city of Mossul. There are still 70,000 Assyrians in Iraq, living between Mossul and Arbil (fig. 4.7.2), and another 100,000 in neighboring countries. Their language is Aramaic (Semitic). They pair loosely in the tree with Jordanians.
6. Jordanians are defined by their country of origin.
7. Druzes (300,000) are an Islamic sect that originated shortly after A.D. 1000; they live in Arabic-speaking countries, mostly Syria, but also in Israel, Lebanon, and Jordan. They pair, not surprisingly, with the Lebanese.
8. Lebanese are defined by their country of origin. Lebanon was once the country of the Phoenicians.
9. Iranians are defined by their country of origin, after exclusion of Kurds, who live in the northwest, and Caspians, who also live in the northwest, but nearer the sea. They pair loosely with the Turks.

10. The Turks are also defined by the country of origin. Turkey, once Asia Minor or Anatolia, has a very long and complex history. It was one of the major regions of agricultural development in the early Neolithic and may have been the place of origin and spread of Indo-European languages at that time. The Turkish language was imposed on a predominantly Indo-European-speaking population (Greek being the official language of the Byzantine empire), and genetically there is very little difference between Turkey and the neighboring countries. The number of Turkish invaders was probably rather small and was genetically diluted by the large number of aborigines. Turks, a political definition, are certainly rather heterogeneous and would deserve a more detailed genetic analysis if the data were available.

Omitting the Arab cluster, which occupies chiefly the southern part of the Middle East and is separated from the northern part by a large desert, there is some linguistic difference between the two subclusters of the major cluster. In the small subcluster are three Indo-European-speaking groups and one North Caucasian one (not well-defined linguistically, but probably speaking North Caucasian languages). In the large subcluster are three Indo-European-speaking populations, one speaking a South Caucasian language, one speaking a Turkic language, and five speaking Semitic languages.

The clustering by geography and, subordinate to geography, by language, is even clearer in the PC map (fig. 4.15.2), which accounts for 54% of the original genetic variation. The Arabian populations form the right pole (as usual, the first PC corresponds to the first tree fission). At the left pole, the second PC tends to segregate the Semitic populations in the upper left quadrant (exceptions are the Kurds and Turks); and in the lower

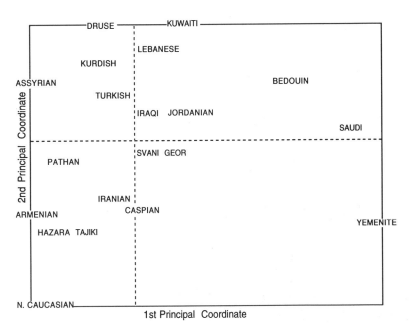

Fig. 4.15.2 Principal-component map of 18 populations from West Asia.

Table 4.15.1. Genetic Distances (in the lower left triangle of the matrix) and Their Standard Errors (in the upper right triangle of the matrix) between Populations in West Asia (all values x 10,000)

	PPU	IRAN	CAS	HTA	KUR	KUW	SAU	YEM	ARM	NCA	SGE	ASY	BED	DRU	IRAQ	JOR	LEB	TURK
Pathan Pushtu	0	4	9	8	6	20	85	124	6	10	8	5	42	10	44	6	17	24
Iranian	76	0	2	10	4	8	15	30	10	8	6	10	15	5	36	5	5	4
Caspian	116	47	0	8	1	10	21	36	18	10	4	7	19	9	4	5	9	7
Hazara Tadzhik	86	169	170	0	13	27	91	132	8	6	11	9	50	14	53	8	30	10
Kurd	117	57	31	198	0	12	61	94	10	14	3	6	24	3	1	4	7	12
Kuwaiti	290	184	141	358	171	0	39	49	22	33	16	24	17	18	7	4	9	9
Saudi	727	321	350	865	683	300	0	13	85	117	26	104	11	113	35	5	8	21
Yemenite	1070	429	390	1121	867	488	100	0	109	147	40	129	19	144	29	11	28	38
Armenian	146	182	237	182	178	423	928	1148	0	7	15	6	29	30	12	10	11	23
North Caucasian	191	105	134	151	208	441	1030	1244	135	0	18	17	56	33	5	17	17	27
Svani Georgian	162	133	61	209	46	179	313	456	240	218	0	5	22	14	9	6	16	14
Assyrian	92	203	160	151	122	332	907	1174	102	271	131	0	45	13	5	6	23	9
Bedouin	468	361	313	581	383	173	145	211	531	690	204	502	0	41	38	8	19	19
Druse	173	133	201	293	73	199	911	1340	326	479	158	169	503	0	6	7	5	6
Iraqi	348	168	61	446	5	76	373	348	146	71	78	49	379	93	0	6	6	15
Jordanian	110	132	95	170	73	81	186	229	147	270	93	78	203	91	116	0	5	4
Lebanese	239	108	170	293	57	135	317	465	174	277	190	202	295	59	76	106	0	10
Turk	256	75	130	158	114	198	335	569	200	279	199	99	397	78	112	77	157	0

left, the non-Semitic ones, Indo-European and Caucasian.

Table 4.15.1 shows F_{ST} genetic distances in West Asia. It is of special interest to consider Turks, the only major group in the region that speaks a language originated at a great geographic distance (probably in the Altai region). They differ very little from all their nearest geographic neighbors, showing distances almost always less than 100, with relatively small standard errors. From data used for building figure 4.15.1, the distance of Turks is 104 ± 55 from Turkmen, 167 ± 67 from Uzbek (not significant because of poor genetic knowledge of the latter tribes), varying between 453 and 735, all highly significant, from all other Turkic languages speaking populations shown in figure 4.12.1 (Tuva, Yakut, Altai). There has been a considerable di-

lution of Mongoloid genes in the peoples of Central Asia (Turkmen and Uzbek), and a practically complete dilution in Turkey, as far as these data show. Especially in Turkey, but to a lesser extent also in Central Asia, language replacement has occurred essentially without, or with very little, gene replacement. Better genetic data would allow the calculation of admixture frequencies, but it is likely that they will differ considerably at a microgeographic level. The relevant historical events have been quite recent, and there has not been much time for a smoothing of differences by migration.

In spite of the complex history of the Middle East and the great number of internal group migrations revealed by history, as well as the mosaic of cultures and languages, the region is relatively homogeneous, as we shall see in the last section of this chapter.

4.16. GEOGRAPHIC MAPS OF SINGLE GENES

Asia is probably the most difficult continent to study from the point of view of genetic geography. The variation in population density is the major factor responsible for this difficulty. Three large areas have very low population densities, below one inhabitant per km^2: Arabia, the western part of China (including Sinkiang, Tibet, and Mongolia), which are mostly deserts, and Siberia. It is not surprising if observations in these regions are scarce; in fact, looking at the most thoroughly investigated locus, *ABO*, one can detect the desert regions simply by looking at the areas in which the density of observations is lowest. For the least investigated genes, there are very large regions in the interior of Asia for which no data were found. As a consequence, they seem to have flat gene-frequency maps. However, the location of data points is always indicated, and it is easy to understand the reason for the apparent flatness. We think several of these maps are still useful because they are informative, at least for areas for which data are available. Another difficulty is generated by two very highly populated regions, India and China, which themselves constitute almost half of the world's population. India is especially rich in data on blood groups, but its great heterogeneity is visible in the high frequency of statistically heterogeneous locations. China was very poorly studied until a short while ago, and the most recent observations, usually published in Chinese journals, came to our attention too late for inclusion in most of our calculations and maps.

Compared with the rest of Asia, Japan has been very intensively studied, especially for genes known for a longer time. We have partially solved the graphic problem thus generated by increasing the relative size of Japan with respect to the rest of Asia by about 60%. For some other islands, rectangles covering the area of

the islands were shaded as appropriate for the island, but only in cases for which data were available. Rectangles are of a size such that the density of the shading should be perceptible even if the gene frequency happens to be very low.

The number of deviations is larger in the *ABO* system than for most other genes. This must be a consequence of the great density of observations in the same location. The concentration of heterogeneities is especially high in India where many different castes may live within the same area defined for map purposes. Moreover, the endogamy of castes inevitably generates considerable drift, as already discussed in section 4.14.

The *O* allele shows a maximum in central-northern Siberia and minima in Korea, Turkmenia, and the eastern Himalayas. The maximum is of potential interest because it may give an indication of the tribe(s) that contributed to the first wave of American settlers. The distance of the location of the maximum from the northeast is not necessarily an impediment to accepting the correlation, given that in the last three centuries there has been much tribal movement in Siberia. More serious is the objection that Siberian tribes are subject to considerable drift because of small population densities.

The *A* allele has maxima in two greatly separated regions, Turkey and southern Japan and Korea. *B* has maxima in the Himalayan region and various parts of Siberia, including one not far from the *O* peak, where there is, as expected, a minimum. *A1* is low where both *O* and *B* are high and also in southern Pakistan and central India, with a maximum around the northern Sea of Japan. The central part of Asia has almost no information for *A1* and *A2*. *A2* is practically absent in East Asia and shows a maximum in Lebanon. It is unfortunate that the *ABO* system, for which there is more information than for any other locus, is perhaps least reliable as an ethnic tracer, given its strong correlation with various infectious diseases.

The variograms of the *ABO* system show intermediate values of the slope (around 0.5×10^{-5} per mile) with a linear portion that sometimes reaches 2000 miles. *A1* is the only allele with a flatter initial portion of the slope.

Acid phosphatase (*ACP1*) has a strong maximum of allele *A* in the extreme northeast. It is highly correlated with climate, but the high frequency does not extend to regions of northwestern Siberia. Here allele *B* has instead a maximum, and another is found in southwestern Arabia. Allele *C* is rare in most areas and reaches a small maximum in southwestern Iran. Variogram slopes are high for the first two alleles.

Adenosine deaminase (*ADA*1*) is essentially unknown in the north. Here the map repeats the patterns observed in the south where there is a fairly regular east-west gradient with a minimum in western India and Pakistan.

Adenylate kinase 1 (*AK1*1*) shows a south-north gradient in the center, the east showing frequencies comparable to those of northwestern Siberia. Variograms of this and the preceding enzyme are fairly regular.

Cholinesterase 1 (*CHE1*U*) is also very poorly known in the north and center. There are minima in India and Turkey. The variogram is invalid because of the very poor geographic distribution.

CHE2 is somewhat better known, but there is little variation of the rare positive (+) allele. The variogram is accordingly flat.

Knowledge of *C3*S* is limited almost exclusively to West Asia, India, and Japan. There seems to be a west-east gradient.

Diego (*DI*A*) is not tested in Central Asia and southern Siberia; the rest of Asia shows a west-east gradient, with a minimum in the west and a maximum around the Sea of Japan. This allele is of special interest because it reaches high frequencies in Amerinds. The geographic distribution in Asia is not sufficiently well tested, however, to tell which East Asian groups contributed most to the peopling of Americas. The variogram seems normal.

Blood-group Duffy (*FY*) is better investigated than the enzymes named so far. Allele *A* has a strong maximum, reaching close to 100% in all the East and along the Arctic Ocean, with a possible minimum in the Kamchatka peninsula and a strong minimum in southern Arabia. Allele *B* has a maximum around the Caspian Sea with a regular decreasing gradient all around, suggesting this is a possible area of origin for the allele. Allele *O* has a maximum in southern Arabia, of possible African origin and in response to malaria. The variograms of *A* and *B* have high slopes; *O* is very irregular.

Esterase D (*ESD*) allele 1 has a maximum in the extreme northeast and a fairly regular gradient toward the south; the minimum is in the Himalayan region. The variogram is fairly flat.

For glucose-6-phosphate dehydrogenase deficiency (*G6PD*), the information is limited to the south and the values interpolated in the upper part merely repeat the southern pattern. There is a strong peak in eastern Arabia, related to malaria, and a moderate incidence of the deficiency in mainland Southeast Asia. Notable is its virtual absence in India.

Glutamate-pyruvate transaminase *GPT*1* has a pattern almost identical to that of *GLO1*1*.

A complement gene for which there is very limited information, *BF*, shows a strong peak in Southeast Asia and a minimum in Arabia of the most common allele, *S*. Allele *F* shows a complementary surface and allele *SO7* has a maximum in Arabia, decreasing regularly toward the east. The regularity is probably the result of poverty of information, especially in the north. Variogram slopes of the common alleles are high.

Glyoxylase *GLO1*1* has maxima in the extreme northwest and in West Asia, with a minimum in the East, but not in insular Southeast Asia. The variogram has a fairly high slope.

Group-specific component or vitamin-D-binding protein (*GC*) has a minimum for allele *1* in Central Asia and maxima in insular Southeast Asia, the Kamchatka Peninsula, and southwestern Arabia. It is difficult to reconcile this pattern with the assumed correlation with solar radiation. Subtype *1F* has an almost regular west-east gradient. Variograms are regular, *1F* with a high slope.

Haptoglobin (*HP*) shows a maximum for allele *1* in several regions, and an apparently single minimum in India with a concentric distribution around it. The variogram is linear to 2000 miles with a fairly large slope.

HLA has a reasonable geographic coverage, even if some areas like the extreme northeast have no data. Two major patterns recur for many alleles and are discussed together.

The most frequent pattern is an almost regular gradient from west to east and occurs in two varieties, higher in the west and lower in the east, and vice versa. The first is found for *A*1*, *A*3*, *A*28*, *A*29*, *A*32*, *B*8*, *B*14*, *B*18*, *B*21*, and *B*35*. The second, opposite, variety is rarer, being found only for *A*2*, *B*13*, and *B*22*. This asymmetry may occur because most antibodies have been selected among Europeans, whose immunological spectrum is therefore better covered than that of Mongoloids. Some of these alleles have additional peculiarities, perhaps caused by drift or local selection: *A*32* has an especially high value in the Philippines, *B*7* has a peak in central Siberia, and *B*18* in Southeast Asia. *B*13* has an additional minimum, and *B*15* a maximum in northeastern and central-northern Siberia, respectively.

The next most frequent pattern is a north-south gradient. It is found in *A*9*, *A*11*, *A*31*, and *B*40*, in all cases with higher values in the north except for *A*11*. Peculiarities are a high value in the Philippines for *A*9* and in Arabia for *A*31*.

Other patterns recur less frequently but show peaks in different regions: *A∗10* inland east of the Caspian Sea, *A∗11*, *A∗33* in Southeast Asia, *A∗19* in southwest Iran and Mongolia, *B∗15* in central Siberia, *B∗16* in the Philippines, and *B∗17* in central-eastern Siberia. *B∗27* and other alleles show little variation or are difficult to define.

The variograms of *HLA* are mostly regular, with slopes varying between 0.1 and 2.0×10^{-5} per mile (median 0.8). The initial linear portion varies from 500 to over 2000 miles. There are 6 variograms out of 29, however, which have negative initial slopes, usually corresponding to alleles with very low average frequency.

For *GM* (*IGHG1G3*), the geographic distribution of the data is reasonably satisfactory, with few gaps compared with most other markers. A strong east-west gradient is shown by *f;b0b1b3b4b5* (a Caucasoid haplotype) with a peak in West Asia, and by *za;g* in the reverse, with a peak in the extreme northeast, possibly the place of origin of the haplotype. Because *fa;b0b1b3b4b5* peaks strongly in Southeast Asia, that area might be the origin of this haplotype, *ZA;B* has a concentric distribution around the Caspian Sea, and *za;b0stb3b5* around central Siberia.

A constant light-chain immunoglobulin gene, *KM* is well studied and alleles *1* & *1,2* seem to have a center of origin in North China. The two alleles cannot be distinguished except with an antibody that is not easily available, but one of them, allele *2*, is much rarer and contributes little to the sum of the two alleles.

The variograms of *GM* and *KM* all have high slopes and reasonable initial linearity, except for allele *za;b0b1b3b4b5*, which has an initial negative slope.

Blood-group Kidd (*JK*) shows two peaks for allele *A*, one in the extreme north, one in insular Southeast Asia, and a minor one in southeastern Arabia, with a flat distribution in the center. Allele *B* is lowest in the north, highest in the east, and intermediate elsewhere.

The allele of the Kell blood group *KEL∗K* is somewhat better studied than Kidd and has a general west-east gradient.

For the Lewis blood group (*LE*), allele *Le(a+)* shows a north-south gradient and allele *Le* an east-west one. The Lutheran blood group (*LU*) (allele *A*) is, like *LE*, poorly studied but seems to show a west-east gradient. The variograms of these four blood-group systems are regular, except for *JK∗A*, and *LU∗A*, which have negative initial slopes.

The *MNS* system, one of the oldest known blood groups, is reasonably well studied. Allele *M* is highest in northeastern Siberia and in the west and south; its two component haplotypes show a clearer distribution, with a definite west-east gradient for haplotype *MS* (peak in Arabia) and a north-south gradient for *Ms* (peak in Southeast Asia). The allele complementary to *M*, *N*, is made of haplotypes *NS*, with a peak in central Siberia

and concentric gradients around it, and *Ns*, with a north-south gradient. Allele *S*, the sum of *MS* and *NS*, peaks in central Siberia. All variograms are regular with initial linear segments over 1000 and even over 2000 miles, with relatively large slopes.

Allele *1* of blood group *P1* has been known for some time and has also been reasonably well studied. It shows a west-east gradient, but in addition to the peak in the west, it is also high in the extreme northeast and in the insular southeast. The variogram is linear to almost 4000 miles, with a relatively large slope.

For phenylthiocarbamide tasting (*PTC*), allele *T* has a fairly regular west-to-east gradient. The variogram is linear to at least 2000 miles, with a moderately high slope.

Phospho-glucomutase 1 (*PGM1*) is one of the best-studied enzymes. Allele *1* shows a high maximum in the extreme northeast and has an overall west-to-east gradient, with a minimum in the west but also in a narrow area of northwestern Siberia.

6-phosphogluconate dehydrogenase (*PGD*) allele *A* has a minimum in Central Asia, and the rarer allele may have originated here. These enzymes have variograms that are linear to 3000 miles with a moderate slope.

The *RH* blood-group system can be conceived as having three biallelic loci *C*, *D*, and *E*. *C* and *E* show opposite north-south gradients, *C* with a maximum in Southeast Asia and *E* with a maximum in the north. *D* shows an east-west gradient with a maximum in the east, but also one in the extreme northwest. Other alleles like C^W and D^U have low frequencies, C^W with maxima in west-central Asia and the extreme northeast and D^U in the western part of Southeast Asia. The most frequent *RH* haplotypes are *CDe*, *cDE*, and *cde*: the first has a strong north-south gradient with a maximum in Southeast Asia where it reaches close to 100%; the second has a similar, but opposite, gradient reaching more than 45% frequency in the north; and the third, the *RH*-negative haplotype, which is characteristically Caucasoid, has a west-to-east gradient, having relatively high frequency (more than 30%) in West Asia and practically zero frequency in Mongoloids. An interesting *RH* negative peak in North Asia needs further analysis. The haplotypes next in importance are *Cde*, *cDe*, *cdE*, and *CDE*. The most common is *cDe*, which has two peaks of more than 25% in the western part of central Asia and on the Kamchatka peninsula, with a secondary peak in Arabia where it is probably of African origin, and is minimal in the east. *Cde* is somewhat higher in southeastern India and is elsewhere uniformly low or absent except around the Caspian Sea. *cdE* is also low everywhere, being slightly higher in the western part of central Asia and around the Sea of Japan. *CDE* is highest on the Kamchatka peninsula and in the western part of central Asia.

Most variograms of *RH* alleles and haplotypes show a long linear portion and medium to high initial slopes.

Three show flat or negative variograms, C^W, cdE, CDE; and all have very low average frequencies.

The secretor gene $SE*Se$ has an approximate north-south gradient with a second maximum in the extreme northeast. The variogram is linear to 2000 miles with a moderately large slope.

The transferrin $TF*C$ allele has 100% gene frequency in most of Asia, except in a few areas in central and Southeast Asia where there is a low percentage of D allele. The variogram is regular, with an initial portion linear to 2000 miles.

In summary, there are a certain number of genes showing west-east and north-south gradients. We can expect them to determine important features of the synthetic maps corresponding to the higher PCs. Other patterns also tend to recur, though less frequently, and are found in synthetic maps corresponding to PCs of levels lower than the first two.

4.17. SYNTHETIC MAPS OF ASIA

The analysis of the variation by PCs was based on 96 genes, the same 98 used for the genetic maps of Asia (see Table of Genetic Maps) and shown in the second part of the book except for $AG*X$, $G6PD*B^+$. Table 4.17.1 shows the relative importance of each of the first seven PCs, which collectively express 80% of the total variance. The genes that contribute most to each of the seven PCs are indicated in table 4.17.2.

The geographic map of the first PC (fig. 4.17.1), accounting for about one-third of the total variation, has a clear west-to-east gradient and may be considered a simple synthesis of the Caucasoid–Mongoloid gradient observed throughout Asia. The gradient is not exactly linear, but is more abrupt around a line that starts in the northern part of the Urals, first descends toward the southeast, then bends toward the south to reach the eastern part of India. This can be taken as the approximate boundary between Caucasoids and Mongoloids. One cannot expect it to be completely abrupt. We know there have been many migrations in the last 2500–3000 years across this boundary, and many populations near it show signs of hybridization. The progress of Mongoloids toward the west has been especially marked in the extreme north; it was less important but still noticeable in the Himalayan region.

The first PC map is in good agreement with the physical data discussed earlier. It is of some importance to note that the extreme western values are not at the boundary between Europe and Asia, but rather in West Asia, particularly at the area of contact with Africa. One might hypothesize that this component indicates the progress of a wave from northeastern Africa or from West Asia. Was this area occupied by other people at these early times, and did important mixture occur with them? Or was the reciprocal dilution of Mongoloid and Caucasoid types entirely a matter of subsequent events? PCs cannot answer these questions and they cannot indicate the direction of a wave of advance, except that, when concentric gradients are observed around a center, as in the present case, it is more likely that the direction of the expansion is centrifugal. They cannot give information on dates, but they may give some indication of the relative demographic importance of the expansions and mixtures.

The high percentage value of the first PC shows that the differentiation between East and West Asia is clearly the most important genetic difference in this continent. There are at least three possible explanations for the expansion. (1) The first expansion of modern humans could have come from Africa eastward via Suez, if there were some humans in East Asia with whom anatomically modern humans mixed. (2) Later expansions from the same origin could have followed the same route. The last such expansion could be that of farmers from the Middle East moving in the same general direction, the western steppes of Asia, Iran, and India. We might very well be seeing the summation of more than one expansion at different times, all of which had their Asian centers of origin in West Asia. (3) There might have been expansions from East Africa via southern Arabia and Iran, or via the southern coast of Asia, which might be difficult to distinguish from each other or from the first and second.

It should be noted that the regularity of the gradient may be deceptive and, in part, artifactual. The central part of Asia is poorly known genetically, and geographic maps of genes inevitably interpolate a smooth surface where there are no data, contributing to the illusion of a regular gradient. Because the region is also partly desert today, few data are available, but known population movements have occurred (e.g., the Uyghur occupied the Xinjiang, but there is little record of them in this data; *HLA* data now available for this population were too late for inclusion).

Table 4.17.1. Percentage of Total Variance Explained by the First Seven Principal Components of Asian Gene Frequencies

Principal Component	% of Total Variance	Principal Component	% of Total Variance
1	35.1	5	5.2
2	17.7	6	4.8
3	7.7	7	3.3
4	6.4		

Table 4.17.2. Genes Showing the Highest Correlations with the First Six Principal Components of Asian Gene Frequencies

P.C.[*]	Range of Correlation Coefficient	Genes
1	1.00 – 0.90	(+) IGHG1G3*f;b0b1b3b4b5, RH*cde
		(–) BF*S, GC*1F
	0.90 – 0.80	(+) ACP1*C, BF*F, BF*S0.7, HLAA*1 HLAA*3, IGHG1G3*za;b0b1b3b4b5, HLAA*28, HLAB*35, HLAB*8, MNS*MS
		(–) ADA*1, C3*S, FY*A, HLAA*2, PTC*T
	0.80 – 0.70	(+) FY*B, HLAA*1, HLAA*2, HLAA*32, HLAB*14, HLAB*18, HLAB*21, HLAB*37, KEL*K, LU*A, P1*1
		(–) AK1*1, HLAB*15, RH*D
2	1.00 – 0.90	(+) —
		(–) HLAA*11
	0.90 – 0.80	(+) —
		(–) LE*Le(a+), MNS*Ms, RH*C
	0.80 – 0.70	(+) ESD*1, HLAA*31, RH*cDE
		(–) IGHG1G3*fa;b0b1b3b4b5, HLAA*33, HLAB*22, RH*CDe
3	0.70 – 0.60	(+) HLAB*5
		(–) ACP1*B, HLAB*27
	0.60 – 0.50	(+) HLAA*10, HLAA*31, HLAB*16, HLAB*22, PEPA*1, JK*B
		(–) HLAB*17, RH*D^u
	0.50 – 0.40	(+) CHE1*U, PGM1*1, TF*C, RH*cdE
		(–) —
4	0.70 – 0.60	(+) ABO*O, HLAA*30, HP*1
		(–) ABO*B
	0.60 – 0.50	(+) FY*O, GC*1, HLAB*21
		(–) G6PD*B-
	0.050 – 0.40	(+) —
		(–) HLAA*10, HLAB*37
5	0.70 – 0.60	(+) ACP1*A, PGD*A
		(–) —
	0.60 – 0.50	(+) JK*A
		(–) IGHG1G3*zax;g
	0.50 – 0.40	(+) MNS*M, P1*1
		(–) MNS*NS
6	0.70 – 0.60	(+) RH*cDe
		(–) —
	0.60 – 0.50	(+) RH*CDE, MNS*NS, MNS*S
		(–) MNS*Ns
	0.50 – 0.40	(+) MNS*M
		(–) ABO*A
	0.40 – 0.30	(+) ABO*O, GC*1, HLAB*7, MNS*MS
		(–) ABO*A1, IGHG1G3*zax;g

Note.– Genes giving positive or negative correlation values are indicated by (+) or (–), respectively.

[*] P.C., Principal component.

The second synthetic map (fig. 4.17.2) shows that the next most important genetic component is the difference between northern and southern Mongoloids. West and South Asia do not participate in the differentiation disclosed by the second PC, showing intermediate values. The separation between northern and southern Mongoloids corresponds approximately to the first isopleth starting from Southeast Asia. Whether this component also absorbs effects caused by correlation with climate is difficult to say without further analysis.

The third synthetic map (fig. 4.17.3) shows a peak in Japan, with rapidly falling concentric gradients. The other extreme PC values are in the extreme north and in the extreme south, posing a difficult problem in interpretation.

It is worth remembering that Japan has been magnified 1.6 times for the purpose of showing more clearly its internal genetic structure, given that it has a higher density of genetic data than any other country in Asia. The third PC explains a nontrivial fraction of the total variance (8%). Taken at face value, one would assume a center of demographic expansion in an area located around the Sea of Japan. Several genes are involved, as we have seen in the single-gene maps, and it is difficult to think of a selective factor of general importance other than climate, involving many genes at once. There seems to be no evidence of nonneutrality from F_{ST} values for these genes. This area was not a center of agriculture, but before 10 kya land was more extended and was later submerged; thus, its archaeology is unknown. Japan, however, was the seat of the earliest known development of ceramics. One can hypothesize that Japan and/or some other region around the inland sea was the seat of a relatively important demographic development, which was not disturbed significantly by later migrations from remote places and therefore has kept a distinctive gene pool. The population dynamics of Jomon Japan (see sec. 4.4) supports this idea because, for at least 6000 years (10,000–4000 B.P.), the population grew steadily and at a high rate. The Japanese population increased in this period from what was probably a very small number, which must have allowed for substantial initial divergence under drift, to a population not far from 300,000, in which gene-frequency change was almost frozen. If there had been no later major immigration to Japan, Jomon gene frequencies might have been similar to modern ones. After 4000 B.P., growth ceased, perhaps under influence from climate change, and resumed around 2000 B.P. under the influence of agriculture.

Jomon population densities are among the highest recorded for a foraging population, although in some areas of the Pacific coast of North America, comparable and even higher figures of population densities have been observed (Hassan 1975). Excellent fishing, in addition to deer hunting, and the gathering of chestnuts and roots, provided resources allowing for substantial population growth in Japan. One possible explanation is that the persistent growth of an initially small population in Jomon Japan (and probably surrounding areas like Korea and Manchuria, which participated in the development

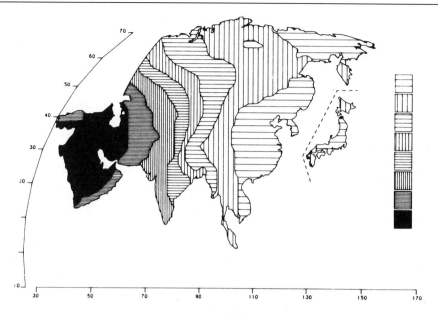

Fig. 4.17.1 Synthetic map of the first principal component of Asia.

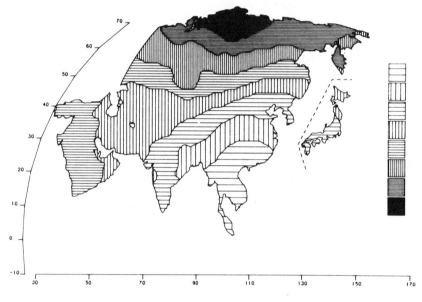

Fig. 4.17.2 Synthetic map of the second principal component of Asia.

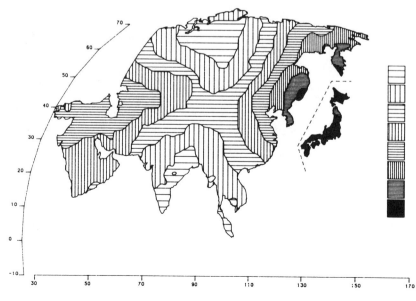

Fig. 4.17.3 Synthetic map of the third principal component of Asia.

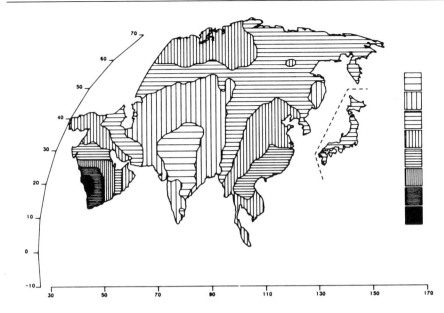

Fig. 4.17.4 Synthetic map of the fourth principal component of Asia.

but are far less well known archaeologically) maintained initial genetic differences generated by drift. Whether this local demographic development also determined a true geographic expansion is hard to say without analysis at a higher resolution.

The fourth synthetic map (fig. 4.17.4) shows a contrast between western Arabia and the eastern part of India and the Himalayas. All the rest of the continent takes intermediate values and is not involved in this contrast. There are no known archaeological interpretations. It is worth recalling that in the Red Sea, opposite the western part of Arabia showing the highest intensity in the figure, we have observed areas of altered PC values. The possibility of an ancient expansion from East Africa, in addition to those discussed above, is worth keeping in mind.

The fifth component (fig. 4.17.5) shows a peak with concentric gradients in the central part of the steppes. A few single-gene maps showed anomalies not far from this region, which corresponds approximately to an early area of development of pastoral nomadism. Given that this economy was responsible for a series of expansions from central Asia, it is tempting to interpret the map of the fifth PC as the result of the radiation of pastoral nomads, probably mostly Caucasoid. Areas of opposite polarity are found in Southeast Asia and in the extreme northeast. The area is very poorly sampled, and it is possible that a more regular sampling grid would shift somewhat the peak of the synthetic map corresponding to this possible center of expansion. It is interesting that similar steppe regions, with slight shifts to the west and east, show peaks in the sixth and seventh components (maps

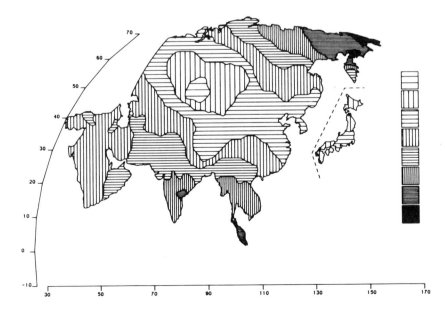

Fig. 4.17.5 Synthetic map of the fifth principal component of Asia.

not given). Different components are uncorrelated, but it is possible that some PCs have one pole in the same position, provided the opposite pole(s) are in different positions. Opposite peaks are found for the sixth component in Armenia and Korea, and for the seventh in south India.

The color picture conveys 60.5% of the original variation and shows three major ethnic zones: pale green with a yellow central tinge in Southwest Asia; blue in north-central Siberia; red in the eastern portion, varying from an extremely dark color that makes Southeast Asia almost invisible, to a very light one among the northern Mongoloids and purple farther northeast.

The expansion of agriculture from the Middle East in light green is superimposed on another expansion responsible for the yellow tinge around the lower half of the Caspian Sea. This is probably associated with Turkic expansions.

In summary, the analysis of synthetic geographic maps shows that the first two formally confirm the major importance of the east-west gradient and secondarily of the difference between northern and southern Mongoloids. It is difficult to distinguish in the first component the effects of two possible radiations from the Middle East, an early one following the first African expansion and a late one of Middle Eastern farmers. They probably have both contributed, and the analysis cannot separate them. The expansions of pastoral nomads seem to characterize lower components, in agreement with the fact that later expansions are likely to explain a smaller fraction of variance. A possible expansion from the Sea of Japan is suggested and has some archaeological support.

4.18. SUMMARY OF THE GENETIC HISTORY OF ASIA

Asia is the largest and genetically most complex continent. The five regions into which we have divided it for convenience of analysis are, if not homogeneous, less heterogeneous than the whole continent, but there are no sharp boundaries between them. They have different degrees of internal heterogeneity, as shown by the following list of the genetic variation internal to each area, measured by the average F_{ST} ($\pm$ standard error) calculated between the 491 populations:

West Asia (Near East)	0.0212 ± 0.0049
Southeast Asia	0.0349 ± 0.0038
South Asia	0.0276 ± 0.0048
Northeast Asia	0.0230 ± 0.0022
Arctic	0.0264 ± 0.0126

The smallest F_{ST} is observed in the Near East, where the highest population densities have existed the longest, especially in the central part (Mesopotamia). Ten thousand years of agriculture, ancient urban developments, and internal migrations are probably responsible for this homogeneity. Many differences between the F_{ST}s above, however, are not statistically significant.

These data alone are not sufficient if taken in isolation for tracing the history of the whole continent. The synthetic maps are obviously more useful: they show that the major separation is on the east-west axis, approximately in the middle. Caucasoids occupy the western half and Mongoloids the eastern half, with a boundary indicated by a slightly sharper gradient in the first synthetic map, running from the northern Urals to eastern India. Mongoloids also separate into northern and southern moieties. We know from the analysis of Chinese data that the separation is approximately between northern Chinese, who are more similar to northern Mongoloids, and southern Chinese who are more similar to southern Mongoloids. The genetic boundary between northern and southern Mongoloids, shown by the second synthetic map, is not very definite, but must have been sharper in Paleolithic and Neolithic times.

There are a few clues that in a more distant past different populations must have occupied the southern part of Asia. "Australoid" Indian tribals, Andamanese, and Negritos of the Malayan peninsula and the Philippines stand out as distinct from their neighbors, even if most of them are poorly known and our judgment is based more on physical anthropology and, to a lesser extent, on genetic data. It is certainly tempting to hypothesize that these populations are relics of early modern humans from Africa, who migrated along the southern coast of Asia all the way to Southeast Asia and Australia. As shown in chapter 7, there is not much evidence of a great similarity of these people to Australians, and therefore this hypothesis currently has very weak support. However, their original separation from Africa must be ancient, perhaps between 100 and 50 kya. These dates are suggested by the earliest times of modern humans in Africa and the time of arrival in Australia. Under these conditions, and with the inevitable later admixtures, little genetic similarity can be expected. In that time period, the passage from Africa to Arabia might have taken advantage of a closer proximity of the continents; we also know that we must credit the early settlers of Australia with skills, however primitive, that enabled them to cross expanses of sea that were not too wide. If these skills were available to the hypothetical first people who left Africa for Arabia, they might have helped them proceed eastward along the coast. Because sea levels are now higher, many possible archaeological clues are under water. Although the thinness of these arguments must be clearly acknowledged, the puzzle of the Australoid relic people in Asia needs to be explained, and it is not unreasonable to think of them as being the remainder of an ancient human bridge between Africa and Australia. Clearly, there is urgent need of further genetic data, possibly using DNA methodology, on these populations.

Not much less hypothetical are the relationships between Caucasoids and Mongoloids. Caucasoids, and more specifically Europeans, tend to be intermediate between Africans and Mongoloids, as we saw in chapter 2. This may have several explanations, to be discussed in the light of the major hypotheses of the origin of modern humans.

As we saw in chapter 2, much evidence favors an African origin for modern humans. The recent discovery of modern humans in the Middle East may seem to have decreased the strength of this interpretation, but the later presence of Neanderthals in the Middle East, the Musterian culture of the early *H. sapiens sapiens* from the Middle East, and, perhaps above all, the existence in Africa of an apparently continuous descent of modern humans from the local archaic *H. sapiens* suggest that the early presence of modern humans in the Middle East does not destroy the hypothesis of an African origin. A hypothesis hard to refute, at this stage, is that West Asia, or some other neighboring part of Asia, was the seat of an important phase of evolution of modern humans, perhaps during the hypothetical 100–50 ky "black" period of the transition from the Musterian culture of 100 kya to the later cultures we find when modern humans started the expansions that took them to settle in the rest of the world. With the present poverty of fossil samples and good dates, it is impossible to make strong statements.

The problem of whether ancient Mongoloid archaic *H. sapiens* or even *H. erectus* contributed to the modern people of East Asia cannot be solved with complete confidence on the basis of present genetic data. Modern humans seem to have appeared in China relatively late and fairly abruptly, but the paleoanthropological sample is even smaller than in the western part of the continent. The claims of similarity between Chinese premodern and modern humans (sec. 2.1), as we have seen, are not considered definitive by several archaeologists who have reexamined the issue. A strictly polycentric origin from *H. erectus* or early archaic *H. sapiens* is not in agreement, as we have seen, with dates from mtDNA. These last estimates do not, however, exclude a partial, though limited, admixture of the first modern humans who arrived in East Asia with local *H. sapiens*.

A view that could explain many of the known facts can be summarized by three points.

1. Ancestors of modern Caucasoids and modern East Asians (let us call them Eurasians) developed either in northeastern Africa, or in West Asia or southeastern Europe from an originally African source during the period between 100 and 50 kya. There was a wide area, difficult to locate in the absence of archaeological evidence, in

which cultural maturation took place until about 50 kya. One speculates that it might have been in West Asia.

2. The migration toward Southeast Asia and Australia may have been independent from the above, possibly even earlier (60–70 kya). It may have originated from East Africa (going across the Red Sea) or from the southern fringe of the speculative West Asian area of maturation. In the first case, these people may have used primitive boats, which must have been also employed by them in the passage to Australia (see chap. 7). This southern stream was directed eastward proceeding through India and Southeast Asia, perhaps along the coasts, and occupied New Guinea and Australia.

3. Whether or not it partially hybridized with local descendants of early archaic *H. sapiens* or *H. erectus*, the Eurasian moiety was ready for an expansion, perhaps about 50–40 kya, and expanded in all directions: north and then east, occupying northeastern Asia, the Arctic, and America; west toward West Asia and Europe; and southeast, where it may have mixed with the descendants of the southern branch of the African migration.

Only for much later times do we have grounds to postulate other important migrations after this (or these) first and major expansions of modern humans. The synthetic maps suggest a previously unsuspected center of expansion from the Sea of Japan but cannot indicate dates. This development could be tied to the Jomon period, but one cannot entirely exclude the pre-Jomon period and that it might have been responsible for a migration to Americas. A major source of food in those preagricultural times came from fishing, then as now, and this would have limited for ecological reasons the area of expansion to the coastline, perhaps that of the Sea of Japan, but also farther along the Pacific coast. According to the analysis of prehistoric demography of Japan by Koyama (1978; see sec. 4.4), the population of Japan peaked in the Paleolithic (20 kya) and again in the Jomon period (4500 years ago). Expansions are likely to occur somewhat before demographic peaks. Therefore, two possible choices for the dates of the expansion from Japan are suggested by the third synthetic map. The number of Paleolithic inhabitants at the peak 20 kya is difficult to estimate but was probably small. An expansion of such a small population may be unlikely, but Japan is perhaps only part of a larger region that enjoyed conditions favoring population increase; it is, however, the part that has been surveyed most accurately both genetically and archaeologically.

Agricultural developments in the Middle East, beginning 9000–10,000 years ago, were responsible for a major expansion of the early Middle Eastern farmers in four directions: a northwestern one, leading to a new settlement of Europe, starting from Anatolia (present-day Turkey); a southwestern one, via Suez to Egypt and North Africa; an eastern direction, from Iran to the Indian subcontinent; and a northern direction, perhaps starting from several different regions, to the central steppes of Asia and eastern Europe. Alternatively the Russian steppes may have been reached from Romania via the Ukraine. The various expansions had different fates, which depended largely on the evolution of the climate and the environment in the various regions and on the technological adaptations that developed locally.

1. The European climate was favorable to agriculture, though the northern part of the continent warmed up sufficiently with some delay, after glaciation. Europe was thus gradually and regularly but slowly populated by Anatolian farmers, with partial admixture with local Mesolithic foragers. High population densities were reached, with the Mediterranean area developing a little earlier (see chap. 5). This expansion probably spread early Indo-European languages from Anatolia.

2. The Iranian plateau, the Pakistan plains, and northern and southern India had a similar, but somewhat later and slower, development (this chapter). This expansion of the farmers from the eastern horn of the Fertile Crescent spread Dravidian languages eastward.

3. North Africa offered more limited opportunities, which were progressively reduced by the desertification of the Sahara (chap. 3). This expansion of farmers from the southwestern horn of the Fertile Crescent may have helped the spread of Afro-Asiatic languages toward Africa.

4. The steppes required a period of adaptation, out of which emerged pastoral nomadism and the local domestication of the horse. These eventually generated a continuously increasing population that was at the center of many geographic expansions, starting about 4000 years ago or earlier in the western and central steppes, and later dominated by Mongolian tribes of the eastern steppes, beginning about 2300 years ago. Pastoral nomads thus started mass migrations and conquests that upset almost all of Eurasia from 2000 B.C. until a few hundred years ago. They also changed the linguistic picture of Asia, extending greatly the ranges of Indo-European and Altaic languages over those that prevailed at the end of the farmers' expansions.

In the nuclear area of the agricultural expansion started in the Middle East, populations increased greatly thanks to sophisticated agricultural developments that permitted the earliest urbanization. Partial desertification, salinification caused by irrigation, and complex political events eventually reduced population numbers in the Middle East, but this area certainly reached a high population density before all others. The eastward agricultural expansion from the Middle East is visible in synthetic maps but cannot be easily distinguished from the earlier and later expansions, summarized by the first synthetic map.

Agricultural developments in East Asia were different. Although the beginnings were not much later than in the Middle East, the local domesticates could not be so easily exported to the whole area. With Middle Eastern crops, the spread was mostly in the same latitude range and met fewer problems of adaptation to different climates. By contrast, northern and southern China had different ecological conditions, and, since the be-

ginning, they were forced to develop different specializations: millet in northern China, rice in southern China and Southeast and South Asia. The relations of southern China with Southeast Asia at the time are obscure but were certainly important. The genetic differences between northern and southern China, which preexisted before these separate agricultural developments, were only partially canceled, given that the cultural segregation of these two areas left them isolated for a long time. The formation of a single Chinese nation took place only in the second part of the first millennium B.C.. Thus began a partial linguistic and cultural integration of the whole region, which had less dramatic genetic consequences.

It seems therefore that the Chinese did not have a major geographic expansion beyond the borders of China in conjunction with the development of agriculture except toward Southeast Asia and Oceania. Fairly early the population grew to large numbers, and the Chinese culture only rarely suffered major disruptions, greatly benefiting from continuity over several millennia. The first census was taken during the Han dynasty about 2000 years ago and has been partly preserved down to the present. The population of China may have increased almost 20-fold since that first census.

In the future, a study of synthetic maps centered on East Asia, or parts of it, may clarify the situation considerably. Although genetic data of parts of this region are currently unsatisfactory, interest in gene geography is increasing in both China and Japan.

Southeast Asia shows enough differences from Northeast Asia, and enough similarity to the inhabitants of the Pacific, that some important separation must have been active. But the basic similarity between northern and southern Mongoloids also suggests that, at some stage, a partial commonality, and hence gradual admixture, must be postulated. The "true" (i.e., extreme) Mongoloids are found in the north, in southern Siberia. The origin of the characteristic traits of these people may be old (extremely old, according to supporters of the polycentric hypothesis). There is a gradient of Mongoloid characteristics from the north to the southeast, well visible in the second synthetic map, which speaks in favor of an admixture that has not, however, completely canceled a strong earlier difference.

The coastal cultures of Southeast Asia were responsible for major expansions toward the islands, corresponding approximately with the spread of Malayo-Polynesian languages. These expansions may have begun 6000 to 7000 years ago. They overlapped and sometimes mixed with probably earlier, more limited Melanesian expansions. The mosaics of languages and people in Southeast Asia and the Pacific are extremely complicated. Enough later migrations, genetic mixtures, and language replacements took place since the first migrations, that a complete resolution of the various expansions is nearly impossible except for some branches. The Pacific is discussed further in chapter 7.

5 EUROPE

5.1. GEOGRAPHY AND ECOLOGY

Europe, a peninsula of Asia, has been promoted to the status of a continent more for historical than for geographic reasons. Its area, 10 million km^2, is less than one-fourth that of Asia and one-fourteenth that of the Earth's landmass, but Europe is second only to Asia in population.

Besides the Alps, which separate Italy from countries farther north, there are other less impressive mountain ranges like the Appennines in the Italian peninsula, the Dinaric Alps along the Yugoslavian coast, the Balkan Mountains, the Carpathian and Transylvanian Alps in the Balkans, the Pyrenees separating Spain from France, the Fennoscandian shield in the western part of the Scandinavian peninsula, and the Urals, which separate European Russia from Asia. The Urals form a modest barrier, but only in the northern part; south of the Urals there is continuous grass steppe from Eastern Europe to China. Otherwise, most of Europe is flat or has modest elevations.

The climate can be roughly described by defining four regions. (1) The Mediterranean climate on the Mediterranean coasts is subtropical, with mild and wet winters and hot summers. (2) The maritime (Atlantic) climate has colder winters and warm to hot summers, with precipitation the year round. (3) The central climate, north of the Mediterranean and east of the maritime one, has colder winters. (4) Farther north, the continental climate is found in most of European Russia and extends to northern Scandinavia. It has the coldest and longest winters with heavy snowfalls. The summers are moderately hot and wet.

Vegetation zones are shown in figure 5.1.1. Europe is heavily agricultural, and the vegetation described is the "natural" one that existed in postglacial times after the last major climatic changes at the beginning of agriculture, around 8000–9000 years ago.

The Mediterranean flora is characteristic of the south and comprises dense vegetation of small evergreen trees (olives, figs, etc.), shrubs, etc.; it is scrubby where the soil is especially poor. The flora is called *macchia mediterranea* and is similar to the flora in other parts of the world at latitudes between 30° and 40° north or south, for example, chaparral in southwestern North America.

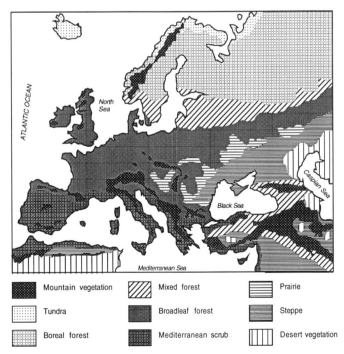

Fig. 5.1.1 Vegetation zones of Europe, western Asia, and North Africa (Times Atlas 1989).

The mixed forest once dominated all of central Europe and the British Isles, including Denmark and part of the Russian plains. Only a small fraction of the area still has forest cover, and much of the change has taken place in the last thousand years.

The boreal forest (alternating birch and conifers) is found in all of Scandinavia (except for small mountainous areas on the northern coasts where tundra predomi-nates) and in the northern half of the Russian plains. It extends to the tree line in the extreme north.

Finally, the southern third of the Russian plains is the steppe, which is mostly grassy, although shrubs occur in some areas. The steppe continues without interruption into Asia; in the last 4000 years it has been the center of developments that have influenced human history enormously.

5.2. PREHISTORY AND HISTORY

The Paleolithic. The earliest humans in Europe were *Homo erectus* and archaic *H. sapiens*; the first well-ascertained date is 700 kya (Gamble 1986). In the last part of the middle Paleolithic, until about 40 kya, all humans in Europe were of the Neanderthal type. Neanderthals were not the first humans found in Europe, but the transition from archaic *H. Sapiens* to Neanderthal is not well understood. The first Neanderthal fossils recognized as distinctive were found in Germany and named after the place of discovery. It later became clear that for a long a time Neanderthals were the only human type that existed in Europe and West Asia. Their skulls were comparable in size to those of modern humans, but they were more robust (see sec. 2.1). They made a stone-tool assemblage called *Mousterian* after the French site where it was first observed and described. Modern humans are believed to have first appeared in the middle Paleolithic in Africa; recently, however, very early dates of appear-ance outside Africa (90–100 kya in Israel) have been re-ported (see chap. 2). The transition from Neanderthal to modern humans is marked by the appearance of people with a new and distinctive skull morphology (called Cro-Magnon after the site of their discovery in France) and new tools (the *Aurignacian* type). On this basis, Euro-pean sites are assigned to Neanderthal or modern hu-mans with some confidence, even when no fossil bones are found. In Europe, the disappearance of Neanderthals and their substitution by modern humans took place be-tween 40 and 30 kya. Recent work seems to support the first rather than the second date (Straus 1989). The situation seems clearer in western Europe, especially in France and southern Germany, where the hypothesis of replacement of Neanderthals by modern humans occur-ring between 35 and 30 kya cannot be rejected (Howell 1984). In Eastern Europe, data are more contradictory and have been interpreted by some paleoanthropologists as showing a morphological continuum. This has sug-gested to some scientists that there was "unidirectional gene flow" from Neanderthals to modern humans (Smith 1984), although the hypothesis of total replacement is not eliminated.

In practice, from a genetic point of view, the choice is between the possibility of replacement without inter-mingling and replacement with intermingling, the latter offering the chance of retrogressive hybridization (taken as gene flow from the type being replaced). Further dis-cussion of the transition may be found in chapter 2.

Modern humans in Europe were also active as artists, as attested to by the rich examples of cave art found in southwestern France and northern Spain. The dating of the most famous French cave, Lascaux, is relatively late (18 kya); direct radiocarbon dates for paintings at the near Niaux caves are even later (12,890 ± 160 years B.P., Valladas et al. 1992), but there are earlier mani-festations of upper Paleolithic European art, especially in southwestern France. Toward 12–10 kya one notes in much of Europe the development of a new style of stone tools, called microliths because of their small size. This period, also called the Mesolithic, concluded the upper Paleolithic. It is regarded by many as a distinct stage from the upper Paleolithic, rather than its last phase. Af-ter the Mesolithic began the Neolithic, which in Europe is strictly associated with the appearance of agriculture.

The Neolithic. The Neolithic period of interest for Europe began in the Near East, more specifically in the most western lobe of the Fertile Crescent, in the southeastern and south-central part of Asia Minor. Here the development of a farming economy may have been a little later than (and perhaps secondary to) that in the southern part (Israel) and northeastern part (northern Iraq and northwestern Iran) of the Fertile Crescent, and around 9000–10,000 B.P. In the Fertile Crescent, local plants—for example, wheat and barley—and animals—sheep and goats, and perhaps slightly later, cattle and pigs—were domesticated. This farming economy spread slowly in all directions. From Turkey it moved to Mace-donia and northern Greece. In the beginning, farmers did not use ceramics (pottery): as mentioned earlier, the first millennium was aceramic both in the Middle East and in the very first European farmers' settlements, but then pottery rapidly spread in the Middle East and from there diffused in an apparently short time to the farmers of southeastern Europe (essentially Greece). From then on, its spread was linked almost exclusively to agriculture. At least in Europe, the finding of pottery is a fairly reliable, though not always perfect, archaeological indicator of the presence of farming. A safer indicator, more difficult to study archaeologically, is the presence of cereal grains, since wheat and barley were common as wild plants only in the Near East and were imported to Europe by the first farmers.

The farming economy allowed an increase in population density over the previous foraging economies. It has been hypothesized that, under the conditions prevailing in Europe, demographic growth stimulated and ensured the spread of farmers, and a specific model has been suggested to explain it (Ammerman and Cavalli-Sforza, 1973). The proposed model (the *wave of advance*) requires no migratory movement other than that which takes place locally in a rural population with primitive "slash-and-burn" agriculture. Only at the front is there a local expansion, with the occupation of new fields in virgin territory, and this process advances the frontier continuously but slowly. Although many archaeologists today are not inclined to favor migrational explanations, the wave-of-advance model is gaining increasing recognition. Its validity has been acccepted by a major authority in the antimigrationist (or "indigenist") field (Renfrew 1974, 1987, 1989a, b). The model can be useful for explaining not only the expansions of Neolithic farmers, but also, with little change, a variety of others (see sec. 2.7).

The early spread of farming from Greece toward the north gave rise to several farming cultures in the Balkans: Starcevo in Yugoslavia, Koros in Hungary, Cris in Rumania, and Karanovo in Bulgaria. In the spread westward to southern Italy, southern France, and eastern Spain, a local cultural adaptation arose, marked archaeologically by a type of pottery known as Impressed or Cordial Ware. Another type of ceramic, Linear pottery, is the archaeo-

logical marker for the extension northward starting from the middle Danube, then along the Rhine and the other rivers going north through the German and Polish plains toward the North Sea (fig. 5.2.1). All these cultures also have other archaeological characteristics, for example, the type of house. The last regions of Europe to be occupied were the British Isles and Scandinavia. Until some time after the beginning of the farming expansion, most of Scandinavia was too cold for growing cereals.

In chapter 2 we described, in a section on expansions, a quantitative analysis of the spread of cereals to Europe by plotting the distance from the nuclear area of agriculture in the Near East versus the time of first arrival in about 100 archaeological European subregions (fig. 2.7.2). The relationship is basically linear, even if it was probably somewhat faster in some areas. The northward expansion was slower than the westward, most probably because early farmers found it easier to advance along the Mediterranean coast. The boats they used cannot be in the archaeological record, but it is unquestionable that they had the necessary technology because they quarried islands in the Aegean and the southern Tyrrhenian for obsidian, a material especially valued for the manufacture of cutting tools. In the German and Polish plains, they advanced faster because they could move and settle along the rich network of rivers. The advance toward Scandinavia was, as already mentioned, delayed by the cold climate determined by the slow retreat of glaciers in that region.

The Middle Neolithic and the transition to the metal age. In the period from 4000 to 3000 B.C., copper mining began in the Balkans. At first, copper was used unalloyed or alloyed with antimony, which produced a much softer metal than bronze, the alloy of copper and tin that

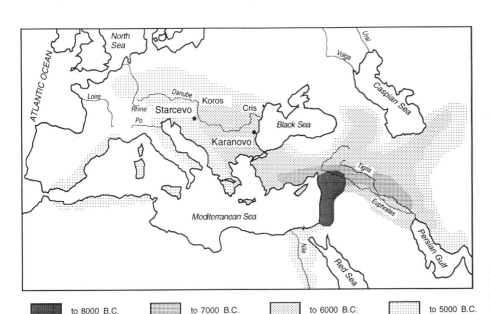

Fig. 5.2.1 Neolithic cultures of the first phase of the spread of farming in Europe (modified from Piggott 1965).

■ to 8000 B.C.	▨ to 7000 B.C.	▨ to 6000 B.C.	▨ to 5000 B.C.

was developed later. In the early period, copper was rare and found in archaeological diggings mostly near ores; it was used for axes and ornaments, but stone tools were still dominant (the Chalcolithic period).

Several new cultures appear in the archaeological record of this period. Among the earliest testimonials are Megalithic monuments, for example, tombs made of large stones and shaped as dolmens (made by a flat stone placed horizontally on top of several other flat stones arranged vertically), menhirs (upright long stones), and alignments or circles of menhirs or other stones. They were sometimes very elaborate; Stonehenge (in Salisbury Plain, England) is the most famous example. They are found over a wide area, the earliest being in Brittany, England (Renfrew 1987), and Ireland and dating from the end of the fifth millennium B.C. until 1000 B.C. They extend to Denmark and northern Germany, the Atlantic coasts of Iberian Peninsula, islands of the western Mediterranean, and Apulia in southern Italy (fig. 5.2.2). It seems likely that the first Megalithic monuments were built by early farmers, who used them as tombs, temples, and astronomic observatories. The pattern of expansion shows that the Megalithic builders were also navigators.

The Bell Beaker group (3500–2500 B.C.) occupied most of western and central Europe in a discontinuous way (see fig. 5.2.3). It is characterized by a distinctive type of pottery that gave the culture its name, and also

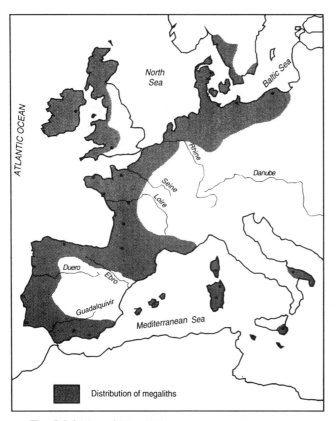

Fig. 5.2.2 Map of Megalithic monuments in Europe (modified from Whitehouse and Whitehouse 1975).

by single graves, copper, and archery. In older archaeology, this culture was considered the result of a migration, whether from west or east it was not clear. In modern times, there seems to be a consensus against population movement associated with the Beaker assemblages, which are instead believed to be "status kits" of the elite. The initially egalitarian Neolithic society was developing stratification, but goods found in Beaker burials are never as rich as those of the later Bronze Age tombs.

Other archaeological objects that appear at this time show a peculiar geographic distribution: the Globular Amphora, Corded Ware, and Battle Axe are distributed in this period in central, northern, and eastern Europe with partial overlap with the Bell Beaker culture.

In the steppes a new economy began, pastoral nomadism, today recognized as a secondary development of the farming economy, which permits a very successful adaptation to these environments (sec. 4.3). Two locations are possible candidates for the nuclear areas of the development of adaptation to the steppes: the Ukraine, and the area north of the Caucasus mountains, between the Black Sea and the Caspian Sea. The first is on the margin of the steppes and could have been settled by the late Neolithic cultures of Cucuteni, in northeastern Rumania, and Tripolye, near Kiev. The Yamnaia culture was located in the second area and gave rise to many other later local cultures. At the moment, radiocarbon data are insufficient to determine which of these cultures came first (Renfrew 1987). In the western part of the steppes, in the third millennium B.C., the horse was domesticated (Anthony and Brown 1991 give earlier dates), two- and four-wheeled chariots for transportation were developed, and the whole culture started spreading, as witnessed by the burial mounds (kurgans) found over the entire steppe by 1500 B.C. (Zvelebil 1980). Around the middle of the second millennium, the Aryan migrations spread southward from Andronovo and other nearby areas (according to several archaeologists but disputed by others; see secs. 4.3 and 4.7) toward Iran, the Indus Valley, and India. Anatolia and Mesopotamia were also occupied by Indo-European-speaking pastoral hordes who started some local dynasties (the Mitanni and others).

Archaeologist Gimbutas (1970) suggested that there was a westward expansion from the Kurgan area above the Caucasus, of pastoral nomads speaking Indo-European languages. She describes (maps are given in her 1991 book) three waves of expansion from the lower Volga-Don region, the first between 4300–3500 B.C., the second around 3500 B.C., and the third around 3000 B.C. Renfrew (1987) criticized these dates as too early. We will discuss in sections 5.3 and 5.12 the linguistic aspects.

The Bronze Age. The early Bronze Age is especially early in western Anatolia (e.g., Troy) and the Greek islands Cyclades and Crete (3200–2000 B.C.; Whitehouse

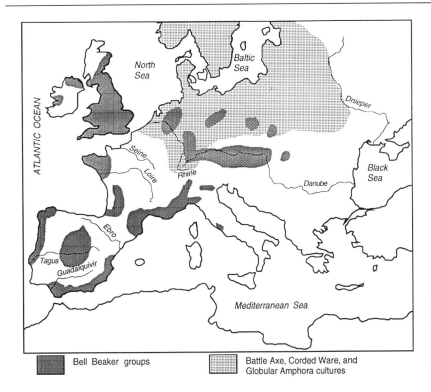

Fig. 5.2.3 The Bronze Age in Europe and the geographic distribution of some cultures: Battle Axe, Bell Beaker, Globular Amphora, Corded Ware (modified from Whitehouse and Whitehouse 1975).

Bell Beaker groups

Battle Axe, Corded Ware, and Globular Amphora cultures

and Whitehouse 1975). After 2000 B.C. Crete and the Greek mainland took over the lead culturally, and probably politically. This gave rise first to the Minoan civilization in Crete, and later to the Mycenean in Greece, which flourished until about 1300 B.C. After this time, there began a dark age for Greece.

In central Europe, the Bronze Age began around 2000 B.C. in southern Hungary and western Rumania and spread

outward. It was earlier thought that bronze technology had diffused to Europe from West Asia, but it is now believed that it was more probably a local development.

The late Bronze Age saw the diffusion of "Urnfield" cultures, beginning in southern Germany, Czechoslovakia, and Austria around the thirteenth century (fig. 5.2.4) and spreading east, south and, to a lesser extent, north and west, to most of Europe by the eighth century

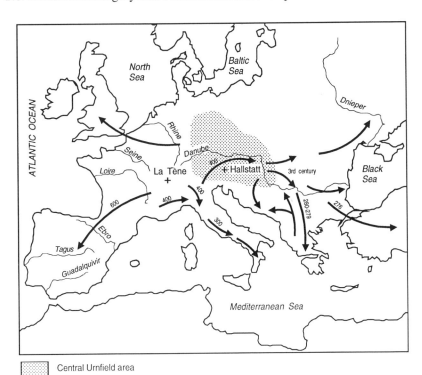

Central Urnfield area

Fig. 5.2.4 Map of Hallstatt and La Tène cultures and Celtic expansions in the first millennium B.C. (modified from Mallory 1989). Numbers are dates in years B.C.

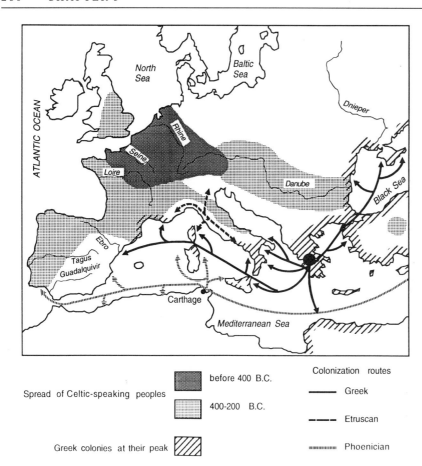

Spread of Celtic-speaking peoples

before 400 B.C.

400-200 B.C.

Greek colonies at their peak

Colonization routes

——— Greek

– – – Etruscan

⁗⁗⁗ Phoenician

Fig. 5.2.5 Europe in the first millennium B.C.: the Celtic community at its peak, and the extent of Greek and Phoenician colonizations (Encyclopaedia Britannica 1972; Alexander 1980).

B.C. Urns were used to contain ashes because cremation had replaced the earlier rite of inhumation.

The Iron Age. Urnfield people in central Europe were successful farmers and had able smiths, who produced beautiful copper objects. They were probably the ancestors of the Iron Age culture named after Hallstatt, in western Austria. A more or less direct descendant of Hallstatt is the culture named after La Tène (in northwestern Switzerland) (fig. 5.2.4). These civilizations flourished in the first millennium B.C. and were unified linguistically through their common Celtic languages. At the peak of its development in the fifth to the third century B.C., the Celtic community included southeastern, central, and southwestern Europe (see also fig. 5.2.5). Its unity was probably based on a common culture and language rather than genetic kinship (see sec. 5.8).

In the first millennium B.C., iron arrived in the Mediterranean, and in the first centuries there developed in central Italy a major civilization of unknown origin, the *Etruscans*, who spoke a non-Indo-European language. Beginning around 750 B.C. and perhaps earlier, Greeks colonized the northeastern coast of Africa, the coasts of southern Italy, and most of the islands of Sicily and Corsica, the northern part of the Iberian peninsula, and the southern coast of France. At about the same time, Phoenicians (and their descendants and political successors in the western Mediterranean, the Carthaginians) colonized the island of Sardinia, that of Ibiza in the Baleares,

the western part of Sicily, the southeastern coast of Spain, and the northwestern part of the African coast (fig. 5.2.5).

Rome and the barbarian invasions. In the last centuries of the first millennium B.C., the eastern part of the Mediterranean fell politically and culturally under the control of Greece; but Roman rule, which began in the western Mediterranean, eventually extended to all of the Mediterranean, the Near East, the Balkans, France, and England (fig. 5.2.6). Roma was originally a small village and Romans had little or no influence on the genetic pattern, but their effect on the social, economic, and linguistic patterns was substantial. In the first five centuries A.D., however, the Roman Empire was under increasing pressure from "barbarians," and in the fifth century, the western moiety succumbed. Many barbarians were originally steppe nomads or their descendants who settled in the northern, central, and eastern parts of Europe. Some, such as the *Huns*, came from as far away as Mongolia, and others from southern Russia or closer. They came mostly for booty, but some also came to settle. Among the latter, the *Goths* (a Germanic people) established an empire in southwestern Russia. Defeated by the Huns, *Ostrogoths* (probably meaning "eastern Goths") established a kingdom in Italy after the fall of the Roman Empire. *Visigoths* ("valiant men" or "western Goths," depending on interpretations), who had been allowed by the Romans to settle in Rumania, sacked Rome in A.D.

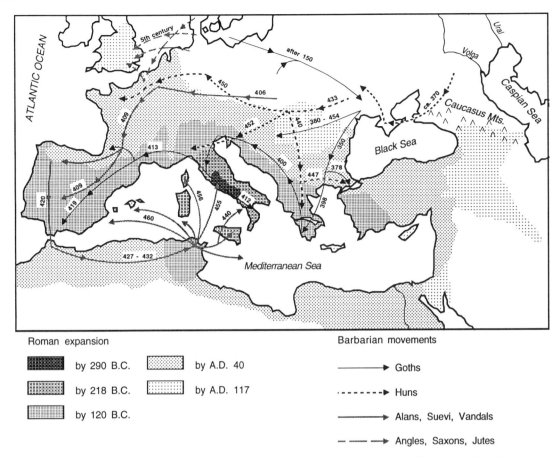

Fig. 5.2.6 Composite map of the Roman Empire and major barbarian movements and invasions. Numbers near arrows indicate dates in years A.D. (Potter 1980; Hammond 1981).

408 and resettled in southwestern France (Aquitania). After the fall of Rome, they were defeated by the Franks in France, and by the Arabs in Spain. *Anglo-Saxon* tribes from northern Germany and the Low Countries settled in England in the fifth century A.D. The *Franks*, another Germanic tribe, left western Germany (near the Rhine) to occupy northern Gaul in the fifth century A.D. Langobards or *Lombards* migrated south from northwestern Germany in the sixth century A.D. and founded a kingdom that extended to much of Italy. These are just a few of the barbarians who settled in Europe, some temporarily, some on a more permanent basis. The numbers of those who settled in the new lands are known only with great approximation. The pathways of barbarians across Europe after the fall of the Roman Empire is shown in figure 5.2.6.

During the early Middle Ages—the second half of the first millennium A.D. and the first years of the second—barbarian kingdoms rose and fell, for example, the Anglo-Saxon kingdom in England. Much of western Europe was unified at the beginning of the ninth century by the Carolingians, descendants of the Frankish rulers who had earlier occupied northern France. In southeastern Europe, the Byzantine empire held sway as far as southern and northeastern Italy. The barbarian settlements (fig. 5.2.7) caused considerable disruption of the European economy. Development came to a stop, the Ro-

man road network disappeared, literacy was largely limited to Catholic monasteries, trade was disrupted, and feudal lords were entrusted with considerable local power. Only at the beginning of the second millennium did local initiatives start to flourish and city states develop under the distant control of two warring powers, the Papacy and the Holy Roman Empire, which formally derived from the Carolingian empire, but was centered east of France.

The eastern moiety of the Roman Empire resisted for another 1000 years after the fall of the western moiety, but eventually it too succumbed to Turkic invaders. The later history of Europe saw other invaders coming from remote distances, none of major importance numerically. However, the *Magyars,* by the ninth century, from an unclear area of remote origin, were in the steppes west of the lower Don, from where they moved to occupy Hungary around A.D. 862. They included perhaps 20,000 horsemen, with a total of 100,000–500,000 people according to different counts (Guglielmino–Matessi et al. 1990). *Arabs* occupied Sicily, Sardinia, southern Italy, and Spain in the seventh to the ninth centuries but were driven out progressively in later centuries; the complete end of Arab rule in Spain did not come until A.D. 1492. *Turks* invaded the Balkans and kept their political control over much of the region until World War I. The demographic contribution of these invaders is extremely

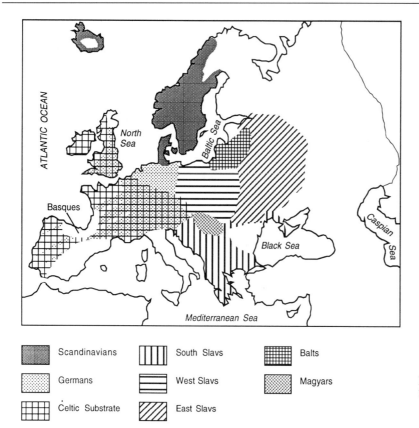

Scandinavians   South Slavs   Balts

Germans   West Slavs   Magyars

Celtic Substrate   East Slavs

Fig. 5.2.7 Europe in ca. A.D. 800, after the barbarian settlement (modified from Kidd 1980).

difficult to assess with any degree of accuracy from historical records. Europe, like China and India, has had a very high population density for a long time, and the genetic impact of invaders during the last two millennia is unlikely to have been profound, given the high numerical ratio of former inhabitants to invaders.

Population densities and numbers. The neolithization of Europe was completed by 4500 B.C. and took almost 4000 years. Farmers spread slowly, in some areas faster than in others. The population density rose considerably with the transition to agriculture; exact figures are not known. There is an estimate of 3000–9000 for the upper Paleolithic population in England, from the analysis of the Star Carr population, based on consumption of game (Clark 1972). This estimate corresponds to a density on the order of 0.02–0.07 inhabitants per km^2. In some other areas, the Mesolithic population may have been somewhat more dense, perhaps in the south of France. The density of Mesolithic populations and of hunter populations in general, is difficult to estimate from archaeological data. Evaluations are based on the number and size of archaeological sites. In particular, the number may be too small unless the search for archaeological sites was exhaustive, or the estimate of population density may be too large because hunters often had various dwellings in different places for different times of the year. Clark's estimate corresponds well, however, with the population density of Tasmania at the time of discovery. If the population density

of England was intermediate with respect to that of Europe, which was practically uninhabited in more northern regions and probably more densely populated in more southern ones, then the late pre-Neolithic population of Europe was 200,000 to 700,000. With the transition to agriculture, the population density rose to 1–5 inhabitants per km^2. An estimate from the Aldenhoven Platte in Germany at the time of the early Bandkeramik suggests a population somewhat above 2 inhabitants per km^2 (Ammerman and Cavalli-Sforza 1984). This was almost certainly a very favorable area for farming; if this density estimate could be extended to the rest of Europe, one would calculate 20 million for the European population, which seems far too high, as most of the north was still scarcely inhabited or uninhabited with glaciers in retreat.

McEvedy and Jones (1978) gave the following European population estimates:

3000 B.C.	more than 2 million
2000 B.C.	5 million
1000 B.C.	10 million
A.D., 200	28–36 million
A.D., 1000	36 million
A.D., 1300	79 million
A.D., 1500	81 million
A.D., 1900	390 million

In the last two millennia B.C., growth was especially sustained in Greece (for more detailed growth curves, see Ammerman et al. 1976), with a dark age between 1200 and 750 B.C. Growth was also marked in Italy after 1000

B.C. and continued throughout the development of the Roman rule, extending to all of the Roman Empire until its collapse. It decreased during the early Middle Ages with the destruction brought about by the barbarian invasions and the collapse of economic and social institutions. It started growing again in the later part of the Mid-

dle Ages, but the population fell markedly when the plague (the Black Death) in A.D. 1348–1353 wiped out perhaps one-third of the total population, in some areas up to two-thirds. By A.D. 1500, the population was again at the pre-Black Death density and from then on grew at a fairly regular, approximately exponential rate to the present numbers.

5.3. LINGUISTICS

The great majority of European languages belong to the Indo-European (or Indo-Hittite) family or phylum (see fig. 5.3.1). Under its more general title, the Indo-Hittite phylum includes two major branches: Anatolian (all extinct: Luwian, Lycian, Lydian, Palaic, Hittite) and Indo-European proper.

Indo-European consists of nine branches, but their reciprocal relationships are unclear. According to Ruhlen (1987) they are:

1. Armenian (a single language)
2. Albanian (a single language)
3. Tocharian (2 languages spoken in western China, extinct)
4. Indo-Iranian (93 languages spoken in Iran, India, Afghanistan, Pakistan)
5. Greek (2 languages)
6. Italic (16 languages: including Latin [extinct] and western and eastern Romance languages)
7. Celtic (4 languages surviving, spoken in Ireland, Scotland, Wales, and Brittany)
8. Germanic (11 languages in north, central, and northwestern Europe, including Yiddish)
9. Balto-Slavic (15 languages, spoken in eastern Europe)

The problem of reconstructing a tree for Indo-European languages has been considered almost insurmountable when approached with rigorous statistical methods, mostly because of the considerable overlap between languages of the different clusters listed above. Today most linguists prefer to list the clusters without any attempt at organizing them in a hierarchy.

Four major languages of the Uralic family are also spoken in Europe: Hungarian, Finnish, Lapp, and Estonian. *Hungarian* was imposed by the Magyars, who conquered the country at the end of the ninth century A.D.; the time of arrival of the other Uralic languages in nothern Europe is not clear but is earlier. The area of origin of these languages is believed to be the central Urals.

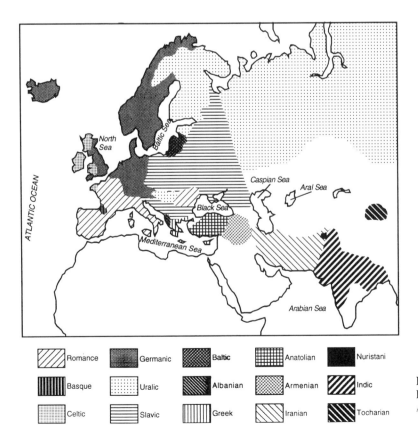

Fig. **5.3.1** Geographic distribution of Indo-European, Basque, and Uralic languages around A.D. 1500 (Ruhlen 1987).

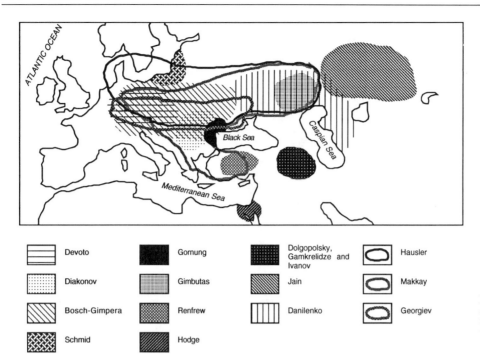

Devoto	Gornung	Dolgopolsky, Gamkrelidze and Ivanov	Hausler
Diakonov	Gimbutas	Jain	Makkay
Bosch-Gimpera	Renfrew	Danilenko	Georgiev
Schmid	Hodge		

Fig. 5.3.2 Hypotheses on the place of origin of Indo-European languages, proposed since 1960 (Mallory 1989).

Another non-Indo-European language spoken in Europe is *Basque*. This is one of nine language isolates (including four extinct ones) that are reasonably well known, but cannot be classified into any of the current phyla. Basque is spoken in the western Pyrenees, with less than 100,000 speakers on the French side (perhaps only 25,000–30,000 according to knowledgeable local sources) and 2–3 million on the Spanish side. There are some similarities indicating a distant relationship with North Caucasian languages (Gamkrelidze and Ivanov 1990). It has also been suggested on tenuous grounds that it is related to Berber and Afro-Asiatic languages (Allières 1979). The simplest hypothesis one can make is that Basque was spoken by the local Mesolithics before the arrival of Neolithics, which may have brought the first Indo-European language into the area (see secs. 5.6 and 5.9).

The origin of Indo-European languages is at the center of the problem of the origin of Europeans. Almost all possible hypotheses have been made (fig. 5.3.2), but two areas of origin seem to deserve greater attention than the others: Anatolia (modern Turkey) and Ukraine.

Anatolia as the area of origin of Indo-Europeans has been suggested a number of times by a variety of linguists, most recently by proponents of the Nostratic superfamily (Dolgopolvsky 1988), chiefly on the basis of early borrowings from neighbors, which were greater and more convincing for the Semitic languages. The Anatolian origin has been forcefully supported in a recent book by the archaeologist Renfrew (1987). The attempt is especially interesting also because traditionally archaeology has refrained from taking a stand on linguistic questions. There is in fact no direct information in archaeological material on language before the advent of writing; but Renfrew argued that language replacement can take place only in specific circumstances that depend on the social structure of the people involved, and archaeology can inform us on this aspect. He has therefore paid special attention to anthropological *process*.

In section 2.6 we summarized three situations (which we reduced to two) that Renfrew listed, in which a change of language can occur. The first situation is exemplified specifically by the movement of early farmers into Europe 9000–5000 years ago. Renfrew (1987) accepted the wave-of-advance model of early farmers (Ammerman and Cavalli-Sforza 1984) and added to it the hypothesis that early farmers in Anatolia (and, in general, in the Fertile Crescent) spoke an Indo-European language. According to him, the expansion of farming to Iran and India also carried Indo-European languages eastward (Renfrew's hypothesis A), but he did not exclude the alternative, traditional hypothesis (hypothesis B) that these languages arrived there with the Aryan pastoral nomads, who occupied these lands around 1500 B.C. (see secs. 4.6 and 5.2).

Under hypothesis B, the diffusion of Indo-European languages to South Asia (including Iran) was a secondary phenomenon. It would then have taken place according to the second of the processes listed by Renfrew, called the *elite-dominance* model, in which a relatively small group of highly organized people, who are militarily and politically efficient, invades a country and successfully imposes its own language. The social organization of the invaders is expected to be highly stratified, or at least to correspond to a "chiefdom" society. Unquestionably, Aryan society fits this description. Ordinarily the conquerors also have some specific military advantage, which was certainly true of the Aryans. In later work, Renfrew (1989a,b,c) dismissed hypothesis. A regarding the origin of Indo-Iranian languages and replaced it with the idea, previously discussed, that the expansion of agriculture from the Middle East diffused three linguistic families: Indo-European toward Europe, Dravidian toward Pakistan and India, and Afro-Asiatic toward Arabia and North Africa. Essentially the same hypothesis was suggested by Cavalli-Sforza (1988).

There is no direct evidence of the language spoken by the early farmers, and the main support for the hypothesis that Indo-European languages were first spread to Europe by Anatolian farmers comes from Renfrew's theoretical considerations. But the alternative hypothesis that Indo-European languages were spread to Europe (as well as to India) by pastoral nomads of the western steppe (Gimbutas 1970) is not incompatible with Renfrew's hypothesis. The Kurgan expansion postulated by Gimbutas is much later than the Neolithic expansion of farmers from the Middle East. Both expansions could well have taken place. Both are supported by genetic evidence (see sec. 5.11). The Kurgan people were possibly descendants of migrants from the Neolithic area of origin. The relationship of their languages, however, is difficult to assess.

To support her hypothesis, Gimbutas also used the notion of the "homeland" of Indo-Europeans, which was identified according to the roots for the names of trees that are common to most Indo-European languages (Friedrich 1970). Many of these trees were frequent in southern Russia 5000 years ago, but they are not characteristic of only that area, nor has the possibility been excluded that many of them were present, say, in Anatolia at earlier times.

According to Renfrew, modern archaeological evidence on the Kurgan culture shows that at the time postulated by Gimbutas the area had not developed the degree of sociopolitical and military sophistication necessary for an expansion of pastoral nomads capable of imposing their language by the process of elite dominance. Only later, perhaps at the time of the Aryan invasion of India, would an elite have been able to expand toward Europe. But at the end of the Neolithic and beginning of the age of metals, the time postulated by Gimbutas, Europe was already fairly heavily populated. It was therefore too late for a process of language replacement by a population expansion in a near vacuum, and too early for a process of elite dominance.

Paraphrasing Renfrew (1987), it is obvious that Middle Eastern farmers carried with them their language, which must have undergone much transformation and local differentiation during the slow colonization of Europe. Because farmers were in greater numbers than hunter-gatherers in most areas, they are very likely to have retained their language, but only some Mesolithics—who had reached unusually high numbers—were able to conserve their own. Basques are the outstanding example of this conservation. It is possible that other non-Indo-European languages—for example, Etruscan—survived

for a long time (or came from other places, according to some classical legends and hypotheses). A few languages spoken in Europe and belonging to other families (Altaic, Uralic) are later introductions, some well-documented historically. For a summary of alternative views of origin of Indo-Europeans and the linguistic criticisms of Renfrew's hypothesis see Mallory (1989) and Diamond (1991a, 1991b) who support Gimbutas' hypothesis.

The Celts, a people and a language subfamily widespread in central and northern Europe at the beginning of history, present a major problem (Renfrew 1987, Mallory 1989). It is unlikely that Celts expanded in the manner of the early farmers; Europe was far too populated at the time and not a near vacuum. How can one explain the spread to Europe of the Celtic language and culture without any expansion of people and without a sociopolitical organization responsible for it? Only a few things seem beyond dispute; the Celtic identity of the Hallstatt culture and the later La Tène culture. The archaeological map shows that the Celtic culture continued its expansion (fig. 5.2.5) until it was finally checked by the Romans in the first century B.C. Northern Italy, France, England, and part of Spain were speaking Celtic languages (but probably not exclusively) at the time of the Roman conquest, around 2000 years ago. Latin replaced Celtic and other languages in all Celtic areas that fell under the Roman rule. Today, Celtic languages survive only at the extreme periphery of their original expansion in areas the Romans did not reach, or in which they did not establish full control, or in which Celts may have sought refuge in their retreat from the Romans.

It would therefore seem that if the Celtic languages spread in the second half of the second millennium B.C. under the influence of a cultural and perhaps also a military elite, with little or no demic component, few genetic traces are likely to be found. Other cultural expansions for which it is difficult to find a genetic or linguistic counterpart are those of the Megalithic monument builders mentioned earlier (sec. 5.2).

We must also consider the spread of other recognized branches of Indo-European languages: Romance, Germanic, Baltic, and Slavic (the latter two are often considered a single branch, Balto-Slavic). Like the Celtic languages, these may have originally been a local dialect that spread at different times and places to a fairly large area by cultural mechanisms, but perhaps also by demic ones or by military conquest. The spread of Romance languages was obviously determined by the military conquest and political control of parts of Europe by the Romans, who imposed their social organization and political control along with the Latin language. This is a clear example of the spread by an elite, with very little, if any, demic component. Latin replaced many languages, but especially other Indo-European ones, like those of the Celtic groups, in all of the Roman Empire (fig. 5.2.6). The introduction of English, Spanish, and

Portuguese to the Americas followed a similar pattern. The spread of Germanic, and Balto-Slavic languages is less well understood for lack of adequate historical information; it may have had more of a demic component than the spread of Latin. The time available for regional differentiation after the spread may have been similar to that for Romance languages, which may have started

diverging more strongly after the fall of the western Roman Empire (about 1500 years ago); the time may have been somewhat less for Slavic languages.

All other Indo-European branches must have arisen at an earlier time and are sufficiently old, or unsuccessful, that they are represented by few, one, or even no living languages.

5.4. PHYSICAL ANTHROPOLOGY

Since the Industrial Revolution, Europeans have undergone a marked change in physique. The oldest series of stature measurements show an increase in average stature that was slow at first during the nineteenth century, but has accelerated during the twentieth. Stature may have reached its final state in some parts of northern Europe and may still be changing in the south, where industrial development had a later beginning (see, e.g., Chamla and Gloor 1987). The causes of this increase are not known. They may be related to improved nutrition, but other, most probably environmental, causes are also at work. As a consequence, it is difficult to attribute a clear genetic meaning to body measurements, especially in Europe.

In general, linear measurements tend to follow stature, even if they need not be simply proportional to it. Because of this, uncertainties concerning the genetic determination of stature carry over to most other linear measurements. There exist many estimates of "heritability" (i.e., the proportion of variation caused by genetics) for metric traits, and they are often very high, but they are not sufficient to indicate that the determination of these traits is mostly genetic. They are obtained from measurements of similarity (correlations) between relatives—like parent-child, sib-sib, or members of twin pairs—but they almost never exclude secular trends, or socioeconomic influences culturally inherited in families, social strata, and countries. There is therefore no guarantee that the observations of anthropometric variation in space or time reflect genetic differences rather than socioeconomic, nutritional, or other environmental and historical factors.

Moreover, ratios between linear measurements—such as the famous or infamous cephalic index by Retzius (1900), which has commanded an enormous and totally unjustified degree of attention for about 100 years—do not guarantee independence from environmental influences. Because, in the last two centuries, European populations, more than other aboriginal groups, have been exposed to recent environmental changes affecting anthropometrics, Europe is probably that part of the world in which it would be most dangerous to rely on genetic conclusions based on anthropometric measurements. Some anthroposcopic traits—for instance, skin color and especially hair and eye color and pattern—are even more difficult to measure, not sufficiently under-

stood from a genetic point of view, and subject to sexual selection of unknown strength, constancy, and direction. It is possible, however, that some of these limitations will be removed in the future when the molecular genetics of these traits will be known. The fact remains that light skin and eyes are almost uniquely characteristic of northern Europeans.

Geographic maps of single physical traits have been studied intensively by early physical anthropologists. Although there undoubtedly can be—and in some cases, there certainly is—at least partial genetic determination, the exact mechanism of inheritance is unfortunately unclear.

The geographic distribution of the pigmentation of hair and eyes in Europe (fig. 5.4.1) is especially characteristic: both hair and eyes are lightest in the south Baltic and their darkness increases fairly regularly and almost concentrically around this region. One may wonder whether this center was the place of origin of mutation(s) to light pigmentation and whether there may have been a cause for selection (natural or sexual) of light pigmentation in this region. Did such mutation(s) recur in other areas? They are found not infrequently in Sicily, where instead of a local recurrence, it is easier to think they were brought by Norman invaders in the eleventh century, and in the Middle East, where crusaders were not dedicated exclusively to the dissemination of the Christian faith.

We have discussed skin color in chapter 2. The evolutionary explanation advanced for skin color is that it represents a selective adaptation to the intensity of solar radiation, whereby the skin is darkest in hot and dry climates and lightest at high latitudes. But white skin has a strong peak only in northern Europe and not in other parts of the world. It has been suggested (Loomis 1967) that light skin color makes it possible to produce the necessary amount of vitamin D in the skin at high latitudes (because the production of vitamin D from precursors contained in food requires that enough ultraviolet radiation reaches the deep layers of the skin), and in the north the intensity of solar radiation is not sufficient unless the skin is unpigmented. The darker skin color of the Lapps and of all the other Arctic populations is explained, in light of the vitamin-D hypothesis, on the basis of their different nutritional customs. In fact,

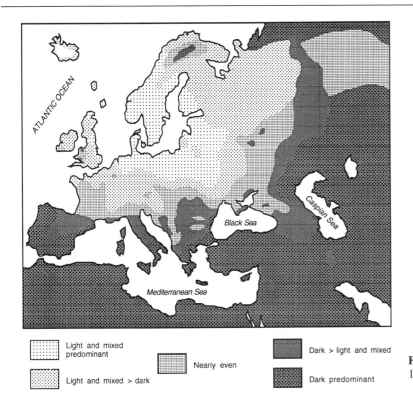

Light and mixed predominant

Light and mixed > dark

Nearly even

Dark > light and mixed

Dark predominant

Fig. 5.4.1 Pigmentation of hair and eyes (Coon 1954).

Scandinavians of non-Lappish origin were farmers and needed unhindered exposure to solar radiation for producing vitamin D from precursors contained in wheat. Arctic inhabitants are herders, hunter-gatherers, or fishermen and have a supply of ready-made vitamin D in their diet. They therefore do not need to have a light skin.

We may also add that all anthroposcopic characters of skin, hair, and eyes are likely to be subjected to strong sexual selection, which, according to Darwin, was the major factor in the ethnic differentiation of humans. Unfortunately, the contribution of sexual selection is difficult to test, preferences are characteristically unstable, and very sensitive to frequency of types in populations. Climate is also believed to be important in the determination of stature and weight, in humans as in all animals. But stature has changed drastically in Europe in the last two centuries, starting in the north, for reasons that are largely unrelated to natural selection and probably depend on artificial modifications of our living environment, diet, etc. (Costanzo 1948; Damon 1968; Chamla and Gloor 1987).

Harding et al. 1989 attempted to use craniometric data of ancient Europeans to test the hypothesis of prehistoric migrations. Results did not support an expansion of farmers from the Middle East but early Neolithic specimens, which would be necessary to test this hypothesis, were tantalizingly few. There have been many attempts at a classification of European peoples based on anthro-

pometric and anthroposcopic characters. As usual, the number of "races" varies greatly with authors; the most influential classification using few races was suggested by the American economist W. Z. Ripley (1899) just before the turn of the century. It recognized three European races: *Nordic* or northwestern Europeans, described as tall, fair, and dolicocephalic (long-headed); *Alpine* or central and eastern Europeans, described as short, stocky, brown-haired with much body hair, and brachycephalic (short-headed); *Mediterranean* (Iberian Peninsula, southern Italy, Greece, North Africa, Near East, Arabia), described as short, slender, dark, often with curly hair, and dolicocephalic.

Alpines were hypothesized to be the result of Neolithic immigration from Central Asia, which separated, like a wedge, the older inhabitants into two groups, Nordics and Mediterraneans. The hypothesis is manifestly at odds with our modern notions on Neolithic colonization of Europe. Seen as a taxonomic effort, Ripley's classification tried to summarize a complex geographic pattern and was inevitably unsatisfactory. Several anthropologists tried to improve on this classification, as usual with divergent results.

Ripley was not to blame (Coon 1954) for the psychological taxonomy of European "races," which became very popular at the beginning of this century and forms one of the most ludicrous confusions among customs, culture, and genetics—in short, a perfect example of "scientific" racism.

5.5. THE GENETIC PICTURE

Europe is the continent with the most genetic information but is also the least easily interpretable on the basis of trees. It owes the first distinction to the high density of universities and medical laboratories per square kilometer and to the great interest in historical problems that also makes Europe the most investigated continent from an archaeological point of view. The second distinction is due to a variety of factors. European populations are classified by countries, which are often heterogeneous, being composed of subgroups on which there is only occasionally enough information for separating them in a clear fashion. Many of these countries have been unified politically or linguistically for a sufficiently long time that internal migration has blurred some of the original differentiation. There is a need, however, to study in detail the internal genetic heterogeneity of these countries. We have done this work so far only in a limited way, but it is rewarding, and we will see examples for Italy, France, and Spain. Moreover, the central part of Europe is fairly homogeneous genetically and has little tendency to form a tree. To some extent, this may be due to the Neolithic diffusion, which has ironed out many initial differences in the central region, leaving a fairly smooth gradient.

A tree analysis (fig. 5.5.1), eliminating several less well tested populations, was made for 26 populations with 26.4% missing data. The average number of genes is 88. The tree shows a number of outliers, a few small clusters, and a larger cluster of central Europeans.

The extreme outliers are Lapps, followed by Sardini-ans; both populations are well known genetically. In a bootstrap analysis for testing the robustness of the tree, Lapps are outliers in 76% of the bootstraps, being replaced as first outliers by Sardinia 18% of the times; Sardinia is the next outlier in 63% of the bootstraps in which Lapps are first.

There is then a group of five less extreme outliers, in the following order: Greeks, Yugoslavs, Basques, Icelanders, and Finns. The rest of the populations form a series of small groups, all of which comprise geographically close neighbors or related populations, in the form of a linear tree. These groups, in order from the most peripheral in the tree to the most central ones, are:

1. Celtic (Scots and Irish)
2. Eastern Europeans (Russians, Hungarians, Poles)
3. Southwestern Europeans (Spaniards, Portuguese, and Italians)
4. Czechoslovaks, who do not join other Slav speakers, but are approximately intermediate between them and the central subcluster of the Germanic group below, to whom they are geographically close
5. Northwestern Scandinavians (Norwegians and Swedes)
6. French, who are related to the Germanic group below
7. Germanic populations, including two subclusters, northern and central: the first subcluster is made up of Dutch, Danish, and English people; the second of Austrians, Swiss, Germans, and Belgians

It is clear that there is a basic linguistic association within these subclusters, with exceptions worth examin-

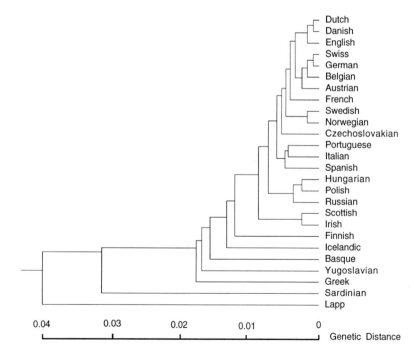

0.04   0.03   0.02   0.01   0   Genetic Distance

Fig. 5.5.1 Genetic tree of 26 European populations. F_{ST} distances are based on an average of 88 genes.

ing. The first small cluster includes the two Celtic populations in the sample. The second includes two of the four Slavic-speaking populations; the third (Czechoslovak) is isolated in the tree, probably because of its intermediate geographic position (it is located between Slavic- and German-speaking people and was part of the Austrian empire for some time). The fourth, Yugoslavian, is an extremely heterogeneous country, historically and genetically, and is not included in the cluster, being an outlier. Hungarians are linguistically intruders in this group, but that is discussed more thoroughly in section 5.6. Swedes and Norwegians are associated both geographically and linguistically: of the other three countries in Scandinavia, two (Iceland and Finland) are outliers for good reasons, also discussed in section 5.6, and the third (Denmark) belongs genetically and geographically to the northern Germanic populations. France is rather heterogeneous genetically and is discussed separately. The most cohesive group consists of people speaking Germanic languages; of these, Belgium and Switzerland are divided linguistically, with Flemish and German speakers being the majority in the two countries, respectively.

This tree most probably has no historical interest in the present case, but other approaches can help disentangle some of the complex historical and prehistorical events that shaped the history of Europe and they will be examined in the next sections. The PC map (fig. 5.5.2) based on an average of 88 ± 0.1 genes accounts for 50% of the original genetic variation and confirms the external position of the outliers Lapps and Sardinians. Other major outliers like Greeks, Basques, and Finns are on the borders. There are two relatively clear clusters, one of northern and one of central Eu-

ropeans, separated by the second PC. Southern Europeans overlap with central Europeans, on the opposite side from the northern ones. Apart from outliers, the PC map reflects to some extent geography, as is often the case. Table 5.5.1 gives F_{ST} genetic distances of Europeans.

An interesting new method of analysis was introduced by Barbujani et al. (1989) and applied to human variation in Europe (Barbujani and Sokal 1990). It formalizes a proposal by Womble (1951) of identifying genetic boundaries as regions of rapid genetic change. The quantities that are used for this analysis are the local slopes of gene-frequency surfaces. The authors visualized *boundaries* as "lines separating two different regions, each one displaying comparatively little variation"; they recognized that no definition permits the unambiguous distinction of boundaries from gradients or clines ("wide areas of gradual biological change," in the authors' words). It seems a boundary is ideally a narrow cline with a recognizably high local slope, but it might also be a local change in the slope of a cline, and it is difficult to give numerical thresholds that might help definitions. It is unlikely that a sharp boundary in a continuously inhabited region is due to change in natural selection across the boundary; it is more likely that it signals a local decrease in genetic exchange. Boundary detection is greatly facilitated by using a multivariate approach and looking for boundaries that are visible with many independent genes, which is what the authors have done in practice. Local decrease could be due to one or more of a great variety of barriers limiting exchange between two regions. The first and most natural are geographic barriers: the sea, mountains, and major rivers. The island condition notoriously predisposes to isolation

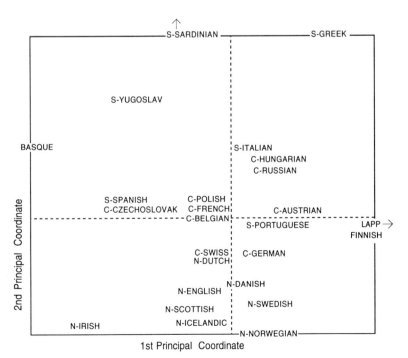

Fig. 5.5.2 Principal-component map of Caucasoids: N, C, and S denote northern, central, and southern Europeans, respectively.

Table 5.5.1. Genetic Distances (in the lower left triangle of the matrix) and Their Standard Errors (in the upper right triangle of the matrix) between European Populations (all values x 10,000)

	BAS	LAP	SAR	AUT	CZE	FRE	GER	POL	RUS	SWI	BEL	DAN	DUT	ENG	ICE	IRI	NOR	SCO	SWE	GRK	ITA	POR	SPA	YUG	FIN	HUN
Basque	0	101	68	39	32	22	37	27	29	37	22	33	22	22	33	27	41	26	25	48	26	30	15	53	38	30
Lapp	629	0	105	55	92	54	49	56	56	64	66	53	55	68	112	109	55	85	63	50	50	59	83	230	36	52
Sardinian	261	667	0	70	77	64	72	74	52	81	71	78	66	70	66	87	88	80	76	30	54	79	58	86	73	69
Austrian	195	308	294	0	14	9	10	39	22	6	4	8	20	26	37	28	34	16	32	13	11	15	17	105	13	9
Czech	159	470	327	14	0	18	15	24	25	8	13	11	17	15	36	30	22	22	24	19	16	10	13	71	41	22
French	93	350	283	9	18	0	5	13	11	6	7	8	7	5	24	20	10	13	16	28	6	14	9	90	13	12
German	169	314	331	10	15	5	0	9	16	2	4	4	7	6	24	21	6	10	18	31	11	15	17	95	17	8
Polish	146	395	282	72	52	13	47	0	8	15	14	16	12	15	36	32	16	21	29	75	24	28	20	125	11	7
Russian	140	323	266	64	64	11	60	30	0	13	13	12	11	16	47	40	18	23	30	49	7	16	29	148	49	9
Swiss	165	375	353	12	31	23	10	60	78	0	5	3	4	9	24	18	10	13	14	27	6	7	13	91	19	14
Belgian	107	333	256	16	43	32	15	40	51	14	0	5	3	6	18	19	6	12	11	27	14	16	11	16	13	17
Danish	184	334	348	27	54	43	16	69	80	19	21	0	2	6	20	15	9	9	15	29	16	18	14	99	25	17
Dutch	118	341	307	38	66	32	16	54	57	16	12	9	0	5	26	18	6	10	16	42	8	9	19	93	31	15
English	119	404	340	55	60	24	22	70	79	28	15	21	17	0	16	7	6	6	10	56	8	9	9	99	21	16
Icelandic	221	494	396	153	173	146	106	144	169	115	78	88	101	76	0	24	18	25	26	68	27	27	34	136	41	44
Irish	145	557	393	115	117	93	84	150	160	86	75	68	76	30	99	0	22	6	21	57	26	27	22	91	50	35
Norwegian	195	317	424	61	76	56	21	58	90	33	24	19	21	25	74	79	0	14	12	62	15	17	18	106	19	18
Scottish	146	447	357	74	104	62	53	121	129	59	59	40	48	27	111	29	58	0	15	42	16	18	18	116	31	26
Swedish	168	333	371	80	90	78	39	82	110	55	34	36	41	37	106	94	18	74	0	57	18	16	15	107	32	31
Greek	231	308	190	86	126	131	144	177	161	148	103	191	199	204	288	289	235	253	230	0	29	25	33	133	16	16
Italian	141	339	221	43	77	34	38	64	75	44	30	72	64	51	143	132	88	112	95	77	0	12	15	88	31	11
Portuguese	145	324	340	48	46	48	51	65	98	53	31	77	60	46	149	115	73	97	78	103	12	0	17	87	19	11
Spanish	104	452	295	69	65	39	69	117	122	43	42	80	76	47	163	113	97	100	99	162	15	48	0	84	16	27
Yugoslavian	176	565	294	110	101	124	118	137	170	120	50	157	136	160	317	272	173	248	213	213	119	139	172	0	141	123
Finnish	236	210	334	77	175	107	77	139	153	112	63	96	123	115	157	223	94	166	82	150	94	119	159	248	0	29
Hungarian	153	338	279	40	69	70	46	25	30	57	52	78	71	70	172	152	77	124	99	88	61	63	118	136	115	0

and hence to genetic difference. Political barriers may be more transient, but they also tend to generate a certain degree of isolation. Language differences are also high on the list of expected barriers. Other social sources of isolation exist, from religion to customs, history, etc., but in general there is a good correlation among many of these barriers.

The results of analyzing Europe by this technique are shown in figure 5.5.3 taken from Barbujani and Sokal (1990), with some simplification, which does not detract from the accuracy of the conclusions. We also omit the description of the significance tests and thresholds imposed by the authors, which are inevitably arbitrary. Using 63 allele frequencies measured at 19 genetic loci, the authors found 33 barriers that passed the tests of significance. They are indicated by bars of different lengths in the figure, and they have all been recognized as such on the basis of a number of different gene-frequency surfaces. These boundaries are listed in the figure legend.

The authors stated that 31 of the 33 boundaries are also linguistic boundaries (more precisely, 26 language differences and 5 dialect differences). There is unquestionably a high correlation of genetic and linguistic boundaries. The authors also count 22 physical boundaries among the 31, 4 montane and 18 marine. The residual nine are linguistic but not physical. They conclude, therefore, that "language barriers may oppose the process of population admixture" (Barbujani and Sokal 1990). This is probably correct, but the proof is not complete because political and socio-historical barriers are not considered. Of the nine genetic boundaries without obvious physical boundaries, three compare Lapps and non-Lapps (see sec. 5.6): an initial genetic difference of some magnitude between Lapps and non-Lapps, an enormous difference in culture and economics, and more than two thousand years of historical separation have created a barrier against cross-mating or marriage. This has been in part overcome (hence some of the observed gene flow among Lapps), but genetic isolation has continued at a level sufficient to maintain a detectable fraction of the initial genetic difference. Some of the other pairs—The Netherlands and Germany, Austria and Hungary, northern and southern Yugoslavia, northern and southern Germany (not linguistically different)—have long had different customs, religions, and political affiliations. Hence, there have been more than linguistic factors of genetic isolation.

In the 22 cases in which there are both physical barriers and genetic boundaries, it is reasonable to postulate that the causal arrow is likely to go more from physical barriers to *both* genetic and linguistic differentiation, rather than in other directions. Once linguistic differentiation has set in, however, it is very likely that it will increase the genetic separation by making marriages across geographic plus linguistic barriers even less likely. An analysis of the power of linguistic differences in deter-

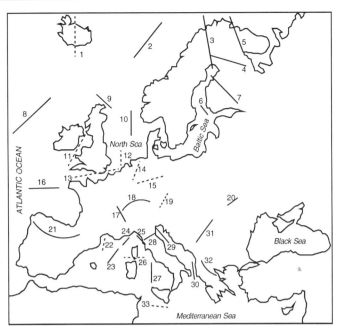

Fig. 5.5.3 Zones of sharp genetic change in Europe and their correspondence with linguistic boundaries (Barbujani and Sokal 1990). 1, western vs. eastern Iceland; 2, Norway vs. Iceland (Norwegian-Icelandic); 3, northern Finland vs. Sweden (Finnic-Germanic); 4, central vs. northern Finland (Finnic-Lappic); 5, Kola Peninsula vs. Finland (Slavic-Finnic); 6, Finland vs. Sweden (Finnic-Germanic); 7, southwestern vs. northern and eastern Finland; 8, Ireland vs. Iceland (English-Icelandic); 9, Scotland vs. Orkney and Shetland Islands (Celtic-Germanic); 10, southern Scandinavia vs. Scotland (Norwegian-English); 11, England and Wales vs. Ireland (Celtic-Germanic, in part); 12, England vs. The Netherlands (English-Dutch); 13, England vs. France and Belgium (Germanic-Romance); 14, Germany vs. The Netherlands (German-Dutch); 15, central vs. southern Germany; 16, Iberia vs. Iceland (Romance-Germanic); 17, France vs. Italy (French-Italian); 18, Switzerland and Austria vs. Italy (Germanic-Romance); 19, Austria vs. Hungary (Germanic-Ugric); 20, Transylvania (Romance-Ugric); 21, northwestern Iberia vs. the rest of Iberia (Basque-Romance); 22, Catalonia vs. Corsica (Catalan-Italian); 23, Catalonia or Baleares vs. Sardinia (Catalan-Sardinian); 24, Corsica vs. France (Italian-French); 25, Corsica vs. Italy; 26, Corsica vs. Sardinia (Italian-Sardinian); 27, Sardinia vs. Italy (Sardinian-Italian); 28, northern vs. southern Italy; 29, Italy vs. Yugoslavia (Romance-Slavic); 30, Italy vs. Albania (Romance-Albanian); 31, northwestern vs. southeastern Yugoslavia; 32, northern vs. central and southern Greece; 33, Sicily vs. Malta (Romance-Semitic).

mining genetic isolation might be more easily attempted at the level of individuals, by comparing the frequencies of marriages in countries like Switzerland or Belgium. Such an atomistic type of research would also suffer from considerable complications, however, because of the confounding effects of propinquity of mates and correlations in social class (geographic and social homogamy), which are not always easy to eliminate. In any case, it seems extremely likely that linguistic differences do help to create genetic isolation and can generate some of it even when they are the only cause of isolation. All things considered, it would be difficult to disagree with Barbujani and Sokal (1990), except for noting that the separation of possible factors of genetic isolation is incomplete in their conditions of study.

It is of interest to compare the specific pattern of genetic differentiation revealed by the map in figure 5.5.3 with that obtained by the tree and PC map analyses of the two previous figures. There is in general very good agreement; of course, figure 5.5.3 cannot indicate, in its present form, the strength and hence the hierarchies of the observed boundaries. It is interesting that the populations we chose quite independently in this section correspond reasonably well with those suggested by figure 5.5.3. We did not include Corsica or Romania among the regions or countries considered in the tree, nor did we separate western from eastern Iceland, or the northern from the southern parts of Italy, Yugoslavia, Greece, or Germany. Some of these were excluded to maintain a high number of genes, a necessity especially when differences are very small. Most of Europe is exceptionally homogeneous.

We were planning to analyze single countries individually, something that has been possible only for a few of them (sec. 5.7–5.9). If we could have chosen, however, to use these genetic boundaries for defining populations to be analyzed by tree or other methods (as suggested by Bateman et al., 1990 a, b, c), we would have had to consider almost all of central Europe, most of Scandinavia, the Baltic countries, and European Russia as a single large population, somewhat against common sense. The approach cannot, by definition, detect slow gradients lacking regions of sufficiently sharp change. With greater resolution, as could be provided by 10 times as many gene frequencies as available now, more boundaries would have been detected and smaller population units created, but with no guarantee that detectable boundaries would form closed units. Europe is, we may remember, by far the continent best known genetically, and inevitably resolution can be expected to be even worse in most other parts of the world.

In the remainder of this chapter we examine the major and best-known outliers, then those European countries for which it has been possible thus far to prepare detailed synthetic maps: Italy, France, and Spain. A few other European countries have enough information that the same work may be extended to them in the near future. Finally, we examine geographic maps of single genes and synthetic maps of the whole continent, which allow us to give a summary of the genetic history of Europe.

5.6. MAJOR OUTLIERS: LAPPS, SARDINIANS, BASQUES, AND ICELANDERS

Four major outliers in Europe on the basis of genetic or historical information, are, in order of divergence, Lapps, Sardinians, Basques, and Icelanders. They have no simple relationship to each other or (with the exception of the last two) to any other population. The greatest distances in the matrix (see table 5.5.1) are those between Lapps and Sardinians (667) and Lapps and Basques (669). The smallest is that between the Dutch and Danes (9). Listed below are the distances ($\times$ 10,000) of all the outliers from their nearest geographic and genetic neighbors.

Lapps: Finns (210), Swedes (333), Norwegians (317)
Sardinians: Italians (221), Greeks (190)
Basques: French (93), Spaniards (104), Italians (141)
Icelanders: Norwegians (74), Irish (99)
Greeks: Italians (77), Austrians (86), Yugoslavs (213)
Yugoslavs: Belgians (50), Czechs (101), Austrians (110), Italians (115), Greeks (213)
Finns: Belgians (63), Austrians (77), Germans (77), Swedes (82), Norwegians (94), Russians (153)

Yugoslavs are lowest in the number of genes sampled (62), with all others having from 75 to 106. This may partly explain their high similarities to nonneighbors, but these similarities are also observed with Finns (96 genes). Likewise, Greeks show greater similarity to Austrians (and to other central Europeans) than to their nearest neighbors, the Yugoslavs. Europe, especially at the center, has low genetic heterogeneity, and distances of peripheral countries from any central European one therefore tend to be similar. Like Greeks and Yugoslavs, Finns show unexpected similarities to central Europeans, greater than those to more immediate neighbors. Many of the distance differences are not significant, and sampling error is responsible for most of these superficially paradoxical results. On the other hand, the similarities of Basques to French and Spaniards, of Greeks to Italians, and of Icelanders to Norwegians are clear, even though all these populations are different enough from the rest of Europe that they are outliers in the tree.

5.6.a. LAPPS AND OTHER URALIC-LANGUAGE SPEAKERS

The origin of the Lapps is obscure. They have been in the same area for over 2000 years, according to archaeological information (Encyclopaedia Britannica 1974). They are found in the northern area of four countries,

Norway (20,000), Sweden (8000), Finland (2500), and the USSR (1900). They call themselves *Saame,* and they speak eight dialects, clustered in three main groups that are mutually unintelligible languages, all belonging to the Uralic family (Hajdee 1975). Unfortunately, it was not possible to separate the groups according to dialects, but only according to the country in which they live. There is clearly considerable variation between samples, as might be expected of a highly fragmented population living in small groups. In the mountains, Lapps are (and were) nomadic herders of reindeer; in the forest, hunters of wild reindeer and freshwater fishermen; on the coast, hunters and fishermen. Today there has been considerable acculturation. Physically they show marked individual differences and vary from a dark to a blonde phenotype.

A genetic analysis has shown that they are an admixture of Mongoloid and Caucasoid people, in approximately equal proportions (47.5% and 52.5%, respectively; standard error, ±4.9%). The ancestral types used were Samoyeds, Komi, and Mari for the Uralic group, and eastern and southeastern Europeans from Poland to northeastern Italy, chosen in order to avoid populations that might have had earlier admixture with Finno-Ugric populations (Guglielmino-Matessi et al. 1990, using Wijsmann's (1984) method of calculating admixture). A second estimate was derived by using the formulas indicated in chapter 1 and the F_{ST} distances given in chapter 2. Here Samoyeds are restricted to people east of the Urals. F_{ST} distance values ($\times$ 10,000) are:

between Samoyeds and Danes	828.5
between Samoyeds and Lapps	857.3
between Danes and Lapps	334.3

Norwegians and Swedes were not used because they may have had admixture with Lapps in the more distant past. The calculated proportion is 82% European and 18% Samoyed. Unfortunately, little is known genetically of western Uralic speakers, who are closer geographically and perhaps might be more satisfactory as an ancestral type. The estimate by Guglielmino-Matessi et al. 1990 incorporated the little genetic knowledge available on them and used a more general European type. Perhaps the 18% estimate sets a lower limit to the admixture proportion, with an upper limit of 47%.

The simplest hypothesis is that Lapps were originally Mongoloids from western Siberia or northern Russia (where some Lapps are still located) who moved west toward Scandinavia in areas where reindeer were common. This animal's optimal habitat was probably at a lower latitude in earlier millennia. The Neolithic came late to Scandinavia (3000 B.C. in the south) because most of the peninsula was still covered by ice or was too cold for the growth of cereals. It is not known where or when contact was first made between Lapps and Scandina-

vian farmers; perhaps in central Scandinavia or southern Finland, or even farther south if Lapps (or more precisely, reindeer) ever reached that far. If that is true, then Lapps must have slowly retreated to the extreme north of Scandinavia where they now live, under pressure from Scandinavian pioneers heading north, but in the process they must have received important gene flow. This might have been reciprocal, but the dominant direction must have been from Caucasoid to Lapp. There is historical evidence of contact for more than 1300 years: Lapps paid tribute to Norwegian chieftains in A.D. 700. In 1500 years, with a gene flow of 1.1% per generation, the final admixture rate expected is 50%; with 2.6%, it is 80%. The time of contact may have been longer, but of course gene flow need not have been so regular, and other scenarios could be hypothesized for the admixture. In the course of the process, Lapps retained their original language, although it certainly underwent profound changes and borrowed heavily from Scandinavian. The estimate of time since genetic admixture is not informative, being inordinately high. The whole problem certainly deserves further examination.

Of the other two major European populations that speak a Uralic language, the Finns have a proportion of 90% European genes and 10% Uralic with a standard error of ±4.1% (Guglielmino-Matessi et al. 1990). An analysis of Hungarians gave similar results. Hungary was occupied by Magyar invaders, herders who spoke a Uralic language and lived on the steppes west of the lower Don in the ninth century A.D. From 100,000 to 500,000 of them, organized in seven major hordes, crossed the Carpathians in A.D. 862 and occupied the territory of modern Hungary. They formed a fraction of 10% to 50% of the total population after the invasion. Subsequent gene flow from neighbors is likely to have decreased this initial proportion if it was higher than 10%. The current estimate of admixture is 13% (±2.3%).

5.6.b. Sardinia

The second major outlier is Sardinia, an island having currently over 1.5 million people in an area of 24,000 km^2, located 200 km from both the Italian coast and the North African coast. It is geographically close to Corsica, but the two populations have had profoundly different histories and are ethnically different. The first inhabitants of Sardinia were pre-Neolithic; a date for human habitation of 9120 ± 380 B.P. in the north-central part of the island (Spoor and Sondaar 1986) is also the earliest date for any island in the Mediterranean. The most important and characteristic archaeological finds are the *nuraghi,* stone houses (sometimes fortresses) of peculiar shape, not dissimilar from buildings found on some other islands or coastal areas in the Mediterranean, from the Baleares to Greece. The nuraghi were built between

1500 and 400 B.C. There were 6000–7000 of them on the island, with a characteristic geographic distribution in which they were always in sight of one another. Some larger nuraghi were centers of villages of 300–500 people. If the nuraghi were all occupied at the same time, the population may have been over 200,000 people.

Around 800 B.C., the Phoenician colonization of the coasts started, especially in the south, and was later continued by a Phoenician colony in Tunisia (Carthage). With the increase of Roman power in the Mediterranean and the defeat of Carthage, Sardinia was occupied by the Romans, who were not, however, able to ever fully conquer the interior. Invasions by Vandals, Byzantines, and Saracens followed each other at the fall of the Roman Empire, with Arab domination lasting until the tenth century. The island then fell under the control of Pisa, followed by the joint forces of Catalonia and Aragon, and finally the Piedmont kingdom. The last three masters left colonies in restricted areas of the island, of which there remain linguistic imprints in three enclaves: a Catalan-speaking area in the northwest, colonists from Piedmont and Liguria still speaking their dialects in the southwest, and Tuscan in the northeast. All settlers in the last two millennia have, however, left limited genetic traces.

By far the most important cause of the considerable genetic differences between Sardinians and any other population known in Europe or Africa is drift, because many genes deviate grossly from regional average gene frequencies. Sardinia has the lowest frequency of RH-negative genes (20%) in the whole Mediterranean, the highest frequency in the world of the *MNS*M* gene (78%), and a unique frequency of the Diaphorase-2 (*DIA2*) gene not included in our analysis. *HLA* and *GM* are also anomalous; the world's highest frequency of *HLAB*18* is present in Sardinia as are a few typical African alleles, although the general distribution is far from that of Africans (Piazza et al. 1985). Thalassemia variants have peculiar frequencies: one molecular variant (β^{39}), has the highest frequency in Sardinia, and is rare in most other places.

An earlier analysis (Piazza et al. 1975) showed that, in the Mediterranean, Sardinians were most similar to Italians, Lebanese, and North Africans (Greeks were not considered). Using the present, larger sample of genes, distances ($\times$ 10,000; with standard errors) of Sardinia from other relevant populations are:

Greeks	190 ± 32
Italians	221 ± 55
Basques	261 ± 67
Lebanese	340 ± 66
North Africans	732 ± 168

Given the magnitude of the standard errors, one can exclude only the North Africans as important contributors. Italy and Greece are likely to have been the sources of the first Neolithic occupants. Neolithics were origi-

nally from the Middle East and Turkey; their genotype, however, was probably already somewhat diluted by possible gene flow from local Mesolithics in the passage through Greece and southern Italy. The modern Lebanese are the most direct descendants of Phoenicians, who also contributed to the Sardinian gene pool. As discussed in Piazza et al. 1988a, there are some similarities between Sardinia and the Caucasus, a very important area, that is unfortunately not very well known genetically. Thus, remote genetic origins still appear recognizable in spite of conspicuous drift that must have occurred at an early time.

The extent of drift expected to have occurred in Sardinia can be approximately estimated by considering that, in the late Paleolithic, the Sardinian population may have been made up of 700–1800 individuals at saturation, assuming a population density equal to that of the aboriginal inhabitants of Tasmania or of England (see sec. 5.2). This small census size would have generated considerable drift over the millennia, and it is unnecessary to invoke a very small number of founders, which is not, however, entirely excluded. Neolithics must have added new genes, but, given the distance from the coasts, they may have been relatively few, adding perhaps new founder drift. Not much is known about the early Neolithic in Sardinia (Lilliu 1983); a clearer demographic picture is available for the later, nuragic population, the density of which was enough to freeze further drift and maintain gene frequencies not far from present levels. Later settlements were limited to the coastal areas, except for those of Phoenician and Carthaginian origins, which were numerous in all the south, but especially in the southwest of the island (Barreca 1986).

Sardinia has been the object of studies on linguistic differentiation and the distribution of surnames. Numerous dialects are spoken on the island; although they are Romance, several words seem to point to a pre-Indo-European substrate of unknown origin (Wagner 1941). Latin is the first Indo-European language known to be spoken in Sardinia. If this statement remains unchallenged either the first Neolithic immigrants to Sardinia were too few to impose their language or they did not speak a language belonging to the Indo-European family. Some dialects are historically known, and late additions are easily identified. The linguistic data base used for this analysis (Contini et al. 1989; Piazza et al., in press) consists of 195 meanings similar but not identical to the Swadesh list, studied in 38 sites distributed over the island. These were grouped into 21 areas. The similarity of areas was studied by trees and principal components. Moreover, surnames in Sardinia have been studied from three sources: consanguineous marriages in the last 150 years, over 30,000 marriages (Zei et al. 1983; Wijsman et al. 1984; Zei et al. 1986); telephone lists, over 180,000 households (Piazza et al. 1987); and electricity bills, over 400,000 households (Zei, pers. comm.).

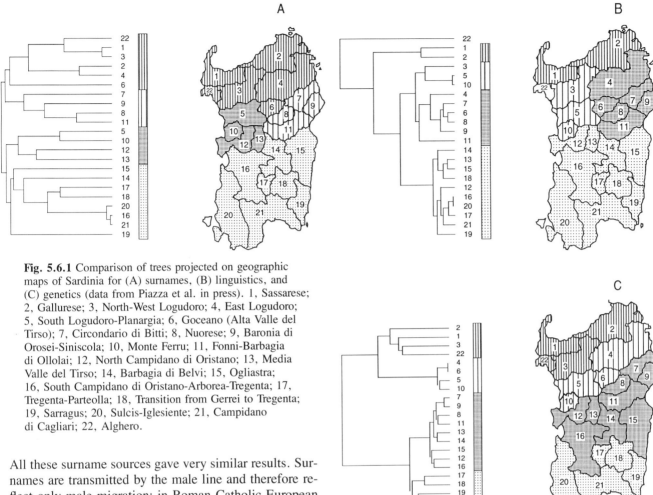

Fig. 5.6.1 Comparison of trees projected on geographic maps of Sardinia for (A) surnames, (B) linguistics, and (C) genetics (data from Piazza et al. in press). 1, Sassarese; 2, Gallurese; 3, North-West Logudoro; 4, East Logudoro; 5, South Logudoro-Planargia; 6, Goceano (Alta Valle del Tirso); 7, Circondario di Bitti; 8, Nuorese; 9, Baronia di Orosei-Siniscola; 10, Monte Ferru; 11, Fonni-Barbagia di Ollolai; 12, North Campidano di Oristano; 13, Media Valle del Tirso; 14, Barbagia di Belvi; 15, Ogliastra; 16, South Campidano di Oristano-Arborea-Tregenta; 17, Tregenta-Parteolla; 18, Transition from Gerrei to Tregenta; 19, Sarragus; 20, Sulcis-Iglesiente; 21, Campidano di Cagliari; 22, Alghero.

All these surname sources gave very similar results. Surnames are transmitted by the male line and therefore reflect only male migration; in Roman Catholic European countries, surnames started officially and on a rather general basis in the late sixteenth century, but they usually represent (1) fathers' names (first names were also transmitted in the family in practically all European traditions) or (2) professions, also transmitted most frequently in the family (Zei, pers. comm.). They are also, more rarely, toponymics (place names, probably specifying early origin) or nicknames, which are of more random nature with respect to genealogy. They all clearly originate with the language, though sometimes they can be carried over from earlier languages spoken in the area. Most surnames are thus older than their official beginning. Sardinia has an environment more favorable to a pastoral than to an agricultural economy and has had very little immigration from the outside. Surname data were treated statistically with procedures similar to those used for genetic data. They give results of considerable precision because of the statistical mass of data covering almost the entire population.

The tree analyses of surname, linguistic, and genetic data are presented also in the form of geographical maps (figs. 5.6.1A, B, C). The genetic data, although abundant, are not as exhaustively representative of the whole island as the surname and linguistic data: they are pre-sented in figure 5.6.1C. One can note a north-to-south gradient in all representations. The darkest area in the geographic presentation of the genetic tree corresponds with the area of highest concentration of pre-Latin place names (Contini et al. 1989). The correlations between linguistic, surname, genetic, and geographic distances are given in the following table:

	Surnames	Genetics	Geography
Language	0.89	0.68	0.76
Surnames		0.66	0.65
Genetics			0.64

These are not Pearson correlation coefficients but similar "congruence" coefficients calculated after the nonlinear transformation by Schönemann and Carrol (1970) has been applied to make the four distance matrices comparable. The Mantel 1967 test modified by Sokal and Wartenberg (1983) gives high significance ($P < 0.001$) for all 6 coefficients.

The correlation is highest between language and surnames, as they might have essentially the same time depth. It is somewhat lower, but still rather high, varying between 0.64 and 0.68, between genetics and all the other variables. Genetics refers to a more distant past and is also less well known than the other three.

Very recently, di Rienzo and Wilson 1991 have tested part of the control region of mtDNA on a number of Sardinians and people from the Middle East. The antiquity of Sardinians is stated to be compatible with a date of settlement from southern Europe between 9000 and 6000 years ago. An analysis of the shape of distributions of distances between all possible pairs of individuals tested for a given population is in agreement with the idea that Sardinians and Middle Easterners (and similarly, Japanese and American Indians of the North Pacific) have been subjected to important demographic growth, whereas Pygmies and Khoisans have not. This is in reasonable agreement with the different demographic patterns of farming populations and of foragers. Important numerical increase was true of farming populations for the last several millennia, and also to some extent of American Indians of the North Pacific, who are fishermen in very affluent areas. By contrast, hunter-gatherers like Pygmies and Khoisans must have been almost stationary demographically for a very long period.

5.6.c. Basques

Basques are today confined to a fairly narrow zone in the western Pyrenees, from southwestern France to northeastern Spain, where the Basque language is still spoken. Basque speakers are few in France (30,000–90,000 according to estimates) and are found only in the extreme southwest corner, although, in the past, their territory was much larger (see sec. 5.8); today they are much more numerous in Spain (see sec. 5.9). It is believed that Basques were in the present area before the arrival of Neolithics some 6000–7000 years ago (Bosch-Gimpera 1943; Cavalli-Sforza 1988; Piazza et al. 1988a) and spoke a proto-Basque language, a pre-Indo-European relic (Collins 1986). If so, they may be direct descendants of upper Paleolithics of the Cro-Magnon type (Cavalli-Sforza 1988; Piazza et al. 1988a). The ancient Basque region is rich in caves with excellent paintings, the best known of which are Lascaux in southwestern France and Altamira in northeastern Spain. The Basque language is of unknown origin and is one of the few language isolates that cannot be classified in any of the known phyla. Some remote resemblances have been found, however, with North Caucasian languages (see sec. 5.3). It has also been claimed that there is some similarity with an extinct language, Iberian, spoken in the Iberian peninsula in pre-Roman times, but Iberian is too poorly known for any clear statement to be possible.

Basques are not completely isolated from their neighbors, and their identity as Basques is mostly linguistic and cultural, but the barriers thus generated are sufficient to maintain a certain level of endogamy. Even so, we have seen that their similarity to neighbors, in particular French and Spaniards, is not small; although their distance from other Europeans is greater than the average inter-European distance, they are not as segregated genetically or geographically as the first two major outliers considered above. One of them, Lapps, lives in an isolated environment where they survived thanks to a very specific cultural adaptation, whereas the other, Sardinians, are isolated on the most remote of the Mediterranean islands. Basques show some genetic similarity with Sardinia (distance 0.0261) and West Asia (distance 0.0247), especially with the area of the Caucasus (Piazza et al. 1988a), as the synthetic maps also show. It would seem as if the linguistic origin is reflected in part by the genetic origin and that the Near Eastern similarity of Basques has antecedents predating the Neolithic expansion from the Near East. Further discussion of Basques is found in sections 5.8 and 5.9.

5.6.d. Iceland

The other isolate that appears as an outlier of Europe is Iceland. This island was settled beginning in A.D. 874, according to tradition, by western Norwegians from the region between Trondheim and Bergen, and the settlement was completed in a relatively short time. Moreover, Icelandic cattle, which were certainly imported by the first settlers on their long boats, are genetically Norse (Kidd and Cavalli-Sforza 1974). From the ninth century until the twelfth century, the Vikings settled widely on the coasts of northern Europe, including the northern coasts of Scotland and Ireland, and other islands in the North Sea; a lively exchange was certainly established among these Viking colonies. The first settlers of Iceland numbered about 20,000 according to tradition (the Landnahmabok), but after the first two centuries there were no further major settlements. The island soon fell under Danish rule, but there probably was very little additional immigration, if any. It is only recently that Iceland became independent of Denmark. The population has multiplied by a factor of about 10 since the beginning, and the language is still more similar to old Scandinavian than to other Scandinavian languages.

In earlier genetic analyses (Bjarnason et al. 1973), it became clear that, for blood-group ABO, Icelanders were more similar to Scots and Irish than to Norwegians; the hypothesis was proposed that the population of Iceland was mostly of Scottish and Irish origin, perhaps because the original settlers made many slave raids in the British Isles. There is also one toponym in Iceland, Papar, which suggests that Irish monks may have come to the island before the Norwegian settlers. Even if this is true, it is unlikely that a few Irish monks could have been demographically as effective as required by the Irish-Icelandic similarity for the *ABO* data. It has become increasingly clear, however, that *ABO* data are not necessarily as reliable as one would like them to be for assessing ethnic origins. They are known to be subject to natural selection for some infectious diseases, many examples of which are summarized by Mourant et al. (1983). *ABO* was extremely popular for a long time for human evolutionary studies, being the first polymorphic gene discovered and having considerable clinical importance. It is still the gene for which the most data exist. When a few other loci were added to the analysis, the conclusions did not change dramatically. However, when a larger number of markers were tested and data from Ireland or Scotland were taken from provinces that were less frequently

settled by other Scandinavians (Wijsman 1984), western Norwegians showed the greatest similarity to Icelanders.

In our data, the population closest to Icelanders is that of Norwegians (distance, 0.0074) followed by the English (0.0076), Belgians (0.0079), and Danes (0.0088). The distance from the Irish is 0.0099, from Scots 0.0112, and from central Europe, a somewhat higher average. The location in the PC map of figure 5.5.2 may seem closer to the British Isles than to Scandinavia, but PC maps give only approximate indications of the distance from nearest neighbors. The standard errors attached to the distances show that the information is not adequate for solving the problem with complete confidence, but at least the data are in the expected direction and in agreement with the traditional information that Icelanders came principally from Norway.

5.7. ITALY

Italy is fairly well known genetically, thanks to the support given to human population genetics by the first three Italian professors of genetics (C. Barigozzi, A. Buzzati-Traverso, and G. Montalenti), even though they were not personally involved in research in this field. Piazza et al. (1988b) published an analysis by principal components of 34 gene frequencies (from the loci *ABO, MNS, KELL, RH, HP,* and 24 *HLA* alleles), including synthetic maps of the first three PCs. The number of locations for which data were available varied from 31 for *HLA* to 319 for *ABO.* The ordinary PC map in two dimensions (not given here) shows that Sardinia is remote from all the other 13 Italian regions represented. Accordingly, Sardinia was not included in PC analyses for building synthetic maps, for it would have used up most of the information of the first PC. Northern Italy shows similarities with countries of central Europe, whereas central and southern Italy are more similar to Greece and other Mediterranean countries. Averages for the whole country are intermediate, putting Italy as a subcluster somewhat peripherally in the major European cluster, together with Spain and Portugal and separate from Greece, which is an outlier.

Figures 5.7.1–5.7.3 reproduce synthetic maps of the first three PCs of Italy, which account for 27%, 18%, and 14%, respectively, of the total variation. The first synthetic map shows a clear gradient from north to south. One pole of the first axis is in the extreme south, in the eastern part of Sicily and the southern part of Calabria, which are separated by the narrow strait of Messina. The opposite pole includes all the north and center. Between Rome—which is centrally located on the peninsula—and the south, there is a progressive gradient of the PC. This corresponds to the well-known differences in physical type (especially pigmentation and general size) between

northern and north-central Italians on one side and southern Italians on the other (Livi 1896–1905).

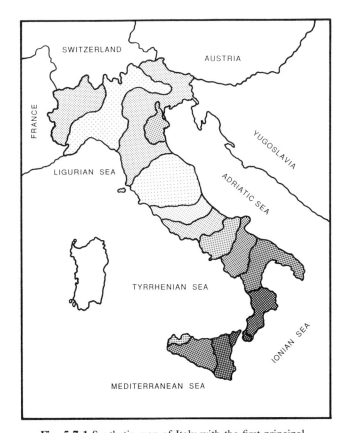

Fig. 5.7.1 Synthetic map of Italy with the first principal component, accounting for 27% of the total variation. The eight classes of proportionally denser shading represent intervals of PC values from lowest to highest (low and high are arbitrary and could be interchanged). Sardinia is excluded; otherwise it would entirely account for the first PC (from Piazza et al. 1988b).

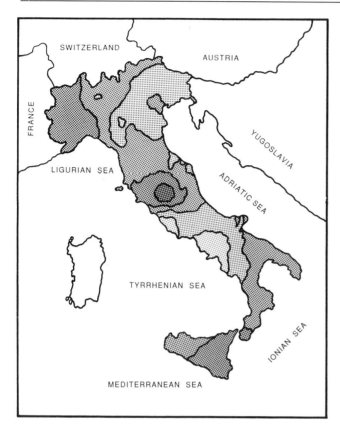

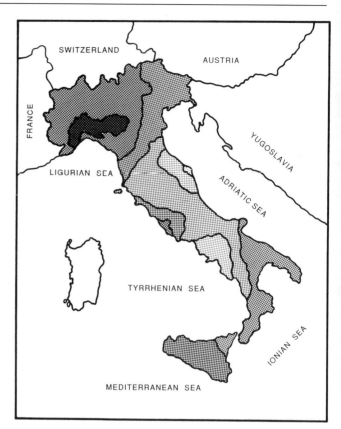

Fig. 5.7.2 Synthetic map of Italy showing the second principal component, which accounts for 18% of the total variation. The five classes of proportionally denser shading represent intervals of PC values from lowest to highest (low and high are arbitrary and could be interchanged). Sardinia is excluded; otherwise it would entirely account for the first PC (from Piazza et al. 1988b).

Fig. 5.7.3 Synthetic map of Italy showing the third principal component, which accounts for 14% of the total variation. The six classes of proportionally denser shading represent intervals of PC values from lowest to highest (low and high are arbitrary and could be interchanged). Sardinia is excluded; otherwise it would entirely account for the first PC (from Piazza et al. 1988b).

Northern Italians are more similar to central Europeans, whereas southern Italians are closer to other Mediterranean people, being darker and smaller. The extreme western portion of Sicily, however, differs from the eastern part. In particular, the northwestern corner, not far from the city of Palermo, is genetically closer to northern Italy. The historical origin of this anomaly can probably be traced to the influx of Norman people in this area; strong artistic evidence of their presence is still visible in much of the architecture of Palermo. Norman conquerors settled here in the early eleventh century and the existence of a nonnegligible proportion of Sicilians with light pigmentation of the skin, hair, and eyes is well known and is generally attributed to the Norman invasion. There is, however, also a prehistoric difference between western and eastern Sicilians. The first inhabitants (Sicani) were pushed west by a wave of immigrants from the peninsula (Siculi). This happened in pre-Greek times (1000 B.C. or earlier), and one cannot exclude the possibility that these earlier inhabitants also contributed to the west-east difference.

Sicily and southern Italy were heavily colonized by Greeks beginning in the eighth to ninth century B.C. (see

fig. 5.2.5). The demographic development of the Greek colonies in southern Italy was remarkable, and in classical times this region was called *Magna Graecia* (Great Greece) because it probably surpassed in numbers the Greek population of the motherland. The Greek language was spoken until the twelfth or thirteenth century A.D. and a proportion of surnames in this area (up to more than 8% in a small region) are still Greek (see fig. 5.7.4), with a geographic distribution very similar to that of the first PC. An area of eight to nine villages in the middle of the heel of Italy still speaks a Greek dialect today, but these might be people of Byzantine origin whose ancestors settled there in the fifth or sixth century A.D. The eastern Roman Empire (Byzantium) held much of southern Italy after the fall of the western Roman Empire, and Greek was its official language. This Greek-speaking enclave shows little or no genetic difference from its neighbors (Modiano et al. 1965).

Much of southern Italy, including Sicily, fell under Arab domination between the seventh and the eleventh centuries A.D. The genetic contribution of Arabs is difficult to assess. In Sicily, the toponomastic pattern of village and town names of Arab origin is spread ho-

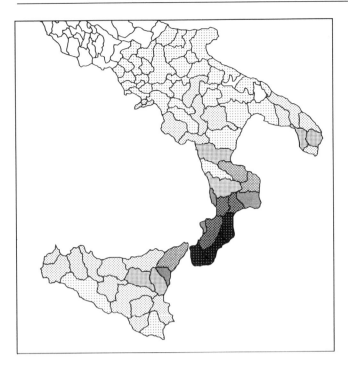

```
< 1
1.0 - 1.9
2.0 - 2.9
3.0 - 3.9
4.0 - 4.9
5.0 - 5.9
6.0 - 6.9
7.0 - 7.9
8.0 or more
```

Fig. 5.7.4 Distribution (%) of Greek surnames in southern Italy (G. Zei, unpubl. data).

mogeneously over the island (Barrai, unpubl.). Interestingly, this finding came about by comparing the names of the Sicilian "comuni" with Arab town names, using a computer program normally employed for detecting homology in DNA sequences.

There is a slightly darker area in the northeast of Italy in the province of Ferrara, near the Adriatic coast, indicative of possible southern admixture. This area was colonized by Etruscans and later, in Byzantine times, by Greeks.

The second synthetic map shows some dark and light areas of interest. The most striking dark area is in a region north of Rome (in southern Tuscany and northern Latium) that is almost exactly congruent with the most ancient area formed by Etruscan cities, beginning before 800 B.C. Before the Romans, the Etruscans were the most developed civilization in Italy. They spread first to northern Tuscany, including Florence (the name Tuscany comes from the old name of the Etruscans, Tusci). Then they extended both north to northern Italy, and south to Rome. According to legend, they came from northwestern Turkey (specifically from the city of Troy). The language is not Indo-European (Pallottino 1978). Among the many hypotheses on its origin, it has been

suggested that, like Basque, it is most closely related to North Caucasian (Gamkrelidze and Ivanov 1990).

One may wonder why and how a people known only through its culture and language—both of which disappeared two thousand years ago, having been obliterated by the development of Rome—may still leave a trace on the genetic map. If the local population of southern Tuscany had a strong demographic development at some early time (in the case of Etruscans, the beginnings of the Iron Age around 3000 years ago), and if later migration from the outside were limited, then the local gene pool would be reasonably resistant to later modification. If any genetic differences with neighbors were present at the beginning, they might thus resist cancellation and persist for a long time; the greater the initial genetic difference, the greater the resistance and persistence. This might indicate therefore that Etruscans were colonists of external origin, but it is difficult to exclude the possibility that they originated from an autochthonous population that diverged genetically from its neighbors because of initial isolation and drift.

The intensity of one PC is usually insufficient to distinguish all possible populations. Thus, Etruscans have about the same intensity of the second PC as people from extreme northwestern Italy, but, on the basis of historical considerations, it is unlikely that there is any special genetic kinship between the two. Etruscans are also relatively similar, for the second PC, to some southern Italians. Only the combination of different PCs can help distinguish these different populations. The color map (Piazza et al. 1988b) superimposing all principal components shows clear differences among the three.

The negative pole of the second map (that is, the light areas) is found around the regions of Rome and Naples. It is probably due to the strong immigration toward these two cities that took place over several periods during the last 2000 years in the case of Rome, and over the past 500 years in the case of Naples. The immigration came from a wide hinterland partly common to both cities and therefore generated some similarities between them.

The third synthetic map shows a dark area corresponding to the core of the region inhabited in pre-Roman times by an ancient, perhaps pre-Indo-European population, the Ligurians. Although the region of Liguria is now limited to the coast south of the Appennine mountain range, it is precisely to such mountainous areas that vanquished populations tend to retreat, because defense is easier. In such areas, they are also least likely to mix with other people and are therefore of special genetic interest. The pole opposite that of the Ligurians, in the third PC, corresponds to another important pre-Roman Italic population of the early Iron Age, the Picenes. The pole is located roughly at the center of the region they occupied, on the coast of the upper Adriatic. They traded with Greece but very little with Etruria. Their number was estimated at 250,000 in the third century B.C.

(Beloch 1886). Linguistically, they were part of an eastern group of Italic languages, Osco-Umbro-Sabellic.

Superimposing the three synthetic maps, printed with different colors (green for the first PC, blue for the second, and red for the third), allows us to distinguish clearly peaks and troughs that are very similar for only one component. In the color map, for instance, Etruscans are pale green, and easily distinguished from southern Italians (dark purple) and from northwestern Italians (brown), although all these populations have a similar intensity for the second PC. Picenes are also more easily distinguishable from neighbors, being white. Rome and Naples are clearly different. As already noted, Sardinia was not included in the analysis.

The Italian population is one of the most dense in Europe and has been so for several millennia. At the time of the Roman Empire it was estimated to be around 6–7 million, about 10% of the total population of the empire. In the Middle Ages, and probably as early as the third century, socioeconomic conditions began to deteriorate. The wars between Byzantines and Goths, and later with Langobards, caused a population decrease, down to a minimum in the seventh century of about 50% of the maximum at the time of the Roman Empire. Numbers started to grow again (McEvedy and Jones 1978) and were 10 million by A.D. 1300. After the serious losses caused by the epidemic of plague known as the Black Death in A.D. 1340, they were down to 7 million and returned to 10 million in A.D. 1500. C. Cipolla (pers. comm.) gave values about 3 million higher with the same trend.

Italian history and later prehistory are reasonably well known for almost 3000 years, especially in the central and southern regions, and genetic data are more complete than in most other countries. It thus becomes possible to interpret the vagaries of the synthetic maps on the basis of plausible historical facts. Naturally, the interpretations are based almost exclusively on the correspondence of the geographic areas showing anomalies in the synthetic map with historically known local developments. Such correlations can hardly be taken as proofs, but frequently the coincidences are sufficiently striking that it would be difficult to ignore them. Whenever other approaches are available, they should be used for corroborating or falsifying the suggestions coming from such correspondences.

5.8. FRANCE

In France, especially in its southwestern corner, the upper Paleolithic period is especially well known and has the further distinction of having produced the most important collection of rock paintings and cave art in the world. The contiguous region of northeastern Spain is also rich in upper Paleolithic art, created at a similar time and ordinarily believed to be closely related. The prototype of modern humans comes from Cro-Magnon, geographically fairly close to the most famous painted cave, Lascaux. The earliest dates of modern humans in this part of Europe (based on artifacts) are around 33 kya, and there are carvings going back to the early period; but the most important artistic contributions are dated to the Magdalenian period, 15–11 kya (Klein 1989a).

Languages related to modern Basque were most probably spoken in this area. Even the name of a southwestern French region, Gascogne, seems related. Today there are only a limited number of French speakers of the Basque language because of strong governmental campaigns in favor of the national language. But toponymy gives evidence that the Basque language was once more widely spread over southwestern France. Figure 5.8.1 shows the location of place names in France and Spain presumed to be Basque on the basis of their endings. It also shows the boundaries of high *ABO*O* blood-group gene frequency. The local high frequency of *RH*-negatives, which gave rise to the belief that Basques might be a proto-European population (Mourant 1954), is more clearly visible in Spanish Basque country. However, no single gene distribution is ever sufficient to give a completely satisfactory picture. We discussed the Basque problem in section 5.6 and further discussion is in section 5.9; suffice it to say here that there is support from many sides to the hypothesis that Basques are a relic upper Paleolithic population. Gene flow from people who arrived later in the area, starting with the Neolithic, did not completely cancel the genetic identity of the group. Despite considerable archaeological research in south-western France, there are no estimates of the population size of the region inhabited by Cro-Magnon people in the Paleolithic, but it might have been comparable to that estimated for England in the Mesolithic (sec. 5.2), that is, perhaps 10,000–20,000 people.

Neolithic immigrants to southern France most probably came via the Mediterranean, whereas those to the northeast and to the Parisian region came from Germany (the Bandkeramik culture, sec. 5.2). This immigration pattern may have established an initial basis for the genetic differentiation of northern and southern France. In the Bronze Age, important megalithic constructions were made, especially between the northwestern corner (Brittany) of France and the center. In the Iron Age, at the time of La Tène (500 B.C.), the majority of the French spoke Celtic languages, as did most people in central Europe and northern Italy; between the fourth and second centuries B.C., Gauls made several raids into Roman territory. It is impossible to say whether the Celtic language was brought by a small group of conquerors or

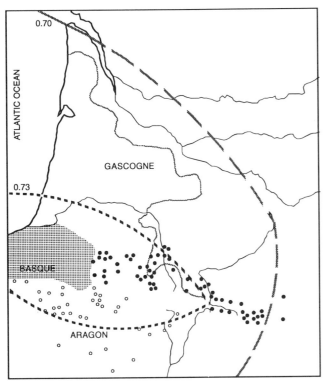

Place names ending in -os, -osse, -ous, -ost, -oz
Place names ending in -ues, -ueste

×ᵐᵐᵐ⁻ 70% limit of blood-type O population
‐ ‐ ‐ ‐ 73% limit of blood-type O population

Fig. 5.8.1 Place names of probable Basque origin in southwestern France and northeastern Spain and blood-group O distribution (Bernard and Ruffié 1976).

a full invasion of Celtic people from central Europe or further East, nor do we know if Celtic dialects were already spoken in Gaul before the Iron Age (see sec. 5.3). Except for the southwestern corner, most of France was under Celtic domination by the end of the third century B.C.

Historical data on Gaul began to be abundant with Julius Caesar's report of his conquest (58–50 B.C.). The southeastern corner (called the Roman Province, hence the modern name of Provence) was conquered by the Romans before Caesar. A wide region including the southwestern corner received the name of Aquitania. Basque was spoken here according to Roman archaeological documents (Allières 1979). The name of Aquitania was later extended north to include most of the wide region between the Garonne and the Loire, the northern part of which was mostly Celtic. North of these two regions, Gaul was divided by the Romans into four administrative zones and included Belgium, French Switzerland, and neighboring parts of Germany. At least 74 tribes, almost all Celtic, can be approximately placed on the map of Gaul (Fierro-Domenech 1986). Tribes lived in areas demarcated by uncleared forest tracts that acted as buffer zones. In Roman times, the population numbered about 5 million people, a high density for that time (though roughly 10 times less than now). The Roman language (Latin) fully replaced the Celtic dialects, of which only a few words survive in languages spoken today. Modern French evolved during the Middle Ages from Latin in this area, as summarized afterward in this section.

Later genetic contributions were probably scarce. German contributions include some German soldiers in the Roman army who were authorized to settle in France as farmers or as auxiliary soldiers. Visigoths were allowed by Rome to settle in Aquitania in A.D. 410, but later moved to Spain and left only a small genetic contribution. Burgundians, originally from the southeastern Baltic, settled in eastern France, but were crushed by the Huns. They gave the name to Burgundy, but survivors settled in Savoy. Another Germanic people, the Alemanni, settled in Alsatia and Switzerland and gave their language to these regions. Yet another Germanic tribe, the Franks, moved from the lower Rhine to northeastern France, which was more sparsely settled at the time. The Franks are also believed to have contributed relatively little genetically to the Parisian area, which was then already heavily settled by Gallo-Romans, but they developed efficient political control of the region. They also imposed the local dialect on most of northern France and finally on the entire country, except for small peripheral enclaves. Altogether the German genetic contribution is estimated to have involved 300,000–350,000 people, or 6%–7% of the population of Gaul (Fierro-Domenech 1986). However, Celtic people were themselves a mixture of central Europeans, and the high homogeneity of central Europe and northern Europe was probably already established, at least in great part, by late Neolithic times.

A Celtic language was reintroduced in the northwest (Brittany) by Celts from the British Isles fleeing Anglo-Saxon invaders in the fifth and sixth centuries A.D. Another important invasion was that of Norwegian and Danish Vikings who settled in Normandy in the ninth century, where some northern Germans had already been resettled by the Romans in the fourth century A.D. In this region, anthroposcopic characters showing a northern contribution (blond hair, blue eyes, pink skin) are somewhat more frequent than in the northeast, where the northern genetic influence was otherwise higher than in the rest of France. From this period until very recent times, there was no further important external genetic contribution to France.

French scientists (Ohayon and Cambon-Thomsen, in press) have recently examined a sample of rural people from 14 French provinces and Corsica. These researchers made a systematic study of a total of 135 alleles—*HLA-A,B,C,DR, Bf; C4A, C4B, GLO-1* on chromosome 6, in addition to *ABO, RH, MNS, P, KELL, GC, HP, PI, TF, GM, KM* genetic loci—that provides a basis for the geographic study of French genes. Part of the genetic analysis was carried out by one of us (Piazza; preliminary results in Piazza 1986). The French provinces, unlike the modern administrative subdivision of "départements" which originated with the French Revolution, are a traditional classification that has changed considerably over time but reflects approximate cultural areas. Basque samples were not included. Unfortunately, for a vast territory like France, 14 points do not permit a high resolution. An important gain in precision is due to the use of the same genes in all locations and the choice of rural populations known to have been historically stable, which guarantees the virtual exclusion of recent immigrants. The use of samples from cities, which is the rule almost everywhere else, causes considerable blurring of the more ancient ethnic dissimilarities.

The analysis by synthetic maps of this material is given in figures 5.8.2–5.8.4. The proportions of variance explained by the first three PCs are 18%, 13%, and 10%.

The first synthetic map shows a north-south gradient. One pole is clearly the northern coast, with an extension inland in the Germanic-speaking regions, and the other in the Basque-speaking region. It contrasts areas of northern Europeans with those of southern Europeans. The bulk

of the northern pole may be the Neolithic component, which entered France from Germany and the Netherlands, with later additions from other northern people (Vikings in Normandy). The southwestern pole includes both the upper Paleolithic component, peaking in the extreme southwest, and the Neolithic component originating in the Mediterranean. The peopling by Neolithics, superimposed in the southwest on the preexisting local descendants of the Cro-Magnon, still forms the most important genetic substrate of the French population.

The second synthetic map shows a gradient with a maximum in the southwest and a regular decrease in the other directions. Ordinarily we tend to interpret this type of pattern as the result of an expansion from an origin placed at either the maximum or the minimum. In the absence of demographic data, it is difficult to choose, but even if the two interpretations are to some extent opposite, they agree in contrasting the center of the upper Paleolithic in the southwest with the Neolithic periphery.

If this is correct, then both the first and second PCs express different aspects of the Neolithic advance: the first PC emphasizes the difference between northern and southern European Neolithic people, whereas the second stresses the differences between the upper Paleolithic

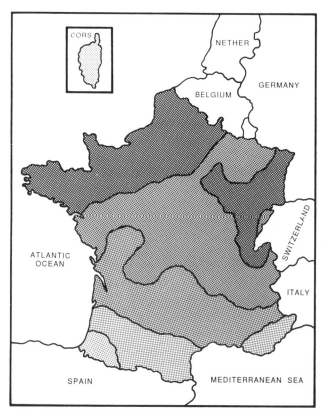

Fig. 5.8.2 Synthetic map of France according to the first principal component, which accounts for 18% of the total variation. The five classes of proportionally denser shading represent intervals of PC values from lowest to highest (low and high are arbitrary and could be interchanged). Cors, Corsica; Nether, Netherlands.

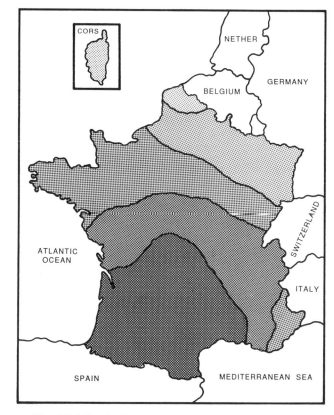

Fig. 5.8.3 Synthetic map of France according to the second principal component, which accounts for 13% of the total variation. The five classes of proportionally denser shading represent intervals of PC values from lowest to highest (low and high are arbitrary and could be interchanged). Cors, Corsica; Nether, Netherlands.

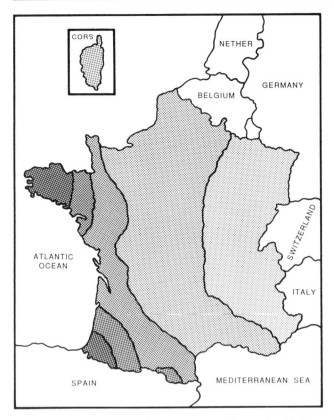

Fig. 5.8.4 Synthetic map of France according to the third principal component, which accounts for 10% of the total variation. The five classes of proportionally denser shading represent intervals of PC values from lowest to highest (low and high are arbitrary and could be interchanged). Cors, Corsica; Nether, Netherlands.

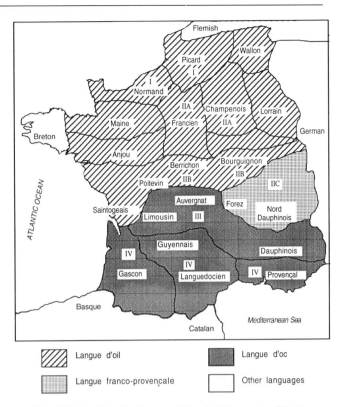

Fig. 5.8.5 A historical map of French dialects (modified from Bonnaud 1981). I, IIA, IIB, IIC, III, IV denote areas of gradual diffusion of the original French.

dominating the southwest and the basic Neolithic pattern, common to both northern and southern Neolithics. Since late migrations into France have been modest, the original ethnic picture may still be determined by the most ancient and most important settlements. One may wonder why the more ancient difference between Paleolithic and Neolithic people is genetically less striking than that between two streams of Neolithic people, as indicated by the fact that the earlier one is expressed by a PC less informative than the later one. The relative importance of PCs is measured by their eigenvalues, which also express the percentage of variance explained by each PC. As shown by simulations (Rendine et al. 1986), the eigenvalues are determined by demographic parameters, like population densities of the various settlers, rates of migration and of admixture, as well as the genetic distances between the various people contributing settlers to the area. Their relative importance does not necessarily reflect the time sequence of settlements.

The third synthetic map shows an east-to-west gradient, with one pole peaking primarily in Brittany (the extreme northwest) and secondarily in the southwest, while the opposite pole forms a light band at the eastern

boundary. It must express a general migration gradient from the east to the west.

The color map (not shown) gives a clear picture of the genetic heterogeneity of France. For instance, the northern band, which appears as a single region in the first PC, is clearly subdivided into at least four zones, from west to east: Brittany, probably colored by an important Celtic settlement coming from the British Isles; Normandy, settled by Scandinavians; a predominantly Frankish region in the extreme north, and, in the extreme northeast, a predominantly Germanic area.

The central zone, a mountainous region (Massif Central), is fairly homogeneous and distinct from the rest. Naturally, the southwest is markedly different from the rest of France, and the Mediterranean coast shows similarity between the western and eastern parts, with some difference from the central part.

The linguistic map of France is well known (Bonnaud 1981), and a geographic-historical reconstruction is shown in figure 5.8.5. The French linguistic community corresponds only approximately to the political territory of France, extending into parts of Belgium, Switzerland, Italy, and the central Pyrenees. Linguistic minorities in political France are the Bretons, the Basques, and Catalans in a small region of the south.

The French language started developing in the early Middle Ages from Latin, first in the Frankish area (northeast), and further evolved as it gradually expanded to the rest of France. In the Middle Ages, there already was a clear distinction between a group of dialects spoken in the north ("langue d'oil") and that in the south ("langue d'oc"), "oil" and "oc" being the equivalent of "yes" in these two groups. The demarcation line between the two is shown in figure 5.8.5, along with major dialects. An east-central area today speaks a surviving "franco-provençal" dialect. Various phases of the Frenchification of the language, and some major modern dialects, are also shown in the figure. One cannot expect a perfect correspondence between genes and languages, but there are some obvious correlations. We will not insist further on the Basque genetic substrate. The exclusion of typically Basque samples from the genetic analysis makes the correlation all the more remarkable. Although Celtic spoken in Brittany was reintroduced from the British Isles, and is therefore not original, the isolation of this part of the country certainly points to a parallel development of genetic and linguistic divergence in this part of France with respect to the rest. The north-south difference is prominent both linguistically and genetically. There is thus some obvious correlation between the genetic and the linguistic mosaics of France. The physical barriers and the lines of communication are common to genes and languages in all times, but inevitably later developments make the correlation less than perfect.

Language is a major expression and determinant, as well as a vehicle of culture. Other cultural phenomena are, however, transmitted over generations like languages. Some show a potentially even greater conservation over time than languages. A study of the geographic distribution of cultural traits in Africa (Guglielmino-Matessi et al. 1987), on the basis of data from the ethnographic atlas by Murdock (1967), has shown significant correlations of differences for some cultural traits with linguistic differences. Traits related to family and kinship, in particular, have the highest correlations with languages. This was interpreted as the consequence of a high conservation of these cultural traits, which tend to reflect the history of population fissions and fusions. However, genetic divergence is also expected to reflect the same history; hence one might expect some correlations between these cultural traits and genes, in the same way as correlations are found between genes and languages. Cultural traits learned early in life are (or were) usually acquired in the family. Thus, they are transmitted similarly to genes and may easily correlate with them (Cavalli-Sforza and Feldman 1981; see also discussion later in this section).

Confusion between causation and correlation from common cause is frequent but also sufficiently widely known that it does not deserve further emphasis. It is interesting to note in France an example which has *not* been misinterpreted. Investigations by a demographer and an anthropologist, Le Bras and Todd (1981) have generated interesting results making use of the wealth of demographic knowledge available in France. A study of data such as age at marriage, joint residence of elders with younger relatives, institutionalization of the old, longevity, and other information related to family structure and customs showed considerable geographic variation in France, and strong correlations with traditional family structures, in particular those described in the classical work by the nineteenth century French sociologist Le Play (1875). Le Play distinguished three types of families.

1. In the strictly nuclear family the children who marry basically rescind their bonds with the parental family and must be financially independent. Later historical work, incidentally, has shown that this makes them free to take residence where job opportunities are best and thus favors industrial development (Laslett 1983).

2. In the extended family of patriarchal type and strictly authoritarian customs, the patriarch is the absolute master of all family decisions.

3. In the extended family of nonauthoritarian type, bonds within the family remain strong, and both the young and the old find support. Here, people can marry young, being supported by the family and often reside with it after marriage; the old are not institutionalized but live with the family and are more likely to die at a later age.

This presentation is inevitably condensed and incomplete, but the important conclusion is that the geographic distribution of the three types of families—the first in northeast France, the second in the northwest, and the third in the southwest—corresponds well with the demographic observations even in modern times.

An interesting finding by Le Bras and Todd is that political opinions are also highly correlated with family structure. For instance, the vote for parties in favor of a liberal economy is highest in the northeast (where the nuclear family prevails); the northwest (in which the authoritarian extended family is common) has a traditional leaning toward absolute power; and the socialist (not the communist) vote comes mostly from the southwest of France (where the benevolent extended family is prevalent). An explanation for this pattern is that political opinions are created in the family, in the sense that the family is a microcosm orienting the young toward a view of the world that is later extended to the macrocosm of social and political life.

Le Bras and Todd have hypothesized that family structures are old, antedating the barbarian invasions. A more cautious attitude, however, is advocated by historians (see, for example, Braudel 1986). We would like to add some theoretical considerations explaining why family structures are likely to be very old, that is, highly conserved. Basic laws of cultural transmission support the Le Bras and Todd hypothesis. In fact, family behavior must be taught by the example of parents, grandparents, and other relatives, a closely knit social circle, especially in extended families. This type of cultural transmission has been demonstrated theoretically to be the most conservative in existence, because it sums the effects of the two more highly conservative mechanisms of cultural transmission: parent-offspring (vertical) transmission and pressure by a social group (Cavalli-Sforza and Feldman 1981; Cavalli-Sforza et al. 1982). Another reinforcing factor is that the family teaching takes place when children are very young and especially, often irreversibly, susceptible. Hence, it is not surprising that cultural transmission of family structures is highly conservative, and they may last for thousands of years, outlasting even languages.

Family structures are so highly conserved, in fact, that their inheritance shows a marked resemblance with the outcome of genetic transmission, which is the most conservative transmission of all. The geographic distribution of the socialist vote, shown by Le Bras and Todd, is similar to our first synthetic map of French genes, describing the difference between Paleolithic and Neolithic populations. It so happens that if one calculated the correlation between the socialist vote in France and, say, the frequency of blood-group *O,* and perhaps even *RH*-negative, one would probably find it very significant. This correlation reflects centuries of common history of genes and certain cultural traits rather than a very unlikely effect of *ABO* and *RH* genes on French family attitudes and on politics.

5.9. IBERIAN PENINSULA

A significant portion of the work published on Iberian gene frequencies appeared in journals not easily available outside of the Iberian peninsula and could not be included in the data base assembled for this book. A full set of gene frequencies published before the end of 1988 was collected by J. Bertranpetit and examined by the method of synthetic maps. The following is a summary of the paper by Bertranpetit and Cavalli-Sforza (1991), in which specific references can be found.

Genetic data for 18 loci (*A1A2BO, MN, RH, K, FY, KEL, LE, HP, PI, GC, TF, C3, ACP1, ADA, PGD, GLO1, AK1, PGM1*) and 34 independent alleles were available for a number of locations varying from 12 to 171 per locus (31.7 average). The first three components accounted for 17%, 14% and 12% of the variation. The synthetic map based on the first PC is given in figure 5.9.1. It

clearly shows that the major genetic influence on Iberian genes is still its Paleolithic component, the Basques. The first isopleth (from the North) corresponds almost exactly to today's Basque region, which still speaks the language. Toponymy and other evidence show that the Basque-speaking country was once much wider, corresponding approximately to the third isopleth. It may have extended even farther, since there are Basque place names in the Ebro River valley in areas where Basque was never spoken in historical times; in the eastern Pyrenees, a Basque dialect was spoken until the Middle Ages.

Rock paintings in caves show that at least cultural influences in the late Paleolithic and Mesolithic extended westward along the north Atlantic coast, being approximately contained within the third isopleth, and have similarities with the cave art of southwestern France. Paleolithic art is seldom accurately dated, but in southern France it is 30 kya old and more, going back to the Magdalenian and even the Solutrean, and continues into the last Paleolithic period, the Azilian.

The archaeological tool kit in the Basque region shows a well-defined Azilian culture; there are local peculiarities in the Mesolithic as well, for example, a microlaminar industry in silex. This has been a continuous local development since the Magdalenian, with a certain degree of success: Mesolithics were certainly not starving, but probably slowly growing in numbers and developing indigenous adaptations. It is clear that in this area hunting-gathering (and fishing along the coast) remained a reasonable alternative to farming for a longer time and may have delayed the full acceptance of agriculture.

The modern morphology of Basques is also somewhat unique; for example, they show a relatively high frequency of increased vertical measurements of the mandible and other traits, which are also found in the older skulls in the region. Although these are not accurately dated, they may go back to the late Paleolithic.

Everything, including genetics, points to a persistence of a pre-Neolithic type in the Basque region that was

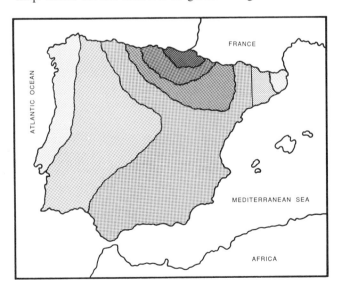

Fig. 5.9.1 Synthetic map of the Iberian peninsula based on the first principal component of gene frequencies, which accounts for 27% of the total variation. The six classes of proportionally denser shading represent intervals of PC values from lowest to highest (low and high are arbitrary and could be interchanged) (from Bertranpetit and Cavalli-Sforza 1991).

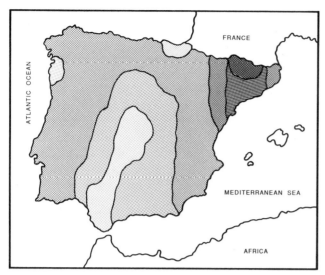

Fig. 5.9.2 Synthetic map of the Iberian peninsula based on the second principal component of gene frequencies, which accounts for 14% of the total variation. The six classes of proportionally denser shading represent intervals of PC values from lowest to highest (low and high are arbitrary and could be interchanged) (from Bertranpetit and Cavalli-Sforza 1991).

once more widespread but is still sufficiently important to determine the first principal component of the region. Undoubtedly, the long conservation of the language must have helped considerably to preserve a local genetic type from external influences by keeping endogamy high, but the local population density must also have been sufficiently high to discourage foreign settlements or to keep them within narrow limits. Basques have always been, and still are, very conscious of their identity and are likely to have efficiently opposed penetration of their territory. Some interesting anecdotal evidence: the powerful army of Charlemagne suffered a serious defeat in A.D. 788 at Roncesvalles, the most important pass in the western Pyrenees, and the best and most celebrated Frankish warriors fell in defense of the rear guard. The defeat is attributed by legend to the Saracens, and a great number of poems were inspired by it; but it is more likely that it was due to ambush and attack by Basque shepherds. Crossing the Pyrenees again 34 years later, one of Charlemagne's successors avoided trouble by taking hostage a number of local women and children during the passage.

The second component (fig. 5.9.2) shows a peak in the northeast of the peninsula, in the region known as Catalonia, and spreads toward the Mediterranean coast, covering almost all of it. The opposite, lowest density is found in the Basque region and in the center. It also seems to extend to the western and the northwestern Atlantic coast.

There is a linguistic correlate, but it is unlikely to be directly relevant. The three main Romance languages of Latin origin spoken in the Iberian peninsula are Galician in the northwest, from which the Portuguese language took its origin; Castilian, originally from the north-central area, from which the Spanish language took its origin; and Catalan, in the northeast. They all came from the north and spread to the south with the successful fight against the Saracens, who settled in Spain in the seventh to eighth centuries A.D. and were eventually chased from the country in 1492. The center of origin of Catalan may superficially seem to correspond with the peak of the second component, but it is actually not in the Pyrenees. In addition, the lowest values of the second PC do not correspond to the areas where the Moorish influence was strongest, which are in the south and east.

The best explanation of the second PC map is the spread of the Neolithic, which must have always involved population growth and is therefore a natural candidate. Its areas coincide well with those of the initially strongest concentration, which also corresponds with a classical area of passage across the Pyrenees. If there was also, as is likely, a major expansion strictly along the coast, it would be more difficult to expect this area to be the darkest one today, since this is an intensely urbanized area and must have attracted immigration from long distance. Also, the dark areas on the Atlantic fringe were heavily settled in Neolithic times and the local, morphologically very distinctive, Mesolithic population is different from the modern one. The latest arrival in the Neolithic was in the central part, which is lightest in the PC geographic map, along with the Basque region, which was probably less receptive to agriculture.

It is of interest that in the Iberian peninsula, the relative demographic importance of the Paleolithic and the Neolithic in determining the genetic map is inverted when compared with that of determining the genetic map of all Europe. In fact, in this region the genetic component bound to the Paleolithic population takes on greater importance than that associated with the Neolithic. In the Iberian peninsula, the Neolithic arrived later than in other parts of Europe, and the Mesolithic lasted longer and was highly developed because it was a more efficient alternative to farming than in other parts of Europe. It is therefore not surprising that the contributions of Mesolithic hunter-gatherers, at both the cultural and the genetic level, may have been more important than those of Neolithic farmers, in agreement with archaeological observations (references in Bertranpetit and Cavalli-Sforza 1991).

The third component (fig. 5.9.3) shows the contrast between the Mediterranean and the Atlantic. This is a well-known theme in Iberian archaeology and history. The time at which this dichotomy finds its fullest expression is in the first millennium B.C., when the Iron Age develops. This culture arrived from the continent through the western and eastern Pyrenees, as already mentioned. At the time, different languages were spoken on the Atlantic and the Mediterranean sides of the peninsula: Celtic (probably imported by Iron Age peo-

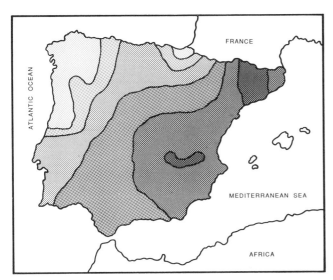

Fig. 5.9.3 Synthetic map of the Iberian peninsula based on the third principal component of gene frequencies, which accounts for 12% of the total variation. The six classes of proportionally denser shading represent intervals of PC values from lowest to highest (low and high are arbitrary and could be interchanged) (from Bertranpetit and Cavalli-Sforza 1991).

ple) on the Atlantic side, and Iberian (a language of non-Indo-European origin) on the Mediterranean side. There are signs of interaction in the area of contact. During this period, population growth was slow, although it was slightly faster between 400 and 200 B.C., approximately doubling the population during the millennium. Growth was somewhat more pronounced in the Mediterranean moiety. There is no information from bones because at this time the dead were cremated. However, considering the modest population increase in this period, the principal component is probably not recording an expansion but just the difference between local populations that were already in situ, aided probably by the

addition of foreign elements, perhaps some immigrants from continental Europe accompanying or carrying the new iron cultures but, more probably, colonizers from the sea. Those historically known are Phoenicians, as well as Punics from the Phoenician colony of Carthage near modern Tunis, who founded many colonies on the Mediterranean coast and extending beyond Gilbraltar, beginning in the eighth to ninth centuries B.C. or earlier. Greeks also founded several colonies in the same general area, but only one of these, Emporyon (in the northeast), has been identified. Greeks and Phoenicians fought for control of the Mediterranean coast of Spain, and the Greeks lost in the sixth century B.C. Eventually, Rome conquered the Phoenician colonies and the rest of Spain.

The fourth synthetic map (not shown) gives peaks in the northwest and southwest and troughs in the west and east; in the fifth component, the same regions (excluding the eastern one) reappear, but in a different combination, the northwestern region is now opposite the western and southwestern ones. There are archaeological developments in these areas that could be associated with the high- or low-density PC areas; for instance the PC pattern observed could be caused by two independent colonizations or strong developments in different times of locally differentiated populations, one in the northwest and west and the other in the west and southwest. The southwestern area, around the mouth of the Guadalquivir River, is now known as Andalusia. In the last millennium B.C. it was the seat of a flourishing civilization called Tartessian after the main city, Tartessus (Tarshish in the Bible) destroyed in 500 B.C. The location of the city has not been identified, but it was an important link in the tin trade with the Greeks. This important metal was mined in Cornwall and was necessary for making the copper alloy known as bronze.

5.10. SINGLE-GENE MAPS

The *ABO* locus in Europe has the highest density of observations, and we have therefore divided the representation of the original data into two maps, one indicating the locations of gene frequencies from which maps were calculated, and the other giving local heterogeneities and discrepancies from the calculated map. We believe the density of available data from the USSR is underestimated, because we have not recorded data from publications in the Russian language, but it is still quite adequate. The *O* gene shows relatively modest variation, with a lower frequency at the central and eastern longitudes. The *A* gene is low in the northern Soviet Union, in the British Isles, in the northern Mediterranean, and in the southern Middle East; it has more heterogeneities than *O*. Gene *B* shows the classic west-to-east gradient

that has traditionally been interpreted on the basis of invasions from the east, where *B* is high. The same trend has been observed by Sokal et al. (1989b). The major subtypes of *A*, *A1* and *A2*, show more interesting patterns, but are based on far fewer data than the classical antigens of more direct interest for transfusion. *A1* has a major peak in central Spain, whereas *A2* shows a gradient from northern Scandinavia to the south. It would seem natural to suspect a correlation with climate; this hypothesis would have to explain exceptions such as high values of *A2* in Israel (Jews excluded) and in the northern British Isles. One cannot rule out the possibility that there were more than one *A2* allele mutations, one of which had a northern origin and spread almost concentrically. All the variograms show a long initial linear increase that often continues for more than 2000 miles, with the exception of *A2*, which has a more pronounced maximum at about 1500 miles. The slope of initial increase is large for *B*, indicating the steepness of the east-west gradient, and is four times larger than that of *A*, three times that of *A1* or *A2*, and twice that of *O*.

The *B* allele of acid phosphatase (*ACP1*) shows an almost concentric gradient from the Middle East, with an unexpected peak in the Basque region. A rare allele, *C*, is high in the northeast. Since, as already noted, acid phosphatase has one of the highest correlations with climate, the geographic distribution in Europe of both alleles may in part follow this general trend. Adenosine deaminase (*ADA*) allele 1 shows a trend that could be due to correlation with climate or to a gradient centered in the Middle East. Variograms are linear, especially for *ACP1*B*.

Adenylate kinase (*AK1*) is, like *ADA*, poorly known, especially in the East. The pattern of allele *1* seems of modest interest. The variogram is regularly linear.

The *M* allele of α-1 antitrypsin, *PI*, has a high frequency in most of the northern Mediterranean. The variogram is linear for 1000 miles.

A β lipoprotein (*AG*X*) is very poorly tested, especially in the western part, but seems to show a strong gradient from the east and north, responsible for the peak of the variogram after 1000 miles. Another β lipoprotein, *LPA*Lp(a+)*, has a central maximum, and the variogram has a clear negative slope.

Cholinesterase 1 *CHE1*U* has a moderate west-east gradient, whereas *CHE2(+)* has a modest north-south gradient. Both have very limited polymorphism, so it is not surprising that the variograms are flat.

A complement protein, *C3*S*, has an interesting gradient from the northeast toward the southwest.

Esterase D *ESD*1* has a definite north-south gradient, in which Lapps do not participate. The variogram has a slow initial slope, then peaks suddenly, very much like that of C3.

The Duffy blood group *FY*A* has a clear east-to-west gradient. Unfortunately, the frequencies of the recessive allele *FY*O* are difficult to estimate when rare, as is true for Europe. Our finding differs from that of Sokal et al. (1989b). They classify the trend of this allele as showing only local patches. The variogram is practically linear for over 2000 miles.

The deficiency of glucose-6-phosphate dehydrogenase (*G6PD*def*) is highest in the east, especially in the Middle East, but shows a peak in Sardinia. The variogram is fairly flat and irregular.

Glutamate-pyruvate transaminase (*GPT*) is a soluble liver enzyme, whose allele 1 shows a fairly regular gradient from the Middle East toward northwestern Europe, which is similar to the gradient shown by another enzyme, *G6PD*. The variogram has a small positive initial slope.

Glycine-rich β glycoprotein (*BF*), also called properdin factor B or C3 proactivator) is linked to *HLA*. Allele *F1* has a peak in the Basque region and Sardinia, echoed by *F* at least for the maximum in the Basques; it also has a minimum in the extreme north. *S* has the opposite pattern, not surprisingly because these two are the more frequent alleles; *S07*, has a very flat distribution. This locus is likely to be important in immunity, but we cannot say if it will help to explain the geographic distribution of the alleles. All variograms increase reasonably linearly.

Glyoxalase 1 (*GLO1*) does not show any regular pattern in its allele 1, merely various peaks in different parts of Europe. The variogram however increases regularly.

The vitamin-D-binding protein, *GC*, has a low *GC*1* frequency in southern Russia and also in the Basque region, with a peak among Lapps. *GC*1F* has a regular west-to-east gradient. *GC* in Europe has little correlation with climate and shows regular variograms.

Haptoglobin (*HP*) has a clear maximum for *HP*1* just north of the Black Sea, and one for the *HP*1S* subtype in southern Scandinavia. Although the variogram of *HP*1* is regular, with a slow initial increase, that of *HP*1S* has a negative initial slope. The subtypes *1S* and *1F* are based on relatively few data.

As usual, the *HLA* system provides the most satisfactory geographic distribution of samples, especially for the oldest known and more frequent alleles. It therefore generates very reliable and informative maps. Four patterns are most easily distinguished on single genes. Some are uncomplicated, and others have added complications caused by the abnormal behavior of the major outliers like Lapps, Sardinians, and Basques. The most frequent patterns show a concentric gradient of either increase or decrease, with a center in the Middle East expanding into Europe. Let us call P_1 the pattern with a maximum in the Middle East and P_2 that with a minimum in the Middle East. They have already been noted in the genes discussed so far, but they are usually clearer in the *HLA* maps.

Alleles with an absolute or relative maximum in the Middle East (including Turkey), are (pattern P_1): *A*9*, *A*19*, *A*28*, *A*31*, *A*33*, *B*5*, *B*14*, *B*21*, *B*35*. Of these, four do not show important maxima other than in the Middle East; others show important secondary, or even primary maxima elsewhere (*A*9* in Lappland; *A*19* among Basques; *A*28* in the north Baltic; *A*31* (which is a subtype of *A*19*) among Basques; *B*5* in northern Greece; *B*14* in Sardinia and southern Spain, but not in the Basque region which has a relative minimum).

Alleles showing a relative minimum in the Middle East and increasing in approximately concentric rings toward Europe are (pattern P_2): *A*1*, *A*31*, *B*8*, *B*12* (*A*1* and *B*8*, typically European, are in tight linkage disequilibrium).

Clearly both patterns strongly indicate the spread of genes from the Middle East toward Europe, accompanying the expansion of farmers as we discuss in the following sections. Alleles higher in the Middle East are those that were more common among Middle Eastern Neolithic farmers than in European Paleolithics; those higher at the periphery of the expansion were more frequent in Paleolithic Europe and were partially replaced by alleles coming from the Middle East. The latter are fewer than the first, probably because Paleolithic Europe had undergone severe population bottlenecks during the last glaciation and had reduced heterozygosity. *HLA* alleles are relatively infrequent, being very numerous, and are more susceptible than alleles at biallelic loci to partial elimination by drift. We therefore expect, for them, more alleles going from the Middle East into Europe than the opposite.

The third pattern, P_3, implies a more or less regular increase or decrease from north to south and may well represent an adaptive response to climate. It is found in a somewhat smaller number of alleles: $A*2$, $A*3$, $A*30$, $B*5$, $B*7$, $B*15$, $B*27$, $B*40$, $B*41$ ($A*3$ and $B*7$, typically European, are in tight linkage disequilibrium).

The fourth pattern shows a maximum or minimum in slightly different locations above the Black Sea (pattern P_4): $A*10$, $A*11$, $A*25$ (which is a subtype of $A*10$), $A*32$, $B*13$, $B*16$, $B*18$.

Other patterns can be observed but not as clearly. Naturally, our recognition of a pattern on single-gene maps is subjective, but patterns that are found repeatedly by this procedure reappear more clearly in synthetic maps. Moreover they are also found with other, non-*HLA* genes, as seen before and as we confirm on the rest of the maps.

Variograms of 15 *HLAA* and 17 *HLAB* alleles showed positive slopes with only one exception ($B*41$), which was also the allele with the lowest average frequency. The linear portions may be somewhat shorter than for other genes, ranging from 500 to 1800 miles, with a median of 800. Initial slopes varied from $-.06$ to 3.5×10^{-5} per mile, with a median value of 0.5 and no differences between A and B.

The polymorphism of the constant heavy chain of G immunoglobulins (*IGHG1G3* or *GM*) shows three major alleles in Europe: $f;b0b1b3b4b5$, $za;g$, and $zax;g$. The first is the most common and has a clear maximum in Rumania, with an approximately concentric decreasing gradient. This may represent the area of origin of this allele, which is typically European, but it may also indicate an area of maximum frequency of an infectious disease to which this allele may have offered protection. Although $za;g$ seems to spread east-to-west, $zax;g$ has a concentric gradient centered in the Middle East, where it has its minimum. $za;b0stb3b5$ is a not so rare allele also centered in the Middle East, and $za;b0b1b3b4b5$ shows a similar behavior. A rarer allele, $fa;b0b1b3b4b5$, has a maximum in Spain. $f;b0b1b3b4b5$, $za;g$, $zax;g$, and $za;b0stb3b5$ all show variograms with extended initial linear portions

with a high slope; $fa;b0b1b3b4b5$ and $za;b$ have flat initial portions.

The κ light-chain immunoglobulin $KM*(1,1\&2)$ shows a predominantly east-west gradient, with a peak in Lappland; the variogram decreases at 1500 miles because northern and eastern sides of the map show areas with equal values.

The Kidd blood group (*JK*) has a wide region of low $JK*A$ gene frequency; the allele K of the *KEL* blood group has several small peaks in central position; and the Le allele of Lewis, another blood group, has a pattern equivalent to P_2 of *HLA*. $LU*A$ (blood-group Lutheran) has a P_2 pattern with a maximum around the eastern rim of the Black Sea. All these blood groups have regular variograms.

The *MNS* blood-group system was analyzed separately for alleles M, S, and the four haplotypes MS, Ms, NS, Ns. M has a P_2-like pattern complicated by several maxima in the northern, eastern, and Mediterranean regions. S has a complicated pattern with a minimum in southeastern Turkey and relative maxima in other parts of the map. The haplotypes have somewhat flatter distributions: MS has a modest maximum in the Middle East, with minor peaks or troughs elsewhere. Ms has a maximum in Finland and Sardinia and a lower one in northwestern Iran. NS has an imperfect P_2 pattern. Ns has a west-east gradient. Most variograms are regular for 1000 or more miles, but MS is initially very flat.

The *P1* blood-group surface is hilly, with no clear dominant gradient, but its variogram is regular for almost 1000 miles.

The phenylthiocarbamide tasting locus (*PTC*) allele T (taster) has two clear maxima in the eastern Aegean and in the Helsinki/Leningrad region, with a minimum not far from the latter, in northern Russia. The variogram is linear for about 1500 miles with a large initial slope.

Phosphoglucomutase 1 $PGM1*1$ has an approximate P_2 pattern, with an added east-west gradient.

6-phosphogluconate dehydrogenase (*PGD*) allele A is rather flat in most of Europe, with a low frequency among Lapps and in northern Russia in general. The variograms are linear for more than 1000 miles; the first enzyme has a median slope and the second a flatter one.

The *RH*-blood group system has also been analyzed by alleles C, D, E, and by haplotypes. It is almost as intensively studied as *ABO* and is more informative, given the wealth of haplotypes that form it. Allele C has maxima in the Mediterranean and among Lapps. Allele D ($RH+$) shows a P_2 pattern, with complications because there are other maxima of $RH+$ besides the Middle East among Lapps, Sardinians, and two regions of North Africa. The Basque region has the lowest frequency in Europe (and in the world). D^u is highest in Turkey and has a P_1 pattern, though not a pronounced one, given the low frequencies of this type observed in Europe. Allele E has a complicated surface, with several maxima and minima.

Variograms are regular, except for D^u, which is flat in the initial portion.

The RH haplotype *CDE* is relatively rare and has a maximum north of the Black Sea. The same is true, in a slightly more western position, for *CDe*, which is also more frequent in other parts of the world. This haplotype is the most frequent in Europe. *Cde* has a P_1 pattern. *cDE*, also numerically important, has a minimum north of the Black Sea and a maximum in northwestern Iran, with a lesser one in the Leningrad region. An allele very frequent among Africans, *cDe*, has a rough P_1 pattern, but also shows a relative maximum in Poland and minima not only in the Basque region, but also in southern Scandanavia and Iceland. The fully *RH*-negative haplotype, *cde*, has the well-known maximum among Basques, with minima not only in the Middle East (a P_2 pattern), but also among the Lapps and in northwestern Africa.

The comparison of alleles and haplotypes in Europe is interesting. "Allele" frequencies of *RH* are the sums of several haplotype frequencies and show complex surfaces, with the exception of D^u, which is more haplotype specific. Haplotypes give simpler surfaces with fewer maxima and minima, which seem easier to interpret in terms of patterns observed with other genes. If there were selection for specific antigens, however, it would show in the alleles and perhaps not so easily in

the haplotypes; but there is no such indication. All variograms are regular with the exception of *CDE*, which is fairly flat, and has a linear portion extending over 1000 miles.

The allele *Se* of the secretor locus *FUT2(SE)* has a north-south pattern with a few exceptions, for example, Iceland. Allele *C* of *TF* (the protein transferrin) has a fairly flat distribution, with a slightly lower frequency in the northeastern region. Variograms are regular.

In summary, there is a dominant pattern showing a nearly concentric gradient between the Middle East and the rest of Europe that is observed in a positive and a negative form: the positive presenting a maximum, and the negative a minimum, in the Middle East. When discussing *HLA* alleles that show these patterns most clearly, they were named P_1 and P_2, respectively. We interpret both of them, more fully discussed in the next section, as resulting from the expansion of farmers from the Middle East at the beginning of the Neolithic. As also discussed by Sokal et al. (1989b), these are not in any way the only patterns observed. The numerically most important among the residual patterns is a more or less precise north-south gradient. The next important pattern is a concentric gradient peaking north of the Black Sea and/or Caucasus. Other patterns are also found, and it is useful to give a synthesis of all these maps by considering their global pattern in the form of PCs.

5.11. Synthetic maps of Europe

Our work on synthetic geographic maps of genes began in 1977 for the purpose of testing the hypothesis that the spread of agriculture from the Middle East to Europe in Neolithic times was a spread of farmers rather than a spread of the farming culture (that is, without human migration). The test assumed that the modern distribution of genes has not been excessively altered by subsequent events. Analysis by principal components offered the hope of separating different patterns. A later demonstration that this is a realistic expectation was given by a simulation of multiple demic expansions in Europe (Rendine et al. 1986). This simulation also showed that the geographic patterns established in Europe by early migration are detectable by principal components in spite of different, later migrations overlapping with them, and that clines established by extended migrations are very resistant to attrition by successive, continuous local migration, which is always present. The simulation also indicated that it is not always easy to separate minor and later expansions by PC analysis and that artifactual results may occasionally arise, but the number of genes (20) in that simulation was small, for reasons of computer economy. Conditions of the simulation were aimed at imitating the expansion of early farmers by a

very precise model (described also in Ammerman and Cavalli-Sforza 1984) and may have to be changed in order to represent more faithfully other types of expansions, especially those that took place after the Neolithic one.

In order to obtain geographic maps of principal components, it was necessary to know gene frequencies in the same geographic positions for a large number of genes and points in space. The distribution of the location of genetic data given in the maps of this book shows their extreme geographic irregularity, except for *HLA* where collection of the information followed, at least in part, a specific design. The calculation of surfaces of gene frequencies in geographic space was a device for allowing interpolation in two dimensions and providing the necessary estimates of gene frequencies at regularly spaced points in space. This was not necessary for *HLA*. In order to test the validity of our approach, we calculated synthetic maps for *HLA* in two different ways (Menozzi et al. 1978a). First, we calculated maps for each allele and used the lattice of points thus generated to evaluate principal components and then their geographic map. Second, we calculated principal components from the original data and then fitted a geographic map to

Table 5.11.1. Percentage of Total Variance Explained by the First Seven Principal Components of European Gene Frequencies

Principal Component	% of Total Variance	Principal Component	% of Total Variance
1	28.1	5	5.3
2	22.2	6	3.2
3	10.6	7	2.6
4	7.0		

Table 5.11.2. Genes Showing the Highest Correlations with the First Six Principal Components of European Gene Frequencies

P.C.*	Range of Correlation Coefficient		Genes
1	1.00 – 0.90	(+)	ACP1*C, HLAB*21
		(−)	C3*S
	0.90 – 0.80	(+)	GC*1F / (GC*1S + GC*1F), HLAA*33, HLAB*7, PTC*T
		(−)	HLAA*2, HLAB*15, HLAB*8, G6PD*A-
	0.80 – 0.70	(+)	RH*Cde, IGHG1G3*za;b0stb3b5, HLAB*35, HLAB*40, LU*A, P1*1, FUT2(SE)*Se
		(−)	ADA*1, PGD*A, HLAB*16, HLAB*27
2	0.90 – 0.80	(+)	MNS*S
		(−)	AG*X, HLAB*12, MNS*NS
	0.80 – 0.70	(+)	LPA*Lp(a+), MNS*Ms, RH*D, IGHG1G3*fa;b0b1b3b4b5, FY*A, CHE1*U
		(−)	BF*F, IGHG1G3*f;b0b1b3b4b5, HLAA*30, LE*(a+), RH*cDe
	0.70 – 0.60	(+)	IGHG1G3*zax;g, JK*A, HLAA*3
		(−)	HLAA*1, HLAA*31
3	0.90 – 0.80	(+)	MNS*MS, HLAA*25
		(−)	—
	0.80 – 0.70	(+)	HLAA*11, HLAA*32
		(−)	HLAA*10, HLAB*13
	0.70 – 0.60	(+)	BF*S, CHE2*(+), PGM1*1, ABO*B
		(−)	—
4	0.60 – 0.50	(+)	HP*1
		(−)	BF*S0.7, HLAA*29, RH*E
	0.50 – 0.40	(+)	ABO*O, HLAB*17, G6PD*A-
		(−)	ABO*A, ABO*A1, RH*CDe, IGHG1G3*za;g, HLAB*5
	0.40 – 0.30	(+)	ESD*1, FY*A, IGHG1G3*zax;g, HLAA*1, HLAA*19, HLAA*28, HLAA*33, HLAB*14, MNS*M, TF*C
		(−)	ADA*1, AK1*1, CHE1*U, HLAA*25, HLAA*30, HLAA*9, HLAB*35, HLAB*40
5	0.70 – 0.60	(+)	MNS*Ns
		(−)	—
	0.60 – 0.50	(+)	HLAA*31, RH*Dᵘ, TF*C
		(−)	—
	0.50 – 0.40	(+)	HLAB*17, HLAB*41
		(−)	GC*2, IGHG1G3*za;g, HP*1, KEL*K
	0.40 – 0.30	(+)	IGHG1G3*zax;g, HLAA*1
		(−)	HLAB*37, RH*cdE, MNS*S, PI*M, RH*E
6	0.50 – 0.40	(+)	ABO*A, RH*CDe, CHE2*(+), LU*A
		(−)	HLAB*18
	0.40 – 0.30	(+)	HLAA*9
		(−)	HLAB*41

Note.– Genes giving positive or negative correlation values are indicated by (+) or (−), respectively.
* P.C., Principal component.

them. There was excellent agreement between the two approaches. There was also reasonable agreement between synthetic maps obtained for HLA and for non-HLA genes analyzed separately, but inevitably halving the number of genes introduced greater noise.

In the 1978 analysis, we used 38 genes. This calculation has now been repeated using 95 genes (32 are of *HLA* alleles). They are the 94 genes whose maps appear in the second part of the book with the addition of *IGHG1G3*fa;b0b1b3b4b5*, the map of which is not shown because its distribution is almost limited to East Asia. Not all the genes considered are fully independent alleles; thus, *A* is not entirely independent of *A1* and *A2*, but the original data are independent. Principal-component analysis eliminates these internal correlations. We give maps of the first seven principal components, which account for variance proportions totaling 80%, as shown in table 5.11.1.

Sardinian data were excluded from this analysis for reasons already given: their inclusion in Europe would have taken a PC by itself.

The genes that give the highest contributions to these PCs are listed in table 5.11.2.

The first three PCs repeat rather accurately the results already published in 1978, though with increased precision. The first synthetic map (fig. 5.11.1) parallels very closely the map of the times of first arrivals of the Neolithic or, archaeologically, of the cultivated forms of cereals domesticated in the Middle East and not present in Europe before the agricultural expansion (fig. 2.7.2). The PC values are given on an arbitrary scale from 0 to 100. The gene map differs from the archaeological one in at least two extreme areas, in Lappland and in the southernmost part of the Middle East. We believe these discrepancies are the result of events external to the main hypothesis and do not cast doubt on the general interpretation. It is unfortunate that the data from the southern coast of the Mediterranean are poorer and cannot be included in the analysis without serious loss of information.

Sokal and Menozzi (1982) repeated this study calculating correlograms limited to the early *HLA* data used by us, and confirmed the existence of a pattern corresponding to that expected for a migration from the Middle East toward Europe. More recent statistical analyses with improved techniques of space correlation fur-

ther confirmed these conclusions; the Sokal et al. 1991 paper in particular extended them to a total of 26 loci, 59 independent alleles. This analysis included a study of

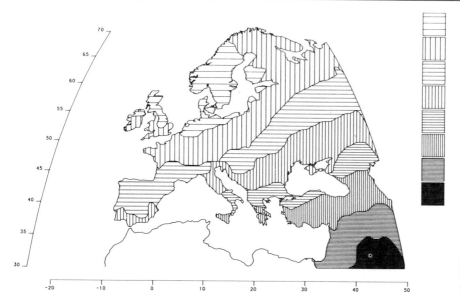

Fig. 5.11.1 Synthetic map of Europe and western Asia obtained using the first principal component.

the correlation with the temporal pattern of migration of Neolithic farmers, which was highly significant for six loci considered individually (*RH, ACP1, PGM1, HLAA, HLAB, TF*). All except the last one were described in section 5.10 as showing P_1 or P_2 patterns visually in agreement with the agricultural expansion. The correlation coefficients between the individual alleles and PC1 shown in table 5.11.2 are in reasonable agreement with the Sokal et al. (1991) statistical analysis.

The second PC map (fig. 5.11.2) shows a concentric gradient like that of the first but centered instead in the Iberian peninsula. The opposite pole of the second PC shows a strong peak among Lapps, which are certainly no candidate for an expansion. There is no known demic expansion from the Iberian peninsula, and an interpretation based on a migration from this area seems unlikely. In general, there is a strong north-to-south gradient that might be interpreted on a climatic or ecological basis; but this interpretation does not take into consideration the

existence of ethnic differences between the populations of northern Scandinavia, like the Lapps and other speakers of Uralic languages who occupied the northeastern areas of Europe, perhaps before the arrival of Neolithics.

The interpretation we gave in Menozzi et al. (1978a) was based on migrations from Asia. Although it would be historically absurd to think in terms of an expansion centered in Lappland, the PC peak in this region may be explained by noting that Lapps have a stronger Mongoloid component than any eastern European population. Thus, the simplest explanation may still be that of Menozzi et al. (1978a), namely, gene flow caused by one or more migrations of Mongoloid Uralic speakers from Northwestern Asia. Migration of steppe nomads or their descendants and the "barbarian" invasions toward the end of the Roman Empire and afterward seem to have no influence on this PC.

The third PC synthetic map of our earlier analysis (Menozzi et al. 1978b) showed a wide peak spread over

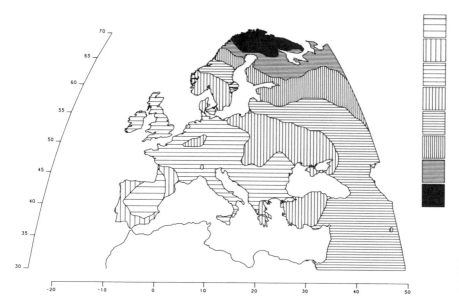

Fig. 5.11.2 Synthetic map of Europe and western Asia obtained using the second principal component.

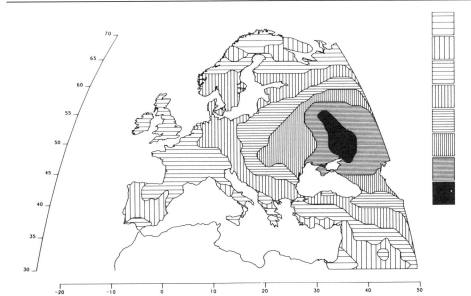

Fig. 5.11.3 Synthetic map of Europe and western Asia obtained using the third principal component.

eastern Germany, Poland, and the Ukraine. This has now been split into at least two components, represented by the third and seventh PC. The third PC (fig. 5.11.3) shows a strong peak in the European steppe north of the eastern part of the Black Sea, with an approximately concentric gradient. The seventh PC shows a peak to the left of that of the third PC, in the western region north of the Black Sea.

An analysis of the statistical robustness of the synthetic maps of Europe (unpublished) indicated that the first five maps are well reproducible under bootstrapping of the 95 genes and the sixth has borderline reliability. Because of its even less satisfactory reliability we do not show the map of the seventh PC. The strongest candidate for an archaeological interpretation of the third (plus, perhaps, the seventh) map is an expansion of the Kurgan culture, which was associated by Gimbutas (1966) with the primary expansion of the Kurgan people. She

describes three separate waves (1991, see sec. 5.2). As we already mentioned, it is possible that this was a secondary expansion of Indo-European speakers. In addition, Scythians were in this area at a much later time and also invaded Europe. Moreover, the same area may have been the original homeland or an intermediate long-term homeland for some other barbarian populations who later invaded central and western Europe. There is no assurance that enough of them remained north of the Black Sea after they moved westward, a necessary condition for finding their genetic traces in our PC maps. It is thus difficult to assign unequivocally the third (or the seventh) PC to a specific migration, but the Gimbutas hypothesis should be given very serious consideration.

Components lower than the third were not tested by Menozzi et al. (1978a) and are also displayed rarely in this book. They are inevitably weaker than the first three, representing a smaller amount of variation. In Europe,

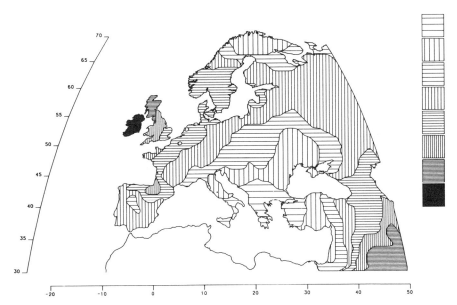

Fig. 5.11.4 Synthetic map of Europe and western Asia obtained using the fourth principal component.

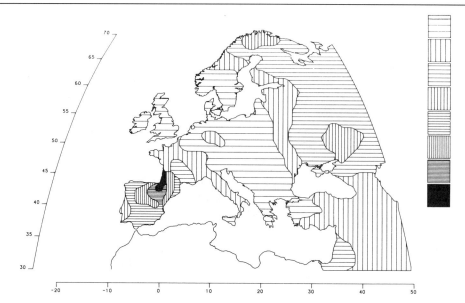

Fig. 5.11.5 Synthetic map of Europe and western Asia obtained using the fifth principal component.

however, the genetic information is more satisfactory than in any other continent, and here we extend our analysis down to the sixth component which shows a reasonable stability under bootstrapping of genes.

The fourth PC synthetic map (fig. 5.11.4) indicates a potential area of expansion centered in Greece. The central area corresponds almost exactly to original Greece and its major colonies, in the Aegean, western Turkey, and southern Italy. A difficulty in interpreting this as a Greek expansion is that the Greek influence in classical times never extended to the large area covered by the two or three rings surrounding Greece, and especially the propagule to the northeast that is seen in the synthetic map. The Aegean region was certainly very active in the second millennium B.C. with cultures of probable Greek origin, and the expansion to southern Italy, Sicily, and the southern and western Mediterranean occupies much of the first millennium B.C. Could the northern area associated in the fourth synthetic map with the Greek expansion and located in the eastern Balkans and western Ukraine represent an origin of "invaders" from the north? This northeastern "propagule," a kind of amoeboid extension of the second ring around Greece, includes regions that were very active in the late Neolithic and Bronze Age. Some archaeological explanation will perhaps be found for this association between the area of the nuclear Greek expansion and its northeastern extension. Although the name "propagule" suggests it was secondary to the development of Greece, one cannot exclude the possibility that the historical sequence was in the opposite direction, with a beginning in the eastern Balkans, a migration to Greece and mixture with local inhabitants, who later took the lead in the expansion and therefore dominate the PC synthetic map (pastoral nomads?).

The fifth PC synthetic map (fig. 5.11.5) has a very wide band of low values across Europe, from the north-west to the southeast. This may represent an old pre-Neolithic relic, since this region corresponds approximately to an area of recolonization of central Europe after deglaciation at 16 kya (Gamble 1986). There are also some similarities with the areas of expansion of cultures in the Bronze and Iron Age, as well as with the area in which Germanic languages were spoken, except for the Balkans. The opposite pole of this PC (indicated as dark) shows an interesting parallel between Basques and a region north of the Caucasus.

The sixth PC synthetic map (fig. 5.11.6) shows a concentric gradient with a maximum in an area located on the eastern coast of the Black Sea, where several North Caucasian languages are spoken. There are similar intensities among Lapps and Basques; interestingly, the fifth component showed a similar phenomenon.

The interpretation of synthetic maps is not always easy. The first question is, how can one hypothesize that a population expansion was involved? If there is a radiation of circular or elliptic clines from a specific area, an expansion is a possible explanation; and its place of origin must clearly be the center of the radiation. The alternative possibility of a centripetal, rather than a centrifugal, population movement may have to be considered. The best way to distinguish them would be on a historical basis, but this evidence is usually lacking. In principle, a centrifugal movement is a priori more likely, but a centripetal migration could be that directed toward an important city, especially a capital.

It is often difficult to link a cline of gene frequencies with a precise historical expansion. Unfortunately, we have only very superficial knowledge of these events, if any. From Herodotus, we know the names of many ethnic groups who migrated to Europe from the east after 500 or 600 B.C. and the approximate areas where they were located before their last movement. We usually have no information on their remote origins or their

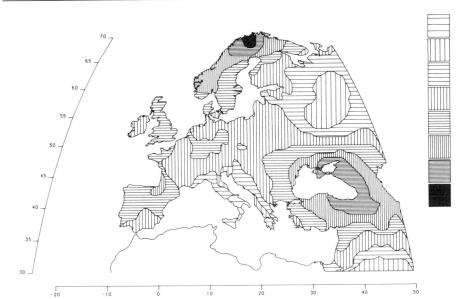

Fig. 5.11.6 Synthetic map of Europe and western Asia obtained using the sixth principal component.

numbers. Most of the groups that came to Europe after the fall of the Roman Empire were probably too small numerically to give a detectable genetic contribution, given that Europe was already fairly densely populated at the time. Consider, for instance, one group already named above: the Scythians, who lived in 500 B.C. on the northern shores of the Black Sea, and were sufficiently important in that their hostile initiatives provoked King Darius of Persia, the most powerful monarch in West Asia, to try and subdue them. He was unsuccessful. The Scythians may have settled relatively late north of the Black Sea, in the area corresponding approximately to the peak of the third PC. They spoke a language, now extinct, of the Iranian subfamily, and their origin is likely to have been from an oasis of the Central Asian steppes, north of the Caspian Sea. They extended also toward the south and the east of Asia, probably having contacts with the Altai, and they also expanded westward to Europe. It is not clear whether their numbers were adequate to create the concentric gradient observed in the third PC. The Sarmatians had an origin similar to that of the Scythians and conquered them; their contribution to the synthetic maps may be undistinguishable from that of the Scythians. Another group, the Goths, were located at some time in their history near the peak of the seventh PC synthetic map. They started applying pressure on the eastern Roman Empire across the Danube in the fourth and fifth centuries A.D., were given land to settle within the boundaries of the Empire, and underwent various resettlements within the Empire in later centuries. Their numbers, even if not precisely known, are likely to have been too small to influence the genetic picture. The same is true of most, if not all, of the barbarian invasions at the end of the Roman Empire, because they settled in very densely populated areas and were therefore genetically diluted. In a few cases, one might be able to recognize their distinct genetic origin,

when they maintained high endogamy, probably an infrequent event. Moreover, small endogamous groups tend to drift away from neighbors and from their original genetic pool, so that their precise origin may be difficult to recognize. Sometimes, however, the remote origin may still be recognizable; see, for example, the Samaritans (Carmelli and Cavalli-Sforza 1979), a very small highly endogamous Middle Eastern group (of the order of 200 people, today) (Bonné-Tamir 1980).

The similarities between Basques and some other populations, in particular in the Caucasus, which appear in two PCs, may be relics of the pre-Neolithic background. An Italian linguist (Trombetti 1923) has first identified a linguistic pre-Indo-European substratum common to Basques and Caucasian populations. Lapps show some relationship in more than one PC synthetic map; this may indicate that part of the Caucasoid background of Lapps is of Paleolithic origin. The mountains of the Caucasus could have been a haven for local aborigines, who might have escaped excessive miscegenation with latecomers. Since mountain populations must have also undergone considerable drift after the Paleolithic, only an intensive genetic study of this area, covering many mountain valleys and many genes, might help trace some of these ancestral relationships more effectively. The Caucasus, the Basque region, other mountains in Europe—including the Sardinian and the Ligurian mountains—and other less accessible and more endogamous parts of Europe and West Asia may still retain some detectable genetic traces of upper Paleolithic Caucasoids. These refuge areas should be a high priority for future investigations.

Improvement in the chances of establishing a correlation between PC geographic maps and archaeological information demands that the relevant archaeological data be subjected to a quantitative analysis not dissimilar to that carried out initially on archaeological material for the spread of the early Neolithic (Ammerman and Cavalli-Sforza 1973, 1984). Even if it is clear that archaeological maps like those of the Kurgan, the Bell Beaker, the Battle Axe, etc. represent mainly cultural diffusions, one cannot exclude a significant demic component, except by rigorous comparisons of accurate archaeological and genetic maps. There may have been a detectable genetic component to some linguistic and cultural diffusions of the past, even if they were determined by elites.

The color picture conveys 60.9% of the original variation and shows five major ethnic zones: blue for Lapps in northern Scandinavia, continuing toward the east to include other Uralic peoples in Finland and northern Russia; dark-red for all Germanic-speaking populations, from Scandinavia to northern Germany to most of England; brown for populations in Celtic areas of the British Isles, excluding Scotland, but including Basque areas in southwestern France and Northern Spain; green for most of the Mediterranean; orange for southern Russia. Very roughly, red and green correspond to northern and Mediterranean regions, respectively, after subtracting the major outliers (Basques and Lapps). Red and green may be the two major streams of Neolithic farmers from the Middle East, one directed to the northwest via the Balkans (but little if anything is left of the red stream in the Balkans); the other directed west through the Mediterranean. One should, of course, remember that the first three components account together for only 60.9% of the total European variation, and the rest is to be found in the lower components. The orange component probably represents the migration from the steppes, in part because of the first pastoral nomads and in part because of their descendants who settled first in southern Russia and then migrated farther west at a later time.

5.12 INTERACTIONS OF GENETIC, ARCHAEOLOGICAL, AND LINGUISTIC INFORMATION

There are many problems in interdigitating genetic, archaeological, and linguistic information. These disciplines supply information on related, but different aspects of the same reality, and data from each of them have different limitations. Genetic knowledge is currently available only for modern populations, and ancient relationships among them must be inferred on the basis of mathematical theories that demand, for being entirely useful, demographic information on population sizes and migratory exchanges. Unfortunately, only guess estimates of the demography of ancient people can be given, in the lack of direct historical information, which is scarce. Archaeological data tell us more about artifacts than about people; they can supply information on certain aspects of demography like population sizes and its dynamics, but are less useful for understanding migration, and in particular for distinguishing between movement of people and of artifacts (which we have often referred to as demic and cultural diffusion). This may explain the oscillations of archaeological thinking between migrationist and indigenist explanations. Languages are very powerful cultural and population markers, the use of which is somewhat limited by the progressive increase of uncertainty as soon as one tries to reconstruct events before writing became available and common. They are also sometimes subject to relatively sudden replacement under conditions that are not entirely predictable. Also genes can be replaced, though with slower kinetics. In spite of these potential limitations, one expects, and usually finds, a strong correlation between genetic and linguistic similarities of different people, but one does observe discrepancies that we are only now learning to predict. Similarly one can find obvious relations of either of them with archaeology and history, or in a more limited way, with paleoanthropology.

The basic principle of genetic data analysis for evolutionary and historical purposes is the need of a great number of genes for robust conclusions. Synthetic maps of gene geography have proved useful because they make full use of this principle, and have proved capable of separating one from the other independent migrations. We have made use of this method for the first time in Europe (Menozzi et al. 1978a), the continent which has the greatest genetic, archaeological, and linguistic information, and therefore lends itself best to this type of analysis. Even so, there inevitably remain uncertainties, tied to knowledge gaps and to the general difficulty of historical investigations, which do not allow experimental tests of hypotheses as easily as in most natural sciences.

The hypothesis put forward by Ammerman and Cavalli-Sforza (1973, 1984) that Neolithic farmers expanded from the Middle East to Europe is today supported by a number of considerations.

1. The strongest evidence is genetic, as summarized in sections 5.10 and 5.11. The major genetic pattern observed in Europe, is a gradient of gene frequencies radiating from the Middle East and partially mixed with preexisting local preagricultural populations (Mesolithics). The original analysis by Menozzi et al. (1978a) has been confirmed by extension to a much greater number of genes in sections 5.10 and 5.11, and with other methods of statistical analysis by Sokal and Menozzi (1982), Sokal et al. (1989b, 1991). The genetic data can be easily interpreted only by assuming that Mesolithics were slowly absorbed into the society of advancing farmers (Ammerman and Cavalli-Sforza 1984) and are in full

agreement with a simulation of the process of expansion of farmers followed by gradual mixing with local Mesolithics (Rendine et al. 1986).

2. The successful acquisition of agriculture is usually accompanied by a radical change in many, if not all, aspects of everyday life that deeply affects the archaeological record so that the patterns of the two cultures are radically different. Agriculture and hunting-gathering take long training and considerable knowledge of an entirely different nature. In particular, the agricultural economy that originated in the Middle East was complex and had reached high efficiency at the time it began to be spread from the nuclear area. It is difficult for a hunting-gathering society to shift to an agricultural way of life; observations on African Pygmies and Bushmen show that the transition is rare, and as long as ecological conditions remain unchanged, the change does not occur, or occurs only partially and slowly (Cavalli-Sforza 1986).

3. In many circumstances, Mesolithics must have continued to live side by side with Neolithics (e.g., in Iberian Peninsula; see sec. 5.9), but were probably slowly absorbed culturally and genetically into the Neolithic culture. This is not different from what happens even today with hunter-gatherers in Africa. Moreover, terrains favorable for agriculture are usually different from those favorable for hunting-gathering, so that, at least initially, some territorial segregation kept the two economies partially separated.

4. Thus, the replacement of an economy by a completely different one is hard to achieve by acculturation on a large scale and in a short time, making it unlikely that the spread of farming was entirely cultural and not determined by migrating farmers. Hunter-gatherers are likely to accept an agricultural economy, only after prolonged genetic and cultural exchanges between the two populations, and only after long periods of observing nearby farmers.

5. The new economy progressed in space without much change except for adaptations to new environments (e.g., in house building), and new cultural fashions (in pottery). These cultural adaptations accompanying the spread of farming were carried forward to farther and farther places, as slowly as expected of a diffusion of people rather than that of cultural innovations, which is ordinarily a much faster process. The rate at which agriculture spread to Europe is compatible with the demographic conditions of early farmers.

Our hypothesis of demic diffusion of Neolithic farmers met with little sympathy for a while until Renfrew (1987, 1989b) examined the whole issue and concluded that it was necessary to accept it. Renfrew's analysis was not based on the support given to the hypothesis by genetic data, which in our view is essential, but was derived merely from considerations of *archaeological process*. As Ammerman (1989) noted, it was somewhat ironical that in the past Renfrew (1969) was the strongest supporter of indigenism. There is no doubt that Renfrew's attitude involved a radical change of thinking: it was unlikely that it would be accepted without resistance by some archaeologists.

Renfrew's interest in the problem came from an independent consideration: the fate of languages spoken by Neolithic farmers, and the likelihood that the Neolithic demic diffusion was responsible for the spread of Indo-European languages to Europe, Iran, and India. In his 1987 book, he pioneered explanations of language replacements which were discussed earlier (sec. 1.8) and advanced the hypothesis that Neolithic farmers in Anatolia spoke ancient Indo-European languages.

M. Zvelebil (1986b, 1986c) also in collaboration with his father, a linguist (1988), has published a long critical analysis of our theory, as well as of Renfrew's hypothesis that Indo-European languages were spread to Europe by Neolithic farmers. He first notes, correctly, that Renfrew's hypothesis is based on the hypothesis of demic diffusion of farmers. From here he goes into long arguments claimed to disprove the hypothesis of demic diffusion, and to conclude that Renfrew's hypothesis must be wrong. Most of Zvelebil's discussion is dedicated to the analysis of demic diffusion, and therefore it seems useful to analyze his arguments in some detail.

It is necessary to note that Zvelebil has been led to believe that we are in error, among other reasons, by a somewhat surprising misunderstanding. In Zvelebil (1986b) he compared our simulation (Rendine et al. 1986) of the expansion to Europe with the "actual" expansion (Zvelebil 1986b, figs. 6–8) and found it lacking, because in certain areas the real expansion has been faster or slower than in our simulation. We were well aware of the local differences in rates of expansion, which are, however, minor (Ammerman and Cavalli-Sforza 1972, 1973, 1984). In fact, Zvelebil's real expansion figures are not dissimilar from our fit of an isochron map (Ammerman and Cavalli-Sforza 1984, fig. 4.5), which he should have used for the comparison. We believe our isochron map is more precise than Zvelebil's figures, because it is based on an exact interpolation procedure of the real data and not on a very approximate rendition by eye of an unspecified set of data perhaps less complete than the one we used and referenced (Ammerman and Cavalli-Sforza 1984).

It is erroneous to compare in detail the real maps with the simulation by Rendine et al. (1986) as done by Zvelebil, because the simulation had other purposes. As we stated clearly, the simulation was set up: (1) to test the possibility of distinguishing different migrations by principal-components analysis; (2) to compare the *average* observed rates of advance with those expected under the model, without any attempt at varying locally the rates of advance; and (3) to test how long the initial genetic gradients could resist before being destroyed by later noise caused by local short-distance migration. The simulation showed that the method responds well to the conditions required for our purposes. There was no attempt to introduce into the simulation different means of transportation by water and by land, as well as local ecological differences, which are presumably responsible for possible irregularities of the advance of the farmer's frontier (probably exaggerated in Zvelebil's figure, Zvelebil 1986b). In fact, in the simulation, only a limited attempt was made to indicate the major mountain chains, given by "holes" of the matrix representing the European map in our figures 7.5 and 7.6 (Ammerman and Cavalli-Sforza 1984). Zvelebil is refusing the model on the basis of an erroneous comparison.

In some areas, however, Mesolithic people may have preserved for a longer time their characteristic way of life side by side with Neolithics. It is interesting that this was observed especially at places most distant from the origin, like Denmark or Spain (both in the Cantabrian region and toward the eastern coast; see sec. 5.9), and therefore at later times. These local peculiarities do not affect the usefulness of the model. Evidence that these situations are as common as claimed by Zvelebil has not been given, but in the area covered by the Bantu expansion, pockets of hunter-gatherers have not entirely abandoned their way of life even after prolonged contact with farmers. They represent, however, exceptions. It is likely that Zvelebil has generalized on the basis of his own very interesting work on an exceptional situation in Spain (see sec. 5.9). In our map of the latest Mesolithic sites in Europe (Ammerman and Cavalli-Sforza 1984) and its comparison with the map of the earliest Neolithic sites there was no important exception to the rule that there were no Mesolithic sites definitely younger than neighboring Neolithic sites.

Other criticisms of Zvelebil (1986c) indicate a lack of understanding of the exact scope of simulations, which can test only one of a few specific factors at a time. In comparing the wave-of-advance theory with qualitative microgeographic models of frontier advance proposed by other authors, Zvelebil failed to see the difference between a macrosimulation, like that of Rendine et al. (1986), and microsimulations, an example of which is described a few lines below. He also confused mathematical theories like the wave of advance, which of necessity assumes the simplest situation (continuous homogeneous space), and qualitative models exemplified by him that represent complex, equally hypothetical, local situations, simply in order to make us understand that geographic reality is complicated. Only a general theory in continuous, isotropic space, as hypothesized in the wave-of-advance model, makes it possible to predict the *average* rate of advance. Although our macrosimulations have not considered, and could hardly consider, local variation of ecology, terrain, and population distribution, we did carry out some microsimulations; for instance, for imitating a model of spread along rivers

of the Linear Band Keramik (Ammerman and Cavalli-Sforza 1984, fig. 7.3). These examples can be useful for understanding differences between a macro- and a microscale of theoretical analysis. However, the more complex spatial models of frontier discussed by Zvelebil (1986c) could be incorporated in simulations and could be tested if there were sufficient data on the greater persistence of Mesolithics in many areas.

A statement by Zvelebil and Zvelebil (1988) would gain considerably by being made explicit, case by case, and quantitative: "in most areas there is continuity of settlement" in the Mesolithic-Neolithic transition; exactly how often and where is this statement correct? There is clearly a considerable increase in population density and number of sites with the Neolithic, and it seems likely that a number of Neolithic sites do not have a Mesolithic antecedent, or at least that there is frequently no good proof of continuity. But again, areas will differ.

Other considerations, such as "continuity of some aspects of material culture (such as lithics) and retention of symbols (bear, waterbird, fishes) across the Mesolithic-Neolithic transition" (Zvelebil and Zvelebil 1988), are also not inconsistent with our model. It is well known that most farmers have retained foraging customs for a long time, together with food production. Ethnographic observations (Cavalli-Sforza 1986a; Hewlett and Cavalli-Sforza 1986) show that African farmers often hunt with local Pygmies. There is some exchange of technical information (e.g., on weapons used, traps, etc.), but the local foragers, the Pygmies, are those who have the superior knowledge of hunting, having been in the business and in the area for a long time, and their know-how is likely to be passed on to the newcomers. The tools used for hunting are in general quite different from those used for agriculture, and farmers might well accept tools that they learn to produce from Pygmies; the reverse has also been observed (e.g., for the crossbow; Hewlett and Cavalli-Sforza 1986).

As to the permanence of symbols, there is also permanence of place names, which continue to persist across many cultural transitions, and undoubtedly there is likely to be local continuity of superstitions; so, why not of symbols, especially those having to do with hunting? The real problem is that archaeology rarely finds direct evidence of the physical arrival of migrants, but only indirect evidence, from the presence of new objects that could always be explained by trade and fashions. Under the conditions, some archaeologists find it easier to deny categorically the possibility of migration. It is true that the hypothesis of migration has been abused in the past, but a total denial of migrations cannot be a rational solution.

Finally, Zvelebil (1986b) did not accept the genetic evidence of migration from the Middle East, which supports farmer migration and is discussed more fully in sections 5.10 and 5.11. He argued that other more recent migrations from the Anatolian region are more likely to be responsible for the genetic gradient observed in the first synthetic map (sec. 5.11). In particular, he cited the expansion of the Turkish people during the growth of the Ottoman Empire. But more recent migrations are less likely to have a detectable genetic impact, because the local population density of the earlier inhabitants is very high in recent times compared with most situations of immigration. Moreover, armies of invaders are usually relatively small minorities, who rarely settle in conquered country. Even in the case of the invasion of Hungary by the Magyars, which was certainly of greater relative demographic weight than the Turkish expansion and was certainly followed by settlement (sec. 5.6), it has been laborious to find specific genetic traces, which turn out to be at the limit of detectability (Guglielmino-Matessi et al. 1990). On the basis of present knowledge, Turks seem to have been relatively unsuccessful in making their genetic presence felt, even when they occupied modern Turkey, coming from the East. By contrast, the genetic gradient in Europe of people originating from the Middle East is of dramatic magnitude and regularity across the whole continent. It is not just limited to the Balkans. The consideration of demographic quantities suggests that the present genetic picture of the aboriginal world is determined largely by the history of Paleolithic and Neolithic people, when the greatest relative changes in population numbers took place.

Renfrew's hypothesis of the Anatolian origin of Indo-European languages has, as mentioned earlier, support from independent linguistic considerations that assign the origin of Indo-European languages to various parts of Anatolia (Dolgopolsky 1988; Gamkrelidze and Ivanov 1990); but it conflicts with a great number of earlier hypotheses putting the homeland of Indo-European languages in many different parts of Europe (Mallory 1989). Major problems encountered in settling linguistic disputes of the kind are the absence of written documents earlier than 5000 years ago, the tenuousness of the arguments used for choosing the homeland of a group of languages, and the high rate of change of languages. Some linguists prefer to consider problems of this kind as insoluble, but others are ready to accept the idea that a Proto-Indo-European language, reconstructed from modern languages, was spoken some 5 or 6000 years ago in a homeland on which there is no complete agreement. Quite a few, however, including Mallory, identify it with the Don-Volga region, essentially accepting Gimbutas' suggestion that the Kurgan culture of pastoral nomads spread Indo-European languages to Europe. We have indicated that our synthetic map of the third genetic PC supports Gimbutas' hypothesis (1970, 1991). In a similar way the first synthetic map supports Renfrew's hypothesis. We have also noted that the two hypotheses are not incompatible; on the contrary they are related and can reinforce each other. We accept both with caution, knowing that it is objectively difficult to reach a high degree of confidence in identifying an archaeologically defined culture with a genetically and/or linguistically defined people. The development of pastoral nomadism in the steppes secondary to that of agriculture in West Asia may well have fostered expansions both west, to Europe, and southeast, to India. These expansions need not necessarily have originated from the same nuclear areas of pastoral nomadism, but there might have been an original relationship between such areas. There is disagreement between Renfrew and Gimbutas regarding the timing of such expansions; but while Renfrew does not accept Gimbutas' migration of the Kurgan people toward Europe, he seems to accept the idea that Iranian and Indic languages originated from an expansion of the pastoral nomads from the steppes. Clearly, improvement of our archaeological knowledge on these expansions of pastoral nomads may be critical for a more convincing test of these hypotheses than is possible today.

5.13. SUMMARY OF THE GENETIC HISTORY OF EUROPE

Europe is the continent that is the best investigated genetically, but it is possible to reconstruct only a small part of its genetic history, which covers a period of at least 40 ky. Written documents give some information only for the most recent 3000 years, a little more in some areas and less in others. The most difficult part is the beginning: we prefer to think that admixture with Neanderthals was not important, but even this is questioned by some archaeologists. The anomaly in eastern Europe shown by the third PC could be due to a number of different events, some of which have been suggested in the last section. One could also add to that list a possible admixture with Neanderthals; synthetic maps cannot speak for or against it, if taken in isolation. The order of importance of PCs and, hence, of the synthetic maps is not determined by order in time but by their potential of explaining as much genetic variation as possible. If a high-resolution analysis of mtDNA were done in Europe, the presence of Neanderthal ancestry might be revealed by independent, old branches. If one accepts, however, as most archaeologists do, that the first development of modern humans (outside Africa) took place in West Asia, it would be surprising if modern humans emerged in western Europe untransformed by the passage through eastern Europe, if there had been important admixture with local Neanderthals in this area. It is, therefore, reasonable to think that there was an essentially undisturbed expansion of modern humans from somewhere in West Asia all the way to western Europe in the period between 40 and 30 kya, unless some less likely explanations are preferred on other grounds.

At least in some parts of Europe, and in particular in the west, living conditions were favorable for the development of strong local cultures, of which archaeologists have found important traces, especially in southwestern France. Whatever the earlier development, at the peak of the last glaciation around 18 kya, the peopling of Europe was certainly limited to the southwest, the south, and the southeast. This may have caused a fairly profound isolation of the Paleolithics of the southwest of France and maybe the Iberian Peninsula from the rest, helping us to understand how some of the local genetic idiosyncrasies may have developed. There clearly was at that time and place plenty of opportunity for strong drift caused by small population size and a high degree of isolation. An example are the very high frequencies of *RH*-negative gene among Basques, who even today have 55% *RH*-negative frequency and show exceptional frequencies also for many other genes. It was practically inevitable that Basques would undergo some admixture with people who arrived later. If about 7,000 years ago, just before contacts with Neolithics, Basques had had an *RH*-negative gene frequency of 100%, a very small gene flow per generation would have been sufficient to lower their gene frequency to present values. This assumes absence of natural selection, which would complicate calculations. Naturally, the evidence from one gene is not sufficient, but Basques differ from their neighbors for many other genes.

In the late Paleolithic, languages spoken in Europe may have been of the type still represented in Basque and Caucasian regions, and it is tempting to speculate that languages of this family were spoken by the first modern humans who arrived in Europe. Most linguists are convinced that languages evolve too fast to allow recognition of relationships of this time depth. Recent preliminary results, however (reviewed in Ruhlen 1990) suggest that this skepticism is unjustified. There is a clear need for a deep investigation of ancient language relationships so that they can inspire general confidence.

It is also difficult to exclude the possibility that the expansion of proto-Caucasian (proto-Basque) speakers was later than the first expansion of anatomically modern humans to Europe, but there is no reason to postulate other radiations until there is evidence for them. If a proto-Caucasian type of language was used by the modern humans spreading to Europe in the period between 40 and 30 kya, its origin need not have been in the Caucasus. It is more likely that the Caucasus is one of the few areas that lends itself, for geographic and ecological reasons, to the survival of relic languages. But, we are clearly asking questions that are very difficult to answer and we do not know whether answers will ever be found. In any case, thorough investigation of the Caucasus populations must be a high priority.

As we come closer to the present, the situation becomes a little less confused. Between 10,000 and 6000 years ago, Europe was deeply transformed by the slow entry of agricultural techniques, introduced by Neolithic farmers from the Middle East, in particular from Anatolia. According to Renfrew (1987), but the hypothesis is hotly debated among linguists, this is also likely to have been the first entry of Indo-European languages into Europe. The slow, gradual spread of Middle Eastern farmers dramatically altered the genetic picture of Europe, determining the most important and most regular multigenic gradient observed there. The persistence of this gradient over a long time, and despite many later phenomena, is not surprising since linear gradients are extremely stable and analysis by synthetic maps can easily extract such latent gradients.

One interpretation of the second synthetic map is that it represents the genetic difference between speakers of Uralic languages and Indo-European languages. One must assume that the Uralic family originated within a group of people who settled not far from the Arctic region at the boundary of Europe and Asia. Its genetic origin may have been Caucasoid or Mongoloid, but there certainly was an admixture of various degrees in different areas: mostly Mongoloid east of the Urals, and at least initially also in northern Scandinavia—with a very substantial Caucasoid component in Finland and in northern and central Russia. The Mongoloid contribution may be responsible for the biological and cultural adaptation to cold climates.

The simplest interpretation of the third and seventh synthetic maps is to associate them with the diffusion of the Kurgan culture, which, according to Gimbutas, spread Indo-European languages to Europe. This is not in contrast with Renfrew's hypothesis that Indo-European languages came from Anatolia with Neolithic farmers. In fact, this first spread could have brought Indo-European languages to the area where the Kurgan culture developed, north of the Caucasus and the Black Sea. Here, new technological developments associated with the use of horses and ox-driven wheel carts were the stimulus to a secondary spread of other Indo-European speakers.

Synthetic maps of lower levels give many indications of other possible expansions from various regions near the Black Sea, the Caucasus, Greece, and elsewhere. There are many possible candidates for these PC patterns, but at the moment it seems difficult to make definite choices for most of them. Better archaeological information is needed; genetic information also needs strengthening, given that synthetic maps of a lower order are statistically less informative.

In spite of considerable linguistic differentiation, central Europe shows substantial genetic homogeneity. Unquestionably, though, there are finer distinctions that can be picked up by new analytical techniques (Barbujani and Sokal 1990). Most European nations are sufficiently old that political and linguistic barriers, which often coincide, are likely to have favored a certain amount of local endogamy within sociopolitical and linguistic boundaries. Language differences may have further contributed to genetic differentiation, by decreasing the probability of exchange across linguistic barriers, which are also often physical and/or political and social. Many linguistic differences in Europe are relatively young, however, and we know very little of those that were present, say, 3000 years ago. By contrast, the major genetic differences in Europe reflect older events, and phenomena that took place in both Paleolithic and Neolithic times determined the major patterns perceptible by synthetic maps.

There are also some real genetic European outliers, some of which may be true genetic relics, others the result of drift taking place with and after their foundation; in both cases, it is necessary and reasonable to assume fairly extreme genetic isolation. Moreover, in some cases, both mechanisms contributed to generating the observed anomalies. The insular condition of Sardinia and Iceland has favored their genetic differentiation. The origin of Iceland is sufficiently recent that the Norwegian derivation is still clear, both historically and genetically; that of Sardinia, which is at least 10 times older and probably had a more important founder effect, is more obscure. Colonization by Phoenicians and Carthaginians may help explain the Lebanese connection, but it

is likely that there are also more remote components: Neolithic, and therefore again connected with the Middle East, and also Paleolithic. Cultural isolation of the Basques, aided by their determination to maintain their language and their culture, has helped preserve some of their presumably original genetic idiosyncrasies (perhaps made more complex by later drift). Basques are the only European people who can aspire to the privileged position of proto-Europeans. The Lapps are the farthest outliers, in part because of their admixture with external populations, and in part because of their extreme ecological and cultural isolation. Less extreme signs of ancient differences persist, for example, in parts of Italy, France, and Iberia, the only countries where we have so far been able to carry our analysis to a greater depth. A more detailed study of such populations, especially descendants of Ligurians, Etruscans, and Osco-Umbrians may be enlightening.

The study of genetic patterns shows that, in spite of the buffering of gene frequencies expected by continuing local migration, there is a remarkable persistence of old differences and a less lively generation of new ones. Usually the only evidence we have for linking these genetic patterns with historical or prehistorical events is limited to a few clues. Sometimes, however, observed coincidences in the geographic distributions of genes and of archaeological observations are sufficiently striking that it would be unreasonable to dismiss them lightly. They are also reproducible. European genetic patterns observed by synthetic maps have been confirmed in at least one case by repetition with a substantially increased number of genes as well as by independent approaches.

6 AMERICA

6.1. GEOGRAPHY AND ENVIRONMENT

The Americas, North and South, form 16% and 12% of the Earth's surface, respectively, and their cumulative area is slightly less than that of the largest continent, Asia, which comprises 29% of the Earth's surface. But today's total population of the Americas is only about 14% (including nonaborigines) of the inhabitants of the world, less than a quarter that of Asia (which is 60%). At the time of discovery, the population level was comparatively much lower, but is not precisely known. At that time, important population densities existed only in Mexico and in the northern and central Andes. Three major demographic changes took place after discovery (McEvedy and Jones 1978). The native population decreased practically everywhere and is now about 5% of the total population (much less in North America); it also underwent considerable admixture in many areas, and the mestizo population may be almost 20%. White immigrants and their descendants became the absolute majority of the population in North America (the United States and Canada) and in the southern part of South America. African slaves were imported for work on the plantations starting in 1650 and grew in numbers in most cases, especially in Brazil. Descendants of slaves now represent 15%–20% of the American population globally, an estimate made very imprecise by the extensive hybridization that took place. As usual, we confine our attention to the native populations living in the Americas before 1492, and begin by describing the environment.

North America. Two chains of mountains of very unequal altitude run along the eastern and western coasts of North America: the Appalachians in the east have been considerably flattened by erosion, whereas the Cordilleras in the west reach altitudes of 6194 m. They extend from Alaska to Mexico and in the region of their maximum width, near the fortieth parallel, they occupy about one-third of the surface of the continent.

The rest of North America is relatively flat; the central shield in the middle is 1400 feet (427 m) high, on the average, but it descends in altitude both in the north toward Hudson Bay and in the south and southeast, toward the lowlands and the Great Lakes region. The northern parts of the lowlands have been marked by moraines accumulated in four major glacial advances; the southern part remained ice free and was molded by rivers, of which the Mississippi is the most important.

Because the continent spans latitudes from 65° to a few degrees above the equator, climate and vegetation are very diverse. The Arctic is mostly a cold desert, with only two months in which temperatures exceed the freezing point. Below the Arctic, in southern Canada, the climate is temperate and cool with frosty winters, short springs, and moderately humid and warm summers. The continental United States has cold to mild winters, depending on latitude, and hot summers with ample rainfall. The western United States is very dry except on the coast, which enjoys, especially in its southern part, a Mediterranean climate. Central America has little variation in temperature with the seasons and has a mild climate with abundant precipitation, except in central areas, which can be very dry.

Two-thirds of North America was once forested, the type of trees depending on temperature and humidity. The rest of the continent is drier, with grassland or desert. In the Great Plains of North America, tallgrass prairies formed the habitat of the bison (often called buffalo) for

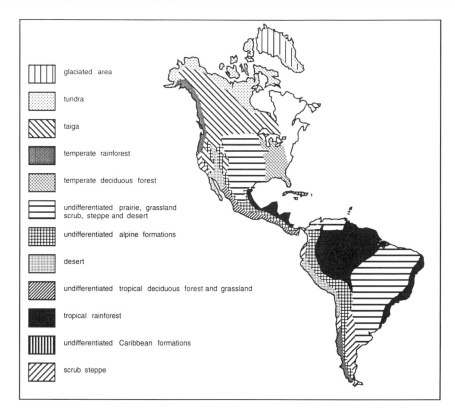

Fig. 6.1.1 Vegetation zones in America (Jennings 1983).

Legend:
- glaciated area
- tundra
- taiga
- temperate rainforest
- temperate deciduous forest
- undifferentiated prairie, grassland scrub, steppe and desert
- undifferentiated alpine formations
- desert
- undifferentiated tropical deciduous forest and grassland
- tropical rainforest
- undifferentiated Caribbean formations
- scrub steppe

many millennia. Tropical savannas are found almost only in parts of Central America; the northern area, however, is mostly desert, whereas tropical forest is extensive in the southern lowlands. The map of vegetation illustrates the climate and ecological conditions (fig. 6.1.1).

South America. To some extent, South America is a mirror image of North America. Here too the western mountains border the Pacific and reach astounding heights; they go from the extreme north to the extreme south and are wider in the middle. Old, flattened highlands occur in the east in northern Guiana and in southern Brazil. Between these highlands is a very wide lowland, the Amazon basin. The Amazon basin occupies all the northeastern part of the continent and is covered by tropical rain forest, having very abundant precipitation and little change in rain or temperature throughout the year. A relatively small fraction, about 10% of the

basin, is excellent for agriculture ("varzea") because it is flooded yearly when the rivers are high, but is not continuously submerged, so that it is naturally fertilized every year; but the rest ("terra firme") lends itself less to agriculture. Where the precipitation is not so heavy, the temperatures are higher and the seasons change, generating tropical savannas common to the Orinoco basin, just northwest of the Amazon in the Brazilian plateau. Farther south is dry forest; and still more to the south lies the basin of another great river, the Parana. Major grassland areas are the Pampas of northern Argentina; farther south lies the Patagonia desert.

The Andes vary in climate and flora, depending on altitude and local conditions, from tropical forest to grasses and plants of small and medium height ("paramos"), to steppe ("puna") that reach the snow line. The extreme south, at a latitude of 56°, has glaciers and mountains, and a frigid climate.

6.2. PREHISTORY: OCCUPATION OF AMERICA

The prehistory of America is shorter than that of any other continent, and its beginnings are more obscure despite enormous interest among scientists who have contributed to the research. Thus, there is considerable uncertainty regarding the origins of native Americans and, as is often the case, uncertainty generates discussion to the point of passion.

There is essential agreement on the idea that the peopling of Americas took place with the passage of nomadic Siberian hunters from Northeast Asia to Alaska (Fagan 1987). Other hypotheses have posited extraordinary journeys—for instance, from Africa to America or from America to Polynesia—but they are not supported by hard evidence (Bellwood 1979). One problem, however, is that among the oldest sites those that are less in dispute (but certainly not entirely accepted) are in South America. Moreover, there are only a few Siberian sites that may have been inhabited by pioneers who later occupied North America. Well-established Siberian sites are more recent than the oldest American sites, which are few and difficult to date. The oldest American sites are not accepted by some archaeologists, whom others accuse of maintaining unreasonably high standards (Bray 1988). Briefly stated, there is strong disagreement between archaeologists who believe that the earliest entry into North America was 30–35 kya (there have even been claims of earlier sites), and those who are prepared to accept, on the basis of present evidence a first date of entry of 15 kya. We briefly review here some of the major finds that are generally accepted and indicate the major controversies.

There is substantial agreement on the lack of evidence of archaic *Homo sapiens* or earlier types in America. All widely accepted American site dates follow the disappearance of Neanderthals in Europe and in Northeast Asia, and there are no finds supporting the migration to America of human types preceding anatomically modern humans (a.m.h.).

The last glaciation occurred 30–13 kya, with a peak at 18 ky; the geography and environment of America and northern Asia when the migration from Siberia to America is believed to have taken place was different from today. In late glacial times (fig. 6.2.1), glaciers occupied almost all of Canada and part of the north-central United States. Temperate and tropical climates were found in North America at much lower latitudes than at present.

The tropical forest had a somewhat smaller extension, especially in South America.

An ice-free corridor is believed to have existed between the eastern edge of the Rockies and the immense glaciers occupying the central and eastern parts of Canada, but the environmental conditions were undoubtedly fairly frigid in the corridor. Perhaps more importantly, at the presumed time of the crossing, the coast line was lower, due to water being retained in the polar ice. This exposed the continental shelf along the coast, causing the temporary disappearance of the Bering strait. A wide and flat land bridge, Beringia, replaced the strait connecting Asia and America, and is believed to have existed between 25 and 15 kya. It is not completely clear what the conditions for life were on Beringia: it was probably a largely treeless land with grasses, dwarf birch, and shrubs, a mosaic of steppe and tundra. It was cold and dry with strong winter winds. Nevertheless, there were mammoth, bison, horse, antelope, and smaller animals (Fagan 1987; Schweger 1990). Certainly the land bridge favored passage between the continents. Without it, the passage would have had to have been made by boat, but direct archaeological evidence of passage by water is difficult to find and, in this case, has not been discovered.

Conditions that permitted crossing from Asia to America by land existed for some time and may have favored the passage of different groups in different periods, some by land and some along the coast. The climate in Beringia was probably never too attractive, although perhaps not very different from that of the Siberian regions of origin, and it may have served as an incentive

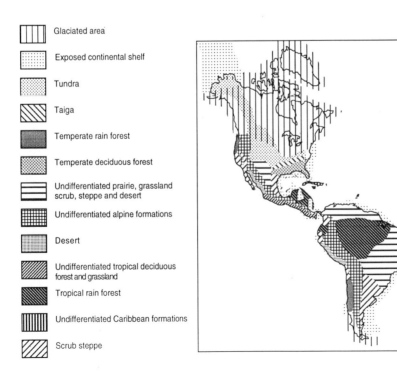

▥	Glaciated area
⬚	Exposed continental shelf
▦	Tundra
▧	Taiga
▨	Temperate rain forest
▩	Temperate deciduous forest
▤	Undifferentiated prairie, grassland scrub, steppe and desert
▦	Undifferentiated alpine formations
▦	Desert
▨	Undifferentiated tropical deciduous forest and grassland
▨	Tropical rain forest
▥	Undifferentiated Caribbean formations
▨	Scrub steppe

Fig. 6.2.1 Glacial environment in the Americas about 15 kya (Jennings 1983).

to continue migration in an eastern and, finally, southern direction.

Several Siberian sites could have been homes of the ancestors of the early Americans.

1. About 20 kya, in Mal'ta and Afontova, in southern Siberia (see fig. 6.2.2), there lived mammoth and reindeer hunters similar to the mammoth hunters of the west-ern Russian steppes north of the Black Sea; among the latter, the best known lived at Mezirich on the Dnieper 18–14 kya (Fagan 1987). Some of their tools are similar to the "microblades" made in Northeast Asia at that time.

2. At the cave situated near Dyukhtai (also spelled Diuktai), near the Aldan River, an affluent of the Lena, a culture was found that was dated at 14–12 kya. By 14 kya, this culture had already spread even farther north, up to the Arctic Ocean where a mammoth burying ground was found at 71° latitude in Berelekh. The discoverer believes its beginning to be earlier and traces the origin of these people to northern China. The Diuktai people used microblades but, unlike Mal'ta people, also made bifacial tools (Fagan 1987). Microblades were used for inset tools and appeared in northern China 30–15 kya; they became common in Japan and perhaps Korea in the later part of this period.

3. A third site is Ushki Lake in Kamchatka where the oldest dates are around 14,000 B.C. The early Ushki cultures used stone-tipped spears, perhaps bows and arrows. The late Ushki culture (12,000–10,000 B.C.) is similar to the Diuktai culture, but more advanced, and has peculiarities of its own. A burial of a husky dated to 11 kya is the oldest northern find of a domesticated dog and may have been connected with the use of dog sleds. Many sites farther north on the Chuckchi peninsula (see fig. 6.2.2) seem to belong to the late Ushki culture and show some intermediacy with Alaskan sites (Dikov 1988).

The earliest archaeological scenario in North America includes sites in central Alaska (fig. 6.2.2) and others in the continental United States (in the parts that were not glaciated at the time) and Mexico (figs. 6.2.2, 6.2.3). Tool finds at Old Crow Flats in the northern

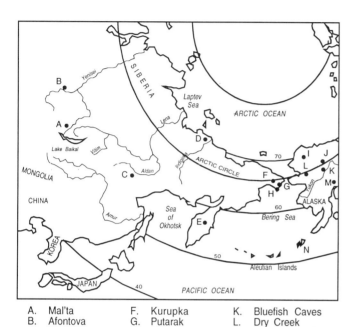

A.	Mal'ta	F.	Kurupka	K.	Bluefish Caves
B.	Afontova	G.	Putarak	L.	Dry Creek
C.	Dyukhtai	H.	Ulkhum	M.	Denali Complex
D.	Berelekh	I.	Akmak	N.	Anangula
E.	Ushki Lake	J.	Old Crow		

Fig. 6.2.2 Archaeological sites in Paleolithic Siberia and Alaska (Fagan 1987; Dikov 1988).

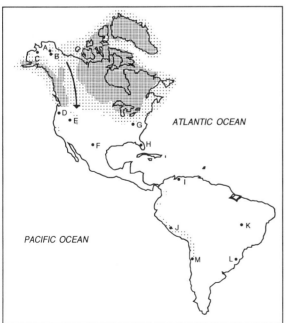

Ice Sheet

ca. 18 kya

ca. 12 kya

A. Old Crow
B. Bluefish Caves
C. Dry Creek
D. Fort Rock Cave
E. Wilson Butte Cave
F. Clovis
G. Meadowcroft Rockshelter
H. Little Salt Spring
I. El Jobo
J. Pikimachay
K. Pedra Furada
L. Alice Boer
M. Monte Verde

Fig. 6.2.3 Paleo-Indian sites in America (Fagan 1987; Guidon 1987).

Yukon (Canada) near the Alaskan border are undoubtedly human, but the date of 27 ky claimed for them is disputed because it comes from animal bones, and there is no consensus that they were "modified" by humans. A human artifact made of bone that had an older date has been redated to 13.9 kya. Another site close to Old Crow, Blue Fish Cave, has bones with dates of 15.5 kya and 12 kya, in addition to human artifacts, including microblades similar to those of the Diuktai caves. A lower layer at Blue Fish has broken bones dated 8,000–10,000 years earlier, but signs of human occupation are not as clear as for the later layers.

Many sites in Alaska have been dated to 12–10 kya; they contain bifaces and/or microblades reminiscent of the Siberian cultures (Denali complex, Dry Creek, Akmak). In summary, there is no evidence on which agreement has been reached that Alaska was occupied by humans before 15 kya.

In the central United States, there was a major explosion of archaeological finds marked by projectile points named after the *Clovis* site, which is dated to the period 11.5–11 kya. There are, however, several finds older than Clovis. A conservative analysis by Fagan (1987) lists a few places that are pre-Clovis and in his view more satisfactory (see location of sites in fig. 6.2.3): Fort Rock Cave, Oregon—13.25 kya; Wilson Butte Cave, Idaho—13–14.5 kya; Meadowcroft Rockshelter, Pennsylvania—more than 12 kya (up to 16.175 kya; Adovasio et al. 1982); Little Salt Spring, Florida—12 kya.

According to MacNeish (1978), Mexico has dates greater than 30 kya associated with chopping-chopper tools, followed by a phase 30–15 kya with bone tools and a unifacial industry (see criticism in Fagan 1987). Other archaeologists also believe dates earlier than 15 kya for Mexico and South America (see also Lynch 1990). Here we cite four major examples of early dates for South American sites.

The Pikimachay Caves in Peru have a more reliable later occupation at 14 kya and an older one at 20 kya considered less reliable.

Dates of 14,200±1150 at Alice Boer Site in south-central Brazil are more reliable than those of earlier tools from a lower layer at the same site, dated to 20–40 kya. Pedra Furada in the northeastern Brazilian plateau (Guidon 1987) has yielded various layers with signs of human occupation, the oldest of which was dated to 32 kya. Monteverde (south-central Chile) is an open settlement with excellent conservation. The people there were mammoth hunters living 12–14 kya.

It is difficult for nonarchaeologists to form a final opinion at this stage, but wide disagreement obviously exists among specialists. It is understandable that there is little tendency to rely on radiocarbon dates especially if they are unique, have high standard error, or come from samples that could have been contaminated with older material. Other often-cited objections are that the strat-

ification is imperfect, or human occupation and use of implements uncertain. The lack of evidence for early, and totally satisfactory, sites in North America is clearly one of the motives for the resistance to accepting sites anterior to 14 kya or 15 kya years in Central and South America. The idea that there is too short an interval of time between occupation of Alaska and that of South America is not a major obstacle, since nomadic hunters could well have covered distances of many thousands of miles in a period of 1000 years. In fact, the whole journey from the extreme north to the extreme south might have taken about that long (Martin 1973). The problems that arise from accepting the hypothesis of this extremely rapid displacement are of two kinds: the hunters had little time to adjust to new environments if they moved so quickly from north to south across such a wide and diverse continent, and they must have reproduced at a high rate in order not to dilute themselves too much in the race toward the south. Approximate calculations indicate, however, that the hypothesis of rapid movement is not unacceptable (Cavalli-Sforza 1986b). Models of genetic consequences of this rapid advance are discussed in the last section of the chapter. The problem of adaptation to new environments must have been simplified by the availability of the same prey (mammoth, mastodon, and probably others) throughout the continent. The idea that South America was occupied before the north, either from the Pacific or the Atlantic Ocean, is more difficult to accept. Whatever trace of African genes are found among living people, it is much more likely to have originated from admixture with African slaves after the sixteenth century. The Pacific islands closest to South America are quite far away and were occupied only very late, in the last two thousand years.

There is no problem with the essentials of the Clovis culture, which developed around 11.5 kya on the Great Plains of North America and lasted for about 500 years. It is marked by mammoth and bison butchering places, where bones of other animals are also occasionally found. The mammoths were killed with spears headed with projectiles that had characteristically fluted stone points and were given additional thrust by using spear-throwers (known as *atlatl*). This culture takes its name from Clovis, one of the important sites; it was supported by a scarce and scattered population. Its origin is uncertain; its end coincided with the disappearance of mammoths from the plains. Shortly thereafter, these animals disappeared from all of America along with several other large mammals that became extinct between 12 and 10 kya, including the mastodon (another elephant), the saber-toothed cat, the horse, several camels, giant sloths, and others (Grayson 1987). One large mammal that survived and was still flourishing on the Great Plains until a few hundred years ago is the bison.

The disappearance of the big mammals has received different interpretations. Martin (1973) suggested that it was due to overkill that started in North America and was continued in South America by hunters that occupied the whole of America in pursuit of this prey. This hypothesis, however suggestive, is certainly simplistic. Pleistocene overkill has been advanced as an explanation for many similar extinctions that happened at about this time in many parts of the world. Although overhunting may have been a partial cause, it seems likely that the change of climate in the postglacial period also had a strong impact by causing profound ecological alterations. Evidence that it affected the fauna comes from the observation that large extinctions of birds also occurred at the same time, whereas small mammals survived and changed their range. Moving to other, more acceptable environments was certainly a mode of adaptation to climatic change (Grayson 1987) that was not equally open to large animals. The bison, however, could survive because it was not bound by its digestive system to eat only the tall grass of the archaic prairies but also the short grass that replaced it in postglacial times. After the disappearance of the mammoths, bison-hunting became the major source of food and other commodities (bones, hides, etc.). Weapons changed somewhat, and new projectile points were developed from the Clovis points. There was some slow evolution in the hunting techniques, but in the Plains the bison remained the major source of food for millennia. Only the introduction of the horse and the gun after the Spanish conquest in the early sixteenth century generated a dramatic change. The bison then came very close to extinction and was saved only by protection in government reserves at the beginning of this century.

Whatever the first date of entry, between 35 and 15 kya, it is clear that there was more than one migration. The linguistic and biological evidence is discussed in sections 6.8 and 6.9–6.13.

The oldest migration from Siberia was that of the Paleo-Indians, to which the above discussion refers, and led to the peopling of the entire continent. There may have been a series of migrational waves, not simply one, or there may even have been a continuous flow. The other two migrations were both later and led to the occupation of more limited and well-defined areas in the north.

Another migration, presumably a second one (15–10 kya) is named after the Na-Dene family of languages spoken by these people. They settled in southern Alaska and on the northwestern coast of North America, perhaps only a little later than the Paleo-Indians. Much more recently, at the beginning of the present millennium, some Na-Dene groups migrated farther south.

The third migration was that of the Eskimo-Aleut (ca. 10 kya), who kept to their Arctic and sub-Arctic habitats, with the Aleuts occupying the Aleutian islands and the Eskimos occupying Alaska and the northern coast of North America, spreading later as far as Greenland. There are still a few Eskimos in the extreme northeast

of the USSR, but Siberian Eskimos are believed to have reentered Asia from the Americas and should not be considered, therefore, an aboriginal Asian group.

The original Asian locations of the Na-Dene and Eskimo-Aleut are not completely clear but are perhaps easier to fit into the general archaeological picture than that of Paleo-Indians, for whom the uncertainty of the time of origin (35–15 kya) is likely to be with us somewhat longer. It is possible that Na-Dene and Eskimo-Aleuts had common origins in Asia.

Dikov (1988) has suggested that the late Ushki culture, dated 10–12 kya and located on the eastern coast of the Kamchatka peninsula, shows similarities with cultures of Alaska and British Columbia and may have contributed to the Eskimo or the Na-Dene populations or both. Dikov also discovered a culture on the southeastern Chukchi peninsula at Puturak Pass, in close proximity to the Bering Sea, that has a technology different from other Asian cultures and similar to that of the Gallagher Flint station in the Brooks Range of northern Alaska. It is dated to 10,549±150 years ago and also has similarities with the culture of Anangula (fig. 6.2.2), a small island in the Aleutians near Umnak Island. The Anangula culture is the oldest known in the Aleutians (dated at 8.7 kya). Laughlin (1980) suggested that Eskimos and Aleuts both come from Anangula and that the occupation of the Aleutian Islands began from it, proceeding both westward and eastward from there. The earliest occupancy of the western and eastern ends of the chain of islands is currently dated to 3000 years ago, but the most interesting early sites of these fishermen and sea-mammal hunters may be submerged. The first known date of occupation of Anangula has also been suggested (Laughlin 1980) as the date of separation of Aleuts and Eskimos. Fagan (1987) indicated more conservatively a date before 4000 years ago. While Aleuts remained on the islands that carry their name and mostly maintained their primary skills in hunting sea mammals, Eskimos developed transportation skills across the Arctic and hunted not only sea but also land mammals (musk ox and caribou). The Dorset culture (Jennings 1983) ranged from the Northwestern Territory in Canada to the Hudson Bay, Labrador, Newfoundland, and Greenland by 1000 B.C., on the average, but there are signs of earlier occupancy of these regions by a pre-Dorset culture.

The difference in origin of Na-Dene and Eskimo-Aleut remains to be clarified. The coast of the Pacific Northwest was colonized by Na-Dene speakers, but the exact time sequence is not clear. Queen Charlotte Island, off the coast of British Columbia, was continuously inhabited between 7000 and 5000 years ago, but the area may have been occupied earlier. The populations of the northwest coast developed a special way of life, reaching high densities especially at the mouths of rivers where salmon was easy to catch. Their cultures at the time of European contact allow us to place them among the world's most successful foragers, and they were the subject of classical research in cultural anthropology.

The eastern coast of Greenland was settled by Vikings coming from Norway and Iceland in the ninth or tenth century A.D., but the Viking settlement lost contact with Europe and disappeared in the fifteenth century. Perhaps in that early time, and probably later after the Danes settled in Greenland (beginning in A.D. 1721), there was some degree of admixture with people of European origin.

In summary, there is little agreement about the first occupation of the Americas; possible dates vary from 35 to 15 kya. There is agreement that this first migration came from Siberia via Beringia and was followed by the rapid occupation of the whole continent by "Paleo-Indians." The next settlement, on the northwestern coast of North America, was between 15 and 10 kya and is attributed to Na-Dene-speaking people. The third, around 10 kya or later, led to the occupation of the Arctic coast by Eskimos. The three-migrations theory has been proposed by Greenberg et al. (1986); see also Greenberg and Ruhlen (1992). It is based on linguistic, dental, and genetic information, as we shall see in the rest of this chapter. A group of linguists (sec. 6.8) vigorously opposes the interpretation of linguistic data proposed by Greenberg (see Ruhlen 1987, 1991; also Rass 1991; Wright 1991).

Other useful references are Kirk and Szathmary (1985), Aikens (1990), and Ruhlen (1990), as well as chapters 54 and 55 of the Cambridge Encyclopedia of Archaeology.

6.3. BEGINNINGS OF AGRICULTURE

The development of human populations was very unequal in the various regions of Americas. The Paleo-Indian hunters occupied the continent with extraordinary rapidity; there later developed local hunting traditions that lasted for millennia in some areas, though inevitably with more or less continuous cultural changes and people displacements. The post-Paleo-Indian period is often called the *Archaic Period* or later hunting-and-foraging period.

The transition to food production from the foraging economy—that is, the hunting-gathering and, near the water, the fishing economy—is sometimes called the *Formative Period;* it occurred at very different times and in different ways in the various regions. In the periods preceding agriculture or in its early development, population density increased somewhat, a stimulus to technological advance in food production. The development of domesticated plants and animals and their adoption as staple food was always a relatively slow process, especially in the Americas, for reasons that depend in part on geography and in part on the nature of the domesticates themselves. Compared with Europe and East Asia, diffusion of agriculture to neighboring regions was slower and more limited. Therefore, at the time of European contact, plants had been cultivated for almost 10 millennia in areas like Mexico and the western part of South America, where important empires with large populations had developed. In many other regions, however, large numbers of American natives were still hunter-gatherers. This was true in particular of the Northwest coast North American Indians, the Na-Dene, and of Californians; but in both regions relatively high population densities had been reached at the time of contact and complex societies had developed, especially among the Na-Dene. The density and, according to some, social complexity of these hunter-gatherers were greater than in other parts of North America that offered only marginal resources and where agriculture, even if it had been adopted as a partial source of food, had only limited development.

The beginnings of agriculture in America are perhaps slightly later than those in the Middle East and in China. By the year 9000 B.P., Middle Eastern agriculture was already a complex economic system using both animal and plant domesticates that could be exported to nearby regions with a somewhat similar ecology. Initial developments in Mexico and the northern and central Andes took place in an environment and with domesticates not widely represented outside the original area. Few if any of the original crops had the potential of being easily exported to a wide area around that of origin before being more fully developed, unlike the Middle Eastern domesticates of wheat, barley, sheep, goats, and cattle. In addition, agriculture in America began in areas like central Mexico and the western part of South America (mostly Ecuador and Peru), which were to some extent unique or isolated. The Mexican plateau enjoyed a temperate climate not found in much drier northern Mexico nor in the tropical forest of the southern part of Central America.

The Andes were another unique environment in which extreme differences in altitude at a short distance provided a great variety of small niches, each suitable for very different types of economic activities. In time, this variety was cleverly used by what is called a "vertical pattern" of exploitation, namely by foraging, cultivating, or breeding very different plants and animals at different altitudes, often very close together, and exchanging these products by a complex network of trade and communications. Systems of seasonal migrations also developed, similar but not entirely comparable to "transhumance" in the Old World. It took time, however, before the social and political conditions of these populations were such that the extraordinary variety of available environments could be turned into a source of wealth.

Native Americans developed a great number of domesticated plants for a variety of uses (Pickersgill and Heiser 1977). Many of them, like maize, potatoes, and tomatoes, were exported to Europe after their discovery in the New World and acquired primary importance as staple food in the Old World. Other American plants like manioc were exported to tropical Africa and radically altered the local food customs. The first plant domesticated in America may have been the bottle gourd (Lathrap 1977), at least 9500 but possibly 11,000 years ago, because of its usefulness as a water container. Maize was domesticated from local plants in Mexico at Tehuacán and Tamaulipas around 9500 years ago, but initially—and for many thousands of years—it remained a small component of the diet. Originally, maize cobs were one-tenth or less the size of modern cobs. Cob size grew with remarkable regularity over the millennia, presumably because of artificial selection exercised consciously or subconsciously by the breeders, who may have been systematically choosing the best cobs for reproduction. At the time of the Spanish conquest of Mexico, agriculture formed an important part of the food supply, which was augmented by the products of hunting and gathering. It is more or less arbitrarily assumed that agriculture became a major source of food supply at a "critical" time about 4000 years ago. At that time, the yield of maize was sufficient to support a sedentary population; pottery made its first appearance then, much later than in Europe and Asia, and almost certainly independently. Beans were also domesticated early in Mexico, with the first examples 9000–10,500 years old; they are a good complement to the maize diet because they supply essential amino acids deficient in maize. Squash was soon added to maize and beans, forming the American Indian triad of staple foods famous for being nutritionally well balanced. Potatoes probably came from Colombia (10 kya). Cotton was grown for use as a textile. Most of these crops could not grow in tropical environments, such as the lowlands of South America, where instead manioc was first domesticated. It later spread to other areas of tropical forest outside the continent.

Few animals were domesticated; however, the use of dog meat for food may be 6000 years old. The turkey is first found in Mexico from 300 B.C. In the central Andes considerable use was made of domestic camelids (llamas, alpacas), which became increasingly common in the last 8000 years for transportation and meat. Guinea pigs were domesticated in Colombia and Peru for meat probably in the last 4000 years. Figure 6.3.1 shows the sites of earliest domestication in America (Bray 1980).

At the time of European contact, American natives were still in the stone age; the only widely used metals, gold and silver, had almost entirely ornamental applications. Some native copper was used for weapons and ornaments. Even so, at the time of contact, two major empires with large populations had developed in Mexico

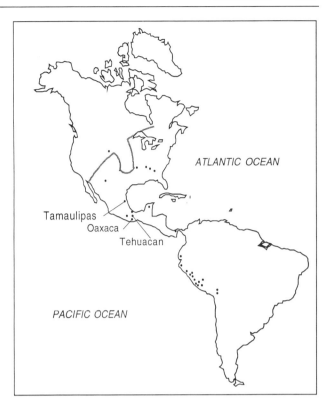

Sites of early domestication of plants and animals

Northern limit of maize cultivation at the time of European contact

Fig. 6.3.1 Distribution of probable places of early domestication (Bray 1980).

and Peru. Elsewhere, population density was still low, although it had increased in the last millennia over the very low densities characteristic of the initial period. The high mobility of the Paleo-Indians allowed them to occupy the whole continent rapidly, but later population growth was slow until the last two or three millennia and increased almost exclusively in areas where previous important agricultural development had occurred. The number of American aboriginals at the time of contact is very imprecisely known and varies greatly with the authors. Early estimates by Kroeber (1939) and Mooney (1928) (whose estimates differ little from Kroeber's) give a total of 1.2 million for all of North America, of which the largest components (in thousands of individuals) come from California (260), Canada (190), the Gulf States (115), and the Plains (100). Later estimates are higher, up to 5 million for the United States (Russell 1987) and 300,000 or more in Canada (Charbonneau 1984). Meso-America was the most densely populated, with perhaps 6–25 million people (McEvedy and Jones 1978). For central Mexico, Cook and Borah (1971) suggested a population of almost 17 million in A.D. 1532, down to 6 million in A.D. 1548 and 1 million in 1608; but Zambardino (1980) corrected the 1548 estimate to

3.6 million. In Peru, the Spanish viceroy estimated 1.3 million in 1572, down to 600,000 in 1620 (Sanchez-Albornoz 1977).

The uncertainty of these estimates should not be surprising. Censuses are difficult even under optimal conditions; at the time of conquest, they were rare even in Europe and there was neither enough interest nor technical skills for carrying them out in the colonies, the occupation of which remained incomplete for a long time. In any case, qualitative evidence shows that population numbers declined rapidly after the conquest, with the spread of epidemics brought by the conquerors and the destruction of the preexisting civilizations. Later censuses are therefore of little use.

Aboriginal population densities largely reflected the degree of development of agriculture and social organization, being higher where the history of agricultural development was older. An exception was North America, where the nonagricultural societies of the west had relatively high densities because of exceptionally favorable environments and advanced sociocultural adaptations. By contrast, central and eastern North America had only a short agricultural history and had not reached high densities at the time of contact.

As noted elsewhere (sec. 2.7, 4.7, 5.2), the onset of agriculture and its successive development is of considerable importance from the point of view of population genetics in that the transition from food collection to food production usually increased population density and thus generally decreased the effect of random genetic drift. It also altered the pattern of migration in many ways, usually reducing individual migration by causing the population to become more sedentary. But migration was always higher in early agricultural times because initial agriculture was of the shifting type (moving to new fields as soil fertility was falling or for other reasons) and in many areas still remains at this stage. Population saturation following initial growth is expected to cause centrifugal migration toward new unexploited fields, when these are available, setting in a slow wave of advance of the agricultural population toward less dense areas. The wave of advance of farmers could begin only when and where cultivated crops had become the major source of food, and where strong physical barriers like mountains or deserts did not impede migration. These conditions occurred relatively late in America, after the Formative Period and therefore after 4000 years B.P. It is unclear whether rapid increases of population density in America caused major demic expansions as they did in Europe or Africa. Mexican agriculture was born in the highlands and expanded late to the north, but it is not clear if there was a demic component; the northern Mexican desert must have acted as a buffer that slowed northward expansions. Agriculture probably spread from Mexico to the south, but there may well have been retrograde flow. The development that took place in Meso-America has much in common with that in the northern Andes. Dates are probably not known in enough detail to allow study of the spread to the south. Lathrap (1977), however, has given tentative dates and directions of expansion for Central and South America. The Andean type of economy was suitable for the particular environment of the Andes, and much of it remained confined to it. However, manioc cultivation, which had an enormous impact on tropical agriculture, may have originated in the forest near the central Andes. The natural way of communication in the South American plains was along rivers, and it is not surprising if spread in this natural network was fast (Lathrop 1977).

As a direct consequence of the economic history we have outlined—mainly the late and limited expansion of agriculture—and its highly localized development, population density remained low in most areas, and the social structure stayed fragmentary, leading to high genetic drift and, with it, high local variation. In the following sections, the development of various regions before and after agriculture is briefly outlined. Surveys of the subject and references can be found in Jennings (1983) and in chapter 56 of the Cambridge Encyclopedia of Archaeology.

6.4. DEVELOPMENT IN NORTH AMERICA

Agriculture arrived late in North America from Mexico, and never reached the western coast during the precontact period. For a general overview of the pre- and post-agricultural development, it is convenient to distinguish four large areas: the West, the Southwest, the central region (the Plains), and the East.

1. The *West* includes for our purposes California, the Great Basin (Nevada and Utah), and the Plateau (Idaho, eastern Washington, and northeastern Oregon). Here, as elsewhere, the more immediate descendants of the Paleo-Indian hunters had to cope with an environment that was becoming warmer and drier. Seven thousand years ago, the climate was already similar to the modern one. But even by 9000 years ago, there was some evidence of a beginning of local differentiation of cultures. A substantial development of the foraging population, accompanied by a trend toward population increase, began only about 3000 years ago but 500 years later in the interior (Aikens 1983). It was once believed that the social system was extremely simple, especially in California, but this view is being corrected. Without increasing sophistication, they would not have eventually reached relatively

high density and local wealth. The foraging peoples in the West were highly sedentary, and there was systematic exchange and trade between local populations.

2. Agriculture from Mexico moved first to its nearest neighbor, the *Southwest*. Defined geographically in various ways, it usually includes Arizona, New Mexico, Colorado, and southern Utah. It is a very dry and almost desert area, but in the Archaic Period, and sometimes even during the Paleo-Indian Period in the eastern moiety of the Southwest, there developed cultures of foragers that lasted for millennia, until the beginning of a sedentary-horticultural mode of living in the Formative Period. The introduction of some cultigens from Mexico, like maize, may be as old as 3000 years B.P. or more; a safer date is 2500 B.P. (Lipe 1983). The beginning of a radically new culture (see fig. 6.4.1) is seen with the *Hohokam* culture in southern Arizona, starting about 2000 years ago. According to some, the Hohokam were migrants from northern Mexico; to others, they were local inhabitants who were under cultural Meso-American influence (Lipe 1983). They grew maize, beans, squash

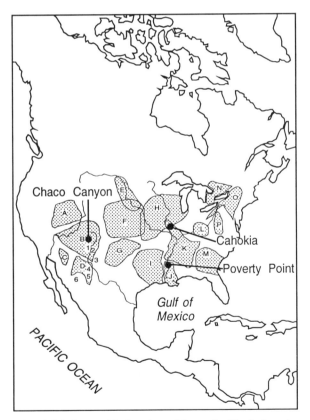

Fig. 6.4.1 Geographic location of the major North American agricultural groups (from Whitehouse and Whitehouse 1975; Griffin 1980). A, Fremont; B, Anasazi; C, Hohokam; D, Mogollon; E, Middle Missouri; F, Central plains; G, Southern plains; H, Onrota; I, Caddoan Mississippian; J, Plaquemine; K, Middle Mississippian; L, Fort Ancient; M, Appalachian Mississippian; N, Hurona; O, Iroquois; P, Monongahela. 1, Hopi; 2, Navaho; 3, Zuni; 4, Apache; 5, Pima; 6, Papago.

and cotton, made ceramics, and with irrigation were able to colonize a vast area. The Hohokam are believed to be ancestral to the Papago and Pima, who still live in the same general region. The case for continuity of culture from the Hokoham to the Pima-Papago is reasonably strong.

North of the Hohokam, the *Anasazi* culture may have developed directly from an earlier Archaic culture that lasted through the millennia (the Oshara), probably with the contribution of migrants. Maize, beans, and squash are well documented by A.D. 600, at which time the population, originally rather diffuse, began to collect in small separate settlements. Between A.D. 900 and 1100, large villages of Pueblo-type appear at Chaco Canyon in northwestern New Mexico. There were cycles in which large villages (Pueblos) were formed, then abandoned collectively when the population moved to other places, often to form larger pueblos. It is believed that the increase in village size made it possible to engage in irrigation works of greater magnitude. Conflicts with immigrants to the area like the Apache and Navajo were earlier believed to have been responsible for the movement of the pueblos, but it is now known that these Na-Dene speakers arrived in the area after A.D. 1200. Many new settlements were built and suddenly abandoned shortly thereafter, at dates that are accurately known thanks to the study of dendrochronology, the sequence of rings in trees. The reasons for movement are less clear. Among the current explanations for the abandonment of pueblos is the recent discovery of cooling and drying of the local climate around A.D. 1100, leading these people to search for areas more suitable for agriculture because more water was available. The descendants of the Anasazi are the modern Pueblo Indians (Hopi, Zuni, etc.; Lipe 1983).

Another culture, the *Mogollon*, started east of the Hohokam and at about the same time, reached its maximum extension around A.D. 900. It was eventually absorbed into the western Pueblo culture under the influence of the Anasazi. Other groups that developed a farming culture in the area, and that are not easily identified with modern descendants, include the Freemont in Utah, the most northern group in the Southwest.

3. Unlike the Southwest, which is dry, the *East* enjoys considerable rainfall, which favored the development of a rich flora and fauna. This area includes the valleys of two major rivers (the Mississippi and the Tennessee), the Appalachian region, and extends farther northeast. In the Paleo-Indian Period, the *Clovis* hunters were the dominant culture, followed by the *Dalton* culture, which clearly derives from the Clovis, but is adapted to a new target, deer. In Paleo-Indian and Archaic times the population was probably scarce and diffuse, made up of small mobile bands with no capacity for food storage.

The transition to a sedentary life was spread over a long period, and domestication of some native plants, like sunflower and amaranth, may have preceded the introduction of cultigens of Mexican origin. An innovation is the building of large mounds as at Poverty Point, Louisiana, with dates ranging from 1700 to 870 B.C. (Jennings 1983). The size of this mound (a diameter of 1200 m) indicates that a degree of social complexity had been reached that made it possible to build such monumental works. Smaller mounds, usually burials, are very common. Domesticated squash is known from the area, but could not have formed an important part of the food supply; maize came somewhat later. Pottery, rare in the beginning, was widespread by 700 B.C. The population clearly became more sedentary during this period, usually called the Hopewellian, but only later (A.D. 700–1000) did clear signs of shifting agriculture appear (the Mississippian period), still combined, as is usual in initial periods, with hunting and gathering. The principal site is Cahokia, near the Mississippi River, almost opposite St. Louis, Missouri. Production of maize and squash increased, and beans were added around A.D. 1000. Communities ranged in size from 100 to 1000, and the larger ones showed indications of social stratification, with chiefs or priests directing ceremonials, mound constructions, and agricultural operations. This culture, *Oneota,* spread north to north-western Illinois and southern Wisconsin after A.D. 1000 and had connections with other nearby cultures. Villages were often fortified (Jennings 1983).

4. Between the Southwest and the East are the *Plains*, which after the disappearance of the forest around 10,000 B.P. became a wide grassland occupied almost since the beginning by large herbivores, particularly bison. The numbers of bison fluctuated over the millennia; there are also fluctuations in the density of occupations and the archaeological record probably for the same reason. At Hell Gap, the archaeological complexes follow one another with few changes from 11,000 to 8000 years ago; after that time, a climate change may have set in. At Mummy Cave, Wyoming, there are 38 distinct fertile levels from 9300 years ago to A.D. 1580, indicating intermittent, perhaps seasonal, occupations for long periods of time (Jennings 1983). Agricultural activity with dependence on maize in the eastern Plains (the Plains Village Tradition) appeared between A.D. 600 and 1000 in South Dakota and nearby regions. The bison remained important, not only as a food source: bison scapulae were used as hoes. There were cultural contacts with the Pueblos and with the Caddoan Mississippian, and many villages were fortified.

In summary, quoting from the Cambridge Encyclopedia of Archaeology (chap. 57, which, along with Jennings [1983], is a good survey of the period), "all North American agricultural developments were related historically and were derived from prior appearances in Central Mexico and further south." The societies of the southeast reached the greatest degree of social complexity and development.

6.5. DEVELOPMENT IN CENTRAL AMERICA

The early development of agriculture at centers like Tehuacán, south of Mexico City, and Tamaulipas, northwest of the capital, has already been described. The slow emergence of an urban civilization reflects the long time necessary to develop an efficient agriculture in a challenging environment where techniques of irrigation were necessary in most of the area. The first indication of water control is in Tehuacán 6000 years ago. Places discussed later are shown in figure 6.5.1.

In the Formative Period (2500 B.C.–A.D. 300), the basis of the Meso-American civilization was laid through the development of intensive irrigation, astronomical observations, ceremonial centers and architecture, and hieroglyphic writing. The first great civilization was the *Olmec* (1200–600 B.C.) which developed its greatest monuments (the colossal stone heads of La Venta, San Lorenzo, and others) in an area of the Gulf coast. But the Olmecs established an exchange system that greatly extended and unified smaller-scale systems existing before in their area and in other areas of Meso-America, thus favoring the spread of cultural diffusion and trade throughout all Meso-America. After

the decline of the Olmecs, important cultures and societies developed in the valley of Mexico (Cuicuilco first, then Teotihuacán) and in the valley of Oaxaca (Monte Albán), where major ceremonial centers were built. In Teotihuacán (200 B.C.–A.D. 800), the population in the later period may have been as high as 100,000 for the whole valley of Mexico, most of it in the capital.

The lowlands of Yucatán and Guatemala were occupied by *Mayas,* who extended also to the highlands in Guatemala. The conditions for agriculture in the Mayan regions were quite similar to those of the Gulf coast where the first urban civilization, that of the Olmecs, had earlier developed. These regions were excellent for sedentary, but not intensive, agriculture with two crops of maize a year. Soil fertility, however, is a serious problem; it is not clear how the Mayas solved it, but they may have employed several different solutions to make slash-and-burn farming more efficient (Jennings 1983). Ceremonial centers like Tikal in the Guatemalan lowlands and Kaminaljuyu in the Guatemalan highlands began developing in 30 and 500 B.C., respectively. The Mayan

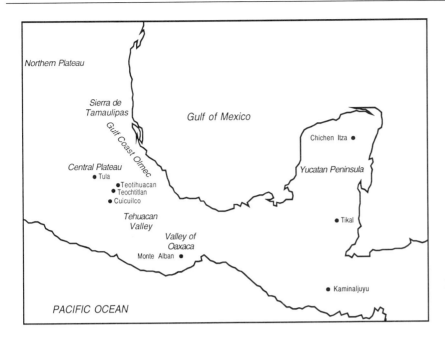

Fig. 6.5.1 Meso-America from Formative Period to European contact (from O'Shea 1980).

culture was strongly influenced by Teotihuacán. It was a multicentric, hierarchical society, with each center having majestic religious and ceremonial monuments. The major center in the Mayan classical period, Tikal, occupied an area of 60 km² (Jennings 1983) and had a population of tens of thousands of people. Outside the center, the population lived in small hamlets and was more diffuse. The classical Mayan period ended abruptly about A.D. 900 for unknown reasons.

The abandonment and destruction of Teotihuacán started a competition between Mexican regional centers, in which the *Toltecs*, from the city of Tollan near Tula in the central Plateau north of Teotihuacán, even-tually gained control and became the first militaristic state of Meso-America. Their influence lasted from A.D. 900 to 1200 and extended as far as northern Yucatán, where Chichén Itzá (ended in A.D. 1224) became the most important center in the so called "Postclassic period" (A.D. 900–1520). Tula had been destroyed a little earlier. Power fell into the hands of the *Aztecs,* who came from the north to found a city at Tenochtitlán, where Mexico City is located. They were in power in 1519 when Hernán Cortés conquered Mexico. A survey of the period and region with additional references can be found in chapter 58 of the Cambridge Encyclopedia of Archaeology.

6.6. DEVELOPMENT IN SOUTH AMERICA

We have already discussed the difficulties associated with the very early dates of some South American sites. Clovis projectiles, indicating the Paleo-Indian Period, are found in most of South America as far south as Patagonia; in the north, at El Jobo, they may even antedate those found in North America. The beginnings of agriculture can be traced to a period between 9000 and 7000 B.P., mostly in the northern and central Andean region (fig. 6.6.1). There is no single, contained nuclear area, but a wide strip all along the northwestern coast that later radiated to other parts of the continent. By contrast, the tropical forest of the Amazon basin had a somewhat later, secondary, and less marked development, but hints of major novelties are already apparent.

There is broad consensus that maize came to the northern Andes from Mexico, along with perhaps squash and beans, but a number of plants were certainly domesti-cated locally in the early period in a variety of different environments. The rich marine fauna remained an important source of food on the coast, but it was later supplemented with agricultural products, and irrigation was developed in arid coastal regions. From the dry highlands came tubers like potatoes, while from higher-altitude forests or the eastern side of the Andes other products emerged, including apparently manioc, which later spread to the Amazon basin. As already mentioned, animal domesticates played a lesser role than in the Old World; however, in the south-central Andes the domestication of camelids provided an important contribution in terms of meat, wool, and animals of burden (for transportation). Around Lake Titicaca in southern Peru there were, at the time of conquest, some 500,000 camelids (llamas, alpacas). These animals had been food for highland hunters since very early times. Their natural range

Location of the Yanomame tribe

Fig. 6.6.1 Map of agricultural settlements in South America. The location of a modern tribe, the Yanomame, is also indicated (from Whitehouse and Whitehouse 1975; Barry 1980; Morris 1980).

is above 3000 m and their domestication may have begun very early (8000 B.P.; Bray 1980).

Agriculture played only a secondary role compared with foraging until about 5000 B.P., but after this date larger settlements supported by agriculture began to appear. Sites like Real Alto and San Pablo, on the Ecuadorian coast, are large stable preceramic farming villages; for example, El Paraiso had a population of 3000–4000 (Bray 1980). Pottery appeared around 5000 years ago at sites as diverse as Puerto Hormiga, near Cartagena, Colombia, and Valdivia, Ecuador.

Irrigation was practiced early and its sophistication increased to remarkable levels. Terracing of the steep Andean slopes was quite common and greatly improved water control and productivity. Cotton (possibly a local domesticate) and the manufacture of textiles soon acquired considerable importance. Improvement in trade networks made it possible to redistribute a variety of materials at long distances, and socioeconomic advances allowed people to make excellent use of the variety of microenvironments present in this region. Through ethnic and kin relations, in addition to trade, it became possible to develop the already mentioned pattern of a "vertical" economy whereby the same people had access to products made in very different environments, from the coast to the highest altiplanos. In the Andes with a day's walk, it is possible to go from one to another of a number of different ecological niches. By wise al-

liances or other social devices, an "archipelago" type of economy was created which gave people and small communities access to, or ownership of, pieces of land in a great variety of areas.

Population density must have risen steadily in this period, and it is not too surprising that the *Inca* empire, which at the time of conquest extended from southern Colombia to south-central Chile, may have been made up of 12 million people. Even if this estimate varies greatly according to sources, the area must have been very densely inhabited, perhaps as much as central Mexico. Complexity of society probably reached a new height about 3000 years ago, as shown by the high rate at which a new sophisticated art form, that of the *Chavin* culture (northern Peru), spread over a vast area, without any evidence of political or military occupations. The Moche site pictures of the north coast of Peru (200 B.C.–A.D. 600) show perhaps the first hints of organized military activity. A major influence was exercised by the *Tiahuanaco* culture, located on the southern rim of Lake Titicaca (1000 B.C. to A.D. 1000). There was progressive development of ceremonial centers and true imperial status was acquired in the last phases. At the peak, the urban population may have been 20,000–40,000 (Jennings 1983). This culture certainly had an important impact on the central Andes, probably initiating or advancing economic innovations later adopted by the Incas. After the collapse of the Moche, the *Huari* culture under Tiahuanaco influence established, probably through military conquest (Morris 1980), an empire that lasted until A.D. 800. Other states (e.g., Chimú, capital Chan Chan; perhaps 25,000 people) existed at the time in the central Andes. The only great South American empire started developing after 1438 when, near Cuzco, the Inca won a battle against a nearby state. They adopted an extremely effective military policy and by building an extensive network of excellent roads (15,000 km) across very difficult terrain, hundreds of road stations and state storehouses, and a well-trained army, they rapidly conquered an extensive territory. Called Tawantinsuyu ("Land of the Four Quarters") it was one of the greatest empires of the world. Inca was the name of the hereditary monarch. The nobility, the priests, and the bureaucrats formed 5%–10% of the population. The rest was a rural population on whom several types of taxes were levied, despite the lack of a currency. Of the agricultural products, roughly two parts of three went for the state and the nonproducing part of the population, and the rest was distributed by the village chief among villagers. Textile products were made by the women for the state. Time in the army and labor for the state were required of the men under the "mit'a" system, which was inherited and perfected from earlier states. It made possible very rapid military conquests and the monumental buildings dedicated to ceremonial and civil purposes for which the Incas are famous. Products taken by the state—food and textiles—

were redistributed to the population according to rank, and individual welfare was assured by an efficient state organization. The "khipu," a system of knotted strings of obscure origin, served the purposes of communication and accounting in lieu of writing.

The enormous Inca empire lasted about a century; at the time of conquest, the empire spanned 36° of latitude from near the present Equador-Colombia border to south central Chile, including much of the Andean region of Bolivia. It was destroyed by 250 *conquistadores* led to Peru by Pizarro in A.D. 1537. The Spaniards were greatly helped by epidemic diseases like smallpox and measles that they involuntarily imported to Peru, and decimated and disorganized the Indian population. They also ably exploited civil unrest.

The remarkable population density and degree of complexity and organization of the Andean states and empires were unmatched in the rest of South America, but a relatively dense population developed in the Amazon forests in spite of the difficulties met by farmers in much of this area. New crops were necessary for the wet soil and climate of the Amazon; the most successful of them was manioc. This plant exists in two varieties, sweet and bitter; the sweet variety was probably domesticated first. The bitter type requires a special fermentation treatment for destroying a poisonous substance that generates cyanide. Manioc cuttings can be easily planted, and propagation is extremely simple. It is especially suitable for tropical environments and provides roots rich in starch but poor in proteins, so it must be coupled with other food. Since manioc seeds are not used, it is difficult to trace it archaeologically; good clues are vats and special bowls employed to make chicha beer from it, or graters. Manioc may have been domesticated at an earlier date farther north, but the earliest well-dated find is from Yarinacocha on the upper Ucayali River in northern Peru (about 4000–3400 years ago). In the same area and time was also found the first pottery, probably derived from the Valdivia types. The Ucayali River is a tributary of the Amazon, and it has been suggested that there were close connections between Amazonia and the Andes during the Chavin culture. This would explain the Chavin paintings of tropical animals and plants that do not exist where this culture developed. The finding of pottery on the lower Amazon, and even at the mouth of the river (island of Marajo, Ananatuba culture, for location see fig. 6.6.1; date 980 B.C.) has suggested that cultural adaptations to the tropical environment, developed on the upper Ucayali River and other tributaries of the Amazon near the Andes, were spread downstream by colonists. There were also later migrations upstream, as in the case of the Omagua and Cocama tribes of the middle Amazon. At the time of European contact, the Omagua had villages of 300–3000 inhabitants, at short distances from each other, and the first visitors were impressed by the quality of pottery. High densities were possible only in areas very favorable for agriculture (varzea), from which natives were soon evicted after conquest, if they were not killed by disease or slave raids (Barry 1980).

Even today in the Orinoco and Amazon basin there exist tribes that have been relatively unchanged by European contact. Several of these—in particular, the *Yanomame* and the *Makiritare*—have been the subject of intensive biological investigations by Neel (1978, 1980; Neel et al. 1977) and his group, including among many others, population geneticists P. Smouse, R. Spielman, and R. Ward, linguist E. Migliazza, and cultural anthropologist N. Chagnon. The bibliography is too extensive for a complete listing, which can be found elsewhere (Chagnon et al. 1970; Smouse 1982; Chagnon 1983). The Yanomame are tropical gardeners who also rely on hunting-gathering activity. Like other hunter-gatherers they have a low number of births, because of long birth intervals. Despite their low fertility, they are at the moment in a period of demographic growth. Their present location is shown in figure 6.6.1. The history of Yanomame villages shows several fissions and fusions. Fissions reflect hostilities between groups and often take place along kinship lines. Although tendentially endogamous, there is migratory exchange between villages of the same tribe and, to a much lesser extent, with other Indian tribes of the region. There were only two documented instances of exchange (Neel, pers. comm.): one was due to the capture of two Makiritare women (Chagnon et al. 1970) and the other was due to the absorption of a few surviving members of a tribe that had come upon hard times (Weitkamp and Chagnon 1968). The genetic exchange between Yanomame villages, in spite of the fusion-fission history, is sufficiently limited that there is considerable genetic heterogeneity between villages, as described in detail in the original papers. In particular, the tendency to fissions following kinship relationships (lineal fission pattern) has the effect of reducing the effective population size of the village and therefore increases the effect of drift over that expected, assuming random fissions. Further strengthening of random genetic drift is due to the high polygamy of village chiefs. The Yanomame move frequently, often under pressure of hostile relationships within the tribe and with other tribes, and are currently drifting slowly southward. They occupied a part of the forest still sufficiently undeveloped at the time of the Neel study that they could keep to their traditional customs, a situation that is rapidly changing now.

The findings in other populations in southern Venezuela or in northern and central Brazil are similar to those of the Yanomame, but there are differences between tribes depending on their economy. Salzano and Callegari-Jacques (1988) have compared groups that they call stage-A tribes (hunters-gatherers with incipient agriculture, like the Yanomame, Trio, Cayapo, Xavante, and others), and stage-B tribes (technologically more advanced agriculturalists and fishermen like the Macushi, Wapishana, Ticuna, Makiritare, Caingang, and many others). Fertility (number of children in completed families) is a little lower, intertribal marriages rarer, and variance of the number of children higher in stage A, but otherwise no major demographic differences were found. It is likely that the average size of villages is greater in stage B.

Contemporary but fragmentary information from other forest people of the Amazon-Orinoco basin shows that most are settling under pressure from governments, but the traditional way of life has been maintained in a few

cases. Movements and admixtures are not uncommon; local economic development, especially mining, farming, and road building, are causes of serious encroachment. Temporary occupation in gold-mining operations and in oil fields is very destructive to traditional Amazon societies and bodes ill for the future of these populations. The extensive destruction of the forest following the opening of roads and modern agricultural and industrial plants create dangers that go well beyond the heavy damage to the local populations.

Surveys relevant to this period and additional references can be found in chapters 59 and 60 of the Cambridge Encyclopedia of Archaeology.

6.7. PHYSICAL ANTHROPOLOGY

Physical anthropologist C. S. Coon (1965) distinguished between Eskimos and Aleuts, on the one side, and American Indians on the other. The first two belong to the Siberian Mongoloids and came by a later migration; American Indians are stated to be Mongoloid in general and more uniform racially, "despite some of their peculiarities in blood groups" and are "more uniform racially than any other group of people occupying an equally vast area, but they are Mongoloids of a particular kind."

The origin of Mongoloids (see chap. 4) is believed to be either in northern China or north of it. According to Alexseev (1979), the maximum development of Mongoloid features is found in central and southern Siberia, especially among: (1) the Tungus-Manchu people of central Siberia, Kamchatka, and the lower part of the Amur Valley; (2) Turko-Mongolic people of southern Siberia and the Yakuts (middle Lena River); (3) the Nivkhs (= Gilyak), a small group in the northern part of Sakhalin and the mainland opposite it; (4) northern Asians like the Nganasan (Taymyr peninsula), Dolgans (a small group south of the Taymyr peninsula), Yukaghir (a small group east of the Lena River), and western Chuckchis. These people have somewhat variable pigmentation in skin and eyes, the lightest being the second group followed by the fourth and then the others. They all have extreme Mongoloid features, mostly reflected in the conformation of the skull and soft parts of the face, which include large cranial and facial dimensions, flattened face, nasal bones, and nasal bridge. It is difficult to give a "nuclear area," especially because the geographic distribution of Siberians has changed considerably in the last three centuries. Although Eskimos and Aleuts have peculiarities of their own, they tend to follow the same general pattern. Like most Mongoloids (with the exception of the Ainu), they have very little, if any, body and facial hair, but abundant and coarse dark hair with rare balding and late, if any, graying. Browridges are small, if any; the eyeballs are wide apart and smaller than in

non-Mongoloids, placed forward in the orbits; the eye opening is narrowed to a slit by eyefolds, with the inner edge of the eye covered by the Mongolian or epicanthic eye fold in a percentage of individuals, which is especially high among Siberians. The lower margin of the orbit lies farther forward and the zygomatic bones protrude forward and laterally, generating the characteristic "high cheek-boned" appearance. The nasal bridge is usually low and flat, but there are also aquiline noses, with little, if any, intermediate forms.

American Indians have less flat faces than Siberians and often prominent, sometimes convex noses. This is perhaps the main difference, but, as just mentioned, the American Indian type of nose is also found in Asia: Coon (1965) cited the Tibetans and the Nagans of Assam. Pigmentation is usually darker among American Indians, but there is also variation among Siberians.

The mean stature of American Indians (Johnston and Schell 1979) varies considerably, being highest at high latitudes (Canada and Patagonia) and lowest in the tropical forests (Guatemala, Brazil). This follows the usual pattern of climate adaptation. In South America, mean stature was mapped for 43 tribes (Salzano and Callegari-Jacques 1988) and there is a slight difference between the northwest and the central-southeast (157 cm vs. 161.3 cm).

Of special interest are the studies of dental characteristics by Turner (1987, 1989). Most northern Mongoloids have shovel-shaped incisors, which are also found in fossil skulls as far back as Chinese *Homo erectus*. This and other cranial peculiarities have been a major reason for claiming independent speciation of Mongoloids (Coon 1965; see also Wolpoff et al. 1984). The genetic exchange at various times and places between local human types, even archaic, and immigrant *H. sapiens sapiens* is a possibility worth considering, but the picture of migrations from Asia to America developed by Turner, and based essentially on dental clues, is unrelated to this question. It is important, however, that on the basis of this evidence, it was stated that a strong difference exists between East Asians from northern China and the Southeast Asian type. For instance, northern Mongoloids ("sinodonts" according to Turner) have 60%–92% shoveling, as against 13%–25% in southern Mongoloids ("sundadonts"). Different percentages refer to different populations sampled. Japanese of the Jomon period (chap. 4) show the lowest percentages, and, together with the Ainu, are classified by dental criteria among the southern Mongoloids, with Thailand, Malay-Java, and Polynesia. Two other traits showing major differences between northern and southern Mongoloids are the number of cusps and the number of roots on molars.

Turner's analysis is based on the premise that dental characteristics are highly inherited, stable in evolution, and not sensitive to evolutionary changes as a function of adaptation to different types of foods. These hypotheses require independent confirmation. Unquestionably, teeth have the advantage of being readable in fossil samples and perhaps also of offering greater detail than bones. Using dental microevolution, Turner calculated 14 kya as the date of the first crossing of the Bering land bridge by the Paleo-Indians. He also postulated that the Na-Dene migration was independent of that of Paleo-Indians and that it occurred 14–12 kya, just before the land bridge of Beringia was completely submerged. In addition, he hypothesized that the Na-Dene may have originated from the late Diuktai culture (fig. 6.2.2), passed along the southern edge of Beringia to Kodiak Island and then to the Northwest coast of the Pacific. He also stated that the third migration, that of the Eskimo-Aleuts, arrived just before the bridge was severed, but after the Na-Dene. These conclusions agree well with other independent sources of evidence (Greenberg et al. 1986) and, apart from dates, with our genetic analysis (sec. 6.9 et seq.).

As we have already briefly indicated in chapter 2, dental data on northern Asia, southeast Asia, and the Americas are generally in excellent agreement with those from single genes. How much further back this agreement will go remains to be seen. The question of how much further back dental data can take us in human evolution is also a matter of conjecture. Apart from the unknown role of natural selection and of dietetic customs—believed to be negligible by Turner—and the unknown level of heritability, an important consideration is the number of independent genes that can be detected by this approach. This is also unknown; only when this number is really large are conclusions insensitive to the addition of further information. Statements based on dental analysis are very interesting, but it would be unwise to rely on them alone until more is known about the problems just mentioned, especially if and when they disagree with other sources of evidence.

6.8. LINGUISTICS

The nonlinguist who approaches the field of the classification of American Indian languages can only be shocked by the segregation of linguists into two groups that hold almost diametrically opposed beliefs: one, more numerous, refuses to recognize unity in these languages and chooses to list a large number of essentially unrelated small families or isolated languages, the interrelationships among which are considered beyond recognition; the other much smaller group proposes three families, corresponding to the three major migrations that are also recognized by other criteria, namely, in time sequence, Amerind, Na-Dene, and Eskimo-Aleut. One cannot fail to see this as the most dramatic example of the usual division between "splitters" and "lumpers," which has been observed repeatedly in almost every classification, be it of living organisms or inanimate objects. To increase the dismay, the group of splitters uses extremely strong language against the author of the unification of Amerind languages, Greenberg (1987) who has earned enormous respect from the whole linguistic community for all his other work. The diatribe has been the subject of articles of popular science (two rather extensive summaries by P. Ross in *Scientific American* and R. Wright in *Atlantic* appeared in April 1991). Another summary of the dispute is in a Postscript to the 1991 edition of Ruhlen (1987).

Ruhlen (1987 and references therein) summarizes the history of classification of Amerind languages, dividing it into three phases. The first was started by the famous anthropologist Alfred Kroeber (1876–1960), who, at the beginning of the century collaborated with R. Dixon to reduce the number of families of North American languages by combining some previously recognized taxonomic units. Edward Sapir carried this effort further, and in 1929 the number of North American families was six, two of which were Eskimo-Aleut and Na-Dene, the languages of the Pacific Northwest. This began a second phase, which can be called a "revolt," and the dismemberment of Sapir's families; after a 1976 conference, the number of independent units of North American languages was back to 63. The list of the results published in 1979 was stated to be "conservative and not very controversial" representing "current received opinion." The third phase was opened by the linguist J. Greenberg, who made the claim that there exist only three families: Eskimo-Aleut, Na-Dene, and Amerind (1987). The Amerind family includes most North American languages and all Central and South American languages, for which there had previously been only limited analysis. For South America, in particular, the information was a list of languages or language clusters rather than a true classification.

The exact meaning of the word "family" (for which some prefer "phylum" or "stock") need not concern us here; it usually refers to the highest "genetic" grouping recognized. Linguists use the word genetic to mean "common descent" similar to "phylogenetic" for geneticists. Today, some linguists have started forming "superfamilies" from the conventional families, hence some of the families are no longer the highest genetic unit.

Nonlinguists, like the authors of the present book, cannot make a contribution to a discussion based on linguistic arguments. From a general scientific point of view, the methodological analysis found in the recent book *Language in the Americas* by Greenberg (1987) is convincing. We accept Greenberg's work as a very serious attempt at a comprehensive classification, which has already achieved some important results by distinguishing the same three major groups found from totally independent sources. Even if this classification changes in the future, it supplies a starting point that is not provided by the extremely fragmentary classifications supported by other authors. As Greenberg's book convincingly shows, the difficulties encountered by the extreme splitters are methodological. They proceed by comparing two languages at a time, with an extremely detailed analysis that makes it impossible to test more than a small fraction of all possible pairs. Their conclusion is limited to the statement that the pair is either "related" or "not related," omitting an estimate of a degree of relationship, without which it is impossible to build a classification that goes beyond the recognition of scattered relationships. The decision on relatedness is based on extremely rigorous criteria, with which, according to Greenberg, it would be impossible to recognize even the unity of the Indo-European family, a step backward by universal consensus. One of these criteria is the belief that "sound correspondences" (rules of change of sounds established on the basis of historical examples) must be followed without exception. Greenberg uses a method of multilateral comparisons, in which many languages are compared for a number of words and other criteria selected for their evolutionary stability. We limit our treatment in the rest of this section to summarizing Greenberg's classification, as given by Ruhlen (1987).

We refer to the three families suggested by Greenberg, called phyla by Ruhlen, as families and to their subdivisions as subfamilies. The geographic distribution of the various subfamilies is shown in figures 6.8.1. A and B.

The ESKIMO-ALEUT family comprises 10 languages and 85,000 speakers; Aleut is presently spoken by 700 people in the Aleutian islands. Three Eskimo languages are spoken by 600 inhabitants of the USSR. The Asian Eskimo languages belong to the Yupik subgroup, found primarily in southwestern Alaska. The Eskimo living on the Arctic coast of North America and Greenland speak three languages: Alaskan Inuit, Canadian Inuit and Greenland Inuit. These are often considered three segments of a dialect chain stretching from northern Alaska to Greenland.

The NA-DENE family is spoken in northwestern North America and consists of two languages, Haida (300 speakers of a total 2000 Haida, living on Queen Charlotte and Vancouver islands) and Tlingit (2000 speakers of Tlingit, out of 10,000 living on the coast north of the Haida), as well as the Athabaskan subfamily made up of 30 languages. The Athabaskan languages are spoken by a northern group of some 70,000 speakers in eastern Alaska and all of western Canada, a few (mostly extinct) groups in California and Oregon, and a southern group of about 130,000 speakers, the Apache and Navajo.

The AMERIND family contains 583 languages, spoken by 18 million speakers. They are subdivided by Greenberg (1987) as follows (see also Ruhlen [1987] and fig. 6.8.1).

I. *Northern Amerind* includes as subfamilies Almosan-Keresiouan, Penutian, and Hokan.
   A.1. *Almosan* consists of Kutenai (a single language), Algic (Algonquian and two isolated languages, Wiyot and Yurok) and Mosan (Wakashan, Salish, and Chimakuan); it covers most of Canada south of the zones occupied by Eskimos (the Arctic) and the Na-Dene (northwestern Canada and central Alaska). It also extends to the Midwest south of the Great Lakes and to New England.
   A.2. *Keresiouan* includes Keres (essentially a single language) and the Siouan, Iroquoian, and Caddoan families; it covers the rest of the Midwest almost to the Atlantic coast.
   B. *Penutian* is a northern group including much of Oregon and California, with outliers (Tsimshian) as far north as Canada; in southeastern North America, a Gulf group includes the Muskogean family and a few isolated languages; in New Mexico, Zuni; a southern group is found in Mexico (Huava, Mixe-Zoque, Totonacan, and the Maya in Yucatán and Guatemala).
   C. *Hokan* is a northern group with small clusters in northern and southern California, Baja California, and parts of Arizona; a southern group in northeastern Mexico and Texas.
II. *Central Amerind* includes three distinct subfamilies: Tanoan, Uto-Aztecan, and Oto-Manguean
   A. *Tanoan* includes Tewa (Arizona and New Mexico) and Kiowa (Oklahoma).
   B. *Uto-Aztecan* is in most of the Southwest, including the Hopi and Pima groups.
   C. *Oto-Manguean* is found in southern Mexico, especially the southwest; also includes the Zapotecan, Chinantecan, Mixtecan, and Mazatecan.
III. *Chibchan-Paezan* includes the Chibchan and Paezan families
   A. *Chibchan* languages are found on the southwestern coast of Mexico and in almost all of Central America

A

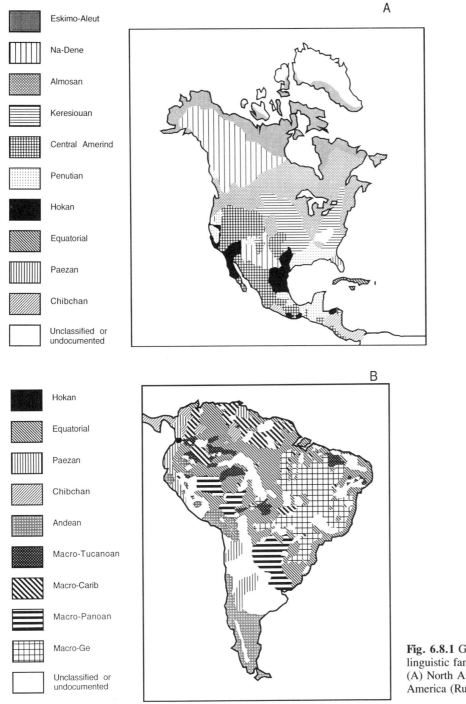

Eskimo-Aleut

Na-Dene

Almosan

Keresiouan

Central Amerind

Penutian

Hokan

Equatorial

Paezan

Chibchan

Unclassified or undocumented

B

Hokan

Equatorial

Paezan

Chibchan

Andean

Macro-Tucanoan

Macro-Carib

Macro-Panoan

Macro-Ge

Unclassified or undocumented

Fig. 6.8.1 Geographic distribution of linguistic families and subfamilies in (A) North America and (B) South America (Ruhlen 1987).

south of the Yucatán; other clusters in Venezuela and Brazil include the Yanomame.

B. The *Paezan* languages formerly found in northern Florida (one language, now extinct), now survive only in South America along the coast of Colombia and Ecuador and farther down in the Chilean Andes; there are also splinter groups in the Brazilian forest and on the northern coast of South America.

IV. The 20 *Andean* languages, of the 583 Amerind languages, account for half of the Amerind population because of the great diffusion of Quechua and Aymara in the central Andes. The Inca empire, and perhaps also the Spanish influence, were responsible for the spread, which is therefore recent. There are also a large number of speakers in the southern Andes, including the Mapuche (= Araucanians). Three small areas in the northern Andes also speak, or spoke, Andean languages.

V. *Equatorial-Tucanoan* includes the Equatorial and the Macro-Tucanoan subfamilies.

A. *Macro-Tucanoan* is found in nine geographic clusters, mostly in western Brazil, with a few in eastern Brazil.

B. *Equatorial* has the greatest number (25%) of all Amerind languages and is widespread from west to east and from the Caribbean islands to Uruguay, in Venezuela, Colombia, Ecuador, Peru, and central and eastern Brazil. The large number of languages is due to the inclusion of two important subfamilies, Arawakan and Tupi-Guarani.

VI. *Ge-Pano-Carib* includes Macro-Ge, Macro-Panoan, and Macro-Carib.
  A. *Macro-Ge* was very widespread, but only a few languages survive, mostly in southern Brazil, in the highlands and farther south. The Kaingang language belongs to this group.
  B. *Macro-Panoan* once extended from Peru to Uruguay; many languages are now extinct.

C. *Macro-Carib* languages were spoken in the northern regions of South America, mostly on the coast of Colombia, Venezuela, the Guianas, and northern Brazil, with outliers farther south.

Geographically, Almosan and Keresiouan are found only in North America; Penutian, Hokan, and Central Amerind are found in North and Central America; Paezan, Chibchan, and Equatorial in Central and South America; and Andean, Macro-Tucanoan, Macro-Carib, Macro-Panoan, and Macro-Ge only in South America. An important point is that the geographic distribution of Amerind languages is extremely fragmentary, especially in South America.

6.9. PHYLOGENETIC ANALYSIS OF AMERICA

Both anthropological and linguistic evidence points to three major groups that may have represented distinct migrations, all from Northeast Asia. The Paleo-Indians were the first, though their date of entry is uncertain, between 35 and 15 kya. There is greater consensus for later dates, but enough uncertainty that an earlier one must be entertained as a possibility. The northwestern American Indians, identified by the family of languages they speak as Na-Dene, were next, as indicated also by their remaining in a more northern area. The Eskimo-Aleut were the latest, and inhabit only the extreme northern region both in America and Asia. The presence of some Eskimos in Asia is believed to be a retrogression from the Americas to Asia, rather than an aboriginal Asian population. The date of entry of the last two groups is probably 15–10 kya.

The question of whether the three migrations can be distinguished on the basis of biological characteristics has recently received some tentative answers, all basically positive. In addition to Turner's (1987, 1989) dental analysis discussed in section 6.7, there is a study of Arctic populations by Szathmary (1981; see also 1985), who used data from 14 genetic loci and found the Athapascan (Na-Dene) are more similar to Eskimos and Chukchi than to northern Algonquians (non-Na-Dene North American Indians). Williams et al. (1985) collected *GM* and *KM* data from the Apache and Navajo (southern Na-Dene), and the Pima, Papago, Hopi, and Walapai (non-Na-Dene from the North American Southwest) and showed that these two groups differ genetically. The difference, however, is not striking and conclusions based on a single genetic system, even one as informative as *GM*, are unsatisfactory. In a more systematic analysis based on data from a larger number of genes and populations, Zegura (Greenberg et al. 1986) tentatively recognized the three migrations, but acknowl-

edged the existence of difficulties for drawing final conclusions.

In our paper (Cavalli-Sforza et al. 1988), which summarizes some of the points made in chapter 2, all the Na-Dene were collected in one group and the rest of the American continent was divided into North, Central, and South America. The Central group was defined on a linguistic basis, taking the Central Amerind subfamily, which is actually partly in North America and does not include all people from Central America. In that analysis, Eskimos clustered with Chukchi and with Turkic-speaking populations of northern Asia, forming a small subcluster of the Northeast Asian cluster, while all American Natives including Na-Dene formed a separate, major subcluster of Northeast Asia. Na-Dene speakers, however, include two major groups, northern and southern. The southern Na-Dene are essentially the Apache and Navajo. Although the exact time of their migration from Canada is not known, it was probably late, and they are believed to have arrived in the Southwest around A.D. 1200. Until recently, there was a splinter Apache group in Kansas.

In the analysis of this section, the major criterion for grouping populations is linguistic. In view of the special linguistic interest, we also added Chukchi and Koriak in order to test possible similarities with Eskimos. Within a few linguistic groups, in particular the Na-Dene, we use a further subdivision on the basis of geography. Because subfamilies are dispersed in widely different areas, it would be especially interesting to distinguish subareas in other subfamilies; but unfortunately, even after the pooling of individual tribes into linguistic groups, there are not enough data to form as many geographic subgroups as would be desirable. Eliminating groups because they take unexpected positions would of course be unacceptable. The procedure adopted was to eliminate systematically groups or subgroups that had fewer markers. Here, as in other chapters, we have tried to limit gaps to not more than 50% in the data matrix. Populations for which there were clear signs of admixture with either Caucasoid or African people, according to the authors who collected the genetic data, were eliminated. We thought it useless to carry out a direct analysis of admixture considering that extreme drift in many American Native groups has generated exceptional gene-frequency variation. There is no assurance for any of the most informative markers, even some *RH* alleles, that they were truly absent in the original American Natives and can therefore be used for inferring admixture. We are reassured by the results of another study that the possible Caucasoid or African admixture of some data we used is not misleading: Salzano and Callegari-Jacques (1988) used 17 non-*RH* alleles potentially useful for evaluating the proportion of non-Indian genes and compared them with results using *RH* alleles, which might be better markers of admixture. There was a correlation, but it was doubtful whether the estimates of admixture could be considered quantitatively valid. Of 58 tribes, only 5 had estimated admixtures of over 25%; 11 between 10% and 25%. Trees from populations believed to have less than 10% admixture gave results very similar to those obtained using the general set. As to our own data, we find there is a clear effect of admixture only in North America, as shown by synthetic maps (sec. 6.13).

The groups for which the number of markers was considered adequate are listed below, together with the names of the major tribes that formed them. In almost every case, however, there were some other, less well-investigated tribes that are not named below but are listed in the tabulations; data from the tabulations were used to calculate the mean gene frequencies of each group. In this way it was possible to increase the representativeness of the data, at least for those genes for which data are more abundant. Such genes, because they are represented in more groups, inevitably have a more important influence on the final conclusions than genes more rarely investigated. Restricting the analysis exclusively to these genes, however, would have reduced its power.

In the list below, the tribes that are named are those that have supplied the most important part of the information, having been tested for more traits. We repeat here that, especially in the Americas, and not only in the southern part, there was enormous drift in many populations, generating great variation from one population to another. This is clearly visible, for instance, in the geographic maps of principal components (Suarez et al. 1985). The averaging over populations can help reduce the effects of drift of individual populations, as already explained in chapter 2.

Figure 6.8.1 shows the geographic distribution of the linguistic groups, and table 6.9.1 the F_{ST} genetic distances among groups. The 23 tribes or groups that contributed most to the genetic data used in the analysis are listed below, with the three-letter symbol used in the table.

I. ESKIMO-ALEUT
　A. Eskimos: U.S.A. other than Inuit (EUS); U.S.A. Inuit (EIN); Canadian Inuit (ECA); Greenland Inuit (EGR)
　B. Aleuts (only U.S.A.; USSR Aleuts had too few markers and tended to associate with Asian populations)
II. NA-DENE
　A. Northern Na-Dene (non-Athabascan): Haida, Tlingit (NDN)
　B. Canadian Na-Dene (Athabascan): Dogrib, Slave, Chipewyan (NDA)
　C. Southern Na-Dene (Athabascan): Apache, Navajo
III. AMERIND (NDS)
　A. *Northern Amerind*
　　1. Almosan (NAL): Blackfoot, Cree, Makah, Montagnais, Micmac + Penobscots, Naskapi, Nootka, Ojibwa, Salish + Mukleshoot + Flathead + Quinault + Okanagan
　　2. Keresiouan (NKE): Caddoan (Caddo + Wichita + Pawnee), Cherokee
　　3. North Penutian: Seminole (= Muskogee), Zuni
　　4. South Penutian: Eastern Maya (Ixil, Kekchi, Cakchiquel, Kiche), Maya, Totonaca, Tzeltalan (Tzeltal + Toztil), Yucatecan
　　Note that Penutian were tested jointly (PEN), and Hokan were eliminated because of strong admixture.
　B. *Central Amerind* (CAN): Chiapaneca, Choluteca, Nahua, Papago, Pima, Tarahumara, Zapoteca
　C. *Chibchan-Paezan*
　　1. Chibchan (MCC): Guaymi, Ica, Misumalpan (Paya, Lenca, Miskito, Sumo), Rama, Talamanca (Cabecar, Bribri, Boruca, Teribe, San Blas), Tarascan, Tunebo, Yanomame
　　2. Paez (MCS): Atacameno (= Kunza), Cayapa (Ecuador), Choco, Colorado, Noanama, Paez, Warao
　D. *Andean* (SAN): Alacaluf, Aymara, Mapuche, Ingano (Colombian Quechua), Quechua
　E. *Equatorial-Tucanoan*
　　1. Equatorial (SEQ): Arawakan (Goajiro, Arawak, Paraujano), Baniwa, Bari, Campa (Maipuran), Chane, Chipaya, Emerillon, Guayaki, Jivaroan (Jivaro, Aguaruna, Yaruro, Cofan, Shuara), Maue, Oyampi, Pacas Novas (Chapacuna), Palikur, Parakana, Piaroa, Piro, Siriono, Wapishana, Zamucoan (Ayore, Imoro, Chamacoco)
　　2. Macro-Tucanoan (SMT): Siona, Ticuna, others
　F. *Ge-Pano-Carib*
　　1. Macro-Carib (SMC): Carib, Galibi, Macushi, Makiritare (= Yecuana), Panare, Pemon, Trio, Wayana, Yupa (= Northern Motilon)
　　2. Macro-Panoan (SMP): Cashinahua, Choroti, Chulupi, Lengua, Mataco, Shipibo, Toba
　　3. Macro-Ge (SMG): Caingang, Cayapo, Craho, Xavante

Table 6.9.1. Genetic Distances (in the lower left triangle) and Their Standard Errors (in the upper right triangle) of American Tribes, Grouped Mostly Linguistically (all values x 10,000)

	PEN	CAN	CKC	CKO	CKR	ECA	EGR	EIN	EUS	ESR*	MCC	MCS	NDA	NDN	NDS	NAL	NKE	SAN	SEQ	SMC	SMG	SMP	SMT
PEN	0	76	181	280	209	234	300	329	224	178	92	70	110	171	50	104	139	40	37	38	98	128	115
CAN	199	0	259	386	246	139	320	265	245	207	85	138	133	124	48	79	39	60	71	88	126	115	126
CKC	896	968	0	170	537	229	198	359	244	84	265	250	313	366	178	165	257	228	180	247	184	281	247
CKO	1241	1367	170	0	678	294	264	627	548	210	413	299	495	480	269	275	333	364	259	362	241	405	326
CKR	1202	1051	980	1414	0	150	376	298	105	201	340	337	325	278	223	193	310	254	308	297	221	312	388
ECA	956	769	658	1070	621	0	86	56	69	66	276	231	191	128	231	75	193	267	197	226	245	257	195
EGR	1286	1152	682	1232	1067	302	0	84	69	156	296	405	314	189	345	247	333	357	298	214	411	307	258
EIN	1438	1211	1370	2015	1494	326	278	0	95	183	404	460	119	300	245	200	318	369	324	360	409	457	288
EUS	1033	977	562	1227	496	259	330	417	0	141	320	360	106	127	210	167	217	263	259	359	285	329	311
ESR	978	1002	246	680	583	407	677	799	351	0	364	385	247	190	238	151	289	289	291	304	237	370	347
MCC	436	359	1389	1871	1732	1267	1312	1548	1372	1669	0	488	211	362	117	208	92	103	55	79	134	81	161
MCS	334	451	1462	1857	1851	1454	1638	2129	1743	1708	488	0	311	531	123	116	68	125	148	96	189	98	288
NDA	744	734	1178	1587	1267	519	1054	492	681	930	1106	1742	0	63	142	111	116	126	129	191	172	213	206
NDN	744	589	1136	1516	1236	645	690	872	595	746	1265	1995	377	0	89	217	289	263	327	401	196	480	451
NDS	256	240	756	1019	892	836	1220	1051	719	787	560	655	426	425	0	52	102	73	68	81	144	145	149
NAL	335	419	618	963	950	367	831	814	669	704	737	634	483	736	217	0	100	167	101	100	111	92	141
NKE	257	146	1002	1136	1406	873	987	1263	703	899	401	237	703	719	319	295	0	41	61	97	153	94	116
SAN	168	280	1151	1586	1437	1144	1387	1374	1367	1335	394	588	965	1094	359	417	204	0	60	62	110	41	140
SEQ	195	230	1089	1522	1673	1072	1293	1450	1292	1437	264	352	906	1109	415	461	219	335	0	17	93	85	87
SMC	174	275	1208	1661	1471	1133	1109	1652	1390	1399	297	340	1067	1193	354	457	300	296	98	0	133	95	137
SMG	381	525	1118	1489	1449	1242	1722	1940	1268	1157	543	578	1118	1054	671	565	524	484	387	496	0	650	811
SMP	505	494	1430	1839	1878	1372	1362	1697	1460	1689	393	393	1037	1410	680	384	403	270	341	410	650	0	734
SMT	524	540	1429	1926	1910	1073	1179	1329	1676	1724	606	739	1075	1234	773	690	468	574	380	495	811	734	0

Note.– PEN, Penutian; CAN, North Central Amerind; CKC, Chukchi; CKO, Koryak; CKR, Reindeer; ECA, Canadian Eskimos; EGR, Greenland Eskimos; EIN, Inupik Eskimos; EUS, Yupik Eskimos; ESR, USSR Eskimos; MCC, Central Macro-Chibchan; MCS, South Macro-Chibchan; NDA, Canadian Na-Dene; NDN, North Na-Dene; NDS, South Na-Dene; NAL, Almosan; NKE, Keresiouan; SAN, Andean; SEQ, Equatorial; SMC, Macro-Carib; SMG, Macro-Ge; SMP, Macro-Parroan; SMT, Macro-Tucanoan. Triangles indicate more compact groups. Tribes included in the groups are listed in the text.
* USSR Eskimos had too few markers and were not used in the tree of Figure 6.9.1; they tend to associate with Chukchi.

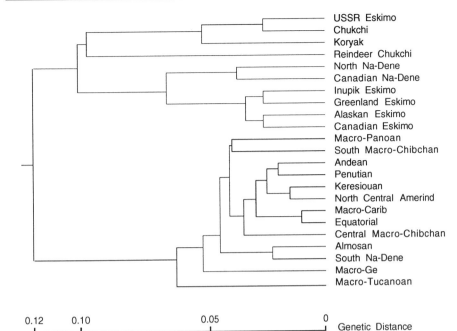

Fig. 6.9.1 Genetic tree of 23 American tribes grouped according to linguistic criteria.

The tree obtained on the linguistic groups formed from subfamilies of North, Central and South America is given in figure 6.9.1 and the PC map in figure 6.9.2. Data include groups with an average number of 72.8 ± 6.8 genes. The PC map accounts for 52% of the original genetic variation.

The genetic tree shows a very clear separation between Eskimo-Aleut and Chukchi-Koryak (in northeastern Siberia, speaking non-American languages) on one side, and all American Indians other than Eskimos on the other. The Na-Dene separate into two groups, the most northern joining the Eskimo and Chukchi cluster and the southern ones the Amerind cluster. These conclusions are in agreement with those reached by studying the matrix of genetic distances. The average distance of southern Na-Dene to the two northern Na-Dene groups is 0.0426, and that between the two northern Na-Dene groups (Canadian and U.S.A.) is 0.0377 (difference not significant); but the northern and southern Na-Dene show average distances of 0.0693 and 0.0957 from the Eskimos. Table 6.9.2 shows the distances between the northern Na-Dene and southern Na-Dene on the one side, and the four most typical Northern Amerind groups on the other.

It is clear from the above distances that the Apache-Navajo, forming the southern Na-Dene, must have had

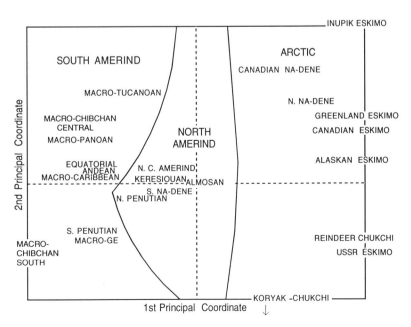

Fig. 6.9.2 Principal-component map of American tribes grouped by linguistic subfamilies.

Table 6.9.2. Genetic distances (x 10,000) among Northern or Southern Na-Dene and Other American Natives from Northern and Central America

	N. Na-Dene	S. Na-Dene
Almosan	609	217
Keresiouan	712	319
Northern Penutian	739	266
North Central Amerindian	662	240

considerable admixture with northern Amerinds. It is also possible that the northern Na-Dene have had some admixture with northern Amerinds, but the data are insufficient to show it.

This observation can also explain why in our earlier world tree (chap. 2) a group made by averaging northern and southern Na-Dene tended to join the Amerinds, splitting from them, however, at an apparently very early time. We know that mixtures tend to attach to an average linkage tree at a higher level than the actual time at which the mixture occurred. The attachment of Na-Dene to other Amerinds indicates that the component in the mixture due to the latter is, on the average, strong enough that it outweighs an original, unknown, component responsible for the difference between Amerind and Na-Dene.

The five Eskimo groups are reasonably clustered in the tree, with Asiatic Eskimos showing greater similarity with their close geographic neighbors, the Chukchi, than with the American Eskimos. USSR Eskimos are a very small group and the separation sufficiently long that this result is not surprising. Furthermore, there are linguistic connections between the Chukchi and Eskimo languages, strengthening the case for a relatively recent common origin of the Eskimo and Chukchi.

Bootstrapping shows that the separation of the two major clusters is clear-cut.

Of 50 bootstraps, 19 show the identical first split of the tree of figure 6.9.1. This may seem a low proportion, but in the other 31 bootstraps, deviations from the tree of figure 6.9.1 are almost always minor.

In 14 bootstraps, the main change is the addition to the Arctic group of the southern Na-Dene; given the strong similarity between the southern Na-Dene and the northern Na-Dene this is not surprising. In 6 of these 14 bootstraps also, the Almosan follow the southern Na-Dene in joining the Arctic cluster. Because Almosan is the Amerind group geographically closest to the Eskimo and northern Na-Dene, the potential for admixture is not negligible.

In 17 bootstraps, one or two populations leave the Arctic cluster; they are, 11 times of 17, the pair of northern Na-Dene and Canadian Na-Dene, which almost always stay together and join the southern Na-Dene in the Amerind cluster. In the other 6 cases, Chukchi or the

Reindeer Chukchi, or, more rarely, the USSR Eskimos join the Amerind cluster.

Even though the USSR Eskimos are today more similar genetically to the Chukchi than to the other Eskimos, the old relationship is still visible in bootstraps. The similarities with Almosan seem modest, and the admixture was probably not a major one, in harmony with the territorial and ecological segregation of Eskimos.

The similarity of northern and southern Na-Dene and their other associations are also clearly visible in the finer details of the bootstrap behavior. The group formed by Haida, Tlingit, and a few Athabascans on the coast is fairly similar genetically to the Canadian Athabascan, and they almost never part. Southern Na-Dene show their affinity with the northern Na-Dene, but they have an even greater affinity with Almosan, which manifests itself in pairing with Almosan in 25 of 50 bootstraps, while they pair with one or the other or both northern Na-Dene in 13 of 50 bootstraps; they show almost no tendency to pair with any other single population. This indicates that the admixture of Navajo-Apache with Amerinds probably happened mostly in earlier times in Canada before the move south.

The Amerind cluster has an internal subcluster of seven North, Central, and South American subfamilies. Two pairs of subfamilies, one central-southern and the other northern are next; Ge and Tucanoan are the outliers.

As mentioned more than once before, an outlier in a tree has several possible explanations. Assuming that evolutionary rates are constant, one can trust the tree structure to correspond to the order of separation of branches, and thus probably to the order of their migration away from the place or places of origin. When an outlier is a very small population that developed in a highly isolated area, the assumption of constant evolutionary rates is difficult to accept, as one would expect it to show a long branch because of high drift. In this case it seems more likely that outliers did not separate particularly early, but being of small size had a very high evolutionary rate because of extreme drift.

In order to avoid the consequences of extreme drift for individual small tribes, we have grouped them, in this case, according to linguistic subfamilies. If linguistic families are formed of groups with greater internal genetic similarity than randomly formed clusters, the pooling of tribes in linguistic groups can help reduce the effects of extreme drift. Although we did not know whether averaging by linguistic family would be truly useful, we attempted it nonetheless. We are currently not aware of better alternatives.

If drift of individual tribes is very high, one may need to average many tribes to obtain a substantial reduction of variation. This has not always been possible here because of a lack of adequate data. In fact, the two worst outliers: Macro-Ge, and Macro-Tucanoan are made up of only four and two populations, respectively. Moreover, the number of individuals in these tribes is small. The Ge are mostly represented by the Caingang (7000 in Brazil), Cayapo (3700 in Brazil), and Xavante (3000 in Brazil). The Tucano are represented by the Ticuna, who number 21,000 in Brazil, Peru, and Colombia combined. Each local population is likely to be a small fraction of the total for the tribe and to have little or no contact with other splinters of the tribe located in other, often distant, regions. These outliers are therefore likely to be cases of very high drift. The next South American outlier, Macro-Panoan, is represented by seven tribes, with numbers of individuals comparable to those above; Central Chibchan-Paezan is represented by 10 tribes. It seems that the greater the number of tribes, the less extreme is the position of the family in the tree. This supports the idea that drift is important in this case; further evidence that high drift is involved comes from geographic multivariate maps, and from other data to be given in later sections, which show extreme differences between geographic neighbors.

A third possible explanation for outliers is an agglomerative origin, with contributions from many groups belonging to very different sources. In urban civilizations, this is often observed in capitals that have received immigrants from widely different regions. They show, therefore, affinity with many other regions without forming close pairs with any particular one. This explanation can be excluded in the present case for forest populations like the Ge and the Tucano, who live (at least today) at a low economic level in isolated areas. The safest general conclusion from the tree, as we discuss later, is that, although the major fissions of the tree are in good agreement with information from other sources, it seems difficult to reconstruct a reasonable genetic history from it as far as the Amerinds of South America are concerned. We see in more detail in section 6.11 that this conclusion is correct. This does not necessarily mean that grouping by linguistic families leads to wrong conclusions, but simply that it was not adequate to improve on a difficult situation.

The PC map (fig. 6.9.2) is more useful, at least in showing the effect of geography: the first axis separates the Arctic populations at the right, puts all northern Amerinds in the center, and the central and southern Amerinds at the extreme left. It is thus in good agreement with basic geography. Arctic, northern Amerinds cluster neatly, whereas southern Amerinds show three major clusters: Tucanoan, Central Chibchan, Panoan; Carib, Equatorial, Andean; southern Penutian, southern Chibchan, and Ge. These results differ somewhat from those obtained with the tree, but they are based on two dimensions only.

At this point, we can ask the most important question, does the proposed three-migration theory agree with the results of genetic analysis? The answer is clearly positive. The two major clusters of the tree, Arctic and Amerind, could certainly be interpreted as separate migrations, and the Arctic cluster does contain a secondary split into Na-Dene and Eskimo, the other two postulated migrations. Thus, the tree is compatible with the three-migrations theory of Greenberg et al. (1987), as is the PC map. The analysis may also support the idea that the two later migrations, Na-Dene and Eskimo, had a related origin in Northeast Asia, in the sense of having come from a common ethnic group in that region. The separation of the Eskimo-Chukchi-northern Na-Dene cluster from the Amerind cluster is also visible in the first principal component of the PC map. The separation of northern Na-Dene from Eskimos is also seen in the second component, though not as clearly. The Na-Dene and Eskimo may have migrated independently to America, or they may have separated in Beringia, or even in Alaska: it is impossible to solve this problem with the present data.

The question of dating these major migrations may be reconsidered again here. In our 1988 paper (Cavalli-Sforza et al. 1988), the divergence between all Amerinds and all northern Mongoloids is in slightly better agreement with the first date of entry proposed, about 35 kya, than with the second. Using the constant calculated in table 2.5.1 we obtain here the date of 31 kya. However, northern Mongoloids are a very diverse population, which underwent considerable internal movement in the last three centuries (Alexeev, pers. comm.). With mixtures and other complications, the divergence between the average Siberian and the average Amerind is likely to be greater than the divergence of Amerinds and their direct Asian ancestors. It is also likely that some of the Siberian populations that remained in Siberia were exposed to more severe environmental conditions and decreased in size, undergoing even greater drift. In any case, our attempts at identifying one Siberian group closer to Amerinds have not been successful. On the basis of relatively few markers (6 loci), Spitsyn (1985) found that among all Siberian peoples, the Tungus, Even, and Yakut located in the northern part of central Siberia are genetically closest to the Athabaskan. The Asian ancestors of Amerinds may have come from a relatively small region, and their Asian descendants may now be diluted by admixture with other less closely related ones, to the point that they are no longer easily recognizable. It is also possible that the majority of the Asian ancestry of the American pioneers has effectively left Asia, as happened, for instance, for Eskimos. All these considerations, and the expectation of high drift in regions of very low density like Siberia, would tend to increase the distance between

Siberians and Americans and thus lead to an overestimate of the time of passage. One may also consider that these are dates of separation, presumably on the Asian mainland, and the date of passage may be later. There are several causes of uncertainty and a dating based on the divergence of Amerind from northern Mongoloids cannot yet be given complete confidence, but we are clearly within the range suggested by archaeology.

From the tree in figure 6.9.1, the genetic separation between Na-Dene and Eskimo is a little more than halfway between the separation of the Arctic group and the Amerind group. If the first is taken as representing the separation between Amerinds and Northeast Asians,

for which we have a not completely convincing estimate (31 kya) discussed above, then the date of separation between Na-Dene and Eskimo, probably still in Asia, is about 18 kya. Note that this is not necessarily a date of entry to America of one, the other, or both; separation may precede entry, which may be later but perhaps not by a large amount.

The tribes were grouped in this section according to a linguistic criterion, modified to some extent by a geographic one. In the next section we consider the tribes that are better known genetically, independent of the linguistic grouping used in this section, for North and South America.

6.10. PHYLOGENETIC ANALYSIS OF INDIVIDUAL TRIBES

The tribes tested for the greatest number of genes in our data files are here individually analyzed. Considering first North and Central America, we have the genetic tree in figure 6.10.1, based on a sample of 17 populations with an average number of 62.7 ± 5.8 markers. Table 6.10.1 shows the F_{ST} genetic distances. Cree and Naskapi, which are very similar linguistically (Voegelin and Voegelin 1977), were pooled; even after pooling, they remain the group with fewest genes.

The Arctic cluster has the same structure as before, with northern Na-Dene (Athabascan and Dogrib) connected to Eskimo, but separating from them in the first split; USSR Eskimos are the most peripheral of the Eskimo cluster.

In figure 6.10.2 the same genetic data are presented as a PC map which accounts for 59% of the original genetic variation. The clusters indicated are linguistic groups and are discussed further in the next section.

The analysis was repeated for 30 populations from South and Central America, including Central Ameri-

can linguistic groups because of the extensive linguistic similarities between some of them. The results of the analysis are shown in figure 6.10.3; distances are given in table 6.10.2. The average number of markers was 61.4 ± 5.7.

Difficult problems arise in the interpretation of this tree. The PC map from the same data (not given) does not bring any clarification. No simple geographic or linguistic correlation is found at first sight, a fact to be discussed further. The amount of genetic drift that has been going on for 10 ky has not abated even today, given that population densities in most of the area are still very small and may even have become smaller in some cases. There has clearly been an extensive geographic movement of tribes, as shown by, among other things, the fragmentation of the linguistic map, and also by modern ethnological observations. There also must have been in the past, and there certainly is at present, a complex network of genetic exchanges within and between tribes, which has been studied in detail only for

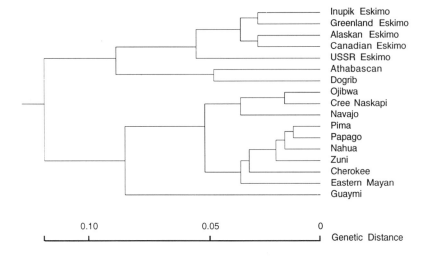

Fig. 6.10.1 Genetic tree of 17 single tribes or geographic groups of tribes from North and Central America.

Table 6.10.1. Genetic distances (in the lower left triangle) and their standard errors (in the upper right triangle) among North American tribes or small groups of them (all values x 10,000). Triangles indicate relatively homogeneous groups.

	ECA	EGR	EIN	EUS	ESR	ANN	ANP	API	MSG	NCD	NAT	NNA	ACN	AOJ	KCH	PZU	PME
ECA	0	81	51	74	63	303	142	139	333	180	111	246	99	140	235	193	322
EGR	302	0	83	77	140	305	315	210	309	308	175	324	291	289	347	236	334
EIN	326	278	0	95	187	420	356	287	544	145	318	288	255	352	371	403	419
EUS	259	330	417	0	133	257	227	241	361	147	129	155	220	175	255	245	327
ESR	407	677	799	351	0	198	247	178	414	302	212	258	201	178	301	213	233
ANN	873	994	1285	675	875	0	40	38	266	334	200	79	148	146	62	47	142
ANP	951	1304	1509	1105	1289	174	0	32	216	211	232	66	80	142	118	55	94
API	814	953	1228	908	865	160	104	0	184	178	192	59	92	106	67	46	145
MSG	1882	1710	2615	2013	1952	714	793	705	0	578	591	300	381	375	317	111	179
NCD	775	1113	734	913	1298	1399	1445	1065	3153	0	66	174	202	308	396	307	131
NAT	754	746	952	676	889	866	810	791	2687	451	0	95	297	322	460	324	189
NNA	872	1253	1072	650	817	297	366	295	1123	629	369	0	55	115	100	123	82
ACN	535	1133	1117	939	930	406	331	340	1191	866	913	220	0	56	100	233	130
AOJ	619	991	1269	717	755	731	611	509	1277	1299	1236	433	157	0	122	221	162
KCH	1177	1168	1726	959	1277	170	266	314	565	2339	1489	540	374	496	0	101	180
PZU	1097	1088	1523	1251	1503	234	210	160	594	1682	1514	441	715	848	422	0	63
PME	1288	1342	1725	1037	848	278	347	317	611	1434	857	384	428	567	399	260	0

Note.– ECA, EGR, EIN, EUS, ESR, Eskimos (see table 6.9.1); ANN, Nahua; ANP, Papago; API, Pima; MSG, Guaymi; NCD, Dogrib; NAT, Athabascan; NNA, Navajo; ACN, Cree, Naskapi, Montaguais; AOJ, Ojibua; KCH, Cherokee; PZU, Zuni; PME, Eastern Maya.

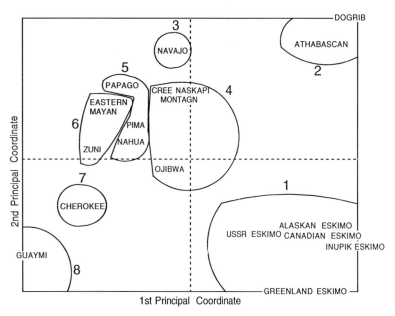

Fig. 6.10.2 Principal-component map of Northern and Central Amerind tribes or geographic groups. The clusters refer to linguistic groupings: 1, Eskimo; 2, Northern Na-Dene; 3, Southern Na-Dene; 4, Almosan; 5, Central Amerind; 6, Penutian; 7, Keresiouan; 8, Chibchan.

two tribes (Yanomame, Makiritare). These investigations are the only ones from which a model can be derived. One wonders how much one can generalize the conclusions reached for these examples, but it is encouraging to have excellent data even for only a few populations, which have not been seriously affected by contact with latecomers, or at least have shown little if any tendency to acculturation. The Yanomame may have originated at a considerable distance from their present location in the upper Orinoco (see fig. 6.6.1), probably in Panama (on the basis of linguistic considera-

tions). They are still moving and expanding (Chagnon 1983). The story that emerges from the Yanomame or Makiritare is one of many scenarios which must exist in South America. It certainly should not be extended to regions with a long history of formation of towns or cities or even villages having a totally different demographic and mating structure. Rather, the Yanomame are a model for populations living as primitive horticulturists in the American tropical forest, which is a significant fraction of Central and South America.

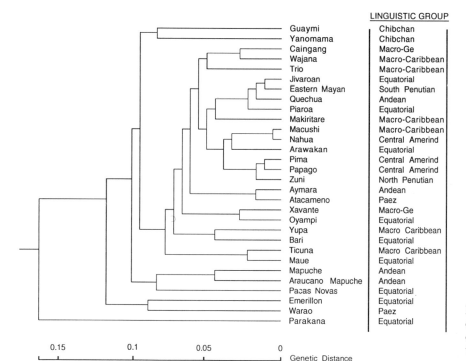

Fig. 6.10.3 Tree based on genetic distances of 30 South and Central or near-Central American Indian tribes.

Table 6.10.2. Genetic Distances (in the lower left triangle of the matrix) and Their Standard Errors (in the upper right triangle) among Central and South American Tribes (all values × 10,000)

	ANN	ANP	API	MCA	MWR	MCY	MSG	PZU	PME	AAM	AAY	AMA	AQU	EAR	EBR	EEM	EJI	EMA	EOY	EPN	EPR	EPI	MMA	MMK	MTR	MWA	MYU	MGC	MGX	MTT
ANN	0	42	36	148	220	217	278	43	161	218	177	247	90	49	178	272	79	125	146	307	328	121	38	142	228	59	115	90	126	154
ANP	174	0	31	207	319	178	196	49	98	323	204	200	74	123	218	276	239	212	129	547	207	292	198	192	135	95	178	148	143	223
API	160	104	0	174	297	169	180	56	154	343	213	193	79	132	127	267	130	154	89	482	320	79	211	186	197	110	138	179	141	178
MCA	478	686	440	0	244	429	428	186	210	368	47	184	73	195	327	328	73	124	220	538	341	68	228	195	397	132	298	226	109	237
MWR	686	1168	1033	1102	0	378	657	250	316	438	146	436	228	228	236	245	333	539	263	538	636	415	259	241	397	239	318	195	458	408
MCY	889	921	701	1436	1368	0	236	121	204	728	267	252	227	343	274	254	194	181	198	463	496	380	175	325	279	292	260	348	732	239
MSG	714	793	705	1341	1529	786	0	110	189	669	192	185	201	333	391	372	180	199	157	267	773	344	142	335	287	302	548	340	484	309
PZU	234	210	160	558	1090	584	594	0	60	328	155	154	54	136	173	255	157	115	108	407	228	223	161	182	108	111	208	184	307	146
PME	278	347	317	565	1093	846	611	260	0	203	267	133	46	71	194	258	27	129	113	247	246	73	90	101	81	137	173	188	72	167
AAM	896	1396	1404	1212	1947	2807	2068	1180	822	0	343	263	390	480	491	447	195	403	553	168	471	265	334	673	192	166	321	380	289	373
AAY	414	598	499	190	578	490	597	537	716	1153	0	165	140	139	253	261	175	183	205	422	396	229	170	175	261	133	246	190	443	182
AMA	604	726	610	642	1422	1160	717	551	486	446	165	0	147	500	457	370	123	155	282	374	423	118	141	224	342	405	458	504	151	225
AQU	326	303	268	403	1045	941	776	208	150	940	140	147	0	112	195	280	71	128	104	358	255	63	135	146	102	97	156	150	141	219
EAR	168	480	548	895	531	854	621	596	375	1399	462	466	112	0	152	227	149	228	115	267	507	228	92	216	111	109	224	89	166	353
EBR	545	639	477	1041	835	714	809	564	697	1874	613	1381	392	152	0	358	162	155	235	304	671	251	157	222	373	316	148	285	221	260
EEM	834	1281	1185	1515	1129	1361	1280	1146	1113	2049	1713	458	732	434	1240	0	358	376	308	447	540	538	258	204	262	230	333	202	271	337
EJI	244	650	329	265	1763	910	970	525	102	674	525	662	446	525	533	1401	0	429	207	244	290	148	82	56	135	160	198	141	490	149
EMA	329	652	593	430	807	912	521	452	446	662	700	1227	533	879	739	1401	429	0	110	103	275	148	128	165	172	217	170	343	318	80
EOY	566	695	578	1077	807	993	521	604	554	1953	1061	1227	416	879	963	854	468	207	0	244	290	227	109	183	140	114	225	68	113	406
EPN	516	1386	1284	1013	1722	1649	1408	1388	778	553	1405	1064	1059	814	1318	1735	643	1065	1257	0	2619	343	230	295	159	398	408	524	353	327
EPR	1188	1146	1496	1554	2644	2094	2731	1315	980	1689	1783	2169	1359	1638	2227	2311	1550	1123	1848	2619	0	1881	118	138	198	185	1643	176	267	327
EPI	760	854	433	342	1887	1580	1195	710	351	698	682	616	239	1112	1082	1738	216	667	845	642	1881	0	363	77	163	185	347	176	267	178
MMA	82	635	394	652	995	1222	960	537	147	1008	625	1024	356	447	423	1047	294	417	497	888	1263	659	0	363	98	105	150	192	222	74
MMK	535	726	708	856	1110	990	960	854	459	2072	922	1227	423	348	842	687	526	776	526	623	1665	469	602	0	296	120	236	102	471	304
MTR	455	682	667	760	1241	1825	1010	500	249	782	802	745	431	328	712	1187	486	565	718	1028	653	872	470	555	0	120	275	230	536	226
MWA	203	458	505	650	1040	850	1026	546	432	415	547	1534	385	328	455	781	505	909	925	823	1329	753	718	497	727	0	196	111	419	337
MYU	489	767	697	1166	812	1359	968	713	558	707	933	1115	547	348	842	1111	526	776	925	731	1643	1009	712	1267	798	624	0	133	443	266
MGC	293	884	872	915	1281	1328	1128	991	621	709	827	1534	749	326	842	687	486	1152	352	738	2105	860	598	555	727	271	437	0	359	478
MGX	294	501	557	682	2051	1825	928	860	267	745	1027	745	494	507	712	915	664	841	352	1496	1541	750	788	1056	1144	913	655	705	0	552
MTT	574	824	718	953	2051	1060	1251	695	825	1321	907	925	820	1052	807	1682	397	240	1572	1267	1487	727	300	1267	835	1058	1043	1371	1311	0

Note.– ANN, Nahus; ANP, Papago; API, Pima; MCA, Atacameno; MWR, Warao; MCY, Yanomame; MSG, Guaymi; PZU, Zuni; PME, Eastern Maya; AAM, Araucano; AAY, Aymara; AMA, Mapuche; AQU, Quechua; EAR, Arawakan; EBR, Bari; EEM, Emerillon; EJI, Jivaroan; EMA, Mane; EOY, Oyampi; EPN, Pacos Novas; EPR, Parakana; EPI, Piaroa; MMA, Macushi; MMK, Makiritare; MTR, Tria; MWA, Wajana; MYU, Yupa; MGC, Cairiguap; MGX, Xavante; MTT, Ticuna.

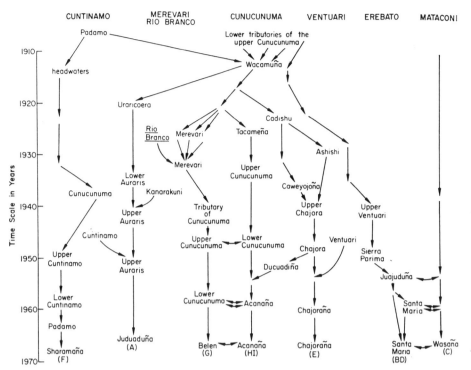

Fig. 6.10.4 History of fissions and fusions of seven Makiritare villages over 60 yr (from Ward and Neel 1970, fig. 1). The phylogenetic tree calculated from genetic similarities between villages can, to some extent, reconstruct the actual history of fissions (but not that of fusions which would demand other approaches). The changing temporal-spatial relationships are indicated by reference to the time scale on the left; headings at the top represent the six main geographical areas.

The history of fissions and fusions for the last few generations of the Makiritare (see fig. 6.10.4) shows a structure of very small groups, on the order of 100 individuals each, which split and reunite, to some extent according to kinship lines. Kinship groups, however, are not necessarily stable entities when viewed over several generations, and the whole picture is one of incomplete randomness of splits and fusions that is not easy to model quantitatively. The genetic variation between villages is about twice what would be expected (Wagener 1973) on the basis of the observed proportions of migration, assuming that migrants are a random sample of the population. Thus, drift is higher than expected from the observed migration and population size, probably because splits and perhaps reunions tend to follow kinship lines and are therefore not random (Smouse et al. 1981; Smouse 1982), as in regular models of population structure.

Another source of amplification of drift effects is strong differential fertility, especially of head men (Neel and Weiss 1975; Neel 1980). The Makiritare are largely endogamous within the village, and even more within the tribe, but give and receive nontrivial genetic contributions to and from neighboring tribes, usually of different linguistic groups since several tribes moved a long distance from their origin. One cannot exclude the possibility that immigrants from other tribes have closer kinship ties with the tribe, decreasing the outbreeding effect caused by mating with members of other

tribes. Considering the frequency with which women are raided from other tribes, a certain amount of random or nearly random outbreeding with neighbors must also occur.

There is only limited information on other intertribal migration. According to a summary of information by Salzano and Callegari-Jacques (1988), genetic exchange is considerable, and it is higher for tribes at a more advanced economic level. Their tabulation does not distinguish between genetic exchange with neighbors and with different tribes. In unpublished data collected with H. Groot and A. Espinel in Colombia, genetic exchanges between different tribes on the upper Orinoco became very high at the end of a long period of intertribal hostilities; in a small area investigated near Puerto Inirida, it was difficult to find marriages where there had not been recent admixture between different tribes. The memory of genetic exchanges in older generations is frequently lost, and such findings make one suspicious about the real isolation of many South American tribes, at least today. Yet, there is enough genetic variation between South American tribes that some degree of isolation must have been maintained in many instances for a long time (Neel and Ward 1970). Our capacity to understand the genetic structure of southern Amerindian tribes can only benefit greatly by an extension of studies like those already cited by the Neel group, before they are made totally impossible by the disruption and disappearance of traditional customs.

6.11. COMPARISON OF GENETICS WITH LINGUISTICS AND GEOGRAPHY

In section 6.9 we have seen that genetic analysis fully confirms the division of American natives into three major clusters, Amerinds, Na-Dene, and Eskimos, which are also clearly distinct linguistically. The hypothesis that they correspond to three major migrations, all from Siberia via the Bering region, is in agreement with current archaeological knowledge, despite present uncertainties on dates. The general picture seems reasonably well established and further analysis given in section 6.10 has clarified possible doubts arising from the ambiguous position of the southern Na-Dene. Their geographic position and the peculiar genetic relationships with other Na-Dene and with Amerinds are best explained by admixtures with the latter that must have accompanied their southern migration. When we come to consider Amerinds, we find greater difficulties in fully reconciling genetic data with information provided by other approaches. In part, this is caused by the poverty of information. At this point however, we must summarize two previous investigations that show without doubt that Amerinds, too, provide good evidence of a strong correlation between genetics and language.

The first is an extensive analysis of the relations between the genetic, linguistic, and cultural similarities of 53 North American Indian tribes carried out by Spuhler (1979). The analysis used a subset of 13 gene frequencies from *ABO*, *MN*, *RH*; Diego blood group was tested only for a subset. Of seven linguistic groups: Arctic-Siberian, Na-Dene, Macro-Algonquian, Macro-Siouan, Hokan, Penutian, and Aztec-Tanoan, 34 (64.2%) of 53 tribes tested were classified correctly using gene frequencies. This indicates a substantial agreement between linguistic and genetic data, but also a number of discrepancies. Most misclassifications in the Spuhler sample are found among Na-Dene, Macro-Algonquian and Macro-Siouan, and in the Hokan group. In Spuhler's analysis by culture areas (Arctic, Subarctic, Northwest Coast, Plateau, California, Plains, Southwest, Northeast, Southeast) 31 of 53 tribes were correctly classified, or 58.5%. Considering that more groups were tested in the latter case, the two approaches gave approximately equivalent results. In conclusion, there is substantial, even if imperfect, agreement between genetic and linguistic or cultural classifications. Some of the discrepancies, especially that of northern and southern Na-Dene are of interest; note, however, that Apache and Navajo are not misclassified in Spuhler's analysis. The statistical approach used by Spuhler (stepwise discriminant analysis) is different from the usual one of calculating correlations between genetic and linguistic (sometimes also with geographic) distances. Moreover, we use more genes and fewer tribes.

In figure 6.10.2, tribes belonging to the same linguistic group are circled. There clearly is a reasonable, though not perfect, agreement between a linguistic and a genetic classification. The small numbers do not permit a completely satisfactory assessment of the correlation. The incomplete agreement indicates that the estimates of genetic and linguistic similarities may need improvement. It may also result from frequent language or genetic replacements. In fact, these explanations are not mutually exclusive, and to some extent, all may have contributed to reduce the correlation without completely destroying it.

Using other more conventional approaches, Spuhler (1972) found no evidence of correlation between genetic and linguistic distance. This negative result may be more of an indictment of the method than of the general correlation between linguistics and genetics. A linear correlation can easily be destroyed by some outliers. The expectation of linearity may be naive when there is a complex fission and fusion pattern; simulations may be appropriate for a comparison of the different methodologies. However, Spuhler (1972) reanalyzed the same data by an analysis of variance, which escapes the strictures of linear-correlation analysis, and found that the variance of genetic distances among linguistic stocks is significantly higher than that within linguistic stocks. This is in line with his result by discriminant analysis. It is worth adding that Spuhler found a moderate but significant correlation between genetic and geographic distances and none between linguistic and geographic distances.

Apart from Spuhler's studies of the genetic-linguistic correlation on the North American continent, there have been many investigations of limited regions or groups of Central and South America. An early one by Spielman et al. (1974) compared the linguistic distances among seven Yanomami dialects and genetic distances among the people occupying the corresponding geographic areas. The matrices of genetic distance, distance calculated from lexical data, and from grammatical data showed in all three cases a significant congruence.

Chakraborty et al. (1976) found no linear correlation between genetic distances and linguistic distances in seven Chilean "highland" Andean populations. Linguistic distances were calculated on a scale based on an early classification by Greenberg. The scale of linguistic distance used may be responsible for the failure.

The same measurement of linguistic distance was used by Murillo et al. (1975) to compare linguistic and genetic distances of the Chipaya of Bolivia to nine South American Indian tribes. They found no correlation.

Salzano et al. (1977) investigated the intra- and intertribal genetic variation within the Ge-speaking Xavante, Kraho, and Cayapo of Brazil. They conclude that the average intertribal genetic distance within this linguistic group is about 63% as great as that between tribes speaking more differentiated languages. They found, however, a weak linear correlation ($r = 0.27$) between genetic distances and cognate percentages in a list of 100 words.

A very thorough and detailed study has been published recently by Barrantes et al. (1990) on the Chibchan-speaking groups of Costa Rica and Panama. Ten such populations were analyzed for 48 genetic loci. The genetic distances between pairs of populations were correlated to the linguistic distances based on cognate percentages. The observed correlation ($r = 0.74$) is high and highly significant, higher than that observed for genetic and geographic distances ($r = 0.49$, not significantly different from zero) and for geography and linguistics ($r = 0.52$, significant at $P = 0.05$).

When we look at figure 6.10.3, we are unable to find a simple interpretation linking genetics and linguistics in the whole of Central and South America. A similar failure is experienced in the related tree given for South America in Salzano and Callegari-Jacques (1988). It seems likely that, in these circumstance, a tree is highly inappropriate for detecting the correlation of interest, but it is also possible that the data are inadequate.

Even the usually strong relation between genetic and geographic distance is blurred in South America. The correlation calculated between the two distances is 0.191 ± 0.048 (standard error calculated by bootstrap). It is positive but low, and confirms the results obtained by plotting the genetic distance between population pairs against their geographic distance (sec. 2.9). Linguistic distance between families showed a negative correlation with genetic distance (-0.139 ± 0.051) and with geographic distance (-0.212 ± 0.051). These results (Minch and Cavalli-Sforza, unpubl.) will need further investigations.

There are many reasons why the correlation of linguistics with genetics and also with geography is especially difficult to study in South America. Part of the problem is tied to the major territorial, economic, and political changes that have taken and are taking place in South America, causing an epidemic of language extinctions that must have been especially dramatic in the last century and earlier. For instance, in Ruhlen's (1987) list, 71 languages of the 117 (61%) that form the Ge-Pano-Carib subfamily are extinct. Similar high percentages apply to many other subfamilies of South and Central America: Equatorial 67/145 (45%), Tucanoan 12/47 (26%), Andean 12/18 (67%), Chibchan 25/43 (58%).

Languages often become extinct when population numbers become too small, or when there is government pressure to expand those of another language, but this does not mean that the people also disappear. In fact, it seems reasonable to assume that in the modern situation, with the continuous shrinking of groups, an increasingly larger proportion of people stop speaking the traditional language and replace it, either with languages imported by the colonial powers or with more widely spoken, traditional languages from other groups. This would certainly contribute to the destruction of the correlation of languages and genes. There may be other important reasons that deserve more research.

One should remember that, as we have already discussed (sec. 2.9), American Natives show an extremely high geographic mobility, as measured by the relationship between genetic distance and geographic distance. Mobility is also detected by studying the distribution of language groups, which is extremely fragmented, with subfamilies forming very complex, interpenetrating patterns. This might be enough to destroy linear correlations between geographic and linguistic distances, and between genetic and linguistic distances. The ecological situation also contributes to this result: the Andean chain forms the backbone of the continent and is very different from the east. It runs from the extreme north to the extreme south and is relatively similar ecologically in spite of the great variation in latitude. It is occupied by people who are also relatively homogeneous genetically, as well as linguistically; only two major subfamilies of the nine spoken in the whole subcontinent occur in the Andean chain today. By contrast, the flatter, eastern part is more heterogeneous genetically and linguistically. Linear correlations are especially unsuitable for measuring the association among geographic, genetic, and linguistic distances in this case. Detailed studies of single linguistic groups that have not undergone too many disruptions and extinctions—for example, the Chibchan (Barrantes et al. 1990)—are best suited for showing the correlation between genetic and linguistic variation. Studies of other groups, that have not been excessively impoverished by extinctions may also be useful.

The studies of correlation between genetics and linguistics in America can give only a very partial answer to the general problem. Of the seven studies we have listed, only one that used linear correlation gave satisfactory results. One can see many reasons why this can happen even if there is a general congruence between the two phenomena. Other methods have given positive results when linear correlation failed. Moreover, even if this is generally overlooked, significant testing of linear correlations between distances calculated between pairs of populations is unsatisfactory because there is usually an internal correlation between the pairs. This does not apply to the sample by Murillo et al. (1977) in which the pairs of populations are independent. For further comments see Cavalli-Sforza et al. 1992.

In summary, three of seven studies favor the hypothesis of congruence between genetics and linguistics but for methodological, theoretical, and historical reasons, one may expect this type of analysis to fail in the Americas, especially using linear correlations. Further work on American data with more refined methods is clearly necessary.

6.12. GEOGRAPHIC MAPS OF SINGLE GENES

The *ABO* system is remarkably different in America from other parts of the world: Amerinds are unique in having almost completely lost the *A* and *B* alleles. By contrast, *A* is conserved among Na-Dene and shows a remarkably high frequency among some Almosan, whereas among Eskimos, the *A* and *B* frequencies are much more similar to those of the rest of the world. Thus, the *ABO* locus is a fairly good, though not a perfect mirror, of the three major postulated migrations.

The reasons for the loss of one or two alleles of this system, which are present at relatively constant frequencies in all other world populations—and to some extent also in many Primates (Socha and Ruffié 1983)—are not entirely understood. The extent to which random variation in gene frequencies affected Amerind populations will be clear from several other examples in this section and suggests that genetic drift played a very important role in America. Did drift determine the irregularities of gene frequencies in America because of a very low number of initial migrants (an initial founder effect), or later bottlenecks, and perhaps persistence of low numbers for long periods? We may anticipate that the behavior of *HLA* loci indicates that the second or third hypothesis may be true, and that many tribes originated from a very small number of founders. Instead of the many alleles of an *HLA* locus commonly found elsewhere, even in small populations, a particular Amerind tribe has only a few alleles at a disproportionately high frequency, with other alleles rare or absent. In another tribe the same rarity of most alleles except a few is observed, but the frequent allele/s are different. This remarkable phenomenon is therefore unlikely to be due to natural selection, given its magnitude, or to the initial founder effect of a small number of first migrants from Asia. *ABO* has far fewer alleles than *HLA*, but in a way there is a somewhat similar phenomenon: an excess of *A* in a few groups, and an excess of *O* (up to 100%) in all the others. A high frequency of *B* is almost never found.

Even if there is a good chance that drift was responsible, at least in part, for the anomalous distribution of *ABO*, it is difficult, if not impossible, to exclude the effects of natural selection. As we have seen in section 2.10, *ABO* phenotypes (or genotypes) react differentially to many infectious diseases, and a popular explanation for the loss of *A* and *B* alleles among Amerinds is differential sensitivity to syphilis, because *O* individuals are more resistant. The origin of the hypothesis is the belief that syphilis was endemic in Central America in the fifteenth century and was spread to Europe by the crew of Christopher Columbus after their return to Spain. The evidence from direct studies of patients (Mourant et al. 1983) showed that *O* individuals heal more rapidly (as judged on the basis of immunological tests) after

treatment with chemotherapeutics. The dates and geography of the European epidemic beginning shortly after the return of Columbus' crew correspond to the expectations of the hypothesis, but others have claimed that the disease originated in Africa from a closely related spirochete responsible for yaws, a nonvenereal disease (McNeill 1976). A search for a correlation between yaws and *ABO* was negative (Cavalli-Sforza 1986b).

The geographic distribution of the *ABO* alleles shown in the maps deserve some comments. Because of the rarity of *A* and *B*, and the omnipresence of *O*, all gene-frequency distributions are very skew. In North America *O* is lower, with allele *A* being high and reaching a peak above 45% (almost all *A1*) in western Canada. Elsewhere, *O* is almost never less than 50%. In the extreme south, there is a small patch with a maximum of *A* greater than 10%, and a corresponding trough in *O*. Greenland is also high in *A (A1)*.

In Eskimos, *B* shows a peak in eastern and southern Canada, where *O* is low and there are also traces of *A*.

Apart from Eskimos, the simultaneous presence of *A* and *B* in proportions of 4:1 is a strong indication of admixture with Caucasoids. This is likely to be the case on the eastern coast of Canada, but the absence of *B* in the western part of Canada, despite the high frequency of *A*, is proof that this is not due to admixture. If Negroids were the donors of *ABO* genes, which is not the case in Canada, the proportion of *A* to *B* would be lower than for white admixture. We have tried to avoid using data from mixed populations but we will see that in the eastern part of the United States and Canada a fair number of mixed groups are present. More intensive contacts with Europeans occurred in this area and, therefore, it is not surprising that it is difficult to find "full-blood" (or even only 3/4 blood) Amerinds.

Variograms of *ABO* alleles have long initial linear segments, with rather small slopes.

Acid phosphatase (*ACP1*B*) shows an almost regular gradient from north to south. The distribution is almost bimodal, reflecting the major difference of Eskimos and Amerinds from the extreme north versus the rest of the continent. The variogram is approximately linear up to 4000 miles, with a fairly large slope.

Adenylate kinase *1* (*AK1*) is, like *ABO*, a marker of Caucasoid admixture. The less frequent allele, *AK1*2* has a frequency of about 5% among Europeans and is essentially absent in other populations. The band of low *AK1*1* (<97%) across the North American continent indicates Caucasoid admixture. It confirms and extends the observations with *ABO*. The variogram is uninformative and is not reported.

The Diego blood group *(DI*A)* is of special significance in America. It was first found in Amerinds, in which, as the map shows, the *A* allele varies from less than 5% to more than 35%; it is also found in some northern Mongoloids but at a lower frequency. It must therefore have originated in Northeast Asia. Its considerable variation in America is most probably due to drift. The maximum is in northern Brazil, but it is rare or absent in North America. The initial slope of the variogram is fairly high, and the linear portion is less than 1000 miles.

The Duffy blood group (*FY*) varies considerably with allele *A*, showing a maximum in the Arctic. The distribution spans almost the complete range, but is concentrated between 40% and 100%. Allele *B* has been studied much less extensively; it peaks with more than 40% frequency between northern Brazil and the Guianas. The variogram of allele *A* is fairly regular, whereas that of *B* has a strongly negative initial slope.

Allele *1* of esterase D (*ESD*1*) shows a maximum in Mato Grosso (southern Brazil) and the Paraguay basin, as well as in Central America; it also shows an absolute minimum in the extreme eastern part of Brazil. The variogram has a large slope and is linear until about 1500 miles.

Glyoxylase-1 allele *1* (*GLO1*1*) has a maximum in Central America and low values in South America; the regular decrease toward the north is artifactual and caused by the near absence of data in North America except in the extreme north. The variogram is approximately linear for almost 2000 miles with a largish slope.

The group-specific component or vitamin-D-binding protein allele *1* (*GC*1*) shows a minimum in central Brazil and a relative maximum farther west; the variogram is irregular, possibly because of the closeness of the minimum and maximum. The electrophoretically fast subtype of *GC*1*, *GC*1F* has two peaks on the western coast of South America, a relative minimum in the extreme south and one in the extreme north. The variogram shows a complex form.

Haptoglobin (*HP*1*) also has a very wide distribution, with gene frequencies ranging from 0% to 100%, with a mean of 55%. The peak is in the extreme south, but there are other secondary peaks in South America; the lowest values are in the extreme north. Basically, there is a north-south gradient, which, in the present case, cannot be attributed to climate. The variogram has a relatively short initial portion with a positive slope.

Antigens specified by *HLA* genes have revealed an unusually narrow range of alleles, especially in South America (Black et al. 1980). Only *HLAA*2, A*9, A*28, A*30, A*31, A*33, HLAB*5, B*15, B*16, B*17, B*27, B*35,* and *B*40* have average frequencies significantly different from zero. This restricted range of polymorphism is expected when the genetic diversity of an ancestral population has been reduced several times by passage through size bottlenecks.

A possible effect of selection should also be considered for *HLA*: in fact, evidence for heterosis in South American Indians had been advocated by Black and Salzano (1981), who found that, in a subpopulation of 122 people whose parents' *HLA* haplotypes were known, there were 56% fewer homozygotes than expected. If this phenomenon is due to differential mortality, it can be efficiently studied only in the few populations still subject to high prereproductive mortality.

The most frequent *HLAA* allele is *A*2*—37%, on the average—and it reaches maxima over 50% in southwestern North America and in Venezuela, with minima in the northern Andes and in eastern Greenland. The distribution is likely to have at least two modes. *HLAA*9* has an average frequency around 31% with a peak over 80% in eastern Greenland and the northwestern Arctic. A secondary peak (over 50%) is found in the northern Andes, whereas the rest of South America has frequencies below 20%. The distribution seems bimodal. Allele *A*19* has an average of 17.5% and a peak of more than 40% in northern Chile, with low frequencies north of Colombia. With an average frequency of 10%, *A*28* has a peak near 40% in the extreme south. Averaging only 1%, *A*30* has a peak of more than 4% in the southeastern United States. A subtype of *A19*, *A*31*, averages 15%, reaching more than 40% in northern Argentina. Again, the distribution seems bimodal. Although it has a maximum above 15% in the southeastern United States, *A*33* averages 1.8%.

With an average frequency of 12% and a peak over 50% in eastern Venezuela, *HLAB*5* has a secondary peak in eastern Greenland. Although its mean frequency is 1%, *B*7* reaches values above 10% in the western Arctic Ocean region. Allele *B*14*, with a 0.8% average, has a frequency greater than 10% in the southern Andes; and *B*15*, average 11.5%, has a peak in northern Chile. *B*16*, average 13%, has a peak in the north-central Andes greater than 50% and minor peaks elsewhere. *B*21*, averaging 1.5%, has a maximum above 10% in the extreme Southwest of the United States. *B*22*, with mean 0.7%, reaches more than 10% among central Eskimos. Well known for its strong association with ankylosing spondylitis, *B*27* has an average frequency of 3.8%, with a maximum above 20% in Alaska. It is interesting to note that the three tribes of the Southwest, the Pima, Papago and Zuni, have similar origins but significantly different frequencies for *B*27*. The most frequent *B* allele, *B*35*, has a 20% average and reaches about 70% in Brazil. With a mean frequency near 19%, *B*40* reaches over 50% among Eskimos of the western Canadian Arctic.

In sum, *HLA* shows great variation, most probably resulting from drift, like the other genetic systems, but as already noted, its multiallelic structure renders variation more evident. This genetic system is ordinarily represented by a great number of alleles in almost every population—even if very small—in the Old World, and all alleles tend to have relatively low frequencies. In the Americas, the situation is different. One or few alleles become definitely dominant in frequency, in one or a few tribes, sometimes reaching values above 50%, and the other alleles are correspondingly rare; but most populations are unique in that the dominant alleles differ from one to the other, sometimes even in neighboring populations. This is exactly what would be expected, at least qualitatively, under drift alone. In fact, in the total absence of cross-migration, drift would eventually lead to the survival of only one allele in each population. The surviving allele is chosen randomly from among those originally present, subject to the rule that the probability of an allele becoming the sole final survivor equals the initial frequency of that allele in the drifting population. Perhaps most alleles were represented at the beginning in Northeast Asia; many are still present in some, but not in all the other tribes.

Some alleles were probably lost, a few among the founders perhaps, but most in the process of evolution of individual tribes, as shown by the very different local patterns of each allele. It seems as if most local populations were started by such small numbers of individuals that they could only maintain two or three alleles at high frequency. Under these conditions, one does not need to postulate a very strong founder effect at the passage from Siberia to America (or even earlier). The remarkable variation among the Indian tribes of South America suggests the existence of a later bottleneck, perhaps more important than the first, if there was a first one. In other words, many alleles may have been present at the beginning and lost later. Only 17 alleles have sufficiently high average frequencies to generate maps of America; this is about half the number of European alleles, but one does not need to conclude that half of the alleles were lost. It is possible that there exist several undetected alleles, because the majority of reagents are of Caucasoid origin and do not necessarily detect all alleles present in other populations.

The variation with distance shows here, as in other *HLA* data, several negative or flat initial slopes: 5 of 17. The initial linear segments of those with positive slopes are in the usual range, and the initial linear portion may sometimes span 2000 miles.

GM (or *IGHG1G3*) also shows considerable local variation. The most common haplotype, *za;g*, varies from 40% to 100%, with several peaks and several minima. The next most important haplotype, *zax;g*, has a maximum in the center of South America and decreases almost regularly around it.

All the other *GM* haplotypes have lower average frequencies, but all show usually single, sometimes extreme peaks in different regions. Thus *za;b0b1b3b4b5*, a Negroid haplotype (very poorly represented in the maps for reasons of reagent availability), has an average frequency near 2%, but peaks at more than 6% in the Guianas

where there is probable African admixture. An Oriental haplotype, *za;b0stb3b5*, has an average frequency of 6% and peaks at more than 20% in Alaska and in Labrador. With an average of 1.6%, *fa;b0b1b3b4b5* has various peaks in the north and south, none too pronounced. A Caucasoid haplotype, *f;b0b1b3b4b5*, has an average frequency of 2.7% and peaks in Greenland and in the northern part of South America.

At first, one might be reluctant to believe that all these maxima and minima for *GM* haplotype frequencies are due to drift. One might hypothesize that this immunoglobulin marker reacts to local infectious diseases, and there is a little evidence for it as discussed earlier. However, drift is expected to operate with the same intensity for all markers. It is therefore likely that many *GM* gene-frequency peaks or troughs in America are due to drift.

The light immunoglobulin constant chain, *KM*(*1&1,2*), has a mean of 37%, with a wide distribution of 0% to 80%, minima in the north, but at least one in the south, and maxima around Panama.

The variograms of immunoglobulins tend to be irregular and uninformative. The Kell blood group (*KEL*K*) is a rare polymorphism almost homogeneously near zero. *KEL*Jsa* is also relatively rare (2% average), but shows a peak above 20% on the northern coast of South America. The Kidd group (*JK*A*) has a distribution of 0% to 80%, with minima in the extreme south and in the Panama region, and various maxima. Its complementary allele, *JK*B* is poorly studied directly; it shows a complementary maximum in Panama. The Lewis blood group *LE*Le* also varies greatly, from 10% to 100%, and has a maximum in Alaska. *LE*Le(a+)* has a maximum in a neighboring region, but has a much smaller range of variation. Almost all these blood groups have irregular variograms.

The *MNS* system shows somewhat less variation than other genes, judging by F_{ST} values, but the range of gene frequencies is not small. The *M* allele varies from 30% to 100%, and the *S* allele from 0% to more than 80%; both frequency distributions are probably unimodal, but both geographic maps are full of relative minima and maxima that span almost the whole range. Of the four haplotypes, only the rarest, *Ns* (6% average frequency), does not have a distribution extending from nearly 0% to nearly 100%; maxima and minima appear in regions already showing strong drift for other alleles, like the north-central Andes or the Arctic, or in new ones, like the coast of southern Brazil. All the variograms have positive initial increases with regular slopes, but with oscillations, except for *Ns* which is fairly flat.

The *P1* blood group, allele *1*, has a distribution varying from 5% to 100%, with a maximum in southern Chile and minima in many places, but mostly among Eskimos. The F_{ST} is elevated, and the variogram increases initially.

Peptidase A (*PEPA*) is poorly studied and shows little variation; allele *2* has an average frequency of only 0.6%. The variogram is uninformative and is omitted. Taster (*PTC∗T*) is poorly known in this part of the world; it varies between 30% and 100%, with maxima in southern Chile and the southwestern part of North America. Minima are among Eskimos. This geographic distribution is in some agreement with an advantage for tasters in an area where antithryoid substances containing plants may be common, at least to the extent that Eskimos, who eat essentially meat and fish, are less exposed to the danger. It is not clear whether the areas with highest frequencies of tasters have a particularly frequent occurrence of edible plants dangerous for thyroid function.

Phosphoglucomutase 1 (*PGM1∗1*) varies from 55% to 100% with a mean of 83.5 for allele *1*; the maximum is in Venezuela, but a secondary peak is found in the Na-Dene region. There are various minima and an irregular variogram, as is almost usual. *PGM2* is less well known and, in any case, shows less variation, being confined to 80%–100% for allele *1*. A minimum is in the extreme south. The variogram of *PGM2* is uninformative.

6-phosphogluconate dehydrogenase (*PGD*) shows a low frequency of allele *B*, with some anomalies in northeastern North America and in northern Chile. Allele *C* is represented on the map, and *B* has the complementary pattern. The variogram has a moderate slope.

The *RH* system is also highly variable. Alleles *C* and *E* span essentially the whole range, while *D* is less variable, having, on the average, 96% frequency. *C* peaks in Panama and is lowest in the Arctic; *D* is universally high everywhere except for minima on the eastern coast of North America (possibly reflecting Caucasoid admixture, since Europeans have the highest world frequencies of the *d* allele [*RH*-]). *E* peaks in the Arctic and in the Andes; it is minimal in Panama.

The most frequent *RH* haplotypes are *CDe* (52%) and *cDE* (36%), and both span almost the entire 0%–100% range. The first peaks in Panama, and the second, in the Arctic. Next in frequency are *CDE* (4%), which also has several relative maxima in North and South America, up to about 30%; and *cDe* (4.6%), which peaks in the Southwest of North America. Ordinarily *cDe* is a good marker of Negroid admixture, which, however, seems very unlikely in the Southwest. The *cde* haplotype is, on the average, 2.5% and can be taken as a good indicator of Caucasoid admixture; not surprisingly, it shows a peak up to 20% on the eastern coast of North America, where we have seen other signs of admixture. It is uncertain if the relative maximum in the extreme northwest of Canada should also be interpreted as a result of Caucasoid admixture, because the other possible markers do not confirm it. Two rare haplotypes, *Cde* and *cdE*, show minor variations. Haplotype *cdE* surpasses 3% in a small area of Mexico and reaches 1%–2% in the extreme south of South America. *Cde* shows very low maxima in Mex-

ico and in the Southwest of the United States. Their maps are omitted. The variograms of *RH* show less extreme oscillations around the curve than most other American alleles, probably because of the greater number of data, and slopes are fairly large on the average.

The secretor locus (*SE*) varies from less than 40% to 100% in frequency of the *Se* allele and has a maximum around the equator. Parts of the map are not supported by data and are unlikely to represent real variation: for example, the maximum in Florida, which is extrapolated from the high Mexican values, and the maximum in the extreme south. The minimum in Brazil seems well supported and is not surprising given the high drift observed throughout America.

Transferrin (*TF*) shows a few troughs of the common allele *C*, where the alternative allele *D* reaches relatively high frequencies, up to 30%: in Panama, northern Venezuela, and Labrador.

The major conclusion is that the Americas, especially South America, show extreme genetic variability. This is also shown by average F_{ST} values, which were calculated for the 491 populations selected for detailed analysis. Below we compare the American average with averages of world groups or regions of interest:

America	0.070 ± 0.006
Caucasoid (no exclusions)	0.043 ± 0.005
sub-Saharan Africa	0.035 ± 0.014
Australia	0.019 ± 0.004
New Guinea	0.039 ± 0.007
Polynesia	0.031 ± 0.004

In the various regions of Asia, F_{ST}s range from 0.021 (Southwest Asia) to 0.035 (Southeast Asia).

Of the various subdivisions of the Americas, South America has the greatest variation of gene frequencies: the average F_{ST} is 0.059 ± 0.006. The gene with the highest variation is *SE∗Se* (0.30), followed by *KEL∗Jsa* (0.19), *PGD∗C* (0.18), and *TF∗C* (0.16). After South America, the extreme North has the greatest variation: 0.051 ± 0.007 (including Eskimo, Aleut, all Na-Dene, and also the Chukchi, who cluster with Eskimos); the most variable genes are *FY∗A* (0.26), *LE∗Le* (0.21), *PCT∗T* (0.13), and *KM∗(1&1,2)* (0.10).

North and Central America combined, including Na-Dene but not Eskimos, has a comparatively low average F_{ST} (0.034 ± 0.004). The most variable gene is *ABO∗A1* (0.17), followed by *A* (0.13), *HLAB∗35* (0.12), and *O* (0.12). Of the various linguistic groups, Chibchan shows a variation comparable to that of South America as a whole: 0.059 ± 0.007, with *DI∗A* being the most variable (0.17), *RH∗cDE* and *CDe* next (0.13 and 0.11), and finally *TF∗D* (0.11).

The impression from the geographic maps and distributions of gene frequencies is thus fully confirmed.

America, in particular South America, is genetically the most variable part of the world. As a consequence, there are extreme oscillations of mean F_{ST} values at various geographic distances around the interpolated variogram curves, that is, of the data points shown in variograms. These oscillations tend to be lower only for genes with high densities of observed frequencies, but even there the strong local geographic variation generates important fluctuations.

The F values indicated in the top right corner of the gene-frequency distributions given in each geographic map are F_{ST} values; but, unlike those given above, they are obtained from the original gene frequencies. They therefore include populations that have been excluded from the 491 selected as genetic references and, more importantly, they were pooled with neighbors. The data from the 491 populations are the basis for the F_{ST} values given above. Pooling neighbors decreases F_{ST} values (Cavalli-Sforza and Feldman 1990), and it is therefore not surprising that the F_{ST} values given in the maps are larger than the F_{ST}s calculated from the 491 populations.

An independent approach that leads to the same conclusions is the study of mitochondrial DNA. With a low-resolution technique, the restriction-fragment-length polymorphisms (RFLPs) of three tribes, Pima, Maya, Ticuna, were studied (Wallace et al. 1985; Schurr et al. 1990), and showed a variation of RFLPs similar to that of genes indicated above. Analyzing DNA markers makes it easier to identify specific mutants and may help us to follow specific migrations more closely. Inferences about the number of migrants to America that have been made in some mtDNA papers, even with techniques allowing higher resolution than those above, seem largely unwarranted at this stage of our knowledge.

6.13. SYNTHETIC MAPS OF AMERICA

Table 6.13.1 shows the partition of the total variation among the first seven PCs, which cumulatively explain 74.3% of the total variation. The seventy-two genes used for the analysis correspond to the 69 genetic maps listed in the Table of Genetic Maps with the addition of $ABO*A2$, $AK1*1$, $GC*1F$. Table 6.13.2 shows correlations of the first six PCs with gene frequencies.

The analysis of single genes shows considerable local variation. Patterns found for different genes are rarely similar. By contrast, in other continents, several geographic patterns of single-gene frequencies were observed repeatedly with different genes. In those continents, one could easily anticipate, on the basis of the repeated patterns, and the number of repetitions of each, the general shape of synthetic maps obtained by PCs and their order of importance. In America we find this occurs clearly only for the first two synthetic maps, which correspond closely to the first two fissions in the genetic tree.

The first PC (fig. 6.13.1) shows a north-south gradient with the greatest slope in Canada, thus emphasizing the distinction between the Eskimos + Na-Dene group and Amerind populations closer to Eskimos on the one side, and the rest of America on the other side. In South America, there is a differentiation between east and west. According to some archaeological dates, not universally accepted (see sec. 6.2), the eastern area may also be the oldest part. There is a good correspondence with the first fission, which separates Eskimos and Na-Dene from all Amerinds. To note: the highest correlation of the first PC axis is with $IGHG1G3*$ $za;b0stb3b5$, a typical marker of Asian origin.

Most of the divergence found in the map of the second PC (fig. 6.13.2) is observed in North America. There is little variation in South America, though the east-west difference is always noticeable. In North America the major divergence is between Eskimos and non-Eskimos, with Na-Dene showing more similarities to the former than to the latter. The peak in the eastern part of North America most likely represents Caucasoid admixture; this is the area in which contact between Caucasoids and Amerinds has been longest. This area has $ABO*B$, relatively high $AK1*2$, $IGHG1G3*f$; $b0b1b3b4b5$ and high $RH*cde$, strongly indicating Caucasoid admixture.

There is an inconsistency between the observations on the frequencies of the Caucasoid markers just indicated, which are drawn directly from the gene-frequency maps, and the correlations of this PC with the gene frequencies shown in table 6.13.2. The reason for this discrepancy is believed to be the existence of inordinate genetic variation in the Americas, which tends to cover other local regularities. The presence of important ethnic heterogeneity—that is, of Eskimos in the North—also tends to alter the meaning of the correlations of a

Table 6.13.1. Percentage of Total Variance Explained by the First Seven Principal Components of American Gene Frequencies

Principal Component	% of Total Variance	Principal Component	% of Total Variance
1	32.6	5	5.7
2	12.7	6	4.8
3	8.6	7	3.9
4	6.0		

Table 6.13.2. Genes Showing the Highest Correlations with the First Six Principal Components of American Gene Frequencies

P.C.*	Range of Correlation Coefficient		Genes
1	1.00 – 0.90	(+)	IGHG1G3*za;b0stb3b5, HLAB*27
		(–)	—
	0.90 – 0.80	(+)	ABO*A, ABO*A1, ACP1*A, AG*X, HLAA*9, LE*Le
		(–)	ABO*O, AK1*1, DI*A, HLAA*1, HLAA*31, HP*1
	0.80 – 0.70	(+)	HLAB*22, HLAB*40, HLAB*7
		(–)	IGHG1G3*zax;g, KM*(1&1,2)
2	0.90 – 0.80	(+)	HLAA*30
		(–)	—
	0.80 – 0.70	(+)	HLAA*33
		(–)	PGD*A
	0.70 – 0.60	(+)	JK*B, GC*1
		(–)	—
	0.60 – 0.50	(+)	HLAA*2, GLO1, HLAB*21
		(–)	TF*C, MNS*Ms
	0.50 – 0.40	(+)	ABO*A2, AG*X, GC*1F, HLAB*16, P1*1
		(–)	FY*B, HLAB*15, LE*Le(a+), RH*D
3	0.80 – 0.70	(+)	—
		(–)	IGHG1G3*za;b0b1b3b4b5
	0.70 – 0.60	(+)	RH*E
		(–)	RH*CDe, RH*C, IGHG1G3*f;b0b1b3b4b5
	0.60 – 0.50	(+)	RH*cdE, RH*cDE, IGHG1G3*fa;b0b1b3b4b5
		(–)	HLAB*5
	0.50 – 0.40	(+)	PTC*T, HLAB*14
		(–)	PGM2*1, LE*Le(a+), JK*A
4	0.70 – 0.60	(+)	HLAA*28
		(–)	—
	0.60 – 0.50	(+)	HLAA*2, HLAB*35, PGM1*1
		(–)	HLAB*14
	0.50 – 0.40	(+)	HLAB*21, PGM2*1
		(–)	HLAB*15
5	0.60 – 0.50	(+)	ESD*1, GLO1*1, JK*B
		(–)	—
	0.50 – 0.40	(+)	GC*1, FUT2(SE)*Se
		(–)	IGHG1G3*za;g, MNS*S, MNS*MS
	0.40 – 0.30	(+)	IGHG1G3*zax;g, HLAB*5, PEPA*1, RH*CDe
		(–)	MNS*Ms, RH*cde
6	0.60 – 0.50	(+)	—
		(–)	FY*B, KEL*K
	0.50 – 0.40	(+)	HLAB*22, RH*E, RH*cDE, FUT2(SE)*Se
		(–)	ABO*B
	0.40 – 0.30	(+)	CHE1*U, GC*1, GC*1F
		(–)	—

Note.– Genes giving positive or negative correlation values are indicated by (+) or (–), respectively.
* P.C., Principal component.

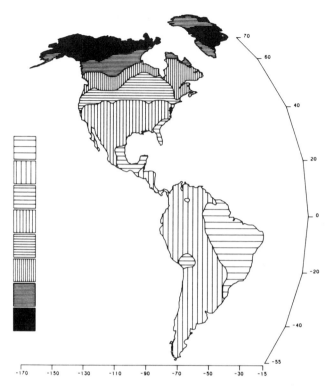

Fig. 6.13.1 Synthetic map of the Americas obtained by using the first principal component.

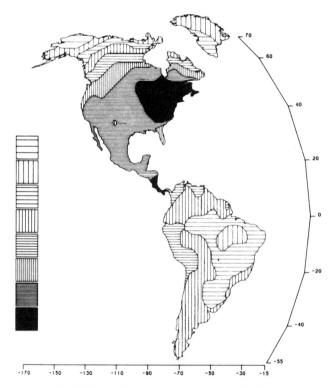

Fig. 6.13.2 Synthetic map of the Americas obtained by using the second principal component.

PC with individual gene frequencies observed in a specific region.

Central America is more similar to North America than to South America. Thus, this map shows approximate correspondence with the fission between Na-Dene and Eskimo, but also with that between South America and the rest of the Americas. It also highlights Caucasoid admixture of the eastern part of North America.

Extreme values for the third PC (fig. 6.13.3) are found especially in South America, the contrast being remarkably strong between the northeastern and the southern Andes. North America also shows some variation between east and west, and in the same direction as in South America. It is possible that the east-west gradients observed in the north and in the south again express Caucasoid admixture which, as we have seen when discussing single genes, is especially prominent in the east-central area of North America, but is not missing in South America. Caucasoid admixture is also probably found among Greenland Eskimos, who were in contact with Vikings, especially on the eastern coast in the ninth to fourteenth centuries A.D.. Eventually, the Vikings died of starvation or were killed by the Eskimos (their fate was never clarified), but there may have been genetic exchange. If this is true, it is not surprising that one finds some similarity in the degree of shading of the three areas that may have had some Caucasoid contribution; some further clarification to this problem comes from the next PC. An admixture of another nature—that is,

with Africans—is likely to have taken place in eastern Venezuela and the Guianas.

The fourth PC (fig. 6.13.4) also has a west-to-east gradient both in North America and in South America, but in contrast to the third PC, the direction of the gradient is inverted in the north and south. The similarity of the third and fourth PCs adds some evidence to the hypothesis that both eastern Greenland and the eastern coast of the United States have had some Caucasoid admixture, but the different behavior of the two components in Guiana may strengthen the hypothesis of admixture with Africans in this region.

The fifth component (fig. 6.13.5) stresses the difference between the Panama area and the rest of America. It is also indicative of migration to the south via Panama. The sixth map (not given) shows very little variation except in the extreme north, where it emphasizes the contrast between the Aleutian islanders and the Yupik Eskimos, occupying the southwestern part of Alaska, with the Eskimos of north-central Canada.

Other authors have used the synthetic map approach in America, O'Rourke et al. in both North (O'Rourke et al. 1986; Suarez et al. 1985) and South America (O'Rourke and Suarez 1986), and Salzano and Callegari-Jacques (1988) in South America. Both groups have found evidence of strong genetic drift in South America as we have, and their maps show less regular patterns than ours, being somewhat more similar to our single-gene maps. Our synthetic maps, however, seem less sensitive

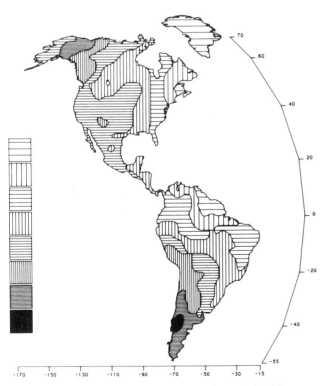

Fig. 6.13.3 Synthetic map of the Americas obtained by using the third principal component.

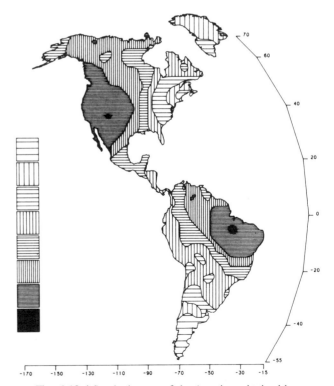

Fig. 6.13.4 Synthetic map of the Americas obtained by using the fourth principal component.

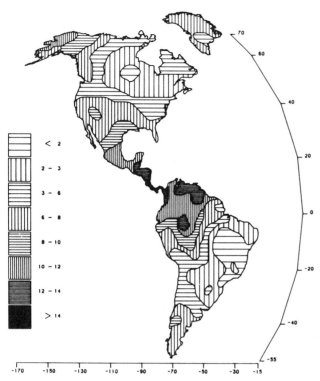

Fig. 6.13.5 Synthetic map of the Americas obtained by using the fifth principal component.

to drift than do individual genes. Our method obtains first maps of single genes and proceeds from them to obtain PCs and then their maps. This tends to smooth maps more than the direct calculation of PCs from original gene frequencies of selected groups or the slightly different mapping method used by O'Rourke and Suarez (1986). Differences in methods inevitably highlight one aspect or another; our synthetic maps are aimed at getting general similarities. Our single-gene maps are more useful than our synthetic maps for seeing highly localized effects of drift.

The conclusions from synthetic maps reinforce previous findings and help visualize major genetic regions. Eskimos, Na-Dene, and Almosan are well characterized and are even further differentiated into subgroups. The Caucasoid infiltration in the eastern United States, in eastern South America, and perhaps in Greenland are clear. The difference between the western and eastern coasts of North America is clear. In South America, several regions can be defined: the Andes show local homogeneity at the level of the higher PCs and always

differ from the eastern part of South America. The lower PCs show differences between northern, and central and southern Andes, with the northern ones more similar to Central America. The fourth PC emphasizes the uniqueness of southern Chile. In the eastern part, one can distinguish a northern region formed by eastern Venezuela and the Guianas (see, e.g., the third PC), probably affected by African gene flow; a central one formed by northern Brazil; and a southern one corresponding to southern Brazil. There are important ecological differences among these areas, and there probably was greater exchange within, rather than between, different ecological regions.

The color map of the Americas conveys 63.9% of the regional variations. In North America there are green and yellow zones, the yellow being Na-Dene speakers and the green areas mostly northern Amerind. The color picture does not supply a clear distinction between these and Inuit (Eskimos), probably because the latter inhabit a very thin area on the coast. The southern part of North America is grayish, and the pink area at the boundary between southern Arizona, New Mexico, and northern Mexico is a sort of average from various local populations: southern Na-Dene (Apache and Navajo, who also have some genetic admixture with Amerinds) and neighboring speakers of Uto-Aztecan languages.

Central America shows a complicated mosaic of colors, as expected of a region that was probably crossed many times by many groups. The area occupied by Chibchan speakers is relatively homogeneous. The Caribbeans are passively stained; there are no aboriginals left.

South America is dominated by two colors, red and blue, neither of which is found in North America. Both colors appear, though not at the same intensity or with the same nuance, in Central America as well, indicating that there are some remnants of the passage across the funnel north of it. Blue extends to the north and northeast and must represent a dominant direction of migration, where languages of Tucanoan, Caribbean, and Ge stocks are spoken preferentially. The other dominant migration in bright red is found in the southern direction along the Andes, but it did expand from the Andes toward the east, mostly into the Amazon plains, as we have seen from archaeology. Is the white spot in the middle of the Andes near Bolivia and Peru, an indication of a possible inverse Thor Heyerdahl (1950) effect, the arrival of Polynesians to South America?

6.14. SUMMARY OF THE GENETIC HISTORY OF AMERICA

The genetic patterns in the Americas fully confirm the three waves of migration suggested by dental and linguistic evidence: Amerinds, Na-Dene, and Eskimo.

Their order in time is strongly suggested by their north-south geographical order. Further refinements may reveal that more than one entry contributed to the first wave, but the archaeological information is contradictory and our understanding of the genetic pattern of Amerinds is incomplete, so that further investigations are required to settle this problem.

Eskimos, the last wave, fairly rapidly settled the Arctic coastal line and rarely occupied the interior. In the extreme east (Greenland) they may have mixed with Caucasoids, most probably because of contact with the Vikings who settled in Greenland and eventually vanished under fairly mysterious circumstances. It seems reasonable to assume that some of that population was reabsorbed by Greenland Eskimos.

The linguistic and geographic split between northern and southern Na-Dene is genetically clear-cut and probably reflects gene flow from other Amerinds, especially in the southern Na-Dene (Apache and Navajo), who had greater opportunities to receive it, because they were in more direct contact.

Amerinds show a much more complex picture. In North America, there is a band across the continent, which is wide in the east, of Caucasoid admixture. This admixture is also found elsewhere but it is less intense than in North America. In general, we have tried to avoid using populations in which admixture of some magnitude was suspected, but it was impossible to avoid mixed populations entirely without introducing an unwarranted bias.

In South America, one can use synthetic maps to distinguish three major genetic regions: the Andes, the Amazon basin, and the southern plateau. They are very different ecologically, and genetic exchange may have been less frequent among them than within. The genetic picture within the regions is so variable that an enormous amount of genetic drift clearly must have occurred. This variation is also found in North and Central America, but it is somewhat less extreme; besides, much of the genetic variation in North America is a direct consequence of major differences among ethnic groups like Eskimos, Na-Dene and Amerind, maintained over the millennia by ecological, behavioral, and social separation among the groups. No such obvious original ethnic differences exist in the rest of the continent. Clearly, fissions of tribes, and probably also fusions, have been numerous. Many tribes have probably originated from a small number of founders, justifying the enormous intertribal and interregional drift; they must also have moved around, as they still do, especially in areas like the Amazon and Orinoco basins. An important testimony to the extensive movements of Amerind tribes is the extreme fragmentation of the linguistic map, especially in South America.

It would be interesting to know whether some of the South American linguistic families existed before the passage through Panama and, if so, in which order they entered. The Andean family is found along the Andes, alternating with Paezan and, in some places, with Equatorial. It is not unreasonable to think that the

Andeans entered before the Paezan, given that they extend farther south. The Paezan family is present in North America (Florida) and is most closely related to Chibchan, which is found mostly in Central, but also in South America. The relationship of Chibchan and Paezan may antedate their entry into South America.

It is very difficult to make inferences about the order of entry of the people who today speak Carib, Equatorial, Ge, and Panoan, on the basis of genetic data. On the basis of the geographic distribution of linguistic families, however, it seems natural to suggest that they entered in the order in which they are found in South America, those located farther south being first. Some subfamilies, however, have a very wide range: the Equatorial family, for instance, is spoken from Venezuela to Uruguay.

	West		East
North (latest)		Chibchan	
			Carib
		Tucanoan	
	Panoan	Equatorial	Ge
South (earliest)	Paezan		
	Andean		

These considerations could have more weight if there was a good correlation between linguistics and genetics in South America. Unfortunately, there is not, or it has not yet been found. Moreover, the considerable genetic noise caused by drift, and probably highly variable from place to place, makes an historical interpretation of the genetic tree less credible in South America than in other parts of the world. With very small populations, of variable size, evolutionary rate from drift is so variable that the length of the branches of the tree is hardly indicative of evolutionary time, using distances based on gene frequencies. It is difficult to say if other approaches—for example, using mtDNA—can be more useful.

At the moment, the simplest hypothesis is that fissions and movements of tribes, their complex gene flows and fusions, and the contrast that can be expected between the genetic and linguistic effects of fusions between tribes all contribute to dissociate genetic and linguistic evolution and to some extent even their relation with geography in this part of the world. Some regularities emerge from the genetic analysis of major geographic regions in South America but, at a microgeographic level, several poor or negative correlations among genetics, geography, and linguistics show the need for more detailed research, perhaps carried out with other methods. The research by Spuhler in North America and that on the Panama Chibchan (Barrantes et al. 1990) reassure us that we are on the right track in assuming a parallelism of genetic and linguistic differentiation in America, that this research model is productive, and some times even more informative than work at a macrogeographic level;

however, not every region will be equally favorable for microgeographic analysis.

In a model designed to test whether the settlement of the Americas could have produced the high genetic variation observed (Cavalli-Sforza 1986), five assumptions were made: (1) demes (tribes) were of census size 500; (2) they produced "buds" 25% of the size of the initial deme; (3) buds doubled in size every generation of 25 years (a rate of growth supported by many observations on populations in free growth; see sec. 2.7); they therefore reached the size of a full deme in 50 years; (4) in a budding cycle (two generations), a deme moved an average of 250 miles (5 miles per year). (5) It is likely that buds advancing in new territory had low mortality, living in environments either not contaminated or less contaminated by previous inhabitants; on the contrary, demes in regions behind the advancing frontier would soon slow down population increase. Perhaps increasing mortality was caused by rapid saturation of local population density. It is a necessary assumption of any expansion that population growth is rapid at the frontier and ceases or slows down considerably back of the frontier (Ammerman and Cavalli-Sforza 1984).

Under these conditions, the occupation of the Americas could be completed in few millennia, and, in the absence of admixture between demes, the final genetic variation between demes would even be too high with $N = 500$. Gene flow between demes would, of course, reduce genetic variation. Tribal fusions are bound to have played an important part because the genetic variation would be excessive if the models above are right.

A demic budding and expansion process in two dimensions would probably be random in direction, certainly unguided except by the search for game, safety, and comfort. The idea that a single band wandered across from Asia to America seems unrealistic. Along coasts and rivers, the process would be closer to unidimensional and unidirectional. The average rate of (random) movement of 5 miles per year is fast, because its randomness means that often, but not always, it would bring the group to new territory. It is, of course, possible that movement was by leaps and bounds, greater than 5 miles per move if people stayed in the same place for several years in a row. This pattern of repeated movement involves a specific behavior that is not typical of present-day hunter-gatherers (e.g., for African Pygmies; Cavalli-Sforza 1986), who move for long distances during the year but on established paths and repetitive, well-known circuits. In the past, Pygmies have certainly moved for long distances, in search of new abodes, but it is difficult to find comparable modern situations.

The model is very approximate, and only an accurate simulation could give more realistic values. Perhaps only at a later stage, closer to saturation of population density, fusion events would become more common. It is difficult to evaluate the saturation density in environments as diverse and poorly known as those in South America. Clearly, population density gradually rose in the Andes to levels much higher than in the rest of the subcontinent. Many urban developments, the skillful exploitation of the variety of ecological niches and astute social management in organized states must have gradually but greatly increased the carrying capacity of the Andean region in the last few millennia.

The most successful civilizations arose in the Andes and in many parts of Central America where the climate was more favorable. No such developments ever took place outside the Andes or other parts of Central America. But in the northern subcontinent, in times before European contact, Plains tribes were probably of relatively large size. More sedentary groups lived in communities that reached numbers in the thousands (sec. 6.4). Thus, wherever population numbers grew, the effects of drift were buffered and, especially where urban communities arose, they were eventually drastically reduced.

7 AUSTRALIA, NEW GUINEA, AND THE PACIFIC ISLANDS

7.1. GEOGRAPHY AND ENVIRONMENT

7.1.a. AUSTRALIA

Australia, the smallest continent, occupies 5% of the earth's land (about one-quarter less than Europe), but its population is only 0.3% of the world total, and the aborigines are only 3% of the present Australian total. Especially from our point of view, therefore, it is certainly the least densely inhabited of the continents. It owes this property to the extreme aridity of its climate in most of its area, and to the consequent scarce economic development of aborigines, who were still hunter-gatherers of the stone age when James Cook's voyages (1768–1779) led to the British settlement of Australia.

Inland Australia is a great plain, almost all less than 600 m (2000 feet) in elevation. Rainfall is extremely low in the central and western parts, increasing somewhat toward the coast, but in the south and west the desert reaches very near the coast. Only 10% of Australia receives more than 1 m of rainfall a year. By contrast, the northern and eastern coasts and the island of Tasmania, located at the southeastern tip, are very humid. The continent extends from 11° to 44° S latitude; accordingly, the north has a tropical climate, with large extensions of mangrove, and the northern half of the east coast (fig. 7.1.1) is covered by tropical rain forest. Somewhat to the interior of the northern coast lies evergreen forest; farther inland are savanna and grassland, and then, in the heartland, the eremian zone (the desert). The southeast, southwest, and Tasmania have a temperate climate. Tasmania was attached to the mainland in earlier times, but may have been separated completely from it by less than 10 kya.

The flora and fauna of Australia are very different from those of Southeast Asia. The vegetation ranges from eucalyptus trees and savannas to acacia scrub in the more arid parts. In the Pleistocene, however, the climate was cooler than now, and a wide freshwater lake system existed in the southeast. Lake Mungo (see fig. 7.2.1), which is located in the southeast but dried up about 16 kya, was one of the oldest inhabited sites (Jones 1989).

7.1.b. NEW GUINEA

New Guinea is the second largest island in the world (after Greenland) and is about one-tenth the size of Australia, to which it was connected before the end of the last glaciation. They are now separated by the Torres Strait, about 100 miles (150 km) wide. Located from 3° to 10° S latitude, New Guinea extends from northwest to southeast. The Central Cordillera crosses the island length and reaches high altitudes (maximum nearly 15,000 feet [5000 m]); the highlands in the center are 4000–6000 feet (1000–2000 m) high. Rainfall is especially high on the coastal plains, where tropical forest is dominant. The highlands are less humid. The total population is over 3 million indigenous inhabitants, a high density for a people that was at the Neolithic stage until this century. Politically it is divided into a western, slightly smaller moiety under Indonesian control (Irian Jaya); and an eastern part belonging to the independent nation of Papua New Guinea.

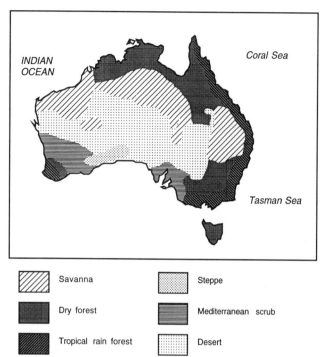

Savanna

Dry forest

Tropical rain forest

Steppe

Mediterranean scrub

Desert

Fig. 7.1.1 Climate and vegetation of Australia (Times Atlas of the World 1989).

7.1.c. PACIFIC ISLANDS

The Pacific islands are traditionally divided into Melanesia (including New Guinea), Micronesia, and Polynesia (including New Zealand). Excluding New Guinea and New Zealand, the total area is about 100,000 square miles (250,000 km²), divided among 10,000 islands (see fig. 7.1.2). Strictly speaking, Melanesia includes New Guinea, which we prefer to treat separately. The total number of inhabitants is in excess of one and a half million (estimated on the basis of speakers of eastern or Oceanic Austronesian languages [Ruhlen 1987]), to which a number of speakers of Indo-Pacific languages other than those living in New Guinea should be added.

7.2. PREHISTORY AND HISTORY

Australia was always separate from Southeast Asia at all times of interest to us, but the sea distance was certainly shorter in the Pleistocene. About 18 kya, Borneo, Sumatra, and Java were connected to the mainland of Southeast Asia, forming the Sundaland peninsula, while Australia, Tasmania, and New Guinea formed a single continent, called Sahulland. The problem of the arrival of *Homo sapiens sapiens* in Australia is not completely understood. There was agreement on setting the date of first arrival at 40 kya or earlier, with direct evidence of settlement about 39 kya, on the extreme southeast, far from the area of entry. Very recently, however, thermoluminescence dates have indicated arrival in southern Australia at 50–60 kya (Roberts et al. 1990). There is (incomplete) consensus that only modern humans settled Australia (see sec. 7.3).

The two oldest archaeological sites with human remains are shown in figure 7.2.1: (1) West of Sydney at Lake Mungo were found the remains of a woman cremated at Lake Mungo I and the burial (without cremation) of a man at Lake Mungo III, 500 m from the first find. The dates are 26–32 kya for the woman (Bowler and Thorne 1976, see also Flood 1983) and somewhat earlier for the man. Other earlier remains suggest that the initial occupation of this area was 35–37 kya (Thorne 1980). (2) The second site is Keilor, west of Melbourne and not far from the sea north of Tasmania, but once certainly located inland. The oldest layer may date from 35 to 45 kya (Thorne 1980).

Most or all the continent was occupied very early, and there is an old date even in the interior; it is believed that the most favorable ecological zones were already occupied by at least 20 kya (Jones 1987). For a more detailed map of dates of occupation of Australia and New Guinea, see Jones (1989). Archaeology thus confirms the hypothesis by Birdsell (1957), based on a simulation of population advance, that the continent was occupied fairly rapidly. Bowdler (1977) has suggested that the first occupation followed the coastline. Given the increase in the water level that has probably occurred since then, most earlier archaeological sites may be under water. The hypothesis that the coastline was occupied first is also made attractive by the inevitable consideration that the earliest settlers must have had considerable practice

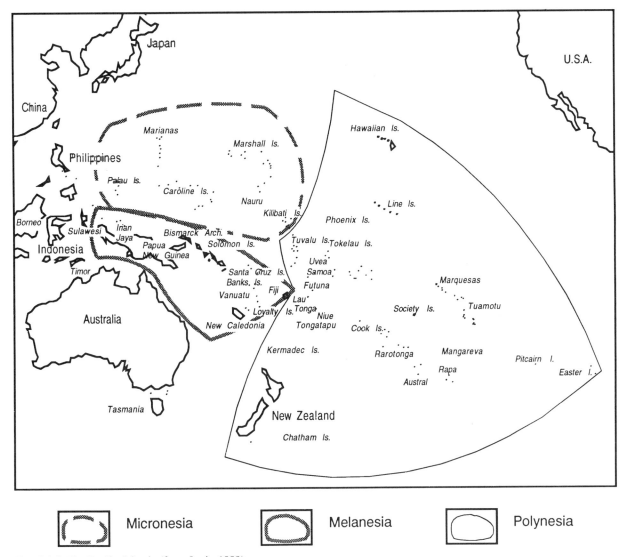

Fig. 7.1.2 The Pacific Islands (from Irwin 1980).

in offshore boating. The type of boats or rafts used is not known.

The only animal brought into Australia by the aborigines was the dingo, one of the few nonmarsupial mammals found in Australia. Most dingoes are wild, but some have been tamed and used for hunting by the aborigines. The oldest dingo remains in Australia are dated to at least 6000 years ago (Bailey 1980). The animal probably came from Southeast Asia; it competed with the Tasmanian wolf and the Tasmanian devil, both of which are now extinct.

Among old dates in New Guinea, Huon peninsula and Kosipe Swamp (fig. 7.2.1) are 40 and 26 kya, respectively (Flood 1983). There is a gap between this date and later ones (Groube et al. 1989, Kirk and Thorne 1976).

It is likely that Australia and New Guinea were occupied by the same people, but some archaeologists have postulated several distinct migrations, despite extremely weak evidence (for a review, see Habgood 1989). There

is agreement on the origin from Southeast Asia; there is no direct evidence of which of the possible routes was, or were, followed. Birdsell (1977) made a thorough analysis of possibilities and suggested two major routes (northern and southern) with a total of five variants. Of two possible southern routes, the more likely one starts from Java and proceeds through the lesser Sunda islands and Timor to northwestern Australia. The most likely northern route is from Borneo through Sulawesi and Seram to the Bird's Head (the extreme northwestern portion of New Guinea). In any case, there were tracts of sea of 80–100 km to be passed, which required better transportation across the sea than anything used by Australian aborigines in these days.

The climate changed repeatedly through the period from first arrival to the present, and there may have been successive waves of migrants showing various biological and cultural adaptations to the changing climate. A new technology of stone artifacts appears in later times and

Fig. 7.2.1 Major archaeological sites in Australia and New Guinea. (From Whitehouse and Whitehouse 1975; Bailey 1980; Thorne 1980).

is called the Australian small-tool tradition, developed primarily after 6000–5000 years ago. This tradition did not reach Tasmania, which was already separated from the mainland. The oldest boomerang was found in South Australia and dated to 10 kya (Bailey 1980).

7.2.a. DOMESTICATION OF ANIMALS AND PLANTS IN NEW GUINEA

Although Australian aborigines never developed farming or the use of metals, in New Guinea agriculture developed early. Until 1960, Neolithic-like stone tools were in common use in the central New Guinea Highlands.

Agricultural developments occurred in the highlands, and one of the major sites east of Mount Hagen is a rock shelter, Kafiavana, which was first occupied at least 11 kya. The earliest strata have a flake industry characteristic of New Guinea, but beginning 9000 years ago there appear axe-adzes, and later also in other parts of New Guinea. Marine shells are found indicating trade with the coast. Swamp-drainage channels dating to 9000 years ago, perhaps used for taro cultivation, were detected in the Western Highlands province. Evidence for swamp drainage in other New Guinea areas is much later (3500 years ago); pollen research, however, shows that deforestation was frequent in the New Guinea highlands by 5000 years ago or earlier. It is believed that these agricultural developments were local and not imported (see Bellwood 1979). The sweet potato, of South American origin, was probably introduced about 300 years ago via the Philippines where it was imported by the Span-

ish. It soon joined taro and yams as a major staple crop, allowing further population growth.

Pigs and fowl came to New Guinea from mainland Southeast Asia, Sumatra, and Java (Bellwood 1979). These are the only animal domesticates in Melanesia other than the dog, but pigs and fowl were never introduced into Australia. The presence of pigs indicates cultivation, because this animal is usually fed on agricultural crops, here probably sugar cane, sago, yams, bananas, and taro (Irwin 1980). Pigs have an especially important role in Melanesian society, much greater than their potential role in meat supply. Women dedicate much time to growing pigs and caring for them individually. This care includes breastfeeding pigs, often along with the woman's babies, so that pigs do not break their milk teeth, greatly valued as ornaments. Pork-eating in special ceremonies is a major use of pigs. There are contradictory reports about the first time of entry of pigs to New Guinea: they might have been introduced by 6000 B.P. or possibly earlier (Spriggs 1984). According to Bellwood (pers. com.) a date of 2000 B.P. would fit the linguistic data better.

7.2.b. OCCUPATION OF THE PACIFIC

Both archaeology and linguistics have shown conclusively (Bellwood 1979) that the settlers of Micronesia had a different origin from those of Polynesia. Western Micronesia (Palau, Marianas, and possibly Yap) was probably settled directly from Indonesia or the Philippines, whereas eastern Micronesia (Nuclear: Caroline, Marshall, Nauru and Gilbert Islands; see Spriggs 1984) may have been settled from Melanesia, possibly Fiji. Fiji shares cultural similarities with Polynesia, with which it may share origins, but the similarities were attenuated in the very long separation that followed.

In the Marianas, the earliest archaeological record shows pottery likely to be of central Philippine origin (1500 B.C.: Bellwood 1979). There are no known B.C. dates for Palau or Yap. The latter island was, at the time of European contact, the head of the Yapese empire.

A major eastward expansion began around 3600 years ago and is tied to a new culture, called *Lapita* from the name of an archaeological site in New Caledonia (Bellwood 1979). From insular Southeast Asia, it was spread first to Melanesia and then to the whole Pacific Ocean in the next 2500 years by very mobile sea colonists with skills in agriculture and pottery. There is reasonable consensus (Bellwood 1979) that Lapita potters spoke Austronesian languages and were responsible for their first arrival in Oceania. The earliest decorated Lapita ceramics have very characteristic motifs, which are found from New Guinea to Samoa. Lapita voyagers also transported obsidian between islands and created characteristic techniques of nonceramic assemblages. On the basis of ce-

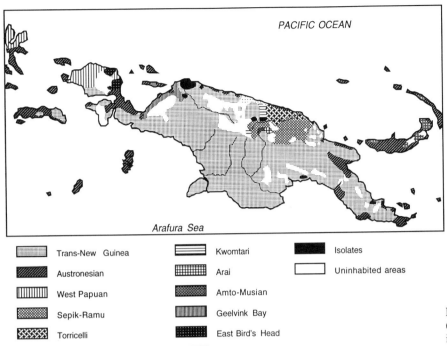

Fig. 7.2.2 Modern distribution of Papuan (New Guinean) and Austronesian languages in New Guinea and nearby islands (after Wurm and Hattory 1983).

ramic features, it has been suggested that Lapita pottery, culture, and the people who carried them around the Pacific had their origin in northeastern Indonesia or the Philippines (if not in China) (Bellwood 1979). In many coastal regions of Melanesia, the Lapita potters lived side by side, and sometimes mixed with, the earlier Papuan inhabitants. In these areas, the two language families are still spoken to this day (see fig. 7.2.2).

The secondary Austronesian expansion from Melanesia to Polynesia was probably made possible by a major invention, the outrigger, which gives considerable stability to a canoe. In insular Southeast Asia a double outrigger is used (one on each side of the hull), but in the open ocean a single outrigger is preferred. In addition to these sailing canoes, skills in navigation and colonization were essential. Other forms of canoes were developed, like the double canoe with two hulls of equal size and the big war canoes, single- or double-hulled, with paddlers (Bellwood 1979).

The following shows the sequence of the occupation of Polynesia (on which there is only incomplete agreement) (Bellwood 1979, 1989); see figure 7.2.3.

Tonga: before 1200 B.C., settlement of Lapita potters
Samoa: settled from Tonga 1000 B.C.
Marquesas: from Samoa (not exclusively), A.D. 200–300
Hawaiian Islands: from Marquesas, A.D. 500–600
Society Islands: from Marquesas, A.D. 800
New Zealand : from Society islands, A.D. 800–1100
Easter Island: the most eastern and most highly isolated island, from Marquesas, A.D. 500.

The Polynesian colonization of the Pacific thus took about 2500 years and was complete by the year A.D. 1000. A suggested scenario (the "express train" to Polynesia, see summary by Diamond 1988) assumes that the ancestors of the Polynesian culture came from insular (or perhaps even mainland) Southeast Asia. In the Philippines and Sulawesi, artifacts resembling Lapita ones were found (see fig. 7.2.3). From there a major step forward may have been through Halmahera, Waigeo (northern Moluccas), Biak, and other islands to the Bismarck islands, where in Mussau the largest known Lapita site was recently discovered, with already developed agriculture and animal husbandry. Unlike other Bismarck islands there is in Mussau no sign of earlier, Melanesian arrivals; this island is relatively isolated from the neighboring ones, indicating that early Polynesians could reach islands farther away and out of sight, thanks to better navigation and sailing techniques. The Mussau site was dated to around 1600 B.C., as early as the other Lapita sites, which are all located farther east (New Caledonia, Fiji, Samoa, and Tonga). These islands are believed to have been first settled by Polynesians, and only later by Melanesians. Figure 7.2.3 also shows the last step of the eastward, northward, and southward Polynesian expansion from the Tonga-Samoa-Fiji triangle, the dates for which were indicated above.

At the time of European contact, the largest populations were those of the biggest islands or archipelagoes, Hawaii (perhaps 200,000) and New Zealand (100,000–150,000; Bellwood 1979). Most Polynesian societies were highly stratified. From the point of view of genetic

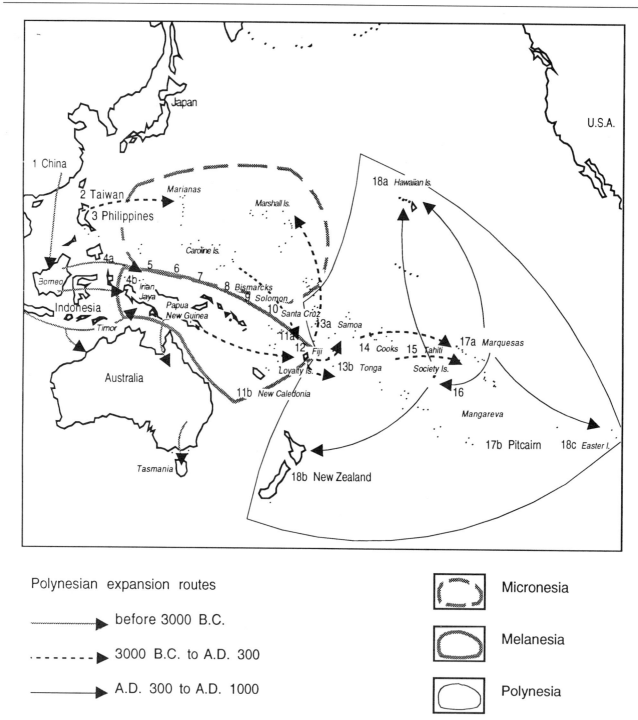

Fig. 7.2.3 A sketch of the "express train to Polynesia" (after Diamond 1988). Lapita pottery flourished between 1600 and 500 B.C. in stations 8–13; possible pre-Lapita pottery was found in 3 and 4a, while 4b–7 are speculative. Polynesian populations are found today in 9–18, with the exception of 17a where Polynesian artifacts were found.

evolution, there are three important considerations: (1) the initial size of founders was probably relatively small in most instances and is not known; (2) there were frequent later bottlenecks in population history, which in some cases were dramatic, especially on smaller islands more exposed to damage by typhoons (see Morton et al. 1972 on the Micronesian island Pingelap); (3) there often were later additions by subsequent migrations.

7.3. PHYSICAL ANTHROPOLOGY

The oldest human remains (sec. 7.2.1) are considered by most archaeologists to belong to anatomically modern humans and are similar to modern aborigines, but a disconcerting difference is found between them and a rich collection of skeletons detected at Kow Swamp, not far from Melbourne (see fig. 7.2.1) dated between 15,000 and 9000 years ago (Kirk and Thorne 1976; Thorne 1980). The Kow Swamp humans are large and more robust, with a marked prognathism. It is somewhat paradoxical that a more rugged type is found at a time intermediate between the early and the modern forms, which are both gracile. Recent dating by ESR (election spin resonance) suggests both robust and gracile were present in the same area 25–30 kya (Mellars and Stringer 1989, p. 11). Thorne (1980) and Wolpoff et al. (1984) postulated that the Kow Swamp remains and two other skulls, one from Queensland and another from Western Australia are descendants of an earlier Indonesian type of *H. erectus*. Brown (1987) hypothesized that both gracile and robust Australian remains are part of a single continuous range.

Birdsell advanced the view (for a summary, see Birdsell 1979) that there were three distinct migrations to Australia. The first was of Oceanic Negritos, who saturated the area, but were replaced by later migrants or mixed with them. Today, Negritos are still found in the Andamans, in Malaysia (the Semang), and in the Philippines (Aeta). They are small, dark-skinned, have peppercorn hair and, at least in the Andamans, steatopygia, the last two characters being found also among African Bushmen and African populations with whom the Bushmen are likely to have exchanged genes. In Australia the aboriginal type is reasonably homogeneous and unlike the Oceanic Negritos. There are, however, a few exceptions: for example, 12 tribes of small aborigines living in a forest area near Cairns (Northeast Queensland) show a high frequency of Negritoid hair forms, also common among Tasmanians.

In Birdsell's model of Australian settlement, the second migration, that of Murrayians, swamped and replaced the first wave of migrants. Murrayians are so named because their physical type would correspond to that of historical aborigines, such as those present in the Murray River basin, in the southeast. The third wave is

named Carpentarian from the Gulf in the north of Australia. It is at the moment difficult to judge the validity of this hypothesis. Perhaps some support for the two latter waves comes from the linguistic divisions of Australian languages, as discussed below.

According to Howells (1989), Tasmanian skulls are more similar to those of Melanesians than to those of Australians. There is again some linguistic support to this claim because the Tasmanian language belongs to the Indo-Pacific family, spoken in New Guinea, and not to the Australian group.

Modern Australians have variable stature with a leaner body build in the northern part and greater hairiness in the south; skin is nearly black in the north, and brown in the south. Eyes are brown and hair is black, but in the desert hair is often blond. Negroid hair is found only in some coastal tribes and was common in Tasmania. All Tasmanians of full descent are extinct today (see Diamond 1991) and the Melanesian phenotype has been altered by intermarriage with ancient Micronesian and Polynesian populations.

In New Guinea, the type is Australoid with some differences. Hair is tightly curled, and the nose is bridged (found in Australia only in the desert). There are some "Pygmy" tribes, especially on the highlands. New Guineans are reminiscent of Oceanic Negritos. It is possible that the Negritic type shows a better adaptation to tropical climates (Omoto 1985). In the other islands of Melanesia, the phenotype is similar to the New Guinean, although it is sometimes diluted. The Papuan curved nose is not the rule, and the skin may be lighter, though still brown.

The Polynesians are brown, of medium stature, usually thickset, with curly to straight hair and facial features with a diluted Mongoloid type. They vary considerably from island to island.

The Micronesians are usually smaller and even more variable than the Polynesians. In Yap the type is Australoid. There is some frizzy hair and some individuals look more Mongoloid than the Polynesians.

These classical anthroposcopic observations are basically confirmed by recent multivariate analyses of anthropometrics and skull measurements (Howells 1989).

7.4. LINGUISTICS

At the time of European contact, there probably lived in Australia some 200,000–250,000 aborigines (Jones 1970). This population size may have been near saturation capacity. In 1980 there were 205,000 aborigines, only a small fraction of whom speak original Australian languages or are considered "full-blood." About 7500

aborigines still live in nearly traditional conditions in an area of the Northern Territory, thanks to a recent initiative of the Australian Government. Until the middle of the present century, the dominant policy was comparable to the traditional one of the United States toward American Indians, perhaps even harsher, but it has become much more enlightened in recent years. In Tasmania, there lived around 4000 people; the last one of full descent died in 1876. Disease and genocide have contributed to the thinning and extinction of many aboriginal groups (Diamond 1991a).

Three different language groups constitute the Pacific languages:

1. Many languages were spoken by Australian aborigines, who are believed to have begun penetrating into this area more than 40 kya and perhaps 55 kya (see sec. 7.2).
2. The Papuan languages form the Indo-Pacific family according to Greenberg (see Ruhlen 1987) and were spoken in most of New Guinea, but also in the Andaman Islands, with the oldest dating back about 15 kya or earlier (see Wurm 1982).
3. The largest and geographically most widely extended group, the Austronesian languages, are spoken from Madagascar to Polynesia. They probably started spreading from or through the Formosa-Philippine region about 5000 years ago, penetrated into Melanesia about 3500 years ago, and into Polynesia 2000 years ago (Tryon 1985).

About 170 Australian languages are still spoken, some by extremely few people, and all except perhaps 50 are likely to be extinct within the next two generations (Ruhlen 1987). Originally there were 400–500 different languages, judged on the basis of the number of tribes, each of which usually had a distinctive language (Dixon

1980). The classification of these languages is not entirely satisfactory, but it is believed by most that they all had a common origin. One large subfamily (*Pama-Nyungan*) is spoken in almost all of Australia except the central north (see fig. 7.4.1), where 14 subfamilies, most of which include only 2 or 3 languages, are present according to Ruhlen.

There is an approximate geographic correspondence between the Carpentarian immigration wave postulated by Birdsell (1977; sec. 7.3) and the non-Pama-Nyungan languages, and between the Murrayian wave and the Pama-Nyungan languages. The extreme fragmentation of the former subfamily, which occupies a relatively small area in the north of the continent, would seem, however, to suggest that the non-Pama-Nyungan speakers settled earlier, in contradiction to Birdsell's model.

New Guinea and the rest of Melanesia speak mostly languages belonging to the Indo-Pacific family, also called Papuan. This large group of more than 700 languages includes Tasmanian and Andamanese, but, as in the case of Australia, the bulk of languages is classified in one large subfamily, the Trans-New Guinea phylum, including over 500 languages (see fig. 7.2.2). Half of these belong to the Main Section, which includes the Highlands and the rest of the mountain chain.

The other New Guinea languages are distributed in 10 subfamilies, of which Sepik-Ramu (about 100 lan-

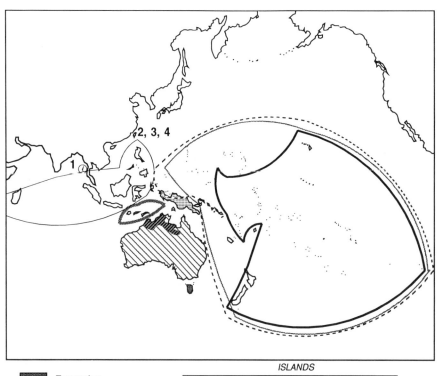

		ISLANDS	
▓ Tasmanian			
▦ Other Indo-Pacific groups	——— West Malayo-Polynesian	1. Andamans	
	▬▬▬ Central Malayo-Polynesian	2. Atayalic	
⧄ Pama-Nyungan	- - - - East Malayo-Polynesian	3. Tsouic	
▨ Non-Pama-Nyungan	········· Oceanic	4. Paiwanic	
	——— Polynesian		

Fig. 7.4.1 Linguistic map of the Pacific Ocean, including the big islands (from Ruhlen 1987).

guages) and Torricelli (about 50) are the richest. Melanesia was, at contact, much more populous than Australia; the lack of obvious riches and the tropical climate made it less attractive to European settlers. It is believed there were about one and a half million inhabitants at first European contact in the sixteenth century, about a million in New Guinea and the rest in the islands. The ecological conditions and the early development of horticulture made it possible for this population to reach much higher densities than in Australia. West Irian or Irian Barat is even now mostly undeveloped. Australian influence in the eastern part of New Guinea and neighboring islands, which began at the end of the last century, was mostly benign. It spread slowly to the interior, and at the time of a visit to the New Guinea highlands by the senior author (1967) many populations in the southwestern part of Papua New Guinea were still not "pacified"; that is, control by Australian authorities had not been established, and traditional customs including head-hunting had not been eliminated. Other Melanesian islands like Fiji and New Caledonia received relatively large numbers of Indian and European immigrants, respectively.

Languages of another family, Austronesian, are also found in Melanesia. Austronesian (almost 1000 languages) is a subgroup of Austro-Tai (itself a subfamily of Austric) and includes two groups.

1. A Western group (with more than 500 languages and over 250 million speakers) ranges from Indonesia, the Philippines, and the Southeast Asia mainland (southern Vietnam, Cambodia) to Madagascar. The languages spoken in western Micronesia (Chamorro, Palauan, Yapese) also belong to this branch.
2. Over 400 Austronesian languages called Eastern, or Oceanic, are spoken by one and a half million people in Melanesia, Micronesia, and Polynesia. They were clearly spread by the Indonesian voyagers who populated the Pacific islands and, in the process, passed their linguistic

heritage to people of the north coast of New Guinea, most other Melanesian islands, eastern Micronesia, and the Central Pacific, that is, Polynesia (including Fiji, which is classified geographically as Melanesia but is on the boundary with Polynesia).

Especially in the smaller Melanesian islands, but also in parts of New Guinea, there is a major linguistic division between Austronesian (AN) and non-Austronesian or Indo-Pacific (NAN) speakers.

Thus, about 1800 languages belong to the Pacific area, close to a third of all the languages in the world. Almost all of these languages, except for the less than 40 Micronesian and Polynesian languages, are in the southwestern and western corners of the Pacific, the area of greatest concentration of different languages in the entire world. The recognition of Austronesian loanwords in Papuan languages located in areas geographically far removed from Austronesian languages today helped to trace language migrations within the New Guinea mainland. It seems plausible that an east-to-west migration of speakers of Papuan languages started about 3500 years ago around the Markham valley in northeastern Papua New Guinea. This migration may have occurred in response to pressure from an aggressively spreading Austronesian population with whom the Papuan speakers had sufficiently long and intensive contact to borrow a considerable number of Austronesian words, including personal pronouns (Foley 1986). Such a migration appears to have been preceded by an earlier, west-to-east "main" migration that spread the ancestral languages of the largest Papuan language group (Trans-New Guinea Phylum) through four-fifths of the New Guinea mainland (about 5000 years ago). This earlier migration is also suggested by the presence of Austronesian loanwords of a more archaic type that seems to form a deep linguistic substratum of the New Guinea mainland (Wurm 1982).

7.5. Genetic population structure in Oceania

Much of our present knowledge of the genetics of Oceania comes from work done in the laboratories of R. T. Simmons (compendia in Simmons [1962] and Simmons and Booth [1971]) and of R. L. Kirk (summaries in Kirk 1965, 1969; Sanghvi et al. 1971). In the multitude of small islands of the Pacific Ocean, there has been extraordinary opportunity for random genetic drift. In addition, Australia and New Guinea are of unique interest in having been, at first European contact, at economic stages of development that had been replaced thousands of years ago in other parts of the world. Although Australians were, in a way, contemporary examples of Paleolithic hunter-gatherers, New Guineans had been through a Neolithic revolution but had not reached the age of metals. They represented (and still do, in part of the island)

modern examples of Neolithic horticulturalists. To the extent that their organization has not been overly changed, they thus are extremely interesting from a genetic and anthropological point of view. Only some populations in the South American tropical forest live in conditions similar to those of New Guinea populations. In no other place can one find almost intact ways of life that disappeared from Europe and most of Asia thousands of years ago, and from Africa centuries ago. Australian aborigines supply instead examples of Paleolithic ways of living, but only a small group is still relatively unacculturated. In Africa one finds hunter-gatherers who cannot, however, be considered perfect examples of Paleolithic life.

From the point of view of population structure, information on tribal customs, demographic properties, and especially migratory exchanges can therefore be extremely interesting. The basic problem, what fraction of genetic variation is due to differential natural selection or random genetic drift, requires knowledge of these factors. No current demographic information can be considered truly satisfactory from a genetic point of view, but at least some approximate estimates exist and are worth considering. There is still a chance of collecting more information, as some thousands of Australian aborigines in the Northern territory live more or less undisturbed, and much of western New Guinea is almost virgin land. The majority of Oceania has, however, been affected from a demographic point of view by European contact and by post-contact migrations. The numbers of inhabitants that we cite below (taken mostly from McEvedy and Jones 1978) refer as much as possible to populations before contact, but are inevitably very approximate. Modern numbers of speakers are taken from Grimes (1984).

The work by Tindale on Australian tribes has generated basic information (summary in 1953), and subsequent work by Birdsell (1957,1977) has clarified some demographic and ecological aspects that are of considerable interest. Birdsell has noted that the Australian tribes, defined on the basis of linguistic and territorial criteria, have a modal census size of about 500. Doubts have been expressed about the validity of the territorial definition of the tribe (see answers in Birdsell 1977), but in any case the linguistic definition seems more important. Tribes are mostly, but not entirely, endogamous; and an estimate indicates that, on the average, 15% of marriages are (were) mixed. Exogamy varies between tribes from 7% to 21% and is equivalent to 3.5%–10.5% migration per generation (Tindale 1953).

Marriage may not be the only cause of intertribal migration, but it is likely to be a major one. Birdsell also found that hunting bands (exogamous in principle) are an average size of 25. The numbers 25 and 500 were called "magic" and "mystical" (see Lee and De-Vore 1968), but they seem to agree, with some variation, with most ethnographic data from other populations (see Cavalli-Sforza 1986b). The effective population size N_e is smaller than the census size by a factor of one third (Cavalli-Sforza and Bodmer 1971a). With this N_e and moderate migration rate, random genetic drift is likely to be important; we noted in chapter 2 that all measurements of genetic variation confirm this hypothesis.

In Melanesia, the number of non-Austronesian speakers was about one million before the demographic increases that followed European contact. With over 700 languages, equating the language to the tribe gives about 1300 individuals per tribe, but there is considerable size variation. Estimates of intertribal migration are scanty (for a detailed summary of biological research in Bougainville, see Friedlaender 1975, 1987). Moreover, tribes do not necessarily have a simple structure. For instance, the Bundi tribe comprises over 4000 people,

subdivided into 14 clans that are exogamous only if they are small, with exogamy decreasing regularly with increasing size of the clan (Malcolm et al. 1971).

In Polynesia and Micronesia, the situation varies considerably with the size of the island. The largest islands, New Zealand (consisting of two major islands) and the Hawaiian archipelago, each had about the same number of inhabitants as the total for Australia. Thus, at European contact, New Zealand had a population density about thirty times higher than that of Australia. In the eighteenth century, however, 80–90% of Maoris were living in the agricultural regions of the northern North Island. The South Island was inhabited basically by hunter-gatherers because of the cold climate (Bellwood, pers. communic.). The Hawaiian Islands had a population density 15–20 times higher than that of New Zealand. New Guinea was probably about as densely populated as New Zealand at the time of contact. These differences reflect ecological conditions, history, and especially the level of development of agriculture, which was completely absent in Australia, but was more advanced in Polynesia than in New Guinea. Tasmania had a population density comparable to that of England in the Mesolithic (see chap. 5).

In the rest of the Pacific islands there are perhaps 100,000 Polynesians; the most populous islands are Tonga, Samoa, and Tahiti. Many islands of the Pacific Ocean have very small populations, some fewer than 100 inhabitants. Even when the number of present-day inhabitants is relatively large, the number of founders must have been very small. After the first settlement, the islanders of many atolls are likely to have gone through a number of population bottlenecks, mostly because of typhoons. In the atoll of Pingelap in Micronesia, the modern population is about 1600 people; one of the many typhoons which plague these islands took place in 1775 and caused a very serious famine, leaving perhaps 30 survivors (Morton et al. 1972).

The populations of the smaller islands are obviously expected to have undergone considerable random genetic drift. Does the same occur for languages? Little empirical information is available on this point. A source of potential confusion is the use of the word "drift," which has different meanings in linguistics and genetics. This word was proposed by Sapir (1921) to describe a trend in linguistic change, especially phonetic, observed to recur in different languages. "Evolution" is another word that has different meanings in the two disciplines; in linguistics, evolution is usually taken to involve "progress," and the word "change" is preferred. In the microevolutionary changes observed during modern human development, there is little chance of observing progress, especially when the major evolutionary factor is random genetic drift. Migration and statistical fluctuations resulting from small population size counterbalance each other in biology; do they do so in linguistic evolution? There may be similar effects of isolation by distance in the two situations, suggesting some analogy. An investigation of the Micronesian islands (Cavalli-Sforza and Wang 1986) showed that linguistic similarities between islands followed the model of isolation by distance developed for genetic variation (summarized in sect. 1.16), with a complication caused by the strong variation of rates of change for different words.

When the population of a small island diverges from that of the population of origin in conditions of complete genetic isolation (no migration), d, the genetic distance between them, increases in direct proportion to the time of separation and inversely proportional to twice N_e, the effective population size:

$$d = t/2N_e,$$

where t is the time of separation in generations. This formula is valid for small distances; otherwise, the relationship is slightly more complicated. In this simple formulation (see sect. 1.4), if N_e is 10 times smaller, genetic distance increases 10 times faster. When there is also migration, the rate of change decreases as migration increases. Furthermore, when N_e is smaller, migration may tend to be larger, partially compensating for the greater drift when N_e is small. Among other complications are later major population movements, which may more or less radically change the ethnic composition of an island. This must have happened repeatedly in the settlement of Pacific islands, making the analysis of genetic differentiation complicated. We have already noted in chapter 2 that the branch of the tree leading to the Polynesian islands is perhaps 10 times too long. Given the small size of many Polynesian islands, there may have been extreme drift which may have been only partly counteracted by taking averages over many islands. Another

factor may have affected the length of the branch leading to Polynesians. Depending on the method of tree reconstruction used, an early admixture (perhaps between Melanesian and Southeast Asian populations) can give rise to a very short branch (with methods allowing variation of branch length) or a too early date of separation with methods forcing constant evolutionary rates.

We have already seen in chapter 2 that New Guinea and Australia are widely separated genetically and that Polynesia, Micronesia, and Melanesia tend to form a small cluster that shows similarities with both New Guinea and Southeast Asia, associating with one or the other depending on methods of analysis. This is a strong indication that important admixtures have occurred before and after the common origin. Perhaps the most important admixture was between Melanesian and Polynesian ancestors. Melanesians were responsible for a first spread from New Guinea to the nearer Pacific islands. Polynesian ancestors probably came originally from the mainland of Southeast Asia through the Philippines and Indonesia and spread much later than the Melanesians, first through islands already occupied by Melanesians and then to the outermost islands of the Pacific Ocean (Pawley and Green 1985).

A recent and very comprehensive book edited by Hill and Serjeantson (1989) on the colonization of the Pacific includes chapters that extensively summarize present genetic knowledge of the area. In particular, Kirk (1989), who is responsible for much population-genetic work in Oceania, analyzed red-cell antigen, serum protein, and enzyme systems of the Pacific, and Serjeantson (1989) analyzed *HLA* data; other authors considered other genetic systems. In the following section we treat separately Australia, New Guinea, and the islands of the Pacific Ocean, with the various methods of analysis used in the earlier chapters.

7.6. Population genetics and synthetic maps of Australia

The genetic picture by tribe is highly fragmentary, given the amount of drift. Moreover, southern areas have almost no aborigines today, but their original location is at least approximately known on the basis of Tindale's (1953) work. It is, however, safer to consider the south of Australia empty and all map values an extrapolation with all the uncertainties tied to extrapolations. Genetic drift was strong in Australia but not as much as in South America, judging from average F_{ST} values, which is, at the original sample level 0.066, 0.050, 0.084 in Australia, New Guinea, and the Pacific islands, respectively. The Pacific islands' high value can be imputed not only to drift but also to the participation of different, and very incompletely admixed, ethnic groups.

Genetic distances from which the tree is calculated are shown in table 7.6.1. The average number of genes in-

Table 7.6.1. Genetic Distances (x 10,000) between Major Geographic-Linguistic Groupings of Australian Aborigines (in the lower left triangle of the matrix) and Their Standard Errors (in the upper right triangle of the matrix)

	NTU	WTN	CPN	NPN	QPN	WPN
NTU	0	76	109	72	200	116
WTN	180	0	146	55	483	79
CPN	293	429	0	75	252	129
NPN	258	223	268	0	448	97
QPN	551	602	872	779	0	422
WPN	354	393	394	381	1154	0

Note.– NTU, Northern Territory, language or tribe unspecified;
WTN, Western Territory, non-Pama-Nyungan language speakers;
CPN, Central Territory, Pama-Nyungan languages;
NPN, Northern Territory, Pama-Nyungan language speakers;
QPN, Queensland, Pama-Nyungan speakers;
WPN, Western Territory, Pama-Nyungan speakers.

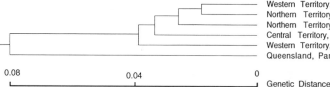

Western Territory, non-Pama-Nyungan
Northern Territory, unspecified
Northern Territory, Pama-Nyungan
Central Territory, Pama-Nyungan
Western Territory, Pama-Nyungan
Queensland, Pama-Nyungan

0.08   0.04   0
|_____| Genetic Distance

Fig. 7.6.1 Genetic tree of six major geographic-linguistic groups in Australia.

vestigated was 58. A reconstruction of the history of the continent was attempted by tree analysis of the major regions for which enough data were available: Northern Territory, Queensland, Central, and Western Australia. For Western Australia, it was possible to separate tribes speaking Pama-Nyungan and non-Pama-Nyungan languages. For the Northern Territory, we separated tribes speaking Pama-Nyungan from tribes speaking unidentifiable languages that are probably non-Pama-Nyungan (fig. 7.6.1).

The tree shows a definite outlier, the Queensland tribes, which speak a Pama-Nyungan language. There are at least two possible explanations for their position. They are not mutually exclusive and may even have some common elements. Being nearest to New Guinea, Queensland people may have had more exchange with this island than was true of other parts of Australia. This hypothesis, however, finds no support in an analysis of contemporary New Guinea data (see table 7.7.3). Another hypothesis is generated by the existence of a group of Negritic tribes in Queensland, which suggests an older heterogeneity. This might agree with the first migration wave proposed by Birdsell (1977), which, however, he assumed was swamped (perhaps not entirely) by later migrations.

The most internal cluster in the tree corresponds to the Northern Territory plus the non-Pama-Nyungan tribes on the northern part of the Western Territory. This is in some contrast to the fact that this area has the greatest linguistic heterogeneity.

The PC map does not add anything to the tree and is omitted. The synthetic maps based on all the 38 genes represented in the second part of the book except G6PD*def and IGHG1G3*fa;b0b1b3b4b5 (see Table of Genetic Maps) are more informative. The fraction of variance explained by the first five PCs is given in table 7.6.2. They account cumulatively for 83.3% of the total variance. The genes contributing mostly to the major PCs are given in table 7.6.3.

The first synthetic map of Australia (fig. 7.6.2.A) shows a difference between north and southwest, with a peak in northern Queensland. The minimum in the southwest is not reliable since there are practically no aboriginal populations left there at their original locations. The lesson from this map is that Queensland is not as sharply different from the northwest as the tree of figure 7.6.1 would simplistically imply. Genes contributing mostly to this PC are ABO, RH, and some enzymes. The poverty or absence of data in the south makes the southern minimum doubtful.

The contrast between the western and the eastern portion is highlighted in the second synthetic map (fig. 7.6.2.B). The western minimum is found in the northern part of the Western Territory, where most non-Pama-Nyungan languages for which data were available are located. Genes contributing here are mostly MNS

Table 7.6.3. Genes Showing the Highest Correlations with the Most Important Principal Components in Australia

P.C.*	Range of Correlation Coefficient		Genes
1	1.00 – 0.90	(+)	ABO*O, RH*CDe, RH*cDe
		(−)	ABO*A, RH*cDE, RH*E
	0.90 – 0.80	(+)	RH*C
		(−)	PGM1*1
	0.80 – 0.70	(+)	ABO*B, CHE1*U, PGK1*1, PGM2*1
		(−)	ABO*A1, PGD*A
2	0.80 – 0.70	(+)	MNS*Ns
		(−)	IGHG1G3*za;b0b1b3b4b5
	0.70 – 0.60	(+)	KM*(1&1,2)
		(−)	MNS*Ms, P1*1
	0.60 – 0.50	(+)	IGHG1G3*za;g, IGHG1G3*zax;g
		(−)	—
3	0.90 – 0.80	(+)	ACP1*B
		(−)	—
	0.80 – 0.70	(+)	—
		(−)	—
	0.70 – 0.60	(+)	CHE2*(+), MNS*S
		(−)	—
	0.60 – 0.50	(+)	RH*CDE, ESD*1
		(−)	—
	0.50 – 0.40	(+)	GC*1, MNS*M
		(−)	ADA*1

Note.– Genes giving positive or negative correlation values are indicated by (+) or (−), respectively.
* P.C., Principal component.

Table 7.6.2. Percentage of Total Variance Explained by the First Five Principal Components in Australia

Principal Component	% of Total Variance	Principal Component	% of Total Variance
1	43.8	4	6.4
2	16.3	5	5.3
3	11.5		

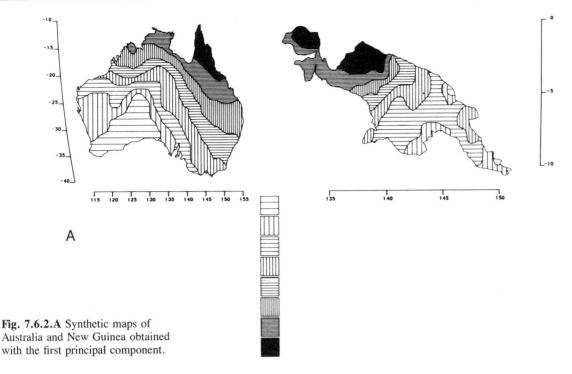

Fig. 7.6.2.A Synthetic maps of
Australia and New Guinea obtained
with the first principal component.

and immunoglobulin genes. The southern minimum is
not considered.

The third map (fig. 7.6.2.C) contrasts the central area
with minima in the southern part of Queensland and
in the central western part. Mostly enzymes determine
this component, in particular *ACP1*B* which, however,
shows in general little variation in Australia, as seen in
the single-gene maps. It may partly reflect a selective
response to climate.

It is difficult to link the synthetic maps with any migration wave, since none is well ascertained. However,
the synthetic map of the second PC is in some agreement with the essence of the linguistic picture and with
the second and third waves hypothesized by Birdsell.

A major problem of interest for the general history of
world migrations is the possible similarity of some relic
populations in South and Southeast Asia with Australian
aborigines on one side, and Africans on the other. These

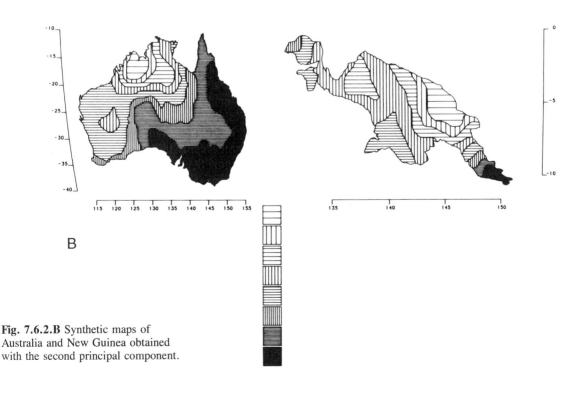

Fig. 7.6.2.B Synthetic maps of
Australia and New Guinea obtained
with the second principal component.

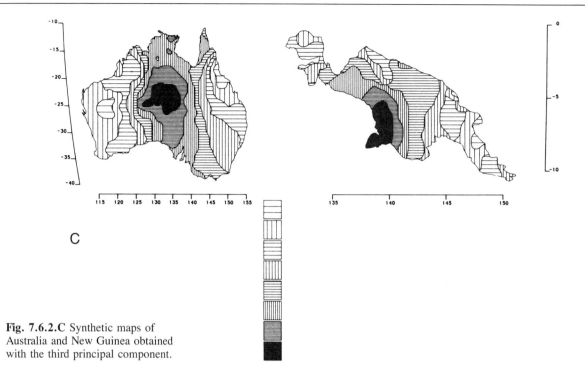

Fig. 7.6.2.C Synthetic maps of Australia and New Guinea obtained with the third principal component.

populations are interesting because they might be evidence for a southern route of migration from Africa to Australia. They are often called Australoid, and some of them are given more specific names like Veddoid, Negritos, pre-Dravidian, etc. (see secs. 4.5, 4.6, 4.13, and 4.14). The resemblance of these populations to Africans or Australians is based on superficial, anthroposcopic characters, skin color, hair or body shape, and there is always the suspicion that they are the product of convergence because of a common climate. Thus far genetic data have not helped to recognize a relationship; see for example, research by Kirk in India (sec. 4.14; also Kirk and Thorne 1976; Kirk and Szathmary 1985) and by Omoto (1985) on Negritos (sec. 4.13). Research using gene frequencies on genes subject to bilinear inheritance and underlying recombination may not be the most suitable means of solving similar problems; genetic systems under unilinear inheritance like mtDNA and segments of the Y chromosome might be more useful.

Table 7.6.4 shows the genetic distances between two "Australoid" populations studied reasonably well genetically: Kadar from Kerala, India, and Negritos from the Philippines. Their potential genetic relatives in Asia and Australia are also given. They clearly show greater sim-

Table 7.6.4. Genetic Distances (x 10,000) (± standard errors) between Three Australoid Populations from South India Kadar and Philippine Negritos, and Other Possible Relatives from Oceania, Africa, and Asia

	Kadar	Negrito
India	554 ± 101	1031 ± 290
Dravidian	610 ± 160	893 ± 192
Indonesia	598 ± 198	1054 ± 362
Malaysia	2176 ± 1031	764 ± 159
Philippines	1169 ± 777	345 ± 82
Australia	1983 ± 535	1714 ± 513
New Guinea	1826 ± 789	1589 ± 367
San	3018 ± 1111	1256 ± 326
West Africa	3247 ± 1190	2210 ± 680
Kadar	0	1550 ± 857

ilarity to their neighbors in India or Southeast Asia than to New Guineans or Australians. Their similarity to the latter two populations is sometimes even smaller than that to representatives of the African continent. The separation is clearly too great and the gene flow from neighboring populations too important to find a significant relationship with the present data by this method.

7.7. POPULATION GENETICS AND SYNTHETIC MAPS OF NEW GUINEA

Thanks to the extensive scientific travels of C. Gajdusek and to the laboratories from Australia and elsewhere collaborating with him, a fairly large number of

New Guinean tribes have been tested for many genes, but genetic knowledge is not homogeneously distributed. The Australian moiety is better investigated than Irian

Jaya, and the Eastern Highlands in particular have been fairly intensively covered. In spite of this remarkable work, our genetic understanding remains incomplete. The linguistic information is also potentially confusing. Interesting cases of rapid language replacements in New Guinea caused by, for example, various types of word taboos have been shown by Wurm (in press). With complex genetic and linguistic patterns, and potentially volatile mechanisms of change, it is not too surprising if it is difficult to correlate them in any simple way. An added complication is that Austronesian languages are also found on the coastal plains, especially in the north.

Several microgeographic studies in Melanesia take into account both genetics and linguistics. Perhaps the most complete is by Friedlaender on Bougainville in the Solomon islands (1971, 1975, 1987). The correlation between genetic distances and linguistics was 0.53 (the average of two measurements of genetic distance, compared with 0.39 between genetics and anthropometrics, 0.38 with geography, and 0.35 with migration). He also concluded "that language boundaries per se do not provide a long-term barrier to marriage migration." In thousands of years of genetic exchange between neighbors speaking different languages, much of the original genetic differences have disappeared, and new ones have been generated by drift. Languages have also changed and may have even been replaced. Such complexity may some times be beyond the analytic means, when there is no historical information to explain unexpected discrepancies.

In an accurate survey of linguistic and genetic differentiation in New Guinea, Serjeantson et al. (1983) have summarized a number of earlier attempts at investigating this problem that suffer from the use of a single gene: GM, the allele AB (for aborigine) of GC, or a Gerbich blood-group allele. Other more recent investigations have used haplotypes of the β-globin region detected by DNA restriction analysis, a relatively powerful genetic system, but potentially subject to incompletely understood natural selection. In general, it is extremely unlikely that a single gene can completely resolve the problem. Clearly, the more complex a situation, the more necessary it is to resort to multivariate analysis using as many genes as possible; even then the number of markers may be critical.

Serjeantson et al. (1983) also reported on their own investigations, carrying out microgeographic comparisons of genetic and linguistic similarity on the north coast of Papua New Guinea, where Austronesian (AN) and non-Austronesian (NAN) languages are found side by side. They found considerable overlap of genetic and linguistic classifications. Five AN linguistic groups are spread almost randomly among 12 NAN groups in the genetic tree formed from these 17 groups. In the region that underwent more detailed study, the Bogia subprovince (east of the mouth of the Ramu river), the correlation

between genetic and linguistic distance (measured on the basis of cognates) was 0.787. The correlations of these two distances to geographic distance are 0.925 and 0.787, respectively. But the correlation between genetic and linguistic distances falls to zero if one partials out the effect of geographic distance. These data are based on only nine languages, one of which is Austronesian, and it is difficult to establish their statistical significance. In the original paper correlations were calculated using genetic and linguistic similarities instead of distances, but we have artificially replaced the term similarity with distance by changing the signs of all correlations, so as to keep all correlations positive for simplicity. The conclusion Serjeantson et al. reached is that Austronesian-speaking groups in Papua New Guinea are only "skin-deep" Austronesians, an expression they borrow from the linguist Capell. A conclusion these authors stress is that linguistic relationships may occasionally give rise to erroneous conclusions if they are used for inferring migrations without also considering genetic relationships.

We believe the major bar to finding clear genetic distinctions between AN and NAN populations in New Guinea (and the rest of Melanesia) is that the AN migration is ancient. Archaeological data set it at 3500–5000 years ago (Bellwood 1979, 1989). In this length of time, we have seen (see section 1.17) that genes can be replaced, perhaps even totally, by gene flow from neighbors. When there is a relatively high probability of gene replacement in addition to language replacement the situation may be very confused, and in the absence of clear history, it may be impossible to understand.

The AN immigration must have been numerically important, given that the linguistic influence was in some areas fairly profound (see fig. 7.2.2). One can expect that, on the average, genetic traces will be clear in the areas more heavily settled by AN immigrants. A trail of migration has been noted by using globin markers, detected by molecular techniques (Hill et al. 1985; O'Shaughnessy et al. 1990). The trail indicated by three relatively rare hemoglobin mutants seems to mark three different segments of the Austronesian migration from the Philippines to north New Guinea, eastern Melanesia, and Polynesia. Rare mutants (or, better, haplotypes) whose frequency is magnified by natural selection cannot be easily used for quantitative migration studies, but can be excellent qualitative markers for it, especially if they are sufficiently complex that it is unlikely that they originated more than once. The data are not inconsistent with a scenario of Polynesian colonization including routes through both Melanesia and Micronesia, perhaps meeting in the melting pot of Fiji-Samoa-Tonga, from which the final later migration to the eastern Pacific took place.

For our analysis of classical polymorphisms, as usual only reasonably well investigated tribes were included. Population numbers are from Grimes (1984). The Highlands are particularly well represented, and data were

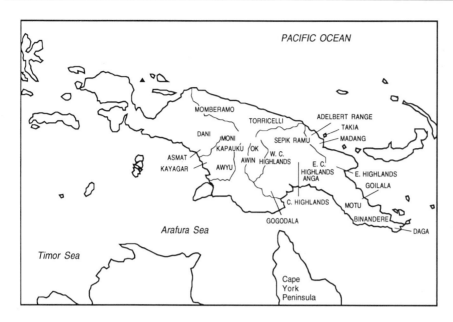

Fig. 7.7.1 Location of the tribes and clusters of tribes in the tree in figure 7.7.2.

grouped by geographic region, as eastern, east-central, central, and west-central Highlands in Papua New Guinea. This corresponds with the linguistic clustering. For instance, the east-central highlands include the Fore (12,000 people, decimated by the kuru epidemic, now ended), the Gimi (16,000), the Gahuku (35,000), the Benabena (12,000), and the Kamano (50,000). The eastern highlands include the Enga (110,000), Gadsup (11,000), Auyana (6500), and Tairora (8000). Other groups are represented in the data by fewer tribes; the census size of the tribes is not always available and tends to be a little smaller, on the average, than for the tribes just given: Gogodala, 10,500; Awin, 8000; Kayagar, 4000; Asmat, 38,000; Goilala (also known as Kunimaipa), 8000; Moni, 25,000; Anga (also known as Kukukuku), around 40,000; Kapauku, 10,000; Awyu, 18,000; Binandere, 3000; and Daga, 5500. The location of these tribes and tribe clusters is shown in figure 7.7.1.

The genetic tree obtained from these data (24 tribes) is shown in figure 7.7.2. Table 7.7.1 gives the F_{ST} genetic distances. The average number of genes investigated was 45 ± 4.3. Three clusters are detected. There is some agreement with geography, but the geographic clustering is not perfect. The first cluster includes tribes like Asmat, Ok, Kayagar, and Gogodala, all from the southern plains or nearby (Dani), but does not include other neighbors like the Moni and Kapauku, who join the second cluster. The second cluster includes all the Papua New Guinea Highland populations and also several populations from the northern coast, from east to west, the Goilala or Kunimaipa, Sepik-Ramu, Torricelli, and even the fairly remote Momberamo. The last (and smallest) cluster is from the extreme southeast (Motu, Binandere) with a group of tribes from the northeast, Madang, Adelbert, Takia. Many of the populations speak AN languages.

Despite a basic agreement with geography, a certain frequency of geographic displacements must be accepted if this picture is correct. The number of genetic markers is low (45 on the average); Torricelli and Kapauku have the lowest numbers (20). In isolation, these data do not add to the problem of correlation with linguistics, which is particularly hard to explore in New Guinea.

In the PC map (fig. 7.7.3), the first PC accounts for 26% of the variation, and the second for 21%. The first PC gives an imperfect split according to an east-west gradient. Populations in Irian Jaya are mostly on the right side of the diagram, and those from Papua New Guinea are at the left. The clearest geographic clustering is that of the Highlands (bottom center).

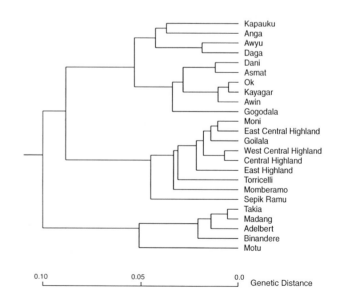

Fig. 7.7.2 Tree of 24 tribes and tribal clusters from figure 7.7.1.

Table 7.7.1. Genetic Distances (in the lower left triangle of the matrix) and Their Standard Errors (in the upper right triangle of the matrix) for New Guinea Tribes (all values x 10,000)

	HGC	HGE	HEA	HWC	EBI	EDA	EGO	NAD	NMA	NSR	NTO	ANG	ASM	AWI	AWY	GOG	KAY	OK	DAN	KAP	MON	MMB	MNM	MNI
Central Highlands	0	44	90	19	382	275	77	161	73	178	139	654	758	667	143	992	168	662	78	188	120	135	1328	823
E. Central Highlands	141	0	82	32	394	345	58	350	60	65	100	106	837	782	123	1062	261	774	83	154	69	141	1455	642
Eastern Highlands	338	240	0	52	163	73	43	104	138	287	243	432	466	401	100	555	332	402	57	172	69	120	799	426
W. Central Highlands	71	140	162	0	360	239	64	33	30	191	215	69	638	427	66	718	185	447	61	177	94	106	1542	855
Binandere	846	956	558	723	0	110	179	99	79	682	459	153	568	552	143	723	209	547	115	176	148	163	292	96
Daga	689	838	295	480	447	0	199	74	140	894	608	70	257	252	81	459	72	244	139	144	188	169	628	286
Goilala	182	153	177	177	515	640	0	79	53	158	79	336	742	729	122	1013	186	692	69	123	60	100	843	451
Adelbert	378	892	452	154	246	241	312	0	33	75	97	425	89	153	129	55	192	112	162	246	131	192	285	57
Madang	135	154	324	64	166	444	147	97	0	59	47	64	103	110	140	77	224	62	100	256	63	118	102	25
Sepik-Ramu	386	317	736	346	1219	1458	413	254	134	0	128	424	1678	1457	121	1846	87	1288	116	351	200	200	1495	1163
Torricelli	235	201	539	400	948	1168	211	383	113	408	0	340	1722	1399	322	2424	28	1267	66	196	140	107	1362	925
Anga	1091	390	745	301	585	357	669	1475	208	1160	837	0	377	271	211	489	51	274	161	128	103	242	768	459
Asmat	1333	1546	808	1051	1315	593	1312	439	309	2678	3086	906	0	66	68	55	62	57	35	252	98	95	1390	947
Awin	1265	1397	699	874	1173	460	1349	454	344	2652	2655	624	216	0	73	76	70	26	174	141	180	283	1620	1178
Awyu	490	399	275	182	332	190	493	437	421	324	474	591	310	254	0	225	310	44	102	92	139	164	758	223
Gogodala	1881	1957	1394	1354	1402	857	1745	170	187	2992	4216	1021	238	438	619	0	103	70	135	332	201	205	1543	1188
Kayagar	494	647	667	491	764	227	462	691	632	492	79	202	181	160	743	289	0	32	89	110	286	199	325	223
Ok	1183	1369	714	833	1095	470	1236	388	232	2435	2357	624	169	52	145	277	51	0	127	69	103	164	1568	1122
Dani	297	291	191	250	592	514	258	616	416	360	187	451	120	509	492	485	309	355	0	340	104	75	207	118
Kapauku	675	419	567	582	615	546	591	577	580	1055	430	377	822	775	245	547	317	256	852	0	249	370	95	310
Moni	262	107	176	189	387	487	133	393	176	455	309	231	287	329	510	413	580	227	377	517	0	134	575	113
Momberamo	305	357	388	352	477	666	318	542	253	611	359	630	386	795	645	490	692	500	308	853	290	0	316	142
Motu	3119	3191	2296	2795	623	1935	2046	803	270	3338	2566	2360	3469	3352	1570	3009	815	2946	706	383	1128	687	0	93
Takia	1792	1795	1248	1391	217	914	1094	187	56	2102	1608	1751	1920	2205	737	2127	849	2009	497	808	418	377	382	0

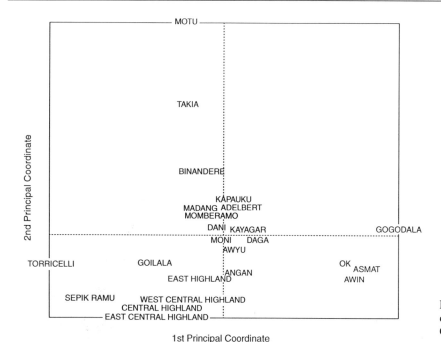

2nd Principal Coordinate

1st Principal Coordinate

Fig. 7.7.3 Principal component map of 24 tribes or tribal clusters from New Guinea.

It is useful to examine the synthetic maps of New Guinea (figs. 7.6.2.A, B, C), noting that they might be somewhat different from the expectations generated by the analysis presented above. Although a selected group of tribes tested for a larger number of genetic markers was shown in figures 7.7.2 and 7.7.3, in synthetic maps all available data were employed to generate single-gene maps (described later), which were then used to build the synthetic maps. Thus, more genetic data are available in these maps than in figures 7.7.2 and 7.7.3, because all data detected in the literature are represented, with the exception of genes for which the information was too scarce to build a single-gene map. The picture may therefore differ in figures 7.6.2, 7.7.2, and 7.7.3.

The amount of information provided by the first six PCs is given in table 7.7.2. The first six PCs express cumulatively 80.4% of the total information. The genes contributing most to the highest PC's are indicated in table 7.7.3. Altogether 38 genes were used, and they correspond to the single-gene maps in this book.

The first synthetic map (fig. 7.6.2.A) shows a longitudinal gradient. In fact we know that there were different migrations west to east and east to west, the last possi-

bly about 4000–3500 B.P. after the Austronesians reached New Britain and New Ireland (see sec. 7.4). This longitudinal gradient can be construed as the major genetic difference in the island of New Guinea.

The second synthetic map (fig. 7.6.2.B) shows a major contrast between the extreme southeast and the northern plains of Papua New Guinea (the Australian moiety). The low values define the Sepik-Ramu area, a language group that appears to share a number of cultural and genetic traits with the Australian aborigenes (Tryon 1985). It is tempting to see the Sepik-Ramu Phylum language speakers as remnants of the original Australian migration which passed through Papua New Guinea.

The third synthetic map (fig. 7.6.2.C) contrasts the southern plains with the western and eastern extremes. It shows some correlation with the AN-NAN settlement pattern.

Questions raised in the discussion above receive a partial answer from a direct analysis of the genetic-distance matrix. In table 7.7.4 we show the distances between regions of New Guinea and Australia to permit the search for possible similarities established at the initial settlement, later migrations, or exchanges between neighboring parts of these two biggest Oceanian islands. There is a weak indication of some exchange between the northern parts of Australia, especially the speakers of non-Pama-Nyungan languages, and New Guineans. One may be tempted to test the hypothesis that there might be some ancient relationship between the non-Pama-Nyungan languages and Indo-Pacific ones. This test may be difficult, because of the presence of very small speech communities who adopt "regional vocabu-

Table 7.7.2. Percentage of Total Variance Associated with the First Six Principal Components in New Guinea

Principal Component	% of Total Variance	Principal Component	% of Total Variance
1	24.0	4	7.7
2	19.8	5	6.7
3	17.0	6	5.2

Table 7.7.3. Genes Showing Highest Correlations with the First Six Principal Components in New Guinea

P.C.[*]	Range of Correlation Coefficient		Genes
1	1.00 – 0.90	(+)	FUT2(SE)*Se
		(−)	—
	0.90 – 0.80	(+)	ESD*1
		(−)	MNS*NS, MNS*S
	0.80 – 0.70	(+)	ADA*1
		(−)	—
2	0.90 – 0.80	(+)	KM*(1&1,2)
		(−)	—
	0.80 – 0.70	(+)	LE*Le(a+)
		(−)	PGM1*1
	0.70 – 0.60	(+)	MNS*Ms, MNS*M
		(−)	PGM2*1
3	0.90 – 0.80	(+)	IGHG1G3*za;b0b1b3b4b5
		(−)	—
	0.80 – 0.70	(+)	HP*1, RH*cDE, RH*E
		(−)	—
	0.70 – 0.60	(+)	—
		(−)	RH*CDe, PGD*A
	0.60 – 0.50	(+)	RH*CDE, GPT*1, MNS*Ns
		(−)	ACP1*B, IGHG1G3*fa;b0b1b3b4b5, RH*C, IGHG1G3*za;g
4	0.70 – 0.60	(+)	ABO*A
		(−)	—
	0.60 – 0.50	(+)	—
		(−)	ACP1*B
	0.50 – 0.40	(+)	LE*Le(a+)
		(−)	RH*CDe, PGD*A, RH*C
5	0.60 – 0.50	(+)	ABO*B, GC*1, IGHG1G3*zax;g
		(−)	ABO*O, RH*cDe
6	0.60 – 0.50	(+)	MNS*N
		(−)	—
	0.40 – 0.30	(+)	RH*cDE, PGK1*1, RH*E
		(−)	ABO*O, RH*CDe, GC*1, HP*1

Note.– Genes giving positive or negative correlation values are indicated by (+) or (−), respectively.
[*] P.C., Principal component.

laries," words particular to certain geographical regions (Capell 1962).

Another attempt was made to detect similarities between potential ancestors of proto-Polynesians, who might have settled in regions of New Guinea, and modern inhabitants of these regions (table 7.7.5). F_{ST} distances between various possible proto-Polynesian ancestors, or their modern descendants, and various New Guinea regions show that, by this criterion, only the extreme western and southeastern parts of New Guinea may have received some gene flow from proto-Polynesians. These are actually the regions where a greater number of Austronesian languages is spoken today (fig. 7.2.2).

A preliminary analysis of mtDNA in New Guinea (Stoneking et. al. 1986) has been considerably extended (Stoneking et al. 1990). From 25 localities of the Australian (the eastern) moiety of New Guinea, 119 individuals were tested by restriction analysis and found to correspond to 65 mtDNA types. They form three major clusters plus other small ones, indicating the possibility of different migrations. There is very little similarity to Australians, no more than to Africans or Caucasoids; this analysis is based on small numbers, but there is definitely a greater similarity to Asians than to all others, and this is probably significant. The authors suggest that the small similarity with Australia may be due to the arrival of later migrations to New Guinea in the last 10 ky that did not extend to Australia. Correlation of genetic distance with geographic distance is present but minimal, and the small correlation with linguistic differences disappears if one corrects for geographic effects. We find mtDNA data are, as far as comparable, in excellent agreement with nuclear DNA data.

The color map includes both Australia and New Guinea. Gene-frequency maps for all alleles were calculated separately for the two so that gene frequencies of one region did not affect the calculations of the other across the sea. This is the same as erecting a total barrier between Australia and New Guinea in the computation of the maps of the gene frequencies. Similarly, principal components were calculated separately for the

Table 7.7.4. Genetic Distances (x 10,000) (± standard errors) between Regions of Australia and of New Guinea

| Australia | New Guinea | | | | |
	East Highlands	Southeast	North	Southwest	West
Northern Territory (non-Pama Nyungan)	1071±243	663±232	806±161	664±217	1249±270
West (non-Pama Nyungan)	801±167	689±120	1031±172	1357±289	1049±307
Central	1184±380	960±274	910±317	2269±654	1241±726
Northern Territory	958±180	852±202	992±158	1828±396	992±273
Queensland	1128±285	1342±395	1291±206	1570±304	1203±296
West	1432±272	1373±272	1431±336	2559±438	1518±427

Table 7.7.5. Genetic Distances (x 10,000) (± standard errors) between Mainland and Island
Southeast Asia, and Regions of New Guinea

Mainland & Island Southeast Asia	New Guinea				
	Southeast	Eastern Highlands	North	Southwest	West
Polynesia	1293±173	1818±366	1843±387	2165±508	1246±251
Southeast Asia	1097±197	1860±350	1718±384	1943±407	1145±240
Indonesia	1196±289	2340±510	2143±593	2403±613	1120±256
Philippines	1161±184	1789±324	1713±366	2213±471	1102±211

two regions. Therefore, any similarity of color between parts of one and of the other is accidental. The color synthetic maps convey 71.6% of the original gene-frequency information in Australia and 60.8% in New Guinea.

The color picture shows four major genetic regions. A pale green found in the north of Australia is also found in the north of New Guinea, turning pale blue in the extreme east; a very dark color in the extreme southwest of Australia is also found in the extreme southwest of New Guinea; red appears in the south of New Guinea and of Australia, as well as in the center of Australia.

7.8. POPULATION GENETICS OF MELANESIA, MICRONESIA, AND POLYNESIA

A selection of 31 better-known populations from the Pacific islands has been subjected to tree analysis in fig. 7.8.1. F_{ST} genetic distances are given in table 7.8.1. The average number of genes investigated is modest: 37 with a standard error of 3.5. (For location of the islands, see fig. 7.8.2.)

One small cluster is separated from the rest and includes all four Eastern Polynesian islands. The major cluster includes two outliers, two populations from islands very close to New Guinea (Tolai and New Britain).

Two other more important components of the major cluster can be easily recognized: (A) one is almost exclusively Melanesian (9 out of 10 populations), in addition to a small, outlying cluster of Samoa and 2 Micronesian islands, and (B) the other component includes the same number (4) of Micronesian, Polynesian, and Melanesian populations.

Despite the incompleteness of the correlation between the genetic data and the standard classification into the three main traditional groups of the Pacific Islands

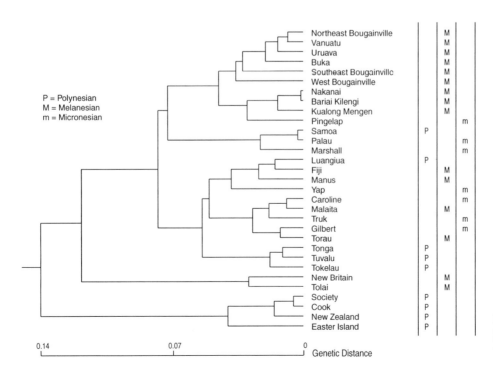

P = Polynesian
M = Melanesian
m = Micronesian

Fig. 7.8.1 Genetic tree of 31 populations from the Pacific Islands.

Table 7.8.1. Genetic distances (in the lower left triangle of the matrix) and their standard errors (in the upper right triangle of the matrix) for the smaller islands of the Pacific Ocean (all values × 10,000)

	MAM	MBB	MBT	MBU	MFF	MMM	MNK	MNA	MNN	MNT	MNE	MIA	MIG	MIM	MIP	MIT	MIL	MIY	NBB	NBE	NBW	NBS	PEC	PEE	PEZ	PES	PSE	PSL	PSS	PSN	PTT
MAM	0	206	93	122	98	63	197	135	191	351	201	155	213	671	353	128	170	222	288	359	303	774	328	535	495	291	128	40	372	137	81
MBB	618	0	142	78	41	47	92	200	155	143	33	116	324	738	467	102	123	72	131	123	67	159	175	346	199	260	516	86	172	98	395
MBT	369	366	0	88	705	49	126	256	181	236	44	51	48	923	126	173	246	180	83	123	139	491	241	579	427	319	257	142	316	114	194
MBU	279	156	163	0	594	39	98	191	119	174	28	43	239	932	137	125	94	132	82	75	195	417	305	674	519	425	472	73	74	188	351
MFF	304	169	1413	1282	0	526	489	131	78	202	54	354	159	426	785	160	60	96	729	580	580	366	171	354	280	145	236	58	210	300	183
MMM	258	180	180	171	907	0	68	173	84	665	201	37	187	587	114	100	120	110	125	163	296	366	206	525	380	276	235	82	83	69	175
MNK	462	203	453	219	988	234	0	35	33	640	178	140	303	65	51	132	42	298	161	189	503	376	326	520	408	343	452	79	146	506	375
MNA	552	504	521	275	343	466	132	0	132	526	118	305	286	168	135	241	43	432	177	259	170	325	238	572	383	304	239	206	500	328	213
MNN	468	288	513	180	328	326	12	142	0	772	201	214	252	215	127	186	263	382	177	100	127	81	354	601	482	417	428	93	34	208	329
MNT	977	344	1242	795	851	1488	995	804	993	0	196	163	406	450	300	307	461	333	292	488	575	21	512	501	564	421	654	342	187	490	547
MNE	550	133	266	105	169	359	402	405	541	532	0	80	120	471	161	88	137	201	25	264	173	166	317	517	449	473	431	246	60	410	386
MIA	475	465	155	195	754	90	281	514	501	799	353	0	129	832	195	61	80	77	145	164	330	329	246	594	437	257	219	44	184	247	158
MIG	790	780	123	654	347	403	1039	1047	885	1643	563	292	0	236	203	70	414	363	223	172	148	370	173	340	265	207	153	264	563	226	121
MIM	1939	1681	1973	1639	989	1216	167	338	502	1166	1122	1639	827	0	1410	111	66	514	837	866	491	774	293	650	474	465	405	979	231	476	365
MIP	1244	1151	608	495	2039	611	239	408	243	761	584	638	1156	2938	0	270	110	388	77	114	309	549	633	1173	919	946	870	148	570	465	748
MIT	421	413	359	444	424	250	532	540	500	1004	419	116	221	676	1055	0	127	135	174	169	330	301	245	502	396	325	158	95	164	125	98
MIL	481	292	456	230	126	269	188	76	590	911	336	208	1128	153	534	566	0	284	117	352	148	540	334	689	475	355	321	243	24	637	281
MIY	551	241	531	490	167	243	727	1062	821	1061	680	298	814	1714	1545	417	566	0	330	238	223	370	360	581	461	351	226	117	44	280	174
NBB	635	313	363	167	1734	490	445	533	246	1056	82	527	951	1633	251	824	354	960	0	40	107	413	511	967	748	892	661	120	138	355	578
NBE	726	419	498	184	1692	572	387	686	220	1274	553	625	1045	1766	420	811	875	988	128	0	96	367	446	667	574	755	785	212	125	416	636
NBW	695	218	603	555	1648	596	762	607	221	1491	358	801	691	1192	804	847	606	660	323	337	0	342	413	497	466	624	640	122	667	224	555
NBS	1513	480	1500	1162	1083	1173	908	688	470	289	446	1119	1452	2696	1343	912	1308	1323	1208	1032	1113	0	857	849	998	1009	1412	429	352	652	1165
PEC	1028	1295	940	1462	623	743	1417	1420	1592	2582	888	1290	642	1273	2228	920	1511	1302	1840	1704	1264	2845	0	110	49	28	134	361	112	255	177
PEE	1690	1445	1293	1807	1222	1218	1385	1651	2005	2389	1704	1387	1164	2040	2420	1357	1794	1819	2042	1967	1356	3093	431	0	84	152	257	644	284	216	260
PEZ	1436	1216	1115	1625	876	870	1436	1252	1946	2626	1453	943	887	1486	2689	1099	1304	1325	2053	1972	1444	3390	145	110	0	57	192	556	106	220	276
PES	1197	1040	886	1536	555	779	1172	1491	1962	2011	1612	716	762	1678	2052	964	1459	1153	2079	1916	1221	2178	53	299	167	0	140	412	192	325	176
PSE	467	1118	781	1136	610	516	1324	1121	1305	2439	1302	535	731	1081	2376	624	1193	741	1842	1787	1388	2947	484	1384	681	478	0	221	66	61	54
PSL	168	252	255	158	151	221	323	613	331	1121	552	134	697	2073	700	269	362	354	402	467	482	1120	1258	1896	1589	1458	645	0	102	160	109
PSS	493	214	897	174	685	191	357	668	80	580	218	391	1032	301	1042	618	29	59	497	390	1129	897	545	679	223	596	131	130	0	114	211
PSN	393	230	487	594	595	218	931	1146	701	1418	901	483	686	949	897	598	1472	591	627	690	363	1017	680	940	768	617	224	337	341	0	55
PTT	357	1008	659	1058	481	425	1129	1085	1042	2147	1090	509	568	1239	1952	477	1167	624	1505	1523	1126	2443	436	1168	735	411	111	437	402	133	0

Note.– MAM, Manus; MBB, Buka; MBT, Torau; MBU, Uruava; MFF, Fiji; MMM, Malaita; MNK, Bariai Kilengi; MNA, Kualong Mengen; MNN, Nakanai; MNT, Tolai; MNE, Vanuatu (New Hebrides); MIA, Caroline; MIG, Gilbert; MIM, Marshall; MIP, Pingelap; MIT, Truk; MIL, Palau; MIY, Yap; NBB, N. E. Bougainville; NBE, S. E. Bougainville; NBW, W. Bougainville; NBS, New Britain; PEC, Cook Islands; PEE, Easter Island; PEZ, New Zealand; PES, Society; PSE, Tuvalu (Ellice); PSL, Luangiua; PSS, Samoa; PSN, Tokelau; PTT, Tonga.

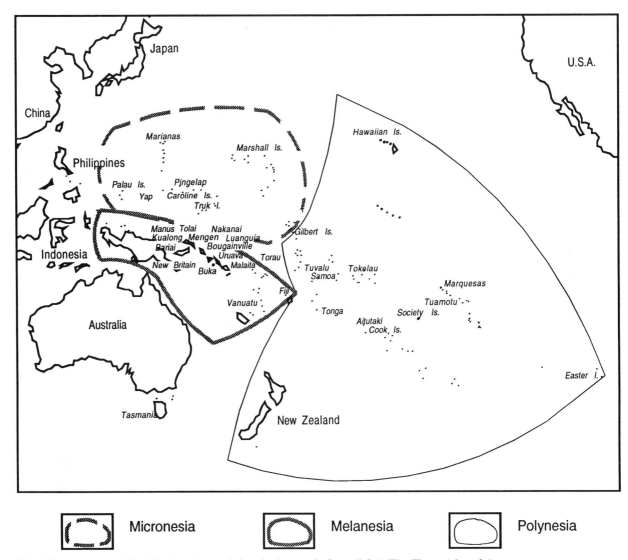

Fig. 7.8.2 Locations of the islands and populations in the tree in figure 7.8.1 (The Times Atlas of the World 1989).

(Melanesian, Micronesian, Polynesian; see fig. 7.8.2), there is a basic order corresponding to this classical nomenclature.

The four eastern Polynesian groups form a cluster that is at a considerable distance from the rest of the Pacific islands. They are: the Society archipelago (including Tahiti, over 100,000 speakers), the Cook Islands (18,000), New Zealand (280,000 Maoris), Easter Island (1600). The other five Polynesians merge with the Melanesian—Micronesian major cluster, indicating admixture with them, but three of these form a small subcluster of the B component mentioned above: Tonga (100,000), Ellice (now called Tuvalu, 7500), and Tokelau (1600). These three have a central geographic position. Two other islands are isolated from these two small Polynesian clusters and show considerable similarities with some Melanesian or Micronesian populations: Samoa (138,000) with Micronesian islands; Luan-

gia (1750, a northern Polynesian outlier close to fully Melanesian islands) with other Melanesians and Micronesians of cluster B. Thus, the Polynesian populations examined fall into three clusters: *eastern, central,* and *northwestern,* with New Zealand belonging genetically to the eastern cluster. The *eastern* group (Easter Island, New Zealand, Cook, Society) shows the greatest genetic autonomy. The *central* group (Tonga, Tuvalu, Tokelau) has internal consistency, but there is some element of Melanesian genetic influence that is definitely stronger for islands closer to Melanesia. We return to the subject at the end of the section when discussing a reconstruction of Polynesian history.

Micronesia is represented by seven islands or groups of islands. Four of these are found in cluster B, one in cluster A, and two form a small outlying cluster with Samoa. This last includes Palau (14,000, one language) in the western part of Micronesia, and the Marshall Islands

(30,000, one language), east of Micronesia. The Micronesian island isolated in the predominantly Melanesian subcomponent A is Pingelap (1300). Truk (26,000), Yap (5000), and the rest of the mostly smaller Caroline islands fall into cluster B, along with the Gilbert Islands (44,000).

The Melanesian populations of cluster A speak mostly Austronesian languages and the prevailingly Melanesian cluster A includes:

1. All populations studied from the Solomon Islands are included except one (Torau, found in component B). The largest island is Bougainville (approximately 80,000 people) where Austronesian (AN) and Indo-Pacific (NAN) languages are spoken.
2. Two populations are from New Britain, the largest island of the Bismarck archipelago (150,000 people): Bariai (1000) and Nakanai (10,000), both Austronesian speakers. Other New Britain inhabitants are with the Tolai of New Ireland (an island of the Bismarck archipelago, north of New Britain, 86,000) as outliers of the major cluster.
3. The New Hebrides, now called Vanuatu, has 86,000 Austronesian speakers.

The four Melanesian populations in the Micronesian-Polynesian cluster B of figure 7.8.1 are from the most northern part of Melanesia: Malaita (30,000), in the northern part of the southern Solomons; Torau, a small island east of Bougainville; Manus, the largest of the Admiralty Islands (23,000); Fiji (260,000 plus many other immigrants from other ethnic origins) at the border of the Melanesian-Polynesian areas. Ancestral Micronesians may have begun their later voyages after having settled for some time on the northern fringe of present-day Melanesia. They may have been early Austronesian settlers to northern Melanesia who mixed to some extent with local Melanesians and proceeded from this area to Micronesian islands.

Figure 7.8.3 shows the PC map of the populations analyzed in the tree of figure 7.8.1. The first PC (27% of the variation versus 11% for the second) clearly separates almost all Melanesian populations (on the right part of the diagram) from Micronesian and Polynesian ones, which fall in the left part. The Micronesians are between Polynesians and Melanesians, indicating that they are, to some extent, the result of an earlier admixture between Southeast Asians and Melanesians. This might be the case if the Micronesians were the first settlers from Southeast Asia in northern Melanesia and radiated later, especially towards the northern Pacific. Some of them may also have reached it directly. The origin of Polynesians would be from the most recent Austronesian migration—also from Southeast Asia, but not from the same region as Micronesians. The ancestors of eastern Polynesians probably arrived directly from Southeast Asia (see sec. 7.2). With very little initial mixing with the local populations in Melanesia, they kept moving east, toward the then uninhabited eastern archipelagoes where they eventually settled permanently. Some tentative dates that may apply to these events and perhaps help to choose between the different scenarios can be found in Pawley and Green (1985); see also Diamond (1988) and section 7.2.

The central and eastern Polynesians, living in areas free from Melanesian settlements, are best suited for a tree analysis. The relevant portion of the F_{ST} genetic distance matrix is shown in table 7.8.2. The matrix gives rise to a tree (fig. 7.8.4) to which locations and arrows have been added on the basis of geographic and historical considerations. The Cook and Society Islands are very

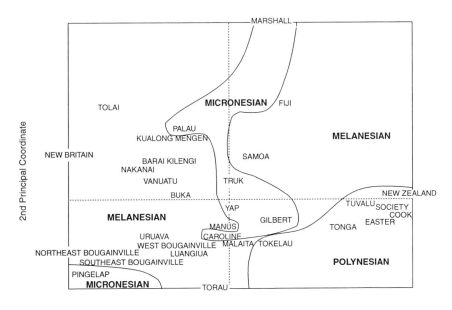

Fig. 7.8.3 Principal-component map of the Pacific Islands.

1st Principal Coordinate

Table 7.8.2. Matrix of Genetic Distances (x 10,000) between Polynesians of the Eastern and Central Areas

	Tonga	Tuvalu	Tokelau	Cook	Society	New Zealand
Tonga	–					
Tuvalu (Ellice)	111	–				
Tokelau	132	223	–			
Cook	435	484	680	–		
Society	411	478	616	52	–	
New Zealand	735	681	767	145	166	–
Easter Island	1168	1384	939	430	477	298

Note.– Boxes indicate genetic distances expected to have the same value (within sampling error) in a satisfactory tree. Note the difference between distances connecting Easter Island to the other islands, and distances among the other islands.

similar, and they were considered as a block, which may be thought of as a secondary center of irradiation from which New Zealand and Easter Island were settled.

As noted earlier, Tonga was probably first settled in 1200 B.C., the Marquesas in A.D. 200–300 from Samoa, the Society Islands from the Marquesas in A.D. 500–600, Easter Island before A.D. 300 and New Zealand before A.D. 1000 (Bellwood 1989). All archaeological dates may obviously be late if traces of the earliest settlers have not yet been unearthed. The order with which the eastern branches detach from the tree is in qualitative agreement with these dates, but the dates are not supported quantitatively. New Zealand has an average genetic distance ($\times$ 10,000) of 155 from Cook and Society, and Easter Island 453. The settlement of Easter Island could hardly be three times earlier than that of New Zealand. The evolution that must have occurred in the Cook and Society islands somewhat inflates the distances given, making the discrepancy a little worse. The example shows that in dealing with very small populations their size can have a disrupting influence on these computations. The population of Easter Island has fluctuated greatly over the centuries. The following demographic information is from Bellwood (1979). The island had undergone a marked environmental decline before its discovery (on Easter Sunday, 1722). It was heavily forested at the time it was settled (as attested by pollen records), but most trees had disappeared by the eighteenth century. At its peak, the population may have been 10,000.

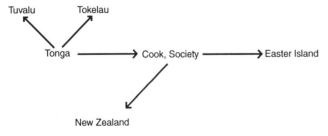

Fig. 7.8.4 Genetic tree from table 7.8.2 and its migrational interpretation.

Cook (1774) estimated 700 people; visiting Spaniards (1770), from 900 to 3000; La Perouse (1786), 2000. In 1862 about 1000 Easter Islanders were taken by slavers to Peru; 900 died within a short time, the remainder were repatriated but only 15 reached home, with smallpox. In 1877 only 110 people remained on the island. No such fluctuations are known to have occurred in New Zealand; in recent times its population is almost 100 times larger than that of Easter Island, and it was probably large at most, if not all, times.

One can calculate approximately the effective population size, N_e, on the assumption that genetic distance d is entirely due to drift, from the formula previously indicated: $d = t/2N$. For Easter Island, t, the time in generations since the beginning, is 70 and for New Zealand, at least 40. The approximation is valid for very short times like these. The effective population size thus calculated is about 500 for Easter Island, whereas that for New Zealand is 1500. These are averages over the whole period, and such averages should be calculated from demographic data, if they were adequately known, by taking their harmonic mean, which is sensitive to low values.

The genetic analysis of Polynesia by Serjeantson (1985) basically agrees with ours, although only *HLA* genes were used. A more complete analysis by Serjeantson (1989) discussed the possibility of a minor Amerind contribution to eastern Polynesia. Although based on tenuous evidence, Serjeantson's note of caution may be added to the observation by Bellwood (1989) that archaeology cannot exclude contact between South America and Polynesia in prehistoric times. Bellwood cited evidence from "the transfer of the Andean sweet potato into prehistoric eastern Oceania," and the "occurrence of an Inca style of stone facing on at least one Easter Island temple platform." Other possible evidence of contact comes from the similarity of fishing gear. The major hope of hard evidence on the question of a South American contribution to eastern Polynesia, as suggested by Heyerdahl (1950), could only come in the future from molecular genetics. With classical markers, the origin of alleles is usually insufficiently clear-cut for this purpose, as shown below. There must be considerable genetic similarity between Amerind ancestors and the Southeast Asian ancestors of modern Polynesians. If one attempts to relate to their descendants the gene frequencies of populations of small islands, which are affected by enormous drift, results can hardly be convincing. But, if one can discover molecular mutants in South Americans that are not present in Southeast Asia, then their detection in Polynesians would build a strong case in favor of Heyerdahl's hypothesis. Variants of hemoglobin studied at the molecular level in Oceania and reviewed before (Hill et al. 1989) have helped show a genetic trail that does not, on the basis of our present knowledge, lead from South America to Polynesia but does not exclude it.

It is difficult to pinpoint the remote origins of Polynesians on the basis of present genetic data. Table 7.8.3 shows genetic distances between Polynesians selected on the basis of a higher number of markers (average 86) and the extant descendants of several possible ancestors.

Eastern Polynesians are most representative of Polynesia, since they had no chance of local Melanesian admixture. Some gene flow may nevertheless have taken place before the final migrations of this very successful group of Oceanic settlers. In the general tree (sec. 2.3), they are placed with Melanesians and Micronesians, whose ancestors seem likely to have made a substantial contribution to Polynesian ancestry in the thousands of years after proto-Polynesians from Asia arrived in Melanesia and before they started, or continued, their migrations eastward. Distances have high standard errors and the conclusions are not all statistically significant, but the best candidates for proto-Polynesians are still in Southeast Asia. The table also shows the distances of the same populations from Easter Islanders. They are always higher, by an approximately constant amount, than

Table 7.8.3. Genetic Distances (x 10,000) (± standard errors) between Selected Polynesians and Descendants of Their Possible Ancestors

	East Polynesia	Easter Island
South American	1185±283	1682±371
Andean	1322±301	1644±312
Southeast Asian	693±185	1289±337
Filipinos	924±269	1251±309
Ainu	921±289	1507±450
Taiwan aborigines	1143±301	1695±444
Average	1031	1511
East Polynesian		207±53

those from eastern Polynesia, indicating that there is no reason to assume a special origin for the inhabitants of Easter Island, despite its location much farther to the east and closer to South America. The great difference between the two relevant columns of table 7.8.3 must be due to drift, which must have been very strong on Easter Island.

7.9. SINGLE-GENE MAPS OF AUSTRALIA AND NEW GUINEA

Maps of single genes for Australia are presented together with those of New Guinea, using a common gene-frequency scale in order to make them comparable. In the map-interpolation procedure, geographic neighbors, even if located across the sea, affect computation results. If Australia and New Guinea were interpolated together, there would be a reciprocal influence. A barrier between them was therefore generated by interpolating them separately, as was done for other continents. Similarities observed between areas of the two islands are therefore real and do not arise from interpolation artifacts.

There are few, if any, aborigines left in southeastern Australia, which is now more densely populated by recent Caucasoid immigrants. The maps of gene frequencies are therefore determined almost exclusively on the basis of information available in the regions farther to the north, and the values interpolated for the southern regions cannot be considered reliable.

There is an obvious difference in the scale at which the maps of Australia and New Guinea are drawn. Data on New Guinea are more numerous, their density per unit of surface is higher, and New Guinea is therefore presented in more detail than Australia. The northern part of Queensland, shaped like a finger pointing to the north, is the part of Australia closest to New Guinea, from which it is separated by the Torres strait. The part of New Guinea nearest to Australia is the tip of the south-central coast.

The number of genes for which maps of these two islands could be drawn is about half that of other continents. This is partly because the number of HLA data is insufficient for drawing maps and partly because of absence, or very low frequency, of some alleles on one or the other of the two islands.

ABO genes show north-south gradients for *A* and *O*, east-west for *B*, and southeast-northwest for A1 in Australia. They are more complex in New Guinea. The *B* antigen is absent in Australia. Variograms have fairly high slopes and are linear for approximately 500 miles.

The *ACP1* enzyme (acid phosphatase), allele *B*, is markedly different in Australia and New Guinea; it is almost fixed in the first, except in Queensland and in a small region to the east. This difference may reflect a difference in climate. Variograms are uninformative.

Adenosine deaminase, *ADA*, allele 1 also shows near fixation in Australia but not in New Guinea, with a flat variogram.

Maps for carbonic anhydrase 1 and 2 (*CA1* and *CA2*) are available only for Australia. In both cases allele 1 is close to fixation; *CA1∗1* shows a northeast-southwest gradient. Variograms are linear for a modest segment, and both have high slopes.

Esterase D allele *ESD∗1* shows a minimum in the center of Australia and a west-east gradient in New Guinea. The variogram is uninformative in Australia and has a high slope in New Guinea.

Glucose-6-phosphodehydrogenase deficiency (*G6PD ∗def*) is present almost exclusively in the southern part of New Guinea.

Group specific component or vitamin-D-binding protein (*GC*) is considerably different in Australia and New Guinea. The *GC*1* allele has a high frequency through most of Australia and peaks in the center of New Guinea. The variogram has a high slope in New Guinea, but only for a short initial segment.

Haptoglobin allele *1* (*HP*1*) again shows a profound difference between Australia and New Guinea; it is at a relatively low frequency and fairly constant in the first, and quite variable with an irregular variogram in the second.

Immunoglobulins, *IGHG1G3*, show considerable variation of allele *fa; b0b1b3b4b5*, which is high in the western and moderately high in the eastern end of New Guinea, with no information for Australia. Allele *za;g* has a peculiar distribution in Australia, with extreme maxima and minima near each other in a small area of the northwestern part, and a strong gradient in New Guinea apparently totally unrelated to the Australian distribution. The variogram is irregular in both. Allele *zax;g* has still another pattern, with maxima in small regions of the eastern part of both Australia and New Guinea. Allele *za;b0b1b3b4b5* has a rather low frequency in Australia, and a strong maximum in the central northwest of New Guinea with a decreasing frequency in all directions. The totally independent immunoglobulin allele *KM(1&1,2)* has, interestingly, a strong similarity with *za;g*, as shown by the presence of highly localized maxima and minima in a region close to that which also shows similar strong variation for *za;g*. Variograms are mostly unreliable, except for a very strong initial slope for *za;b0b1b3b4b5* in New Guinea, an indicator of the strong and regular gradient observed for that allele.

The exceptional behavior of *GM* alleles in Southeast Asia and Melanesia was noticeable in the geographic maps drawn by Steinberg and Cook (1981). All nonimmunoglobulin genes do not show such drastic variations in this general region. The conclusion seems inescapable that the phenomenon cannot be due to drift, but must be the result of strong local selection for one allele or the other. There is also a correlation in northern Australia between an allele of the *KM* and one of the *GM* locus. Both genes contribute parts of antibody molecules, but *KM* does not show such extreme variation in any other part of the world. The antibodies in question serve functions of defense against infectious diseases and it is legitimate to hypothesize that at some time in these areas there occurred major epidemics to which the various immunoglobulin alleles have shown differential resistance. These epidemics may have since totally disappeared, leaving us with no clues as to their nature. The health situation in these areas is not well enough known, however, that one can reject the possibility that some such diseases are still present. The *GM-KM* correlation in Australia is based, however, on observations with very small numbers and could be a statistical fluke.

An allele of the Lewis blood-group gene, *LE*Le*, shows a moderate gradient in New Guinea that has little effect over the initial part of the variogram.

The *MNS* blood-group system shows reasonable variation of allele *M* in both Australia and New Guinea, without much correlation between the two islands. By contrast, allele *S* is practically absent in Australia but present with an average frequency of 9.7% in New Guinea, where it shows a maximum in the northern coastal plains. Consequently, only haplotypes *Ms* and *Ns* have a geographic distribution worthy of note, which tends to imitate very closely that of allele *M* and of its counterpart *N*. The map of *MS* is not given because it would be identical to that of *M* with gene frequencies equal to 100 minus those of *M*, but there is enough *NS* in New Guinea that the map of *NS* is worth showing. Most variograms have a high initial slope with linearity usually restricted to a fairly small initial segment.

The *P1* blood group, allele *1*, shows some modest variation without major gradients.

Allele *1* of enzymes *PGM1* and *PGM2* (phosphoglucomutases) show reasonably small variations with gradients not unique to these genes. The first has a high initial slope, the second a negative one.

Phosphogluconate dehydrogenase *PGD*A* has little variation in Australia and more in New Guinea. The latter has an irregular variogram.

Phosphoglycerate kinase *PGK1*1* is practically not polymorphic in Australia and has little variation in New Guinea.

The *RH* blood-group gene system shows polymorphism only for genes *C* and *E*, while allele *d* is absent. *C* is more frequent in New Guinea where it reaches close to 100% in some northern regions, with fairly large initial slopes. In New Guinea *E* is rare, but it increases in Australia from north to south. Variograms are fairly regular with a rather large slope. Only four haplotypes are possible, given the absence of allele *d*. In Australia, *CDe* is present with appreciable frequencies in only some parts, especially the west and east. The highest frequencies of *CDe*—often near 100%—are in New Guinea; lower ones in New Guinea have a northeast-southwest gradient. *cDE* shows behavior opposite to that of *CDe*, and *cDe* is rather low all over: 10.5% on average in Australia, with a higher frequency in the east, and 2.3% in New Guinea. Variograms have moderate slopes, when positive.

Secretor (*FUT2*Se*) has a rather irregular distribution, as shown by the relatively large number of heterogeneous data points and the very negative initial slope of the variogram. It is also poorly known in western New Guinea, and there is no distribution for Australia.

Transferrin (*TF*) allele *D* shows a gradient in Australia that is determined entirely by one southern value showing a high *D*, of doubtful significance given the rarity of aborigines in the south. There is only modest variation in New Guinea.

7.10. SINGLE-GENE MAPS OF THE PACIFIC ISLANDS

The Pacific islands are presented in the context of the rest of the populations facing the Pacific: Australia, New Guinea, eastern Asia, and the Americas. Because they are almost all too small for indicating their gene frequencies within their actual physical boundaries, the shading has been extended to a rectangle of area sufficient to express unambiguously the gene frequency. Whenever there was more than one observation for the island or group of islands, their average is indicated. For Pacific islands, given their isolation and in analogy with the procedure used for small islands of all continents, no interpolation was used. As a consequence principal components were not computed. For all non-Pacific islands and continents the computations performed for single continents were used. The great variation of interisland distances and the ethnic heterogeneity of the Pacific islands made the calculation of variograms potentially very confusing. It is already known (for a summary, see Friedlaender 1971) that when the interisland distance is very high and islands are small—as, for instance in Micronesia—the genetic similarity observed per unit of distance is much greater than when villages on land are compared. Clearly the smallness of the island populations and the distance from other islands are compensated by covering much greater distances for contacting other groups and thus widening the circle of potential marriage partners. The genetic maps of the Pacific Ocean are of special interest for comparing centers of origin of migrants and their destinations.

As is well known, the *ABO* system shows very strong differences between most of America and the rest of the world. Because of the suspicion that it may have selective origins, it is not especially useful for understanding ethnic similarities.

In most of Melanesia and Micronesia, *ACP1*∗*A* and *ACP1*∗*B* show considerable variation, possibly due to drift. On a larger scale, the gradient with latitude seems in agreement with the general view of a selective response to climate.

FY∗*A* shows much difference between Melanesia and Polynesia. Taken in isolation, it would seem to support Heyerdahl's hypothesis that Polynesia was settled from the Americas, but few other genes give similar evidence.

Glucose-6-phosphodehydrogenase deficiency (*G6PD*∗*def*) shows pockets of high frequencies in parts of Southeast Asia and New Guinea, probably because of a local reaction to malaria. Glutamic pyruvate transaminase (*GPT*) allele *1* has a high frequency in Australia, parts of New Guinea, and a few other locations.

For *HLAA*∗*2*, some drift is evident in Polynesia, but not as strong as, for instance, South America (see chap. 6). In southern New Guinea and in South America

HLAA∗*9* has major peaks. Although *A*∗*10* has a maximum in Australia, in the rest of the Pacific it is relatively flat at mostly low frequencies. By contrast, *A*∗*11* is almost absent in Australia and has a maximum in Southeast Asia. Absent in eastern Polynesia, *B*∗*13* has a maximum in Australia. The peak for *B*∗*15* is in South America (southern Andes) and for *B*∗*16* near the northern Andes. With peaks in Australia, New Guinea, southern Melanesia, and central Polynesia, *B*∗*22* is absent in America. The maxima for *B*∗*27* gradient with latitude lie especially in North America, and *B*∗*40* is highest in the extreme north, with secondary maxima in the southern Pacific. Many *HLA* alleles which are common in Caucasoids such as *A*∗*1*, *A*∗*3*, *B*∗*5*, *B*∗*7*, and *B*∗*8* do not occur in Pacific populations. The *A*∗*1*, *A*∗*3*, *B*∗*7*, and *B*∗*21* alleles are also missing in American natives, but the last two are present in North America.

Figure 7.10.1 gives a summary of the *HLA* relationships among Pacific populations, the Americas included. Several facts are worth noting. The Australian *HLA* polymorphism is restricted in comparison with that in New Guinea. Austronesian speakers carried *HLAA*∗*2* and *HLAB*∗*18* into Melanesia but not into the New Guinea interior. The number of Lapita potters who left the Samoa area for eastern Polynesia about 2000 years ago may have been small enough to have caused the loss of *B*∗*13* and *B*∗*27* by genetic drift. Micronesia has a non-Melanesian *HLA* gene, *B*∗*35*, which is absent everywhere else in the Pacific but is present in the Philippines. Finally the *HLA* profile of American natives has no relation to any Pacific Group, militating against Heyerdahl's hypothesis (but see the Duffy and Rhesus genes below).

Haptoglobin (*HP*∗*1*) peaks in parts of New Guinea, on Easter Island, and shows a fair amount of variation in America.

GM (*IGHG1G3*) has considerable variation, as already noted. Peaks for *fa;b0b1b3b4b5* (map not given) are found in Southeast Asia, Borneo, and some parts of Melanesia and Micronesia, and *za;b0b1b3b4b5* shows a peak in western New Guinea. Although *za;g* is es-

Population	A2	A9	A10	A11	B13	B15	B22	B40	B27	B16	B18	B35
Australia												
Papua New Guinea												
Austronesia	∗											
West Polynesia												
East Polynesia												
Micronesia												
America												

∗ Austronesian-speaking Melanesians

Fig. 7.10.1 Presence of *HLA* antigens in Pacific populations (adapted from Serjeantson 1985).

pecially frequent in parts of America, it shows highly localized peaks in Australia and New Guinea. The allele *zax;g* shows very heterogeneous behavior in South America and somewhat higher frequencies in a few other regions.

The light immunoglobulin chain *KM* allele (*1&1,2*) shows a heterogeneous distribution in South America and a highly localized irregularity in northern Australia, as already noted.

Lewis (*LE*Le*) shows clinal variation in the Pacific Ocean; *Le(a+)* has considerable variation among islands of Southeast Asia and the Pacific.

The *MNS* system shows high frequencies of allele *M* in America, the extreme north and extreme south of East Asia, and mostly low frequencies in the Pacific. The *S* allele is very high in southern Siberia and in eastern South America, but is rare in the Pacific. Haplotypes *MS* and *Ms* have higher variation; *MS* and *NS* are rare in the Pacific islands.

Blood group *P1* allele *1* is also relatively rare in most places except America, and extreme frequencies are found as usual in South America.

Phosphogluconate dehydrogenase (*PGD*) allele *A* is relatively low in Melanesia. It has a very high frequency in the Americas.

*PGM1*1* shows great variation in South America and a relatively high one also in Oceania. The same is true of *PGM2*1*.

Phospho-glycerate kinase (*PGK1*) allele *1* is variable, especially in the western Pacific islands, including New Guinea.

For *RH*, gene *C* shows considerable variation in Melanesia, whereas *E* is uniformly low except in southern and eastern Polynesia. Again, *C* and *E* like *FY*A* would favor Heyerdahl's hypothesis. However, when the analysis is carried out at the level of haplotypes, one finds *CDE* much more favorable to a Southeast Asian origin, *cDE* and *CDe* to a South American origin, and *cDe* indifferent. It is clear that tests based on single genes may be unreliable for studying ethnic origin.

The secretor gene (*FUT2*Se*) shows much variation, especially in New Guinea.

The *TF*C* allele is especially low in Australia and New Guinea.

7.11. SUMMARY OF THE GENETIC HISTORY OF THE PACIFIC

The settlement of Australia and New Guinea goes back at least 50 kya and was probably not a single event. Several migration waves were postulated by the various scholars cited earlier; the first wave perhaps went through New Guinea to northeast Australia but may have been partly swamped by the later ones, especially in western New Guinea. The genetic evidence is not in disagreement with this idea, because northwestern and northeastern areas differ, but there is no sharp gradient and it is also possible that a cline arose later. In Australia the coast had chances of reaching higher population densities, at least in recent times; in earlier times the internal regions were not, however, as arid as today, and the center was settled relatively early. The center now shows some genetic difference from the coast, possibly an indication of adaptations to extremely arid or humid conditions.

On the whole, the genetic data provide little information about the possible sequence of migrations and successive settlement patterns. The population structure of Australia consisted of hundreds of small tribes that filled the whole continent practically continuously. This structure lends itself to generating a fair amount of local drift such that even nearest neighbors could differ from one another to a nontrivial degree in their gene frequencies. No major heterogeneities among regions are expected, in the absence of internal geographic barriers, and they are not found. A few extreme variations are observed for very few genes that are known to be potentially subject to strong local selective pressures.

Australian languages are very diverse, but considerable diversity could be expected if their separation is very old. It is possible that the structure in small tribes, each of which speaks a different language, may help local random differentiation of languages in a manner analogous to random genetic drift. It should be acknowledged, however, that there is no strong empirical or theoretical basis for expecting populations showing great genetic drift to also show considerable local linguistic variation. The only clear linguistic differentiation in Australia is that between Pama-Nyungan and non-Pama-Nyungan languages. The latter occupy a relatively small region in the northwest, and the inability to classify them in one or a few subfamilies indicates that there is high local linguistic variation in this small area. There is also some correlation between the geographic map of genes and the distinction between the two major subfamilies. The hypothesis may come to mind that Pama-Nyungan speakers spread at a relatively late time to most of the continent, replacing older languages; but the reasons for success of Pama-Nyungan speakers remain unexplained in the light of current evidence.

The relations between New Guinea and Australia seem to be one of remote kinship, slightly greater perhaps for the northern and northwestern part of Australia. According to Greenberg (1971), Tasmanians spoke an Indo-Pacific, and not an Australian, language, but in the absence of genetic information it is impossible to say whether they had greater genetic similarity to New Guineans or to Australians.

New Guinea shows an east-west gradient and some differences between the coastal plains and the mountainous interior. There may have been many migrations from Southeast Asia, of which the latest, the Lapita potters, was most probably responsible for the introduction of Austronesian languages to the area. Before that, most nearby Melanesian islands had already been settled by people speaking Indo-Pacific languages. Austronesians settled in various parts of the New Guinea coast and on Melanesian islands. A complex mosaic of villages of people speaking languages from the two different linguistic families must already have formed at the beginning and still exists today. Austronesian speakers, however, tend to be limited to the coasts. In many areas of Melanesia, it is not easy to find a good correlation between languages and genes. The time elapsed since the first migration may be 3000 years or more and, at the usual rate of genetic exchange with neighbors, there would be, in any case, practically complete gene replacement (sec. 1.17). We do not know how often language substitution occurs; it might contribute further to our inability to find a clear correlation. It is therefore not at all surprising that in Melanesia AN and NAN speakers ordinarily show little genetic difference. Unless there is unusual resistance to exogamy or very poor opportunities for cross marriages or other forms of gene flow, the boundaries of an initially complex genetic mosaic are most likely to be blurred after this length of time. Moreover, with small tribes and much drift, another mosaic may replace the original. New Guinea reached the agricultural stage fairly early; thousands of years ago, its population density and, to some extent, tribe size must have already been much greater than those more characteristic of hunter-gatherers. European contact caused further population increase by the introduction of new crops but did not substantially alter the tribal structure of New Guinea, at least until a few decades ago. Australian aborigines remained at the hunting-gathering stage until European contact. Only a few of them still practice their original way of life in a small, protected area.

The first migrations to more remote islands, leading to the peopling of Micronesia to the north and Polynesia to the east, began later. Micronesians have a greater Melanesian component than Polynesians, as confirmed also by genetic data with single classical and molecular markers (Kirk 1989; Serjeantson 1989; Serjeanston and Hill 1989). Descendants of the Polynesian migrants to the central and eastern Polynesian islands are clearly genetically distinct from populations of Oceania located farther to the west. It is possible that the original proto-Austronesians did not proceed beyond Tonga at first, but, when there were additional eastward migrations from Melanesia, they were pushed to expand farther to the east, north, and south. With this hypothesis, the heavy dilution of Polynesian genes by Melanesian genes in islands west of Tonga was a secondary event. Another possibility suggested by Bellwood (pers. communic.) is that intermarriage of Polynesian settlers and previous Melanesian inhabitants was only a later event, say, of the last millennium.

Genetic analysis is in agreement with the idea that Polynesians proceeded first from Tonga to nearby islands and to the east. Data are insufficient to determine whether the first steps were to the Cook Islands or the Society Islands. From there, New Zealand and Easter Island were occupied. Easter Island shows especially conspicuous drift, not only because of a founder's effect, but also because of a contribution from later bottlenecks. It is difficult at this stage to completely exclude Heyerdahl's (1950) hypothesis of a contribution of Amerind genes to Polynesians, but it does not seem likely that it was an important component.

8 EPILOGUE

8.1. THE MULTIDISCIPLINARY APPROACH

One message that we would like to convey is the need for a multidisciplinary approach to historical problems like the evolution of modern humans. One can considerably enrich our understanding of this difficult problem by bringing in as many relevant disciplines as possible, from historical demography to archaeology, paleoanthropology and linguistics, and perhaps ethnography, together with population and molecular genetics. The use of all available evidence from various fields is the only way to increase our confidence in our results. If we neglect some of it, we may discover later that it deeply affects our conclusions; it is therefore better to include from the beginning whatever information is available. But we should not delude ourselves into thinking that the desirable evidence is abundant and can be easily acquired. The opposite is true: in practically any of the disciplines that we need to tap, the information relevant to our aims is extremely scanty and usually imprecise. There is almost always a multitude of possible interpretations, and opinions are almost always divided. Yet, it seems to us that a multidisciplinary approach is a priori preferable, even if one must recognize a serious disadvantage; it is easier to make mistakes in fields that are not our own. It is also possible to make mistakes in one's own field, of course, and we will be grateful to people who will point out our errors and correct them. It also seems important to clarify which information is useful for our purposes, for example, in the archaeological and historical sciences. This increases the chances that it will accumulate in the future.

Some possible misunderstandings need to be avoided. It is important to distinguish between genetic information (on characters that are inherited according to strict Mendelian rules; these vary from observations on DNA to those on proteins or other macromolecules, e.g., by immunological tests as for blood groups, etc.)

and evidence from physical anthropology in the narrow sense (observations of anthropometric and anthroposcopic characters like stature, bone and tooth measurements, and qualitative traits such as skin color, etc.). Physical anthropology has permitted us to dig into the past and is, in this sense, superficially more attractive than genetic information, which has so far enlightened us concerning only living populations. This situation is in the process of changing, however, and great expectations about the possibility of analyzing "fossil" DNA are being generated. It should be stressed, however, that physical anthropology data are *not* entirely reliable for purposes of reconstructing history; they are susceptible to short-term change because of either natural selection or direct phenotypic modification. Natural selection causes convergence, or divergence in ways and directions that have nothing to do with coancestry, and can cause serious errors in historical interpretation. Direct phenotypic modification introduces even worse misunderstandings. Genetic data are entirely free from the second type of error and are largely free from the first type. The main reason for the lack of deviations caused by natural selection is that many—probably the great majority—of the genetic traits we study are selectively neutral. Especially when a wide range of genetic data is considered, it seems extremely unlikely that selective convergence or divergence would cause serious deviations.

It is therefore essential to understand that evidence from physical anthropology in the narrow sense and that from genetics are subject to different constraints and provide different information. It is also important to keep in mind the distinction between genetic evidence from gene frequencies for polymorphic genes, which have formed the bulk of the material considered in this book, and mutational changes observed between individuals as studied in some approaches to mitochondrial DNA, which have been less extensively investigated. The interpretation of the two types of genetic information relies on different methods of study and leads to different types of conclusions: the first considers populations, and the second considers single individuals or groups of them, which cannot be equated to populations.

8.2. THE USES OF GENETICS IN HUMAN EVOLUTIONARY HISTORY

Having introduced these caveats, the use of genetic data for reconstructing what is, in a wide sense of the word, human history, has the attraction of bringing in hard data where there is only softer evidence or no information at all. Genetic data are strongly reproducible, and, if need be, the size of samples can usually be increased to decrease statistical error. Difficulties may, however, arise later, at the level of interpretation. Sometimes, the evolutionary understanding of the genetic data is straightforward. But this is not always true; and when one is interested—as we are here—in historical interpretations, one may still be overwhelmed by the complexity and the multiplicity of possible explanations. Hard data are nice, but their interpretation may still suffer from the weakness of historical research: unlike ordinary experimental data, the events in which one is interested cannot be reproduced at all. Thus, interpretation of hard data may be soft. Nevertheless, the collection and use of genetic evidence on populations of historical interest is a healthy trend that should be further encouraged. It is fortunate that a fair number of scientists enjoy collecting genetic information on populations for the purpose of understanding their origins. Genealogical interest is common and may correspond to some deep-rooted human custom or drive. Otherwise, the thousands of articles that make books like this possible would not exist. Those of us who have collected data know that it is usually demanding work. We all owe considerable gratitude to these thousands of scientists, some of whom, in particular A. E. Mourant and his collaborators, have truly dedicated their lives to it.

Naturally we have to be humble enough to recognize the limitations and the gaps that will probably remain forever in our attempts to reconstruct the genetic history of humans. The laboratory scientist, who is used to asking problems answerable in a firm way by setting up appropriate experiments, may find historical problems unattractive because in this case it is much more difficult to choose without reasonable doubt between a variety of possible explanations. The historian cannot devise strictly controlled experiments for unambiguously testing different possible hypotheses. Yet he can often find clues that allow him to give a reasonable backing to his conclusions. There is also considerable intellectual stimulation from trying to make some headway toward understanding a difficult puzzle by using pieces from very different sources of knowledge and introducing new informative ones.

Some of the specific conclusions we reach in this book are new or differ in a number of points from those that others have expressed before. It has been difficult to make sure that all previous opinions and data were duly acknowledged, and we apologize to those authors who, for one reason or another, were not cited or whose ideas were not considered. Our effort was truly one big investigation in which we have tried to reexamine without preconceptions an enormous amount of material. We have given priority to data rather than to earlier interpretations; even so, we may have lost a fraction of existing data. We have given as much weight as we could to the need for using as many genes as possible. This has led us to the innovative procedure of making comparisons that are not, or not always, based on exactly the same genes. In practice, most or many of the genes are nevertheless the same, because some polymorphisms have been studied much more frequently than others. This approach is entirely justified if the sample of genes used each time is large and can be considered random. It is difficult to test in a completely rigorous way how close we come to this requirement, and this may generate a limitation to our conclusions that future research may test. Had we tried to compare populations strictly on the basis of the same genes, the number of available genes would have shrunk beyond acceptability in a great number of cases. However, the data are there and can be tested by other approaches.

We obviously do not claim that all our interpretations are correct. At least some of them are provocative, and we offer them for further testing by other techniques or, perhaps more usefully, by a larger number of data. Some more genetic information has appeared in the literature since our data collection was finished, and more will appear in the future. No new hypothesis is usually accepted in science unless it is confirmed by at least another independent laboratory. Other scientists may want to retest the new and nontrivial interpretations we suggest before accepting them; this is the normal course of science. It is difficult to predict how many or which of our hypotheses will be confirmed in the long run. Even if, in the worst hypothesis, many of them were to fall under the axe— and we have obviously tried to use statistical procedures that should minimize the probability of such an event— we would feel these incorrect interpretations have benefited science in that they are clearly stated and have stimulated others to test them and set the record straight if necessary. One great advantage of genetic data is that they can be increased almost at will, at least as long as individuals of the populations of interest or samples of their hereditary material are available. Another advantage is that the statistical significance of conclusions can be tested. Few other historical sciences, including archaeology and linguistics, enjoy this advantage to the same extent.

When it seemed necessary, we have clarified the uncertainties tied to some statements, and we have chosen to stress rather than to ignore earlier wrong conclusions, including our own, when this could be instructive.

8.3. COMPARISON OF DIFFERENT METHODS OF GENETIC ANALYSIS

We have used various types of methods for genetic analysis: trees, PC maps, geographic maps of single alleles, or synthetic maps summarizing the information from many alleles, the correlation between geographic and genetic distances under the form of standardized variograms, and others. Other methods have been developed too recently to use here. One of them is the analysis of genetic boundaries (Barbujani and Sokal 1990), which has promise and has already been useful for showing correlations with physical and linguistic boundaries. The method is not programmed to form close boundaries and therefore cannot delimit regions. This demand may be hard to satisfy, given that boundaries are sharp clines and sharp clines do not necessarily exist in all directions around any region. Another method that has now reached a certain degree of development (Sokal and Oden 1978; Sokal 1979; Sokal and Menozzi 1982; Sokal and Wartenburg 1983; Sokal et al. 1986; Sokal et al. 1989) is the study of autocorrelation in space. We have limited our study to the correlation of genetic and geographic distance by using standardized variograms. In principle, autocorrelation in space is a statistical study of geographic maps. The maps themselves are always easier to read and carry more detail. Autocorrelation analysis is a useful supplement for providing tests of significance of theoretical geographic trends (Sokal et al. 1991).

It may be of some interest to summarize the interpretive logic of the methods we have used on the basis of our comparative experience. No method is completely adequate by itself, and one can only gain by looking at the data in a variety of ways, which may supplement each other very usefully. Even so, some methods are more commendable than others, depending on the nature of the questions we ask.

In spite of their serious flaws, trees remain very useful. There are problems in choosing methods. The type of genetic distance is not too critical; more important is the choice of the tree method. We have described, however briefly, several ways of constructing trees. We have heavily relied on average linkage (UPGMA sec. 1.12b) because it is also very fast with large numbers of populations and gives results close to maximum likelihood. It therefore corresponds to a specific evolutionary model: independence of branches (no selective convergence or divergence, no admixtures) with a constant evolutionary rate. Both hypotheses may be wrong in specific cases, hence some of the flaws of the method. But using other techniques of tree reconstruction does not necessarily help, and methods of maximum parsimony really have no strong theoretical backing in the study of gene frequencies. Minimum-evolution methods, which give similar but not identical results, may be better, but the method that is simplest to use in practice (neighbor-joining; Saitou and Nei 1987) needs to be tested by a variety of simulations before its usefulness can be evaluated in full. In any case, it seems to suffer from the same drawbacks as maximum parsimony: it does not correspond to a specific evolutionary hypothesis that can be tested.

Other difficulties are the need of substantial data for obtaining significant results, and the problems of testing significance. In recent years, the bootstrap has been a very valuable addition to the arsenal of tree analysis. Not only can it test the significance of tree segments and nodes, but it can also point to potential admixtures. The rather lengthy procedure of comparing trees from many bootstrapped distance matrices is simplified by a computer algorithm developed by E. Minch (unpubl.), which has, incidentally, confirmed that our data on the world tree leave uncertain the order of separation in the tree of New Guinea, Australia, and Southeast Asia. This problem requires further analysis. As to the other potential problem in interpreting trees, selective convergence, it is likely to generate fewer difficulties than do admixtures. By providing easy access to knowledge of ancestral alleles, molecular genetics is starting to give us some clues, but much more work will be necessary before an adequate set of data is generated.

Two-dimensional displays of the first two principal components are a useful supplement to trees. Especially (but not only) in the way in which we have calculated them, via genetic distances and not via correlations between gene frequencies, PCs tend to give results similar to those of trees, and theory (Cavalli-Sforza and Piazza 1975) explains that this is, indeed, expected. But PCs are more flexible (relying on a greater number of parameters) and seem to provide interpretations that are occasionally more sensible—in situations made difficult by the poverty or complexity of data—than those given by trees. Inevitably, this is a subjective impression, but experience teaches us to use common sense in accepting conclusions from statistical methods.

It seems reasonable to say that trees are more an expression of history. Indeed, in their evolutionary interpretation, they do represent a temporal sequence of fissions. By contrast, two-dimensional PC displays express geography. Frequently, in fact, two-dimensional PC maps resemble geographic maps of the populations, though with characteristic distortions. One may think there is a paradoxical situation, even a contradiction, when we note a high similarity between trees and PCs. This is only apparent because, at this level of approximation, history and geography are themselves highly integrated.

We have not explored in any detail the correlation between two-dimensional PC maps and geography, in spite of well-known methods for carrying out this job (which has been aptly called "Procrustean"; Schönemann and Carroll 1970; Lalouel 1973). Frequently the similarity to geography is self-evident and, if it is not, there is not much to be gained in forcing it on the data and measuring it. The correlation of genetic and geographic distances does a similar analysis in a more synthetic way. In almost every geographic map of gene frequencies, we have given the "variogram" analysis, which shows the increase of genetic distance with geographic distance. It replaces the display introduced by Morton (1982; Morton et al. 1971, 1982) of an inverse relationship between genetic similarity and geographic distance. Both relations have an initially linear portion (in one case increasing, in the other decreasing) and an asymptote. We insisted on calculating "standardized" variograms gene-by-gene for detecting possible heterogeneities, which exist and agree with other findings on the same genes but are less important than we thought. Many variograms are not linear, either because the information for the particular gene is insufficient, or because the pattern of geographic variation is far from that expected at equilibrium. In a majority of cases, however, there are enough data, or enough time has elapsed since major migrations of the past, that genetic distance in many variograms increases monotonically with geographic space.

In difficult cases, analyzing the matrix of genetic distances can provide some help. Distance matrices, if large, may be threatening. One can reduce the threat enormously by ordering populations according to a tree. A "good" matrix, with good treeness (no admixtures, no large discrepancies in evolutionary rates) will then show a structure of rectangular and triangular blocks that are self-explanatory, and which we have exemplified. Peculiarities because of admixtures or other complications may then pop up as relatively simple deviations from this block structure. Smaller values in the row and column belonging to the mixed population will usually appear upon inspection, but the accurate analysis of admixtures is demanding. It is simplest when restricted to the two putative ancestors and the mixed population, but most existing methods have tried to account very accurately

for the least important sources of errors, ignoring the serious ones (e.g., the occurrence of more than two putative ancestors), and there is the need for more work along these lines.

It is commonly believed that, unlike trees or principal components, the matrix of distances carries all the information. This is not entirely correct, because different genes may lead to different evolutionary conclusions, and genetic distances calculated as averages of distances for single genes may hide such facts. Fortunately, there is only limited evidence that this phenomenon is really important, but we have not systematically looked for it.

Geographic maps of single alleles provide a visual aid to appreciating the extent and the details of genetic variation. We chose to extend the use of maps as much as possible, even if parts of the area are not well known genetically and had to be extrapolated from nearby genetically known areas. We were not worried about extrapolating even long distances from known points, because the locations of available data are always indicated and give a visual feel for the hazards of relying on the map value for locations at a great distance from observed gene frequencies. Local heterogeneities of gene frequencies that are statistically significant (at the expected rate of one per map, if everything is random) are also displayed on maps, and they are certainly more common than one would expect by chance alone. There is considerable heterogeneity, in some regions more than in others, and we have stored statistical information while building maps that we decided not to use here for reasons of book space.

Gene-frequency geographic maps were built as the necessary basis for calculating synthetic maps, and it is interesting to note that one can anticipate genetic patterns found in synthetic maps by inspecting the geographic maps of single alleles. This is obviously not surprising. Synthetic maps are much more efficient than maps of single genes in giving a visual impression from which one can derive a fairly accurate knowledge of the origin and direction of population movements or location of isolates.

Unfortunately, synthetic maps do not give direct indications with respect to the time of the population phenomena involved. Some indirect information can be gained. Simulations are still limited (Sgaramella-Zonta and Cavalli-Sforza 1973; Rendine et al. 1986), but those in existence show that for major migrations, the most important factor is the ratio of numbers of migrants to that of earlier settlers. This is reflected in the rank of the principal component: the first is the most important one, and its relative importance is determined by demographic factors such as the final population densities of migrants versus those of earlier settlers. Local genetic anomalies (absolute or relative maxima and minima) observed by synthetic maps are those that affect many genes simultaneously and are therefore the consequence of local genetic drift; the clines observed in synthetic maps frequently must be the result of major migrations (often, population expansions). These clines have not been completely smoothed out by later local migration, which usually takes many millennia to level major clines, as shown by simulation. The accuracy of the location of maxima or minima depends on the resolution of the map, that is, the unit used for map interpolation, which is usually of one or more degrees.

Three major phenomena are detectable by synthetic maps.

1. Expansions from a center of genetic diffusion can be observed from synthetic maps if, initially, the expanding population differed genetically from that living in the area into which the expansion occurred. If the center of diffusion is not too large, its geographic location can be distinguished fairly accurately. The majority of "demic" (population) expansions detected in the synthetic maps of higher order seem to be associated with patterns of expansion of Paleolithic or Neolithic people, or both. Population densities were lowest during the Paleolithic, and therefore drift was most extreme; the greatest genetic differences were established during this time, that is, before 10 kya. In Neolithic times, the introduction of agriculture caused a substantial increase in population density, the numerical importance of which was rarely equaled later: increases by factors of 10 or 100 times occurred in a matter of millennia, sometimes in a few centuries. It is therefore fitting that the genetic patterns observed in synthetic maps of higher rank were largely determined in Paleolithic and Neolithic times, keeping in mind that these patterns are relatively resistant to later change.

In any case, major demographic phenomena must be invoked to explain highest principal components. In terms of ratios of increase of population densities over previous settlers, clearly the most ancient epochs are those most likely to be involved, and these are the ones usually found in synthetic maps. One can paraphrase this conclusion by saying that the most readily observed genetic patterns are determined by the most ancient events. This is only a rough guide to the problem of timing the phenomena underlying the genetic patterns found in synthetic maps, but is better than no guide at all.

2. Dark or light local areas observed in a synthetic map are probably the consequence of drift differentiating a local population from its neighbors. If this initial local divergence was followed by some population increase— not enough perhaps to determine a large expansion— and not followed by immigration from the outside, there might be a local maximum or minimum, usually secondary, and not necessarily accompanied by evidence of expansion from it. The examples of Basques and Lapps in Europe are the most striking. Other examples are local dark or light areas in the second and third synthetic maps of Italy. The dark area north of Rome corresponding geographically to ancient Etruria may reflect a local, minor population increase determined by the high level of agriculture and social organization of Etruscans in the first half of the first millennium B.C. The initial genetic difference may have been generated either by local inbreeding and drift, or by the arrival of migrants from elsewhere. The latter origins are suggested by ancient legends; more substantive information may come from genetic data yet to be obtained on populations from the possible areas of origin, or from typings of Etruscan bones. Some population increase at the time of Etruscan rise, which is very likely, accompanied and followed by relative genetic isolation, which is possible, would be sufficient to allow the conservation of traces of this genetic isolate. In other cases—for example, the Ligurians in the northern Appennines— conservation of the genetic differences may simply be the result of the original population having remained less mixed in the refuge offered by high valleys in relatively inaccessible mountains. Demographic considerations may help distinguish this interpretation from that of a later differentiation caused by local drift. Every case may require a special analysis, remembering that population sizes and migration patterns are the determining factors.

3. Although expansions are usually the consequence of population explosions, the opposite, "implosions," can also occur. A capital may attract immigrants from a large area. In earlier synthetic maps of Europe (Menozzi et al. 1978a)—but not in the present ones, which use a smoother method of interpolation—in the approximate locations of Paris or London, one can detect areas of peculiar density and color that are similar in density and color to the area of origin of migrants. The migration to capitals does not usually or necessarily take place from immediately neighboring areas, but rather from poorer areas farther apart: the north of the British Isles for London, and the south of France for Paris. Thus, when the resolution is adequate, the capitals may show as an island of different color from the immediate background. In the maps of Italy, the present capital, Rome, and the old capital of the south, Naples, are different from the rest of the country.

The detailed interpretation of synthetic maps requires historical or archaeological information. Archaeological maps of artifacts that indicate contacts, and perhaps also migrations, do exist but are usually of poor quality in terms of completeness or detail. They are badly needed for our purpose. The distribution of relevant artifacts in innumerable museums, and the rarity of exact origins and dates for such artifacts, make the construction of improved maps a demanding but important job.

8.4. THE FUTURE OF THIS RESEARCH

The summaries in chapters 2 to 7 make a general summary redundant. It seems more constructive to dedicate some space to discussing the future of research on this topic and, in particular, the type of data that should be collected.

A great hope is that some crucial paleoanthropological material may still contain enough unfragmented DNA to be analyzed by the most recent techniques (Pääbo et al. 1989). Unfortunately, material that can be examined in this way is probably very limited. At the time of writing, the most hopeful example refers to an unknown Indian tribe that lived near Windover, Florida seven to eight millennia ago. The bodies, probably the result of ceremonial sacrifice or burial, belong to 91 individuals who died over a period of 1000 years. They were found in a peat bog under water of high saline content and nearly neutral pH; these conditions are responsible for the excellent conservation of the DNA of several brains, which could be amplified and shown to contain various HLA reactivities (Doran et al. 1986; Pääbo et al. 1988). However, DNA may be fragmented under these conditions, and the reliability of genetic information obtained from very old material is uncertain at this time. Other similar studies may be of very high value, but it is too early to assess the chances that similar material in good conditions of conservation will be found from several different places, times, and individuals.

The methodology of statistical analysis is still far from perfect; no single method gives exhaustive answers. Confidence in results from a single method is therefore almost as dangerous as confidence in conclusions from a single gene. There are a number of preconceived ideas about which methods are preferable; they are usually not expressed, and they are mostly unjustified. We have stated our reasons for preferring some methods, but we also believe it essential to try several different ones and compare the results, keeping in mind the idiosyncratic properties of each method. The major advances that are desirable are the development of satisfactory methods of reconstruction of networks—trees with admixtures—and of further statistical testing along the lines of bootstrapping.

We are now at a crossroads in the genetic analysis of human populations for two important reasons. The first is that a number of populations of considerable interest are rapidly disappearing. They are usually not physically endangered, but changes in living conditions and the opening of large areas to exploitation and/or development are rapidly and irreversibly altering the tribal worlds that still survive on every continent. The disappearance of the old ways of life destroys the old communities. Their dispersal makes it practically impossible to sample them; it is only for a few markers (from mtDNA and part of the

Y chromosome) that cross-marriages with members of other communities do not provoke a profound disruption of their genetic identity. The number of linguistic communities that are on the verge of extinction is appalling, as a perusal of Grimes (1984) shows. With the disappearance of language, that of tribal identity is likely to follow sooner or later.

The second reason is that our perspectives on methods of laboratory analysis have changed fairly radically in the last few years. The data used in this survey are mostly those obtained with the "classical" polymorphisms. These methods are still being used, but their popularity is likely to decline in favor of new techniques, which also have the advantage of being highly streamlined, so that they will be automated in part or in full within the next few years. The number or type of genes that one could use is limited only by the extent of the financial support, a crucial consideration. The number and type of genes employed are important now that DNA analysis allows so many more choices. As to the number of genes to be used: our empirical results indicate that 50 polymorphisms (independent alleles, possibly from many loci) are a minimum for applications to simpler trees but not to bigger ones. One hundred or more may not solve all the problems of trees with many populations but are clearly better and, in fact, necessary for large trees. For research on complicated trees with admixtures, more than 200 may be necessary. Whenever we have used a small number of polymorphisms (50 or less), results seem shakier. The degree of resolution depends on this number.

With the new molecular techniques, one can choose special polymorphisms with a higher number of alleles, but the number must not be so high that the distinction between different alleles is impaired (see a discussion of this point later). A reasonable balance of the numbers of loci and alleles in making up the total number of polymorphisms should also be reached. Too few loci with too many alleles—as, for example, when one uses only *HLA* or just one locus like mtDNA—leave room for the feeling of not being guaranteed against possible idiosyncrasies in these genetic systems. Moreover, adding another allele to a multiallelic system is not statistically as efficient as adding a new biallelic locus (Pamilo and Nei 1988).

Even where one can reach satisfactory results with 100 polymorphisms, it is obviously better to use more. Fortunately, the new molecular methodologies lend themselves to automation, which is in the process of developing especially under the impetus of the Human Genome Project. This and the chance of using a number of new classes of DNA polymorphisms (summary in Bowcock and Cavalli-Sforza 1991) open up totally new vistas. One may also choose different types of genes, depending on the populations being compared. For our purposes (distant populations), we prefer genes that have very low mutation rates because they are expected to have fewer alleles. This decreases the probability of confusion between alleles that might seem identical because of the testing technique in use, but are actually the result of different mutations. With DNA, one can sequence, and in some cases distinguish, mutations that appear similar by techniques of lower resolution but are not. If the same mutation appeared more than once, it may still be possible to distinguish different mutational events by studying neighboring polymorphisms (the haplotype) in which they are found. This method, however, is not as foolproof as it might seem, because gene conversion may transfer a mutation or a segment containing more than one mutation to another haplotype.

Rather than having too frequent ambiguities because of recurrent mutation, it is preferable to use polymorphisms with low mutation rates, as indicated by the small number of alleles. This preference of loci with fewer alleles does not exclude *HLA*, which owes part of its multiplicity of alleles to selection for diversity and not necessarily to high mutation rates. It excludes many "satellite" DNA systems (Jeffreys et al. 1990), which show an enormous variety of alleles, most of them difficult to identify precisely. Some of this has been shown to be due to high mutation rate. But for certain comparisons—for example, between very similar populations—one may need genes that have a higher number of alleles. Demands in this case are not unlike those of forensic problems (search of paternity, identification of sources of spots of blood or other material) or special research problems (identity of twins, linkage in pedigrees). In all of these studies, not only in those of populations, one wants to use genetic systems in which the number of alleles is not so high that the identity of spots shown by a probe in different individuals can be in doubt. The location of spots may be subject to small variation, and when different spots are too numerous, identification may be difficult. The mutation rate may be high enough to generate many mutants with similar or identical appearance. Somatic mutation may also further undermine the interpretation of results (Bowcock and Cavalli-Sforza 1991).

The new methodologies of analysis at the DNA level make it possible to store hereditary material from individuals and examine it at a later time, with less constraints than those encountered for storing and examining classical markers. By "immortalization" of B lymphocytes (infection by Epstein-Barr virus causing these cells to multiply indefinitely in culture), one can also assure a potentially infinite supply of DNA from an individual. The only known limitations to the use of this DNA are the rearrangements of DNA regions synthesizing immunoglobulins. The main advantage of immortalization is that it provides a potentially infinite amount of material for testing. More recently the amounts of needed DNA have decreased; and it is possible to store DNA for long times in relatively small amounts and use it for many tests. The conditions are therefore entirely favorable for collecting blood samples from disappearing aborigines *now*, by storing preferably, but not exclusively, lymphoblastoid cultures.

One cannot overemphasize the need for not waiting to strengthen the effort of collecting blood samples from populations of importance and storing them in DNA banks. An initiative in this direction was taken by one of the authors (Cavalli-Sforza) in collaboration with professor K. Kidd of the Yale Genetics Department (Bowcock et al. 1987; Cavalli-Sforza et al. 1987), but because of limited financial support the number of populations immortalized so far is small (less than 20). Immortalization is fastidious and relatively expensive, but not too elaborate, and could be done in many laboratories around the world including developing countries, after appropriate training and some addition of equipment. The main limitation is the need for a liquid nitrogen supply, not easily available outside developed countries. If immortalization were carried out in some laboratories in suitable locations—for example, in Asia and in South America—the problem of transporting blood samples would be greatly simplified. For a high probability of success, blood samples have to reach a laboratory in which the procedure can be carried out within 48–72 hours after collection. The great majority of critical populations are far from international airports, and they are almost all in developing countries where equipped laboratories are rare, but not nonexistent. Immortalizations are expensive; the minimum cost for 50 individuals is about $10,000, plus collecting and shipping costs. But there is no need to immortalize every population in existence. Moreover, the more expensive procedure of immortalization may be reserved for the exceptionally interesting and unique populations, of which there exist only a few hundred. The effort could be shared by a number of laboratories (Cavalli-Sforza et al. 1991). The cost could be perhaps 2% of that of the Human Genome Project which, in its original format, did not include any study of individual variation. An initiative called the Human Genome Diversity is being developed to meet these needs (see Cavalli-Sforza et al. 1991 and reports by Roberts 1991 and 1992). We know that an important part of the genome varies greatly from one individual to another and that this information is important.

From all these considerations, it follows that one should not wait any longer to start a collaborative study that could save the endangered information about the genetic history of *Homo sapiens sapiens* from disappearing. We have developed many ideas on criteria for collection and it may be worth listing at least some of them. Several considerations are of importance in choosing individuals and/or groups.

1. Direct information on gene flow into the group, the more detailed the better, is extremely important. Usually, this information is restricted to the most recent period. Groups that had notorious and frequent acceptance of neighbors as marriage partners, especially of entirely different origin, are obviously less good candidates for sampling.

2. Except for rare circumstances, children of exogamous matings (one parent external to the group) should not be sampled. The rationale for this exclusion is that, in most parts of the world, exogamy has probably increased in the last decades. Demanding that all known remote ancestors be from the group is usually unrealistic, but as a minimum, one should include only people who have both parents born in their same group (be it tribe, village, or whatever unit is used). In some cases, this rule can be extended to include the four grandparents.

3. Marked acculturation indicates extensive contacts with neighbors at a different level of development, which may be old and have favored genetic exchange for a long time; such samples should be avoided if possible.

4. Very small populations (one to a few hundred individuals) may have been exposed to drift so profound that their gene frequencies may have been altered capriciously and dramatically; they are therefore less interesting for these purposes, but some such groups—for example, from the Andaman Islands—are so unique that all their adult members should be immortalized even if there are only 100 or fewer.

5. Cities should be avoided as much as possible as sources of samples. They usually represent immigrants from a poorly defined area and a poorly defined earlier period. At least until recently, most cities have never reproduced their population number, which has been maintained or increased mainly through immigration. Cities are therefore a poor source of samples, except for first-generation immigrants or their immediate descendants.

6. Rural areas are better sources than urban ones, and an analysis of local history is always desirable and frequently possible.

7. Areas that have had the least immigration over the centuries are those that are economically poor and geographically isolated. They may have been occupied as refugia in early times by people who were chased from their original abode by latecomers. Most mountainous areas fall in this category and are therefore of great interest; the main limitation is that they are subject to drift as in (4). For this reason, it is better to sample, not from a single village, but from a larger, fairly homogeneous region, like a valley with a number of villages and inhabitants, taking only a few individuals (or even only one) from each village. Here demographic and surnames studies are especially useful for identifying the patterns of gene flow.

8. Important developments in the strategy of population sampling may be introduced when sequencing of highly variable segments of DNA is more common. Currently, there exist data for only some regions of mitochondrial DNA and, to a more limited extent, for some *HLA* genes. The extension of this practice will probably revolutionize the field when automation of the procedure is more advanced.

It would be useful to prepare a list of populations that seem to deserve detailed study. This list will demand considerable work in consultation with anthropologists. What follows is not a list but a series of loose considerations on the theme, organized by continent.

Africa. In Africa all populations of the Sahara should be studied (some have been almost entirely inaccessible for many years, like those of southern Lybia and northern Chad). Ethiopia and, in general, East Africa also need thorough studies; the forested area in western Ethiopia, and the Omotic-speaking tribes deserve much attention. Populations from southern Sudan are in considerable distress from civil war but are of great interest. The possible traces of Khoisan and/or Asian populations in East Africa are worth considerable attention. In South Africa, the Khoisan are slowly disappearing and have not been immortalized in sufficient numbers; some Pygmy groups might be further studied. The great number of tribes in the west, center, and south of Africa offer unlimited challenge. They are fairly homogeneous genetically, but there are still enough differences among them that may help in understanding the different agricultural expansions that took place in the area. A number of other small groups, a few of which are still hunting and gathering, are probably residues of older, preagricultural populations. The possible connections because of the putative Saharan origin of pastoral tribes of sub-Saharan Africa are worth exploring further.

Asia. In Asia there exist the two largest populations on Earth: Chinese and Indians, who are together responsible for almost a half of the world population. Both of them are highly heterogeneous genetically. In India there are a number of potentially very interesting "Australoid" populations, so-called because of their exterior phenotype, which are possible remnants of migrants from Africa to Australia during the early spread of *H. sapiens sapiens*. They are few and far between, but they are well worth a very detailed analysis. They may be similar to other related groups (often called Negritos) in the islands (the Andamanese, in particular) in the Malayan peninsula and in insular Southeast Asia. The great number of ethnic minorities in China, almost all in the south and a few in the west, are now being studied systematically at Academia Sinica in Beijing. These minorities are characterized by having kept their language, but they may be related to many neighbors, who have been absorbed in the Han culture over the centuries. An interesting problem here is the genetic boundary between

Northeast and Southeast Asia and its correlation with archaeological and paleoanthropological observations. Not only Negritos, but many other populations of mainland and insular Southeast Asia have been poorly studied and are potentially very interesting.

The movements that have taken place in Siberia in the last few centuries may make it more difficult to trace the connections between Siberians and early Americans, but all tribes that have not been excessively acculturated should be immortalized. The possible participation of people from the Japanese archipelago in expansions are worth being considered further; islands around Japan and the low-caste Japanese, the Etas, are potentially very interesting, if there truly is hereditary transmission of this social condition.

West Asia is normally considered a hodge-podge of populations that have been completely altered by too many wars, invasions, and successions of empires. The effect of conquest by armies is often trivial in terms of genetic demography, unless truly large-scale genocides took place and mass migrations—forced or spontaneous—of the vanquished people occurred. In any case, mountainous parts like the Caucasus and several regions of Turkey should be explored carefully. If the populations of southern Iraq (the "marshland people") can be traced in spite of the recent wars that have deeply affected the areas they occupied, their historical connections with Sumerians make them particularly interesting.

Europe. Lapps have distinct dialects, and the different groups have also different degrees of accultura-tion and mixture with other Scandinavians and should be studied separately. Basques from the core of the Basque area (Guipuzcoa) have not been intensively studied; isolated populations in the Pyrenees, Alps, Appennines, Carpathians, and, again, the Caucasus deserve further analysis. In general, all these populations would gain from parallel studies of demography, linguistics, genetics, and surname distribution as done, for instance, in Sardinia. There are many other other groups of interest to be studied in Europe.

America, Australia, New Guinea and the Pacific Islands. These areas simply have too many populations to be analyzed. Criteria for choice are: size not too small; least acculturation; least, possibly no admixture, especially with Caucasoids and Africans, but also with other tribes. It would be best to choose populations representing different linguistic groups. There are serious limitations for reaching many populations; the most interesting ones are the most difficult to sample. Much single-handed work was done in New Guinea by Carleton Gajdusek, unfortunately at a time when nobody knew that white cells could and should have been preserved; otherwise, there would be an incredibly good collection of New Guinea samples already in cold storage.

In the above, we have considered the tribal world, but modern populations should also be studied even though they are in lesser danger of disappearing, and it is not necessary to immortalize them. Many of them could be sampled most easily through blood banks, by carefully selecting donors whose origins are better known.

8.5. GENETIC AND LINGUISTIC EVOLUTION

One of the most intriguing results of this inquiry was the finding of important correlations between the genetic tree and what is understood of the linguistic evolutionary tree. It is well known that languages evolve so fast that the tracing of ancient relationships is difficult. No linguist doubts the existence of phylogenetic relationships between languages of the Indo-European family; and even if there is some disagreement over the exact boundaries of many other linguistic families, most of them are widely accepted. The major disagreement arises over the American languages, as we have discussed. A fairly large group of Americanists from the United States would like to recognize no less than 50–60 families among which no relationship can be discerned (see chap. 6). Greenberg (1987), developing ideas that originated in part with Sapir and Kroeber, grouped all these languages in one family, Amerind, and distinguished it from two other families of North-American languages, Na-Dene and Eskimo.

There is clearly a profound chasm between these two views, and agreement may be difficult to reach because methods are widely different on the two sides. Our observations go a long way in supporting the major statements by Greenberg.

The central question is, *why should there be any congruence between genetic and linguistic evolution?* The main reason is that the two evolutions, in principle, follow the same history, which can be represented, in a simplified or sometimes oversimplified way, as a sequence of fissions. In two or more populations that have separated, there begins a process of differentiation of both genes and languages. One does not need to assume that the genetic process has evolutionary rates that remain strictly constant in the various branches, as long as there exists a very rough proportionality (more generally, a monotonic relationship) of the accumulated evolutionary changes to elapsed time, so that the greater the separation in time, the more likely that the differentiation is also greater. We are not postulating a rigid genetic clock; though we found it valid in first approximation for long periods, it is likely to be less useful for shorter periods. A similar assumption is to be made for linguistic evolution, but again it is not at all necessary that a real "linguistic clock" be rigorously valid. The average rates and modes of change can be quite different for genes and languages (and indeed they are). Of course, it is reasonable to expect that later events, like language replacements and/or gene substitutions, may blur the picture; but our conclusion was that they do not blur it entirely. Language replacements are more likely to be historically known, and we have discussed some examples. Gene substitutions caused by gene flow are more subtle and not so easily proved historically, but can be tested genetically on modern populations.

One can also use another, complementary explanation for the congruence of genetic and linguistic evolution. We know that genes are transmitted from parents to children. We have mentioned earlier (sec. 5.8) that culture (including language) is transmitted in a great variety of ways, one of which (called *vertical*) is very similar to genetic transmission, being directed from parents to children. Cultural traits thus transmitted tend to be almost as conservative as genetic ones. Although there is very little or no information on how the teaching of language takes place, especially in traditional societies where no schools exist and age-peer groups are small because population groups are small, it is very likely that it is a task largely undertaken by parents. For one traditional society, African Pygmies, there is evidence that parents are responsible for close to 90% of all cultural transmission investigated (which did not, however, include a direct study of language; Hewlett and Cavalli-Sforza 1986). It seems inevitable that cultural traits transmitted vertically will mimic closely the pattern of historical variation of genes, and thus will language. In modern society, the correlation will inevitably diminish, with the increased importance of other mechanisms of cultural transmission which, unlike the vertical one, follow very different paths and make it easier for traits to spread horizontally between unrelated individuals. Limiting the analysis to aboriginal people has increased the relative emphasis on traditional societies and, therefore, on vertical cultural transmission.

Language is not the only culturally transmitted trait that will closely parallel the transmission of genes under such conditions. In modern society, parents influence their children in unexpected ways, for example, in the transmission of religion and political opinions (Cavalli-Sforza et al. 1981; see also sec. 5.8). Religions that do not proselitize (i.e., are not transmitted horizontally, but essentially only vertically), like the Jewish religion, are more likely to be transmitted with genes. It is therefore not surprising that Jews have maintained considerable genetic similarity among themselves and with peo-

ple from the Middle East, with whom they have common origins (Carmelli and Cavalli-Sforza 1979; Cavalli-Sforza and Carmelli 1979; Motulsky 1980).

Regions where we found little or no agreement between genes and languages are New Guinea, Australia, and South America. They all share a long permanence in Paleolithic and Neolithic conditions, but other factors may be involved. It is easiest to explain the failure to find a precise correlation in New Guinea for Austronesian versus non-Austronesian languages. The Austronesian penetration was limited to relatively few coastal groups and happened sufficiently long ago that gene replacement from genetic exchanges with Melanesian neighbors had time to blur the picture. We have given indications of the length of time necessary for gene replacement to be sufficiently advanced that it may be undetectable, and the timing of the Austronesian migrations certainly falls beyond those limits. Difficulties in South America are likely to arise from an extreme genetic differentiation that is hard to reconcile even with geographic distance, a very rare phenomenon. In other parts of America or of Melanesia, the correlation of genetics and linguistics reappears. It is likely that the use of linear correlation coefficients is not ideal (Cavalli-Sforza et al. 1992).

The discrepancies we found in the correlation between languages and genes in the general tree have been useful in allowing us to detect reasonable historical or genetic explanations of the exceptions. Historical knowledge is not, however, always sufficient to explain all possible discrepancies that may be found in the future. Even so, the general correlation we have studied, and the regional ones found by several authors in a number of regions are well established. We found the correlation between the European-Northern Asian-American branch of the genetic tree and the Nostratic-Eurasiatic-Amerind linguistic superfamily especially illuminating.

Once the general principle of a correlation between genetic tree and linguistic families and superfamilies is established, it opens up the possibility of making predictions on times and perhaps places of origin of linguistic families and superfamilies, and finally of completing the linguistic evolutionary tree. We have given a few tentative examples in this book, but a rigorous application of the method demands specific investigations. All this proves basically valid a prediction by Charles Darwin, even if it does not foresee the complications from genetic admixtures and language substitutions. It is well known that Darwin was greatly inspired by considerations on linguistic evolution and analogies with biological evolution, as is clear from many comments in *The Descent of Man* (1871). In chapter 14 of *On the Origin of Species* (1859), he stated specifically:

If we possessed a perfect pedigree of mankind, a genealogical arrangement of the races of man would afford the best classification of the various languages now spoken throughout the world; and if all extinct languages, and all intermediate and slowly changing dialects, were to be included, such an arrangement would be the only possible one.

This optimistic statement is largely correct today, even if admixtures set restrictions to the validity of trees for genes and also for languages, demanding the refinement of the trees we now have or make, with more complicated networks showing connections between tree branches. It would of course be more rigorous, and interesting, if linguistic research could independently establish its own complete phylogenetic tree or network, because then the comparison between the two trees or networks would be even more convincing and revealing.

LITERATURE CITED

Adovasio, J. M., J. D. Gunn, J. Donahue, and R. Stuckenrath. 1982. Meadowcroft Rockshelter: A synopsis. In *Peopling of the New World*. J. E. Ericson, R. E. Taylor, and R. Berger, eds. pp. 97–103. Los Altos, Calif.: Ballena Press.

Aikens, C. M. 1983. The far West. In *Ancient North Americans*. J. D. Jennings, ed., pp. 149–201. New York: W. H. Freeman.

——. 1990. The first peopling of the new world. *Prehistoric Mongoloid Dispersals* spec. issue 9:1–34.

Aird, I., H. H. Bentall, and J. A. F. Roberts. 1953. A relationship between cancer of stomach and the ABO groups. *Br. Med. J.* 1:799–801.

Akbari, M. T., S. S. Papiha, D. F. Roberts, and D. D. Farhud. 1984. Serogenetic investigations of two populations of Iran. *Hum. Hered.* 34:371–377.

Alexander, J. 1980. First-millennium Europe before the Romans. In *The Cambridge Encyclopedia of Archaeology*. A. Sherratt, ed., pp. 222–226. New York: Crown.

Alexseev, V. P. 1979. Anthropometry of Siberian peoples. In *The First Americans: Origins, Affinities, and Adaptations*. W. S. Laughlin and A. B. Harper, eds., pp. 57–90. New York: G. Fischer.

Allières, J. 1979. Manuel Pratique de Basque. *Roman Archaeological Documents*. Paris: Picard.

Allison, A. C. 1954a. The distribution of the sickle-cell trait in East Africa and elsewhere, and its apparent relationship to the incidence of subtertian malaria. *Trans. R. Soc. Trop. Med. Hyg.* 48:312–318.

——. 1954b. Protection afforded by sickle-cell trait against subtertian malarial infection. *Br. Med. J.* 1:290–294.

——. 1961. Genetic factors in resistance to malaria. *Ann. NY Acad. Sci.* 91(3):710–729.

Ambrose, S. H. 1982. Archaeology and linguistic reconstructions of history in East Africa. In *The Archaeological and Linguistic Reconstruction of African History*. C. Ehret and M. Posnansky, eds., pp. 104–157. Berkeley: University of California Press.

Ammerman, A. J. 1989. On the Neolithic transition in Europe: a comment on Zvelebil and Zvelebil (1988). *Antiquity* 63:162–165.

Ammerman, A. J., and L. L. Cavalli-Sforza. 1973. A population model for the diffusion of early farming in Europe. In *The Explanation of Culture Change*. C. Renfrew, ed., pp. 343–357. London: Duckworth.

——. 1984. *Neolithic Transition and the Genetics of Populations in Europe*. Princeton, N.J.: Princeton University Press.

Ammerman, A. J., L. L. Cavalli-Sforza, and D. K. Wagener. 1976. Toward the estimation of population growth in Old World prehistory. In *Demographic Anthropology: Quantitative Approaches*. E. Zubrow, ed., pp. 27–61. Albuquerque: University of New Mexico Press.

Ananthakrishnan, R., and H. Walter. 1972. Some notes on the geographical distribution of the human red cell acid phosphatase phenotypes. *Humangenetik* 15:177–181.

Anderson, S., A. T. Bankier, B. G. Barrel, M. H. L. de Bruijn, A. R. Coulson, J. Drouin, I. C. Eperon, D. P. Nierlich, B. A. Roe, F. Sanger, P. H. Schreier, A. J. H. Smith, R. Staden, and I. G. H. Young. 1981. Sequence and organization of the human mitochondrial genome. *Nature (Lond.)* 290:457–465.

Antonarakis, S. E., S. H. Orkin, H. H. Kazazian, S. C. Goff, C. D. Boehm, P. G. Waber, J. P. Sexton, P. Ostrer, V. F. Fairbanks, and A. Chakravarti. 1982. Evidence for multiple origins of the beta E-globin gene in Southeast Asia. *Proc. Natl. Acad. Sci. USA* 79:6608–6611.

Antonarakis, S., C. D. Boehm, G. R. Sergeant, C. E. Theisen, G. J. Dover, and H. H. Kazazian. 1984a. Origin of the α^S globin gene in Blacks: the contribution of recurrent mutation or gene conversion or both. *Proc. Natl. Acad. Sci. USA* 81(3):853–856.

Antonarakis, S., S. H. Irkin, T. C. Cheng, A. F. Scott, J. P. Sexton, S. P. Trusko, S. Charache, and H. H. Kazazian. 1984b. Beta-thalassemia in American Blacks—Novel mutations in the tata box and an acceptor splice site. *Proc. Natl. Acad. Sci. USA* 81(4):1154–1158.

Anthony, D., and D. Brown. 1991. The origins of horseback riding. *Antiquity* 65:22–38.

Archie, J. W. 1989. Homoplasy excess ratios: new indices for measuring levels of homoplasy in phylogenetic studies and a critique of the consistency index. *Syst. Zool.* 38:253–269.

Astolfi, P., A. Piazza, and K. K. Kidd. 1978. Testing of evolutionary independence in simulated phylogenetic trees. *Syst. Zool.* 27:391–400.

Astolfi, P., K. K. Kidd, and L. L. Cavalli-Sforza. 1981. A comparison of methods for reconstructing evolutionary trees. *Syst. Zool.* 30:156–169.

Bachman, C. G. 1961. *The Old Order Amish of Lancaster County*. Lancaster, Pa.: Franklin and Marshall College.

Bailey, G. N. 1980. Holocene Australia. In *The Cambridge Encyclopedia of Archaeology*. A. Sherratt, ed., pp. 333–341. New York: Crown.

Barbujani, G. 1987a. Autocorrelation of gene frequencies under isolation by distance. *Genetics* 117:777–782.

——. 1987b. Diversity of some gene frequencies in European and Asian populations. III. Spatial correlogram analysis. *Ann. Hum. Genet.* 51:345–353.

————. 1988. Diversity of some gene frequencies in Europeans and Asian populations. IV. Genetic population structure assessed by the variogram. *Ann. Hum. Genet.* 52:215–225.

————. 1991. What do languages tell us about human microevolution? *Trends Ecol. Evol.* 39:151–156.

Barbujani, G., and R. R. Sokal. 1990. Zones of sharp genetic change in Europe are also linguistic boundaries. *Proc. Natl. Acad. Sci. USA* 87 (5):1816–1819.

Barbujani, G., N. L. Oden, and R. R. Sokal. 1989. Detecting regions of abrupt change in maps of biological variables. *Syst. Zool.* 38(4):376–389.

Barrantes, R., P. E. Smouse, H. W. Mohrenweiser, H. Gershowitz, J. Azofeifa, T. D. Arias, and J. V. Neel. 1990. Microevolution in lower Central America: Genetic characterization of the Chibcha-speaking groups of Costa Rica and Panama, and a consensus taxonomy based on genetic and linguistic affinity. *Am. J. Hum. Genet.* 46(1):63–84.

Barreca, F. 1986. *La Civiltà Fenicio-Punica in Sardegna.* Sassari, Italy: C. Delfino.

Barry, I. 1980. Tropical forest cultures of the Amazon basin. In *The Cambridge Encyclopedia of Archaeology.* A. Sherratt, ed., pp. 398–402. New York: Crown.

Bastin, Y., A. Coupez, and B. de Halleux. 1983. Classification lexicostatistique des langues bantoues (214 relevés). *Bull. Seances Acad. R. Sci. Outre-Mer* 27(2):173–199.

Bateman, R., I. Goddard, R. T. O'Grady, V. A. Funk, R. Mooi, W. J. Kress, and P. F. Cannell. 1990a. Speaking of forked tongues: the feasibility of reconciling human phylogeny and the history of language. *Curr. Anthropol.* 31(1):1–13.

————. 1990b. On human phylogeny and linguistic history—reply to comments. *Curr. Anthropol.* 31(2):177–183.

————. 1990c. Reply to comments. *Curr. Anthropol.* 31(3):315–316.

Bayard, D. 1980. East Asia in the Bronze Age. In *The Cambridge Encyclopedia of Archaeology.* A. Sherratt, ed., pp. 168–173. New York: Crown.

Bellwood, P. S. 1979. *Man's Conquest of the Pacific: The Prehistory of Southeast Asia and Oceania.* New York: Oxford University Press.

————. 1989. The colonization of the Pacific: some current hypotheses. In *The Colonization of the Pacific: A Genetic Trail.* A. V. S. Hill and S. W. Serjeantson, eds., pp. 1–59. Oxford: Clarendon Press.

Beloch, J. 1886. *Die Bevölkerung der Griechisch-römischen Welt.* Leipzig: Duncker and Humblot.

Bengston, J. D., and M. Ruhlen. In press. Global etymologies. In *On the Origin of Languages: Studies in Linguistic Taxonomy.* M. Ruhlen, ed. Stanford, Calif.: Stanford University Press.

Bernard, J., and J. Ruffié. 1976. Hematologie et culture: le peuplement de l'Europe de l'ouest. *Ann. Ecole Sci. Prat. Hautes Etud.* 4:661–676.

Bernini, L., C. Borri-Voltattorni, and M. Siniscalco. 1966. Studies on dissociation and reassociation of haptoglobins. Hybridization between Hp1 Hp1 and Hp2 Hp2. *Atti Accad. Naz. Lincei Rend. Cl. Sci. Fis. Mat. Nat.* 40:279–289.

Bertranpetit, J., and L. L. Cavalli-Sforza. 1991. A genetic reconstruction of the history of the population of the Iberian peninsula. *Ann. Hum. Genet.* 55:51–67.

Bhattacharrya, A. 1946. On a measure of divergence between two multinomial populations. Sankhya 7:401–406.

Biasutti, R. 1959. *Le Razze e: Popoli della Terra.* Turin, Italy: UTET.

Biraben, J.-N. 1980. An essay concerning mankind's evolution. *Population* 4:1–13.

Birdsell, J. B. 1957. Some population problems involving Pleistocene man. *Cold Spring Harbor Symp. Quant. Biol.* 22:47–69.

————. 1977. The recalibration of a paradigm for the first peopling of Greater Australia. *In Sunda and Sahul. Prehistoric Studies in Southeast Asia, Melanesia and Australia.* J. Allen, J. Golson, and R. Jones, eds., pp. 113–167. London: Academic Press.

Bjarnason, O., V. Bjarnason, J. H. Edwards, S. Fridriksson, M. Magnusson, A. E. Mourant, and D. Tills. 1973. The blood groups of Icelanders. *Ann. Hum. Genet.* 36:425–458.

Black, F. L., and F. M. Salzano. 1981. Evidence for heterosis in the HLA system. *Am. J. Hum. Genet.* 33:894–899.

Black, F. L., L. L. Berman, and Y. Gabbay. 1980. HLA antigens in South American Indians. *Tissue Antigens* 16:368–379.

Blumenbach, J. F. 1775. De generis humani varietate nativa. M. D. thesis, University of Gottingen.

Boas, F. 1940. *Race, Language, and Culture.* New York: Collier Macmillan.

Bodmer, W. F., and L. L. Cavalli-Sforza. 1968. A migration matrix model for the study of random genetic drift. *Genetics* 59:565–592.

————. 1976a. *Genetics, Evolution, and Man.* San Francisco: W. H. Freeman.

————. 1976b. Genes in populations. In *Genetics, Evolution, and Man.* W. F. Bodmer and L. L. Cavalli-Sforza, eds., pp. 229–258. San Francisco: W. H. Freeman.

————. 1976c. Genetic applications of probability and statistics. In *Genetics, Evolution, and Man.* W. F. Bodmer and L. L. Cavalli-Sforza, eds. pp. 709–754. San Francisco: W. H. Freeman.

Bonnaud, P. 1981. *Terres et Langages, Peuple et regions.* Clermont-Ferrand, France.

Bonné-Tamir, B. 1980. The Samaritans—A living ancient isolate. In *Population Structure and Genetic Disorders.* A. W. Eriksson et al., eds., pp. 27–41. London: Academic Press.

Bonné-Tamir, B., A. Zoossmann-Diskin, A. Ticher, A. Oppenheim, S. Avigad, and S. Kleiman. 1992. Genetic diversity among Jews reexamined: preliminary analyses at the DNA level. In *Genetic Diversity among Jews: Diseases and Markers at the DNA Level.* B. Bonné Tamir and A. Adam, eds., pp. 80–94. New York: Oxford University Press.

Bosch-Gimpera, P. 1943. El problema de los origines vascos. *Eusko-Jakintza* 3:39.

Boserup, E. 1965. *The Conditions of Agricultural Growth: The Economics of Agrarian Change under Population Pressure.* Chicago: Aldine.

Bowcock, A. M., C. Bucci, J. M. Hebert, J. R. Kidd, K. K. Kidd, J. S. Friedlaender, and L. L. Cavalli-Sforza. 1987. Study of 47 DNA markers in five populations from four continents. *Gene Geogr.* 1(1):47–64.

Bowcock, A., and L. L. Cavalli-Sforza. 1991. The study of variation in the human genome. *Genomics* 11: 491–498.

Bowcock, A. M., J. R. Kidd, J. L. Mountain, J. M. Hebert, L. Carotenuto, K. K. Kidd, and L. L. Cavalli-Sforza. 1991. Drift, admixture, and selection in human evolution: A study with DNA polymorphisms. *Proc. Natl. Acad. Sci. USA* 88:839–843.

Bowcock, A. M., J. M. Hebert, J. L. Mountain, J. R. Kidd, K. K. Kidd, and L. L. Cavalli-Sforza. 1992. Study of an additional 58 DNA markers in five populations from four continents. *Gene Geogr.* 5:151–173.

Bowdler, S. 1977. The coastal colonisation of Australia. In *Sunda and Sahul: Prehistoric Studies in Southeast Asia, Melanesia and Australia.* J. Allen, J. Golson, and R. Jones, eds., pp. 205–236. London: Academic Press.

Bowler, J. M., and A. G. Thorne. 1976. Human remains from Lake Mungo: Discovery and excavation of Lake Mungo III. In *The Origin of the Australians.* R. L. Kirk and A. G. Thorne, eds., pp. 127–138. Canberra: Australian Institute of Aboriginal Studies.

Bowles, G. T. 1977. *The People of Asia.* New York: Charles Scribner.

Boyd, M. F., ed. 1949. Malariology. 2 Vols. Philadelphia: Saunders.

Boyd, W. C. 1950a. *Genetics and the Races of Man.* Boston: Little, Brown.

———. 1950b. Use of blood groups in human classification. *Science (Wash., D.C.)* 112(2903):187–196.

———. 1963. Four achievements of the genetical method in physical anthropology. *Am. Anthropol.* 65:243–252.

Braudel, F. 1986. L'identité de la France. Vol. 1, pp. 88–94. Paris: Arthaud-Flammarion.

Bräuer, G. 1989a. The ES-11693 hominid from West Turkana and Homo Sapiens evolution in East Africa. In *Hominidae.* G. Giacobini, ed., pp. 241–245. Milan: Jaca Books.

———. 1989b. The evolution of modern humans: A comparison of the African and non-African evidence. In *The Human Revolution: Behavioural and Biological Perspectives on the Origins of Modern Humans.* P. Mellars and C. Stringer, eds., pp. 123–154. Princeton, N.J.: Princeton University Press.

Bray, W. 1980. Early agriculture in the Americas. In *The Cambridge Encyclopedia of Archaeology.* A. Sherratt, ed., pp. 365–374. New York: Crown.

———. 1988. The Paleoindian debate. *Nature (Lond.)* 332:107.

Brock, D. J. H., and O. Mayo, eds. 1978. *The Biochemical Genetics of Man.* London: Academic Press.

Brooks, A. S., and B. Wood. 1990. Palaeoanthropology: the Chinese side of the story. *Nature (Lond.)* 344: 288–289.

Brown, P. 1987. Pleistocene homogeneity and Holocene size reduction: the Australian human skeleton evidence. *Archaeol. Oceania* 22:41–67.

Brown, W. M. 1980. Polymorphism in mitochondrial DNA of humans as revealed by restriction endonuclease analysis. *Proc. Natl. Acad. Sci. USA* 77(6):3605–3609.

———. 1983. Evolution of animal mitochondrial DNA. In *Evolution of Genes and Proteins.* M. Nei and R. K. Koehn, eds., pp. 62–88. Sunderland, Mass.: Sinauer.

Campbell, L., and M. Mithun, eds. 1979. *The Languages of Native America: Historical and Comparative Assessment.* Austin: University of Texas Press.

Camps, G. 1974. *Les Civilisations Préhistoriques de l'Afrique du Nord et du Sahara.* Paris: Doin.

Cann, R. L., and A. C. Wilson. 1983. Length mutations in human mitochondrial DNA. *Genetics* 104(4): 699–711.

Cann, R. L., W. M. Brown, and A. C. Wilson. 1982. Evolution of human mitochondrial DNA: A preliminary report. In *Progress in Clinical and Biological Research.* Vol. 103A. B. Bonné-Tamir, T. Cohen, and R. M. Goodman, eds., pp. 157–165. New York: Alan R. Liss.

Cann, R. L., M. Stoneking, and A. C. Wilson. 1987. Mitochondrial DNA and human evolution. *Nature (Lond.)* 325:31–36.

Cao, A., M. Gossens, and M. Pirastu. 1989. Annotation: Alpha thalassaemia mutations in Mediterranean populations. *Br. J. Haematol.* 71:309–312.

Capell, A. 1962. *A Linguistic Survey of the Southwestern Pacific.* Oceanic Linguistic Monograph. Vol 7. New Caledonia: South Pacific Commission.

Carmelli, D., and L. L. Cavalli-Sforza. 1979. The genetic origin of the Jews: A multi-variate approach. *Hum. Biol.* 51:41–61.

Casals, T., V. Nunes, P. Maña, M. Chillon, C. Lazaro, C. Vasquez, and X. Estivill. 1990. Cystic fibrosis in the Basque Country: European origin of the 4F508 mutation (Abstract). *Am. J. Hum. Genet.* 47(suppl.):A129.

Casals, T., C. Vasquez, C. Lazaro, E. Girbau, F. J. Gimenez, and X. Estivill. 1992. Cystic fibrosis in the Basque Country: high frequency of mutation F508 in patients of Basque origin. *Am. J. Hum. Genet.* 50:404–410.

Casanova, M., P. Leroy, C. Boucekkine, J. Weissenbach, C. Bishop, M. Fellous, M. Purello, G. Fiori, and M. Siniscalco. 1985. A human Y-linked DNA polymorphism and its potential for estimating genetic and evolutionary distance. *Science* (Wash., D.C.) 230: 1403–1406.

Cassero, C., R. Montana, M. Arena, and G. Modiano. 1986. The distribution of human genetic polymorphism in India with special reference to the Hindus. *Atti Accad. Naz. Lincei* 18:3–252.

Cavalli-Sforza, L. L. 1958. Some data on the genetic structure of human populations. *Proc. 10th Int. Congr. Genet.* 1:388–407.

———. 1963. The distribution of migration distances, models and applications to genetics. In *Human displacements; measurement, methodological aspects.* J. Sutter, ed., pp. 139–158. Entretiens de Monaco en Sciences Humaines. Monaco: Editions Sciences Humaines.

———. 1966. Population structure and human evolution. *Proc. R. Soc. Lond.* 164:362–379.

———. 1973. Some current problems in human population genetics. *Am. J. Hum. Genet.* 25:82–104.

———. 1974. Genetics of human populations. *Sci. Am.* 231(3):81–89.

———. 1986a. *African Pygmies.* Orlando, Fla.: Academic Press.

———. 1986b. Population structure. In *Evolutionary Perspectives and the New Genetics.* H. Gershowitz, D. L. Rucknagel, and R. E. Tashian, eds., pp. 13–30. New York: Alan R. Liss.

——. 1988. The Basque population and ancient migrations in Europe. *Munibe* 6:129–137.

——. 1990. How can one study individual variation for three billion nucleotides of the human genome? *Am. J. Hum. Genet.* 46(4):649–651.

Cavalli-Sforza, L. L., and W. F. Bodmer. 1971a. *The Genetics of Human Populations*. San Francisco: W. H. Freeman.

——. 1971b. Mendelian populations. In *The Genetics of Human Populations*. L. L. Cavalli-Sforza and W. F. Bodmer, eds., pp. 39–70. San Francisco: W. H. Freeman.

Cavalli-Sforza, L. L., and D. Carmelli. 1979. The Ashkenazi gene pool: Interpretations. In *Genetic diseases among Ashkenazi Jews*. R. M. Goodman and A. Motulsky, eds., pp. 93–102. New York: Raven Press.

Cavalli-Sforza, L. L., and A. W. F. Edwards. 1964. Analysis of human evolution. *Proc. 11th Int. Congr. Genet.* 2:923–933.

——. 1967. Phylogenetic analysis: Models and estimation procedures. *Am. J. Hum. Genet.* 19:223–257.

Cavalli-Sforza, L. L., and M. Feldman. 1981. *Cultural Transmission and Evolution: A Quantitative Approach*. Princeton, N.J.: Princeton University Press.

——. 1990. Spatial subdivisions of populations and estimates of genetic variation. *Theor. Popul. Biol.* 37:3–25.

Cavalli-Sforza, L. L., and A. Piazza. 1975. Analysis of evolution: evolutionary rates, independence and treeness. In *Theor. Popul. Biol.* 8:127–165.

Cavalli-Sforza, L. L., and W. S. Wang. 1986. Spatial distance and lexical replacement. *Language* 62(1):38–55.

Cavalli-Sforza, L. L., M. W. Feldman, K. H. Chen, and S. M. Dornbusch. 1982. Theory and observation in cultural transmission. *Science* (Wash., D.C.) 218:19–27.

Cavalli-Sforza, L. L., J. R. Kidd, K. K. Kidd, C. Bucci, A. M. Bowcock, B. S. Hewlett, and J. S. Friedlaender. 1987. DNA markers and genetic variation in human species. *Cold Spring Harbor Symp. Quant. Biol.* 51:411–417.

Cavalli-Sforza, L. L., E. Minch, and J. Mountain. 1992. Co-evolution of genes and languages revisited. *Proc. Natl. Acad. Sci. USA* 89:5620–5624.

Cavalli-Sforza, L. L., A. Piazza, P. Menozzi, and J. Mountain. 1988. Reconstruction of human evolution; bringing together genetic, archaeological, and linguistic data. *Proc. Natl. Acad. Sci. USA* 85:6002–6006.

——. 1989. Genetic and linguistic evolution. *Science (Wash., D.C.)* 244:1128–1129.

——. 1990. Comments following Bateman et al. 1990. *Curr. Anthropol.* 31(1):16–18.

Cavalli-Sforza, L. L., A. C. Wilson, C. R. Cantor, R. M. Cook-Deegan, and M.-C. King. 1991. Call for a worldwide survey of human genetic diversity: a vanishing opportunity for the Human Genome Project. *Genomics* 11:490–491.

Chagnon, N. A. 1983. *Yanomamo: The Fierce People*. G. Spindler and L. Spindler, eds. New York: Holt, Rinehart, Winston.

Chagnon, N. A., J. V. Neel, L. Weitkamp, H. Gershowitz, and M. Ayres. 1970. The influence of cultural factors on the demography and pattern of gene flow from the Makir-

itare to the Yanomama Indians. *Am. J. Phys. Anthropol.* 32(3):339–349.

Chakraborti, D. 1980. Early agriculture and the development of towns in India. In *The Cambridge Encyclopedia of Archaeology*. A. Sherratt, ed., pp. 162–167. New York: Crown.

Chakraborty, R. 1986. Gene admixture in human populations: models and predictions. *Yearb. Phys. Anthropol.* 29:1–43.

Chakraborty, R., and A. K. Roychoudhury. 1978. Is there a pattern of gene differentiation in the Indian populations? *Hum. Genet.* 43:321–328.

Chakraborty, R., R. Blanco, F. Rothhammer, and E. Llop. 1976. Genetic variability in Chilean Indian populations and its association with geography, language, and culture. *Soc. Biol.* 23:73–81.

Chamla, M.-C., and P.-A. Gloor. 1987. Variations diachroniques depuis trois siècles. Données et fracteurs responsables. In *L'Homme, Son Evolution, Sa Diversité: Manuel d'Anthropologie Physique*. D. Ferrembach, C. Susane, and M.-C. Chamla, eds., pp. 463–490. Paris: Doin.

Chang, K.-C. 1977. *The Archaeology of Ancient China*. 3rd. ed. New Haven, Conn.: Yale University Press.

Charbonneau, H. 1984. Trois siècles de dépopulation amérindienne. In *Les Populations Amérindiennes et Inuit du Canada*. L. Normandeau and V. Piche, eds., pp. 28–48. Montreal: Presses de l'Université de Montréal.

Chard, C. S. 1974. *Northeast Asia in Prehistory*. Madison: University of Wisconsin Press.

Chebloune, Y., J. Pagnier, G. Trabuchet, C. Faure, G. Verdier, D. Labie, and V. Nigon. 1988. Structural analysis of the 5' flanking region of the beta-globin gene in African sickle-cell anemia patients: further evidence for three origins of the sickle-cell mutation in Africa. *Proc. Natl Acad. Sci. USA* 85:4431–4435.

Chen, K.-H., and L. L. Cavalli-Sforza. 1983. Surnames in Taiwan: Interpretations based on geography and history. *Hum. Biol. Oceania* 55:367–374.

Chen, K.-H., H. Cann, T. C. Chen, B. Van West, L. Wang, and L. L. Cavalli-Sforza. 1985. Genetic markers of an aboriginal Taiwanese population. *Am. J. Phys. Anthropol.* 66:327–337.

China Handbook Editorial Committee, comp. 1985. *Life and Life Styles*. Beijing: Foreign Language Press.

Christiansen, F. B., and M. W. Feldman. 1986. *Population Genetics*. Palo Alto, Calif.: Blackwell Scientific.

Cipriani, L. 1966. *The Andaman Islanders*. D. T. Cox, ed., transl. London: Weidenfeld and Nicolson.

Clark, J. D. 1970. *The Prehistory of Africa*. New York: Praeger.

——. 1989. The origins and spread of modern humans: A broad perspective on the African evidence. In *The Human Revolution: Behavioural and Biological Perspectives on the Origins of Modern Humans*. P. Mellars and C. Stringer, eds., pp. 565–588. Princeton, N.J.: Princeton University Press.

Clark, J. G. D. 1965. Radiocarbon dating and the expansion of farming culture from the Near East over Europe. *Proc. Prehist. Soc.* 31:57–73.

——. 1972. Star Carr: a case study in bioarcheology. In *An Addison-Wesley Module in Anthropology*. Vol. 10. J. B. Casagrande, W. H. Goodenough, E. A. Hammel, and R.

F. Heizer, eds., pp. 1–42. Menlo Park, Calif.: Addison-Wesley.

Cleveland, W. S. 1979. Robust locally weighted regression and smoothing scatterplots. *J. Am. Stat. Assoc.* 74:829–836.

Cliff, A. D., and J. K. Ord. 1981. *Spatial Processes: Models and Applications.* London: Pion.

Cohen, M. N. 1977. *The Food Crisis in Prehistory: Over Population and Origins of Agriculture.* New Haven, Conn.: Yale University Press.

Collins, R. 1986. *The Basques.* Oxford: Blackwell.

Constans, J., H. Cleve, D. Dykes, M. Fischer, R. L. Kirk, S. S. Papiha, W. Scheffran, R. Scherz, M. Thymann, and W. Weber. 1983. The polymorphism of the vitamin D-binding protein (Gc): isoelectric focusing in 3M urea as an additional method for identification of genetic variants. *Hum. Genet.* 65:176–180.

Contini, M., N. Cappello, R. Griffo, S. Rendine, and A. Piazza. 1989. Géolinguistique et géogénétique: une démarche interdisciplinaire. *Geolinguistique* 4:129–197.

Cook, S. F., and W. Borah. 1971. *Essays in Population History. Mexico and the Caribbean. Vol. I, chap. VI.* Berkeley: University of California Press.

Coon, C. S. 1954. *The Races of Europe.* New York: Macmillan.

———. 1963. *The Origin of Races.* New York: Alfred A. Knopf.

———. 1965. *The Living Races of Man.* New York: Alfred A. Knopf.

Costanzo, A. 1948. La statura degli italiani ventenni nati dal 1854 al 1920. *Ann. Stat.* 2 (serie VIII):64–123.

Count, E. 1950. *This Is Race.* New York: Shuman.

Crick, F. H. 1988. *What Mad Pursuit: A Personal View of Scientific Discovery.* New York: Basic Books.

Crow, J. F., and M. Kimura. 1970. *An Introduction to Population Genetics Theory.* New York: Harper & Row.

Crystal, D. 1987. *The Cambridge Encyclopedia of Language.* Cambridge: Cambridge University Press.

Dahlberg, G. 1943. *Mathematical Methods for Population Genetics.* London: Karger.

———. 1947. *Mathematical Methods for Population Genetics.* New York: Karger.

Dahlberg, G., and S. Wahlund, eds. 1941. The race biology of the Swedish Lapps. II. Anthropometrical survey. Uppsala: The Swedish State Institute for Race Biology.

Damon, A. 1968. Secular trend in height and weight within Old American families at Harvard, 1870–1965. *Am. J. Phys. Anthropol.* 29:45–50.

Darlington, C. D. 1947. The genetic component of language. *Heredity* 1(1):269–286.

Darwin, C. 1859. *On the Origin of Species.* London: J. Murray.

———. 1871. *The Descent of Man, and Selection in Relation to Sex.* London: J. Murray.

Das, S. K., B. N. Mukherjee, K. C. Malhotra, P. P. Majumdar, and P. Partha. 1978. Serological and biochemical investigations among five endogamous groups of Dehli, India. *Ann. Hum. Biol.* 5(1):25–31.

Dausset, J., and A. Sveigaard. 1977. *HLA and Disease.* Baltimore: Williams and Wilkins.

Davis, K. 1973. Cities and mortality. *International Population Conference.* Liege: International Union for the Scientific Study of Population.

Dawson, W. R. 1936. Pygmies and dwarfs in Ancient Egypt. *J. Egypt. Archaeol. (Lond.)* 24(2):185–189.

Delfiner, P. 1976. Linear estimation of non-stationary spatial phenomena. In *Advanced Geostatistics in the Mining Industry.* M. Guarasio, M. David, and C. Haijbregts, eds., pp. 49–68. Dordrecht: Reidel.

Denaro, M., H. Blanc, M. J. Johnson, K.-H. Chen, E. Wilmsen, L. L. Cavalli-Sforza, and D. Wallace. 1981. Ethnic variation in HPA I cleavage patterns of human mitochondrial DNA. *Proc. Natl. Acad. Sci. USA* 78(9):5768–6772.

Denbow, J. 1984. Prehistoric herders and foragers of the Kalahari: the evidence for 1500 years of interaction. In *Past and Present in Hunter-Gatherer Studies.* C. Schrire, ed., pp. 175–193. Orlando, Fla.: Academic Press.

———. 1989. From the Congo to the Kalahari: data and hypotheses about the political economy of the western stream of the Early Iron Age. In African Studies 16th Annual Spring Symposium, Urbana, Ill.: 1–21.

Denbow, J. R., and E. N. Wilmsen. 1986. Advent and course of the pastoralism in the Kalahari. *Science (Wash., D.C.)* 234:1509–1515.

Desmarais, J.-C. 1977. Idéologies et races dans l'ancien Rwanda. Thèse présentée à la faculté des études supérieures en vue de l'obtention du Philosophiae Doctor (Anthropologie). Université de Montréal, Canada.

Devoto, M., G. Romeo, J. L. Serre, G. Meerman, N. Cappello, and A. Piazza. 1990. Gradient of distribution in Europe of the world CF mutation and of its associated haplotypes. *Hum. Genet.* 85:436–441.

Diamond, J. A. 1990. The talk of the Americas. *Nature (Lond.)* 344:589–590.

———. 1991a. *The Rise and Fall of the Third Chimpanzee: Evolution and Human Life.* London: Radius.

———. 1991b. The earliest horsemen. *Nature (Lond.)* 350:275–276.

Digombe, L., P. R. Schmidt, V. Mouleingui-Boukosso, J. Mombo, and M. Locko. 1988. The development of an Early Iron Age prehistory in Gabon. *Curr. Anthropol.* 29(1):179–184.

Dikov, N. N. 1988. On the road to America. *Nat. Hist.* 1:12–15.

di Rienzo, A., and A. C. Wilson. 1991. Branching pattern in the evolutionary tree for human mitochondrial DNA. *Proc. Natl. Acad. Sci. USA* 88:1597–1601.

Dixon, R. M. 1980. *The Languages of Australia.* Cambridge: Cambridge University Press.

Doe, B. 1980. The emergence of Arabia. In *The Cambridge Encyclopedia of Archaeology.* A. Sherratt, ed., pp. 212–215. New York: Crown.

Dolgopolsky, A. B. 1970. Gipoteza drevnejshego rodstva jazykov severnoj evrazii. *Proc. 7th Int. Congr. Anthropol. Ethnol. Sci. (Moscow).* 5:620–628.

———. 1988. The Indo-European homeland and lexical contacts of Proto-Indo-European with other languages. *Mediterr. Lang. Rev.* (Harassowitz) 3:7–31.

Doran, G. H., D. N. Dickel, W. E. Ballinger, O. F. Agee, P. J. Laipis, and W. W. Hauswirth. 1986. Anatomical, cellular and molecular analysis of 8,000-yr-old human brain tissue from the Windover archaeological site. *Nature (Lond.)* 323:803–806.

Dow, M. 1989. Categorical analysis of cross-cultural survey data: effects of clustering on chi-square tests. *J. Quant. Anthropol.* 1:335–352.

Du, R., Y. Yuan, J. Hwang, J. Mountain, and L. L. Cavalli-Sforza. 1991. *Chinese Surnames and the Genetic Differences between North and South China.* Paper 0027. Stanford, Calif.: Morrison Institute for Population Resource Studies.

Dutta, P. C. 1984. Biological anthropology of Bronze Age Harappans: new perspectives. In *The People of South Asia.* J. R. Lukacs, ed., pp. 59–75. New York: Plenum.

Edwards, A. W. F., and L. L. Cavalli-Sforza. 1964. Reconstruction of evolutionary trees. In *Phenetic and Phylogenetic Classification.* V. E. Heywood and J. McNeill, eds., pp. 67–76. London: The Systematics Association.

Efron, B. 1982. *The Jackknife, Bootstrap, and Other Resampling Plans.* Philadelphia: Society for Industrial and Applied Mathematics.

Embleton, S. M. 1986. Statistics in historical linguistics. *Quantitative Linguistics* (Bochum) 30:1–194. Studienverlag Dr. N. Brockmeyer.

Encyclopaedia Britannica. 1974. 15th ed. Chicago: Helen Heningway Benton.

Erlich, H. A. (ed.) 1989. *PCR Technology: Principles and Applications for DNA Amplification.* New York: Stockton Press.

Excoffier, L. 1990. Evolution of human mitochondrial DNA: evidence for departures from a pure neutral model of populations at equilibrium. *J. Mol. Evol.* 30:125–139.

Excoffier, L., and A. Langaney. 1989. Origin and differentiation of human mitochondrial DNA. *Am. J. Hum. Genet.* 44:73–85.

Excoffier, L., B. Pellegrini, A. Sanchez-Mazas, C. Simon, and A. Langaney. 1987. Genetics and history of sub-Saharan Africa. *Yearb. Phys. Anthropol.* 30:151–194.

Fagan, B. M. 1987. *The Great Journey: The Peopling of Ancient America.* London: Thames and Hudson.

Falconer, D. S. 1960. *Introduction to Quantitative Genetics.* New York: Ronald Press.

Farris, J. S. 1972. Estimating phylogenetic trees from distance matrices. *Am. Nat.* 106:645–668.

———. 1973. On the use of the parsimony criterion for inferring evolutionary trees. *Syst. Zool.* 26:77–88.

Feldman, M. W., and L. L. Cavalli-Sforza. 1989. On the theory of evolution under genetic and cultural transmission with application to the lactose absorption problem. In *Mathematical Evolutionary Theory.* M. W. Feldman, ed., pp. 145–173. Princeton, N.J.: Princeton University Press.

Feldman, M. W., M. Nabholz, and W. F. Bodmer. 1969. Evolution of the Rh polymorphism: a model for the interaction of incompatibility, reproductive compensation, and heterozygote advantage. *Am. J. Hum. Genet.* 21(2):171–193.

Felsenstein, J. 1973. Maximum-likelihood estimation of evolutionary trees from continuous characters. *Am. J. Hum. Genet.* 25:471–492.

———. 1982. Numerical methods for inferring evolutionary trees. *Q. Rev. Biol.* 57(4):379–404.

———. 1985. Confidence limits on phylogenies: an approach using the bootstrap. *Evolution* 39:783–791.

Fierro-Domenech, A. 1986. *Le Pre 'Carre': Geographie Historique de la France.* Paris: Robert Laffront.

Fisher, R. A. 1930. *The Genetical Theory of Natural Selection.* New York: Dover.

———. 1937. The wave of advance of advantageous genes. *Ann. Eugen. (Lond.)* 7:355–369.

Fitch, W. M., and M. Margoliash. 1967. Construction of evolutionary trees. *Science (Wash., D.C.)* 155:279–284.

Flannery, K. V. 1973. The origins of agriculture. *Ann. Rev. Anthropol.* 2:271–310.

Flatz, G. 1967. Hemoglobin E: distribution and population dynamics. *Hum. Genet.* 3:189–234.

———. 1987. Genetics of lactose digestion in humans. *Adv. Hum. Genet.* 16:1–77.

Flatz, G., and H. W. Rotthauwe. 1977. The human lactase polymorphism. Physiology and genetics of lactose absorption and malabsorption. *Prog. Med. Genet.* 2:205–250.

Flight, C. R. 1988. The Bantu expansion and the SOAS network. *Hist. Africa* 15:261–301.

Flood, J. 1983. *The Archaeology of Dreamtime.* Honolulu: University of Hawaii Press.

Foley, W. A. 1986. *The Papuan languages of New Guinea.* Cambridge: Cambridge University Press.

Foley, R. A. 1991. The silence of the past. *Nature (Lond.)* 353:114.

Fraser, G. R. 1976. *The Causes of Profound Deafness in Childhood.* Baltimore: Johns Hopkins University Press.

Freeman, L. 1980. The development of human culture. In *The Cambridge Encyclopedia of Archaeology.* A. Sherratt, ed., pp. 79–86. New York: Crown.

Friedlaender, J. S. 1971. Isolation by distance in Bougainville Island. *Proc. Natl. Acad. Sci. USA* 68:704–707.

———. 1975. *Patterns of Human Variation: The Demography, Genetics, and Phenetics of Bougainville Islanders.* Cambridge, Mass.: Harvard University Press.

———., ed. 1987. *The Solomon Islands Project: A Long-term Study of Health, Human Biology, and Culture Change.* 1987. Oxford: Oxford University Press.

Friedlaender, J. S., L. A. Sgaramella-Zonta, K. K. Kidd, L. Y. C. Lai, P. Clark, and R. J. Walsh. 1971. Biological divergences in South-Central Bougainville: an analysis of blood polymorphism gene frequencies and anthropometric measurements utilizing tree models, and a comparison of these variables with linguistic, geographic, and migrational "distances." *Am. J. Hum. Genet.* 23:253–270.

Friedrich, P. 1970. *Proto-Indo-European Trees: The Arboreal System of a Prehistoric People.* Chicago: University of Chicago Press.

Gabriel, K. R. 1971. The biplot geographical display of matrices with application to principal component analysis. *Biometrika* 58:453–467.

Gadgil, M., and R. Thapar. 1990. Human ecology in India: some historical perspectives. *Interdiscip. Sci. Rev.* 15(3):209–223.

Gamble, C. 1986. *Cambridge World Archaeology. The Palaeolithic Settlement of Europe.* Cambridge: Cambridge University Press.

Gamkrelidze, T. V., and V. V. Ivanov. 1990. The early history of Indo-European languages. *Sci. Am.* 262(3):110–116.

Garlake, P. 1980. Early states in Africa. In *Cambridge Encyclopedia of Archaeology*. A. Sherratt, ed., pp. 348–354. New York: Crown.

Garn, S. M. 1971. *Human Races. 3*. Springfield, Ill.: Charles C. Thomas.

Ghosh, A. K., L. Kirk, S. R. Joshi, and H. M. Bhatia. 1977. A population genetic study of the Kota in the Nilgiri Hills, South India. *Hum. Hered.* 27:225–241.

Giacobini, G., ed. 1989. *Hominidae*. Milan: Jaca Books.

Giacobini, G., and F. Mallegni. 1989. Les Néandertaliens Italiens. Inventaire des restes et nouvelles découvertes. In *Hominidae*. G. Giacobini, ed., pp. 379–385. Milan: Jaca Books.

Giblett, E. R. 1969. *Genetic Markers in Human Blood*. Oxford: Blackwell Scientific.

Gillespie, J. H. 1987. Molecular evolution and the neutral allele. *Oxf. Surv. Evol. Biol.* 4:10–37.

Gimbutas, M. 1970. Proto-Indo-European culture: The Kurgan culture during the fifth, fourth, and third millennia B.C. In *Indo-European and Indo-Europeans*. G. Cardona, H. M. Hoenigswald, and A. Senn, eds., pp. 155–195. Philadelphia: University of Pennsylvania Press.

——. 1991. *The Civilization of the Goddess*. San Francisco: Harper.

Glass, B., and C. C. Li. 1953. The dynamics of racial intermixture, an analysis based on the American Negro. *Am. J. Hum. Genet.* 5:1–20.

Glover, I. C. 1980. Agricultural origins in East Asia. In *The Cambridge Encyclopedia of Archaeology*. A. Sherratt, ed., pp. 152–161. New York: Crown.

Goodenough, W. H. 1966. The evolution of pastoralism and Indo-European origins. In *Indo-European and Indo-Europeans*. G. Cardona, H. M. Hoenigswald, and A. M. Senn, eds., pp. 253–265. Philadelphia: University of Pennsylvania Press.

Goodman, M. 1985. Rates of molecular evolution: the hominid slowdown. *BioEssays* 3:9–14.

Gould, S. J. 1989. Grimm's greatest tale. *Nat. Hist.* 2:20–23.

Gower, J. C. 1966. Some distance properties of latent root and vector methods used in multivariate analysis. *Biometrika* 53(3,4):325–338.

Grayson, D. K. 1987. Death by natural causes. *Nat. Hist.* 5:8–13.

Great Britain Meteorological Office. 1968. *Tables of Temperatures, Relative Humidities, and Precipitation of the World*. London: Air Ministry Meteorological Office.

Greenberg, J. H. 1949–50, 1954. Studies in African linguistic classification. *Southw. J. Anthropol.* 8 parts, 5 (1949), 6 (1950), 10 (1954).

——. 1955. *Studies in African Linguistic Classification*. Bradford, Conn.: Compass Publishing.

——. 1963. *The Languages of Africa*. Bloomington: Indiana University Publication.

——. 1971. The Indo-Pacific hypothesis. *Curr. Trends linguistics* 8:807–871.

——. 1976. Lecture, Stanford University, Stanford, Calif.

——. 1987. *Language in the Americas*. Stanford, Calif.: Stanford University Press.

Greenberg, J. H., and M. Ruhlen. 1992. Linguistic origins of native Americans. *Sci. Am.* 267 (5):94–99.

Greenberg, J. H., C. G. Turner II, and S. L. Zegura. 1986. The settlement of the Americas: a comparison of the linguistic, dental, and genetic evidence. *Curr. Anthropol.* 27(5):477–497.

Griffin, J. B. 1980. Agricultural groups in North America. In *The Cambridge Encyclopedia of Archaeology*. A. Sherratt, ed., pp. 375–381. New York: Crown.

Grimes, B., ed., 1984. Languages of the World: Ethnologue. 10th ed. Dallas, Tex. Wycliffe Bible Translators.

Groube, L. M., J. Chappell, J. Muke, and D. Price. 1986. A 40,000 year-old human occupation site at Huon Peninsula, Papua New Guinea. *Nature* 324:453–455.

Groves, C. P. 1989. A regional approach to the problem of the origin of modern humans in Australasia. In *The Human Revolution: Behavioural and Biological Perspectives on the Origins of Modern Humans*. P. Mellars and C. Stringer, eds., pp. 274–285. Princeton, N.J.: Princeton University Press.

Guerin, P., P. Rouger, and G. Lucotte. 1988. A new TaqI BO variant detected with the p49 probe on the human Y chromosome. *Nucleic Acids Res.* 16:7759.

Guglielmino-Matessi, C. R., P. Gluckman, and L. L. Cavalli-Sforza. 1979. Climate and the evolution of skull metrics in man. *Am. J. Phys. Anthropol.* 50:549–564.

Guglielmino-Matessi, C. R., C. Viganotti, and L. L. Cavalli-Sforza. 1983. Spatial distributions and correlations of cultural traits in Africa. Pavia, Italy: Consiglio Nazionale delle Richerche.

Guglielmino-Matessi, C. R., A. Piazza, P. Menozzi, and L. L. Cavalli-Sforza. 1990. Uralic genes in Europe. *Am. J. Phys. Anthropol.* 83(1):57–68.

Guidon, N. 1987. Cliff notes. *Nat. Hist.* 8:6–12.

Guthrie, M. 1967. *Comparative Bantu: An Introduction to the Comparative Linguistics and Prehistory of Bantu Languages*. Vol. 1. Farnborough, England: Gregg Press.

Habgood, P. J. 1989. The origin of anatomically modern humans in Australasia. In *The Human Revolution: Behavioural and Biological Perspectives on the Origins of Modern Humans*. P. Mellars and C. Stringer, eds., pp. 245–273. Princeton, N.J.: Princeton University Press.

Hajdu, G. F. 1975. Finno-Ugrian languages and peoples. G. F. Cushing, transl. London: Andre Deutsch.

Haldane, J. B. S. 1949. Disease and evolution. *Ricerca Sci.* 19(Suppl. 1):3–10.

Hammond Medallion World Atlas. 1981. Maplewood, N.J., Hammond.

Hanihara, K. 1985. Origins and affinities of Japanese as viewed from cranial measurements. In *Out of Asia: Peopling the Americas and the Pacific*. R. Kirk and E. Szathmary, eds., pp. 105–112. Canberra: The Journal of Pacific History.

Harding, R. M., F. W. Rosing, and R. R. Sokal. 1989. Cranial measurements do not support Neolithization of Europe by demic expansion. *Homo* 40:45–58.

Harlan, J. R. 1971. Agricultural origins: Centers and noncenters. *Science (Wash., D.C.)* 174:468–474.

Harris, H. 1966. Enzyme polymorphisms in man. *Proc. R. Soc. B* 164:298–310.

——. 1980. *The Principles of Human Biochemical Genetics*. 3d rev. Amsterdam: Elsevier/North-Holland Biomedical Press.

Harrison, G. A., and J. J. T. Owen. 1964. Studies on the inheritance of human skin colour. *Ann. Hum. Genet.* 28:27–37.

Hartl, D., and A. G. Clark. 1989. *Principles of Population Genetics.* 2d ed. Sunderland, Mass.: Sinauer.

Hassan, F. A. 1975. Determination of the size, density, and growth rate of hunting-gathering populations. In *Population, Ecology, and Social Evolution.* S. Poloyor, ed., pp. 27–52. The Hague: Mouton.

Heincke, F. 1898. *Naturgeschichte Des Herings.* Berlin: Salle.

Heine, B., H. Hoff, and R. Vossen. 1977. Neuere Ergebnisse zur Territorialgeschischte der Bantu. In *Zur Sprachgeschichte und Ethnohistorie in Afrika.* W. J. G. Möhlig, F. Rottland, and B. Heine, eds., pp. 57–72. Berlin: Dietrich Reimer.

Hennig, W. 1966. *Phylogenetic systematics.* Urbana: University of Illinois Press.

Henrici, A. 1973. Numerical classification of Bantu languages. *Afr. Lang. Stud.* 14:82–104.

Herodotus. 1964. History IV. In *Historiens Grecs: Bibliotheque de la Pleiade.* Paris: Editions Gallimard.

Hewlett, B. S., and L. L. Cavalli-Sforza. 1986. Cultural transmission among the Aka pygmies. *Am. Anthropol.* 88:922–934.

Heyerdahl, T. 1950. *The Kon-Tiki Expedition.* London: Allen and Unwin.

Hiernaux, J. 1968a. Bantu expansion: the evidence from physical anthropology confronted with linguistic and archaeological evidence. *J. Afr. Hist.* IX(4):505–515.

———. 1968b. *La Diversité Humaine en Afrique Sub-Saharienne: Récherches Biologiques.* Bruxelles: Institut de Sociologie, Université Libre.

———. 1975. *The People of Africa.* New York: Scribner's.

Hill, A. V. S., and S. W. Serjeantson, eds., 1989. *The Colonization of the Pacific: A Genetic Trail.* Oxford: Clarendon Press.

Hill, A. V. S., D. F. O'Shaughnessy, and J. B. Clegg. 1989. Haemoglobin and globin gene variants in the Pacific. In *The Colonization of the Pacific: A Genetic Trail.* A. V. S. Hill and S. W. Serjeantson, eds., pp. 246–285. Oxford: Clarendon Press.

Hillis, D. M., and C. Moritz, eds. 1990. *Molecular Systematics.* Sunderland, Mass.: Sinauer.

Hirszfeld, L., and H. Hirszfeld. 1919. Essai d'application des méthodes au problème des races. *Anthropologie* 29:505–537.

Hostetler, J. A., and G. E. Huntington. 1980. *The Hutterites in North America.* New York: Holt, Rinehart and Winston.

Hotelling, H. 1933. Analysis of a complex of statistical variables into principal components. *J. Educ. Psychol.* 24:417–441, 498–520.

Howell, F. C. 1984. Introduction. In *The Origins of Modern Humans: A World Survey of the Fossil Evidence.* F. H. Smith and F. Spencer, eds., pp. xiii–xxii. New York: Alan R. Liss.

Howells, W. W. 1973. Cranial variation in man: a study by multivariate analysis of patterns of difference among recent human populations. *Pap. Peabody Mus. Archaeol. Ethnol. Harv. Univ.* 67:1–259.

———. 1981. Origins of the Chinese people: Interpretation of the recent evidence. In *The Origins of Chinese Civilization.*

D. N. Keightly, ed., pp. 297–319. Berkeley: University of California Press.

———. 1986. Physical anthropology of the prehistoric Japanese. In *Windows on the Japanese Past.* R. J. Pearson, G. L. Barnes, and K. L. Hutterer, eds., pp. 85–99. Ann Arbor, Mich., Center for Japanese Studies.

———. 1989. Skull shapes and the map: craniometric analyses in the dispersion of modern Homo. *Pap. Peabody Mus. Archaeol. Ethnol. Harv. Univ.* 79:1–189.

Illich-Svitych, V. M. 1971. Opyt Sravnenija Nostraticheskix Jazykov. Moscow: Nauka.

Imaizumi, Y., N. E. Morton, and D. E. Harris. 1970. Isolation by distance in artificial populations. *Genetics* 66(3):569.

Ingram, V. M. 1957. Gene mutations in human haemoglobin: the chemical difference between normal and sickle cell hemoglobin. *Nature (Lond.)* 180:326–328.

Irwin, G. J. 1980. The prehistory of Oceania: colonization and cultural change. In *The Cambridge Encyclopedia of Archaeology.* A. Sherratt, ed., pp. 324–332. New York: Crown.

Isaac, G. L. 1972. Chronology and the tempo of cultural change during the Pleistocene. In *Calibration of Hominoid Evolution: Recent Advances in Isotopic and Other Dating Methods Applicable to the Origins of Man.* W. W. Bishop and J. A. Miller, eds., pp. 381–430. New York: Scottish Academic Press.

———. 1976. Stages of cultural elaboration in the Pleistocene: possible archaeological indicators of the development of language capabilities. In *Origins and Evolution of Language and Speech.* Vol. 280. S. R. Harnald, H. D. Steklis, and J. Lancaster, eds., pp. 275–288. New York: New York Academy of Sciences.

Jacob, F. 1977. Evolution and tinkering. *Science (Wash., D.C.)* 196:1161–1166.

Jaeger, G. 1974. Étude hémotypologique d'une communauté sara centrafricaine. *Cah. Anthropol. Ecol. Hum.* II(2):19–124.

Jakubiczka, S., J. Arnemann, H. J. Cooke, M. Krawczak, and J. Schmidtke. 1989. A search for restriction fragment length polymorphism on the human Y chromosome. *Hum. Genet.* 84:86–88.

Jeffreys, A. J., R. Neumann, and V. Wilson. 1990. Repeat unit sequence variation in minisatellites: a novel source of DNA polymorphism for studying variation and mutation by single molecule analysis. *Cell. Mol. Neurobiol.* 60(3):473–485.

Jenkins, T., A. Zoutendyk, and A. G. Steinberg. 1970. Gammaglobulin groups (Gm and Inv) of various southern African populations. *Am. J. Phys. Anthropol.* 32:197–218.

Jennings, J. D. 1983. Origins. In *Ancient North Americans.* Vol. 1. J. D. Jennings, ed., pp. 25–67. New York: W. H. Freeman.

Johanson, D. C. 1989. The current status of Australopithecus. In *Hominidae.* G. Giacobini, ed., pp. 77–96. Milan: Jaca Books.

Johnson, M. J., D. C. Wallace, S. D. Ferris, M. C. Rattazzi, and L. L. Cavalli-Sforza. 1983. Radiation of human mitochondria DNA types analyzed by restriction endonuclease cleavage patterns. *J. Mol. Evol.* 19:255–271.

Johnston, F. E., and L. M. Schell. 1979. Anthropometric variation of Native American children and adults. In

The First Americans: Origins, Affinities, and Adaptations. W. S. Laughlin and A. B. Harper, eds., pp. 275–291. New York: G. Fischer.

Jones, F. L. 1970. The Structure and Growth of Australia's Aboriginal Population. In *Aborigines in Australian Society.* 1. Canberra: Australian National University Press.

Jones, R. M. 1987. Paleolithic archaeology—Pleistocene life in the dead heart of Australia. *Nature (Lond.)* 328(6132):666.

——. 1989. East of Wallace's line: issues and problems in the colonization of the Australian continent. In *The Human Revolution: Behavioural and Biological Perspectives on the Origins of Modern Humans.* P. Mellars and C. Stringer, eds., pp. 743–782. Princeton, N.J.: Princeton University Press.

Jorde, L. B. 1980. The genetic structure of subdivided human populations: A review. In *Current Developments in Anthropological Genetics.* Vol. 1. J. H. Mielke and M. H. Crawford, eds., pp. 135–208. New York: Plenum.

——. 1985. Human genetic distance studies: present status and future prospects. *Annu. Rev. Anthropol.* 14:343–373.

Jukes, T. H., ed. 1987. Molecular evolutionary clock. *J. Mol. Evol.* 26:1–171.

Karlin, S., R. Kenett, and B. Bonné-Tamir. 1979. Analysis of biochemical genetic data on Jewish populations. II. Results and interpretations of heterogeneity indices and distance measures with respect to standards. *Am. J. Hum. Genet.* 31:341–365.

Karve, I., and K. C. Malhotra. 1968. A biological comparison of eight endogamous groups of the same rank. *Curr. Anthropol.* 9(2/3):109–124.

Kendall, D. G. 1965. Mathematical models of the spread of infection. In *Conference on Mathematics and Computer Science in Biology and Medicine,* Oxford, pp. 213–225. London: Medical Research Council.

Kennedy, K. A. R., J. Chiment, T. Disotell, and D. Meyers. 1984. Principal-components analysis of prehistoric South Asian crania. *Am. J. Phys. Anthropol.* 64:105–118.

Khazanov, A. M. 1984. *Nomads and the Outside World.* Cambridge: Cambridge University Press.

Kidd, D. 1980. Barbarian Europe in the first millennium. In *The Cambridge Encyclopedia of Archaeology.* A. Sherratt, ed., pp. 295–303. New York: Cambridge University Press.

Kidd, K. K. 1973. *L'Origine dell'Uomo.* Accad. Naz. Lincei. Quaderno No. 182.

Kidd, K. K., and L. L. Cavalli-Sforza. 1974. The role of genetic drift in the differentiation of Icelandic and Norwegian cattle. *Evolution* 28(3):381–395.

Kimura, M. 1968. Evolutionary rate at the molecular level. *Nature (Lond.)* 217:624–626.

——. 1983. *The Neutral Theory of Molecular Evolution.* Cambridge: Cambridge University Press.

Kimura, M., and G. H. Weiss. 1964. The stepping-stone model of population structure and the decrease of genetic correlation with distance. *Genetics* 49:561–576.

Kingman, J. F. C. 1982a. On the genealogy of large populations. *J. Appl. Probab.* 19A:27–43.

——. 1982b. The coalescent. *Stochastic Processes Their Appl.* 13:235–248.

Kirk, R. L. 1965. *Genetic Markers in Australian Aborigines.* Canberra: Australian Institute of Aboriginal Studies.

——. 1976. Serum protein and enzyme markers as indicators of population affinities in Australia and the Western Pacific. In *The Origin of the Australians.* R. L. Kirk and A. G. Thorne, eds., pp. 329–346. Canberra: Australian Institute of Aboriginal Studies.

——. 1979. Genetic differentiation in Australia and the Western Pacific and its bearing on the origin of the first Americans. In *The First Americans: Origins, Affinities, and Adaptations.* W. S. Laughlin and A. B. Harper, eds., pp. 211–237. New York: G. Fischer.

——. 1989. Population genetic studies in the Pacific: Red cell antigen, serum protein, and enzyme systems. In *The Colonization of the Pacific: A Genetic Trail.* A. V. S. Hill and S. W. Serjeantson, eds., pp. 60–119. Oxford: Clarendon Press.

Kirk, R., and E. Szathmary, eds. 1985. *Out of Asia: Peopling the Americas and the Pacific.* Canberra, Australia: The Journal of the Pacific History.

Kirk, R. L. and A. G. Thorne, eds., 1976. *The Origin of the Australians.* Canberra: Australian Institute of Aboriginal Studies.

Klein, R. G. 1980. Later Pleistocene hunters. In *The Cambridge Encyclopedia of Archaeology.* A. Sherratt, ed., pp. 87–95. New York: Crown.

——. 1989a. *The Human Career: Human Biological and Cultural Origins.* Chicago: University of Chicago Press.

——. 1989b. Biological and behavioural perspectives on modern human origins in southern Africa. In *The Human Revolution: Behavioural and Biological Perspectives on the Origins of Modern Humans.* P. Mellars and C. Stringer, eds., pp. 529–546. Princeton, N.J.: Princeton University Press.

Kluge, M., and J. S. Farris. 1969. Quantitative phyletics and the evolution of anurans. *Syst. Zool.* 18:1–32.

Koyama, S. 1978. Jomon subsistence and population. *Senri Ethnol. Stud.* 2:1–65.

Kreitman, M. 1987. Molecular population genetics. *Oxford Surv. Evol. Biol.* 4:38–60.

Kroeber, A. L. 1939. *Cultural and Natural Areas of Native North America.* Berkeley: University of California Press.

Kruskal, J. B. 1971. Multi-dimensional scaling in archaeology: time is not the only dimension. In *Mathematics in the Archaeological and Historical Sciences.* F. R. Hodson, D. G. Kendall, and P. Tautu, eds., pp. 119–132. Edinburgh: Edinburgh University Press.

Kruskal, J. B., I. Dyen, and P. Black. 1971. The vocabulary and method of reconstructing language trees: innovations and large-scale applications. In *Mathematics in the Archaeological and Historical Sciences.* F. R. Hodson, D. G. Kendall, and P. Tautu, eds., pp. 361–380. Edinburgh: Edinburgh University Press.

Kuzolik, A. E., J. S. Wainscoat, G. R. Sergeant, B. C. Kar, B. Al-Awamy, G. J. F. Essan, A. G. Falusi, S. K. Hague, A. M. Hilali, S. Kate, W. A. E. P. Ranasinghe, and D. J. Weatherall. 1986. Geographical survey of the beta S-globin gene haplotypes: evidence for an independent Asian origin of the sickle-cell mutation. *Am. J. Hum. Genet.* 39: 239–244.

Lal, B. B. 1954. Excavations at Hastinapura. *Ancient India* 10:5–151.

Lalouel, J. M. 1973. Topology of population structure. In *Genetic Structure of Populations*. N. E. Morton, ed., pp. 139–149. Honolulu: University of Hawaii Press.

Landsteiner, K. 1901. Uber Agglutinationserscheinungen normalen menschlichen. *Wiener Klin. Wochenschr.* 14: 1132–1134.

Landsteiner, K., and A. S. Wiener. 1940. An agglutinable factor in human blood recognized by immune sera for rhesus blood. *Proc. Soc. Exp. Biol. Med.* 13:223.

Langley, C. H., and W. M. Fitch. 1973. Appendix to the constancy of evolution: a statistical analysis of the alpha and beta hemoglobins, cytochrome *c*, and fibrinopeptide A. In *Models and measures of population structure*. N. E. Morton, ed., pp. 259–262. Honolulu: University of Hawaii Press.

Laslett, P. 1983. Demographic and microstructural history in relation to human adaptation: reflections on newly established evidence. In *How Humans Adapt: A Biocultural Odyssey*. D. J. Ortner, ed., pp. 343–370. Washington, D.C.: Smithsonian Institution Press.

Lathrap, D. W. 1977. Our father the cayman, our mother the gourd: Spinden revisited, or a unitary model for the emergence of agriculture in the new world. In *Origins of Agriculture*. C. A. Reed, ed., pp. 713–751. The Hague: Mouton.

Lathrop, G. M. 1982. Evolutionary trees and admixture: Phylogenetic inferences when some populations are hybridized. *Ann. Hum. Genet.* 46:245–255.

Laughlin, W. S. 1980. *Aleuts: Survivors of the Bering Land Bridge*. New York: Holt, Rinehart and Winston.

Le Bras, H., and E. Todd. 1981. *L'Invention de la France: Atlas Anthropologique et Politique*. Paris: Livre de Poche.

Lee, D. E. 1982. On K-nearest neighbor Voroni diagrams in the plane. *IEEE (Inst. Electr. Electron. Eng.) Trans. Comp.* C-31:478–487.

Lee, R. B., and I. DeVore, eds., 1968. *Man the Hunter*. Chicago: Aldine.

Leff, G. 1975. *William of Ockham: The Metamorphosis of Scholastic Discourse*. Manchester: Manchester University Press.

Le Play, F. 1875. *L'Organisation de la Famille*. Tours, France: Mame.

Levine, P., and R. E. Stetson. 1939. An interesting case of intragroup agglutination. *J. Am. Med. Assoc.* 113:126–127.

Lewontin, R. C. 1972. The apportionment of human diversity. In *Evolutionary Biology*. Vol. 6. T. H. Dobzhansky, M. K. Hecht, and W. C. Steere, eds., pp. 381–398. New York: Appleton-Century-Crofts.

Lewontin, R. C., and J. L. Hubby. 1966. A molecular approach to the study of genetic heterozygosity in natural populations. II. Amount of variation and degree of heterozygosity in natural populations of *Drosophila pseudoobscura*. *Genetics* 54:595–609.

Lewontin, R. C., and J. Krakauer. 1974. Distribution of gene frequency as a test of the theory of the selective neutrality of polymorphisms. *Genetics* 74:175–195.

Li, W. H., and D. Graur. 1991. *Fundamentals of Molecular Evolution*. Sunderland (Massachussetts), USA: Sinauer.

Lieberman, P. 1989. The origins of some aspects of human language and cognition. In *The Human Revolution: Behavioural and Biological Perspectives on the Origins of Modern Humans*. P. Mellars and C. Stringer, eds., pp. 391–414. Princeton, N.J.: Princeton University Press.

Lilliu, G. 1983. *La Civilta dei Sardi. Dal Neolitico All'eta dei Nuraghi*. Torino, Italy: ERI.

Lipe, W. D. 1983. The Southwest. In *Ancient North Americans*. J. D. Jennings, ed., pp. 421–493. New York: W. H. Freeman.

Lisker, R. 1981. *Estructura Genetica de la Poblacion Mexicana (Aspectos Medicos y Anthropologicos)*. Barcelona: Salvat.

Livi, R. 1896–1905. *Antropometria Militare*. Rome: Giornale del Regio Esercito.

Livingstone, F. B. 1976. Hemoglobin history in West Africa. *Hum. Biol. Oceania* 48:487–500.

———. 1984. The Duffy blood groups, vivax malaria, and malaria selection in human populations: A review. *Hum. Biol. Oceania* 56:413–425.

———. 1985. *Frequencies of Hemoglobin Variants: Thalassemia, the Glucose-6-Phosphate*. New York: Oxford University Press.

———. 1989. Simulation of the diffusion of the Beta-globin variants in the old world. *Hum. Biol. Oceania* 61(3): 297–309.

Livshits, G., R. R. Sokal, and E. Kobyliansky. 1991. Genetic affinities of Jewish populations. *Am. J. Hum. Genet.* 49:131–146.

Loomis, W. F. 1967. Skin-pigment regulation of vitamin-D biosynthesis in man. *Science (Wash., D.C.)* 157:501–506.

Lucotte, G., and K. Y. Ngo. 1985. A highly polymorphic probe that detects TaqI RFLPs on human Y chromosome. *Nucleic Acids Res.* 13:8285.

Lucotte, G., K. R. Sriniva, F. Loirat, S. Hazout, and J. Ruffié. 1990. The p49/TaqI Y-specific polymorphisms in three groups of Indians. *Gene Geogr.* 4:21–28.

Lundborg, H., and S. Wahlund, eds. 1932. *The race biology of the Swedish Lapps. I. General survey. Prehistory. Demography*. Uppsala: The Swedish State Institute for Race Biology.

Luzzatto L., E. A. Usanga, and G. Modiano. 1985. Genetic resistance to *Plasmodium falciparum*: studies in the field and in cultures in vitro. In *Linnean Society Symposium Series*. Vol. 11, *Ecology and Genetics of Host-Parasite Interactions*. D. Rollinson and R. M. Anderson, eds., pp. 205–214. London: Academic Press.

Lynch, T. F. 1990. Man in South America? A critical review. *American Antiquity* 55:12–36.

MacNeish, R. S. 1978. Late Pleistocene adaptations. *J. Anthropol. Res.* 34:475–496.

Mahalanobis, P. C. 1936. On the generalized distance in statistics. *Proc. Natl. Inst. Sci. India* 12:49–55.

Malcolm, L. A., P. B. Booth, and L. L. Cavalli-Sforza. 1971. Intermarriage patterns and blood group frequencies in the Bundi people of the New Guinea highlands. *Hum. Biol. Oceania* 43(2):187–199.

Malécot, G. 1948. *Les Mathematiques de L'Hérédité*. Paris: Masson.

———. 1950. Quelques schémas probabilités sur la variabilité des populations naturelles. *Ann. Univ. Lyon Sci. A* 13: 37–60.

———. 1965. Les covariances dans un milieu en équilibre statistique. *Cah. Rhodaniens* 14:1–29.

———. 1966. *Probabilité et Hérédité*. Paris: Presse Universitaire de France.

———. 1967. Identical loci relationships. *Proc. Fifth Berkeley Symp. Math. Statist. Prob.* 4:317–332.

———. 1969. *The Mathematics of Heredity*. San Francisco: Freeman, Cooper.

Mallory, J. P. 1989. *In Search of the Indo-Europeans: Language, Archaeology and Myth*. London: Thames and Hudson.

Mantel, N. 1967. The detection of disease clustering and a generalized regression approach. *Cancer Res.* 27:209–220.

Marshall, J. C. 1989. The descent of the larynx? *Nature (Lond.)* 338:702–703.

Martin, P. 1973. The discovery of America. *Science (Wash., D.C.)* 179:969–974.

Martin, R. 1957. Lehrbuch der Anthropologie. 3. Stuttgart: G. Fischer.

Masson, V. M., and V. I. Sarianidi. 1972. Ancient Peoples and Places. Vol. 79. In *Central Asia: Turkmenia before the Achaemenids*. R. Tringham, ed., New York: Praeger.

Mather, K. 1949. *Biometrical Genetics*. London: Methuen.

Mather, K., and J. L. Jinks. 1977. *Biometrical Genetics*. Ithaca, N.Y.: Cornell University Press.

Matheron, G. 1971. *The theory of regionalized variables and its applications*. Cahiers du Centre de Morphologie Mathematique de Fontainebleau. 5. Paris: Ecole Nationale Superieur de Mines.

McAlpin, D. W. 1981. *Proto-Elamo-Dravidian: The Evidence and Its Implications*. Philadelphia: University of Pennsylvania Press.

McEvedy, C., and R. Jones. 1978. *Atlas of World Population History*. New York: Penguin Books.

McKusick, V. A. 1992. *Mendelian Inheritance in Man: Catalogs of Autosomal Dominant, Autosomal Recessive, and X-Linked Phenotypes*. 10th ed. Baltimore, Md.: Johns Hopkins University Press.

McNeill, W. H. 1976. *Plagues and Peoples*. Garden City, N.Y.: Anchor Books.

Mellars, P., and C. Stringer, eds. 1989. *The Human Revolution: Behavioural and Biological Perspectives in the Origins of Modern Humans*. Princeton, N.J.: Princeton University Press.

Menges, K. H. 1977. Dravidian and Altaic. *Anthropos* 72:129–179.

Menozzi, P., A. Piazza, and L. L. Cavalli-Sforza. 1978a. Synthetic maps of human gene frequencies in Europe. *Science (Wash., D.C.)* 201:786–792.

———. 1978b. Synthetic gene frequency maps and an application to the analysis of the spread of the neolithic in Europe (Abstract). *Am. J. Hum. Genet.* 30:125A.

Merimee, T. J., B. Hewlett, L. L. Cavalli-Sforza, and J. Zapf. 1987. The riddle of pygmy stature. *N. Engl. J. Med.* 317(11):709–710.

Miller, L. H., S. J. Mason, O. F. Clyde, and M. H. McGinniss. 1976. The resistance factor to *Plasmodium vivax* in blacks. *N. Engl. J. Med.* 295:302.

Minch, E. 1992. (Unpubl.) Estimating reliability of phylogenetic trees: bootstrapping and the lattice representation.

Modiano, G., A. S. Benerecetti-Santachiara, F. Gonano, G. Zei, A. Capaldo, and L. L. Cavalli-Sforza. 1965. An analysis of ABO, MN, Rh, Hp, Tf, and G6PD types in a sample from the human population of Lecce Province. *Ann. Hum. Genet.* 29:19–31.

Mooney, J. 1928. The aboriginal population of America north of Mexico. *Smithson. Misc. Collect.* 80(7).

Moorey, P. R. S. 1980. Mesopotamia and Iran in the Bronze Age. In *The Cambridge Encyclopedia of Archaeology*. A. Sherratt, ed., pp. 120–127. New York: Crown.

Moran, P. A. P. 1950. Notes on continuous stochastic phenomena. *Biometrika* 37:17–23.

Mori, F. 1974. The earliest Saharan rock engravings. *Antiquity* 48:87–92.

Morris, A. G. 1986. Khoi and San craniology: A re-evaluation of the osteological reference samples. In *Variation, Culture and Evolution in African Population*. R. Singer and T. Lundy, eds., pp. 1–12. Johannesburg: Witwatersrand University Press.

Morris, C. 1980. Andean South America: From village to empire. In *The Cambridge Encyclopedia of Archaeology*. A. Sherratt, ed., pp. 391–397. New York: Crown.

Morton, N. E., ed. 1973. *Genetic Structure of Populations*. Honolulu: University of Hawaii Press.

———. 1982. *Outline of Genetic Epidemiology*. Basel: Karger.

Morton, N. E., J. F. Crow, and H. J. Muller. 1956. An estimate of the mutational damage in man from data on consanguineous marriages. *Proc. Natl. Acad. Sci. USA* 421: 855–863.

Morton, N. E., S. Yee, and R. Lew. 1971. Bioassay of kinship. *Biometrics* 27 (1):256.

Morton, N. E., R. Lew, I. E. Husserl, and G. F. Little. 1972. Pingelap and Mokil Atolls: historical genetics. *Am. J. Hum. Genet.* 24:277–289.

Morton, N. E., R. Kenett, S. Yee, and R. Lew. 1982. Bioassay of kinship in populations of Middle Eastern origin and controls. *Curr. Anthropol.* 23 (2):157–167.

Motulsky, A. G. 1980. Ashkenazi Jewish gene pools: admixture, drift, and selection. In *Population Structure and Genetic Disorders*. A. W. Eriksson, H. Forsius, H. R. Nevanlinna, P. L. Workman, and R. K. Norio, eds., pp. 353–365. London: Academic Press.

Mountain, J. L., A. A. Lin, A. M. Bowcock, and L. L. Cavalli-Sforza. 1992. Evolution of modern humans: evidence from nuclear DNA polymorphisms. *Phil. Trans. R. Soc. Lond. B.* 337:159–165.

Mourant, A. E. 1954. *The Distribution of the Human Blood Groups*. Oxford: Blackwell Scientific.

———. 1978. *The Genetics of the Jews*. Oxford: Clarendon Press.

Mourant, A. E., A. C. Kopec, and K. Domaniewska-Sobczak. 1976a. *The Distribution of the Human Blood Groups and Other Polymorphisms*. London: Oxford University Press.

Mourant, A. E., D. Tills, and K. Domaniewska-Sobczak. 1976b. Sunshine and the geographical distribution of the alleles of the Gc system of plasma proteins. *Hum. Genet.* 33:307–314.

Mourant, A. E., A. C. Kopec, and K. Domaniewska-Sobczak. 1978. *Blood Groups and Diseases*. Oxford: Oxford University Press.

Murdock, G. P. 1959. *Africa: Its Peoples and Their Culture History*. New York: McGraw-Hill.

———. 1967. *Ethnographic Atlas*. Pittsburgh, Pa.: University of Pittsburgh Press.

Murillo, F., F. Rothhammer, and E. Llop. 1977. The Chipaya of Bolivia: Dermatoglyphics and ethnic relationships. *Am. J. Phys. Anthropol.* 46:45–50.

Murphy, R. 1969. *The Population of China: An Historical and Contemporary Analysis.* Stanford, Calif.: Stanford University Press.

Myres, J. L. 1953. *Geographical History in Greek Lands.* Oxford: Clarendon Press.

Nakahori, Y., T. Tamura, M. Yamada, and Y. Nakagone. 1989. Two 47z [DXYS5] RFLPs on the X and the Y chromosome. *Nucleic Acids Res.* 17:2152.

Neel, J. V. 1978a. Rare variants, private polymorphisms, and locus heterozygosity in Amerindian populations. *Am. J. Hum. Genet.* 30:465–490.

——. 1978b. The population structure of an Amerindian tribe, the Yanomama. *Annu. Rev. Genet.* 12:365–413.

——. 1980. On being headman. *Perspect. Biol. Med.* 23:277–294.

Neel, J. V., and R. H. Ward. 1970. Village and tribal genetic distances among American Indians and the possible implications for human evolution. *Proc. Natl. Acad. Sci.* 65:323–330.

Neel, J. V., and K. M. Weiss. 1975. The genetic structure of a tribal population, the Yanomama Indians. XII. Biodemographic studies. *Am. J. Phys. Anthropol.* 42 (1):25–52.

Neel, J. V., M. Layrisse, and F. M. Salzano. 1977. Man in the tropics: the Yanomama Indians. In *Population Structure and Human Variation.* Vol. 2. International Biological Programme. G. A. Harrison, ed., pp. 109–142. Cambridge: Cambridge University Press.

Nei, M. 1978. The theory of genetic distance and evolution of human races. *Jpn. J. Hum. Genet.* 23:341–369.

——. 1987. *Molecular Evolutionary Genetics.* New York: Columbia University Press.

Nei, M., and G. Livshits. 1989. Genetic relationships of Europeans, Asians and Africans and the origin of modern *Homo sapiens. Hum. Hered.* 39:276–281.

Nei, M., and A. K. Roychoudhury. 1972. Gene differences between Caucasian, Negro, and Japanese populations. *Science (Wash., D.C.)* 177:434–435.

——. 1974. Genic variation within and between the three major races of Man, Caucasoids, Negroids, and Mongoloids. *Am. J. Hum. Genet.* 26:421–443.

——. 1982. Genetic relationship and evolution of human races. *Evol. Biol.* 14:1–59.

——. 1988. *Human Polymorphic Genes: World Distribution.* New York: Oxford University Press.

Nei, M., and F. Tajima. 1983. Maximum likelihood estimation of the number of nucleotide substitutions from restriction sites data. *Genetics* 105:207–217.

Ngo, K. Y., G. Vergnaud, C. Johnsson, G. Lucotte, and J. Weissenbach. 1986. A DNA detecting multiple haplotypes of human Y chromosome. *Am. J. Hum. Genet.* 38:407–418.

Nijenhuis, L. E., and I. M. Hendrikse. 1986. Blood group studies in Pygmies and in a few Bantu populations. In *African Pygmies.* L. L. Cavalli-Sforza, ed., pp. 181–199. Orlando, Fla.: Academic Press.

Nurse, G. T., J. S. Weiner, and T. Jenkins. 1985. *The peoples of southern Africa and their affinities.* Research Monographs on Human Population Biology. 3. Oxford: Clarendon Press.

Oakey, R., and C. Tyler-Smith. 1990. Y chromosome DNA haplotyping suggests that most European and Asian men are descendent from one of two males. *Genomics* 7:325–330.

Oates, J. 1980. The emergence of cities in the Near East. In *The Cambridge Encyclopedia of Archaeology.* A. Sherratt, ed., pp. 112–119. New York: Crown.

The Odyssey World Atlas. 1966. New York: Odyssey Press.

O'Grady, R. T., I. Goddard, R. M. Bateman, W. A. DiMichelle, V. A. Funk, W. J. Kress, R. Mooi, and P. F. Cannell. 1989. Genes and tongues. *Science (Wash., D.C.)* 243:1651.

Ohayon, E., and A. Cambon-Thomsen. In press. *Les Marquers Génétiques dans les Provinces Françaises.* Vols. 1,2. Paris: Inserm-Doin.

Omoto, K. 1972. Polymorphisms and genetic affinities of the Ainu of Hokkaido. *Hum. Biol. Oceania* 1 (4):278–288.

——. 1973. The Ainu: a racial isolate? *Israel J. Med. Sci.* 9 (9–10):1285–1290.

——. 1985. The Negritos: genetic origins and microevolution. In *Out of Asia: Peopling the Americas and the Pacific.* R. Kirk and E. Szathmary, eds., pp. 123–145. Canberra: The Journal of Pacific History.

O'Rourke, D. H., and B. K. Suarez. 1986. Patterns and correlates of genetic variation in South Amerindians. *Ann. Hum. Biol.* 13:13–31.

O'Rourke, D. H., B. K. Suarez, and J. D. Crouse. 1985. Genetic variation in North Amerindian populations: covariance with climate. *Am. J. Phys. Anthropol.* 67:241–250.

O'Shaughnessy, D. F., A. V. S. Hill, D. K. Bowden, D. J. Weatherall, and J. B. Clegg. 1990. Globin genes in Micronesia: Origins and affinities of Pacific Island peoples. *Am. J. Hum. Genet.* 46:144–155.

O'Shea, J. 1980. Mesoamerica: From village to empire. In *The Cambridge Encyclopedia of Archaeology.* A. Sherratt, ed., pp. 382–390. New York: Crown.

Pääbo, S., J. A. Gifford, and A. C. Wilson. 1988. Mitochondrial DNA sequences from a 7000-year-old brain. *Nucleic Acids Res.* 16 (20):9775–9787.

Pääbo, S., R. G. Higuchi, and A. C. Wilson. 1989. Ancient DNA and the polymerase chain reaction: The emerging fields of molecular archaeology. *J. Biol. Chem.* 264 (17):9709–9712.

Pacciarini, M. L., A. M. Bowcock, K. K. Kidd, and L. L. Cavalli-Sforza. (Unpubl.) Variation and evolution of an endogenous retrovirus (ERV3) in humans and primates. 1–10.

Pagnier, J., J. G. Mears, O. Dundabelkodja, K. E. Schaeferego, C. Beldjord, R. L. Nagel, and D. Labie. 1984. Evidence for the multi-centric origin of the sickle-cell hemoglobin gene in Africa. *Proc. Natl. Acad. Sci. USA* 81 (6):1771–1773.

Pallottino, M. 1978. *The Etruscans.* New York: Penguin Books.

Pamilo, P. 1990. Statistical tests of phenograms based on genetic distances. *Evolution* 44:689–697.

Pamilo, P., and M. Nei. 1988. Relationships between gene trees and species trees. *Mol. Biol. Evol.* 5 (5):568–583.

Pampiglione, S., and M. L. Ricciardi. 1986. Parasitological surveys of pygmy groups. In *African Pygmies*. L. L. Cavalli-Sforza, ed., pp. 153–165. Orlando, Fla.: Academic Press.

Pandit, X., and M. Chattopadhayay. 1991. A note on bio-anthropological profile of the Onge of Little Andaman. *Symposium on Genetic and Cultural Diversity in Indian Populations*. Bangalore: Indian Institute of Science (January 29, 1991).

Papiha, S. S., S. M. S. Chahal, D. F. Roberts, and I. P. Singh. 1980. Genetic studies among Kanet and Koli of Kinnar District in Himachal Pradesh, India. *Am. J. Phys. Anthropol.* 53:275–283.

Parr, P. 1980. The Levant in the early first millennium B.C. In *The Cambridge Encyclopedia of Archaeology*. A. Sherratt, ed., pp. 196–199. New York: Crown Publishers.

Pauling, L., H. A. Itano, S. J. Singer, and I. C. Wells. 1949. Sickle-cell anemia, a molecular disease. *Science (Wash., D.C.)* 110:543–548.

Pawley, A., and R. C. Green. 1984. The proto-oceanic language community. In *Out of Asia: Peopling the Americas and the Pacific*. R. Kirk and E. Szathmary, eds., pp. 123–146.

Pedersen, H. 1931. *Linguistic Science in the Nineteenth Century: Methods and Results*. Cambridge, Mass.: Harvard University Press.

Phillipson, D. W. 1980. Iron Age Africa and the expansion of the Bantu. In *The Cambridge Encyclopedia of Archaeology*. A. Sherratt, ed., pp. 342–347. New York: Crown.

———. 1982. The Later Stone Age in sub-Saharan Africa: Physical anthropology. In *The Cambridge History of Africa: from the Earliest Times to c. 500 B.C.* J. D. Clark, ed., pp. 410–477. Cambridge: Cambridge University Press.

Piazza, A. 1986. The genetic data from the French provinces: a tentative summary. In *Génétique des Populations Humains*. E. Ohayon and Cambon-Thomsen, eds. Paris: Inserm.

Piazza, A., and P. Menozzi. 1983. Geographic variation in human gene frequencies. In *Numerical Taxonomy, Proceedings of a NATO Advanced Study Institute*. J. Felsenstein, ed., pp. 444–450. Berlin: Springer.

———. 1984. HLA-A, B genes in the world and the effect of climate. In *Histocompatibility Testing*. E. D. Albert, M. P. Baur, W. R. Mayr, eds., p. 602. Berlin: Springer.

Piazza, A., L. Sgaramella-Zonta, P. Gluckman, and L. L. Cavalli-Sforza. 1975. Fifth histocompatibility workshop gene-frequency data—phylogenetic analysis. *Tissue Antigens* 5:445–463.

Piazza, A., P. Menozzi, and L. L. Cavalli-Sforza. 1980. The HLA-A, B gene frequencies in the world: migration or selection? *Hum. Immunol.* 4:297–304.

———. 1981a. The making and testing of geographic gene-frequency maps. *Biometrics* 37 (4):635–659.

———. 1981b. Synthetic gene frequency maps of man and selective effects of climate. *Proc. Natl. Acad. Sci. USA* 78 (4):2638–2642.

Piazza, A., W. R. Mayr, L. Contu, A. Amoroso, I. Borelli, E. S. Curtoni, C. Marcello, A. Moroni, E. Olivetti, P. Richiardi, and R. Ceppellini. 1985. Genetic and population structure of four Sardinian villages. *Ann. Hum. Genet.* 49:47–63.

Piazza, A., S. Rendine, G. Zei, A. Moroni, and L. L. Cavalli-Sforza. 1987. Migration rates of human populations from surname distributions. *Nature (Lond.)* 329 (6141):714–716.

Piazza, A., N. Cappello, E. Olivetti, and S. Rendine. 1988a. The Basques in Europe: a genetic analysis. *Munibe (Antropologia-Arqueologia)* 6:168–176.

———. 1988b. A genetic history of Italy. *Ann. Hum. Genet.* 52:203–213.

Piazza, A., R. Griffo, N. Cappello, M. Grassini, E. Olivetti, S. Rendine, and G. Zei. In press. Genetic structure and lexical differentiation in Sardinia. In *Language Change and Biological Evolution*. A. Piazza and L. L. Cavalli-Sforza, eds. (in preparation).

Pickersgill, B., and C. B. Heiser, Jr. 1977. Origins and distribution of plants domesticated in the new world tropics. In *Origins of Agriculture*. C. A. Reed, ed., pp. 803–835. The Hague: Mouton.

Piggott, S. 1965. *Ancient Europe from the Beginnings of Agriculture to Classical Antiquity*. Edinburgh: Edinburgh University Press.

Pope, G. G., and J. E. Cronin. 1984. The Asian hominidae. *J. Hum. Evol.* 13:377–396.

Postgate, N. 1980. The Assyrian empire. In *The Cambridge Encyclopedia of Archaeology*. A. Sherratt, ed., pp. 186–192. New York: Crown.

Potter, T. W. 1980. Rome and its empire in the west. In *The Cambridge Encyclopedia of Archaeology*. A. Sherratt, ed., pp. 232–238. New York: Crown.

Presland, G. 1980. Forest cultures of South and South-East Asia. In *The Cambridge Encyclopedia of Archaeology*. A. Sherratt, ed., pp. 272–276. New York: Crown.

Quack, B., P. Guerin, J. Ruffié, and G. Lucotte. 1988. Mapping of probe p49 to the proximal part of the human Y chromosome. *Cytogenet. Cell Genet.* 47:232.

Quinton, P. M. 1982. Abnormalities in electrolyte secretion in cystic fibrosis sweat gland due to decreased anion permeability. In *Fluid and Electrolyte Abnormalities in Exocrine Glands in Cystic Fibrosis*. P. M. Quinton, R. J. Martinez, and U. Hopfer, eds., pp. 53–76. San Francisco Press.

Race, R. R., and R. Sanger. 1975. *Blood Groups in Man*. Oxford: Blackwell Scientific.

Rackham, H., trans. 1979. Books I, II. In *Pliny: Natural History*. 10 vols. Cambridge, Mass.: Harvard University Press.

Rao, C. R. 1948. The utilization of multiple measurements in problems of biological classification. *J. R. Stat. Soc. Ser. B* 10(2):159–193.

Reddy, B. M., and D. C. Rao. 1989. Phenylthiocarbamide taste sensitivity revisited: complete sorting test supports residual family resemblance. *Genet. Epidemiol.* 6: 413–421.

Reed, C. A., ed. 1977. *Origins of Agriculture*. The Hague: Mouton.

Reed, T. E. 1969. Caucasian genes in American Negroes. *Science (Wash., D.C.)* 165:762–768.

Rendine, S., A. Piazza, and L. L. Cavalli-Sforza. 1986. Simulation and separation by principal components of multiple demic expansions in Europe. *Am. Nat.* 128 (5):681–706.

Renfrew, C. 1969. Trade and culture process in European prehistory. *Curr. Anthropol.* 10:151–169.

——. 1973. *Before Civilisation: The Radiocarbon Revolution and Prehistoric Europe.* London: Jonathan Cape.

——. 1987. *Archaeology and Language: The Puzzle of Indo-European Origins.* London: Jonathan Cape.

——. 1989a. Models of change in language and archaeology. *Trans. Philolog. Soc.* 87(2):103–155.

——. 1989b. The origins of Indo-European languages. *Sci. Am.* 261(4):106–114.

——. 1989c. Reply to comments. *Trans. Philolog. Soc.* 87(2):172–178.

——. 1992. Archaeology, genetics and linguistic diversity. *Man.* 27:445–478.

Retzius, G. 1900. *Crania Suecica Antiqua.* Stockholm: Aftonbladts Druckerei.

Reynolds, J., B. S. Weir, and C. C. Cockerham. 1983. Estimation of the coancestry coefficient: basis for a short-term genetic distance. *Genetics* 105:767–779.

Rice, T. T. 1957. *The Scythians.* London: Thames and Hudson.

Rightmire, P. G. 1989. Middle Stone Age humans from Eastern and Southern Africa. In *The Human Revolution: Behavioural and Biological Perspectives on the Origins of Modern Humans.* P. Mellars and C. Stringer, eds., pp. 109–122. Princeton, N.J.: Princeton University Press.

Ripley, W. Z. 1899. *The Races of Europe.* New York: Appleton.

Roberts, D. F. 1973. Climate and human variability. In *An Addison-Wesley Module in Anthropology.* Vol. 34. P. T. Baker, J. B. Casagrande, W. H. Goodenough, and E. A. Hammel, eds., pp. 1–38. Menlo Park, Calif.: Addison-Wesley.

Roberts, L. 1991. A genetic survey of vanishing people. *Science (Wash., D.C.)* 252:1614–1617.

——. 1992. Anthropologists climb (gingerly) on board. *Science* 258:1300–1301.

Roberts, R. G., R. Jones, and M. A. Smith. 1990. Report of thermoluminescence dates supporting arrival of people between 50 and 60 k.y. in southern Australia. *Nature (Lond.)* 345:153.

Robertson, A. 1975. Gene frequency distributions as a test of selective neutrality. *Genetics* 81:775–785.

Ross, P. E. 1991. Hard words. *Sci. Am.* 264(4):138–147.

Rouhani, S. 1989. Molecular genetics and the pattern of human evolution: plausible and implausible models. In *The Human Revolution: Behavioural and Biological Perspectives on the Origins of Modern Humans.* P. Mellars and C. Stringer, eds., pp. 47–61. Princeton, N.J.: Princeton University Press.

Roychoudhury, A. K. 1974. Gene differentiation among caste and linguistic populations of India. *Hum. Hered.* 24:317–322.

——. 1977. Gene diversity in Indian populations. *Hum. Genet.* 40:99–106.

——. 1983. Tribes of India. In *Peoples of India: Some Genetical Aspects.* G. V. Satyavati, ed., pp. 1–30. New Delhi: Indian Council of Medical Research.

Roychoudhury, A. K., and M. Nei. 1988. *Human Polymorphic Genes: World Distribution.* New York: Oxford University Press.

Ruhlen, M. 1975. *A Guide to the Languages of the World.* Stanford, Calif.: Stanford University Press.

——. 1987. *A Guide to the World's Languages.* Stanford, Calif.: Stanford University Press.

——. 1990. Phylogenetic relations of Native American languages. *Prehistoric Mongoloid Dispersals Spec. Issue* 9:75–96.

——. 1991. Postscript. In *Guide to the World's Languages.* M. Ruhlen, pp. 379–407. Stanford, Calif.: Stanford University Press.

Russell, T. 1987. *American Indian Holocaust and Survival.* Norman: University of Oklahoma Press.

Saha, N., R. L. Kirk, S. Shanbhag, S. H. Joshi, and H. M. Bhatia. 1974. Genetic studies among the Kadar of Kerala. *Hum. Hered.* 24:198–218.

——. 1976. Population genetic studies in Kerala and the Nilgiris (South West India). *Hum. Hered.* 26 (3):175–197.

Saitou, N., and M. Nei. 1987. The neighbor-joining method: a new method for reconstructing phylogenetic trees. *Mol. Biol. Evol.* 4 (4):406–425.

Salzano, F. M., and S. M. Callegari-Jacques. 1988. *South American Indians: A Case Study in Evolution.* Oxford: Clarendon.

Salzano, F. M., J. V. Neel, H. Gershowitz, and E. C. Migliazza. 1977. Intra- and intertribal genetic variation within a linguistic group: the Ge-speaking Indians of Brazil. *Am. J. Phys. Anthropol.* 47:337–347.

Sánchez Albornoz, N. 1977. La población de América Latina desde los tiempos precolombinos al año 2000. Madrid: Alianza Editorial.

Sanchez-Mazas Cutanda, A. 1990. Polymorphisme des systèmes immunologiques Rhésus, *GM* et *HLA* et histoire du peuplement humain. Ph.D. diss. University of Geneva.

Sanderson, M. J., and M. J. Donoghue. 1989. Patterns of variation in levels of homoplasy. *Evol. Biol.* 43 (8):1781–1795.

Sanghvi, L. D., R. L. Kirk, and V. Balakrishnan. 1971. A study of genetic distance between populations of Australian aborigines. *Hum. Biol.* 43:445–458.

Sankoff, D. 1970. On the rate of replacement of word-meaning relationships. *Language* 46:564–569.

Sapir, E. 1921. *Language: An Introduction to the Study of Speech.* New York: Harcourt, Brace and World.

——. 1958. Time perspective in aboriginal America culture. In *Selected Writings of Edward Sapir in Language, Culture, and Personality.* D. G. Mandelbaum, ed., pp. 389–462. Berkeley: University of California Press.

SAS Institute. 1985. *SAS User's Guide: Basics, Statistics.* Version 5. SAS Institute, Cary, N.C.

Schanfield, M. S., T. E. Alexeyeva, and M. H. Crawford. 1979. Studies on the immunoglobulin allotypes of Asiatic populations. VIII. Immunoglobulin allotypes among the Tuvinians of the U.S.S.R. *Hum. Hered.* 30 (6):343–349.

Schepartz, L. A. 1988. Who were the later Pleistocene eastern Africans? *Afr. Archaeol. Rev.* 6:57–72.

Schönemann, P. H., and R. M. Carroll. 1970. Fitting one matrix to another under choice of a central dilation and a rigid motion. *Psychometrika* 35:245–255.

Schurr, T. G., S. W. Ballinger, Y.-Y. Gan, J. A. Hodge, D. A. Merriwether, D. N. Lawrence, W. C. Knowler, K. M. Weiss, and D. C. Wallace. 1990. Amerindian mitochondrial DNAs have rare Asian mutations at high frequencies, suggesting they derived from four primary maternal lineages. *Am. J. Hum. Genet.* 46 (3):613–623.

Schwarcz, H. P., R. Grun, B. Vandermeersch, O. Bar-Yosef, H. Valladas, and E. Tchernov. 1988. ESR dates for the hominid burial site of Quefzeh in Israel. *J. Hum. Evol.* 17:733–737.

Schweger, C. E. 1990. The full-glacial ecosystem of Beringia. *Prehist. Mongoloid Dispersal Bull.* 7:35–51.

Serjeantson, S. W. 1985. Migration and admixture in the Pacific. In *Out of Asia: Peopling the Americas and the Pacific.* R. Kirk and E. Szathmary, eds., pp. 133–145. Canberra: The Journal of Pacific History.

———. 1989. HLA genes and antigens. In *The Colonization of the Pacific: A Genetic Trail.* A. V. S. Hill and S. W. Serjeantson, eds., pp. 120–173. Oxford: Clarendon.

Serjeantson, S. W., and A. V. S. Hill. 1989. The colonization of the Pacific: the genetic evidence. In *The Colonization of the Pacific: A Genetic Trail.* A. V. S. Hill and S. W. Serjeantson, eds., pp. 286–294. Oxford: Clarendon.

Serjeantson, S. W., D. P. Ryan, and A. R. Thompson. 1982. The colonization of the Pacific: the story according to human leukocyte antigens. *Am. J. Hum. Genet.* 34: 904–918.

Serjeantson, S. W., R. L. Kirk, and P. B. Booth. 1983. Linguistics and genetic differentiation in New Guinea. *J. Hum. Evol.* 12:77–92.

Sgaramella-Zonta, L., and L. L. Cavalli-Sforza. 1973. A method for the detection of a demic cline. In *Genetic Structure of Populations: Proceedings of a Conference Sponsored by the University of Hawaii and Dedicated to Sewall Wright.* N. E. Morton, ed., pp. 128–135. Honolulu: University of Hawaii Press.

Shaffer, J. G. 1984. The Indo-Aryan invasions: cultural myth and archaeological reality. In *The People of South Asia.* J. R. Lukacs, ed., pp. 77–90. New York: Plenum.

Shannon, C. E., and W. Weaver. 1972. *The Mathematical Theory of Communication.* Urbana: University of Illinois Press.

Shaw, T. C. 1980. Agricultural origins in Africa. In *The Cambridge Encyclopedia of Archaeology.* A. Sherratt, ed., pp. 179–184. New York: Crown.

Shepard, D. 1968. A two-dimensional interpolation function for irregularly spaced data. In *Proceedings of the 1968 ACM National Conference,* pp. 517–524. Princeton, N.J.: Brandon Systems.

Sherratt, A. 1980. The beginnings of agriculture in the Near East and Europe. In *The Cambridge Encyclopedia of Archaeology.* A. Sherratt, ed., pp. 102–111. New York: Crown.

Shevoroshkin, V. 1989. Methods in interphyletic comparisons. *Ural-Altaisch Jahrb.* 61:1–26.

Shinnie, P. L. 1978. The Nilotic Sudan and Ethiopia, c. 660 B.C. to c. A.D. 600. In *The Cambridge History of Africa: from c. 500 B.C. to A.D. 1050.* J. D. Fage, ed., pp. 210–271. London: Cambridge University Press.

Sibley, C. G., and J. E. Ahlquist. 1984. The phylogeny of the hominid primates, as indicated by DNA-DNA hybridization. *J. Mol. Evol.* 20:2–15.

Simmons, R. T. 1962. Blood group genes in Polynesians and comparisons with other Pacific peoples. *Oceania.* 32: 198–210.

Simmons, R. T., and P. B. Booth. 1971. *A Compendium of Melanesian Genetic Data. Parts I–IV.* Melbourne: Roneoed, Commonwealth Serum Laboratories.

Singer, R. 1978. The biology of the San. In *The Bushmen: San Hunters and Herders of Southern Africa.* P. V. Tobias, ed., pp. 115–129. Cape Town: Human and Rousseau.

Skellam, J. 1951. Random dispersal in theoretical populations. *Biometrika* 38:196–218.

———. 1973. The formulation and interpretation of mathematical models of diffusionary processes in population biology. In *The Mathematical Theory and Dynamics of Biological Populations.* M. Bartlett and R. Hiorns, eds., pp. 53–85. London: Academic Press.

Smith, F. H. 1984. Fossil hominids from the upper Pleistocene of Central Europe and the origin of modern Europeans. In *The Origins of Modern Humans: A World Survey of the Fossil Evidence.* F. H. Smith and F. Spencer, eds., pp. 137–209. New York: Alan R. Liss.

Smith, P. E. L. 1982. The Late Palaeolithic and Epi-Palaeolithic of Northern Africa. In *The Cambridge History of Africa.* Vol. 1. *From the Earliest Times to c. 500 B.C.* J. D. Clark, ed., pp. 342–409. Cambridge: Cambridge University Press.

Smouse, P. E. 1982. Genetic architecture of swidden agricultural tribes from lowland rain forests of South America. In *Current Developments in Anthropological Genetics.* Vol. 2. M. H. Crawford and J. H. Mielke, eds., pp. 139–178. New York: Plenum.

Smouse, P. E., V. J. Vitzhum, and J. V. Neel. 1981. The impact of random and lineal fission on the genetic divergence of small human groups: a case study among the Yanomama. *Genetics* 98:179–197.

Sneath, P. H. A., and R. R. Sokal. 1973. *Numerical Taxonomy.* San Francisco: W. H. Freeman.

Sober, E. 1988. *Reconstructing the Past: Parsimony, Evolution and Inference.* Cambridge, Mass.: Massachusetts Institute of Technology Press.

Socha, W. W., and J. Ruffié. 1983. *Blood Groups of Primates: Theory, Practice, Evolutionary Meaning.* Monographs in Primatology, Vol. 3. New York: Alan R. Liss.

Sokal, R. R. 1979. Ecological parameters inferred from spatial correlograms. In *Contemporary Quantitative Ecology and Related Ecometrics.* G. P. Patil and M. Rosenzweig, eds., pp. 167–196. Fairland, Md.: International Co-operative Publishing House.

———. 1988. Genetic, geographic, and linguistic distances in Europe. *Proc. Natl. Acad. Sci. USA* 85:1722–1726.

Sokal, R. R., and P. Menozzi. 1982. Spatial autocorrelations of HLA frequencies in Europe support demic diffusion of early farmers. *Am. Nat.* 119:1–17.

Sokal, R. R., and C. D. Michener. 1958. A statistical method for evaluating systematic relationship. *Univ. Kansas Sci. Bull.* 38:1409–1438.

Sokal, R. R., and N. L. Oden. 1978. Spatial autocorrelation in biology. 2. Some biological implications and four applications of evolutionary and ecological interest. *Biol. J. Linn. Soc.* 10:229–249.

Sokal, R. R., and P. H. A. Sneath. 1963. *Principles of Numerical Taxonomy.* San Francisco: W. H. Freeman.

Sokal, R. R., and D. E. Wartenburg. 1983. A test of spatial autocorrelation analysis using an isolation-by-distance model. *Genetics* 105:219–237.

Sokal, R. R., and E. M. Winkler. 1987. Spatial variation among Kenyan tribes and subtribes. *Hum. Biol. Oceania* 59:147–164.

Sokal, R. R., P. E. Smouse, and J. V. Neel. 1986. The genetic structure of a tribal population, the Yanomama Indians. XV. Patterns inferred by autocorrelation analysis. *Genetics* 114:259–287.

Sokal, R. R., N. L. Oden, and B. A. Thompson. 1988. Genetic changes across language boundaries in Europe. *Am. J. Phys. Anthropol.* 76:337–361.

Sokal, R. R., N. L. Oden, P. Legendre, M. Fortin, J. Kim, B. A. Thomson, A. Vaudor, R. M. Harding, and G. Barbujani. 1989a. Genetics and language in European populations. *Am. Nat.* 135:157–175.

Sokal, R. R., R. M. Harding, and N. L. Oden. 1989b. Spatial patterns of human gene frequencies in Europe. *Am. J. Phys. Anthropol.* 80:267–294.

Sokal, R. R., N. L. Oden, and C. Wilson. 1991. Genetic evidence for the spread of agriculture in Europe by demic diffusion. *Nature (Lond.)* 351:143–145.

Southern, E. M. 1975. Detection of specific sequences among DNA fragments separated by gel electrophoresis. *J. Mol. Biol.* 98:503–517.

Spielman, R. S., E. C. Migliazza, and J. V. Neel. 1974. Regional linguistic and genetic differences among Yanomama Indians. *Science (Wash., D.C.)* 184:637–644.

Spitsyn, V. A. 1985. *Biokimicheskii Polimorfism Cheloveka: Antropologicheskie Aspekti*. Moscow: University of Moscow Press.

Spoor, C. F., and P. Y. Sondaar. 1986. Human fossils from the endemic island fauna of Sardinia. *J. Hum. Evol.* 15:399–408.

Spriggs, M. 1984. The Lapite cultural complex: origins, distribution, contemporaries, and successors. In *Out of Asia: Peopling the Americas and the Pacific*. R. Kirk and E. Szathmary, eds., pp. 202–223. Canberra: The Journal of Pacific History.

Spuhler, J. N. 1972. Genetic, linguistic, and geographical distances in native North America. In *The Assessment of Population Affinities in Man*. J. S. Weiner and J. Huizinga, eds., pp. 72–95. Oxford: Clarendon Press.

——. 1979. Genetic distances, trees, and maps of North American Indians. In *The First Americans: Origins, Affinities, and Adaptations*. W. S. Laughlin and A. B. Harper, eds., pp. 135–183. New York: G. Fischer.

Spurdle, A. B., D. J. Morris, and J. Jenkins. 1989. Y-chromosome probe p49f detects new TaqI variants and complex PvuII haplotypes. *Cytogenet. Cell Genet.* 51:1084.

Starostin, S. A. 1989. Nostratic and Sino-Caucasian. In *Exploration in Language Macrofamilies*. V. Shevoroshkin, ed., pp. 42–66. Bochum: Studienverlag Dr. N. Brockmeyer.

Steinberg, A. G., and C. E. Cook. 1981. *The Distribution of the Human Immunoglobulin Allotypes*. Oxford: Oxford University Press.

Stern, C. 1973. *Principles of Human Genetics*. San Francisco: W. H. Freeman.

Stoneking, M., and R. L. Cann. 1989. African origin of human mitochondrial DNA. In *The Human Revolution: Behavioural and Biological Perspectives on the Origins of Modern Humans*. P. Mellars and C. Stringer, eds., pp. 17–30. Princeton, N.J.: Princeton University Press.

Stoneking, M., K. Bhatia, and A. C. Wilson. 1986. Rate of sequence divergence estimated from restriction maps of mitochondrial DNAs from Papua New Guinea. *Cold Spring Harbor Symp. Quant. Biol.* 51:433–439.

Stoneking, M., L. B. Jorde, K. Bhatia, and A. C. Wilson. 1990. Geographic variation in human mitochondrial DNA from Papua New Guinea. *Genetics* 124:717–733.

Straus, L. G. 1989. Age of the modern Europeans. *Nature (Lond.)* 342:476–477.

Stringer, C. B. 1978. Some problems in middle and upper Pleistocene hominid relationships. *Evolution (Lond.)* 3:395–418.

——. 1988. Paleoanthropology—the dates of Eden. *Nature (Lond.)* 331: 565–566.

——. 1989a. The origin of early modern humans: A comparison of the European and non-European evidence. In *The Human Revolution: Behavioural and Biological Perspectives on the Origins of Modern Humans*. P. Mellars and C. Stringer, eds., pp. 232–244. Princeton, N.J.: Princeton University Press.

——. 1989b. Neanderthals, their contemporaries, and modern human origins. In *Hominidae*. G. Giacobini, ed., pp. 351–355. Milan: Jaca Book.

Stringer, C. B., J. J. Hublin, and B. Vandermeersch. 1984. The origin of anatomically modern humans in Western Europe. In *The Origin of Modern Humans: A World Survey of the Fossil Evidence*. F. H. Smith and F. Spencer, eds., pp. 51–135. New York: Alan R. Liss.

Stringer, C. B., R. Grun, H. P. Schwarcz, and P. Goldberg. 1989. ESR dates for the hominid burial site of ES Skhul in Israel. *Nature (Lond.)* 338:756–758.

Stronach, D. 1980. Iran and the Achaemenians and Seleucids. In *The Cambridge Encyclopedia of Archaeology*. A. Sherratt, ed., pp. 206–211. New York: Crown.

Suarez, B. K., J. D. Crouse, and D. H. O'Rourke. 1985. Genetic variation in North Amerindian populations: The geography of gene frequencies. *Am. J. Phys. Anthropol.* 67(3):217–232.

Sulimirski, T. 1970. *The Samaritans*. New York: Praeger.

Swadesh, M. 1971. *The Origin and Diversification of Language*. J. Sherzer, ed. Chicago: Aldine/Atherton.

Swafford, D. L. 1985. *Phylogenetic Analysis Using Parsimony (PAUP), Version 2.4*. Champaign, Ill.: Illinois Natural History Survey.

——. 1989. *Phylogenetic Analysis Using Parsimony (PAUP), Version 3.0g*. Champaign, Ill.: Illinois Natural History Survey.

Szathmary, E. J. E. 1981. Genetic markers in Siberian and Northern North American populations. *Yearb. Phys. Anthropol.* 24:37–73.

——. 1985. Peopling of North America: clues from genetic studies. In *Out of Asia: Peopling the Americas and the Pacific*. R. Kirk and E. Szathmary, eds., pp. 79–104. Canberra: The Journal of Pacific History.

Tattersall, I., E. Delson, and J. Van Couvering, eds., 1988. *Encyclopedia of Human Evolution and Prehistory*. New York: Garland.

Thapar, R. 1980. India before and after the Mauryan empire. In *The Cambridge Encyclopedia of Archaeology*. A. Sherratt, ed., pp. 257–261. New York: Crown.

Thomas, J. M. C. 1963. *Les Ngbaka de la Lobaye.* Paris: Mouton.

Thomas, J. M. C., and L. Bouquiaux. 1967. La détermination des catégories grammaticales dans une langue à classes. In *La Classification Nominale dans les Langues Négro-Africaines.* G. Manessy and M. Houis, eds., pp. 27–44. Paris: Centre National de la Recherche Scientifique.

Thompson, E. A. 1975. *Human Evolutionary Trees.* Cambridge: Cambridge University Press.

Thompson, G. 1988. HLA disease associations: models for insulin-dependent diabetes melletus and the study of complex human genetic disorders. *Annu. Rev. Genet.* 22: 31–50.

Thorne, A. 1980. The arrival of man in Australia. In *The Cambridge Encyclopedia of Archaeology.* A. Sherratt, ed., pp. 96–100. New York: Crown.

Tills, D., A. C. Kopec, and R. E. Tills. 1983. *The Distribution of the Human Blood Groups and Other Polymorphisms.* Oxford: Oxford University Press.

The Times Atlas of the World. 1989. 7. Edinburgh: John Bartholomew & Son.

Tindale, N. B. 1953. Tribal and intertribal marriage among the Australian aborigines. *Hum. Biol.* 25:169–190.

Tiwari, J. L., and P. I. Terasaki. 1985. *HLA and Disease Associations.* Heidelberg: Springer.

Tobias, P. V., ed. 1978. *The Bushmen: San hunters and herders of Southern Africa.* Cape Town: Human and Rousseau.

——. 1989. Hominid variability, cladistic analysis and the place of Australopithecus Africanus. In *Hominidae.* G. Giacobini, ed., pp. 119–127. Milan: Jaca Books.

Torgerson, W. S. 1958. *Theory and Methods of Scaling.* New York: Wiley.

Torroni, A., O. Semino, R. Scozzari, G. Sirugo, G. Spedini, N. F. M. Abbas, and A. S. Santachiara-Benerecetti. 1990. Y chromosome DNA polymorphisms in human populations: Differences between Caucasoids and Africans detected by 49a and 49f probes. *Ann. Hum. Genet.* 54: 287–296.

Tringham, R. 1971. Hunters, fishers and farmers of eastern Europe: 6000–3000 B.C. London: Hutchinson University Library.

Trinkaus, E. 1984. Western Asia. In *The Origins of Modern Humans: A World Survey of the Fossil Evidence.* F. H. Smith and F. Spencer, eds., pp. 251–293. New York: Alan R. Liss.

Trombetti, A. 1923. *Le Origini della Lingua Basca.* Bologna: Forni.

Tryon, D. T. 1985. Peopling the Pacific: a linguistic appraisal. In *Out of Asia: Peopling the Americas and the Pacific.* R. Kirk and E. Szathmary, eds., pp. 147–159. Canberra: The Journal of Pacific History.

Tucker, A. N. 1969. Sanye and Boni. In *Wort und Religion: Kalima Na Dini.* H.-J. Greschat and H. Jungraithmayr, eds., pp. 67–81. Stuttgart: Evangelischer Missionsverlag.

Turner, C. G., II. 1987. Late Pleistocene and Holocene population history of East Asia based on dental variation. *Am. J. Phys. Anthropol.* 73 (3):305–321.

——. 1989. Teeth and prehistory in Asia. *Sci. Am.* 260(2):88–96.

Turri, E. 1983. Gli Uomini Delle Tende: I Pastori Nomadi Tra Ecologia e Storia, Tra Deserto e Bidonville. Milan: Edizioni di Comunita.

Undevia, J. V., V. Balakrishnan, R. L. Kirk, N. M. Blake, N. Saha, and E. M. McDermid. 1978. A population genetic study of the Vania Soni in western India. *Hum. Hered.* 28(2):104–121.

Central Statistical Board of the USSR, eds. *The USSR in Figures for 1986: Brief Statistical Handbook.* 1987. Moscow: Finansy i Statistika Publishers.

Valladas, H., J. L. Joron, G. Valladas, B. Arensburg, O. Bar-Yosef, A. Belfer-Cohen, P. Goldberg, H. Laville, L. Meignen, Y. Rak, E. Tchernov, A.-M. Tillier, and B. Vandermeersch. 1987. Thermoluminescence dates for the Neanderthal burial site at Kebara in Israel. *Nature (Lond.)* 330:159–160.

Valladas, H., J. L. Reyss, J. L. Joron, G. Valladas, O. Bar-Yosef, and B. Vandermeersch. 1988. Thermoluminescence dating of Mousterian 'Proto-Cro-Magnon' remains from Israel and the origin of modern man. *Nature (Lond.)* 331:614–616.

Valladas, H., H. Cachier, P. Maurice, F. Bernaldo de Quiros, J. Clottes, V. Cabrera Valdés, P. Uzquiano, and M. Arnold. 1992. Direct radiocarbon dates for prehistoric paintings at the Altamira, El Castillo and Niaux caves. *Nature* 357:68–70.

Vandermeersch, B. 1989a. The evolution of modern humans: Recent evidence from Southwest Asia. In *The Human Revolution: Behavioural and Biological Perspectives on the Origins of Modern Humans.* P. Mellars and C. Stringer, eds. pp. 155–164. Princeton, N.J.: Princeton University Press.

——. 1989b. Homogénéité ou hétérogénéité des Néandertaliens. In *Hominidae.* G. Giacobini, ed., pp. 311–317. Milan: Jaca Books.

——. 1989c. L'origine de l'homme moderne: le point de vue du paléoanthropologie. In *Hominidae.* G. Giacobini, ed., pp. 415–421. Milan: Jaca Books.

Vansina, J. 1984. Western Bantu expansion. *J. Afr. Hist.* 25:129–145.

Vansina, J., and H. Brunschwig. 1982. *Études Africaines: Offertes à Henri Brunschwig.* Paris: Editions de l'Ecole des Hautes Etudes en Sciences Sociales.

Vark, G. N. van. 1985. Multivariate analysis in physical anthropology. In *Multivariate Analysis IV.* P. R. Krishnaiah, ed., pp. 599–611. New York: Elsevier Science.

Vidyarthi, L. P. 1983. Tribes of India. In *Peoples of India: Some Genetical Aspects.* G. V. Satyavati, ed., pp. 85–103. New Delhi: Indian Council of Medical Research.

Vidyarthi, L. P., and B. K. Rai. 1985. *The Tribal Culture of India.* 2. New Delhi: Concept.

Vigilant, L., R. Pennington, H. Harpending, T. D. Kocher, and A. C. Wilson. 1989. Mitochondrial DNA sequences in single hairs from a southern African population. *Proc. Natl. Acad. Sci. USA* 86:9350–9354.

Vishnu-Mittre. 1977. Origins and history of agriculture in the Indian sub-continent. *J. Hum. Evol.* 7:31–36.

Voegelin, C. F., and F. M. Voegelin. 1977. *Classification and Index of the World's Languages.* New York: Elsevier North Holland.

Vogel, F., and A. G. Motulsky. 1986. *Human Genetics: Problems and Approaches.* Berlin: Springer.

Wagener, D. K. 1973. An extension of migration matrix analysis to account for differential immigration from the outside world. *Am. J. Hum. Genet.* 25:47–56.

Wagner, M. L. 1941. *Historische Lautlehre Des Sardischen.* Halle (Saale): Max Niemeyer.

Wahba, G. 1982. Constrained regularization for ill-posed linear operator equations, with applications in meteorology and medicine. In *Statistical Decision Theory and Related Topics III. Vol. 2.* S. S. Gupta and J. O. Berger, eds., pp. 383–418. New York: Academic Press.

———. 1984. Surface fitting with scattered, noisy data on Euclidian d-spaces and on the sphere. *Rocky Mt. J. Math.* 14:281–299.

Wahba, G., and J. Wendelberger. 1980. Some new mathematical methods for variational objective analysis using splines and cross validation. *Mon. Weather Rev.* 108:36–57.

Wainscoat, J. S. 1987. Out of the garden of Eden. *Nature (Lond.)* 325:13.

Wainscoat, J. S., A. V. S. Hill, A. L. Boyce, J. Flint, M. Hernandez, S. L. Thein, J. M. Old, J. R. Lynch, A. G. Falusi, D. J. Weatherall, and J. B. Clegg. 1986. Evolutionary relationship of human populations from an analysis of nuclear DNA polymorphisms. *Nature (Lond.)* 319:491–493.

Wainscoat, J. S., A. V. S. Hill, S. L. Thein, J. Flint, J. C. Chapman, D. J. Weatherall, J. B. Clegg, and D. R. Higgs. 1989. Geographic distribution of alpha- and beta-globin gene cluster polymorphisms. In *The Human Revolution: Behavioural and Biological Perspectives on the Origins of Modern Humans.* P. Mellars and C. Stringer, eds., pp. 31–38. Princeton, N.J.: Princeton University Press.

Wallace, D. C., K. Garrison, and W. C. Knowler. 1985. Dramatic founder effects in Amerindian mitochondrial DNAs. *Am. J. Phys. Anthropol.* 68:149–155.

Wang, W. S.-Y., and L. L. Cavalli-Sforza. 1986. Transport proteins in pygmies. In *African Pygmies.* L. L. Cavalli-Sforza, ed., pp. 291–303. Orlando, Fla.: Academic Press.

Wang, W. S.-Y. 1991. Explorations in language. Taipei: Pyramid Press.

Ward, J. H. 1963. Hierarchical grouping to optimize an objective function. *JASA (J. Am. Stat. Assoc.)* 58:236–244.

Ward, R. H., and J. V. Neel. 1970. Gene frequencies and microdifferentiation among the Makiritare Indians. IV. A comparison of a genetic network with ethnohistory and migration matrices; a new index of genetic isolation. *Am. J. Hum. Genet.* 22:538–561.

Wayne, R. K., B. Van Valkenburgh, and S. J. O'Brien. 1991. Molecular distance and divergence time in carnivores and primates. *Mol. Biol. Evol.* 8(3):297–319.

Weidenreich, F. 1939. Six lectures on *Sinanthropus pekinensis* and related problems. *Bull. Geol. Soc. China* 19:1–110.

———. 1945. *Apes, Giants, and Man.* Chicago: University of Chicago Press.

Weir, B. S. 1989. *Genetic Data Analysis: Methods for Discrete Population Genetic Data.* Sunderland, Mass.: Sinauer.

Weir, B. S., and C. C. Cockerham. 1984. Estimating F-statistics for the analysis of population structure. *Evolution* 38:1358–1370.

Weiss, K. M. 1988. In search of times past: gene flow and invasion in the generation of human diversity. In *Biological Aspects of Human Migration.* C. G. N. Mascie-Taylor and G. W. Lasker, eds., pp. 130–166. Cambridge: Cambridge University Press.

Weiss, K. M., and T. Maruyama. 1976. Archeology, population genetics and studies of human racial ancestry. *Am. J. Phys. Anthropol.* 44:31–50.

Weitkamp, L. R., and N. A. Chagnon. 1968. Albumin Maku: a new variant of human serum albumin. *Nature (Lond.)* 217:759–760.

Whitehouse, D. 1980. The expansion of the Arabs. In *The Cambridge Encyclopedia of Archaeology.* A. Sherratt, ed., pp. 289–294. New York: Crown.

Whitehouse, D., and R. Whitehouse. 1975. *Archaeological Atlas of the World.* San Francisco: W. H. Freeman.

Wijsman, E. M. 1984. Techniques for estimating genetic admixture and applications to the problem of the origin of the Icelanders and the Ashkenazi Jews. *Hum. Genet.* 67:441–448.

———. 1986. Estimation of genetic admixture in pygmies. In *African Pygmies.* L. L. Cavalli-Sforza, ed., pp. 349–358. Orlando, Fla.: Academic Press.

Wijsman, E. M., and L. L. Cavalli-Sforza. 1984. Migration and genetic population structure with special reference to humans. *Annu. Rev. Ecol. Syst.* 15:279–301.

Wijsman, E., G. Zei, A. Moroni, and L. L. Cavalli-Sforza. 1984. Surnames in Sardinia. II. Computation of migration matrices from surname distributions in different periods. *Ann. Hum. Genet.* 48:65–78.

Williams, R. C., A. G. Steinberg, H. Gershowitz, P. H. Bennett, W. C. Knowler, D. J. Pettitt, W. Butler, W. Baird, L. Dowda-Rea, T. A. Burch, H. G. Morse, and C. Smith. 1985. GM allotypes in Native Americans: evidence for three distinct migrations across the Bering land bridge. *Am. J. Phys. Anthropol.* 66:1–19.

Wilson, A. C., and V. M. Sarich. 1969. A molecular time scale for human evolution. *Proc. Natl. Acad. Sci. USA* 63:1088–1093.

Wilson, A. C., R. L. Cann, S. M. Carr, M. George, U. B. Gyllensten, K. M. Helm-Bychowsky, R. G. Higuchi, S. R. Palumbi, E. M. Prager, R. D. Sage, and M. Stoneking. 1985. Mitochondrial DNA and two perspectives on evolutionary genetics. *Biol. J. Linn. Soc.* 26:375–400.

Wilson, A. P., and I. R. Franklin. 1968. The distribution of the Diego blood group and its relationship to climate. *Caribbean J. Sci.* 8:1–13.

Wobst, H. M. 1974. Boundary conditions for paleolithic social systems: a simulation approach. *Am. Antiq.* 39:147–178.

Wolpoff, M. H. 1989. Multiregional evolution: The fossil alternative to Eden. In *The Human Revolution: Behavioural and Biological Perspectives on the Origins of Modern Humans.* P. Mellar and C. Stringer, eds., pp. 62–108. Princeton, N.J.: Princeton University Press.

Wolpoff, M. H., X. Wu, and A. G. Thorne. 1984. Modern *Homo sapiens* origins: A general theory of hominid evolution involving the fossil evidence from East Asia. In *The Origins of Modern Humans: A World Survey of the Fossil Evidence.* F. H. Smith and F. Spencer, eds., pp. 411–483. New York: Alan R. Liss.

Womble, W. H. 1951. Differential systematics. *Science (Wash., D.C.)* 114:315–322.

Wright, R. 1991. Quest for the mother tongue. *Atlantic* 267(4):39–68.

Wright, S. 1943. Isolation by distance. *Genetics* 28:114–138.

——. 1946. Isolation by distance under diverse systems of mating. *Genetics* 31:39–59.

——. 1951. The genetical structure of populations. *Ann. Eugen.* 15:323–354.

Wu, X. 1988. Comparative study of early Homo sapiens from China and Europe. *Acta Anthropol. Sinica* 7:287–293.

Wurm, S. A. 1982. Papuan languages of Oceania. *Ars Linguistica* 7. Tübingen: Gunter Narr.

——. (Submitted). Migrations and languages in the Pacific area. In *Language Change and Biological Evolution.* A. Piazza, and L. L. Cavalli-Sforza, eds. (in preparation).

Wurm, S. A., and S. Hattory, eds. 1983. *Language Atlas of the Pacific Area.* Canberra: Australian Academy of the Humanities.

Yamamoto, F., H. Clausen, J. White, J. Marken, and S. Hakamori. 1990. Molecular genetic basis of the histoblood group ABO system. *Nature (Lond.)* 345:229–233.

Zambardino, R. A. 1980. Mexico's population in the sixteenth century: demographic anomaly or mathematical illusion? *J. Interdiscipl. Hist.* 11(1):1–27.

Zei, G., C. R. Guglielmino-Matessi, R. Siri, A. Moroni, and L. L. Cavalli-Sforza. 1983. Surnames as neutral alleles: observations in Sardinia. *Hum. Biol. Oceania* 55(2):357–365.

Zei, G., A. Piazza, A. Moroni, and L. L. Cavalli-Sforza. 1986. Surnames in Sardinia. III. The spatial distribution of surnames for testing neutrality of genes. *Ann. Hum. Genet.* 50:169–180.

Zhimin, A. 1990. The proto-handaxe and its tradition in China. *Acta Anthropol. Sinica* 9(4):311–320.

Zohary, D., and M. Hopf. 1988. *Domestication of Plants in the Old World.* Oxford: Clarendon.

Zubrow, E. 1989. The demographic modelling of neanderthal extinction. In *The Human Revolution: Behavioural and Biological Perspectives on the Origins of Modern Humans.* P. Mellars and C. Stringer, eds., pp. 212–231. Princeton, N.J.: Princeton University Press.

Zvelebil, M. 1980. The rise of the nomads in Central Asia. In *The Cambridge Encyclopedia of Archaeology.* A. Sherratt, ed., pp. 252–256. New York: Crown.

——., ed. 1986a. *Hunters in Transition: Mesolithic Societies of Temperate Eurasia and Their Transition to Farming.* Cambridge: Cambridge University Press.

——. 1986b. Mesolithic prelude and neolithic revolution. In *Hunters in Transition: Mesolithic Societies of Temperate Eurasia and Their Transition to Farming.* M. Zvelebil, ed., pp. 5–15. Cambridge: Cambridge University Press.

——. 1986c. Mesolithic societies and the transition to farming: Problems of time, scale and organization. In *Hunters in Transition: Mesolithic Societies of Temperate Eurasia and Their Transition to Farming.* M. Zvelebil, ed., pp. 167–188. Cambridge: Cambridge University Press.

Zvelebil, M., and K. V. Zvelebil. 1988. Agricultural transition and Indo-European dispersals. *Antiquity* 62:574–583.

INDEX

(Note: page numbers in italics refer to figures or tables)